Urbanization
in a Global Context

Urbanization in a Global Context

Edited by Alison L. Bain and Linda Peake

OXFORD
UNIVERSITY PRESS

Oxford University Press is a department of the University of Oxford.
It furthers the University's objective of excellence in research, scholarship,
and education by publishing worldwide. Oxford is a registered trade mark of
Oxford University Press in the UK and in certain other countries.

Published in Canada by
Oxford University Press
8 Sampson Mews, Suite 204,
Don Mills, Ontario M3C 0H5 Canada

www.oupcanada.com

Copyright © Oxford University Press Canada 2017

The moral rights of the author have been asserted

Database right Oxford University Press (maker)

First Edition published in 2017

All rights reserved. No part of this publication may be reproduced, stored in
a retrieval system, or transmitted, in any form or by any means, without the
prior permission in writing of Oxford University Press, or as expressly permitted
by law, by licence, or under terms agreed with the appropriate reprographics
rights organization. Enquiries concerning reproduction outside the scope of the
above should be sent to the Permissions Department at the address above
or through the following url: www.oupcanada.com/permission/permission_request.php

Every effort has been made to determine and contact copyright holders.
In the case of any omissions, the publisher will be pleased to make
suitable acknowledgement in future editions.

Library and Archives Canada Cataloguing in Publication
Urbanization in a global context / edited by Alison Bain and Linda Peake.

Includes bibliographical references and index.
ISBN 978-0-19-902153-6 (paperback)

1. Urbanization–Case studies. I. Bain, Alison L., 1974-, editor II. Peake,
Linda, 1956-, editor

HT361.U65 2017   307.76   C2016-907285-1

Cover images images (clockwise from top, right): Slums in Lima, Peru, © iStock/FrankvandenBergh;
Chinatown in Bangkok, Thailand, © iStock/aluxum,; Alberta suburban housing, © iStock/Daniel Barnes;
Dubai Skyline,
© iStock/Owen Price.

Oxford University Press is committed to our environment.
This book is printed on Forest Stewardship Council® certified paper
and comes from responsible sources.

Printed and bound in the United States of America

1 2 3 4 — 20 19 18 17

Alison L. Bain

This book is for my inquisitive and inspiring undergraduate Urban Geography students who will hopefully continue to explore cities and become part of the next generation of urban scholars.

Linda Peake

This book is for my much loved family, Karen de Souza, Essie B. and Barbara Peake, and the memory of my much missed mother and father, Marjorie and Frederick Peake.

Contents

Contributors   ix
Tables and Figures   xi
Preface   xv
Acknowledgments   xxii

1 Introduction: Urbanization and Urban Geographies   1
Linda Peake and Alison L. Bain

I ✢ Urban Development   17

2 Shifting Urban Contours: Understanding a World of Growing and Shrinking Cities   19
Kenneth Cardenas and Philip Kelly

3 National Urban Systems in an Era of Transnationalism   36
André Sorensen

4 Globalizing Cities and Suburbs   52
Richard Harris and Roger Keil

5 Incremental and Instant Urbanization: Informal and Spectacular Urbanisms   70
Grace Adeniyi Ogunyankin and Michelle Buckley

II ✢ Urban Policy and Planning   87

6 Urban Policy and Governance: Austerity Urbanism   89
Betsy Donald and Mia Gray

7 Land Use and Creativity in Post-industrial Cities   103
Alison L. Bain and Rachael Baker

8 Socialist and Post-socialist Cities in the Twenty-First Century   120
Lisa B.W. Drummond and Douglas Young

9 Urban Planning, Indigenous Peoples, and Settler States   136
Ryan Walker and Sarem Nejad

10 Urban Policy and Planning for Climate Change   155
Daniel Aldana Cohen

III ✢ Urban Forms   171

11 Gentrification, Gated Communities, and Social Mixing   173
Nicholas Lynch and Yolande Pottie-Sherman

12 Unequal and Volatile Urban Housing Markets   190
Alan Walks and Dylan Simone

13 Urban Public Spaces, Virtual Spaces, and Protest   209
Ebru Ustundag and Gökbörü S. Tanyildiz

14 Urban Geopolitics: War, Militarization, and "The Camp" 227
Nicole Laliberté and Dima Saad

IV Urban Lives 243

15 Placing the Transnational Urban Migrant 245
Harjant S. Gill and Margaret Walton-Roberts

16 The Urban Poor: The Urban Majority and Everyday Life 260
Sabin Ninglekhu and Katharine Rankin

17 Women in Cities 276
Linda Peake and Geraldine Pratt

18 Urban Governance, Ethnicity, Race, and Youth 295
Beverley Mullings and Abdul Alim Habib

19 Disabling Cities 309
Nancy Worth, Laurence Simard-Gagnon, and Vera Chouinard

20 Cities, Sexualities, and the Queering of Urban Space 326
David K. Seitz and Natalie Oswin

V Urban Infrastructure and Livability 345

21 Plants, Animals, and Urban Life 347
Laura Shillington and Alice Hovorka

22 Healthy Cities 361
Godwin Arku and Richard Sadler

23 Urban Water Governance 377
Rebecca McMillan, Sawanya Phakphian, and Amrita Danière

24 Delivering and Managing Waste and Sanitation Services in Cities 394
Carrie L. Mitchell, Kate Parizeau, and Virginia Maclaren

25 Global Convergence and Divergence in Urban Transportation 409
Craig Townsend

26 Conclusion: Envisioning Global Urban Futures 426
Alison L. Bain and Linda Peake

Glossary of Key Terms 436
Index 459

Contributors

Godwin Arku
Associate Professor, Department of Geography, University of Western Ontario

Alison L. Bain
Associate Professor, Department of Geography, York University

Rachael Baker
Doctoral Candidate, Department of Geography, York University

Michelle Buckley
Assistant Professor, Department of Human Geography, University of Toronto Scarborough

Kenneth Cardenas
Doctoral Candidate, Department of Geography, York University

Vera Chouinard
Professor, School of Geography and Earth Sciences, McMaster University

Daniel Aldana Cohen
Assistant Professor, Department of Sociology, University of Pennsylvania

Amrita Danière
Professor, Department of Geography and Planning, University of Toronto

Betsy Donald
Associate Professor, Department of Geography and Planning, Queen's University

Lisa B.W. Drummond
Associate Professor, Urban Studies, Department of Social Science, York University

Harjant S. Gill
Assistant Professor, Department of Sociology, Anthropology and Criminal Justice, Towson University

Mia Gray
Senior Lecturer, Department of Geography, University of Cambridge

Abdul Alim Habib
Instructor, Department of Geography and Planning, Queen's University

Richard Harris
Professor, Department of Geography and Earth Sciences, McMaster University

Alice Hovorka
Professor, Department of Geography and Planning, Queen's University

Roger Keil
Professor, Faculty of Environmental Studies, York University

Philip Kelly
Professor, Department of Geography, York University

Nicole Laliberté
Assistant Professor, Teaching Stream, Department of Geography, University of Toronto at Mississauga

Nicholas Lynch
Assistant Professor, Department of Geography, Memorial University of Newfoundland

Virginia Maclaren
Associate Professor, Department of Geography and Planning, University of Toronto

Rebecca McMillan
Doctoral Candidate, Department of Geography and Planning, University of Toronto

Carrie L. Mitchell
Assistant Professor, School of Planning, University of Waterloo

Beverley Mullings
Associate Professor, Department of Geography and Planning, Queen's University

Sarem Nejad
Doctoral Candidate, Department of Geography and Planning, University of Saskatchewan

Sabin Ninglekhu
Postdoctoral Research Fellow, Earth Observatory of Singapore, Nanyang Technological University

Grace Adeniyi Ogunyankin
Assistant Professor, Women's and Gender Studies, Carleton University

Natalie Oswin
Associate Professor, Department of Geography, McGill University

Kate Parizeau
Assistant Professor, Department of Geography, University of Guelph

Linda Peake
Professor, Urban Studies, Department of Social Science, York University

Sawanya Phakphian
Doctoral Candidate, Department of Geography and Planning, University of Toronto

Yolande Pottie-Sherman
Assistant Professor, Department of Geography, Memorial University of Newfoundland

Geraldine Pratt
Professor, Department of Geography, University of British Columbia

Katharine Rankin
Professor, Department of Geography and Planning, University of Toronto

Dima Saad
Master of Arts Candidate, Department of Anthropology, University of Toronto

Richard Sadler
Assistant Professor, Department of Family Medicine, Division of Public Health, Michigan State University

David K. Seitz
Visiting Scholar in Sexuality Studies, Centre for Feminist Research, York University

Laura Shillington
Lecturer, Geosciences Department, John Abbott College

Laurence Simard-Gagnon
Doctoral Candidate, Department of Geography and Planning, Queen's University

Dylan Simone
Doctoral Candidate, Department of Geography and Planning, University of Toronto

André Sorensen
Professor, Department of Human Geography, University of Toronto Scarborough

Gökbörü S. Tanyildiz
Doctoral Candidate, Department of Sociology, York University

Craig Townsend
Associate Professor, Department of Geography, Planning and Environment, Concordia University

Ebru Ustundag
Associate Professor, Department of Geography, Brock University

Ryan Walker
Associate Professor, Department of Geography and Planning, University of Saskatchewan

Alan Walks
Associate Professor, Department of Geography and Planning, University of Toronto

Margaret Walton-Roberts
Professor, Department of Geography, Wilfrid Laurier University

Nancy Worth
Assistant Professor, Department of Geography and Environmental Management, University of Waterloo

Douglas Young
Associate Professor, Urban Studies, Department of Social Science, York University

Tables and Figures

Tables

1.1 One hundred years of growth in the world, urban, and rural populations and percentage population increase, 1950–2050 (billions) 3
1.2 Annual average rate of change in total, urban, and rural populations, selected periods, 1950–2050 4
1.3 Milestones in world and urban populations 5
2.1 Key variables employed to define urban areas 21
5.1 Gulf-focused real estate and infrastructure development funds, 2007–2009 77
6.1 Ontario municipalities operating expenses by function, as percentage of total operating expenses 2009–2012 99
17.1 Feminist research on gender in the urban global South 278
17.2 Feminist research on gender in the urban global North 279

Figures

0.1 Map of case study cities   inside cover
2.1 Global urbanization, 1950–2050 22
2.2 Dense, low-rise buildings in Delhi, 2013 23
2.3 The global cluster of financial services in London 26
2.4 Map of Manila, the Philippines 30
2.5 Workers outside a new call centre in Manila 31
2.6 Taguig, Metro Manila 31
3.1 Commuters near Osaka Station 42
3.2 Map of Japan, the Tokaido Megalopolis, and the Pacific Belt 44
3.3 Fish market in Tokyo 46
3.4 Club district near Ueno Station, Tokyo 46
3.5 Restaurants under the Bullet Train tracks, Tokyo 46
4.1 Pudong skyline, Shanghai, viewed from the Bund 54
4.2 Informal arrangements for tapping electrical wires in Kolkata, 2012 56
4.3 Map of the Los Angeles Region, California 61
4.4 View of Bunker Hill from the Bonaventure Hotel, Los Angeles 62
4.5 Map of the social geography of Tehran, Iran 63
4.6 An informally built house in Tehran 64
5.1 Four prominent narratives framing understandings of contemporary urbanization and urbanisms 72
5.2 The Burj Khalifa, the world's tallest building 73
5.3 An architectural maquette for the Lagoons megaproject in Dubai 74
5.4 Map of Dubai, United Arab Emirates 75
5.5 Map of Ibadan, Nigeria 81
5.6 A house in Foko, Ibadan 81
5.7 Houses in Ibadan, Nigeria, that contravene planning regulations 82
6.1 Map of Stockton, California 93
6.2 Stockton, California, foreclosure 94
6.3 The city of Stockton goes into bankrupty 94
6.4 View from Woodward and Garfield in midtown Detroit 95
6.5 Location of Detroit, Michigan 96

6.6 Poster advertising "Day against Austerity (Detroit)" 97
7.1 Map of market, community, school, and family gardens in Detroit, Michigan 109
7.2 Brother Nature Organics farm located in Detroit's North Corktown neighbourhood 110
7.3 Farnsworth Street community garden on Detroit's east side is used all year 110
7.4 Map of the Lindenau-Plagwitz "creative cluster" in Leipzig, Germany 111
7.5 A *Wächterhaus* guardian house program project on Karl-Heine Straße in the Lindenau and Plagwitz neighbourhoods 112
7.6 The Spinnerei, a 125-year-old former cotton-spinning mill and now a commercially successful arts district in West Leipzig 113
7.7 Paint-bombing of a new middle-class residential development in Leipzig's central city 114
8.1 Map of Hanoi, Vietnam, showing the city's larger collective housing estates 127
8.2 Resident-adapted but derelict *Khu Tap The* (collective housing block) in Hanoi, Vietnam 128
8.3 Hanoi residents light incense at the mausoleum of Vietnam's first president and founder of the Communist Party, Ho Chi Minh, at the Lunar New Year 2015 128
8.4 Map of Vällingby, Sweden 129
8.5 Vällingby in 1960 130
8.6 Vällingby in 2015 131
9.1 Map of "the Block" in Sydney, Australia 140
9.2 "the Block," Aboriginal Housing Company (Redfern), Sydney 142
9.3 Map of Ōrākei in Auckland, New Zealand 145
9.4 Development on Ngāti Whātua Ōrākei Trust land, Auckland 145
9.5 Yellow Quill First Nation urban reserve, Saskatoon 147
9.6 Muskeg Lake Cree Nation urban reserve, Saskatoon 147
9.7 Aberdeen Square Algonquin basket-weave pattern, Lansdowne Park, Ottawa 149
9.8 Algonquin teaching circle, Lansdowne Park, Ottawa 150
10.1 Infographic of planned changes to São Paulo's transportation system, Brazil 161
10.2 Housing movements marching in central São Paulo to demand more affordable housing in the city core 162
10.3 Central São Paulo 162
10.4 Map of flooded areas in New York, USA 163
10.5 The remains from a home in Staten Island, New York 164
11.1 Map of London, UK, and its boroughs 175
11.2 Luxury church conversions in London's gentrified and gentrifying neighbourhoods 180
11.3 Graffiti along a new housing development in Brixton, London 180
11.4 Reclaiming Brixton: Community members and local activists attend Another Lambeth Is Possible 181
11.5 Map of New York City, USA, and its boroughs 181
11.6 Poor doors in New York City 182
11.7 Protesters during "Turning the Tide" march following Hurricane Sandy 183
12.1 House prices in real terms, selected countries, 2000–2014 194
12.2 The unfinished "Red Arches" estate in Baldoyle, Dublin Region, Ireland 2013 197
12.3 Map of unfinished (ghost) estates in the Dublin region, Ireland, 2010 198
12.4 Map of neighbourhood-based housing clusters in Barcelona, Spain 199
12.5 The heart of Barcelona: Catalonia Square 200
12.6 House Value Index, selected Canadian metropolitan areas, 1996–2013 202
12.7 Total household debt as a proportion of disposable income, by census tract, Montreal CMA 2012 203
13.1 Map of Gezi Park in Istanbul, Turkey, in 2013 216
13.2 Map of Gezi Park in Istanbul, Turkey, after the government's Urban Transformation Plan 217

13.3   A protestor being showered with tear gas by the police in Gezi Park in 2014   218
13.4   Map of the Bronx neighbourhood in New York City, USA, used in the Morris Justice "critical mapping project" to investigate residents' experiences with police   220
14.1   Map of Awere IDP Camp, Uganda   233
14.2   An IDP camp in Northern Uganda where the inhabitants had to build their own homes   234
14.3   ADO members in 2010   235
14.4   Map of al-Baqa'a Refugee Camp, Jordan   236
14.5   In one of al-Baqa'a's main intersections, street art portrays symbols of Palestinian claims to the homeland   237
14.6   A depiction of al-Baqa'a's unfinished infrastructure   237
15.1   Billboard advertising study-abroad opportunities dominate the provincial Punjabi landscape   247
15.2   IELTS Institutes and migration-related service industry established in Chandigarh's modernist city centre, Sector 17   249
15.3   Mastering English-language skills is regarded as the essential step in successful student migration   249
15.4   Map of Punjab District, India, locating Chandigarh   250
15.5   New office development, Ram Darbar Colony, Chandigarh   251
16.1   Map of Kathmandu Metropolitan City, Nepal   266
16.2   A slum settlement in Sinamangal, western end of Kathmandu   267
16.3   A slum settlement on the banks of the Bagmati River in Kathmandu   267
16.4   Residents of Acharya Tole rebuilding their homes after road expansion   269
16.5   Acharya Tole in the aftermath of road expansion   270
17.1   Woman gives side glance to man-spreading   280
17.2   "Housewives please finish travelling by 4 o'clock": a WWII poster   282
17.3   Bagong Barrio West in Manila, the Philippines   284
17.4   Map of Greater Georgetown, Guyana   286
17.5   Georgetown city centre on the banks of the Demerara River   287
17.6   Edmonton SlutWalk, 2011   291
18.1   Map of the Kingston Metropolitan Area, Jamaica   299
18.2   A typical "downtown" West Kingston neighbourhood   300
18.3   Passa Passa street dance event in West Kingston, Jamaica   300
18.4   Map of Accra, Ghana   301
19.1   The medical model of disability   310
19.2   The social model of disability   310
19.3   Map of Greater Georgetown, Guyana   317
19.4   Map of Liverpool, UK   319
19.5   Independent mobility: A visually impaired young person travels through the city   319
19.6   Saint-Roch Mall, 1974   321
20.1   Map of Boystown, Chicago   331
20.2   Boystown rainbow pylons designating the presence of the LGBTQ community   332
20.3   "Pink Dot": An LGBTQ gathering in Singapore   333
20.4   Foreign-worker dormitory in Serangoon Gardens neighbourhood, Singapore   334
20.5   Map of La Zona de la Paz, Ciudad Juárez, Mexico   336
21.1   Map of Managua, Nicaragua   353
21.2   Landscape of large fruit trees in patios of an informal barrio in Managua   354
21.3   Map of Greater Gaborone, Botswana   355
21.4   Female-owned urban poultry farm   356
21.5   City Farm School solar house and gardens, Montreal, Quebec   357
21.6   Urban chickens in Guelph, Ontario   357
22.1   Informal activity in Ekwendeni Trading Centre near the city of Mzuzu, Malawi   366

22.2 Cycling culture juxtaposed against the provision of fast foods shows the challenge of providing healthy urban amenities, Vienna, Austria   367
22.3 Map locating Accra, Ghana, and Flint, Michigan, USA   369
22.4 Original Afrikiko Park in Accra   370
22.5 The "Occupy Flint" movement encampment sits alongside recent downtown redevelopment in Flint, Michigan   371
23.1 Map of water service areas in Pathum Thani, Thailand   384
23.2 Pathumthani Water Company   384
23.3 A utility official explains the water system to community members in Pathum Thani   385
23.4 Map of municipalities and parish boundaries of Caracas, Venezuela   386
23.5 The barrios of Antímano in the west of Caracas   386
23.6 Public MTA meeting in Antímano, Caracas   387
23.7 Residents of Barriolita, Caracas, explain their water problems to Hidrocapital engineer Daniel Pereira   387
24.1 Informal recyclers in Buenos Aires, Argentina   395
24.2 Informal recycler in Hanoi, Vietnam   396
24.3 Open dump outside of Vientiane, Lao PDR   397
24.4 Indiscriminate dumping of waste in canal, Jakarta, Indonesia   398
24.5 Map of Agbogbloshie e-waste recycling in Accra, Ghana   401
24.6 E-waste incineration at the Agbogbloshie metal scrapyard in Accra   402
24.7 Map of Delhi, India   403
24.8 Woman collecting water at community tap, outskirts of Delhi, India   404
25.1 Motorcycle-covered sidewalk in Ho Chi Minh City, Vietnam, 2004   412
25.2 Motorized three-wheel taxis in Bangkok, Thailand, 2011   413
25.3 Map of railways in Tokyo metropolis, Japan   415
25.4 Dense Tokyo suburbs, view facing southeast from Shinjuku, 2013   416
25.5 Subway station exit, Bangkok, Thailand, 2011   417
25.6 Map of Singapore's urban passenger railways and HDB public housing   418
25.7 View north towards Singapore's high-rise suburbs from the central area, 2009   420
25.8 Electronic road pricing gantry, Singapore, 2009   420
25.9 Transit-oriented development in suburban Vancouver, 2013   421

Preface

A Guide to the Text

This textbook engages with dimensions of global urbanization, introducing Canadian undergraduate students to cities outside Canada across the global North and South. Such a global reach emphasizes the interconnectedness of urban places and speaks to an approach to urban analysis that accentuates the similarities and differences between cities in different world regions. Each chapter focuses on different contemporary urban issues that are central to current urban scholarly debates and grounds them in international case studies. How these issues resonate with the Canadian urban context is discussed through descriptive accounts that draw on examples from a selection of small-, mid-, and large-sized Canadian cities. Activities and questions at the end of each chapter are designed as concrete exercises to prompt students to collaborate with peers to further reflect critically upon how these urban issues could relate to their lived experience in, and growing knowledge of, Canadian cities. In this preface we outline the foci of the chapters as well as the structure of the textbook, ending with a brief description of how readers can productively engage with this textbook to actively deepen their knowledge of the urban.

Our critical examination of how cities work—the roles they play and the meanings they engender—and of how and why urbanization is important, is structured into five sections: (i) urban development; (ii) urban planning and policy; (iii) urban forms; (iv) urban lives; and (v) urban infrastructure and livability. This organizational structure permits analysis of contemporary cities in a conceptual sequence that builds from theoretical foundations about patterns and processes of urban change within national and global urban systems, to urban planning and policy frameworks that govern the production of urban space, to the contemporary built form and appearance of cities, to lived experiences within cities, and finally to the material infrastructural challenges for sustainable and just urban futures. To capture some of the diversity of lived urban experiences, each chapter engages with cities in the global North, the global South, or both, offering distinctive perspectives as well as providing an urban case study of one, two, or three different cities (see the map on the inside front cover for a complete list of the cities discussed in the case studies). We journey with our readers in and beyond North America and Europe so that they might appreciate how the urban is differently manifested in Africa, Latin America and the Caribbean, the Middle East, South and Southeast Asia, and Australasia.

The authors of the chapters comprise a mix of late-, mid- and early-career Canadian scholars, ranging from full professors to graduate students, permitting a breadth of critical reflection that is both cutting-edge and rooted in rich scholarly experience and training. The majority of chapters are co-authored and multi-authored, which not only serves to increase the range of voices and empirical examples provided, but also emphasizes the value of dialogue between scholars. And while the majority of contributors teach in geography, other urban studies–related disciplines are represented, not least by those individuals whose affiliations cross disciplinary boundaries to anthropology, development studies, health sciences, sexuality studies, sociology, urban planning, and women's and gender studies. Given the cross-disciplinary reach of contributors, the range of research methods used to collect the data that inform each city case study is many and varied, including a combination of quantitative and qualitative analytical techniques: statistical analysis of censuses and other large data sets; archival research; participant observation; questionnaire surveys; interviews; and discourse analysis. It should also be clear from the range of images and maps incorporated into each chapter, that visual methods—particularly photography and cartography—are important tools for researchers seeking to understand and represent the complexity of the urban.

The goal of this textbook is to challenge users to understand urbanization in the global South and global North as relational rather than as analytically and empirically separate. In Chapter 1, we, the editors, describe urbanization both as a *level* and as a *process*, providing a geographical and historical overview of the current dynamic pace of urbanization at the global scale. The chapter evaluates the extent to which the Urban Age thesis, which stipulates that the majority of the world's population now lives in urban places, is of value. Counteracting this Urban Age–view of the urban as a container and the city as a bounded and knowable entity, the chapter subsequently turns to a theoretical consideration of "the urban question." Addressing both foundational and contemporary urban theory, attention is then given to different ways in which the urban can be understood as a socio-spatial process. The chapter concludes by asking what can be gained from the adoption of a geographical approach to the urban. Concepts of urban space and urban places and the many philosophical paradigms used by geographers to explain them help the reader to better appreciate what is at stake in how the city is studied.

Part I: Urban Development

The first section of the textbook gives readers an overview of urban development at global and national scales, covering both causes and consequences of urbanization, its multiple forms, and its links to urbanism as a way of life. Chapter 2 explores how understandings of urbanization are contested, both in terms of definitions and measurements, and it raises questions about what it means to be urban. The authors, Kenneth Cardenas and Philip Kelly, examine how the global drivers of urbanization operate within the context of uneven development to produce patterns of urban growth and decline, and how these patterns have changed between the twentieth and twenty-first centuries. They also explore the challenges and opportunities that these changing contours of urbanization present, illustrating these issues through a case study of the megacity of Manila in the Philippines. Spatial scales are switched in Chapter 3 in order to explore how processes of globalization and transnationalization affect urbanization within nations. Through a historical-geographical examination of urbanization in Japan, André Sorensen illustrates the ways in which one country and its city systems, in particular the Tokyo–Nagoya–Osaka corridor, have experienced and responded to globalization. Different conceptual approaches to understanding national urban systems are explained and an argument laid out for the importance of engaging with the national scale in a comparative context.

The remaining two chapters in this section return to a global scale of analysis. Chapter 4 provides an overview of the built forms of urbanization. Richard Harris and Roger Keil argue that processes of urbanization have been characterized by both a spatial concentration of activities, seen in urban forms such as the megacities of the global South and the world (or global) cities of the global North and, by urban extension, witnessed through suburbanization, exurbanization, and post-suburbanization. Considering the differentiated nature of suburbanization, they provide case studies of Los Angeles, United States, and Tehran, Iran, to examine questions of suburbanism as a way of life, of suburban development and governance, and of the many different forms that suburban developments can take. Chapter 5 further develops an understanding of different forms of urbanization and urbanism in the global South with case studies on the cities of Dubai, United Arab Emirates, and Ibadan, Nigeria. The authors, Grace Adeniyi Ogunyankin and Michelle Buckley, highlight two forms of urbanism that characterize life for large numbers of urban residents: informal urbanism, which engages in securing resources in quasi-legal or illegal ways, or in a manner that lies outside of formal governmental regulation and spectacular urbanism, in which expressions of a "global" sense of place are developed involving grand architecture or spectacular events. Two common forms of urbanization are also investigated: instant urbanization, which refers to rapid processes involving the development of large areas of land, capital and labour in the production of ambitious development projects; and incremental urbanization, which can be either rapid or gradual in nature, whereby forms of urban growth take place by accretion.

Part II: Urban Policy and Planning

The chapters in the second section explore some of the differing global contexts in which urban policy and planning are contextualized (e.g., capitalist, socialist, and indigenous), the differing ways in which planning and policy have been employed in the economic contexts of post-industrialism and austerity in cities of the global North, and specific applications of planning and policy within cities, namely in relation to climate change and land use. These chapters also focus attention on how urban planners and policy-makers can perform particularly influential roles in controlling urban development and ensuring that the different functional requirements of cities—for production, distribution, reproduction, and exchange—are appropriately located and connected. Chapter 6, by Betsy Donald and Mia Gray, examines urban policy and planning in the context of capitalism, introducing the neoliberal concept of austerity urbanism as a theoretical lens through which to consider the state policies imposed on American cities, especially since the global financial crisis of 2007–2008, that seek to reduce budget deficits. The case studies of the municipal bankruptcies of Stockton, California, and Detroit, Michigan, probe the complex interplay between global capitalist forces, particular state policies, and historically contingent urban circumstances. The topic of declining cities in the global North is also addressed by Alison Bain and Rachael Baker in Chapter 7. Using case studies of Leipzig, Germany, and Detroit, United States, they explore the extent to which post-industrial decline, with its associated social and land-use implications, has shaped urban development opportunities within cities. Particular attention is directed to the creative and cultural turn in neoliberal urban policy and to the power of the urban planning profession and urban policy discourses in shaping and controlling property development.

Chapter 8 turns to the operationalization of urban policy and planning in the economic context of socialism. It examines how urban life has been shaped in significant ways by the urban planning and policy legacies of twentieth-century socialism. Within the literature on socialist and post-socialist cities, Lisa Drummond and Douglas Young identify two core questions that have been debated by urban scholars: was or is there a distinctly socialist city; and what was or is socialist about socialist urbanism? Case studies from Hanoi, Vietnam, and Stockholm, Sweden, are used to illustrate different visions of socialist urbanism and different contemporary fates for socialist urban projects. Chapter 9 turns from global systems of power to address urban planning from the "bottom-up" perspective of a global group of marginalized people, Indigenous people, the descendants of the original inhabitants of settler countries around the world, many of whom now live in urban areas. This chapter, by Ryan Walker and Sarem Nejad, provides an overview of some key features of the academic literature on Indigenous peoples living in cities, as well as case studies of Sydney, Australia, and Auckland, New Zealand, that examine the historic and contemporary contours of what urban co-existence between Indigenous and settler peoples looks like. The urban planning and policy focus of Chapter 10, by Daniel Aldana Cohen, addresses climate change—what some see as the greatest global challenge of the twenty-first century. Urban-based activities are responsible for between half to three quarters of the world's greenhouse gas (GHG) emissions. This chapter asks what role do urban activities play in causing GHG emissions and what are the impacts of climate change on cities and their residents. It features two case studies that consider the social and political dimensions of urban climate policy: São Paulo's efforts to reduce GHG emissions and New York City's experience of Hurricane Sandy in 2012.

Part III Urban Forms

The third section turns to the scale of the urban itself to address material aspects of urban form—that is, the physical characteristics that constitute built-up areas. A consideration of urban form includes the shape, size, density, and configuration of settlements, the layout of which underscores processes of power that produce patterns of inequality across the urban landscape. In Chapter 11, Nicholas Lynch and Yolande Pottie-Sherman explore how the urban landscape speaks to inequality through an examination of specific urban forms that represent spaces of division in contemporary cities: gentrified areas; gated communities; and socially mixed neighbourhoods. Gentrification and gating are both processes by which exclusion is mapped onto the built environment of cities either through displacement or by physical exclusion, whereas social mixing is a planned attempt to reduce division, albeit not always a successful one. Case studies of London, UK and New York City, USA, focus on specific aspects of the built environment—adaptive reuse and poor doors—that highlight the challenges in bridging socio-economic divides in the city. In their examination of housing markets in Chapter 12, Alan Walks and Dylan Simone also highlight the ways in which housing serves to increase social and spatial inequalities in wealth accumulation among different social groups and neighbourhoods. Their chapter discusses the evolution and structure of residential housing systems and markets in countries in the global North, including the globalization of housing finance and its pivotal relationship to the post-2008 global financial crisis. The effects of the crisis are explored in relation to the volatility experienced in housing production and prices, particularly in the United States, which has the deepest and most globalized housing and mortgage markets, but also through case studies of the Irish and Spanish housing bubbles in relation to the metropolitan regions of Dublin and Barcelona.

Residential land use occupies the greatest amount of land in cities, combining "public" spaces (e.g., streets, parks, community centres, libraries, and schools) and "private" spaces (e.g., homes). Chapter 13 addresses one aspect of public space, that of protest movements and the role of the urban built environment in the formation of democratic polities. Ebru Ustundag and Gökbörü Tanyildiz examine the dynamic relationship between public space, protests, citizenship, and the new digital technologies of the twenty-first century, using case studies of Istanbul, Turkey, and New York City, USA, to demonstrate how marginalized social groups mobilize physical and virtual resources in order to claim new urban rights and to expand upon existing ones. The final chapter in this section, Chapter 14, co-authored by Nicole Laliberté and Dima Saad, considers the way in which the built fabric of cities has been destroyed and reconfigured through warfare and political violence. Employing an urban geopolitical analysis of militarization and forced displacement, they use case studies of Al-Baqa'a refugee camp in Jordan and the Awere internally displaced persons camp in Uganda to question fundamentally what constitutes the urban.

Part IV: Urban Lives

The fourth section considers how urban built environments consist of more than the materiality of buildings and public spaces; these are actively transformed and re-imagined by the many different and ordinary types of people who live in them, who through their difference give vitality and a unique character to individual cities. The chapters in this section engage with the everyday lives of specific marginalized groups of urban inhabitants—the poor, women, youth, the disabled, and the queer—engaging with processes of social inclusion and exclusion. Those in positions of relative privilege are also considered, as in Chapter 15 in which Harjant Gill and Margaret Walton-Roberts consider how people move between cities, with a specific emphasis on how transnational migration connects multiple urban regions across national borders. Their focus is on the specific example of migration for educational purposes and how the globalization of higher education has shaped Indian cities.

The case study of Chandigarh in northern India analyses the pathways of students who travel abroad, assessing how the city's urban form and function services this migratory desire.

The following two chapters focus on the majority social groups to be found in cities of the global South. Chapter 16 addresses the urban poor, a social group that is increasing in size with the urbanization of poverty in the twenty-first century. Sabin Ninglekhu and Katharine Rankin ask, what do we know of the places the urban poor occupy, the social and economic practices in which they engage, and their forms of social belonging? They draw on postcolonial and Marxist literature on "the right to the city" to draw attention to the resourcefulness of this majority group in adapting to ever-changing social and economic conditions. A case study of a squatter settlement in Kathmandu, Nepal, illustrates the ways in which development interventions aimed at city rebuilding inevitably contain within them the threat of dispossession and displacement for the urban poor, revealing as well how the poor navigate the complex sociopolitical and bureaucratic terrains of the cities they inhabit to make claims about citizenship and belonging. In Chapter 17, Linda Peake and Geraldine Pratt turn to another specific social group, that of women, to reveal that urbanization is a deeply gendered process and that democratic urban transformations have at their heart issues of gendered equity, belonging, and justice. A review of the dominant analytical framings employed in the study of women's urban lives is undertaken, particularly in relation to two of the longest-standing analyses of women in cities—issues of gendered fear, safety, and violence, including the right to the city, and women's engagement in activities of production and social reproduction. These issues are further explored through two case studies that address the similarities of the challenges of everyday life facing working poor women in very different cities in the global South, namely Manila, Philippines, and Georgetown, Guyana.

The remainder of the chapters in this section also engage with specific marginalized social groups. Chapter 18 addresses youth, who constitute approximately a quarter of the world's population and most of whom live in the urban global South. In this chapter, Beverley Mullings and Alim Habib consider how social constructions of "race" and ethnicity are implicated in neoliberal forms of urban restructuring in the context of the urban global

South. Using case studies from Kingston, Jamaica, and Accra, Ghana, they analyze how ethno-racial forms of inequality, modes of exclusion, and regimes of violence have come to reassert themselves with the embrace of the market as the primary mechanism for organizing and managing urban space. Focusing on the survival experiences and strategies of young people located at the intersection of marginalizing systems of oppression, this chapter invites imaginative thinking about the ways in which young people potentially offer new modes of creativity and interaction in cities. Identity markers of age, racialization, and disability intersect in Chapter 19, wherein authors Nancy Worth, Laurence Simard-Gagnon, and Vera Chouinard focus on what makes cities more or less disabling for persons with physical and mental health impairments. In this chapter, "disabling" refers to processes of physical and social exclusion arising from physical and social barriers to full participation in city life. Over 80 per cent of people with disabilities live in the global South, and a case study of Georgetown, Guyana, explores some of the ways in which this city further disables persons with physical impairments. A second case study of visually impaired young people living in Liverpool, UK explores how they negotiate social space in the city and how the aids they use or choose not to use change the way they are treated by others. This chapter argues that while persons with physical and mental health impairments continue to be marginalized in urban places, it is important to recognize that disabled people, at least sometimes and in some places, find ways of resisting these processes of marginalization. The last chapter in this section, Chapter 20 by David Seitz and Natalie Oswin, engages the city as a central site for the regulation of and resistance to notions of sexual propriety and impropriety. The chapter outlines some of the multiple meanings of the term "queer," both as an umbrella term to represent a wide range of identities on the LGBTQ spectrum and as a critical approach to the shifting and exclusive ways in which sexual norms work in distinct urban contexts. Following this conceptual discussion, they present three case studies—from Chicago, United States; Ciudad Juárez, Mexico; and Singapore—to showcase the relevance of queer theory's view of gender and sexuality as discursive and social forces for understanding urban life. The chapter highlights how a queer approach helps to put LGBTQ and other sexual experiences and politics on the agenda of critical urban scholarship.

Part V: Urban Livability and Infrastructure

The final section bridges critical reflections upon the diverse lived experience of cities with a consideration of how urban infrastructure is a fundamental component of urban livability. Infrastructure allows the relationship between urban form and urban flows of people, capital, information, and goods to function. All cities contain areas of residential, commercial, industrial, and public space, connected through public and private systems of infrastructure—transportation, information and communications technology, energy, water, sewage, waste disposal, health, education, and cultural facilities—to enable systems of service delivery. Discussion in this section focuses upon these elemental infrastructures, both their hard physical networks and their soft frameworks (e.g., the organizational structure of software, rules and regulations, and finance), in relation to the provision of basic needs: health, food, water, sanitation and waste, and transportation. It also explores the notion of both human and non-human actors as infrastructure. Chapter 21 argues that a sustainable urban future is premised upon recognizing all human and non-human actors and acknowledging their roles in urban infrastructures and urban life more broadly. Laura Shillington and Alice Hovorka discuss the ways in which scholars have theorized nature and cities, highlighting the concept of socio-nature; they reference recent work on urban agriculture and urban animals to argue that cities must be viewed as more-than-human. They provide specific case studies of how plants and trees in Managua, Nicaragua, and chickens in Gaborone, Botswana, are critical elements of urban infrastructures in relation to people's food security.

The remaining chapters in this section explore infrastructure in terms of meeting basic needs. Chapter 22 by Godwin Arku and Richard Sadler provides an overview of the ways that urban areas have approached health issues in order to frame how "healthy cities" are thought about. Key areas of concern are introduced regarding urbanization and health: communicable and non-communicable disease; poverty and economic inequality; and the delivery of urban amenities. The two case studies of Flint, Michigan, and Accra, Ghana, highlight that effective and socially responsible governance is one of the most important macro-level determinants of the creation of infrastructure for healthy cities. Chapter 23, by Rebecca McMillan, Sawanya Phakphian, and Amrita Danière, engages with water-supply challenges in cities in the global South, focusing on the merits and limitations of service delivery models. Case studies are provided of recent water-governance reforms in two very different contexts: Bangkok, Thailand, where the provision of water has been privatized; and Caracas, Venezuela, where a public utility operates with community participation. Both examples show how actors' interests and activities interact in practice and the important roles played by local communities in struggles for reliable, affordable, and healthy infrastructure in relation to water.

Water infrastructures, like sanitation and waste, are critical urban planning concerns. Chapter 24 focuses on sanitation and waste as urban planning issues with important environmental and public health dimensions. Carrie Mitchell, Kate Parizeau, and Virginia Maclaren examine how access to, delivery of, and management of sanitation and waste services in cities in the global South are connected to broader patterns of urbanization and globalization. They survey current global trends in the provision of sanitation and discuss common themes between sanitation and waste infrastructure, including health and environmental impacts, social inequity in the provision of service, and service delivery mechanisms. A case study of sanitation infrastructure in Delhi, India, investigates how lack of access to toilets affects men and boys and women and girls differently, while a study in Accra, Ghana, on the repairing and refurbishing of electronic equipment contributes to our understanding of the growing problem of electronic waste to the urban economy. Shifting from sanitation to transportation infrastructure, Craig Townsend in Chapter 25 describes features of urban transport infrastructures. This chapter emphasizes conditions in the global South, including the fast pace of transport-system motorization, which often includes extraordinarily high levels of motorcycle use, the high human health impacts of motorization, the expansion of informal transport, and the flows of foreign investment in transport infrastructure. Case studies from Tokyo, Japan, and Singapore highlight aspects of two of the most technologically advanced, innovative, and productive urban transport systems in the world. Together the chapters in this section show how the day-to-day experience and perception of urban infrastructure is a powerful lens through which to examine the urban.

Infrastructure, like the city, is also not just an "object" or a "system" or an "output," but rather it is a complex socio-spatial process that enables—or disables—particular kinds of urban lives and futures.

In the conclusion, Chapter 26, we address the future of cities. Utopian thinking of alternative futures has played an important political role in imagining different urban arrangements, or socio-spatialities, that address how power relations can play out differently across urban space. Both urban theorists and practitioners have made significant contributions not only to utopian thinking but also to enacting urban transformations on the ground. Yet forces of urban change are more commonly driven through crises. Hence, the major challenges facing cities in the twenty-first century are addressed: the rapid rate of urban growth in the global South in relatively poor postcolonial countries; the persistent spread of urban neoliberalism as the dominant mode of urban development and regulation; the rise of inequality; and the detrimental effects of stress, including climate change, on urban environments. The textbook ends with a consideration of some of the major discourses addressing urban futures that are circulating in urban scholarly, planning, and policy circles—urban sustainability, the role of electronic data, and resilience—as well as the emergence of urban policy on a global scale. As the potential next generation of urban scholars, we conclude by inviting you to help imagine a global urban future.

How to Engage with This Textbook

We hope that you, the reader, will take this book as more than just a collection of facts and examples of urban scholarship about a number of diverse geographical locations. Our intent is to help you to engage critically—intellectually and politically—with its material. Being a critical reader is about questioning assumptions and discovering where you stand on issues that you read about, not only in relation to, in this volume, the case studies on individual cities but also on the material covered in the disciplines addressed, namely geography and urban studies. Being critical involves not only learning about the underlying social relations and political and economic processes that lead to oppressions, inequalities, injustices, but also using theory and analysis to take a stance against them. The activities and questions at the end of each chapter are a starting point for you and your classmates to engage in discussion, employing different modes of learning: visual, auditory, kinesthetic, and reading and writing. We do not contend that taking a critical stance enables one to know the "right" answers or be able to grasp the full picture of what is happening in cities. Indeed, engaging in critical knowledge production requires the recognition of its partial nature; this is why talking to and listening to others is a necessary part of critical learning, of becoming part of an epistemic community of learners. In this sense, the text is interactive; the questions are learning tools to explore further how you and other readers interpret the material in each chapter and to recognize and articulate positions in relation to it. The Glossary is a further aid, highlighting concepts and approaches used by the critical urban scholars in this textbook.

Each chapter contains one to three city case studies that are also valuable tools for learning about the urban. Not only do they illustrate and enrich the material discussed, but they also present realistic, complex, and contextually rich situations that investigate a contemporary urban phenomenon within its real-life setting. The case studies are probably best understood as stories that can help you to understand and interpret a spatially and temporally bounded set of events within an urban area through a particular theoretical lens. In other words, they serve to show the application of a theory or concept to real situations and allow you to ask "how" and "why" questions.

The case studies also allow you to consider whether there are aspects that apply to Canadian cities. The Canadian text boxes illustrate how the issues discussed in relation to cities outside Canada also have relevance to the Canadian context in which you may live.

The Canadian urban content is not, however, the focus of the textbook; thus, it is deliberately neither accentuated nor intended to function as direct comparison to the more detailed international case studies. What the text boxes bring to their chapters is only the gesture of a comparative approach, allowing you to think about the case studies in relation to the experience of Canadian cities. The interest in learning by thinking about similarities and differences among different cities has grown substantially with globalization. Just as you can learn from discussion with others, you can also learn from thinking across different urban contexts and relating those examples to urban environments and lived experiences that may be more familiar to you.

After reading this textbook, you will be able to fulfil the following learning objectives: you will be able to explain different theories, concepts, and issues of concern to critical urban study; to understand processes and outcomes of urbanization; to understand cities as socio-spatial expressions of processes of inequality and difference and as built environments that can perpetuate these expressions but that are also subject to change; to appreciate the complexity of problems confronting cities around the world and the ways in which urban planners and urban policy-makers respond; to recognize relationships between city structures, socio-economic organization and conflict and urban protest; to identify urban ways of being in terms of how the residents of cities organize their daily lives, the issues they face, and how they may work collectively to affect who gets what and where; to understand the relationship between infrastructure provision and the livability of cities; and to critically interpret contemporary cityscapes through photographs, maps, and texts.

Acknowledgments

It is fitting that we find ourselves finalizing this book in 2016, the centennial of the birth of the Canadian-based urban scholar and maverick Jane Jacobs. A vast range of other critical urban scholars have influenced us, too many to list here, so we reserve mention to those powerful feminist urban scholars, past and present, who have shaped our thinking, including Dolores Hayden, Janet Lippman Abu Loghod, Linda McDowell, Suzanne Mackenzie, Susan Parnell, Ananya Roy, and Gerda Wekerle.

We wish to wholeheartedly thank all the contributors to this volume who worked so obligingly to exacting deadlines. Anyone who has edited a book with such a large number of contributors relies tremendously on their "being on side" with the project and on their joie de vivre in enthusiastically taking on multiple drafts. Both were very much in evidence. We appreciate all those people who have kindly donated photographs, including those of the creative commons. Another important thank you needs to be extended to the Social Sciences and Humanities Research Council of Canada (SSHRC), who funded much of the research reported in this book.

We also want to thank those whose presence is not obvious but without whom we would not have engaged in this project. We owe a particular shout out to our undergraduate and graduate students who have shown such a love of learning about cities and all things urban, who remain excited at the prospect of discovering new authors and new texts, who give us an engaged platform for sharing knowledge, and whose engagement in city life and politics gives us hope for the future. And we would be remiss not to mention our "urban" colleagues at York University in the Geography Department and the Urban Studies Program, Department of Social Sciences, in the Faculty of Liberal Arts and Professional Studies, the Faculty of Environmental Studies, and the City Institute with whom we work, play, and interact on a daily basis: Teresa Abbruzzese, Ranu Basu, Jon Caulfield, Lisa Drummond, George Fallis, Jenny Foster, Liette Gilbert, Shubhra Gururani, Laam Hae, Jin Haritaworn, Peggy Keall, Roger Keil, Stefan Kipfer, Sara Koopman, Ute Lehrer, Lucia Lo, Sara Macdonald, Janine Marchessault, Valerie Preston, Barbara Rahder, Roza Tchoukaleyska, Steven Tufts, Gerda Wekerle, Patricia Wood, and Doug Young. A special thanks goes to Carolyn King, the recently retired cartographer in the Department of Geography at York University, who created a number of the maps with great care and attention to detail.

Finally, we wish to thank the anonymous reviewers who read the whole manuscript and provided valuable comments and Oxford University Press, whose interest generated this book. We have had the pleasure over the last few years of working with the super efficient, helpful, and friendly Caroline Starr, acquisitions editor, and Peter Chambers, supervising developmental editor. It was their belief that there was no textbook that addressed urban issues and urbanization outside the Canadian context and yet could also address a Canadian audience that was the inspiration for this book. We hope that we have delivered.

Alison L. Bain and Linda Peake

Introduction
Urbanization and Urban Geographies

Linda Peake and Alison L. Bain

Introduction

Whatever their location in the global process of **urbanization** that has led the twenty-first century to be declared "the century of the city," cities are complex and multifaceted places that ignite the imagination and inspire intellectual curiosity. They have found expression not only through academic texts, but also through a range of media, including novels, films, poems, video games, and television shows. Cities are also, of course, experienced first-hand. The embodied experience of walking through a city, for example, individually or in a crowd, with friends or with strangers, has further inspired scholars (De Certeau, 1984; Debord, 1996), philosophers (Gros, 2014), public intellectuals (Self, 2003; Solnit, 2001) and poets to contemplate the nature of urban life and its **psychogeographies**—the effects of the geographic environment on the emotions and behaviours of individuals. Whether through first- or second-hand modes of encounter, the city fascinates as a site that communicates messages about power, war, inequality, poverty, violence, surveillance, and **loneliness**, but also possibility, hope, freedom, citizenship, **anonymity**, peace, and many other relations and domains of human life. It would be surprising if you, the reader, did not already have your own knowledge based on experiences of and feelings about urban places. In this book we aim to build on your lived experience and knowledge by introducing you to academic scholarship about the world's cities and to the processes of urbanization that have produced them.

As complex social, economic, ecological, and political systems, cities require analytical scrutiny in order for us to understand how they change. This textbook provides a critical introduction to the conceptual framings, theoretical frameworks, planned dynamics, and lived experiences of contemporary cities around the world as products of capital investment, as collective **urban commons** or public goods, and as socio-cultural-natural spaces. The Canadian urban scholars who have contributed to this textbook and the international scholars with whom they have collaborated offer different perspectives on the changing forms and geographies of urbanization. Collectively, we recognize that urbanization processes produce variegated socio-spatial forms, conditions, and urban ways of life (**urbanisms**) that necessitate contextually specific analyses and theorization. The now global reach of the economic, political, and social processes that produce cities also demands that we situate the study of urbanization in a global context. Hence, a geographical approach to the urban dimensions of space and place also speaks to addressing urbanization at different spatial scales, from the macro-scale factors that influence urban development, to the micro-scale contingent local dynamics that produce places we call urban.

Cities have changed through history—expanding and shrinking, extending and concentrating—and will continue to change in response to technological innovation, migration, population growth and decline, the demands of businesses and households, and crises and disasters. The nature of these changes is also expressed through networks of relations to other places and spatial scales. In this book we unpack these relational complexities in order to render the everyday life of cities legible as unique representations of the societies that build, inhabit, and maintain them. We seek to answer fundamental questions about how knowledge of the urban is constituted; we ask, through the study of what material sites and socio-spatial relationships is such knowledge gained? In academic study the elevation of some places as exemplary models of the urban (e.g., London, Manchester, Paris, Berlin, New York City, Chicago, and Los Angeles) has produced an uneven comparative framework of scholarly analysis and theorization in which cities in the global North are repeatedly positioned as leaders and cities in the global South as followers (Brenner and Keil, 2006; Jacobs, 2012; Robinson, 2006). By deliberately focusing on a range of cities from the global North and global South that may not be conventional exemplars in urban scholarship, we challenge our readers to assume different vantage points in order to reframe and reimagine what constitutes the urban in both theory and practice.

This introductory chapter explores the tension between understanding the city as an empirical container (an inert, pre-existing product or thing) and perceiving it as a socio-spatial process (an ongoing production of spatialized relations of power). It starts by addressing the city as container, by outlining the main features of urbanization and concerns that critical urban scholars have with the concept of the **New Urban Age**. The chapter then turns to the development of urbanization, from the emergence of ancient cities to the current globally connected urban fabric. This discussion is followed by a consideration of the city as socio-spatial process in order to raise questions about how the urban has been defined theoretically and what the value-added is of a geographical perspective.

Urbanization in a Rapidly Transforming Urban World

The mid twentieth century to the contemporary period has been witness to a rapidly transforming urban world, marked by geographically uneven change that has resulted in the global redistribution of urban growth. In some places the pace of change has been dynamic, with dramatic changes to the urban landscape (e.g., London, Hong Kong, and Dubai) and the creation of entirely new cities (e.g., Brasilia, Chandigarh, Canberra, Abuja, and Shenzhen). In other places the pace of urban change seems glacial, either tied up in red-taped regulatory frameworks or trapped in a state of decline resulting from decay and neglect. Cities such as those of the North American Rust Belt, the slowly shrinking cities of the former Soviet Union and Eastern Europe fall into this latter category. As low rates of population growth now barely allow some cities and towns in the global North to secure sustainable urban futures, in the global South, natural increase and rural to urban migration has fuelled urban growth, resulting in what UN-Habitat refers to as the New Urban Age. This "age" is estimated to have arrived at some point in 2007–2008, when half of the world's population became urban, an estimate that by 2050 is expected to have risen to two-thirds (Table 1.1).

From the mid-twentieth century onward, demographers, sociologists, and historians engaged in international urban scholarship and policy research have sought to statistically investigate urban population growth in different national contexts so as to determine the size of the world's urban population. Since the 1960s, the United Nations (UN), particularly the UN Population Division, has played a central role in collecting and representing worldwide urban data on city-size distributions and city populations (their most recent publication is *World Urbanization Prospects: The 2014 Revision*, from which the figures below are taken (UN-DESA-PD 2015a, 2015b). What the figures in Table 1.1 show, apart from the fact that urbanization is taking place on a global scale, is that urbanization can be something of a slippery concept, understood in at least three different ways as level, process, and rate.

- The *level* of urbanization refers to a condition at a specific time (usually depicted as the percentage of population living in urban areas of a country).
- The *process* of urbanization is one that occurs over time as a function of the respective rates of change and relative sizes of the urban and rural populations in a country. This process can include reference to migration from rural areas to urban areas, the reclassification of rural settlements into cities and towns, the demographic process of absolute growth in the urban population, and relative growth (i.e., urban growth that is faster than rural growth).
- The *rate* of urbanization refers to the change (increase/decrease) in the proportion of urban population over time, calculated as the rate of change

(growth/decline) of the urban population minus that of the total population.

Intuitively, it may seem that urbanization is the result of urban population growth and rural population decline (which is occurring in many European countries, in North America, as well as in parts of Asia and Latin America and the Caribbean). But urbanization can also occur in situations of both urban population growth and rural population growth if the absolute increase in the urban population exceeds the absolute increase in the rural population (as in most of the countries of Africa and Asia). In rare instances, urban population decline can also be associated with urbanization when the absolute decline in the urban population is less than a concurrent absolute decline in the rural population (as in parts of Eastern Europe) (UN-DESA-PD 2015a).

The large size of the absolute numbers in Table 1.1 can be difficult to grasp. The 100-year period they cover starts in 1950, when the urban transition in the global South was starting, and ends in a few decades, in your not-too-distant future. In 1950 the world's population of 2.53 billion was largely rural (70.4 per cent), with the world's urban population constituting less than 1 billion. By 2050 the world's urban population is estimated to rise to approximately 6.34 billion, a sevenfold increase (66 per cent of the world's total population). This is a reversal of the global rural–urban population distribution of the mid-twentieth century.

With respect to location, since the 1970s the majority of urban growth has taken place in the global South. In the first five decades of the twenty-first century it is estimated that over 90 per cent of urban growth will take place in Asia and Africa alone. The population of the urban global South will have increased from 0.30 billion people in 1950 to 5.23 billion in 2050 (a 17-fold increase). By contrast, in the urban global North the urban population will have increased less than threefold, from 0.44 billion in 1950 to 1.11 billion in 2050. Hence, by 2050 the global South is projected to have 82 per cent of the world's urban population and 86 per cent of the total world population. Correspondingly, the urban areas of the global North will account for only 18 per cent of the urban population of the world and 14 per cent of the world population (UN-DESA-PD, 2015a).

These broad comparisons serve to hide wide-ranging national and regional differences in urbanization, with just over half the world's urban population living in only seven countries in 2014: China, India, the United States, Brazil, Indonesia, Japan, and Russia. The largest absolute increases in urban population between 2014 and 2050 are estimated to continue to be in China, India, and Nigeria, the latter speaking to the fact that although Africa is currently the least urbanized region, its urban

Table 1.1 One hundred years of growth in the world, urban, and rural populations and percentage population increase, 1950–2050 (billions)

Population (billions)	1950	2014	2050 (estimated)
World population	2.53	7.24	9.4–10.0
Global North	0.81	1.26	1.30
Global South	1.71	5.99	8.25
World urban population	0.75 (29.6%)	3.88 (53.6%)	6.34 (66.4%)
Urban global North	0.44 (54.6%)	0.98 (78.0%)	1.11 (85.4%)
Urban global South	0.30 (17.6%)	2.90 (48.4%)	5.23 (63.4%)
World rural population	1.78 (70.4%)	3.36 (46.4%)	3.21 (33.6%)
Rural global North	0.37 (45.4%)	0.28 (22.0%)	0.19 (14.6%)
Rural global South	1.41 (82.4%)	3.09 (51.6%)	3.02 (36.6%)

Adapted from UN-DESA-PD, 2015a, 2015b.

population will triple in this period. By 2050 most of the world's urban population will be concentrated in Asia (52 per cent) and Africa (21 per cent) (UN-DESA-PD, 2015a).

The rate of urbanization in a country tends to be associated with the level of urbanization, with more urbanized countries urbanizing more slowly than less urbanized countries. As Table 1.2 shows, the rate of world urbanization is slowing from a high of 2.96 per cent from 1950 to 1970 to an annual estimated rate of 1.66 per cent from 2014 to 2030 and 1.13 per cent from 2030 to 2050. These figures mask substantial differences, however, between the urban global North and South. Whereas between 1950 and 1970 the rate of urbanization in the global South (4.04 per cent) was nearly double that of the global North (2.08 per cent), between 2030 and 2050 the rate in the urban global South (1.33 per cent), albeit much reduced, is estimated to be five times that of the urban global North (0.27 per cent) (UN-DESA-PD, 2015a). Moreover, while urban population growth is virtually global (only Japan and Russia are expected to experience declines in urban growth), rural growth is not. Around two-thirds of countries will experience reductions in the size of their rural populations in the first five decades of the twenty-first century (especially in China and India), while the remainder will see their rural populations stay the same or grow (especially in Nigeria, Ethiopia, and Uganda), with rural growth peaking in 2020. The global rural population will have increased from less 1.78 billion in 1950 to nearly 3.21 billion in 2050, a near twofold increase, but will experience a decline in absolute numbers between 2014 to 2050 of 0.15 billion. This means that all world growth after 2030–2035 is expected to be urban growth (UN-DESA-PD, 2015a).

A temporal overview of these figures is provided in Table 1.3, which further illustrates features of urbanization by highlighting dates at which major landmarks have been reached. These figures not only show that the world's population continues to urbanize, but also that the number of years taken to increase by 1 billion rapidly decreased in the nineteenth and twentieth centuries (although the number of years is slowly increasing in the twenty-first century). There is agreement that the world's population will increase up to 2050 while after this its trajectory is largely dependent upon fertility rates. In recent years fertility rates have declined in all major areas of the world (except for Europe) as a result of aging populations (especially the increase in the number of people over the age of 60). Hence the rate of urbanization is expected to slow in the future, with both the absolute size and the proportion of the urban population likely to grow less rapidly.

Table 1.2 Annual average rate of change in total, urban, and rural populations, selected periods, 1950–2050

Population (billions)	Average Annual Rate of Change (%)				
	1950–170	1970–90	1990–2014	2014–30	2030–50
World population	1.90	1.83	1.29	0.94	0.63
Global North	1.08	0.65	0.37	0.18	0.04
Global South	2.24	2.21	1.50	1.09	0.73
World urban population	2.96	2.63	2.21	1.66	1.13
Urban global North	2.08	1.06	0.69	0.45	0.27
Urban global South	4.04	3.82	2.88	2.02	1.33
World rural population	1.37	1.30	0.43	0.01	0.23
Rural global North	−0.47	−0.28	−0.58	−0.87	1.18
Rural global South	1.76	1.52	0.53	0.08	0.17

Adapted from Table II.1, UN-DESA-PD, 2015a.

Table 1.3 Milestones in world and urban populations

Total Population			Urban Population		
Population billions	Year when reached	No. of years it took to increase by 1 billion	Population billions	Year when reached	No. of years it took to increase by 1 billion
1	1804		1	1959	
2	1927	123	2	1985	26
3	1960	33	3	2002	17
4	1974	14			
5	1987	13			
6	1998	11			
7	2011	13	4	2016	14
8	2024	13	5	2029	13
9	2039	15	6	2045	16

Adapted from Table II.2, UN-DESA-PD, 2015a.

Troubling the Urban Age Thesis

Why have scholars, governments, and policy-makers attempted to categorize cities as measureable objects? For many reasons it can be essential to know urban "facts." For planners and policy-makers, for example, it can be important to know population size and location when planning services and infrastructure and deciding if a settlement should expand spatially or grow (usually vertically) through the intensification of the existing urban area. Quantifying cities through various rankings has also proved popular with city officials aiming to market their city through its inclusion on such rankings, including those of "most livable city" and the "most prosperous city." But to what extent can these figures of existing and projected urban population levels and rates of urbanization be taken as truths?

There are systematic problems with cross-national data comparability and compatibility, which cast into doubt the claim that we now live in the "New Urban Age" in which over 50 per cent of the world's population resides in cities (Brenner and Schmid, 2014). One of the most obvious problems in comparative analysis is that there is no global definition of what constitutes an urban settlement. The United Nations relies on state-centric data from nationally specific censuses. Yet nation states define what is urban in different ways. For most countries what constitutes an urban area is defined by administrative criteria or *political boundaries* (e.g., an area within the jurisdiction of a municipality or town committee), a threshold *population size* (where the minimum for an urban settlement is typically approximately 2,000 people, although this varies globally between 200 and 50,000), or *population density* (or combinations of these factors). Less common defining features relate to *economic function* (e.g., where a significant majority of the population is not primarily engaged in agriculture or where there is surplus employment), the presence of *infrastructure* related to urban functions (e.g., paved streets, electric lighting, piped water or sewerage), or the presence of *education* or *health facilities*.

Apart from varying definitions across countries, comparison can also be difficult within countries in which definitions have changed over time. Another problem with definitions that are dependent on administrative boundaries is that urban populations often over-spill them, as is the case with informal settlements. Rapid growth can also lead to these boundaries being quickly out of date. The usual data collection and publication period for national censuses (normally 10 years) can also quickly render both boundaries and data obsolete. A further problem arises from the "trustworthiness" and completeness of the compiled data. In some countries the experienced and qualified staff needed to collect and analyze data are lacking, lending little credence to the figures. In others, periods of political upheaval can prevent data from being collected. In China, for example, no census data were collected during the Cultural Revolution (1966–1976). Consequently, the Chinese census of 1982 provided the first comprehensive source of data on the urban population of China since the 1950s. Reclassification of urban and rural areas has also played an important role in estimates of urbanization trends in China since 1980, with huge implications for the measurement of the world's urban population (Wu and Gaubatz, 2013).

Such caveats serve to dampen the dramatic predictions that have been repeatedly made about the worldwide explosion of urbanization. Brenner and Schmid (2014: 740) argue that such widely varying criteria of the urban produces, "an extremely blurry vision of the global urban condition." They reveal and question three assumptions that underlie the Urban Age thesis; namely, that

1. the world can be neatly classified into different settlement types;
2. the urban transition is produced by the redistribution of population from the rural to the urban; and
3. the urban and the rural are distinct and oppositional.

The third of these assumptions is particularly contested with many critical urban scholars arguing that the urban and the rural are intimately connected (Roy, 2015). Households can now not only stretch across continents but also, especially in cities of the global South, maintain strong links with rural areas, with members of the same household strung out along a rural–urban continuum. Furthermore, as many a car bumper sticker reads in Canada's largest anglophone cities, "Farmers feed cities"—a reminder that cities are not autonomous places, but rather they depend upon agricultural practices and resource extraction in rural areas, near and remote, for their existence.

While urbanization is undoubtedly taking place and urban places are essential to the political, economic, social, cultural, and ecological life of the planet, for Brenner and Schmid (2014), the metanarrative of the Urban Age thesis is misleading. Although the thesis captures the ideas of a worldwide shift towards an urban way of life, it uncomfortably combines many different conditions of settlement, population, infrastructure, and administrative organization that may have little in common. The notion of a new urban ascendancy is misleading on many levels, not least because it does not herald "a golden age of human prospect" for everyone (Gleeson, 2015: 1). It is important, then, to reflect critically upon how processes of urbanization emerged on a global scale.

The Development of Global Urbanization

While the Urban Age thesis is contested there is agreement that urbanization is a relatively new phenomenon in the history of humanity. The social and economic changes that brought about the development of the world's first cities were identified by the archaeologist V. Gordon Childe as the first "urban revolution." The origins of these ancient urban settlements, however, are still debated. Scholars disagree on whether some early large settlements constituted cities. They also disagree on whether agricultural activity predated cities, allowing nomadic hunter-gatherers to settle permanently in areas where an agricultural surplus was available to support trade and a relatively large population (the commonly held view) or whether city formation predated agriculture (the view held by, for example, Jane Jacobs [1969] and Peter Taylor [2013]). The population sizes of ancient cities are also debated. The historian Lewis Mumford (1961) claims it is doubtful whether ancient cities had populations much greater than one million.

Ancient cities are commonly considered to date back to 4500 and 3100 BCE in the region known as Mesopotamia (largely corresponding to modern-day Iraq, Syria, and Kuwait). Here the Sumerian and Babylonian civilizations gave rise to prosperous urban settlements in the fertile areas of the region between the Euphrates and Tigris Rivers, home to the ancient cities of Babylon, Eridu, Lagash, Nineveh, Ur, and Uruk (Mark, 2014). Uruk is considered to be the first city in the world for which there is physical evidence, including written documentation, although it is likely that cities such as Jericho (Palestine) and Çatal Hüyük (Turkey) developed as trading cities as early as 7500 BCE. Other major city systems arose in the first century CE: the Mayan cities of the Andes (e.g., Tiwanaku and Wari), the Aztec, Olmec, and Zapotec cities of Mesoamerica (e.g., Tenochtitlan and Teotihuacan), the Kymer sacred cities of Southeast

Asia (e.g., Angkor), and those of sub-Saharan Africa (e.g., Great Zimbabwe and Jenné-Jeno). Growing independently of each other, urban states developed in at least six world areas: Mesopotamia, Egypt, Mesoamerica, the Indus Valley, north China, and the Andes. There was significant diversity found in these ancient cities. "The two best documented ancient urban traditions—Mesopotamia and Mesoamerica—each included small city-state capitals, huge imperial capitals, port cities, industrial towns, and cultural centers" (Smith, 2010: 29). Many cities in these areas became the seats of powerful empires and civilizations (the root of the word "civilization" is derived from the Latin word for "city"—*civitas*). Many of the problems that beset these early civilizations and led to their demise—growing populations, mounting environmental pressures, and potentially unsustainable development—still pose problems today, albeit these are problems now encountered on a global scale.

The seat of the Roman Empire, Rome, was the largest city by the first century CE with over a million inhabitants (Aldrete, 2010). With the exception of Rome in Italy, Baghdad in Iraq, and some cities in China, until the nineteenth century there were no other cities to reach a population size of 1 million. It was with the emergence of bourgeois society in the medieval period in Europe, along with mercantile capitalism, that cities started to develop as commercial centres (Parker, 2015). And yet, even by the start of the nineteenth century only London, UK, and Beijing, China, had populations of over 1 million, and only 3 per cent of the world's population lived in cities. The **urban transition** that was to take place in Europe and North America over the late nineteenth and twentieth centuries was linked closely to the economic development and demographic growth of the industrial revolution that spawned the industrial cities of Manchester, Liverpool, and Birmingham in United Kingdom the Ruhr cities of Duisburg, Essen, Bochum, and Dortmund in Germany, and Chicago, Pittsburgh, and others in the United States.

By the start of the twentieth century 15 per cent of the world's population was urban and the century was to be witness to an increasingly urban world. Building on the urban growth of the previous century in Europe and North America, London and New York City, as well as Tokyo, came to dominate the world stage as **global cities**. The rapid pace of the urban transition observed since 1950 in many countries of the global South, resulted, however, in the urban population in the global South surpassing that of the global North within a mere 20 years (UN-DESA-PD, 2015a). By the latter decades of the twentieth century, some countries of the global North started to experience economic and demographic decline, while urbanization continued in countries of the global South.

The twentieth century also saw the emergence of very large cities, those with populations measured not in millions but in tens of millions, the **megacities** and **metacities** of Southeast and East Asia, Africa, and, to a lesser extent, Latin America. By 2014 there were 28 megacities (with 6 in China), the largest—Tokyo-Yokohama—with a population of over 30 million. Cities familiar to Canadian students—such as Beijing, Cairo, Delhi, Istanbul, Jakarta, Karachi, Lagos, London, Manila, Mexico City, Moscow, Mumbai, New York, Osaka, Paris, Sao Paulo, Seoul, and Shanghai—now compete in size with those far less familiar—such as Abidjan, Ahmedabad, Bengaluru, Chongqing, Dongguan, Faisalabad, Shenyang, and Yangon. In some cases these cities have formed super-sized clusters of cities, giving rise to huge conglomerations of cities (**megalopolises** or **megaregions**). Although urban growth has been marked in academic circles by attention to the growth in size of these megacities, it is important to remember that people live in a huge diversity of settlements. Only 12 per cent of the world's urban population lives in megacities. Nearly 50 per cent of the world's urban population lives in relatively small settlements of less than 500,000 inhabitants, and one in five urban dwellers lives in a medium-sized city with 1 to 5 million inhabitants (e.g., Toronto, Montreal, and Vancouver in Canada) (UN-DESA-PD, 2015a).

Whether in small-, middle-, or large-sized cities, in the twenty-first century, urbanization appears unstoppable. Urbanization (and the increasing levels of **inequality** that it engenders) has led the United Nations, in 2015, to develop the first global goal in relation to cities. It forms one of the **Sustainable Development Goals (SDGs)**: SDG 11 aims to "[m]ake cities and human settlements inclusive, safe, resilient, and sustainable") (ICSU, ISSC, 2015). Furthermore, **Habitat III**'s New Urban Agenda (announced in late 2016) will determine how cities will be analytically located in relation to the SDGs and has the potential to change the normative base of how urban issues are understood and acted upon by national and international bodies (Parnell, 2016). These new global **urban policy** arenas, developed because of the large range of problems that cities pose for sustainable futures, raise the question of whether we now live in an epoch, an **Urban Anthropocene**, in which humans, living in cities, are the dominant drivers of global environmental change and in which cities as the locus of humanity can create tipping points of global sustainability.

In such an epoch, what does it mean to talk about urban settlements? Can the city still function as its normative ideal? Can urban theory still account for the increasing number and diversity of forms of urban settlements, of **urban villages**, **informal settlements**, **suburbs**, city agglomerations, and megalopolises? For example, what about the rapidly developing urban places in the global South for which the catalyst is often human social reproduction, and not the standard late twentieth-century factors used to theorize the shaping and growth of cities—urban **agglomeration economies** and land markets (Roy, 2003)? If it no longer makes sense to continue to analyze urban places as discrete units with distinct boundaries, lying within nested scales, do we need to abandon twentieth-century modes of analysis in favour of radically new ways of thinking about the urban? These are profound and complex questions that have no straightforward answers. Any preliminary answers demand a theoretical engagement with the urban—a consideration of how foundational and contemporary urban theorists have grappled and continue to grapple with the city as a microcosm of larger society (Parker, 2015).

The Nature of the Urban Question

For the most part, urban theorists have shunned knowing about cities as quantifiable objects, so that they might understand cities in terms of their form and socio-spatial processes. This section explores how foundational and contemporary theorists have thought about and defined the urban. Although cross-cut with dubious ethnocentric, androcentric, and Eurocentric understandings of the city, foundational arguments about the urban have been and continue to remain highly influential (Peake and Rieker, 2013).

Foundational Theory

European cities became objects of fascination in the academy in the late nineteenth century during the urban transition, and since the early twentieth century, urban theorists have engaged in foundational debates regarding the nature of modern urban life. They were intrigued by the ways in which the movement of people from rural into urban areas changed habitual ways of life. The German sociologist Ferdinand Tönnies (1855–1936) defined these in terms of the ideal types of **Gemeinschaft** (**community** groupings based on feelings of togetherness and mutual bonds, such as families, in rural areas) and **Gesellschaft** (groups based on instrumental, rational, impersonal, and voluntary forms of interaction typically found in urban areas) (Tönnies, 1887; Tonkiss, 2005). For the German sociologist Georg Simmel (1858–1918), the city was of interest for its effects on the minds of individuals and their **subjectivity**. In an essay entitled "The metropolis and mental life" Simmel (1971: 324) claimed, "[t]he deepest problems of modern life flow from the attempt of the individual to maintain the independence and individuality of his existence against the sovereign powers of society." Simmel also theorized city life as an escape from stifling rural traditions and **surveillance**, as it provided the opportunity to live among strangers. Simmel (1971: 148, emphasis in original) defined the social type of the **stranger** as a synthesis of nearness and remoteness, someone who is "near and far *at the same time.*" Although he was ultimately ambivalent about the effect of the city on individuality, his work is most commonly remembered for his view that city life leads to a **division of labour** and increased financialization, as money becomes the key to exchange. As money takes on increasing importance, he claimed that societal focus shifted to what the individual could do, instead of who the individual was, leading to the development of a blasé (indifferent) attitude to urban life.

The work of Tönnies and Simmel (and that of other sociologists, such as Emile Durkheim and Max Weber) was taken up by members of the **Chicago School**, which since the turn of the twentieth century until the 1930s was most influential in defining the city and its social dimensions, such as **alienation**, "inequality, loneliness in crowds, mixing and segregation, and immigration" (Jonas, McCann, and Thomas, 2015: 18). Inspired by Darwinian theories of natural selection, and by the economic powerhouse of Chicago itself, a small group of scholars at the University of Chicago established what has proved to be an enduring legacy for urban theory. Two of its most renowned members, Ernest Burgess (1886–1966) and Robert Park (1864–1944) developed the idea of the city as an ecological system in which "survival of the fittest" was used to explain the location of the wealthiest groups in prime locations and those most disadvantaged in the most undesirable residential locations. Urban growth was understood to be the result of the ecological concepts of "invasion" and "succession."

In their book *The City* (Park, Burgess, and McKenzie, 1925), Burgess developed the **Concentric Zone Model**, a model that remains the best-known visual representation of the city, although it is important to remember that this model was based upon urban development in one city (Chicago) at one point in time. In it, the city was depicted as growing outward from Zone I to Zone V, from the **Central Business District (CBD)** to the Commuter Zone, in a pattern of radial expansion. Surrounding the CBD was Zone II, a Zone of Transition, where industry encroached upon declining neighbourhoods containing houses that were divided into apartments. In Zone III, the Zone of Independent Workingmen's Homes, working-class families lived in close proximity to factory jobs, while the outlying Residential Zone IV and Commuter Zone V contained higher-quality, single-family homes. Each zone was characterized by the mobility of its residents. Those who lived closest to the CBD were perceived as having the highest levels of mobility as well as the greatest need for affordable housing and social services. The central city was also the first destination of wave upon wave of new immigrants; as the material conditions of immigrants improved, they moved further out into more spacious, better-quality, and newer homes, their places taken by new rounds of immigrants. This model was based upon a number of fallacious assumptions including those of private property ownership and the relative absence of urban planning constraints (Pacione, 2005). It assumed, for example, that property owners were free to develop land and that only the wealthy could desire and afford the cost of residential locations away from the central city. Over time, the Concentric Zone Model was adapted to account for the impact of transportation (Hoyt, 1939) and polycentric urban forms (Harris and Ullman, 1945). Although much critiqued for its understanding of social processes as somehow "natural" (Saunders, 1981) and for its racist and sexist underpinnings (Kobayashi, 2016; Sibley, 1995), the Chicago School's work did emphasize understanding the city as a socio-spatial process, albeit one largely devoid of the role of **capitalism** in urban development.

Although other members of the Chicago School adopted a range of analytical approaches to the study of the city, they shared a belief in the value of collecting empirical data and undertaking ethnographic fieldwork using Chicago as their urban laboratory. One of these members, the German-American sociologist Louis Wirth (1897–1952), in his article "Urbanism as a Way of Life" produced a now classic definition of the urban as a settlement type based on the spatial coexistence of three properties: *population size*, *density*, and *demographic heterogeneity* (Wirth, [1903] 1938). The combination of these three properties allowed for the proliferation of social **encounters** but also the weakening of social ties.

Like the Chicago School scholars, the Marxist theorist Walter Benjamin (1892–1940) was also a keen observer of urban life. His observations focused on life in European rather than American cities, and he emphasized the possibilities the city offered for reinventing itself and its citizens (Parker, 2015). In his unfinished study of the commercial passageways of Paris in the nineteenth century, commonly known as the Arcades Project, Benjamin addressed broad questions of modernity. He approached modernity through a study of urban types, the best known of which is the bourgeois male figure of the **flâneur**, who engages in aimless wandering of the city streets. In a nostalgic search for the hidden wonders of the city, Tonkiss (2005: 125–126) suggests that the "concept of *flanerie* provides a simple, if not always very precise shorthand for cruising and perusing the city . . . a version of subjectivity produced out of the modern city." This character of the flâneur helped to define the modern urban subject. For Benjamin, the city was also a complex site of perception and memory (Tonkiss 2005). His important notion of the history of the present, of "time as filled by the presence of the now" (Parker 2015: 19), was expressed in his view of cities as living archives of the past. For Benjamin, city buildings and spaces form a palimpsest of past meanings, no longer visible but exposable through excavation of the experiential memories they hold.

Other Marxist urban scholars have drawn attention to the public geographies of city life and the role they play in democratic formations and **social mobility** (Brenner et al., 2011). The French urbanist Henri Lefebvre (1901–1991), whose work has been incredibly influential in and beyond Urban Geography and Urban Studies, advocated for a radical urbanism that has encouraged urbanites to rethink what is possible or desirable in and for cities (Lefebvre, 1968). With Paris as his conceptual laboratory, Lefebvre asserted that the city is humanity's greatest creative achievement—it is a work of art that is created and recreated every day by the actions of its inhabitants. Lefebvre posited that all urban spaces—abstract, absolute, or social—are historically constructed. They become and are organized, represented, and given meaning by collective human action; for Lefebvre, **space** and the **spatialities** it produced, was *the* human project. He argued that the social production of space is a top-down conceptual imposition by those in power that can be resisted from the bottom-up through lived experience and the power of the imagination. For Lefebvre, urban space is produced abstractly in plans and policies by urban planners, architects, engineers, scientists, and urbanists—those who have a regulatory role and can over-code space in an effort to maintain authority. But he reminds us that the "textures of the city" are also produced through the socio-spatial practices of the everyday lives of city inhabitants (Hubbard, 2006).

In the late twentieth century, **everyday life**, with its daily routines and rhythms of movement and activity, is where capitalism survives and reproduces itself, but it also creates urban space and, in so doing, becomes the ground for legitimizing urban rights claims. In this zone of everydayness, the critique of the everyday realities of boredom versus societal promises of free time and leisure could lead to people understanding and then revolutionizing their everyday lives. Thus, Lefebvre argued, city residents individually and collectively earn the **right to the city** by living in it. The right to the city is a particular form of **urban citizenship** that gives a political community of urban inhabitants two distinct rights: to appropriate urban space and to produce urban space (Mitchell, 2003). The right of appropriation challenges both the ability of capital to manipulate property rights regimes that validate the exchange value of urban space and the ability of neoliberalism to understand freedom as individual choice. In its place, the right of appropriation elevates use value over exchange value.

Lefebvre's theoretical work also involved rethinking the city as the locus of the urban. In *The Urban Revolution* (2003) he presented a picture of urban modernity in crisis. The industrial city was dismantled and replaced with "complete urbanization." Urban society transcended industrial society as the engine of capitalism in what he referred to as a process of "implosion-explosion." By this he was referring to processes of concentration and dispersion in which cities could be understood as zones of agglomeration that themselves implode, fragment, and destruct while also extending their infrastructural reach deep into previously remote areas (Brenner, 2014). Lefebvre is describing a new global urban imaginary of an urban fabric that has thickened and extended its borders, one that has been recently taken up by contemporary theorists of the urban.

Contemporary Theory

Although not all contemporary urban theory is Marxist, it does pay a large debt to it. Within a Marxist conceptual framework, concern is much less with the ways in which the urban is measured than with how urbanization is understood. Urbanization is considered a complex process of continual socio-spatial transformation that unfolds in contextually specific ways in relation to the socio-cultural and political-economic dynamics of capitalism, producing a variegated geography that extends unevenly across significant portions of the world. The work of the Marxist geographer David Harvey has been extremely influential in this understanding of the city as a socio-spatial process in which capitalism annihilates space in order to ensure its own reproduction. In *Social Justice and the City* (1973), he argues against approaches such as that of the Chicago School or liberal economic analyses. He works instead to understand how the production and reproduction of urban space not only expresses wider capitalist relations but also materially shapes how these relations are played out in society. Urban land markets, he asserts, are connected to the ways in which capital circulates in order to secure the highest profits (Harvey, 1985). Building on this Marxist appreciation of the ways in which capital's search for profit creates and exacerbates inequality in cities, Neil Smith has explored urban growth as a process of **uneven development** (Smith, 1984). His work on the **rent gap**, a process whereby a gap emerges between the value of a property (ground rent) and its potential value should it be redeveloped, in relation to **gentrification** shows urban social change to be a class-based process with winners and losers. Urban development at the scale of the neighbourhood is shown to engage a large number of urban actors—builders, developers, landlords, mortgage lenders, government agencies, and real estate agents (Smith, 1979)—and to be a conflictual process.

A critical investigation of class processes has also been central to the work of the Spanish Marxist Manuel

Castells (1983), who initially defined the city as a "spatial unit of collective consumption," i.e., as the site of the location of goods and services (e.g., transportation, housing, health care, and education) provided by the state to reproduce **labour**. As the state periodically pulls back from the investment of capital in the built environment, it provides opportunities for the development of "pluri-class alliances" (in which members of different classes work together) to protest against such withdrawals. These alliances, labeled **urban social movements**, when aligned with radical left-wing political parties had the opportunity to revolutionize larger society. Feminist scholars critiqued Castells's work; collective consumption, for example, could not account for the pivotal role that women's unpaid labour played in the **reproduction of the labour force** (Peake, 2016). There is no denying, though, that Marxist theorizations of the urban have opened up new (and hopeful) conceptual spaces, challenging inhabitants of cities to collectively produce and govern alternative non-commodified urban spaces, institutional forms, and political frameworks (Harvey, 2012). These demands for access to social wealth that are not mediated by competitive market relations can be seen on the global stage. Such demands manifest in movements such as Occupy, opposition to the privatization of public space, support for affordable housing and basic public infrastructures, and development of the urban commons and **commoning practices** (that include non-capitalist, non-proprietorial independent spaces in cities from food co-operatives to community gardens to alternative living arrangements to alternative currencies and non-monetary trading, amid many other practices) (Bresnihan and Byrne, 2015).

Most recently, Neil Brenner and Christian Schmid (2015) have taken up Lefebvre's thesis of complete urbanization in their theorization of **planetary urbanization**. They argue that new forms of urbanization are emerging that challenge older conceptions of the urban as a fixed, bounded, and universally generalizable settlement type and that a new epistemology of the urban is needed to decipher the rapidly changing geographies of urbanization and urban struggle under early twenty-first-century capitalism. They see the formation of megacities and megalopolises as only one important expression of the ongoing reconstitution of urbanizing landscapes. Its other dimensions include the densification of infrastructural networks, stretching across territories and continents as well as oceanic and atmospheric environments; the restructuring of traditional "hinterlands" through new economic spaces such as new export-processing zones and data-processing facilities; the spatial extension of large-scale land-use systems devoted to resource extraction, energy, and water and waste management; the transformation of vast "rural" areas through the expansion of large-scale industrial agriculture and forms of land grabbing and territorial enclosure; and the commodification of "wilderness" spaces, including the atmosphere itself, to serve the profit imperatives of a planetary formation of capitalist urbanization. Thus Brenner and Schmid argue that these planetary formations mean there is no longer an "outside" to the urban world. Critiques of planetary urbanization have challenged what appears at face value to be the folding of the rural into the urban as well as the seemingly totalizing nature of the theory. For many scholars, Brenner and Schmid's epistemological understanding that the urban is now planetary is an empirical question that remains to be determined (Peake 2015).

Postcolonial scholars have also opened up important new ways of thinking about the city that challenge **developmentalism**. A developmentalist discourse is one in which most cities in the global South are assessed as lacking in qualities of cityness and, in terms of urban theory, can therefore only be understood by their difference from cities in the global North, which form the objects of urban theorizing. Postcolonial scholars argue that the Euro-American experience of capitalist urbanization has been unreflexively used to interpret processes of urban development across the global South and it now needs to be "provincialized," or seen as just one way of understanding the contextually specific urbanization of the global South (Parnell and Robinson, 2012; Sheppard, Leitner, and Maringanti, 2013). The work of Jennifer Robinson (2006, 2011, 2014) and Ananya Roy (2011, 2014) has been particularly important in showing how all theories come from specific contexts. The geographer Jennifer Robinson has called for comparative urban research based on an understanding that all cities are **ordinary cities**. She eschews urban hierarchies for their privileging of those in the global North, the ones at the supposed pinnacle of the hierarchy. The urban theorist Ananya Roy has drawn on the concept of **"worlding"** to explain how cities and their inhabitants in the global South are positioned within a global power regime in which they are understood primarily only through reference to dominant global North perspectives. The sociologist AbdouMaliq Simone (2010) adds to this body of work though his conceptualization of people as infrastructure, investigating ways in which "peripheralized" peoples make livings in ways that most urban theorists have not yet begun to investigate. Scholars working within and across the disciplines of sociology, anthropology, cultural studies, political science, environmental studies, urban studies, and geography have all had valuable roles to play in deepening postcolonial understandings of the urban, questioning the geographical sources of urban theory, and reinforcing a claim for southern urban knowledge. It is to the intellectual distinctiveness of a geographical approach to the urban that the next section now turns. What is the scholarly value-added that urban geographers can provide?

A Geographical Approach to the Urban

Human geographers approach the study of the urban through the conceptual lens of space and **place**, seeking to investigate processes of relational connectedness in terms of both similarities and differences across a variety of scales. Space and place are two of the most important concepts in the social sciences, but it is the disciplines of geography and urban studies that have most robustly theorized their application to the urban. Space is a complex term with multiple understandings. Sometimes it is depicted in absolute terms as a "container" or as unchangeable. Sometimes it is a metaphor (e.g., points, hubs, doorways, gateways, room, edge, boundary, and territory). And other times it is a concept (e.g., the stretching out of power relations). Scholars have largely now departed from a "container" or absolute view of space to a relational understanding of the underlying political, economic, and social processes that create spatial patterns and relationships. That society and space are mutually constitutive is conveyed through the concept of **spatiality**.

Place is commonly thought of in terms of the dimensions of location, locale, and sense of place. While location equates place with position in space, locale consists of "the settings for everyday routine social interaction provided in a place" (Agnew and Duncan, 1989: 2). Locales can exist at a range of scales, from the corner in a room to a local market to a neighbourhood to a city. A third understanding of place, that of a sense of place, engages with human occupation with place, referring to a form of "identification with a place engendered by living in it" (Agnew and Duncan, 1989: 2).

An urban geographical approach is immersed in understanding how economic, political, social, and environmental processes play out across space and in the formation of places. The formation of **neighbourhoods**, for example, will differ depending on their historical geographies as national and global processes (e.g., the international division of labour) manifest themselves differently in local places. An urban geographical approach is considered by many as being primarily concerned with explaining both the locations of urban places relative to one another and the socio-spatial similarities and differences within and between them (Pacione, 2005). This understanding makes a distinction between two important traditions within the study of cities: **systems of cities** and **cities as systems**. The former tradition considers the linkages and interdependencies that connect cities to one another within regional, national, and global urban systems through flows of spatial interaction. The latter tradition interrogates the inner workings and socio-spatial processes that operate within and between cities to influence the locational arrangements of humans, activities, and institutions. It aids an understanding of how space is organized in the **built environment**—across the scales of the body, the **household**, the **street**, and the neighbourhood—in ways that speak to power relations in wider society and the economy. For both traditions, an emphasis on process also allows the dynamism of urban transformation to be investigated, not only in terms of local character but also through the global connections of urban places.

The study of the urban within geography is concentrated within the sub-discipline of urban geography, which has been, and continues to be, practised differently depending on what philosophical paradigm(s) (frameworks for thinking about and viewing the world) the researcher identifies with. Within urban geography a number of dominant paradigms have broached the study of cities differently. From a **positivist** standpoint, cities are studied using **spatial analysis** and the **scientific method** (e.g., hypothesis, data collection to test hypothesis, and theory or model development) to explain or predict phenomena. **Behaviouralism** seeks to modify the intent of spatial analysis by applying it to the study of human behavior and particularly geographical decision-making, with attention to attitudes towards and expectations of places. Just like behaviouralists, Marxists find spatial analysis to be too consumed with the geometry of location and emphasize instead capitalist production and labour relations within land and **labour markets** as key to understanding the underlying causes of urban processes. Where Marxists study contradictions within the structural socio-economic contexts of urban phenomena, humanists focus on how people shape and change cities, paying particular attention to **human agency**, perceptions, and experiences. Building off, while simultaneously critiquing, **humanism**, **structuralism** directs attention to the social, economic, and political structures that produce cities. **Feminism** extends this critique still further, accentuating the gendered relations that underpin how cities are built, managed, and experienced. Where feminists have emphasized how one's **positionality** (e.g., gender, sexuality, age, class, racialization, [im]mobility, health, and religion) changes the lived experiences of cities, postmodernists have accentuated these differences, illustrating how cities are the product of contestation and differing perceptions and representations. Within **postmodernism** and **poststructuralism** there is no singular experience of the city or universal understanding of what the city is; instead it is understood as constructed through **discourses**, **representations**, and varied urban experiences. And, finally, within **postcolonialism**, a Western hegemonic perspective is deconstructed and decentred; imperial legacies and disjunctures of economic, political, and cultural control are critically examined to reveal how historical forms of colonialism have influenced the nature of urbanization and the development of cities in the global South and global North.

Urban geographers cannot all be neatly located within one intellectual paradigm. We, for example, identify as critical urban scholars, as urban geographers engaged in both feminist and queer analyses, with one also deeply committed to critical race and postcolonial scholarship. Each of the contributors to this textbook has a different philosophical position. As you read each chapter, we invite you to consider the potential intellectual alignments of the author(s) and to critically reflect upon how different philosophical and political commitments might influence the arguments made and the content and perspective of the urban snapshot provided.

Key Points

- The twenty-first century has been declared "the century of the city" in which urbanization is occurring on a global scale.
- Cities are complex social, economic, ecological, and political systems.
- Urbanization processes produce variegated socio-spatial forms, conditions, and urban ways of life that require contextually specific analyses and theorization.
- Scholars, governments, and policy-makers have attempted to categorize cites as measureable objects, but there is no global definition of what constitutes an urban settlement and administrative boundaries change, making quantification and comparison within and between cities difficult.
- Urban theorists seek to understand the city as a microcosm of the larger society, both in terms of its form and how this relates to socio-spatial processes.
- Foundational theories of the urban were inspired by Darwinian theories of natural selection and used Chicago as a scholarly laboratory to explore patterns of residential settlement and land use.
- Like the Chicago School scholars, Marxist theorists keenly observed urban life but also emphasized the role of capitalism in urban development.
- Lefebvre's theorizations about the production of urban space and rights of urban citizenship have inspired contemporary Marxist, feminist, and postcolonial scholars to reflect critically upon how urban space is produced and reproduced as an expression of wider capitalist relations, colonial legacies, and social difference.
- A geographical approach to the urban dimensions of space and place is concerned with explaining the locations of urban places relative to one another and the socio-spatial similarities and differences within and between them.

- Urban geographers cannot easily be categorized within one intellectual paradigm: they utilize positivism, humanism, behaviouralism, structuralism, Marxism, feminism, postmodernism, poststructuralism, and postcolonialism.
- The reader is invited to consider what the philosophical position might be of each of the chapter contributors and how that might impact upon the arguments made and the urban vantage point provided.

Activities and Questions for Review

1. In your own words, explain to a classmate what urbanization is.
2. In a small group, create a table that summarizes the advantages and disadvantages of understanding "the urban" as an empirical container versus a socio-spatial process.
3. Make a list of the different urban theories discussed in this chapter. In a second column, identify the key concepts associated with each theory.
4. Debate as a class whether life in urban Canada significantly differs from life in rural areas of the country.

References

Adams, P.C. (2016). Place. In Richardson, D. Castree, N., Goodchild, M., Kobayashi, A., Liu, W., and R. Marston (eds), *The AAG encyclopedia of geography*. Oxford: John Wiley and Sons.

Agnew, J.A., and Duncan, J.S. (Eds.) (1989). *The power of place: Bringing together geographical and sociological imaginations*. Oxford and New York: Routledge.

Aldrete, G.S. (2010). Rome, Italy. In R. Hutchison (Ed.), *Encyclopedia of urban studies*. London: SAGE, pp. 675–679.

Archer, K. (2013). *The city: The basics*. New York: Routledge.

Brenner, N. (2014). *Implosions/Explosions: Towards a study of planetary urbanization*. Berlin: Jovis Verlag.

Brenner, N., and Keil, R. (Eds.) (2006). *The global cities reader*. New York: Routledge.

Brenner, N., Marcuse, P., and Meyer, M. (Eds.) (2011). *Cities for people, not for profit: Critical urban theory and right to the city*. New York: Routledge.

Brenner, N., and Schmid, C. (2014). The "urban age" in question. *International Journal of Urban and Regional Research*, 38(3), 731–755.

Brenner, N., and Schmid, C. (2015). Towards a new epistemology of the urban? *City: Analysis of Urban Trends, Culture, Theory, Policy, Action*, 19(2–3): 151–182.

Bresnihan, P., and Byrne, M. (2015). Escape into the city: Everyday practices of communing and the production of urban space in Dublin. *Antipode*, 47(1), 36–54.

Castells, M. (1983). *The city and the grassroots: A cross-cultural theory of urban social movements*. Berkeley: University of California Press.

Debord, G. (Ed.) (1996). *Guy Debord presente Potlatch*. Paris: Folio.

De Certeau, M. (1984). *The practice of everyday life*. Berkeley: University of California Press.

Gleeson, B. (2015). *The urban condition*. London: Routledge.

Graham, S., and McFarlane, C. (2015). *Infrastructural lives: Urban infrastructure in context*. London and New York: Routledge.

Gros, F. (2014). *A philosophy of walking*. London: Verso.

Harris, C., and Ullman, E. (1945). The nature of cities. *Annals of the American Academy of Political and Social Sciences*, 242: 7–17.

Harvey, D. (1973). *Social justice and the city*. Athens: University of Georgia Press.

Harvey, D. (1985). *The urbanization of capital*. Baltimore, MD: The Johns Hopkins University Press.

Harvey, D. (2012). *Rebel cities: From the right to the city to the urban revolution*. London: Verso.

Hoyt, H. (1939). *The structure and growth of residential neighbourhoods in American cities*. Washington: Federal Housing Administration.

Hubbard, P. (2006). *The city*. London and New York: Routledge.

ICSU, ISSC (2015). *Review of targets for the sustainable development goals: The science perspective*. Paris: International Council for Science.

Jacobs, J. (1969). *The economy of cities*. New York: Random House.

Jacobs, J.M. (2012). Urban geographies 1: Still thinking cities relationally. *Progress in Human Geography*, 36(3), 412–422.

Jonas, A., McCann, E., and Thomas, M. (2015). *Urban geography: A critical introduction*. Oxford: Wiley Blackwell.

Knox, P. (Ed.) (2014). *Atlas of cities*. Princeton, NJ: Princeton University Press.

Kobayashi, A. (2016). Spatiality. In Richardson, D. Castree, N., Goodchild, M., Kobayashi, A., Liu, W., and R. Marston (eds), *The AAG encyclopedia of geography*. Oxford: John Wiley and Sons.

Lefebvre, H. (1968) 1996. "The Right to the City." In E. Kofman and E. Lebas (Eds.), *Writings on cities by Henri Lefebvre*. Oxford: Blackwell, pp. 63–182.

Lefebvre, H. ([1970] 2003). *The urban revolution*. Translated by R. Bonnono. Minneapolis: University of Minnesota Press.

Mackenzie, S. (1989). Women in the city. In R. Peet and N. Thrift (Eds.), *New models in geography: The political economy perspective*. Vol. 2. London: Unwin Hyman, 109–126.

Madden, D. (2012). City becoming world: Nancy, Lefebvre, and the global-urban imagination. *Environment and Planning D: Society and Space, 30*: 772–787.

Mark, J. (2014). The ancient city. In *Ancient History Encyclopedia*. Retrieved from http://www.ancient.eu/city

Mitchell, D. (2003). *The right to the city: Social justice and the fight for public space*. New York: Guildford Press.

Mumford, L. (1961). *The city in history*. San Diego: Harcourt.

Pacione, M. (2005). *Urban geography: A global perspective* (2nd ed.). New York: Routledge.

Park, R.E., Burgess, E.W., and McKenzie, R.D. (1925). *The city*. Chicago: University of Chicago Press.

Parker, S. (2015). *Urban theory and the urban experience: Encountering the city* (2nd ed.). Abingdon, UK.: Routledge.

Parnell, S. (2016). Defining a global urban development agenda. *World Development, 78*: 529–540.

Parnell, S., and Robinson, J. (2012). (Re)theorizing cities from the global south: Looking beyond neoliberalism. *Urban Geography, 33*(4): 593–617.

Peake, L. (2016). The twenty-first century quest for feminism and the global urban. *International Journal of Urban and Regional Research, 40*(1): 219–227.

Peake, L. (2015). On feminism and feminist allies in urban geography. *Urban Geography, 37*(6): 830–838.

Peake, L., and Rieker, M. (Eds.) (2013). *Rethinking feminist interventions into the urban*. London: Routledge.

Robinson, J. (2006). *Ordinary cities: Between modernity and development*. New York: Routledge.

Robinson, J. (2011). Cities in a world of cities: The comparative gesture. *International Journal of Urban and Regional Research, 35*(1), 1–23.

Robinson, J. (2014). New geographies of theorizing the urban: Putting comparison to work for global urban studies. In S. Parnell and S. Oldfield (Eds.), *The Routledge handbook on cities of the global South*. New York: Routledge, pp. 57–70.

Roy, A. (2003). *City requiem, Calcutta: Gender and the politics of poverty*. Minneapolis; University of Minnesota Press.

Roy, A. (2009). The 21st century metropolis: New geographies of theory. *Regional Studies, 43*(6), 819–830.

Roy, A. (2011). Slumdog cities: Rethinking subaltern urbanism. *International Journal of Urban and Regional Research, 35*(2), 223–238.

Roy, A. (2014). Worlding the South: Towards a post-colonial urban theory. In S. Parnell and S. Oldfield (Eds.), *The Routledge handbook on cities of the global South*. New York: Routledge, pp. 9–20.

Roy, A. (2015). What is urban about critical urban theory? *Urban Geography, 37*(6): 180–823.

Saunders, P. (1981). *Social theory and the urban question*. London: Hutchinson.

Self, W. (2003). *Psychogeography*. London: Bloomsbury Publishing.

Sheppard, E., Leitner, H., and Maringanti, A. (2013). Provincializing global urbanism: A manifesto. *Urban Geography, 34*(7): 893–900.

Sibley, D. (1995). *Geographies of exclusion: Society and difference in the West*. London: Routledge.

Simmel, G. (1971). The metropolis and mental life. In D.N. Levine (Ed.), *Simmel: On individuality and social forms*. Chicago: Chicago University Press.

Simone, A.M. (2010). *City life from Jakarta to Dakar: Movements at the crossroads*. New York: Routledge.

Smith, M.E. (2010). Ancient cities. In R. Hutchison (Ed.), *Encyclopedia of urban studies*. Los Angeles: SAGE, pp. 25–29.

Smith, N. (1979). Toward a theory of gentrification: A back to the city movement by capital, not people. *Journal of the American Planning Association, 45*(4): 538–548.

Smith, N. (1984). *Uneven development*. Oxford: Blackwell.

Solnit, R. (2001). *Wanderlust: A history of walking*. New York: Penguin.

Taylor, P. (2013). *Extraordinary cities*. Cheltenham, UK: Edward Elgar.

Tonkiss, F. (2005). *Space, the city and social theory*. Cambridge, UK: Polity Press.

Tönnies, F. (1887). *Gemeinschaft und Gesellschaft*. Translated in 1957 as *Community and Society*. Leipzig, Ger.: Fues's Verlag.

United Nations, Department of Economic and Social Affairs, Population Division (2015a). *World urbanization prospects: The 2014 revision* (ST/ESA/SER.A/366).

United Nations, Department of Economic and Social Affairs, Population Division (2015b). *World population prospects: The 2015 revision*.

Wirth, L. ([1903] 1938). Urbanism as a way of life. *American Journal of Sociology, 44*: 1–24.

Wu, W., and Gaubatz, P. (2013). *The Chinese city*. London and New York: Routledge.

I Urban Development

Shifting Urban Contours
Understanding a World of Growing and Shrinking Cities

Kenneth Cardenas and Philip Kelly

Introduction

Sometime in 2007–2008 the world crossed a threshold. From that point onward, over half of humanity was living in urban areas (United Nations, 2008). This shift in how, and where, people live has perhaps been the defining transformation in twenty-first-century history, and it is accelerating sharply. It has not, however, been a uniform movement towards city life, as patterns and speeds of urbanization have varied significantly around the world. Amid a general trend towards urbanization, some cities have declining rather than growing populations. Nor is urbanization necessarily an unproblematic shift to a better way of life—for many people, cities involve drudgery and hardships that are barely an improvement on a rural existence, and the expansion of urban areas has far-reaching social and environmental consequences.

Nowhere is urbanization more dramatically on display than in China. In 1990 just 26 per cent of the country's population lived in urban areas, but by 2014 this had risen to 54 per cent. It is estimated that 76 per cent of China's population will be urbanites in 2050. The ways in which this process is unfolding give us some clues about how we can approach urbanization in general in a critical and analytical manner.

Canadian writer and journalist Doug Saunders (2010) examines the process of urbanization at the level of lived experiences. He describes a Chinese village called Liu Gong Li in Sichuan province, just south of the megacity of Chongqing. His story begins in the mid-1990s, when the village was still, in some ways, the same as it had been for generations, with land cultivated using basic (and unmechanized) techniques by the same families. But the village was far from being the peaceful rural idyll that one might imagine for an Asian village. It had seen economic hardship and political repression, including many deaths from famine and starvation in the 1950s and 1960s.

With the construction of a road and bridge in 1995 to link Liu Gong Li with the booming city of Chongqing (just six kilometres [4 miles] away) and the opening up of land markets so that agricultural land could be converted to other uses, the village was transformed. By 2010 Liu Gong Li had become a "forest" of apartment blocks and a vast landscape of seemingly random self-built concrete shops and houses. In the intervening 15 years, the village population increased from just 70 to over 10,000, blending together with other villages to form part of Chongqing's mega-urban region. It is no longer a village at all.

The problems of rapid urbanization experienced in Liu Gong Li are numerous—open sewers and inadequate garbage disposal, congested roads, unsafe construction, and a hive of low-paid and unregulated workshops and factories. But at the same time, this newly minted urban space is also a foothold in the city for rural–urban migrants, many of whom live in buildings or neighbourhoods alongside other migrants from their former villages in the rural hinterland. Liu Gong Li represents a pathway out of rural life and into the possibilities—if not for migrants, then for their children—of upward mobility in the city.

The example of Liu Gong Li provides a useful starting point for several issues that will be discussed in this chapter. The main question concerns global patterns of urban growth and decline. While China's rate of urbanization is impressive and far exceeds that of most other countries in terms of the absolute numbers involved, rapid rates of urban growth are a global phenomenon. And yet, while the overall trend is towards increasing urbanization, some towns and cities (referred to as **shrinking cities** especially in North America, Eastern Europe, and central Asia) have seen declining populations in recent years. It is important, therefore, to recognize the dynamic geography of urban population change around the globe. Inherent to this concern is the question of how we define urbanization. At what point does a village engulfed by new growth become urban rather than rural? How do we classify households in more distant villages that are being supported by migrants to the city—are they rural or urban? Given that almost every country has its own unique definition of what constitutes an urban space, urbanization statistics are peculiarly uneven and idiosyncratic and notoriously difficult to compare (Brenner and Schmid, 2014).

This chapter examines the processes that lie behind urban growth (and decline). Increases or decreases in population due to demographic trends or migration flows are key processes causing urban growth or decline. In the case of Liu Gong Li, nearly all growth is due to internal migration within China. In a country such as Canada, most net population growth in cities is caused by international in-migration. But in all cases, behind these population changes are underlying economic and political contexts that drive them in a process commonly understood as **uneven development**. It is therefore important to ask why cities grow and attract new residents in the first place. This chapter also reflects upon some of the negative consequences and planning challenges that result from urbanization and the economic, social, and environmental changes that it entails.

Many of the challenges of urbanization are evident in the case study of Manila. While 12 million people live within Manila's metropolitan borders, the capital of the Philippines sits at the centre of an extended region that is home to twice as many inhabitants. The city and its region exhibit many of the driving forces, and consequences, of rapid urbanization in the twenty-first century. The chapter ends with an examination of issues of growing and shrinking cities within the Canadian context.

Patterns of Urban Growth and Decline

What we know about urbanization processes around the world is subject to inconsistent definitions. When levels of urbanization are discussed, the usual basis for measurement is the proportion of a population that lives in urban areas. But what exactly constitutes an urban area and therefore an urban population? The definition varies across different countries and has also shifted over time. There are, however, a few key variables that tend to be used, outlined in Table 2.1. Only the last of these measures begins to hint at a definition of the urban that relates to the lifestyles and social interactions of people who live in urban or non-urban areas. The lifestyles that are possible in many "non-urban" areas of Canada, for example, with telecommunications, transport links, and retail services, allow many people to pursue lifestyles that are quite urban in every way except for their physical location.

These conceptual difficulties should be kept in mind whenever one reads and interprets claims regarding urbanization. Any standard for defining what is "urban" likely includes populations, landscapes, and lifestyles that other standards might not include, and vice versa. Setting these parameters is very much a social, and thus political, exercise. For instance, designating a place as urban often carries fiscal, administrative, and planning implications; a legally defined "city" might have taxation powers that a "county" does not, and different interests might line up behind either classification.

The *World Urbanization Prospects* report (United Nations, 2015a) is perhaps the best data set available for tracking urban population trends from 1950 to the present and then projecting these trends up to 2050 (http://esa.un.org/unpd/wup/). Among the most important trends that can be seen in this data are the following (also illustrated in Figure 2.1):

- *A greater share of the world's population is living in cities.* The share of the world's population living in cities has steadily increased over the past century, and it is expected to keep increasing in the decades to come. In 1950 about 30 per cent of people worldwide lived in cities. Around 2008 the balance tipped so that for the first time in human history, more than half of the world's population was classified

Table 2.1 Key variables employed to define urban areas

Jurisdictional Boundaries. In some countries, an urban area is taken to be the same as the administrative unit that forms the city government. This can, however, be very misleading and arbitrary. For example, the administrative area of Beijing in 1999 was estimated to be 11 times larger than the actual built-up area (Angel et al., 2011). In many cases, though, the situation is the opposite, with administrative city governments at the core of an urban region representing only a fraction of the actual urban area. The city of Vancouver, for example, had a 2011 population of just over 600,000, but the wider Vancouver region, represented by a Census Metropolitan Area, was home to nearly four times that number (Statistics Canada, n.d.).

Population Criteria. Regardless of where the boundaries are placed around a jurisdiction, there is still the question of whether it constitutes an urban area or not. This is decided on the basis of criteria that vary from country to country, usually based on the total population and the population density (or people per square kilometre). In Canada, an urban area is one with 1000 inhabitants or more and a population density of at least 400 inhabitants per square kilometre. The United Kingdom simply requires a settlement to have at least 10,000 people, while in France a commune is urban if it has 2000 inhabitants or more who live in houses separated by no more than 200 metres (219 yards) (United Nations, 2015a).

Built-Up Area. One way of going beyond arbitrary thresholds is to use satellite images to generate precisely delimited zones of built-up areas comprising pavements, rooftops, and compacted soil (Angel et al., 2011). While such techniques allow the physical urban landscape to be defined, it is also important to note that urban forms can be quite different in various contexts. For example, in parts of Asia with dense rural populations, there are regions around major cities that have intensely mixed landscapes. They juxtapose both urban and rural land uses and are hard to define as one or the other. Geographer Terry McGee called these zones "*desakota*"—combining the Indonesian terms for "village" (*desa*) and "city" (*kota*) (McGee, 1991).

Functional/Sectoral. Another set of criteria for defining the urban relates to the categories of economic activity undertaken and the existence of certain types of services. Urbanization is generally taken to include a shift from resource-based economic activities to those involving manufacturing and service provision. The provision of certain kinds of services such as health care, education, and transportation may also be used as a basis for determining urban status. Finally, deciding the extent of an urban area may be based on its functional integration. In Canada, for example, a Census Metropolitan Area is defined as an urban agglomeration with at least 100,000 people and incorporating areas that have a high degree of integration with the central urban area as measured by commuting flows.

Kenneth Cardenas and Philip Kelly

as urban. By 2050, it is estimated that 66 per cent, or two-thirds of humanity, will be living in urban areas (United Nations, 2015b).

- *The rate of urbanization is accelerating.* While the trend towards ever-greater urbanization can be traced back to the Industrial Revolution in the nineteenth century, what will set the next half-century apart is the pace and scale at which global urbanization will proceed. From 2000 to 2050 the world's urban population is expected to grow by 3.48 billion people. To put this figure in perspective, the world's urban population only reached the 3.5-billion mark in 2010. By 2050 some 6.3 billion people are expected to live in urban areas, a number equivalent to the world's whole population in 2003.
- *The process of global urbanization is uneven.* Countries around the world are not urbanizing at the same rate. The period until 2050 will see some countries become "more urban" much faster than the overall pace. Haiti, Thailand, and China, in particular, are expected to zoom from 35 to 75 per cent urban between 2000 and 2050. Other countries, meanwhile, will see little change in the proportion of their population living in urban areas. This is typical of countries that already have fairly high levels of urbanization, such as high-income countries in Europe. Canada, for instance, is expected to see the share of its urban population grow by only a little over 8 per cent, from 79.5 to 87.6 per cent. Likewise, within national contexts, some cities and regions are growing faster than others, even as other cities and regions see their populations shrink. And, at the scale of individual cities themselves, certain neighbourhoods and districts experience growth or decline that may defy the trend for the city as a whole. Urbanization, in other words, is a geographically uneven process.
- *Most urban growth will be in the global South.* A little more than 93 per cent of urban growth from 2000 to 2050 is expected to take place in the global South, what the United Nations refers to as less-developed regions. Urban areas in these parts of the world will grow from 1.97 billion to 5.22 billion people (a staggering 165 per cent increase) in the first half of the twenty-first century. Urban populations in

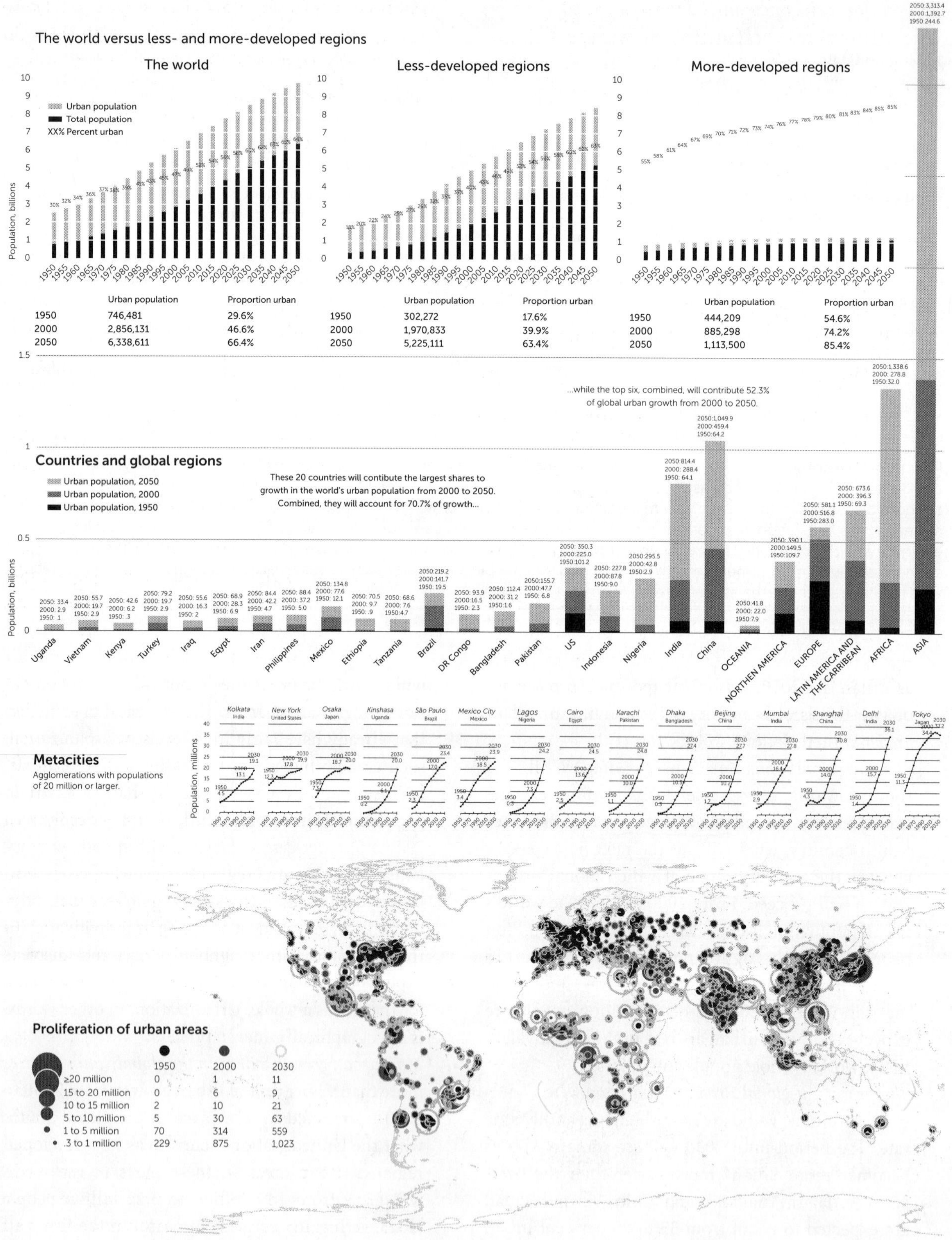

Kenneth Cardenas

Figure 2.1 Global urbanization, 1950–2050

more-developed regions, on the other hand, are projected to increase by only 25 per cent, from 885 million in 2000 to 1.11 billion in 2050. When these figures are broken down by continent, Asia's dominance is clear. Of the total urban population growth, 55 per cent (or 1.92 billion) will be in Asia; 1.06 billion (30 per cent) will be in Africa; and 277 million (8 per cent) will be in Latin America and the Caribbean. North America and Europe will contribute only 4 per cent and 1.8 per cent respectively to global urban growth. Both China and India will add more than 500 million people to their urban populations between 2000 and 2050, meaning that both countries will individually contribute a larger share to global urbanization than Latin America and the Caribbean, North America, and Europe combined.

- *Some large cities will become megacities.* Population projections foresee some cities becoming truly enormous (a **megacity** is usually defined as a city with a population over 10 million). From 2000 to 2030, Delhi in India is expected to register the largest growth in its population (Figures 2.1 and 2.2). It will grow by over 20 million people, to reach a size of 36 million by 2030. This means that almost as many inhabitants will be added to Delhi's population as will be added to the entire urban population of Western Europe (which is projected to grow by over 22 million). Nine other cities are expected to add more than 10 million to their populations from 2000 to 2030. All 10 cities in this particular club are in less-developed regions: Beijing, Shanghai, and Guangzhou (China); Delhi and Mumbai (India); Dhaka (Bangladesh); Kinshasa (Democratic Republic of the Congo); Cairo (Egypt); Lagos (Nigeria); and Karachi (Pakistan).
- *The fastest growing urban areas are smaller and newer cities.* Despite the prominence of a few megacities, most of the fastest-growing cities in the twenty-first century will be smaller cities in less-developed regions. There are 208 cities that are expected to at least triple in size from 2000 to 2030. Of these, 180 had populations below 1 million in 2000. The city that is expected to register the fastest growth rate over this period—Myanmar's new capital, Nay Pyi Taw—did not even exist in 2000. The proliferation of small, rapidly growing cities in less-developed regions has important implications for **infrastructure**, planning, and growth management, as many such cities will not be well equipped to handle such growth.
- *Some cities are shrinking.* Between 2000 and 2030, 65 cities are expected to see their populations decline. Of these, the vast majority (39) are in the former Soviet Union, with Russia (22) and Ukraine (11) containing the most. Other shrinking cities are in western and southern Europe (8), East Asia (7), the United States (2), and the Caribbean (2). They include, for example, Leipzig (Germany), Athens (Greece), Vilnius (Lithuania), Donetsk (Ukraine), and Detroit and New Orleans (the United States). In most cases, these cities are declining either because of overall demographic declines in the total national population that are not being offset by in-migration, or because of out-migration due to long-term economic decline in specific cities. In the case of the United States, for example, Detroit's total population will fall slightly because of a declining manufacturing sector.
- *Cities are spreading to become mega-urban regions.* As larger, more-established cities continue to grow and as smaller urban areas proliferate along transportation corridors, the world is seeing a growing number of **megalopolises**—areas of continuous urbanization that have grown to encompass what were previously distinct conurbations. The most obvious feature of these megalopolises are vast

Figure 2.2 Dense, low-rise buildings in Delhi, 2013

expanses of built-up areas, but they may also be connected together by frequent and high-speed intercity passenger rail or air service. Examples include the Bosnywash corridor in the northeastern United States, extending from Boston to Washington and including New York City, Philadelphia, and Baltimore, and the Taiheiyō Belt in central Japan. In other cases, very dense "rural" populations may constitute large **peri-urban** regions around established cities (e.g., **desakota** in densely populated, lowland, rice-producing areas of Asia). On the island of Java in Indonesia, for example, with an area of over 120,000 square kilometres, (46,332 square miles) rural population densities are so high that the entire island would constitute an urban population by many measures. Such areas tend to develop in places with forms of agriculture that support high levels of population, especially wet rice cultivation.

Drivers of Urban Growth and Decline

The causes of urban growth and decline can be examined at many different levels. The growth and decline of cities occurs in sync with the demographic patterns of the societies in which they are embedded. Hence birth rates and death rates are key drivers of growth or decline. Even more importantly in many cases, urban growth and decline is also the product of inward and outward migration, both within and between countries. While demographic trends take shape over a generational time scale, it is migration that largely accounts for dramatic changes in urban population. However, neither demographics nor migration really identify the root causes of urbanization—the processes that cause people to leave rural areas and move to urban centres or to move from one urban centre to another. To understand these processes, we need to examine the ways in which urban economies work. In this section, we will examine all of these drivers: demographic changes, migration, and urban economies.

Demographic Change

The patterns described in the previous section are partly a reflection of the pace and scale of population growth from the mid-twentieth century onwards. The 20 countries that are expected to account for 70 per cent of total urban population growth from 2000 to 2050 are also expected to account for 68 per cent of total global population growth (United Nations, 2015b). The fact that the largest and fastest-growing cities and urban populations can be found in Asia and Africa is partly a consequence of the "**youth bulge**" found in national populations on these continents. Many countries in Asia and Africa have large cohorts of people born during periods of **demographic transition**, when reduced mortality rates, owing to improvements in maternal and infant care, better nutrition, and public health, are coupled with high birth rates to cause rapid population growth.

Conversely, negative rates of natural increase where prevailing death rates are higher than birth rates will be important drivers of city shrinkage over the next half-century in some countries. Of the 18 countries that are expected to see their urban populations shrink from 2000 to 2050, 16 will see declines in their overall populations as well. The populations of these societies are characterized by low fertility rates and aging demographic structures.

Nonetheless, rates of natural increase do not completely explain country-level trends in urbanization. A comparison of China and India provides a good illustration of this point. As a consequence of a rapid reduction of fertility rates in the late twentieth century—partly due to the recently rescinded one-child policy but also to a pre-existing trend towards lower birth rates linked to economic development—China's population is forecast to hit a peak at some point between 2030 and 2035 and then to decline slightly (Whyte, Feng, and Cai, 2015). China's urban population, however, is set to continue rising, even as the country's total population stabilizes. Indeed, China's urban population will grow by 590 million between 2000 and 2050, while its total population will only grow by 70.5 million. This suggests that China's urbanization is taking place alongside a fairly aggressive depopulation of its rural places (or, as the example of Liu Gong Li suggested, the urbanization of some rural places).

India, on the other hand, is expected to see both its total and urban populations increase steadily over the first half of the twenty-first century. While about 80 per cent of its population growth is expected to take place in its cities, there will still be significant growth in rural areas. India's rural population will grow from 754 million in 2000 to 893 million in 2030 before declining to 805 million by 2050. Despite contributing the second-largest share to global urbanization over this half-century, India is only expected to hit the 50 per cent urban mark by around 2050 (United Nations, 2015b).

These differences indicate that population growth—or decline, for that matter—does not reliably and predictably correspond to a pattern of urban growth. Important differences at the scales of countries, and of individual cities, matter. This suggests that other factors, apart from demographic transitions, need to be considered.

Rural–Urban Migration

In countries where urbanization is proceeding rapidly, the bulk of new urbanites are migrants from areas still designated as rural. This is a worldwide phenomenon but one that is often difficult to measure accurately for several reasons. First, in some cases new migrants to urban areas are undercounted by census surveys because they live in informal settlements and do not have documented ownership or rental rights to their homes. Such residents will also often avoid detection by census enumerators or other forms of outreach by the state. Second, some residents may not be formally recognized by the state as residents because they lack a required formal permit to live in the city. In China, such formal residency permits, or *hukou*, are required to access a variety of social services but are difficult to obtain for low-skilled migrants from rural areas (Chan, 2009). Finally, rural–urban migrants may circulate frequently between rural and urban areas, so their movement is sometimes better characterized as short-term mobility rather than long-term migration (Fan, 2008). New urbanites might return to rural communities on a regular basis, for weekends, public holidays, or extended sojourns in between jobs. China again provides the most dramatic example of this phenomenon. Around the time of Chinese New Year or Spring Festival, the largest recorded human migration takes place annually, as many of China's 140 million internal migrants return to their home villages from industrial cities in the south and east of the country.

As noted above, the places in a city where new migrants from rural areas are able to reside are often informal and characterized by low incomes. These areas have been given many labels, including "arrival cities" (Saunders, 2010) and "**gateway cities**" (Montsion, 2009), as they provide an accessible entry point into the social, economic, and political life of the city. Examples range from the improvised informal settlements of Rio de Janeiro's "**slums**" to the multicultural high-rise neighbourhood of Thorncliffe Park in Toronto. What all have in common is that they represent a staging ground in between migrant families' places of origin and their integration into a new urban space.

Another feature of rural–urban migration is that it is predominantly undertaken by young people. In Indonesia, for example, in the period 2005–2010, 65 per cent of migrants within the country were aged 15–34 (compared to only 39 per cent of the non-migrant population) (Sukamdi and Mujahid, 2015). While this creates a youthful and energetic population for growing cities, it also means that rural areas are increasingly home to the elderly segments of the population. The result is that even in countries with growing populations, agricultural labour can be hard to find and the social context of rural areas is profoundly changed. In the long term, as small-scale farmers become too old to cultivate the land and their children are absent, the basis for rural life and production is often threatened.

The Economic Drivers of Urban Growth

Processes of demographic expansion or contraction explain only a part of the complex processes of urban growth or decline. More important is the existence or absence of economic opportunities that will lead to migration into, or out of, urban areas. But what leads to expanding economic opportunities? This question needs to be broken up into two parts. First, why does economic growth tend to occur in cities? Second, why do some cities grow economically more than others?

The question of why economic growth occurs in cities has three answers. First, economic advantages in the size of population and the concentration of enterprises are found in urban areas. Some of these advantages relate to the size of a market for products and services or to the presence of suppliers—both helping new enterprises to develop. These economic advantages are called **agglomeration economies**—meaning that there are cost savings or other benefits that result from being part of a cluster of activities. The agglomeration of people and firms also means that urban areas are where new ideas develop and spread. This happens because concentrations of people working in the same industry will spread knowledge and innovations among themselves. This spread of information is often in the form of **tacit knowledge**, meaning it comprises skills and ways of doing things that are not easily transmitted or codified. These kinds of processes make cities especially conducive to new ideas and innovative ways of doing things and therefore to economic growth (Wolfe, 2014). Specific cities may also become specialized in particular forms of economic activity—usually called industrial clusters—and further growth will be spurred by their reputations. Such cities may also have a large pool of discerning consumers who will force improvements and innovations in the products and services being made. A classic example of this process would be Los Angeles, where the film industry represents a globally important cluster of economic activity that has contributed to the city's growth for almost a century (Scott, 2004).

The second reason for economic growth in cities is the concentration of human talent that they represent. Skilled workers and innovators tend to be disproportionately found in urban centres, and their presence attracts other talented individuals. In some cases, these concentrations lead to growth in particular industries. In Bangalore, India, the concentration of software engineers and other high-tech workers means that growth has been driven by IT-related firms. Mumbai, on the other hand, has become a huge magnet for Indian talent drawn to its media cluster, encompassing films, television, and digital media, collectively known as Bollywood (Punathambekar, 2013). For some authors, the ability of urban centres to attract talented individuals who generate new ideas—the so-called **creative class**—is the fundamental basis for urban economic growth (Florida, 2012). The idea that creative professionals drive growth for everyone else does, however, have a number of critics, who point out that when translated into urban policies it privileges the needs of already privileged professionals (Peck, 2005).

The third reason for urban economic growth is that cities represent concentrations of infrastructure and institutions. These facilities range from international airports and seaports to universities to public or private sector financial institutions. For many economic activities, being close to these facilities or services is important and so location in urban areas is logical. Economic growth in the financial sector of London, UK, for example, is driven in part by the previous two factors—clusters of similar activities fostering more of the same and concentrations of talent—but proximity to key financial sector regulators, international transport hubs, media outlet and other forms of infrastructure has also facilitated its growth and dominance (Figure 2.3).

Figure 2.3 The global cluster of financial services in London

The factors mentioned here all explain why cities tend to be sites of innovation and clustering for economic activities and new growth and therefore attract in-migration both nationally and internationally. They do not, however, fully explain why some cities may grow or decline relative to others. That phenomenon relates to a wider process of uneven development in a capitalist system (Scott, 2012). Whether at a national scale or across global space, cities increasingly represent sites that mobile investment or financial flows can enter or leave according to the opportunities for profit that exist in a given place. Although some cities have remained consistently at the forefront of global development for many decades (e.g., New York and London), others have grown rapidly in importance as capital has sought emerging possibilities for profit. Such a process has been referred to by some geographers as a **spatial fix**: over time, as one urban centre becomes less suited to future profitability, economic growth starts to shift somewhere else. This explains why some cities decline over time. Liverpool, UK, for example, was a major centre for shipping in the nineteenth and early twentieth centuries, but was superseded by newer, larger, and therefore more profitable ports that could handle larger container traffic. Another example is provided by Detroit, United States, which depended heavily on a large auto manufacturing base but suffered as employers have found it more profitable to relocate their production to new places—for example, in countries of the global South, the southern United States, or in Mexico. In this way, the relative importance of different centres in an urban system at one point in time can shift as new centres of profitability open up. At a global scale, the rise of Chinese cities could therefore be viewed as just the latest frontier in an ever-shifting capitalist landscape of uneven development.

In sum, while population growth and rural–urban migration might be indicators of urbanization, it is the underlying drivers of economic growth that propel the process. These drivers include the agglomeration benefits of clusters of activity, the development of concentrations of talented workers, and the creation of urban infrastructure to support economic activity. The economic lives of cities are, however, never static. They are always in a dynamic process of development and decline as the shifting landscape of capitalism is made and then remade. Nor, as the next section reveals, does growth come without significant social and environmental costs that are usually borne unevenly by different social groups.

Challenges and Opportunities of Urban Growth

What truly sets current urbanization apart is the sheer pace and scale of urban growth that often pose significant challenges. Many issues would be familiar to big cities all over the world; dealing with air and water pollution, providing services, ensuring efficient transportation, and implementing land-use planning are key features of urban life and politics everywhere. At the same time, however, these issues can take on a very different character in megacities (populations above 10 million) and **metacities** (populations above 20 million). At present, the busiest rapid transit systems in the world have annual ridership figures that run into the billions, the largest landfills in the world process in excess of 10,000 tons of trash a day, and the largest urban police departments have employees numbering the size of small national armies. The cities whose populations are now reaching 10 and 20 million are in largely uncharted waters: the first megacity only came into existence in the mid-twentieth century, and the largest cities of the global South have only been metacities since around 2010 (United Nations, 2015b). The very wide wealth disparities among these cities, from Tokyo at one end to Kinshasa at the other, will make it difficult to create solutions that can apply to all of them.

These issues will play out very differently for the large number of newer, smaller, and rapidly growing cities whose infrastructures, governments, and citizens are being pressed to respond, within just a few years, to the same level of urban growth that took decades in their older counterparts. In a lot of cases, official designations matter here: a place that is officially classified as a "village" might have to deal with issues in the near future that are beyond the fiscal and planning capabilities laid out in its official mandates. When Toronto inaugurated its first subway line in 1954, it had a population of about 1.5 million. Most of the cities that will see their populations grow to this size over the next few decades have neither the local revenue base nor the support of national governments to invest in expensive rail-based rapid transit systems. Consequently, established planning and urban governance practices that were developed in response to urbanization in the nineteenth and twentieth centuries might not be particularly useful for these cities.

Other issues and challenges will relate to *where* the urban growth and decline is taking place. The massive growth in urban populations in the global South will take place alongside other significant social transformations. The urbanization of their populations will also entail the **urbanization of poverty** (Ravallion, Chen, and Sangraula, 2007); for much of the twentieth century, most poverty was located in poorer rural areas. As more people move to and are born in their cities, these countries will also see a shift in the demographics of poverty incidence away from their rural areas and towards their cities. While there is some overlap between urban and rural poverty, the two pose very different challenges and may involve very different causes and consequences. For instance, being cash-poor in an agrarian community is very different from being cash-poor in the city, where more aspects of social life, from rent to food and water, are subsumed within the monetized market economy. The growth of slums is another distinctly urban phenomenon that will be a significant feature of urbanization in the countries of the global South. As most of these countries were predominantly rural until recently, their economic and development policies are often geared towards addressing rural development. These policies will have to be retooled, if not redesigned from the ground up, to address urban poverty.

Perhaps counterintuitively the urbanization of poverty is accompanied in many cities by growth in the number of middle-class residents, who in turn are experiencing increased buying power. However, in other cities, such as in the Anglo-Caribbean, there is a decline in the number of middle-class residents as many fall into poverty themselves or embark on migration to join their country's diaspora. But as the overall population of middle-class urbanites in cities in the global South grows, problems such as sedentary lifestyles and obesity, a greater dependence on private vehicles for transportation, higher levels of solid-waste generation, and a larger carbon footprint will present themselves at a much larger scale.

Some attention must be paid to how global urbanization relates to broader twenty-first-century trends. The juxtaposition of glitzy malls with shantytowns found in many cities in the global South hints at one such trend: that fundamental differences in how economies are organized and wealth is distributed set this phase of urbanization apart. During the twentieth century, the experience of developed societies showed that economic growth and urbanization correlated with broad improvements in the quality of life for large numbers of new urbanites. It was expected that the same will take place in the global South, and indeed, this was once a very influential idea. Yet instead of converging on the paths that were taken by cities in the global North, the economic growth and urbanization in the global South that we are seeing have not necessarily produced the same outcomes in well-being. This indicates that growth and urbanization are taking place alongside higher levels of **inequality**, with much of the new wealth and opportunity concentrated in a smaller portion of the population (Shatkin, 2007).

Demographically, the youth bulge that underpins the growing populations of global South countries can yield a "demographic dividend," in which high employment levels and a young population combine to deliver high economic growth rates. If opportunities for education and employment are not available, however, a youth bulge can instead turn into a "demographic bomb," in which high levels of youth unemployment and poverty feed into wide dissatisfaction and political instability. In either scenario, cities will play an important role, either as engines of growth or as focal points for dissent and upheaval.

The role of information and communications technology (ICT) in urbanization is both difficult to predict yet hard to overstate. At the least, ICT created new sunshine industries in cities that serve as hubs for technology companies (such as Kitchener-Waterloo, Canada), as concentrations for startups (such as Silicon Valley, USA), or as sites for IT-enabled services **offshoring** (such as Bangalore, India). While it created significant and sometimes unexpected sources of growth and jobs for such cities, the role of ICT in these examples is hardly revolutionary and in fact can be seen in terms of well-understood processes. Technology and startup hubs are fairly straightforward examples of industrial clusters. Services offshoring, meanwhile, applies the same logic of the spatial fix, originally applied to shifting manufacturing centres to countries with cheaper production costs, to forms of service work that can be performed remotely: tech support, animation, software development, and even legal and accounting services.

What remains to be seen is how ICT could fundamentally alter the ways in which cities themselves are built and organized and how people live in them. The investment in so-called **smart cities** by companies such as IBM seeks to apply developments in real-time remote monitoring and control, embedded computing, connectivity in physical infrastructure (the "Internet of Things"), and the collection and analysis of big data into urban planning and management. Similarly, developments spearheaded by Google in self-driving cars, navigation technologies, and ride-sharing apps are some examples of what may turn out to be an ongoing revolution. While these technologies may promise cleaner, more efficient cities, the experiences of many cities over the twentieth century show that top-down technological solutions cannot be exclusively relied on to "solve" inherently complex social and political processes and often give rise to their own set of unintended consequences. The self-driving car, for instance, might lead to less turbulent traffic flows and lower overall demand for parking space, even as it tracks and compiles the movements of all of its subscribers.

Climate change will also shape both patterns of urbanization and the ways in which cities themselves are built. Some of the largest cities in the world are on low-lying coastal areas and river floodplains: in cities like Dhaka in Bangladesh, Jakarta in Indonesia, and the Pearl River Delta in China, rising sea levels, storm surges, and flash floods are fast becoming important concerns. Rising sea levels and unreliable growing seasons might give rise to climate refugees as another factor behind urban growth. Extreme weather, such as tropical storms that are either stronger or take unusual paths, heat waves, and droughts will come to be important considerations for cities—especially cities in which large numbers live in vulnerable conditions or where the infrastructure is not built in anticipation of extreme weather. We explore a number of these issues in a case study of Manila, in the Philippines, a rapidly growing megalopolis in the global South that illustrates the patterns of global urbanization in the twenty-first century, the processes that underpin them, and the problems and opportunities that arise.

Manila: The Challenges Facing a Rapidly Growing Megacity

Having reached a population of 10.2 million in 2005, Manila is one of 22 megacities in the global South today and 1 of 16 in Asia (Figure 2.3). Its population is expected to continue growing well into the twenty-first century.

Creating jobs and providing services for young populations are ongoing challenges for all countries with demographic profiles similar to the Philippines. As their young populations enter the workforce, there is great potential for these countries to enjoy gains in wealth and well-being, but only if opportunities for education, employment, and socio-economic inclusion are made available. Cities in such countries play a key role in determining whether this potential is realized. The disparities between rural and urban areas often mean that the population's ability to pursue higher education, find work, and avail themselves of important services entails moving to cities. These disparities can be particularly acute in countries where urbanization is dominated by a **primate city**. In such cities, standards of living for its elites can equal and even exceed those seen in cities of the global North, even as other regions of the country struggle to generate wealth and provide basic services. This is certainly the case with Manila: its **extended metropolitan region** has consistently accounted for between 50 and 60 per cent of the entire country's gross domestic product, despite accounting for only about a quarter of the national population (Philippine Statistics Authority, 2012, 2015). Infrastructure, higher education, advanced medical care, and government services are likewise concentrated in this region.

The primacy of Manila has continued, and in some aspects has even deepened, as economic activity in the Philippines shifted to manufacturing, services exports, and the deployment of overseas labour. **Export processing zones** (EPZs), set up in the 1990s on the southern flank of the Manila region, now generate 45 to 55 per cent of the country's total merchandise exports (Kelly, 2013). Proximity to two container ports, an international airport, and a pool of skilled labour were key factors to the success of these zones in this area. An ongoing boom in IT-enabled services offshoring that began in the early 2000s is likewise concentrated in the city's business districts, with 75 per cent of the jobs in this industry being in Manila (Figure 2.4). Most of these, in turn, are concentrated in a few privately developed enclaves that provide reliable, high-quality uplinks to communications networks (Kleibert, 2015). Finally, Manila has served as the primary staging point for the large numbers of Filipinos who go overseas to work or settle; these migrants, in turn, play important roles in offsetting labour shortages in economically expanding (but demographically shrinking) cities around the world, from the Canadian prairies, to the Persian Gulf states, to Singapore. For prospective overseas workers, Manila and its adjoining regions host the largest concentration of higher-education, language-training, and testing facilities that provide the accreditations necessary for work placements; the government and consular offices that process their work permits and visas; and the hotels, factories, and hospitals that are often their first workplaces before they embark on their overseas contracts. Meanwhile, a sizable proportion of the remittances they send to the Philippines end up in Manila and its peripheries, whether in the form of investments in condominiums in its booming property sector, consumption in its malls, or expenditures on health care and education in its hospitals and schools (Pernia, 2007).

As a consequence, sharp differences in opportunities and outcomes between Manila and the rest of the Philippines persist: average per capita incomes in the city are 3 times that of the Philippines' national average and 6 to 10 times that of its poorest regions (Philippine Statistics Authority, 2015). From 2000 to 2010, average per capita incomes in the city grew 80 per cent, compared to a national average of 32 per cent. These disparities go a long way to accounting for both push and pull factors for rural–urban migration to Manila and its continued growth in the twenty-first century.

This type of urban growth in primate cities, driven by the concentration of wealth in one place and the impoverishment of other places, is often interpreted as

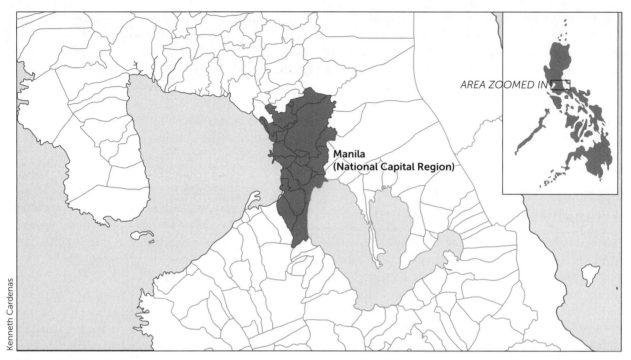

Figure 2.4 Map of Manila, the Philippines

a negative process that produces crowded environments with strained infrastructures. Cities like Manila were once widely viewed as examples of **over-urbanization**, involving rates of urban growth that were higher than what would be expected based on levels of industrial growth (Sovani, 1964). Seen in this light, such patterns of urbanization need to be corrected, either by encouraging growth in secondary cities and rural areas or by dissuading people from moving into the primate city in the first place. But cities such as Manila are also sites of possibility and mobility, and it is important to recognize this for two reasons. First, the simplistic narrative of "congestion" and "overcrowding" can lead to an exclusionary view of the city, implying that only some people, and not all, have a claim on the benefits of urban living. In Manila such exclusionary rhetoric has long been used as the pretext for discriminatory practices, policies, and attitudes towards the urban poor, ranging from forced slum clearances done in the name of "city beautification" to "*balik probinsya*" (back to the countryside) schemes based on the idea that the urban poor would be better off living in rural areas (Arn, 1995; Lakha and Pinches, 1977). Such attitudes have been especially evident after flooding episodes, which have been blamed on slum dwellers living along the city's waterways and drainage canals (Bordadora, 2011). Yet, it is precisely their marginalization that has left the city's poorest families most vulnerable to the effects of extreme weather events, climate change, and natural disasters.

A second reason to reject narratives of over-urbanization is that cities provide unique opportunities for upward mobility for migrants from rural areas and other systematically marginalized groups (as noted at the start of this chapter). In the city these groups are sustained by a range of practices that often take place within familial networks and that are unplanned, illegal, or informal, such as living in slum housing or engaging in informal work, unregistered micro-businesses, or market gardening. In Manila a negative view of rural–urban migrants also fails to appreciate the impact they may have on development in poorer regions. Up to 8 per cent of households in poorer regions of the Philippines receive remittance income from other parts of the country (Kelly, 2009), and remittances have been shown to raise incomes by up to 21 per cent for receiving households in the poorest income quintile (Pernia, 2008).

Similar themes are at play when we look at another aspect of urban life in Manila: the persistence of slums alongside an ongoing real estate boom and the resultant socio-economic segregation. At first glance, these issues may present themselves as merely a symptom of over-urbanization. A closer look, however, will reveal that they are bound to many of the processes at the heart of global urbanization in this century, and that in some cases, what might intuitively appear as a problem might actually be an opportunity for creatively rethinking how, and for whom, cities grow, are built, and are transformed.

As in many Asian cities, Manila's skyline is presently being remade by numerous construction cranes, and a property boom is in full swing. From 2000 to 2014, 40.6 million square metres (48.6 million square yards) of new residential construction was built within Manila and its extended metropolitan region, accounting for 51 per cent of all residential construction in the country (Housing and Land Use Regulatory Board, n.d.; Philippine Statistics Authority, n.d.). Much of this new construction is driven by demand for self-contained enclaves that host the inward flows of foreign investment and international remittances from Filipinos living and working overseas.

Yet it is also fairly obvious that Manila has a housing crisis: about 40 per cent of its population lives with some form of housing deprivation. The ongoing property boom has done little to address this issue, as within the core areas of the city, most of the new construction targets affluent market segments, such as expatriates

Figure 2.5 Workers outside a new call centre in Manila

working in the offshore services industry (e.g., call centres) and households receiving remittances from overseas (Figure 2.5). The lifestyles and skylines found within these globally oriented enclaves, such as in the financial district of Makati and the newly developed Bonifacio Global City in Taguig (Figure 2.6), might invite favourable comparisons with more prosperous Asian cities, such as Singapore—indeed, their developers often invoke aspirations to this mode of "global city"-hood in their designs and their marketing materials. Meanwhile, the population living in Manila's slums rose by 7.3 per cent from 2007 to 2011, some of the best years of the ongoing

Figure 2.6 Taguig, Metro Manila, with the new Bonifacio Global City and the financial district of Makati rising in the background

property boom (LGI Consultants, 2015). Moreover, the demand for urban land has created new problems, as the upward pressure on the price of land has made it too expensive for organizations working with Manila's urban poor to acquire it for socialized housing or in-city relocation. Low-income groups are thereby excluded from the core and displaced to the peripheries of the city, where commutes from new suburbs to workplaces in the core may take up to three hours (Kleibert and Kippers, 2015). In the absence of investment in rapid transit infrastructure, intensified development in outer areas threatens to lock the city into a car-dependent transportation system, even as demand for peri-urban land spurs the dispossession of land from farmers (Kelly, 2003; Ortega, 2012).

Economic exclusion in Manila is also often, but not always, matched by exclusion from planning and decision-making processes. In Manila, people's participation in urban planning is a highly uneven process with unpredictable outcomes, and some communities have understandably grown weary of rhetorical commitments to these concepts. Nonetheless, there are success stories that offer lessons for mobilization and engagement. Most recently, the story of Alyansa ng Mamamayan sa Valenzuela at Caloocan (Alliance of Citizens of Valenzuela and Caloocan, AMVACA), an organization of inner-city slum dwellers, demonstrates how these challenges can provide an opportunity. Its members originally lived on a high-risk site along a river that experiences flashfloods after heavy rain, and their community was often the target of forced evictions. After being served another notice of eviction in 2012, the community mobilized and activated a network of non-government organizations. These organizations, in turn, connected the community with a national government fund for relocating communities in high-risk zones, which itself was put in place as a response to the 2009 floods. The involvement of the community in the formulation of a "people's plan" was a key component in the design, site-selection process, and selection of a contractor. As a direct result, it was successful in winning in-city relocation to high-density housing and in setting up a co-operative ownership structure that could serve as a model for other communities (Garcia, 2015).

This snapshot of Manila in 2015 illustrates some of the issues and open questions with respect to how global urbanization is understood. When the numbers are considered in isolation, this phase of urban growth, concentrated as it were in the global South and involving unprecedented scales of movement, size, and need, can often take on a nightmarish, apocalyptic tone. It is, of course, vitally important to recognize that life in these cities can be difficult, and to analyze how new forms of deprivation arise. Yet it is also important to pay attention to *why* people choose to move to and live in cities in the first place, and to appreciate the role of cities in enabling possibility and opportunity. The following text box considers similar issues of growth and decline at the national scale in the context of the Canadian urban system.

Growth and Decline in the Canadian Urban System

Canada has been predominantly urban for almost a century, having reached the 50 per cent urban threshold in the 1920s. By 2011, 81 per cent of Canadians were designated as urbanites (Statistics Canada, n.d.). Nevertheless the Canadian urban system continues to change, with several interesting trends apparent in recent years.

International in-migration to Canada has become the most important source of population growth increase for the country as a whole. The decade between censuses in 2001 and 2011 was the first in Canadian history in which national population growth due to net in-migration was greater than growth due to natural increase. In the coming decades, nearly all net population increase is expected to be due to in-migration (Statistics Canada, 2015). The result is that major immigrant-receiving metropolitan areas such as Toronto and Vancouver have grown rapidly. Between the censuses in 2006 and 2011, both metropolitan areas saw their populations grow by just over 9 per cent in five years.

The highest rates of growth between 2006 and 2011 were not, however, in the very largest cities, but in smaller and medium-sized urban areas (Statistics Canada, 2012). Furthermore, the four areas with the fastest population growth were all in western Canada: Calgary, Edmonton, Saskatoon, and Kelowna. Each of these cities grew by over 10 per cent in just five years, reflecting a buoyant economy in these regions, driven to a large extent by intensified oil and natural gas extraction. The same pattern was true among smaller centres (with populations below 100,000). In this group, 10 of the 15 fastest-growing municipalities were in Alberta. In just five years Okotoks grew by 42.9 per cent and Wood Buffalo (containing Fort McMurray in Alberta's oil patch) by 27.1 per cent. Nearly all of this growth was due to net in-migration, as workers and families were drawn to employment opportunities. In Fort McMurray, until devastating forest fires destroyed around 10 per cent of the city in May 2016, the result had been runaway house prices and difficulties in providing the infrastructure and urban services (such as community centres, health care, affordable housing, and public transit) for a burgeoning population (Dorow and O'Shaughnessy, 2013).

The driving force of economic restructuring is also apparent in (the relatively few) urban centres with declining populations. Between 2006 and 2011, the only Census Metropolitan Areas in Canada to see falling populations were Windsor and Thunder Bay (both in Ontario) (Statistics Canada, 2012). In the case of Windsor, reduced manufacturing employment in the auto sector (especially following the financial crisis that started in 2008, but also because of the relocation of production to low-wage regions in the United States and Mexico) meant that people left for opportunities elsewhere. The deep recession in neighbouring Detroit also affected Windsor's economy. Between 2006 and 2011, Windsor's population declined by just over 4,000 people, or about 1.3 per cent. Thunder Bay, sitting in northern Ontario at the head of Lake Superior, presents a similar scenario but in a different sector. The municipality saw a population decline of 1.1 per cent between 2006 and 2011 (Statistics Canada, n.d.). This reflected a collapse in the forestry sector, as home construction across North America slowed during the recession. Both Windsor and Thunder Bay illustrate the vulnerability of urban centres that are heavily dependent on a single economic sector. The problems they face are, of course, the opposite of boom towns. As opportunities dry up, young people tend to leave, and a spiral of decline can set in. Over the last century, this has been a phenomenon in single-industry **resource towns** across Canada (Norcliffe, 2005).

Conclusion

Perhaps the defining feature of global urbanization in the twenty-first century is that it is an array of very different processes, masquerading as one. It is important to keep in mind that more people are living in cities today than at any other point in the past, but our understanding cannot end there. Different wealth levels, demographic profiles, patterns of migration, and historical trajectories are producing very diverse experiences of urbanizations and vastly different conditions, ways of life, and landscapes across cities. These, in turn, translate into a very diverse set of challenges and opportunities: what might work for one city might not work for another, and there likely is no single set template for responding to what are inherently place-specific challenges.

Understanding and responding to these challenges also require us to appreciate how urbanization relates to other important processes. The demographic and economic factors behind urban growth and decline are themselves undergoing tremendous shifts; forms of urban life that came into being in the twentieth century and that we now take for granted will likely have to be rethought as populations, economies, and climates change. At the same time, cities have long been engines of creative thought and innovation; as cities (and city dwellers) proliferate, new ways of thinking about, and responding to, city life also blossom. The potential for learning from and across the diversity of new urban experiences—what the urban theorist Ananya Roy (2009) refers to as the establishment of new "roots and routes" for thinking about the city—will be greater in the coming decades.

Key Points

- Definitions of what constitutes the "urban" vary around the world, with every country adopting its own criteria and thresholds in terms of administrative boundaries, population numbers, physical landscape, and economic activities.
- Urbanization is accelerating globally, but it is an uneven process with both smaller cities growing rapidly and mega-urban regions emerging, especially in the global South.
- The key drivers behind urban growth relate to demographic trends, migration patterns, and the processes behind concentrations of economic wealth in cities.
- The challenges of big city growth around the world include housing, infrastructure, employment, environmental quality, and infrastructure; but the opportunities presented by big cities for upward mobility and innovation must also be recognized.
- The Manila case study shows that these challenges and opportunities are intertwined in complex and sometimes unexpected ways: economic growth can pose a challenge for equitable access to the city, and communities marked as living in "vulnerable" slums can craft solutions from the ground up.
- Nearly all of the population growth or decline in the Canadian urban system is attributable to internal or international migration, much of it driven by dynamic patterns of economic development in different regions and sectors.

Activities and Questions for Review

1. Why do some cities grow faster than others? In pairs, discuss the answer to this question in relation to a specific national and regional context.
2. With a classmate, develop a list of the key challenges faced by governments in the expanding megacities of the global South. In your estimation, which are the top challenges and why?
3. Collect three different e-media sources on Manila. What positive and negative features of Manila are highlighted in these representations?

References

Angel, S., with Parent, J., Civco, D.L., and Blei, A.M. (2011). *Making room for a planet of cities*. Policy Focus Report 27, Lincoln Institute of Land Policy. Cambridge, MA: Lincoln Institute.

Arn, J. (1995). Pathway to the periphery: Urbanization, creation of a relative surplus population, and political outcomes in Manila, Philippines. *Urban Anthropology and Studies of Cultural Systems and World Economic Development*, 24(3/4): 189–228.

Bordadora, N. (2011). Aquino bares land distribution plan to relocate squatter-families. *Philippine Daily Inquirer*, August 8.

Brenner, N., and Schmid, C. (2014). The "urban age" in question. *International Journal of Urban and Regional Research*, 38(3): 731–755.

Cardenas, K. (2014). Urban property development and the creative destruction of Filipino capitalism. In W. Bello and J. Chavez (Eds.), *State of fragmentation: The Philippines in transition*. Bangkok: Focus on the Global South.

Chan, K.W. (2009). The Chinese *hukou* system at 50. *Eurasian Geography and Economics*, 50(2): 197–221.

Dorow, S., and O'Shaughnessy, S. (2013). Fort McMurray, Wood Buffalo, and the oil/tar sands: Revisiting the sociology of "community." *Canadian Journal of Sociology/Cahiers canadiens de sociologie*, 38(2): 121–140.

Fan, C. (2008). *China on the move: Migration, the state, and the household*. London and New York: Routledge.

Florida, R. (2012). *The rise of the creative class—Revisited: 10th anniversary edition—Revised and Expanded*. New York: Basic Books.

Garcia, G.P. (2015). Former OFW leads neighborhood housing project. *Rappler.com*, March 19.

Housing and Land Use Regulatory Board (various years). *License to sell statistics*. Quezon City: Housing and Land Use Regulatory Board.

Kelly, P.F. (2003). Urbanization and the politics of land in the Manila region. *The Annals of the American Academy of Political and Social Science*, pp. 170–187.

Kelly, P.F. (2009). From global production networks to global reproduction networks: Households, migration, and regional development in Cavite, the Philippines. *Regional Studies*, 43(3): 449–461.

Kelly, P.F. (2013). Production networks, place and development: Thinking through global production networks in Cavite, Philippines. *Geoforum*, 44, 82–92.

Kleibert, J.M. (2015). Islands of globalisation: Offsore services and the changing spatial divisions of labour. *Environment and Planning A*, 47: 884–902.

Kleibert, J.M., and Kippers, L. (2015). Living the good life? The rise of urban mixed-use enclaves in Metro Manila. *Urban Geography*, pp. 1–23.

Lakha, S., and Pinches, M. (1977). Poverty and the "new society" in Manila. *Australian Journal of International Affairs, 31*(3): 371–378.

LGI Consultants (2015). Land and related constraints in socialized housing provision in Metro Manila. Report presented at the 2nd session of the Technical Working Group on Land and Housing, Housing and Urban Development Summit, House of Representatives, Republic of the Philippines, August 17.

McGee, T.G. (1991). The emergence of desakota regions in Asia: Expanding a hypothesis. In N. Ginsburg, B. Koppel, and T.G. McGee (Eds.), *The Extended Metropolis: Settlement Transition in Asi*a. Honolulu: University of Hawaii Press, pp. 3–26.

Montsion, J.-M. (2009). Relocating politics at the gateway: Everyday life in Singapore's global schoolhouse. *Pacific Affairs, 82*(4): 637–656.

Norcliffe, G. (2005). *Global game, local arena: Restructuring in Corner Brook*. St John's: ISER Books.

Ortega, A.A. (2012). Desakota and beyond: Neoliberal production of suburban space in Manila's fringe. *Urban Geography, 33*(8): 1118–1143.

Peck, J. (2005). Struggling with the creative class. *International Journal of Urban and Regional Research, 29*(4): 740–770.

Pernia, E.M. (2007). Diaspora, remittances, and poverty. In R.C. Severion and L.C. Salazar (Eds.), *Whither the Philippines in the 21st century?* Singapore: Institute of Southeast Asian Studies.

Pernia, E.M. (2008). Migration, remittances, poverty and inequality in the Philippines. UP School of Economics Discussion Paper No. 2008–01, University of the Philippines, Quezon City.

Philippine Statistics Authority (2012). *2010 Census of population and housing.* Manila, Phil.: National Statistics Office.

Philippine Statistics Authority (2015). *Gross Regional Domestic Product. Base year: 2000; 2012–2014.* Quezon City: Philippine Statistics Authority.

Philippine Statistics Authority (various years). *Construction Statistics from Approved Building Permits*. Quezon City: Philippine Statistics Authority.

Punathambekar, A. (2013). *From Bombay to Bollywood: The making of a global media industry*. New York: New York University Press.

Ravallion, M., Chen, S., and Sangraula, P. (2007). New evidence on the urbanization of global poverty. *Population and Development Review, 33*(4): 667–701.

Roy, A. (2009). The 21st-century metropolis: New geographies of theory. *Regional Studies, 43*(6): 819–830.

Saunders, D. (2010). *Arrival city: The final migration and our next world*. Toronto: Alfred A. Knopf Canada.

Scott, A. (2004). *On Hollywood: The place, the industry*. Princeton, NJ: Princeton University Press.

Scott, A. (2012). *A world in emergence: Cities and regions in the 21st century*. Cheltenham, UK: Edward Elgar.

Shatkin, G. (2007). Global cities of the South: Emerging perspectives on growth and inequality. *Cities, 24*(1): 1–15.

Sovani, N.V. (1964). The analysis of "over-urbanization. *Economic Development and Cultural Change, 12*(2): 113–122.

Statistics Canada (2012). *Census dictionary*. Catalogue no. 98-301-X2011001. Ottawa: Statistics Canada.

Statistics Canada (2015). Population growth: Migratory increase overtakes natural increase. http://www.statcan.gc.ca/pub/11-630-x/11-630-x2014001-eng.htm

Statistics Canada (n.d.). Population and dwelling count highlight tables, 2011 Census. Ottawa: Statistics Canada.

Sukamdi, S., and Mujahid, G. (2015). Internal migration in Indonesia. UNFPA Indonesia Monograph Series No. 3. Jakarta, Indonesia: UNFPA.

United Nations, Department of Economic and Social Affairs, Population Division (2008). *World urbanization prospects: The 2007 revision* (ESA/P/WP/205).

United Nations, Department of Economic and Social Affairs, Population Division (2015a). *World urbanization prospects: The 2014 revision* (ST/ESA/SER.A/366).

United Nations, Department of Economic and Social Affairs, Population Division (2015b). *World population prospects: The 2015 revision*.

Whyte, M.K., Feng, W., and Cai, Y. (2015). Challenging myths about China's one-child policy. *The China Journal*, No. 74 (July 2015): 144–159.

Wolfe, D. (Ed.) (2014). *Innovating in urban economies: Economic transformation in Canadian city-regions*. Toronto: University of Toronto Press.

3 National Urban Systems in an Era of Transnationalism

André Sorensen

Introduction

A **national urban system** can be defined as the network of cities within a particular country that are linked together by flows of people, goods, money, and information. The fundamental conditions for such networks of cities are the transportation and communications systems that link varied cities together. Over the course of the nineteenth and twentieth centuries, a series of transportation and communications revolutions—from canals to railways, steamships, telegraph and telephone, automobiles, containerization of shipping, and the Internet—continually reduced the costs and time associated with connectivity between cities, producing ever-denser linkages within and between national urban systems. This in turn is understood to have encouraged specialization of economic activity in the cities and regions where they had the greatest comparative advantage—whether because of location, natural resource endowments, or local skills—and this served to increase the overall efficiency and productivity of the system. Thus, national urban systems evolved into networks of functionally differentiated cities within a particular country.

As is well known, the last half-century has seen an acceleration of **globalization** (the process of increasing linkages and flows of goods, money, information, and people between cities and countries in different parts of the globe) and transnationalization (the process of increasing linkages and more complex ties and networks between people in different places around the world). It is clear that different countries and city systems experience and respond to globalization and transnationalization in different ways. This chapter suggests that an examination of the evolution of national urban systems provides important insights about these differences and why urbanization processes and outcomes are so varied in different countries. In these processes, geography and history are important, but so too are government policies, economic structure, and the timing of urbanization in relation to technological change, global events, and global norms and ideas.

This chapter argues that an examination of national urban systems provides a valuable perspective on the ways in which globalization and transnationalization are having an impact on different countries and cities. Particular attention is paid to the city system as an aggregation of investment over many decades in land and property and to the Japanese urban system. Japan is an interesting and distinctive case, where modernization and urbanization processes have been compressed into a relatively short period. The Japanese urban system is also old and well established and has long been tightly integrated with the global economy. Japan is one of the few countries in the world that was never colonized, retaining its own legal systems, urban culture, and traditions. In particular, the Japanese case underlines the point that ideas of national urban systems are closely related to the policies designed to shape them. The Japanese case is also valuable because of the significant regional planning machinery it has created since the 1960s. These highly politicized regional policies invested significant resources into attempts to reshape the Japanese urban system. Also, a recurring cycle of land booms and busts that crippled the financial system in the early 1990s and devastated the Japanese economy was closely related to the government policies designed to encourage investment in real estate. The resulting collapse of the bubble of the late 1980s led to over two "lost decades" of stagnation since the early 1990s. Japan also provides a clear contrast with Canada, as the Japanese urban system is older and has a much more constrained geography, a unitary government system rather than a federal system with strong provincial governments, and a much more activist approach to economic and urban system management.

The first section of this chapter introduces the concept of national urban systems. This is followed by a review of the "Varieties of Capitalism" conceptual framework for understanding the very different systems of state–society relations in different national contexts, and of ideas associated with New Institutionalist approaches to international comparative research. The third section provides a brief history of the development of the Japanese urban system and discusses the major attempts to intervene in shaping urban system development through regional policies in the postwar period. It is followed by a brief discussion of the Canadian urban system that highlights three points of comparison with the Japanese urban system, namely differences in models of national government, of development policies, and of geographies. These specific examples of national urban systems demonstrate the importance of national specificity within the context of an increasingly globalized urban world.

National Urban Systems

Research on national urban systems flourished from the 1960s to the 1980s, motivated in part by the dramatic processes of urbanization and structural urban system transformations occurring in the postwar period (Bourne, 1975; Pred, 1977). Bourne (1975) argued that research on national urban systems was encouraged by the high visibility of a range of urban problems and challenges during the 1960s combined with the fact that some countries showed demonstrable success in managing rapid urban growth. This provided a clear rationale for attempts to better understand patterns of urban system change during a period when, in the global North, state resources and capacity to intervene were growing exponentially and rapid urban economic growth in the global South suggested that it too would need to develop policies and capacity to manage its urban systems.

As Bourne (1984: 2) put it, most urban problems boil down to misallocations in the spatial and social distribution of societal resources, including

> differences in growth rates, income, employment opportunities, housing, public services, and the like. It has been extensively argued in the literature that such disparities represent a direct source of social injustice, a misallocation of national resources, a measure of structural inefficiency in the economic order, and, in some instances, pose a direct threat to national unity.

Market processes tended in most urban systems to encourage greater growth in some locations at the same time as either lower growth or decline occurred in other parts of the network, in patterns described by Marxist geographers as inherent to capitalism and labeled as **uneven development**, in which growth and accumulation in the centre is always dependent on underdevelopment in the periphery (Smith, 1984). Such core–periphery relationships can occur either within a nation or at the international or global scales. Such disparities prompted many efforts at regional economic development planning aimed at promoting more balanced growth throughout a national territory by promoting growth in lagging regions. Bourne states the central goal of the regional planning project concisely: "Regulating urban systems, and shaping the consequences of redistributive mechanisms operating through those systems, is also a means of achieving greater equality in society as a whole" (Bourne, 1975: 3). The fundamental goal of policies designed to shape national urban systems, in Japan and elsewhere, has been to prevent excessive growth in already-large cities and to encourage growth in lagging or declining regions. Such efforts are a major part of the story of the Japanese urban system discussed below.

Research on national urban systems contributed a number of important insights, including analysis of city-size distributions and the **rank-size rule** and **urban primacy** concepts. These found regular size distribution patterns for cities within a particular system and sought to explain why these would exist and why there were differing patterns in various countries. Other work, such as on optimal city size, seemed important when new towns were actively being built, but determining such optimal sizes eventually proved to be impossible, as while particular functions or infrastructures might have optimal population-size thresholds, any given city will have potentially hundreds of different functions with different thresholds (Begovic, 1991; Richardson, 1973). By the 1980s the limits of national urban systems research were commonly noted, including the problem that while robust empirical patterns were found, no adequate theoretical explanation for the rank-size rule emerged. City-size distributions were accused of being too reliant on population as a metric, and it was suggested that the nation-state is not necessarily the best unit of analysis, as regions within countries or cross-national networks may be more relevant.

Perhaps more important, however, was the political shift beginning in the 1980s towards more laissez-faire, market-based policies. Because national urban systems research was so closely linked with regional economic development policies that were abandoned in many countries with the rise of neoliberalism in the early 1980s, it began to seem less relevant. Major investments and subsidies in support of shifting industrial location patterns to revitalize declining regions lost favour, and the larger project of using government policies to contribute to greater social and spatial equality was incrementally abandoned in countries around the world. Furthermore, with increasing globalization, the assumption of the nation-state as the primary unit of analysis became increasingly untenable, as cities were increasingly tied in to global urban networks and influences and focus shifted to world or global city systems (Friedmann, 1986; Sassen, 2000). The same factors of cheaper transportation and communications that had initially strengthened linkages and integration within national urban systems began to expose them to increased international competition and linkages to cities in other countries, and in many cases actually weakened links within countries. For all these reasons the detailed description and analysis of national urban systems have declined greatly since their peak in the 1960s and 1970s, but this chapter asserts that there is still a lot to be learned from analysis of national urban systems, particularly in comparative perspective and in the context of increasing transnationalization.

Approaches to Understanding National Urban Systems: Varieties of Capitalism

Two important contributions to understanding the sources of variability in national economic and developmental trajectories are the "**New Institutionalist**" (NI) and "**Varieties of Capitalism**" (VOC) research agendas, which emerged in the 1990s in part as a way of explaining persistent differences in economic development between nations. The new institutionalism is a broad research agenda in social sciences that focuses on the institutional structures that shape economic, social, and political outcomes. Institutions are commonly defined as the laws and regulations, shared norms and understandings, and standard operating practices in any given society. Major questions concern the ways in which institutions shape social processes, how they are created, and how they change over time.

Whereas neo-classical economic theory assumes that over time differences in levels of development between countries will decrease in response to normal market processes, in fact disparities in levels of development have persisted and in some cases increased. Nobel Prize–winning economist Douglass North (1990), in particular, convincingly showed that a major source of differentiation between countries can be found in the institutional structures that regulate economic activity, including legal, financial, and governance institutions, as well as shared norms and ideas.

VOC researchers developed these ideas further, arguing not only that there are profound and enduring differences in the institutional infrastructures of different countries and groups of countries, but that these differences amount to significantly different forms of capitalism (Hollingsworth and Boyer, 1997; Streeck and Yamamura, 2001). An influential work by Hall and Soskice (2001) divided capitalist countries into two main groups: liberal market economies (LME) such as the UK, Canada, and the USA, and coordinated market economies (CME) such as Germany and Japan. They suggested that whereas in LMEs firms coordinate activities primarily through competitive markets, CMEs allow a much greater role for non-market relationships in intra- and inter-firm coordination, largely through state guidance and "steering" of investment decisions. Crucially, such differences tend to persist or become path dependent because each system is composed of multiple interlocking institutions, and while individual institutions may change, it is much more difficult to change the overall orientation of the system. The suggestion here is that the VOC thesis provides useful insights for understanding national urban systems in comparative perspective and that different capitalist systems work with differing assumptions about the appropriate relationships between state, capital, and society in the production of urban space.

Japan is considered to be a prime case of a CME: it had a relatively autonomous state bureaucracy that actively engaged in guiding private investment decisions so as to stimulate investment in priority economic sectors, encouraging cooperation and managing competition between firms, influencing exchange rates and wage levels, and shaping firm–labour relations. This set of Japanese market coordination institutions is often referred to as the "developmental state." In the seminal formulation of Johnson (1982), the goal of the developmental state was to help push the country's industrial structure towards higher value-added industries and balance the costs and benefits of industrial restructuring between firms and sectors in pursuit of more rapid economic development. The effectiveness of this approach is supported by the fact that Japan transformed itself from a poor country devastated by war in 1950 to the world's second-largest economy by 1970.

Different capitalisms tend to produce significant differences in national urban systems in terms of their urban governance structures, regulatory landscapes, and institutional capacities. Responses to processes of transnationalization are equally varied. The focus here is on the distinctive ways that private investment in urban real estate is regulated and supported in different countries. This perspective highlights the centrality of land, infrastructure, and regulation of the built environment as components of contemporary capitalism, as discussed below.

Attempts to Shape National Urban Systems

In Japan the role of the state in shaping and reshaping the urban system was central, as even though Japan had a weak city-planning system with few land-development controls, it did have a relatively robust regional planning system and the state has always made exceptionally large investments in infrastructure, to the extent that some described Japan as shifting from a "developmental state" to a "construction state" in the 1990s (Woodall, 1996; Sorensen, 2002). In its strong regional policies, Japan largely conformed with dominant policy ideas about urban systems planning from the 1950s through to the 1970s, including attempts to use infrastructure investments to encourage economic development, attempts to restrain the growth of the largest cities and redistribute development to more peripheral areas, and attempts to develop new economic activities in targeted growth centres (see Glickman, 1979; Hall, 1992). The rationale for such policies was the high cost of "excessive" growth and congestion in the largest cities and the desire to encourage development in lagging regions. Such peripheral growth centres were seen as a way to both raise overall economic growth rates and prevent migration to core regions. Large state investments in promoting economic growth in peripheral regions were also often seen to be politically advantageous for the ruling party (Calder, 1988). Similar policies were pursued in many countries, from the early postwar period through to the 1980s, including Canada, although Japanese regional policies were often envied for their expansive ambition, scope, and resources.

After the Second World War, major urban rebuilding projects were underway, especially in Europe but also in Japan and elsewhere, and the spatial distribution of industry was being transformed by new transportation modes and new industries, so the development of policies to shape such change is understandable. The idea of national-scale spatial planning was also supported by the major role of planning during the war, and it was widely assumed that states should play a central role in managing processes of reconstruction and economic development. In the United Kingdom the question of regional policy had been prominent since the 1930s when the decline of the industrial regions of the north began to accelerate. A turning point was the Barlow Report on the "distribution of the industrial population" (Barlow Report, 1940), which recommended measures to ensure investment in declining regions and prevent overconcentration in the south, especially around London. Regional policy in the UK has taken a variety of forms, including nationalization of industries, the building of new towns, and the creation of development corporations and enterprise zones (Hall, 1992). Strong state interventions in regional planning were also encouraged by the French achievement of rapid economic growth managed in a top-down manner (Hansen, 1968), in contrast to the enduring failures of Italy's attempts to encourage development in its southern regions. Attempts to reshape national urban systems are not a simple matter, and the Japanese case itself is quite mixed in its impacts.

The basic logic of regional policies is clear: to achieve national development and prosperity, many agree that states should encourage balanced development over the whole of their national territory. This seemed most obvious where one region was declining and experiencing significant unemployment or population loss and other regions were growing so fast that they experienced capacity constraints on infrastructure, congestion, pollution, or land price inflation. In such cases it seemed obvious that states should attempt to redirect some of the growth to the declining region.

Such interventions can be justified on the basis of both equity and efficiency considerations. The equity argument is clear: where one region is becoming richer while another is becoming poorer, some equalization payments or investments may help mitigate the pain of the region in decline. This argument is most convincing in cases where growth of the growing region directly or indirectly imposes costs on the declining region, such as by driving up inflation, attracting firms to move from the declining region, or attracting skilled workers and driving population loss. The efficiency argument is also straightforward: if the growing region is experiencing capacity constraints, pollution, or inflation, then moving some investment to the declining region will increase overall economic efficiency and problems in the growing region may be reduced.

In democratic systems such policies can easily become major political priorities as a result of pressure from the declining regions for a share of the prosperity and wealth being generated in the growing regions. Whether such redistribution occurs on any meaningful scale will depend on the particular political balance and conjuncture, on the effectiveness of advocates for regional policies, on the availability of appropriate policy tools and institutions, on the institutional capacity of governments and their agencies, and on prevailing policy norms and ideas. In many countries today such attempts at regional planning have been dying away in the face of neoliberal development and its lack of concern for equity issues.

Historical Development of the Japanese National Urban System

Even if a national urban system is being reshaped by transnationalization, the legacies of earlier development patterns and linkages are always profound. Japan had a fully developed urban system before the advent of modernization in the middle of the nineteenth century. The Japanese case therefore allows an analysis of the impacts of the earlier round of globalization in the late nineteenth century, of the urban-industrial transformation of Japan during the twentieth century, and of the more recent phase of transnationalization, economic and urban system maturity, and urban and economic decline.

A brief history of the evolution of the Japanese city system is necessary at this point. The first truly national urban system emerged during the Edo period (1600–1867), which began with the unification of the country under the Tokugawa Shogunate. The shoguns were military leaders who—under the nominal leadership of the emperor, who was entirely dependent on them for security and finances—were the effective rulers of Japan during the Edo period. The Tokugawa family were the feudal lords of the eastern Japan region centred around what is now Tokyo, and they maintained control of the shogunate government until 1867. Peace and national unification after a long period of civil wars allowed growing prosperity and rapid population growth, the development of better transport and communications, and increasing integration of the economy.

A few rules imposed by the shogunate were key in structuring the emerging city system. First, in reaction to the increasing influence of Christianity in the south, where Dutch and Portuguese traders and missionaries had bases, the new government, during its first decades, virtually eliminated contact with the rest of the world, banned Christianity, and prohibited all manufacture and possession of guns. The carrying of weapons was restricted to the samurai warrior class, which was between 10 and 15 per cent of the population. Japan was effectively sealed off from the world for the next two and a half centuries, it became entirely self-sufficient, and thus ports for international trade were not a significant part of the city system, as they were in most of the rest of the world.

Second, in an edict of 1630 the shogunate allowed only one castle for each feudal domain, or *Han*, of which there were about 300 in the seventeenth century. This led to the demolition of most of the existing castles and the building of new castle-towns. The primary difference between Japanese castle-towns and those of medieval Europe was that in Japan the castle was designed only to house the samurai warrior class and the towns that grew up outside were never protected by an encircling fortification, whereas most European castle walls enclosed a town. Most domains chose new locations for their single castle-town, which instead of being located in strategic defensive positions on hills or mountains were now mostly located in the centre of the largest agricultural areas, as taxes were almost entirely on farmers and were paid in rice. These new castle-towns became the administrative centres of each domain and grew quickly, as all samurai were now required to live in the town with their lords. The second half of the seventeenth century thus saw one of the world's greatest periods of new town building and the creation of the basic framework of the contemporary Japanese urban system. As Kornhauser (1982) notes, the fact that many of the largest cities of contemporary Japan originated from these castle-towns is strong evidence that their locations were chosen well. Also, as domains were of roughly similar sizes and based primarily on watersheds and river basins and as each domain had one major town, a relatively evenly distributed city system emerged.

Third, all feudal lords (*Daimyo*) were required to spend every second year living in the shogun's capital, Edo (renamed Tokyo in 1868), and their families lived there permanently as hostages. This was a key part of the Tokugawa strategy to unify the country and to achieve hegemony over the formerly relatively independent domains, which had been embroiled in over a century of brutally destructive civil wars. The Tokugawa directly controlled only about a third of the country and had close alliances with domains controlling another third, belonging to the *Daimyo* who fought with them in the decisive battle at Sekigahara in 1600 and were considered loyal to the Tokugawa. The remaining domains were seen as potential enemies, and so requiring their families to live year-round in Edo was considered an effective strategy to prevent open rebellion. This "alternate residence system" (*Sankin-kotai*) had several major consequences. As each lord had numerous retainers and servants, they established large households in the capital, often consisting of several hundred people. The high costs of maintaining large establishments in Edo and journeying back and forth each year also consumed huge resources and limited the lords' ability to develop significant military forces. As in other castle-towns, the presence of large numbers of the samurai ruling class in Edo supported a growing population of townspeople who provided goods and services to them.

Edo quickly grew to become one of the largest cities in the world, with over a million residents by the beginning of the eighteenth century, when London and Paris each had populations of only about 600,000 (Rozman, 1973). This giant city represented a huge market for food and products of all sorts, as much of the income of every domain was spent to support their establishments in the capital. A major factor contributing to the integration of the national economy was the resulting creation of national supply chains in response to the enormous demand for goods and services in Edo (Hayashi, 1994). This fostered the development of well-established ocean shipping routes and highways, and differentiated roles in economic activity emerged in different cities.

Edo was the political capital and a huge centre of consumption and cultural production. Osaka was the merchant and financial capital, while Kyoto was the imperial capital, the centre of aristocratic and court life, and also a major textile and luxury goods production centre. Most other cities were regional service centres, although many were known for particular products, such as silk, metallurgy, coal mining, and pottery production. Annual travel between domains and the capital contributed to the development of the transport system, particularly the five great pedestrian highways, as most people travelled on foot and most goods by boat along the coasts. Of these the two most important were the highways that connected Edo with Kyoto and Osaka, the *Tokaido* (literally East Sea Road) along the Pacific coast and the *Nakasendo* (Inner Road), which ran through the mountains from Edo to Nagoya via Nagano. Famous temples and shrines also frequently gave rise to larger towns, as did some ports.

After a period of rapid population growth and the creation of the new city system in the seventeenth century, Japan quickly reached the limits of its food production capacity, and thus the eighteenth century saw repeated famines when harvests were poor, and population stabilized. The feudal-era city system therefore achieved considerable stability, supported by its particular set of structuring rules. With the revolution of the 1860s in response to the threat of the European powers, which had colonized much of the rest of the world, however, the Tokugawa order was overthrown and the emperor was "restored" to power. The alternative residence system was abolished, Edo (renamed Tokyo in 1868) lost almost half its population during the 1870s as the *Daimyo* withdrew both their large households and their spending to their provincial capitals, and the economy of the city collapsed. Restrictions on movement and on farmers were lifted, title deeds to land were issued to farmers and a new capitalist market in land created, and central government efforts to promote modern industries, especially heavy industry, railways, and shipping, were launched. This new set of economic, political, and social factors worked to reshape the settlement system in Japan's first bout of globalization (Sorensen, 2002; 2010).

Of these factors, the most important forces shaping the urban system were rapid population growth, industrialization, rural to urban migration, and the development of the railway system. The national population doubled between the 1890s and 1940 at the same time that rural areas were either stable in population or declining. The food supply increased and food prices declined in part because of an increased supply from Japan's colony Taiwan, which Japan had occupied after winning the Sino-Japanese war of 1894–1895. Declining food prices and rural poverty provided a push of population from rural to urban areas, while the growth of industry pulled migrants from increasingly impoverished rural areas.

Population growth was therefore concentrated in urban areas and particularly in the industrial centres along the Pacific coast, including Tokyo-Yokohama, the Nagoya area, Osaka-Kyoto-Kobe, and in northern Kyushu in the west. Urban industrialization was initially led by light industries such as cotton and silk spinning and weaving, but during and after the First World War, which Japan joined on the Allied side, production increasingly shifted to heavy industries such as steel, shipbuilding, and machinery. At the same time as Allied purchases grew rapidly, imports from Europe were largely cut off, particularly from Germany, which had been a major supplier, and Japanese industry grew rapidly in response to virtually unlimited demand.

Another key factor shaping the restructuring of the urban system at this time was the growth of railways. As Japan is compact and mountainous, with poor road systems, the new central government decided early on that railways were the priority in terms of investments in internal transportation, and it invested large amounts to create a national railway system (Ericson, 1996; Yamamoto, 1993). Location on a railway line soon became a decisive factor in local economic development, and local elites competed to ensure that new train lines and stations were located near their towns. Murayama (2000) argues that railway investment strengthened the integration of the Japanese urban system with a steady and thorough building of the national rail system from 1900 to about 1960. Most of this was accomplished by the state-owned Japanese National Railways, with private commuter railways operating primarily in the Tokyo and Osaka metropolitan regions (Figure 3.1).

Figure 3.1 Commuters near Osaka Station

As in other countries, an increasingly reliable, frequent, and fast railway transportation system tended to favour concentration of new investment in the core area, as companies based in the centre could easily serve markets throughout the country and benefited from agglomeration economies of co-location with other large companies. During the period of rapid economic growth from the 1950s to the 1970s, the central government prioritized government infrastructure investment in the core Pacific Belt (Tokaido) area from Tokyo through to Osaka as a way of promoting economic development through industrial concentration and agglomeration economies. These investments in the core region are widely understood to have been important contributions to the extraordinary expansion of the Japanese economy, with an average GDP growth of over 9 per cent from 1955 to 1973 (Glickman, 1979; Harris, 1982; Johnson, 1982).

Growth in the Pacific Belt region of Japan was spectacular during the period of rapid economic growth of the 1950s and 1960s. Most of that growth was concentrated in the core region from Tokyo to Osaka (Gottmann, 1980; Nagashima, 1981; Rimmer, 1986). Although the Japanese economy is capitalist, with large private corporations organized into giant vertically organized conglomerates called *Keiretsu*, the role of the state in coordinating industrial investment, providing infrastructure and industrial land supply, preventing over-competition, and winding down declining industries was significant, particularly during the rapid growth period (Johnson, 1982). Investments were made in railways, electric power, water supply, port infrastructure, industrial land reclamation from coastal waters, and highways.

Concentration of both public and private investment in the Pacific Belt region also meant that this area was the main destination for rural to urban migrants seeking work. As Glickman (1979) showed, the population of the Pacific Belt area grew much more rapidly than the rest of Japan during both the 1950s (32 per cent versus 12.3 per cent) and the 1960s (31 per cent versus 10.4 per cent), with similar disparities in employment growth. This concentration of population represented a profound transformation of the Japanese urban system, to the extent that the core region began to be referred to as the Tokaido **megalopolis,** following the publication of the French regional geographer Jean Gottmann's famous study of the vast urban-industrial agglomeration in the northeastern seaboard of the United States, from Boston to Washington (Gottmann, 1961). This period also saw the first widespread losses of population in rural and peripheral regions of Japan in the north and west and along the Japan Sea coast, owing to migration towards the Pacific Belt, a pattern that has continued until the present (Figure 3.2).

From the mid-nineteenth to the late twentieth century, therefore, industrialization and economic growth associated with integration into the global economy caused a transformation of the Japanese urban system from one that was relatively evenly distributed across the archipelago, with the largest city of Edo with less than 3 per cent of the national population in 1850, to a situation where the much larger Tokaido core region from Tokyo to Kobe now holds over two-thirds of the national population in a continuously urbanized corridor along the Pacific coast and all other regions have seen several decades of continuous population decline (Sorensen, 2002).

Although concentration in the core area was rational as economic policy, it had a number of unwanted consequences. These included significant increases of land and housing prices in core regions, urban sprawl as households sought affordable housing ever farther from city centres, and significant commuting and congestion problems due both to sprawl and to absolute growth in the major cities. But these were all relatively minor consequences compared to the biggest problem, the tragic environmental pollution crisis that was experienced throughout Japan in the 1950s and 1960s, but was most severe in the Tokaido megalopolis because of the concentration of heavy and chemical industries there. Weak and unenforced pollution regulations, the emphasis on growth above all other priorities, and the government's stubborn disregard of the risks of environmental pollution all combined to produce one of the world's worst-ever environmental crises. Hundreds died from polluted air, water, and food, and thousands were crippled or suffered enduring illnesses because of exposure to toxins (Ui, 1992; Tsuru, 1999). Environmental opposition movements proliferated and eventually forced the government to enact stricter laws and forced corporations to change some practices (Broadbent, 1998).

Concentration in the core area and decline in the periphery also led to sustained political lobbying in favour of regional policies that would redistribute economic opportunities throughout the rest of the country. In particular, the significant role of the developmental

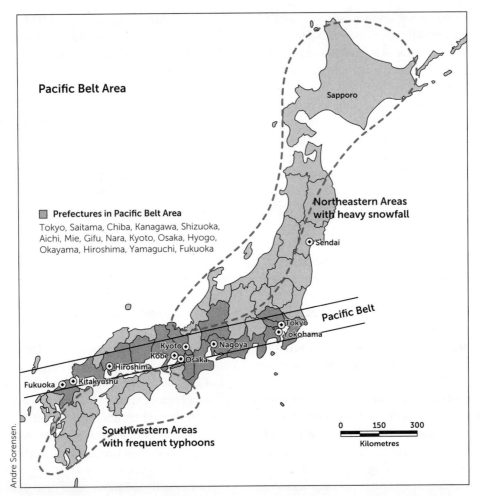

Figure 3.2 Map of Japan, the Tokaido megalopolis, and the Pacific Belt

state in facilitating economic growth and agglomeration in the core areas exposed the government to criticism from those in peripheral regions who saw national tax revenues invested in promoting the booming Tokyo–Osaka corridor, even while their areas were declining. This analysis was linked to the particular political context of postwar Japan, where the long-ruling Liberal Democratic Party (LDP), in power continuously from its formation in 1955 to 1993 and most of the years since, enjoyed its most secure electoral base in the rural and less-developed periphery. As Calder (1988) explains, a major political dynamic of postwar Japan was that whenever the LDP saw serious political crises challenging its continuing power, it expanded distributive policies of increased central government investment in the regions. For individual politicians, the "pipeline" to central government resources was a powerful political asset that, as long as the LDP was in power, was enjoyed only by LDP members. Although many countries have seen such corrupting politics, in Japan they were particularly important, and the Japanese central government has long spent a much higher share of GDP on infrastructure than most other countries (Woodall, 1996).

In Japan the full range of possible policy tools were used at different times to support growth in declining peripheral regions. These included infrastructure investments in railways, roads, dams, ports, and airports; subsidies to local governments to maintain comparable levels of education and services as in the core regions; industrial land development, tax incentives, and subsidies for firms that relocated to designated areas: high-technology "**technopoles**," new towns, and research institutes; and decentralization of government agencies and offices to regional centres (Castells and Hall, 1994). These policies had significant impacts, at least in the medium term, but the really interesting aspect is the politics underlying these policies. Regional policies have been particularly contentious in Japan.

An important element of state planning and guidance for investment, particularly for the national urban system, was the Comprehensive National Development Plan (CNDP) system based on a law first passed in 1951. This was developed explicitly to encourage investment in peripheral regions, but did not work out as initially intended. The first CNDP was passed in 1962, pushed forward by politicians in the peripheral regions of Japan who saw the linked trends of concentration in the Pacific Belt and decline everywhere else as a consequence primarily of an unfair bias in government spending. While business elites pushed for investment in two to three new **growth poles** in the core areas, the plan established 15 "New Industrial Cities" that were scattered around peripheral regions (Calder, 1988; Sorensen, 2002). Before these cities could be fully established, business interests succeeded in lobbying for the establishment of six "Special Areas for Industrial Concentration," all in the core area of the Pacific Belt, and these ended up receiving the majority of new industrial investment. Glickman shows that in fact the central government did continue to disproportionately focus investment in the Pacific Belt, at least through to the end of the 1960s (Glickman, 1979).

The second CNDP, passed in 1969, focused primarily on increasing investment in transportation networks, including the Shinkansen bullet trains and highways. Glickman argues that this too contributed primarily to further concentration in core areas. In particular, Shinkansen-linked cities converged as the core of the national economy, while those not linked were marginalized (Murayama, 2000: 151). The third CNDP passed in 1977, however, marked a significant shift towards major investments in peripheral regions, with its theme of comprehensive development of projects for human habitation. Calder (1988) argues that this real shift towards increased spending in the periphery was in response to the perceived electoral vulnerability of the LDP after the serious recession of the early 1970s, the growing environmental movement's response to the pollution crisis, and the growing success of socialist and communist politicians at the local, state, and national level in the core regions of Tokyo, Osaka, and Kyoto.

The 1970s are seen as the beginning of the *Chihô no Jidai* or "Era of the Regions" (Steffensen, 1996), which celebrated local autonomy, regional distinctiveness, and quality of life over GDP growth. While the associated citizen-based movements continued, this direction in CNDP policies was short-lived, as the fourth CNDP of 1989 made an about-face and controversially focused on the importance of the National Capital Region of Tokyo as central to the Japanese economy. More investment was focused on Tokyo as Japan's global city and as an important competitor in international trade and financial markets. The current fifth CNDP of 1998, labeled the "Grand Design for the Twenty-first Century: Promotion of Regional Independence and Creation of Beautiful National Land," was generally seen as a "something for everyone" approach to spreading public investment widely in the effort to revitalize the national economy, which had seen sluggish growth since the collapse of the **bubble economy** of 1990.

By the 2000s the nexus of the LDP, infrastructure investment, and construction industry was widely denounced as the "construction state" approach to government spending and economic stimulus (Kerr, 2002; McCormack, 2002; McGill, 1998). Huge economic stimulus spending on construction projects was a major factor behind Japan's enormous public debt, which at over USD$10 trillion by 2013 was more than double Japan's annual GDP, easily the highest rate in the world, higher even than Greece today. Indiscriminate pouring of concrete for roads, dams, landslide prevention, river re-engineering and flood protection, coastal defences, and the rest was also increasingly seen as a major cause of environmental destruction throughout Japan.

Although Japan never had a primate city structure because of the balancing effects of the giant Osaka-Kobe-Kyoto metropolitan area and the Nagoya region, in recent decades only the Tokyo region has seen population growth, a process described as "uni-polar concentration." This process continues to create serious challenges for declining regions in Japan, as the total national population is now projected to decline, based on medium-mortality and fertility assumptions, from 128 million in 2010 to 97 million in 2050 (Japan National Institute of Population and Social Security Research, 2012, Table 1.1). As Tokyo continues to attract population from the rest of the country (Figures 3.3, 3.4, and 3.5), peripheral towns and villages are seeing accelerated

Figure 3.3 Fish market in Tokyo

Figure 3.4 Club district near Ueno Station, Tokyo

Figure 3.5 Restaurants under the Bullet Train tracks, Tokyo

decline and aging, and increasing numbers of smaller villages are disappearing entirely. This has prompted a range of efforts to attract new residents, but none have been successful in countering decline (Sorensen, 2006).

One final aspect of government intervention in regional development and the national urban system concerns land prices and attempts to shape them. As part of the effort to grow the Japanese economy and capital assets, the development of the urban system and support of land investment have long been a priority. Rapid economic growth and investment in urban areas were accompanied, however, by a series of land-asset inflationary bubbles in the early 1960s, the early 1970s, and the late 1980s. By far the most damaging of these was the last one, which crashed the Japanese economy and financial system in 1989–1991; bankrupted thousands of firms, with many more individuals losing their life savings; prompted trillions of dollars of government bank bailouts to prevent bank collapses; and was the major factor in the last two "lost decades" of economic decline. The bubble economy period from 1985 to 1990 was also damaging in itself, as asset price inflation, particularly of land, led to predatory real estate development practices, the destruction of many inner-city neighbourhoods for redevelopment, the employment of gangsters to intimidate tenants into moving, and the rapid decline of housing affordability and growth of homelessness in urban areas (Douglass, 1993; Sorensen, 2003).

Many have argued that attempts to control land prices represented a gross failure of land policy in Japan, as the state consistently acted to increase incentives to invest in urban land and to hold land assets through preferential land and inheritance taxes, banking system regulation, and land policy (Haley and Yamamura, 1992; Noguchi, 1990; Noguchi and Poterba, 1994; Yamamura, 1992). This reflects a long tradition of collusion between the LDP and real estate interests, both at the local and the national level, which led to repeated scandals. The main point here is that central government support for real estate investment, combined with weak land policies and weak land development controls, contributed significantly to environmental damage, urban sprawl, housing problems, financial losses for both the individual and the corporate sectors and sucked capital out of productive and otherwise profitable industries and into real estate speculation (Sorensen, 2002: 289). With the long, slow collapse of the bubble many people became a lot poorer, and not just those who had speculated. The fact that this collapse ended the growth of the formerly dynamic Japanese economy for over two decades since 1990 is an indication of how important careful policies to manage investment in urban systems and urban property can be.

This examination of the Japanese experience of planning and intervention in the development of its national urban system will now be considered in relation to the Canadian urban system. Three issues are put into sharp relief: the significant differences between the unitary national government of Japan and the federal divided-powers system of Canada, the impact of Japanese "developmental state" policies on spatial development patterns, and the over-arching role of differing national geographies.

The Canadian Urban System

Compared to Japan's approach, Canadian policies that were intended to reshape the Canadian national settlement system have been tentative, short-lived, and piecemeal, at least if we ignore the nineteenth-century colonization policies of national unification, railway building, and settlement of the Prairies and the West, which were certainly robust. But they occurred long before the period when national urban systems became a policy target, and were much more about seizing and settling a national territory than about redistributing economic activity or preventing the growth of the largest cities. During and after the Second World War, the Canadian federal government did intervene to promote new settlements, such as Ajax, Ontario, for munitions factories, and Shipshaw-Saguenay in Quebec and Kitimat in British Columbia for aluminum smelting. But all of these projects are best understood as opportunistic efforts to promote military capacity or economic growth and were not really attempts to reshape the national urban system.

Efforts to reshape Canada's national urban system began in the 1960s, much like Japan's, and with a very similar basic analysis of overconcentration in core areas and decline in rural peripheries. In 1940 the Rowell-Sirois Royal Commission (officially named the Royal Commission on Dominion-Provincial Relations) had suggested that settlement in Canada was highly concentrated in the Windsor to Quebec City urban corridor, but little action was taken to intervene until the 1960s. Then, inspired in part by regional planning efforts in other countries, including the United States, Britain, and France, the Canadian government passed the Agricultural Redevelopment and Rehabilitation Act in 1961, which along with the Atlantic Development Board and the Fund for Rural Economic Development, were brought together in the federal Department of Regional Economic Expansion (DREE) in 1969 (Hodge and Gordon, 2008: 232). The idea that overconcentration in core areas could be detrimental to economic growth and could contribute to decline in peripheral areas as well as the idea that the growth and change of national urban systems could and should be planned and managed were arguably at the height of their influence in the mid-1970s. These ideas contributed greatly to the creation of a federal Ministry of State for Urban Affairs (MSUA)(1971–1979), which favoured policies of deconcentration, and to the work of the United Nations Conference on Human Settlements (Habitat) in Vancouver in 1976 (Robinson, 1981).

Canada never did develop a national urban system policy. The agency that might have initiated such a national policy existed for only a few years before it was dismantled, in response to strong provincial objections to the creation of a federal department and policies in an area—municipal affairs—that was constitutionally a provincial responsibility (Spicer, 2011). This division of powers made it very difficult to develop a national urban policy, and virtually impossible to implement one. In contrast, Japan (like the UK and France) is a unitary state, not a federation, and can make national

continued

laws that affect urban policy without interference from other levels of government. Comparison with the case of Japan shows clearly that the Canadian multi-level government system inhibits certain kinds of policy approaches and encourages others.

This point is also illustrated by Japan's "developmental state" approach to encouraging rapid economic growth in the 1950s and 1960s. Japan's interventionist approach to economic planning, including huge investments in infrastructure to support economic investment—industrial water supply, electricity, ports and railways, huge coastal landfills for industrial complexes, export credits and coordinated access to capital, and state-directed cartels to prevent over-investment and maintain profits—was wildly successful in economic terms, helping Japan to grow from a devastated country on the brink of starvation after the Second World War to the second-largest economy in the world by 1970, despite its almost complete lack of raw material resources. Such economic intervention has never been a part of the Canadian approach to economic policy, which conforms much more closely to the liberal market economy approach of relatively limited direct involvement in industrial policy or in promoting particular industrial sectors.

Finally, the comparison between Canada and Japan points to the significance of Canada's vast and varied geography in shaping approaches to national-level urban policies. Japan's small size and high population densities, its limited arable land with over 75 per cent of the country covered with forested mountains, its long settlement history, its relatively homogeneous population, and its unitary government system all make a national urban policy much more likely there than in Canada. Here, our huge low-density areas make it harder to convince the public that a crisis of overconcentration in a particular region should be a national concern, even though Toronto, Montreal, and Vancouver each do have serious growth-management issues. The huge diversity of Canadian regions, from the Atlantic provinces to Quebec, Ontario, the Prairies, the West, and the North, each with distinctive urban challenges and priorities, means that crafting a national policy that works for all is very difficult.

As a result of all three of these factors, Canada's constitutional division of powers, its tradition of limited government intervention in economic policy, and its great diversity and variety of regions and regional policy challenges have meant that there is a much stronger tendency to allow room for individual cities and regions to manage their own affairs. Canadian municipal governments have more autonomy and greater legal authority than their counterparts in Japan. Canadian urban governance arrangements therefore more closely approach the ideal of subsidiarity or encouraging decision-making to occur at the lowest level of government that can effectively manage that policy level, in contrast to the highly top-down governance approaches of Japan.

Conclusion

Globalization, transnationalization, and increasing global economic integration suggest a decreasing significance of national boundaries and a disruption of national urban systems and economies, which are increasingly re-oriented towards international networks. But the degree to which this occurs, the speed of change, and the precise consequences are very different for each country, depending on its economic structure, geography, and the nature of existing linkages to the global economy. National urban systems that were already very open to the global economy may be affected less, and existing patterns may even be reinforced in some cases, so that outcomes are extremely varied between countries and cities. This variability is important and is potentially revealing of deeper structures and differences between countries and regions.

The Japanese case in particular shows the degree to which global integration and transnationalization are not new phenomena, as repeated transformations of the Japanese urban system—from the Edo period when borders were sealed to block Christian missionaries, guns, and ideas, to the nineteenth-century opening and industrialization, to the rapid economic-growth period of integration into the global economy, to the promotion of Tokyo as a global city in the 1990s—were all driven by international influences. In the most recent period, the most visible impact of transnationalization has been the hollowing out of the Japanese economy since the mid-1980s, as high land prices and wages drove more and more manufacturing offshore, especially to Asia.

Japan also shows the significant impact of transnational flows of ideas, as it borrowed heavily from internationally current best practice ideas, whether in terms of championing the Tokaido megalopolis, or of attempting to prevent urban sprawl, or of attempting to analyze and shape the national urban system through regional policies, technopoles, and new towns. Japan has been a continual borrower and adaptor of ideas from abroad and, to some extent, an exporter of ideas as well.

The comparison of the Japanese and Canadian cases suggests two main points. First, the very active policy interventions and investments designed to shape the Japanese national urban system make it clear how limited such efforts were in Canada. Although Canada has arguably much stronger city-planning and land-development control systems and much stronger provincial governments than in Japan, the power of the Japanese national government to shape national-scale changes is impressive, even if not always leading to the outcomes intended. Second, the comparison of Japan and Canada illustrates that different countries do indeed have very different political, economic, social, and cultural systems and practices, which some describe as amounting to different forms of capitalism. The very different institutional and legal capacities of different countries mean that even when policy ideas are shared, the ways in which they are implemented are often very different. Over time this leads to the emergence of divergent institutional and policy structures in different places. International comparison is therefore a valuable way of understanding the essential characteristics of different countries and different institutional geographies.

Key Points

- A key premise of many studies of globalization is that increased economic integration among countries, regions, and cities is producing more similarity both between and within national urban systems; yet this is often not the case.
- National urban systems are the product of the intersection of geography, economic development patterns, demographic changes, and transportation systems, and provide a useful window for comparing different national and regional patterns of development.
- Two important approaches to understanding national urban systems include the new institutionalism and varieties of capitalism research agendas.
- Within the Japanese urban system, cities have been profoundly shaped by government-led investment in railway networks both for inter-city and daily commuter travel.
- The Japanese and Canadian urban systems can be most usefully compared in relation to differences in models of national government, development policies, and geographies.

Activities and Questions for Review

1. Developing policies to shape something as large as a national urban system is a complicated and expensive undertaking. Work in a team to draft a policy brief arguing for such interventions in Canada, listing both the main reasons for such a policy and the main policy instruments to be employed.
2. The Japanese case reveals some of the benefits and costs of active intervention in attempts to reshape national urban systems. List some of the major unanticipated consequences of the Japanese urban policies discussed in this chapter.
3. Based on the case of Japan and your own knowledge of Canada, set up a debate with one team arguing that the "Varieties of Capitalism" conceptual framework is useful in helping us understand differences between countries, and the other arguing against this interpretation.

References

Barlow Report (1940). *Report of the Royal Commission on the Distribution of the Industrial Population*. London: HMSO.

Begovic, B. (1991). The economic approach to optimal city size. *Progress in Planning*, 36: 93–161.

Bourne, L.S. (1975). *Urban systems: Strategies for regulation: A comparison of policies in Britain, Sweden, Australia, and Canada*. Oxford: Clarendon Press.

Broadbent, J. (1998). *Environmental politics in Japan*. Cambridge: Cambridge University Press.

Calder, K.E. (1988). *Crisis and compensation: Public policy and political stability in Japan, 1949–1986*. Princeton, NJ: Princeton University Press.

Castells, M., and Hall, P. (1994) *Technopoles of the world: The making of twenty-first-century industrial complexes*. London and New York: Routledge.

Douglass, M. (1993). The "new" Tokyo story: Restructuring space and the struggle for place in a world city. In K. Fujita and R.C. Hill (Eds.), *Japanese cities in the world economy*. Philadelphia, PA: Temple University Press, pp. 83–119.

Ericson, S.J. (1996). *The sound of the whistle: Railroads and the state in Meiji Japan*. Cambridge: Council on East Asian Studies, Harvard University.

Friedmann, J. (1986). The world city hypothesis. *Development and Change*, 17: 69–83.

Glickman, N. (1979). *The growth and management of the Japanese urban system*. New York: Academic Press.

Gottmann, J. (1961). *Megalopolis: The urbanized northeastern seaboard of the United States*. Cambridge, MA: MIT Press.

Gottmann, J. (1980). Planning and metamorphosis in Japan: A note. *Town Planning Review*, 51(2): 171–176.

Haley, J.O., and Yamamura, K. (Eds.) (1992). *Land issues in Japan: A policy failure?* Seattle, WA: Society for Japanese Studies.

Hall, P. (1992). *Urban and regional planning*. London: Routledge.

Hall, P.A., and Soskice, D.W. (2001). *Varieties of capitalism: The institutional foundations of comparative advantage*. Oxford: Oxford University Press.

Hansen, N.M. (1968). *French regional planning*. Bloomington: Indiana University Press.

Harris, C.D. (1982). The urban and industrial transformation of Japan. *Geographical Review*, 72: 50–89.

Hayashi, R. (1994). Provisioning Edo in the early eighteenth century: The pricing policies of the shogunate and the crisis of 1733. In J.L. McClain, J.M. Merriman, and K. Ugawa (Eds.), *Edo and Paris: Urban life and the state in the early modern era*. Ithaca and London: Cornell University Press, pp. 211–233.

Hodge, G., and Gordon, D.L.A. (2008). *Planning Canadian communities: An introduction to the principles, practice and participants*. Toronto: Thomson/Nelson.

Hollingsworth, J.R., and Boyer, R. (1997). *Contemporary capitalism: The embeddedness of institutions*. Cambridge: Cambridge University Press.

Japan National Institute of Population and Social Security Research (2012). Population projections for Japan: 2011 to 2060. Tokyo: National Institute of Population and Social Security Research, http://www.ipss.go.jp/index-e.asp

Johnson, C.A. (1982). *miti and the Japanese miracle, the growth of industrial policy, 1925–1975*. Stanford: Stanford University Press.

Kerr, A. (2002). *Dogs and demons: Tales from the dark side of Japan*. New York: Hill and Wang.

Kornhauser, D. (1982). *Japan: Geographical background to urban-industrial development*. London and New York: Longman.

McCormack, G. (2002). Breaking the iron triangle. *New Left Review*, 13(1): 5–23.

McGill, P. (1998). Paving Japan—the construction boondoggle. *Japan Quarterly*, 45(4): 39–48.

Murayama, Y.J. (2000). *Japanese urban system*. Dordrecht and Boston, MA: Kluwer Academic Publishers.

Nagashima, C. (1981). The Tokaido megalopolis. *Ekistics*, 289 (July/August): 280–300.

Noguchi, Y. (1990). Land problem in Japan. *Hitotsubashi Journal of Economics*, 31: 73–86.

Noguchi, Y., and Poterba, J.M. (Eds.) (1994). *Housing markets in the United States and Japan*. Chicago and London: University of Chicago Press.

North, D.C. (1990). *Institutions, institutional change, and economic performance*. Cambridge: Cambridge University Press.

Pred, A. (1977). *City-systems in advanced economies: Past growth, present processes and future development options*. London: Hutchinson.

Richardson, H.W. (1973). *The economics of urban size*. Lexington, MA: Saxon House.

Rimmer, P. (1986). Japan's world cities: Tokyo, Osaka, Nagoya or Tokaido Megalopolis. *Development and Change*, 17: 121–158.

Robinson, I.M. (1981). Canadian urban growth trends: Implications for a national settlement policy. In *Human settlement issues*. Centre for Human Settlements. Vancouver: University of British Columbia Press.

Rozman, G. (1973). *Urban networks in Ch'ing China and Tokugawa Japan*. Princeton, NJ: Princeton University Press.

Sassen, S. (2000). *Cities in a world economy*. Thousand Oaks, CA: Pine Forge Press.

Smith, N. (1984). *Uneven development: Nature, capital, and the production of space*. Oxford: Blackwell.

Sorensen, A. (2002). *The making of urban Japan: Cities and planning from Edo to the 21st century*. London: Routledge.

Sorensen, A. (2003). Building world city Tokyo: Globalization and conflict over urban space. *Annals of Regional Science*, 37(3): 519–531.

Sorensen, A. (2006). Liveable cities in Japan: Population ageing and decline as vectors of change. International Planning Studies, 11(3–4): 225–242.

Sorensen, A. (2010). Land, property rights and planning in Japan: institutional design and institutional change in land management. *Planning Perspectives*, 25(3): 279–302.

Spicer, Z. (2011). The rise and fall of the Ministry of State for Urban Affairs: A re-evaluation. *Canadian Political Science Review*, 5(2): 117–126.

Steffensen, S.K. (1996). Evolutionary socio-economic aspects of the Japanese "Era of Localities" discourse. In S. Metzger-Court and W. Pascha (Eds.), *Japan's socio-economic evolution: Continuity and change*. Folkestone, UK: Curzon Press.

Streeck, W., and Yamamura, K. (2001). *The origins of nonliberal capitalism: Germany and Japan in comparison*. Ithaca, NY: Cornell University Press.

Tsuru, S. (1999). *The political economy of the environment: The case of Japan*. London: Athlone Press.

Ui, J. (Ed.) (1992). *Industrial pollution in Japan*. Tokyo: United Nations University Press.

Woodall, B. (1996). *Japan under construction: Corruption, politics and public works*. Berkeley: University of California Press.

Yamamoto, H. (Ed.) (1993). *Technological innovation and the development of transportation in Japan*. Tokyo: United Nations University Press.

Yamamura, K. (1992). LDP dominance and high land price in Japan: A study in positive political economy. In J.O. Haley and K. Yamamura (Eds.), *Land issues in Japan: A policy failure?* Seattle, WA: Society for Japanese Studies, pp. 33–76.

4

Globalizing Cities and Suburbs

Richard Harris and Roger Keil

Introduction

In this chapter, we sketch the major contours of the "urban revolution." The process of **urbanization** has been characterized by a tremendous spatial *concentration* of activities (in urban forms such as **megaregions, megacities,** and **global cities**) accompanied by a vast urban *extension* (with ubiquitous **edge cities**, **suburbanization**, and **post-suburbanization**); these are the core characteristics of what some have called "the urban revolution" (Lefebvre, 2003) or even the "suburban revolution" (Keil, 2013).

Cities have indeed been recast as not just the sites of but also the producers of tremendous demographic and economic change as more people are becoming urbanized in absolute and relative terms (Saunders, 2010). This process of concentration and extension at the beginning of the twenty-first century is not entirely new. Almost 50 years ago, Henri Lefebvre (2003) speculated that the world was on an inevitable path towards total urbanization in which little to no "outside" to the urban would remain. Lefebvre's proposal was not just an admission of the quantitative growth of the urban fabric across the surface of the world, it implied that the urban would became *the* organizing societal principle where sprawl and density would exist simultaneously, where parts of cities would be networked into global streams of urban economies and lifestyles, and where fragmentation would be the rule (Lefebvre, 2003: 14, 169). These complex incarnations of the dialectics of urban implosion and explosion (Brenner, 2014) are characterized in equal fashion by massive processes involving the concentration of capital and decision-making power in global cities and in the deconcentration of regional economies in, for example, vast 100-mile zones or 60-mile circles around city centres (Soja, 1986). These sprawling urban regions have become more diverse in population and economic activities. Infrastructural pressures have created new geographic dynamics of urbanization as airports, highways, and high-speed commuter transit cut through residential suburbs, office centres, and protected greenspaces like greenbelts. This chapter considers global cities and megacities as specific concrete urban forms developed in the urban revolution.

While the chapter discusses processes of worldwide suburbanization, contemporary expressions are different from the building of picket-fenced peripheries of much of the twentieth century. And, to a considerable extent, the blurring of boundaries between city and **suburb** has been a consequence of progressive suburbanization, a process driven by rising incomes, innovations in transportation technology, and the decentralization of employment. In the global North, for many decades suburbanization was favoured by experts, who viewed the suburbs as a solution to urban congestion and ill-health. Many urban residents agreed, and moved to the suburbs as soon as they could afford to do so. This was especially true of heterosexual families with children, who were attracted to the promise of greater space and privacy.

Some observers spoke of a suburban way of life, centred on the home, property ownership, patriarchy, and, often, conservative politics (Whyte, 1956). Others disagreed with the physical determinism of this point of view, noting social-class differences and suggesting that those who chose the suburbs did so because they preferred the domesticity that this environment enabled

(Gans, 1968). Lately, researchers have developed new arguments about the effects of suburban living. They suggest that auto-dependency is unhealthy and environmentally irresponsible (Frumkin, Frank, and Jackson, 2004). Far from being the solution to urban problems, the suburbs are now being seen as their main source. Such arguments are relevant in both the global North and global South, a distinction that itself is blurry. At the global, national, and urban scales, many now believe that it is more useful to think in terms of a global continuum than to think of binary categories.

The first part of this chapter focuses critically on two processes of spatialization that have shaped metropolitan areas everywhere and that have received particular attention: the emerging large super-conurbations of 10 million or more inhabitants called megacities, invariably found in the global South; and global or world cities that are command centres in the global economy, usually situated in the global North. The second part of the chapter introduces suburbanization as a key element in the global process of urbanization. Two closely related types of suburban difference correlate closely with national income levels: the forms of suburban development, notably the prevalence of informality, and the effectiveness of suburban governance. Where governance is weak, development often fails to conform to regulatory guidelines and may occur where it is supposedly prohibited. The Tehran metropolitan area provides an example of such issues, one that contrasts with the situation typical in Los Angeles. A third type of difference among suburbs—the meaning that suburban *residence* has for local people—varies more by culture than by wealth or income. Case studies of Tehran, Iran, and Los Angeles, USA, illustrate the first two forms of suburban difference and follow up with a consideration of urban form and city types in the Canadian context. The Canadian urban system is shown to be highly globalized and suburban yet without megacities or first-rank global centres.

Global Cities and Megacities

Changes in urban form in response to, in support of and as a consequence of, globalization have a long history. As early as in the beginning of the twentieth century the emergence of world cities, the **megalopolis** and the **conurbation,** was acknowledged as an expression of a generalization of industrial and post-industrial forms of settlement (Gottmann, 1961; Hall, 1966). With roots in these earlier debates, in recent years the concept of the megaregion has gained prominence once again among critical scholars of globalized urbanization. But as the editors of a definitive new collection on the topic assure us, "megaregions are just one of an increasingly large number of competing spatial imaginaries which purport to reflect globalization's new urban form" (Harrison and Hoyler, 2015: 236). The megaregional imaginary can sometimes be associated with extensive and expensive infrastructures typical for landscapes of high-speed rail, autoroutes, international airports, logistics, and ports (Cidell and Prytherch, 2015). Megaregions are associated with both global cities and megacities.

The Emergence of Megacities

At some point in the 1980s, we entered "a world of giant cities" (Dogan and Kasarda, 1988). The birth of megacities as a general phenomenon of our emerging globalized planet was also seen as a shift of the focus and dynamics of urbanization from the global North to the global South and as an indication of high urbanization rates in countries of the global South (Figure 4.1). While in the mid-twentieth century most of the largest cities in the world were still to be found in Europe and North America, at the twentieth century's end the majority of urban giants were in Africa, Asia, and Latin America (Dogan and Kasarda, 1988; Elledge, 2015). In many ways, then, after the 1980s the debate in urban studies, and even more clearly in the popular media, on the spread of urban settlements in the global South centred around the idea of the megacity in particular. The general euphoria about the megacity, however, eclipsed its initially rather marginal significance. Hardoy, Mitlin, and Satterthwaite (2001: 30) note:

> Most of the urban population in Africa, Asia, Latin America and the Caribbean (and in other regions) live in urban areas with less than 1 million inhabitants. Megacities had less than 3 per cent of the world's population in 1990. Megacities can only develop in countries with large non-agricultural economies and large national populations; most nations have too small a population and too weak an urban-based economy to support a mega-city. Close to 50 independent countries in Africa, Asia, and Latin America and the Caribbean had no urban centre that had reached 500,000 inhabitants in 1990.

Figure 4.1 Pudong skyline, Shanghai, viewed from the Bund

Nonetheless, megacities are here to stay. While the larger, well-known metropolises and capitals of the global South like Mexico City, Sao Paulo, Shanghai, Delhi, and Seoul are often mentioned in this group, it is "cities like Ghaziabad, Surat or Faridabad in India, or of Toluca in Mexico, Palembang in Indonesia or Chittagong, the Bangladeshi port" that give the phenomenon its name today (Webster and Burke 2012). There are now almost 30 megacities, with a total population of more than half a billion people, making up about 8–10 per cent of the urban population of the earth (UN-DESA, 2014; United Nations, 2014). Their economic weight is remarkable, both inside individual countries and in the world economy, with cities such as São Paulo and Bangkok producing almost half of the national GDP in their respective countries (Suri and Taube, 2013: 196). But size is not everything, as some of the most economically powerful global cities remain in the global North (Elledge, 2015).

There is no doubt that the sheer size and scale of urbanization in this current period is unprecedented. It poses a number of problems that need to be urgently addressed, as awareness is growing that they may not be easily solved. Among them are infrastructure, health, education, and public safety as well as rising vulnerability to disasters and threats caused by climate change, especially for coastal cities (Von Glascow et al., 2013). Moreover, megacities are characterized by huge social and economic disparities, segregations, and fragmentations that call for new forms of governance (Suri and Taube, 2013). While the massive urban expanses of the global South provide opportunities not available to country dwellers, environmental, health, and infrastructural conditions such as basic access to water and wastewater networks are lacking in poor megacities and lead to an overall reduction of expectations of what level of services would be considered standard in urban settings in the twenty-first century (Kotkin, 2014).

In dealing with these intractable issues, we can differentiate two camps that cut across ideological lines. Some believe that the urbanization of the world and especially its hyper-urbanized form of megacity development provide the conditions from which human and economic growth spring (Saunders, 2010). Some urban enthusiasts have even portrayed poverty as a wellspring of human development (Glaeser, 2011). But observers have critiqued this enthusiasm from various perspectives. Prominent progressive intellectual Mike Davis sounds the alarm in his book *Planet of Slums* (Davis, 2006) and points to the inherent dangers of an urbanization process that is mostly based on poverty and massive exclusion. At the other end of the ideological spectrum, libertarian Joel Kotkin (2014: 12) posits that

> the megacity is increasingly a phenomena [*sic*] of countries that are struggling to find their way in the modern world economy. Size used to be more correlated with economic and political success and dominance on a global scale. Today, some of the largest cities are disproportionately poor, and seem likely to remain that way for the foreseeable future.

Indeed, the runaway success of megacity economies can hardly hide the misery and poverty, and even hunger, that their formation also produces.

The Emergence of World Cities

Based on the early work of writers like Geddes and Mumford at the beginning of the twentieth century, the British planner Peter Hall coined the term "**world city**" in the 1960s (Hall, 1966). He was fascinated by the growing role of large urban centres in the large leading states of the world economy, from North America to Europe and the Soviet Union. Influenced more by the changing international division of labour during the 1970s, a new generation of world and global city researchers pointed to the emergence of an entirely novel urban form that would serve the international (and later global) economic order that was in the making. When the world city thesis was first developed, the world was imagined as one of national state systems that engaged in international relations. This was reflected in the common use of the term "internationalization" to express the intensification of worldwide relationships. It was during the 1980s and 1990s that the term "globalization" gained ground as the prevailing concept, testament to the perforation of national economies, to pervasive state rescaling, and to changing intellectual preferences. John Friedmann, in particular, systematized the field's speculations into a fully fledged world city hypothesis (Friedmann and Wolff, 1982; Friedmann, 1986, 1995) that led to theoretical debate and a rich practice of empirical study. In his view, global cities were seen as basing points of the global economy, control centres of a new international division of labour in which processes of proletarianization (such as occurred in the garment industry in Los Angeles in the 1980s) stood alongside the building of financial citadels of money and power (Keil, 1998). Viewed hierarchically for the most part, global city networks were considered a practical expression of (what counted as) the world economy at the time: a network of articulated urban and regional economies that defied the logic of the state system as it gave expression to a world system of global capital and labour, the former ever more concentrated, the latter ever more diverse (Knox and Taylor, 1995). Later, the idea of the global city was especially closely associated with the work of Saskia Sassen (1991, 2000), whose detailed study of the leading global financial centres—London, Tokyo, and New York City—defined the pinnacle of the hierarchized and layered network that Friedmann had already identified. A large variety of research methodologies and approaches has been deployed in the study of global cities, most prominently perhaps the Global and World Cities research network in Loughborough University in the UK (www.lboro.ac.uk/gawc/) under the original leadership of Peter Taylor (Taylor, 2004; Taylor et al., 2013). The global city literature itself has been changing under challenges to its limited view of what constitutes the global and global control, and attention has shifted to "globalizing cities" more generally (Keil, Ren, and Brenner, 2017).

The emergence of globalizing cities is not a natural occurrence, but a contested one. Governance and politics are critical to the conceptualization of how to chart a course for future urbanization. The megacities of the global South, for example, have been subject to projections by global development organizations and corporations interested in investment in construction and infrastructure and in designing better urban environments (Burdett and Sudjic, 2007). But they also have been seen as sources of mega-problems by global agencies such as the World Bank (Desmet and Rossi-Hansberg, 2014). Creating participatory structures among the millions moving to and building the cities of this age and, at the same time, allowing for large-scale strategic investment into urban structure and infrastructure are seen as major challenges in a context where fluid boundaries of formal and **informal urbanization** exist (Suri and Taube, 2013) (Figure 4.2). The governance and politics of the megacity, therefore, have taken centre stage for many scholars and policy-makers as urbanism itself becomes a prime medium for the negotiations of postcolonial futures. In this process, the size and scale of the megacity (and the megaregion) have become subject to much concern in terms of governance.

Global cities have also been seen as contested terrain. For Friedmann and Wolff (1982), they were originally primarily a landscape of sharp distinctions in political power, each divided into a citadel and a ghetto whose political trajectories rarely met and overlapped. Warren Magnusson (2011: 25) critically points to the constraints that come with the striving for global city status, especially from the perspective of the urban elites: "Once globalization emerged as the theme for business development, these nascent urban regions or city-regions were faced with a new challenge: how to organize themselves to take advantage of the new opportunities." Yet we must also view the global city as the outcome of struggles between different local trajectories of world city formation and as a strategic result of state action.

Suburbanisms and Suburbanization

Researchers debate over how to describe metropolitan areas in part because the distinction between city and suburb has become blurred. Suburbs grew with industrial urbanization in the nineteenth century. They emerged as residential communities at the urban fringe, having quite affluent residents whose male heads of households commuted to city jobs. To many, these are still the quintessential characteristics of suburbs (Forsyth, 2012). In time, however, the distinctive features of suburbs grew blurry as suburbanization continued in the global North and suburbs became more like cities in terms of land use, social composition, and density. This is also true in the global South, where fringe development has often taken the distinctive form of low-income informal settlement.

Physically, most cities grow primarily through suburbanization rather than through redevelopment and densification: they expand at their periphery, always continuously and sometimes also in a leapfrog fashion that invites the label **sprawl**. This term is also used, loosely, to refer to any type of low-density development at the urban fringe. The global urban trend of the past two centuries—first in Britain and Western Europe, soon after in North America, then in Latin America, more recently in east and South Asia, and lately in Africa—has involved suburbanization on a massive scale (Nicolaides and Wiese, 2016).

Indeed, because urban densities have declined over the twentieth century, the suburban trend has been greater than the growth of urban centres might suggest (Angel et al., 2012). Even cities that grew a little, or not at all, have suburbanized as their residents have chosen

Figure 4.2 Informal arrangements for tapping electrical wires in Kolkata, 2012

to live further from the centre at lower densities. Several developments have enabled this. Rising incomes, coupled with improvements in transportation technology, are the most important. The railroad and the streetcar (or tram) enabled people to move out along particular routes; from the 1920s in North America and increasingly after the Second World War elsewhere, the automobile permitted dispersal and sprawl. The failure of municipalities to charge developers and residents enough to cover the costs of providing suburban infrastructure has exacerbated the problem of sprawl (Blais, 2010). The decentralization of employment, first of manufacturing in the late 1800s and then of offices after 1945, also played a role. As people moved out of the city, they were followed by retail and service activities. These brought more jobs, encouraging still more people to move to the suburbs in a process that became a wholesale decentralization of metropolitan areas.

The distinction between city and suburb became increasingly unclear. A century ago, most jobs were concentrated in or near the downtown, and except for some industrial communities, most suburbs were essentially residential. Today, suburbs contain the full range of urban land uses, including many that have abandoned the city (most manufacturing) as well as land-intensive uses that never located there (e.g., airports). Moreover, the ubiquity of the automobile and truck has allowed suburban areas to develop in ways that appear incoherent and formless. Except where city plans decide otherwise, or where zoning prevents it, very different types of land use can exist side-by-side, while at the metropolitan scale only the most major of activities, such as airports, create broad elements of coherence in the pattern of land use.

The Suburban Way of Life, and Its Consequences

In many countries, one of the reasons that suburbanization has proceeded more rapidly than urban growth has been a widespread preference for suburban living. A century ago, cities were dense and often unhealthy, above all for low-income households that were compelled to live in crowded and unsanitary conditions. Suburbs were commonly healthier because they were lower in density and blessed with gardens and parks. In Britain and North America, although not so much in Europe, planners and other urban experts viewed suburbanization as the solution of urban problems (Nicolaides and Wiese, 2016). Many urban residents agreed, and with encouragement from developers and then government policies that favoured automobility and homeownership, they moved to the suburbs when they could afford to do so. Yet, it was not all about choice. In Europe, the suburban peripheries increasingly became the areas where the poor and unwanted land uses ended up in landscapes dominated by large-scale housing and production facilities.

Even in cosmopolitan centres such as Paris and New York, where the attractions of city living have always been appreciated, the suburban option has often been preferred by families with children and increasingly by immigrants (as seen in the increasing ethno-racialized suburbs and the production of **ethnoburbs**). The suburbs ideally offered more domestic space for family privacy and public space for recreation; they were perceived to be quieter, cleaner, and safer, although the true contrast with city neighbourhoods has often been more modest than many suppose, especially in the towers in the parks at the outskirts of Canadian and European cities that have become the home of many families driven from gentrified inner-city neighbourhoods. Because suburbs contain many families with children, they have been associated with an automobile-dependent way of life focused on the home, with homeownership, and with a broadly conservative political outlook. In Britain and North America in the 1950s and 1960s, some writers suggested that it was the physical character of the suburbs that produced this lifestyle. They contrasted it with the urban way of life, recently characterized by Louis Wirth (1938). Others, notably Herbert Gans (1968), disagreed. Noting that people in middle-class and working-class suburbs acted differently, he argued that class differences mattered more than physical setting and that any commonalities reflected the prior choices that residents had made, most typically with children in mind. If a suburban way of life existed, then, it was because residents had chosen to live in the environment that made it possible (Nicolaides and Wiese, 2016; Walks, 2006).

Debates about the effects of the built environment revived in the 1990s and are currently a major subject of research. Although many still discuss the socio-political aspects of suburban living, the main focus of debate concerns people's health and environmental consequences. Improvements in sanitation and public health over the twentieth century largely eliminated the hazards of high-density city living, at least in the global North, so that one advantage of the suburbs has been nullified. Indeed, suburban environments that discourage walking are now criticized as a cause of the rising incidence of obesity, heart disease, and diabetes (Frumkin, Frank, and Jackson, 2004). Although the health benefits of exercise are unquestioned, the precise significance of the built environment remains unclear. Many researchers suggest that diet and the sedentary character of modern work have more substantial effects on health than where we live (Biswas et al., 2015).

Clearer connections have been made between auto-dependence, greenhouse gas emissions, and climate change. Much more energy is required to transport people in cars than on public transit or by bicycle, especially because most drivers travel alone. Suburbs are commonly poorly served by transit, while long distances, coupled with the fact that roads are designed primarily for vehicles, discourage cycling. This exacerbates regional inequalities in access to mobility options. In many modern suburbs, almost all trips—to work, to stores, and even to school, and, ironically, to exercise class—must be made by car. This does not have to be the case. As long as suburbs are not built at very low densities or through **leapfrog development**, they are compatible with transit. For example, after 1945 the city of Stockholm developed mass transportation in tandem with suburbs that were focused on transit stops. Their residents are still able to manage many daily tasks without a car. In parts of the global South, too, suburbs and whole new towns, such as Gurgaon outside New Delhi, have been developed at high densities and with public transit. But everywhere—from Sao Paulo to San Francisco, Nairobi to Winnipeg—many suburbs are now built around the car.

Changing patterns of suburban development is not easy. Planners everywhere are encouraging redevelopment at higher densities. In some cases **densification** is feasible: a struggling mini-mall can be refashioned as a mixed-use condominium. But such changes are often resisted by local residents, who worry about parking and traffic. More generally, because ownership of suburban land is typically dispersed among many individuals, anything other than piecemeal redevelopment is difficult. The wheel, then, has turned full circle. Instead of being seen by experts as the solution to urban problems, the suburbs are now viewed as their main source. But many—perhaps the majority—of residents see things differently. They appreciate the personal flexibility that cars offer, the space for personal and family privacy. In Stockholm's planned suburbs, levels of automobile ownership approach those in North America. The threat of climate change can seem remote in time and space, while transit presents itself as crowded and inconvenient. Suburban living, despite rush hour congestion, often seems the best option. Changing popular opinion is the planner's biggest challenge.

North and South

These observations about the causes and effects of suburbanization apply everywhere, but in different degrees. As with the academic and popular literature on cities, suburbanization has been framed in terms of the experience of the global North and, above all, of English-speaking nations: the United States, Britain, Canada, Australia, and New Zealand. In several respects the experience of the global South is different. So different, in fact, that until recently it was treated separately: researchers focused on cities in one or the other world region; textbooks spoke about suburbs in the global North but emphasized **slums** and shantytowns in the global South. But the world, and the way researchers think about it, is changing.

The once-clear line between the global North and South is now a continuum, and as discussed earlier, there has been a corresponding globalization of urban forms and a convergence in suburban experiences. In Shanghai, Jakarta, and Mumbai, many suburbs are being designed around the automobile. Although not all are directly modeled on American predecessors, gated communities are sprouting in Bangalore, Buenos Aires, and Beijing (Glasze et al., 2006). And the rapid pace of urbanization in the global South means that this region contains a rapidly growing share of the world's suburbs, indeed more now than the global North. China is the prime case in point, and one without historical precedent (Wu and Shen, 2015). Between 1990 and 2010, the level of urbanization in China jumped from 26 per cent to 50 per cent, as the number of people living in urban areas more than doubled, to just over 600 million. Out of necessity, most of this growth was accommodated through development at the urban fringe, commonly in high-rise developments, although also through the informal redevelopment at higher densities of rural villages as they were absorbed into the expanding urban fabric. The scale and scope of this change defy description. In the face of this growth and of increasing commonalities, it becomes more fruitful and indeed necessary to survey urban and suburban developments as a whole.

But if, as many believe, there is a global convergence in the urban and suburban experience, there are still important differences. Although the binary categories of global North and South no longer work well, there are several distinctions that correlate approximately with average incomes. The most important of these concern suburban development and its governance. In the South, development often does not conform to state regulation, a clear sign of weak governance. There is also a less tangible cultural difference pertaining to the meaning of the suburbs, a difference that is apparent in the words that are used to describe the urban fringe and also in the connotations of those words.

Suburban Development and Governance

Suburban development involves the conversion of rural or vacant land to urban use. In some countries and for certain periods of time, a branch of the state has played a leading role in this process. Prime examples are the countries of the Communist bloc between the late 1940s and the disintegration of the Soviet Union in 1989. More typically, state agencies initiate specific developments—housing projects, for example, although even then the actual work of construction is usually subcontracted. But with these exceptions, the great majority of suburban development is undertaken by private land developers (Rybczynski, 2008). These typically aim to make their profit by acquiring land cheaply, ideally at its value for agriculture, and then selling it for urban use, for which it is much more valuable.

In the global North, land developers play, or at least appear to play, by the rules. They only build or arrange for construction where allowed; they attend to zoning regulations, which govern land use, and building regulations, which define methods and materials; and these days, typically, they are required to install hard services (e.g., water, sewers, and roads) prior to the sale and occupancy of homes, and this they do. That is not to say that everything is above board. In most countries, the development and construction industries are rife with corruption, because large profits can be made through the rezoning of land. But, nominally, modern suburban subdivisions usually conform to state regulation, although attempting to stretch the planning rules is common.

But this is not true elsewhere. The most visible suburban difference between countries lies in the prevalence of what are still referred to as slums, that is, areas seriously deficient in their physical character (Davis, 2006). Slums take various forms but share two characteristics: their presence indicates poverty as well as an *informal* manner of land development. The two are connected. The poor can rarely afford accommodation that conforms to state regulations, and so they must compromise and improvise.

Informality refers to economic activity that is not in principle illegal, but that is carried out in a place or manner that contravenes one or more laws (Davis, 2006). Although all types of people are associated with informal activity, it is most commonly associated with the poor. **Squatting**, where people occupy land that they do not own, rent, or otherwise have rights to, is the most striking example. It happens most frequently on publicly owned land, because governments find it politically difficult to evict large numbers of potential voters—although many, such as Zimbabwe's President Mugabe, have been willing to face that challenge. Squatting is also common around cities like Cairo and Tehran, where fringe land has little or no agricultural value. Squatting is a feature of many cities in the global South, and nowhere more than in Latin America in the 1950s and 1960s, but the supply of valueless or publicly owned land is finite.

More common these days are various types of unauthorized or **pirate settlements** in which residents have legal right to land but where development has been prohibited, where dwellings violate building standards, or where developers have not installed required services (Davis, 2006). In all types of informal settlements, especially those that are squatted, residents often build their own homes. The style, dimensions, and quality of the results typically vary, sometimes wildly. In poorer areas, conditions may be appalling, with flimsy structures that offer little shelter and open sewers that compromise health. More than any other feature, such settlements symbolize the distinctiveness of cities in the global South.

The existence of informal settlements points to a weakness of suburban governance. To the extent that state regulations are well designed, this is unfortunate. In practice, however, some have done more harm than good, especially when they have demanded unreasonably high standards. This is especially true of those regulations that were inherited from the colonial era. Governance refers to the manner in which suburban development is regulated and serviced through the modalities of state action, market initiative, and often authoritarian forms of privatization (Hamel and Keil, 2015). Regulation is usually seen as the domain of the state, but other agents, including residents and especially developers, are also significant. For example, although squatter settlements may appear anarchic, even they are shaped by forms of regulation. These may include verbal agreements between neighbours as well as bylaws enforced by associations of residents.

In the global North, the state plays a dominant though never exclusive role in governance. Cities provide a range of services and enforce various regulations that govern development. Except for affluent enclaves, suburban areas are often less well serviced and less tightly regulated. Although we think of state controls being imposed on resentful developers, the latter often take the initiative in establishing land-use and building regulations. This makes their subdivisions more marketable. Today, those who develop gated communities include many provisions that constrain what residents can do with their property, down to the colour they can paint their doors (McKenzie, 1994). Developers suppose that buyers will tolerate such constraints because they limit what neighbours can do, thereby protecting property values.

The existence of informality underlines the limits of state power. Governments may tolerate it, knowing that they cannot stop it, but they lack the ability to enforce regulations or to provide alternative accommodation. Indeed, they often welcome informal settlements, which provide accommodation, however inadequate, thereby reducing the political pressure to address the housing shortage. Governments may also welcome them as a potential political resource: at election time they may buy votes by promising to regularize or service settlements that are nominally illegal. Here, informal settlements are not so much a sign of the weakness of the state as they are an indicator of a distinctive mode of governance (Roy, 2011).

The Meaning of the Suburbs

Researchers have lately referred to suburbanization as the combination of non-central population and economic growth with urban spatial expansion. Qualitatively distinct suburban ways of life have been labeled **suburbanisms** (Hamel and Keil, 2015; Keil, 2013). One aspect of the suburbanization process that continues to resist global convergence is its local meanings. In Canada, as in the United States and Britain, suburbs have always had particular connotations. These were once overwhelmingly positive but are now mixed. In speaking about suburbanization as a global phenomenon, we should not assume that its meaning is universal. On this issue, more than on the subjects of development and governance, precise comparisons are difficult. Many peoples, of course, have words in their own language for the type of area that we call a suburb. Among the more notable words are *banlieue* in France, *suburbio* in Italy, *periferia* in Spain and Mexico, *Stadtrand* in Germany, *jiaoqu* in China, and *prigorod* in Russia. But, although they all gesture towards the urban fringe, these words do not denote precisely the same phenomena. And, to the extent that they do, they do not have the same connotations as "suburb." In many European countries, for example, their social connotations have long been quite negative (Harris and Vorms, 2017).

Elsewhere, more strikingly, many societies simply do not have a word for the places we call suburbs. India and Italy are examples, as is, to a lesser extent, Iran. Local experts, trained in the West, familiar with Western discourse, and wishing to communicate internationally, may use the language of suburbs, new towns, and so forth, but their language has not filtered down to the streets. Unless local people learn to see a significant difference between city and suburb, they may never have reason to adopt such terminology. We should be careful, then, in imposing our words and meanings on their experience. But perhaps researchers will follow their example. As the boundaries between cities and suburbs have become blurred, English-speaking urban experts have become uneasy with such terminology. Many have tried to develop new terms, such as "post-suburban" (Phelps, Wood, and Valler, 2010). None has yet caught the popular imagination, but perhaps someday one will. For the present, we have only imperfect terms to grapple with a bewilderingly complex global reality. Two metropolitan areas, Los Angeles and Tehran, provide a glimpse of that global diversity.

Los Angeles: Post-suburban Megacity in the Global North

Los Angeles is a prime example of the kinds of sub/urbanization processes discussed in this chapter (Figure 4.3). Born as a colonial outpost in the Spanish and later American empires, the city had a high proportion of white, native-born Americans until the Second World War. The war years brought a new and more diverse population to the Southland. By the 1980s, finally, the Southern California metropolis had become the ultimate symbol of a globalized immigrant region that, for a while, had the highest percentage of non-native born residents in the United States.

Los Angeles also typifies many of the conundrums currently faced in urban research. It starts with the problem of scope and definition. Is it a city, a region, a city-region? Is the region of 15 million a global city or a megacity? Is it urban or suburban? As early as the 1920s, the city had grown out of scale compared to all other cities in the world. Sprawling across Southern California, with an annexed port in San Pedro to the south and a hydrocephalic San Fernando Valley to the north, the city of Los Angeles had exploded the original pueblo, fanning out along railway lines at first and later along the emerging systems of roads and freeways into residential neighbourhoods that spread from the sea to the mountains. Fed by imported water through the Owens Valley aqueduct, the city was able to sustain itself beyond its local desert metabolisms; the port and the railway kept it provisioned when there was no longer any regional wealth to be exploited once

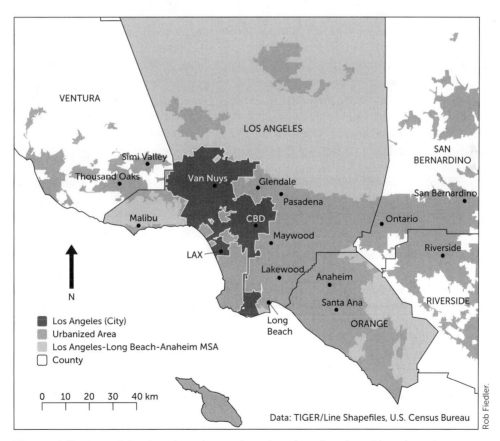

Figure 4.3 Map of the Los Angeles region showing the city of Los Angeles, Los Angeles, California and surrounding counties and various points of interest.

the ranchos and orange groves had receded into subdivisions. The land rush of the 1880s, which had created real estate fiefdoms out of agricultural land, laid a pattern of developer-driven, railroad-supported, media-hyped development that allowed for the highest degrees of fragmentation possible. Each developable patch of land seemed to become incorporated into yet another city. The founding of cities merely for functional purposes—industry, commerce, and the "black-gold suburbs" in the South Bay oilfields—established a pattern of segregated land use and ultimately population that was cast in stone with the Lakewood Plan of 1954. This allowed unincorporated lands that could not provide their own services to become municipalities and have their public duties picked up by the County of Los Angeles. An incorporation boom ensued that made for a process of "privatizing with class" whereby each taxpayer could theoretically live with only those neighbours that fit their social expectation and income. But it also made initially invisible those communities, like the African Americans of Watts before the 1965 riots, the Chicanos of East Los Angeles, and the gays of West Hollywood, that lived in pockets of the municipal puzzle.

Talking about Los Angeles, then, is always fraught with the problem of definition. Born disparate, dispersed, and suburban, Los Angeles grew into one of the densest carpets of urbanized humanity anywhere in the world, tied together by a federally financed and globally fueled infrastructure of the military, media, and mobility. The Southern California metropolis was at once the largest industrial city in the West and anyone's post-industrial dream, sweatshop nightmare, and postmodern phantasmagoria all in one. In the 1980s and 1990s, when Los Angeles enjoyed its moment in the sunshine of the urban studies universe, the city and its region were rethought by the influential LA School as both a palimpsest of possibilities and dystopian hell, depending on which one of their representatives you believed (Davis, 1990; Dear, Shockman, and Hise, 1996). The Southland had megacity ambitions, as it outpaced most economies in the United States and elsewhere. It served as the ultimate model of the "luxurious, splendid" world cities whose very "splendor obscures the poverty on which their wealth is based" (Friedmann and Wolff, 1982: 319) (Figure 4.4). And it provided anyone seeking to study megaregions a good place to start. At its functional core, it is what it had been from the start: a horizontal city,

a suburb that had grown into post-suburbia, the first city of the twenty-first century to believe and confirm the hype of the Urban Age. Today, Los Angeles is one of the densest urban areas in North America, perpetually growing economically and demographically, although not at the clipped speed of previous decades, remaining international and culturally diverse in population, although the region's foreign-born population takes up a smaller percentage of the total than in the recent past. The region has also become a site of significant transit development based on an ultimately extensive rail network and environmental innovation, as the once smoggy air is cleaned up and its namesake river begins to be a green spine rather than a concreted flood-control channel. Los Angeles remains the place where the world's dreams are produced and televised in motion pictures while it continues to provide opportunity for millions of

Figure 4.4 View of Bunker Hill from the Bonaventure Hotel, Los Angeles. These buildings were once considered the "citadel" of postmodern world city Los Angeles and still represent the core of the city's revitalized downtown.

its inhabitants as well as newcomers to help define the notion of living the urban revolution.

Tehran: A Globalized Megacity with Extensive Informality but No "Suburb"

If Los Angeles still typifies many people's image of the suburban metropolis, Tehran is a place where no one speaks of suburbs in any language. No place can typify cities of the global South, but Tehran, the capital of Iran, illustrates several key themes: globalizing forces, informality, and the varied meanings of the urban periphery. A rapidly growing global centre, Tehran boasts many suburbs that developed informally and that locals view in ways unfamiliar to a Canadian ear (Figure 4.5).

Iran, or Persia as it was known until 1935, has always had international connections. For centuries its urban settlements were important stops on the Silk Road,

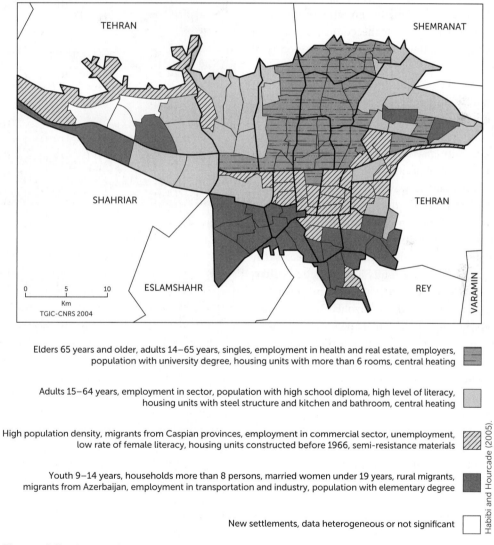

Figure 4.5 Map of the social geography of Tehran, Iran based upon "all population and housing data" shows the affluent northern sector, its much poorer southern sector, and an outer ring of informally developed and poorly documented "new settlements."

the trade routes that linked Europe with Asia. Tehran was a late addition to these, growing steadily from the 1700s and rapidly since the interwar period of the twentieth century. In 1931 its population reached 334,000, similar to that of Vancouver; by 1951 it drew level with Toronto, at 1.5 million; today, 8.5 million in the city alone, it rivals New York City (Angel et al., 2012). Rural–urban migration has fueled much of its growth, especially during the Iran-Iraq war (1980–1988). Since the Revolution of 1979, which overthrew the US-backed Shah and rejected Western, secular values, Iran has been subject to international sanctions of varying severity. Despite this, Tehran's growth and modest prosperity have depended on the country's role as a major exporter of oil, the leading global commodity of the past century.

In recent decades, rapid growth has meant extensive suburbanization, but its process, form, and meaning would be unfamiliar to Canadians (Habibi and Hourcade, 2005). The very rapidity of growth defied planning. In 1962 the deputy mayor commented that "buildings and settlements have been developed by whomever has wanted in whatever way and wherever they have wanted" (quoted in Madanipour, 2006: 435). Various types of informality have been common, although less so than elsewhere in the global South. Normally, even where tolerated, squatting is illegal, but not necessarily in Iran. Much of the land around Tehran is desert-like, unclaimed, and without value. Under Islamic law, such land can legally be settled by anyone. Millions have done so, typically by erecting their own modest dwellings, with the help of friends or family. Other forms of informality include unauthorized construction where official plans prohibit settlement, the addition of prohibited extra storeys to existing dwellings, and the evasion of taxes through a failure to report property transfers (Figure 4.6).

As it has expanded, Tehran has absorbed pre-existing settlements. Now multi-centred, it encompasses 22 municipalities, each with its own mayor, while 4–5 million people live beyond city limits in a poorly serviced fringe. This pattern of poor outer suburbs is similar to that of European cities like Rome and Paris. Similar, too, is the city's sectoral pattern: in the affluent northern sector, reaching towards attractive foothills, land prices are double the city-wide average; to the south they are half.

Figure 4.6 An informally built house with a flat roof in Tehran, where even middle-class families in quite substantial dwellings have found it convenient and economical to work on their own homes, often in stages, by adding storeys

What is most distinctive is the way locals think about their city and suburbs, whether the latter lie within or beyond city limits. "Suburb" has no Persian equivalent. The closest term, *houmeh,* does not imply a particular social class, distance from the city centre, or even an area that is primarily residential. Residents do use the term *haashieh* when referring to the poorer, informal "new settlements" in the outer ring, but their mental maps of the city, including areas recently developed, are dominated by the north–south contrast and not by location or density. Tehran, then, has been experiencing extensive suburban development, but locals do not think of it in such terms. In Canada, however, the concept of suburbanization is widely accepted and used by residents, planners, policy-makers, and theorists alike.

Canada: Highly Globalized and Suburban

There are few superlatives in Canada's urban system: we have no megacities, no first-rank global centres. But a trend towards centralization is apparent. By 2001, 62.7 per cent of Canada's urban population lived in large urban areas—those with a population in excess of half a million—and by 2011 this proportion had risen to 63.9 per cent (Statistics Canada, 2011). By contrast, the proportion of those living in towns and small cities under 50,000 fell from 8.1 per cent to 7.6 per cent. And the largest centres grew most rapidly of all, as they attracted and retained a disproportionate share of immigrants. But in varying degrees all of Canada's cities show aspects of most of the processes we have discussed in this chapter. Canada made the statistical transition to urban society long before it became fashionable to talk about an "urban revolution." Rapid urbanization, by any measure, made the majority of its population urban by 1931, although Canadians' sense of identity remained unimpressed by that fact (Kipfer and Keil, 2003). The country really does have one discontinuous megacity along the American border with tentacles spreading north along resource routes, real estate corridors, and rivers. The Calgary–Edmonton corridor in Alberta and cottage country in Ontario are examples of that spread.

Canada is a country in which global city formation has been an important feature of its overall urban process. More than most nations and from the beginnings of European settlement, Canada has always depended on transnational flows of goods, capital, and people. Today, Toronto and Vancouver, and to a lesser extent Montreal, are most obviously tied into such global flows, but no city escapes this influence.

There clearly is a Canadian school of suburban and post-suburban research (for an overview, see Hamel and Keil, 2015; Harris 2004, 2015). Canada is a country where the term "suburb" still carries much of its original meaning: residential areas, recently developed, relatively affluent, and lower in density than the city (Harris, 2004: 18–33). The fact that the residents of many suburbs have been poor and racialized has dented this image only slightly. Historically, many immigrants built their own modest homes at the fringe; in the 1960s, many were assigned to new public housing projects in suburban locations, and these help to define the character of what are now called inner suburbs, notably in Toronto (Harris, 2015). Unlike in the United States, but similar to the United Kingdom, a place in Canada does not have to be a municipality to be a suburb, although there is a sense that areas lying beyond city limits are the most suburban.

In recent decades, the stereotype that suburbs are lower in density has been challenged. During the 1960s and 1970s, many apartment buildings—both privately and publicly owned—raised suburban densities. Since then, the size of the average suburban lot has shrunk while that of the average dwelling has increased. Modern Canadian suburbs are much more densely built up than those of the early postwar decades or than their American counterparts. This is something that governments are encouraging. In 1973 the government of British Columbia established the Agricultural Land Reserve, which has constrained urban development ever sense. In 2005 Ontario established a green belt that has helped to restrain sprawl and redirect development inwards. The effect has been to change the geography of land values, the extent of infrastructure planning, and the landscape of politics in the region.

Typical of cities in the global North, Canada's suburbs have been developed in an essentially legal, formal manner. Since 1945, cities have enacted comprehensive plans that define what types of development are permitted in each area, and such plans have generally been followed. Squatting has been very rare, and few property owners evade taxes for long. But modest informality has always existed, especially during boom periods. During the housing shortage of the late 1940s, building inspectors turned a blind eye to buildings that violated

continued

> the letter of the law. More recently—in Hamilton for example—professional builders may begin construction before receiving permits, assuming that the city's red tape will catch up in due course. But these are viewed as exceptions to the rule.
>
> Like their counterparts almost everywhere, Canadian cities and suburbs have suffered from weak metropolitan or region-wide planning (Keil et al., 2015). Patterns of commuting in modern metropolitan areas reveal complex connections between city and suburb, and among suburbs. Effective planning of highways and public transit requires coordination on a regional scale, but political inertia and local resistance often prevent this. The problem has been compounded by the reluctance of provincial governments to raise the taxes and make the investments necessary to maintain, still less to improve, such infrastructure. The result, most apparent in the largest centres, has been a growing level of congestion that frustrates commuters and limits productivity.

Conclusion

The portrait painted in this chapter has revealed a kaleidoscopic image of globalized urbanization. No stark division into global and ordinary cities can be discerned; the old city-suburb distinction is largely a thing of the past (if it was not already always a fiction). In the more and more tightly woven fabric of the urban world, we find a broad spectrum of forms, functions, and uses that make up the urban relationality that Lefebvre describes as urban society. Los Angeles, once the ultimate example of postwar suburbanization, has now become a city of dense sprawl, a powerful globalizing megacity-region that continues to present a kaleidoscope of the urbanization trends we have portrayed in this chapter.

The hybrid megaregions, both city and network (Wachsmuth, 2015), resemble tapestries of remarkable diversity. As Xuefei Ren (2013: 104) reminds us,

> on the Chinese periphery, one can find residential new towns of massive scale, exclusive European themed villas, migrant villages, brand-new university campuses, military-style manufacturing facilities and workers' dorms, and often times, agricultural fields in the midst of urban construction as well. Heterogeneity of socio-economic composition, high population density, and dependency on public mass transit characterize the nature and process of urban territorial expansion.

Migrant villages are the Chinese version of informal settlement, a common feature of suburban growth in cities of the global South such as Tehran, featured here. With the notable exception of informality, most of the same features, albeit not at the same scale, can be found in Europe and North America, where in-between urbanization (Sieverts, 2003) has now become the norm and large urbanized regions are inevitably fragmented in appearance and function.

This is the landscape that Phelps and Wu (2011) have characterized as typically "post-suburban." There are many parts of the world where this phrase has little or no meaning for local people. In Tehran, as in Indian cities and indeed many parts of Europe, residents have never used a single term to refer to the urban fringe, and certainly not "suburb." But urban experts, for whom English is the global language, together with the residents of Britain, Australasia, and North America, arguably need to go beyond the old categories of city and suburb. The global and the post-suburban are terrains of articulation that give form to this present period of urbanization. In this new, amorphous, emerging form, there exists "a field of negotiation between multiple agents at stake, with different positional backgrounds and different resources that condition new processes of urbanization via asymmetric positions of power" (Savini, 2013: 15). Political governance, steering, and planning are difficult in this landscape, as power is not easily localized and centralized (Young and Keil, 2014) and defies conventional explanation and access. The perceived tension between formal and informal modalities of governance is also crucially in need of attention as states seek control over spaces they have not created because so-called informal city-building practices outpace other modalities of governance (Hamel and Keil, 2015).

The urban landscape that is taking shape in the first half of the twenty-first century is revolutionary, as it transforms a largely rural human experience into a largely urban one (Saunders, 2010). In the process, our understanding of "the urban" itself is challenged by far-reaching suburbanization and post-suburbanization that do away both with the notion of the city as a limited container and with suburbanization as a linear process that starts in the centre. Yet despite its revolutionary appearance, the process that brings us worldwide global and megacities and suburban constellations also broadly confirms the existing network of cities that has been around since the beginning of human settlement. The future will tell whether these tremendous and fascinating globalized sub/urban landscapes will be, as some make them out to be, the fields of opportunity for a new generation of urban dwellers rising out of poverty or marginality, or whether they will become the supercharged reincarnation of the hopeless Dickensian cities of the early industrial period.

Key Points

- With over 50 per cent of the world's population now urban, we can speak of an urban revolution that refers to major processes of concentration and extension in urban form and function.
- Among the many forms of urbanization we find today around the world are megaregions, megacities, and global cities.
- Rather than looking at suburbanization and suburbanisms as processes of novel urbanization, we have now begun to see them as permanent features of a post-suburban landscape.
- As the case of Los Angeles illustrates, suburbanization is now ubiquitous, blurring the line between city and suburb. It is possible to talk about post-suburbanization as a general experience.
- As the case of Tehran illustrates, remaining differences concern the extent of informality in land development and weaknesses of governance.
- Canadian suburbanization, high-density by North American standards, could support better transit but requires investments in infrastructure.

Activities and Questions for Review

1. Create a chart with a column titled "global North" and a column titled "global South" and in each column identify the distinguishing features of suburbanization. How is suburbanization similar and/or different in the global North and global South?
2. How relevant are the categories of city and suburb to your own family's lived experience? Compare your experiences to those of a classmate and discuss how your experiences converge and diverge.
3. With reference to social, economic, and environmental influences, illustrate in a diagram the points at which a megacity develops into a megaregion.
4. As a class, debate whether "globalizing cities" is a more appropriate concept than "global cities" for capturing contemporary urban realities in both the global North and global South.

Acknowledgments

This research was partly supported by the SSHRC-funded Major Collaborative Research Initiative on Global Suburbanisms: Governance, Land and Infrastructure in the 21st Century. We gratefully acknowledge the research assistance of Jenny Lugar, Ali Madanipour, and Mehrnaz Nazari, together with the constructive suggestions of the editors.

References

Angel, S., Parent, J., Civco, D.L., and Blei, A.M. (2012). *Atlas of urban expansion*. Cambridge, MA: Lincoln Institute for Land Policy, http://www.lincolninst.edu/subcenters/atlas-urban-expansion

Biswas, S., et al. (2015). Sedentary time and its association with risk for disease incidence, mortality, and hospitalization in adults: A systematic review and meta-analysis. *Annals of Internal Medicine, 162*(2): 123–133.

Blais, P. (2010). *Perverse cities: Hidden subsidies, wonky policies, and urban sprawl*. Vancouver: University of British Columbia Press.

Brenner, N. (Ed.) (2014). *Implosions/Explosions*. Berlin: Jovis.

Brenner, N., and Keil, R. (Eds.) (2006). *The global cities reader*. London: Routledge.

Burdett, R., and Sudjic, D. (Eds.) (2007). *The endless city*. London: Phaidon Press.

Cidell, J., and Prytherch, D. (Eds.) (2015). *Transport, mobility, and the production of urban space*. London: Routledge.

Davis, M. (1990). *City of quartz: Excavating the future in Los Angeles*. London: Verso.

Davis, M. (2006). *Planet of slums*. London: Verso.

Dear, M., Shockman, H.E., and Hise, G. (Eds.) (1996). *Rethinking Los Angeles*. Thousand Oaks, CA: SAGE.

United Nations Department of Economic and Social Affairs (UN-DESA) (2014). World's population increasingly urban with more than half living in urban areas. United Nations (July 10). Retrieved from http://www.un.org/en/development/desa/news/population/world-urbanization-prospects-2014.html

Desmet, K., and Rossi-Hansberg, E. (2014). Analyzing urban systems: Have megacities become too large? World Bank: Policy Research Working Paper 6872. May.

Dogan, M., and Kasarda, J.D. (1988). *The metropolis era*. Vol. 1: *A world of giant cities*. Newbury Park, CA: SAGE.

Elledge, J. (2015). Where are largest cities in the world? *CityMetric*, May 19. Retrieved from http://www.citymetric.com/skylines/where-are-largest-cities-world-1051

Forsyth, A. (2012). Defining suburbs. *Journal of Planning Literature, 27*: 270–281.

Friedmann, J. (1986). The world city hypothesis. *Development and Change, 17*: 69–83.

Friedmann, J. (1995). Where we stand: A decade of world city research. In P. Knox and P. Taylor (Eds.) (1995), *World cities in a world system*. New York: Cambridge University Press, pp. 21–47.

Friedmann, J., and Wolff, G. (1982). World city formation: An agenda for research and action. *International Journal of Urban and Regional Research, 6*(3): 309–344.

Frumkin, H., Frank, L., and Jackson, R. (2004). *Urban sprawl and public health: Designing, planning and building for healthy communities*. Washington, DC: Island Press.

Gans, H. (1968). Urbanism and suburbanism as ways of life. In H. Gans (Ed.), *Essays on urban problems and solutions*. New York: Basic Books.

Glaeser, E. (2011). *Triumph of the city*. Oxford: Pan Books.

Glasze, G., Webster, C., and Frantz, K. (Eds.) (2006). *Private cities: Global and local perspectives*. London: Routledge.

Gottmann, J. (1961). *Megalopolis: The urbanized northeastern seaboard of the United States*. New York: Twentieth Century Fund.

Habibi, M., and Hourcade, B. (2005). *Atlas of Tehran Metropolis*. Vol. 1: *La terre et les hommes—Land and people*. Tehéran/Paris: Tehran Geographic Information Center and CNRS. Retrieved from http://www.irancarto.cnrs.fr/volume.php?d=atlas_tehran&l=en

Haddad, E.A., and Teixeira, E. (2015). Economic impacts of natural disasters in megacities: The case of floods in São Paulo, Brazil. *Habitat International, 45*(2): 106–113.

Hall, P. (1966). *The world cities*. New York: McGraw-Hill.

Hamel, P., and Keil, R. (Eds.) (2015). *Suburban governance—A global view*. Toronto: University of Toronto Press.

Hardoy, J.E., Mitlin, D., and Satterthwaite, D. (2001). *Environmental problems in an urbanizing world*. London: Earthscan.

Harris, R. (2004). *Creeping conformity: How Canada became suburban, 1900–1960*. Toronto: University of Toronto Press.

Harris, R. (2015). Using Toronto to explore three suburban stereotypes. *Environment and Planning A, 47*(1): 30–49.

Harris, R., and Vorms, C. (Eds.) (2017). *What's in a name? Talking about the urban periphery*. Toronto: University of Toronto Press.

Harrison, J., and Hoyler, M. (2015). Megaregions reconsidered: Urban futures and the future of the urban. In Harrison and Hoyler (Eds.), *Megaregions: Globalization's new urban form?* Cheltenham, UK: Edward Elgar, pp. 230–256.

Keil, R. (1998). *Los Angeles: Globalization, urbanization, and social struggles*. Chichester, UK: Wiley.

Keil, R. (2011). Suburbanization and global cities. In B. Derudder, M. Hoyler, P.J. Taylor, and F. Witlox (Eds.), *International Handbook of Globalization and World Cities*. London: Edward Elgar.

Keil, R. (Ed.) (2013). *Suburban constellations*. Berlin: Jovis.

Keil, R., Hamel, P., Chou, E., and Williams, K. (2015). Modalities of suburban governance in Canada. In P. Hamel and R. Keil (Eds.), *Suburban governance—A global view*. Toronto: University of Toronto Press: 80–109.

Keil, R., Ren, X., and Brenner, N. (Eds.) (2017). *The globalizing cities reader*. London: Routledge.

Kipfer, S., and Keil, R. (2003). The urban experience. In L. Vosko and W. Clement (Eds.), *Changing Canada: Political economy as transformation*. Montreal and Kingston: McGill-Queen's University Press, pp. 335–362.

Knox, P., and Taylor, P. (Eds.) (1995). *World cities in a world system*. New York: Cambridge University Press.

Kotkin, J. (2014). *The problem with megacities*. Orange, CA: Chapman University Press.

Lefebvre, H. (2003). *The urban revolution*. Minneapolis: University of Minnesota Press.

McKenzie, E. (1994). *Privatopia: Homeowner associations and the rise of residential private government*. New Haven, CT: Yale University Press.

Madanipour, A. (2006). Urban planning and development in Tehran. *Cities, 23*(6): 433–438.

Magnusson, W. (2011). *Politics of urbanism: Seeing like a city.* London and New York: Routledge.

Nicolaides, B., and Wiese, A. (Eds.) (2016). *The suburb reader* (2nd ed.). New York: Routledge.

Phelps, N.A., Wood, A.M., and Valler, D.C. (2010). A post-suburban world? An outline of a research agenda. *Environment and Planning A, 42*: 366–383.

Phelps, N.A., and Wu, F. (Eds.) (2011). *International perspectives on suburbanisation: A post suburban world?* Basingstoke, UK: Palgrave-Macmillan.

Ren, X. (2013). *Urban China.* Cambridge, UK: Polity.

Roy, A. (2011). Slumdog cities: Rethinking subaltern urbanism. *International Journal of Urban and Regional Research, 35*(2): 223–238.

Rybczynski, W. (2008). *Last harvest: How a cornfield became New Daleville.* New York: Scribner.

Sassen, S. (1991). *The global city.* Princeton, NJ: Princeton University Press.

Sassen, S. (2000). *Cities in a world economy.* Thousand Oaks, CA: Pine Forge Press.

Saunders, D. (2010). *Arrival city: The final migration and our next world.* Toronto: Vintage.

Savini, F. (2013). *Urban peripheries: The political dynamics of planning projects.* Amsterdam: Faculteit der Maatschappij-en Gedragswetenschappen.

Sieverts, T. (2003). *Cities without cities: An interpretation of the Zwischenstadt.* London and New York: Routledge.

Soja, E.W. (1986). Taking Los Angeles apart: Some fragments of a critical human geography. *Environment and Planning D: Society and Space, 4*(3): 255–272.

Statistics Canada. 2011. *Census of Population.* Ottawa: Statistics Canada.

Suri, S.N., and Taube, G. (2013). Governance in megacities: Experiences, challenges, and implications for international cooperation. In F. Kraas et al. (Eds.), *Megacities: Our global urban future.* Netherlands: Springer.

Taylor, P. (2004). *World city network: A global urban analysis.* London and New York: Routledge.

Taylor, P., Beaverstock, J.V., Derudder, B., Faulconbridge, J., Harrison, J., Hoyler, M., Pain, K., and Witlox, F. (Eds.) (2013). *Global cities.* London: Routledge.

United Nations (2014). *World urbanization prospects [highlights].* United Nations. Retrieved from http://esa.un.org/unpd/wup/Highlights/WUP2014-Highlights.pdf

Von Glascow, R., et al. (2013). Megacities and large urban agglomerations in the coastal zone: Interactions between atmosphere, land and marine ecosystems. *Ambio, 42*: 13–28.

Wachsmuth, D. (2015). Megaregions and the urban question: The new strategic terrain for US urban competitiveness. In Harrison and Hoyler (Eds.), *Megaregions: Globalization's new urban form?* Cheltenham, UK: Edward Elgar, pp. 51–74.

Walks, R.A. (2006). The causes of city-suburban political polarization: A Canadian case study. *Annals of the Association of American Geographers, 96*: 390–414.

Webster, P., and Burke, J. (2012). How the rise of the megacity is changing the way we live. *Theguardian.com*, January 21. Retrieved from http://www.theguardian.com/society/2012/jan/21/rise-megacity-live

Whyte, W. (1956). *The organization man.* New York: Simon and Schuster.

Wirth, L. (1938). Urbanism as a way of life. *American Journal of Sociology, 44*(1): 1–24.

Wu, F., and Shen, J. (2015). Suburban development and governance in China. In P. Hamel and R. Keil (Eds.), *Suburban governance: A global view.* Toronto: University of Toronto Press, pp. 303–324.

Young, D., and Keil, R. (2014). Locating the urban in-between: Tracking the urban politics of infrastructure in Toronto. *International Journal of Urban and Regional Research, 38*(5): 1589–1608.

5

Incremental and Instant Urbanization

Informal and Spectacular Urbanisms

Grace Adeniyi Ogunyankin and Michelle Buckley

Introduction

Consider and compare the following magazine articles. In 2002, *National Geographic* ran a profile on the incremental growth of the city of Lagos, Nigeria, and the informal economy that has developed alongside the city. In the article, the author writes that

> "Struggling to Cope" would look very well inscribed as the motto on the city's coat of arms.... Not only are the markets a chaos of commerce, the streets are lined with vendors offering delicately peeled oranges or golden brown smoked fish, and at every intersection hordes of young men and boys weave among vehicles offering passengers every conceivable object: magazines, used shoes, mousetraps, envelopes of starch, bags of fruit juice, ... even toilet seats. (Zwingle, 2002: 73)

In 2007, another article in the same magazine—"Sudden City" (Molavi, 2007: 94)—chronicled the rapid and spectacular development of the city of Dubai. The city is, in the author's words, "the world capital of living large; the air practically crackles with a volatile mix of excess and opportunity. It's the kind of place ... where diamond-encrusted cell phones do a brisk business at $10,000 apiece; where millions of people a year fly in just to go shopping."

That same year, *National Geographic* published a third, similarly titled article about "instant cities." This one focused on China's rapid urbanization, bearing a subheading claiming that "China is in the fast lane, ignoring every speed limit. Cities spread like a cartographic contagion" (Hessler, 2007).

Stories like these seem to be everywhere these days—breathless accounts of the "explosion" of urban centres, the "flood" of migrants into cities bursting at the seams, or the chaotic and unregulated growth of megacities across the global South. The urban centres in these stories are often depicted either as epicentres of human deprivation and unruly development across the planet or as unmodern global city contenders trying to leapfrog their economies into the twenty-first century through the ruthless, relentless pursuit of unbridled consumerism and globally marketed megaprojects. Underpinning these narratives, however, are often troublesome colonial and Eurocentric assumptions about what cities *should* look like and how they *should* grow. Popular narratives like these tend to frame urbanization in Asia, the Middle East, or Africa in pathological terms. In the quotation above regarding China's urban growth, for example, the country's urbanization is portrayed as both deviant and rule breaking; it is an epidemiological disaster, a viral "contagion," unstoppable, unplanned, and chaotic. What stories like these demonstrate is that narratives often implicitly compare urban growth and development in parts of Asia, Latin America, Africa, or the Middle East with the histories—ones largely imagined, as this chapter explores later on—of gradual, "orderly" growth in cities across North America and Europe (Roy 2011a).

Focusing primarily on the cities of Dubai, United Arab Emirates (UAE), and Ibadan, Nigeria, this chapter deconstructs and unpacks popular and academic narratives associated with instant and incremental urbanization and spectacular and informal urbanisms. While the political economy, colonial histories, and current social relations shaping contemporary urbanization in Dubai and Ibadan are very different, we argue that thinking about them together is useful because it attunes us to the ways that very different forms of urbanism and processes of urbanization in the global South tend to be framed in rather similar ways in Anglo-American discourse—as deviant, perverse, excessive, and ultimately illegitimate. In order to move beyond these stereotypes, the questions this chapter explores are: What kinds of *relations*—economic, political, social, and spatial—underpin these processes of urban change? Why are they happening now? And who wins and who loses in their realization?

The chapter begins by reviewing perspectives on what kinds of urbanization have taken place in the past several decades and discussing the implications of this for low-income urban dwellers as well as alternative narratives for understanding urbanization and urbanism in the global South. Following this, the case studies of Dubai and Ibadan are examined to highlight some of the specific relations that have enabled urban growth in these two locales to take place. The Dubai case study emphasizes the ways that international markets for finance and labour have sustained processes of rapid, large-scale urbanization. The Ibadan case study highlights the gendered relations of urban settlement and development and the everyday politics of class and income that shape growth and development. The urban trajectories of Dubai and Ibadan find parallels in Canada where large cities, in particular, are characterized by spectacular urban development and where cities of all sizes contain diverse forms of unauthorized, quasi-legal forms of settlement and occupancy urbanism. The chapter concludes with a critical reflection upon the limitations of the concepts of instant and incremental urbanization and spectacular and informal urbanisms to adequately capture the complexities of urban transformation; the forms and processes of urbanization will always exceed the attempts of conceptual efforts to fully capture them.

Placing Urban Growth and Change in Perspective

While the claim is often made that urbanization has become a global phenomenon, this process is deeply uneven in its pace and form, as well as in the social, economic, and environmental relations that drive it. In 2014, by some estimates, more than 54 per cent of the world's population lived in cities, and this number is projected to continue to grow (United Nations, 2014). Other estimates about the global urban population are that by 2050, an additional 2.5 billion people will live in cities and over 90 per cent of this urban growth will take place in the global South (UN-Habitat, 2014). Within the global South, this process is profoundly geographically concentrated: for example, the urban population of the world's two poorest regions, sub-Saharan Africa and South Asia, is expected to double over the next two decades (UN-Habitat, 2013). Meanwhile, as of 2013, Asia held 7 out of the world's 10 most populous urban centres (UN-Habitat, 2013). An overwhelming amount of this growth, moreover, will happen through the cumulative, minute, and independent actions of a vast number of people.

In teasing apart these different processes and relations, it is important to distinguish between the concepts of **urbanism** and **urbanization**. Urbanism generally relates to ways of *being* in the city, such as urban residents' practices of living and interacting with each other, or the production and reproduction of place-based imaginaries among those who live within and outside of the city. Urbanization, on the other hand, very broadly refers to a broad set of *processes* taking place to transform space; to material, economic, and demographic growth and development; to socio-material transformations in the built environment; and to the connections—economic, ecological, social, or cultural—that settlements are forging with places and processes well beyond their own borders.

This chapter considers two forms of urbanism, *informal* and *spectacular*, and two processes of urbanization, *instant* and *incremental* (Figure 5.1). These terms are conceptualized not simply as concrete phenomena that are taking shape "out there" in cities, but equally as imperfect concepts that the popular media and academic discourses tend to use in order to make sense of complex processes of urban growth and change. **Informal urbanism** generally refers to particular ways of living and interacting in cities that can involve garnering or distributing resources (e.g., water, food, jobs, and housing) in quasi-legal or illegal ways or in a manner that lies outside of formal governmental regulation, monitoring, and taxation. **Spectacular urbanism**,

Forms of Urbanism	Processes of Urbanization
Informal urbanisms Refer to ways of living and interacting in cities that can involve garnering or distributing resources (e.g., water, food, jobs, and housing) in quasi-legal or illegal ways. These kinds of relations transect the lives of both poorer and richer urban residents across the global North and South	**Instant urbanization** Urbanization that appears to be happening rapidly, often involving the fast development of very large parcels of land and involving complex, multi-use developments on a grand scale; a process often involving centralized forms of urban development undertaken a small cadre or state or market actors.
Spectacular urbanisms Refers to ways of being or relating in cities that rely heavily on the promulgation of a "global" sense of place; on the development and interaction with urban built environments that are architecturally striking, grand or novel, and/or the production of urban lifestyles that rely heavily on cosmopolitan forms of urban consumption.	**Incremental urbanization** A term often used to describe forms of urban growth and change that happen "by accretion" (i.e., bit by bit). This form of urbanization can be rapid or gradual, but generally entails the production of urban space through a large number of small, cumulative and independent actions undertaken by many different people over time.

Figure 5.1 Four prominent narratives framing understandings of contemporary urbanization and urbanisms
Michelle Buckley.

meanwhile, relates broadly to ways of being or relating in cities that rely on expressions of a "global" sense of place. These relations can often involve grand architectural gestures or spectacular events that create new meanings of place. The phrase **instant urbanization** typically refers to processes of urbanization that appear to be happening very rapidly and that often involve the development of large parcels of land, capital, and labour in the production of ambitious development projects. The final concept explored in this chapter is **incremental urbanization**. This term is often used to describe forms of urban growth and change that happen "by accretion" (i.e., bit by bit). This form of urbanization can be rapid or gradual, but generally entails the production of urban space through a large number of small, cumulative, and independent actions undertaken by many different people over time.

These kinds of relations transect the lives of both poorer and richer urban residents across the global North and South, albeit at times in very different ways. Moreover, the concepts above do not exist in isolation, but can be connected to each other. For example, much incremental urbanization can take place through the enactment of informal forms of urbanism, whether by large numbers of individuals building their own houses on illegally settled lands, without official building permits, or without meeting municipal standards. Overall, this chapter highlights them here to draw attention to the complex ways that diverse practices of living and interacting in cities can intersect with different forms of urban growth.

Urbanization, Globalization, and Spectacle

Grand and spectacular forms of development, especially those focused around the attention-catching gesture of a signature built edifice, are nothing new; from the pyramids of Ancient Egypt to the Louvre Palace in Paris, France, bold and centralized interventions into the urban built environment have long been a fundamental strategy of expressing national, imperial, dynastic, colonial, authoritarian, or totalitarian power throughout history. In the late twentieth and early twenty-first centuries, however, urban scholars have noted how globalization and economic change are prompting state and market actors with a stake in local economies to repurpose local urban identities in spectacular ways. This is in part due to the fact that municipalities in the twenty-first century have become increasingly dominated by **urban entrepreneurialism**—forms of governance and global place-making that aim to attract both regional and international investment. In a world where metropolitan centres are increasingly competing for investment, professional talent, and tourist dollars, the imperative to demonstrate that one's locale is an attractive place to live, work, and play has led to strong imperatives of **urban imagineering** (van den Berg, 2015). Marketing the city as a desirable destination for business elites, or as a decadent weekend stopover for globally mobile professionals, has become an important facet of efforts to attract tourist dollars by rebranding the urban everyday as exciting, cosmopolitan, and out of the ordinary. Such efforts frequently include self-conscious forms of place-branding through sophisticated advertising campaigns, often coupled with a constellation of policies on the ground (e.g., policing strategies and transit planning) meant to produce and preserve these locales as an urban playground for those who are good consumers, such as tourists, business travelers, and affluent locals. As a result, the performing and projecting of the spectacle of specific forms of "urban success" (Ong, 2011: 206)—economic, cultural, architectural, or infrastructural—has become a key imperative of local development strategies internationally (Figure 5.2).

At the same time, urban scholars have traced the growing prevalence of specific *kinds* of urban development projects that are sometimes associated with efforts to engineer these spectacular urbanisms. For example, Gavin Shatkin (2011: 77) has documented the rise of what he calls **urban integrated megaprojects** throughout Asia. These are large-scale development projects typically built on a for-profit basis and often by a single developer. While they can span from a few dozen to thousands of hectares in area, Shatkin (2011: 78) suggests that what these developments have in common is how their development has heralded the privatization of urban and regional planning. In his words, private and often transnational actors (ranging from architects to urban planning consultants) have come to play a "central role in the core functions of urban planning—the visioning of urban futures, and the translation of these visions into the planning, development and regulation of urban space in the network infrastructures that connect them on an urban and regional scale." In these cases, not only has the *scale* of building projects in some locales become much larger, but states have also taken on a more entrepreneurial role as *facilitators* of private-sector development.

Figure 5.2 The Burj Khalifa, the world's tallest building

While scholars have noted the growing role of private capital and interests in state-led spectacular megaprojects, some spectacular forms of urbanization are also undertaken directly by governments. Aihwa Ong (2011: 205), for example, describes the development of the China Central Television (CCTV) building in Beijing, China, as an assertion of the state's sovereignty through the built environment. In her view, signature buildings like these are intended to radiate *outward* to the world as a gesture of "national arrival on the global stage," but also inwardly, as a demonstration of state control aimed at local residents. Reworking Rem Koolhaas's notion of **hyperbuilding**, Ong (2011: 206) emphasizes that we must take care not to assume that all "showy and flamboyant urban architecture" is merely the hand of powerful market actors colonizing urban space. Nor should we assume that speculative real estate markets are built only by the private sector; state actors in many cities have been some of the chief architects behind speculative property bubbles of the late twentieth and early twenty-first centuries.

The Finance–Real Estate–Construction Nexus of Fast Development

The growing place of internationalized property markets has accompanied the rise of global city economies whose key economic drivers have been the classic "**FIRE**" sectors—finance, insurance, and real estate. The entrepreneurial turn in large-scale urban development is also connected to the growing transformation of local real estate and infrastructure holdings (e.g., office towers, toll roads, private housing, or condominium units) into "global" asset classes that can be bought and sold internationally. Geographers such as David Harvey (2012) argue that the entrepreneurial marketization of urban property has become a driving force of contemporary urbanization processes. Although some countries place severe restrictions on non-nationals buying land and property (for example, in Sri Lanka and Saudi Arabia, non-citizens are permitted to buy property but are very heavily taxed), others, such as Britain, have relatively few restrictions. Since the late 1980s, the urban built environment and the value associated with it have been increasingly subject to financialization through the growth of international demand for local mortgage debt by global investment banks, the growing use of financial instruments like real estate investment trusts (**REIT**s) in property markets across the Middle East and Asia, or the development of special development companies for particular megaprojects that then sell shares of the company on local stock markets to raise funds to build.

Large-scale building projects at the scale of the Lagoons megaproject (Figure 5.3), for example, require very large infusions of financing into a development that will take years to build and even longer to turn a profit; such projects have prompted banks to develop innovative ways to secure such large amounts of development capital and manage the perceived risks of doing so in so-called emerging markets like Saudi Arabia, the UAE, or Indonesia. In the Gulf region, this has led to the development of a host of new financial products and strategies of pooling capital, including the growth of syndicated loans between local Gulf banks and European banks, and the rapid growth of Islamic banking sectors offering Shari'a-compliant financial products catering to construction financing, real estate leasing, and build-operate-transfer (BOT) partnerships (Figure 5.4).

Figure 5.3 An architectural maquette for the Lagoons megaproject in Dubai

The unprecedented amount of wealth being accumulated in the global economy, coupled with financial markets' growing ability to gather ever-larger amounts of this wealth together across borders, is one key reason that such "instant," large-scale forms of urbanization can take place. Moreover, because property speculation often requires a commitment of funds for a fairly significant period of time—months if not years—funneling these sources of global capital into local real estate markets effectively *urbanizes* this form of finance by tying it to the built environments of these cities. It is also worth noting that these relationships between real estate and finance also work in the opposite direction; for example, the search for large amounts of financing for infrastructure and real estate projects in Dubai fuelled one of the most successful waves of investment on the Dubai financial market between 2004 and 2008 (Buckley and Hanieh, 2014).

The Waged Labour of Spectacular and Instant City-Building

Another important facilitator of rapid and large-scale urban development has been the supply of large amounts of low-cost and often exploitable labour. Urban development always mobilizes a vast assemblage of professional (and frequently male) labour—that of developers, financiers, architects, planners, engineers, and politicians. However, a crucial sector implicated in these processes is the construction industry. Indeed, the very cities often

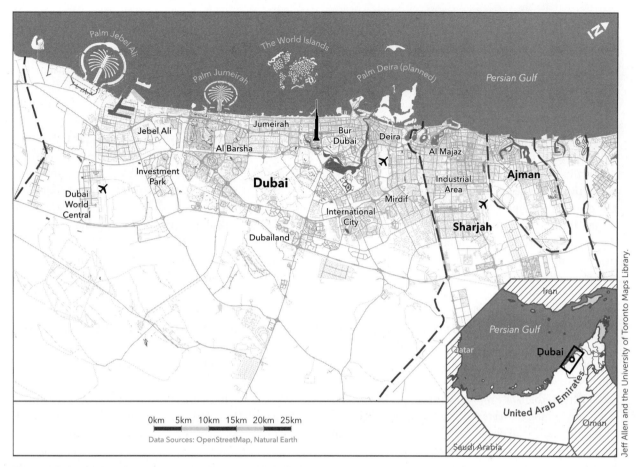

Figure 5.4 Map of Dubai, United Arab Emirates

scrutinized for their purportedly "instant" character, like Dubai, UAE, Shenzhen, China, or Doha, Bahrain, often rely heavily on a huge number of labourers undertaking painstaking manual work such as masonry, carpentry, cement forming, steel fixing, drywall installation, or window glazing.

Despite decades of innovation in construction, the industry remains one of the most labour-intensive worldwide, particularly in cities of the global South. As a result, construction sectors tend to rank among some of the largest employers of many national economies in *both* the global South and the North. It has been a key source of blue-collar and professional jobs for local workers in many cities. Throughout history, moreover, newly arrived migrants to cities have played a crucial role in building them. Irish and Italian immigrants to New York, for example, formed the backbone of the construction workforce that produced the city's skyline throughout the nineteenth and early twentieth-centuries. In addition to bringing crucial skills and expertise to the construction labour markets of many cities, immigrants and temporary migrants employed in the trades have often been important to city-building because they have taken jobs that no one else wants to do; a great deal of construction has long been, and continues to be, dirty, dangerous, poorly paid, and low-status work. Across China, for example, rural migrants in rapidly urbanizing cities have comprised a construction workforce numbering in the tens of millions. However, these migrants are not technically permitted to settle permanently in these cities because of the country's *hukou*, or housing registration system, which designates every citizen as either a "rural" or "urban" resident. The *hukou* system has effectively enabled local governments to treat internal migrants like foreign citizens, because without an urban *hukou* designation, rural migrants in the city are denied access to an array of social-welfare services, such as unemployment insurance, while their children can be denied access to education (Chan, 2010).

Although urban construction labour markets also tend to be highly segmented on the basis of race, ethnicity, nationality, and religion, construction as an industry is also typically gendered. In North America, while women are growing as a part of the workforce in the skilled trades and are also increasingly represented in the industry's professional ranks, the industry has tended to be overwhelmingly male. Women are also frequently overrepresented in some of the most difficult, low-status, and low-paid jobs in the industry. In India, where women have been estimated to make up 30 per cent of the construction workforce, many tend to be paid less than men for the same work and also to be employed at the bottom end of the labour market as brick kiln workers, cement mixers, and material carriers (Vaid, 1999). Similarly, in cities across Thailand, women from Burma or Cambodia also tend to be overrepresented as general labourers or in other low-paid jobs in the industry (Human Rights Watch, 2010). For a closer look at the role of transnational labour in the process of urbanization, this chapter now focuses on the city of Dubai.

Spectacular Urbanism and Instant Urbanization: Constructing Dubai

Curving along the southern coast of what in North America is called the Persian Gulf but is sometimes referred to as the Arabian Gulf, the UAE possesses some of the largest oil reserves in the world. These reserves have dominated both the country's exports and its economy since the early to mid-twentieth century. Within the UAE, however, the emirate of Dubai, a heavily urbanized federated state that is one of seven semi-autonomous territories that make up the UAE, has had comparatively little of this oil within its own borders. Since the mid-twentieth century, thus, Dubai's ruling family has looked to other sources of revenue for their economy. The emirate is dominated by the city of Dubai, which, with a population of about 2.2 million people (Government of Dubai, 2013), is the largest city in the UAE. With a very small population of national citizens, moreover, the UAE (and especially Dubai) has been structurally dependent on foreign workers across the public and private sector since the middle of the twentieth century, with the vast majority of the country's population comprising non-citizens.

While much of Dubai's economic development activities over the past 50 years has focused on its becoming a bustling port city with a major export/re-export industry, the emirate has also pursued an array of strategies aimed at expanding its urban built environment through the rapid development of tax-free industrial zones, information technology parks, elite residential enclaves, spectacular retail landscapes, and other tourist attractions. All of these activities have propelled a massive construction market involving some of the largest development sites in the world, some of which span kilometres. Alongside the "bricks-and-mortar" activities associated with constructing new spaces for economic activity, the emirate has also worked rapidly to create speculative markets for real estate; this required the rewriting of property laws to allow foreign ownership, establishing new legal frameworks for governing construction activities, and instituting new rules regarding bank lending and borrowing to finance very large development projects. By the mid-2000s, billions of dollars from speculative and state sources circulated through thousands of active construction sites across the Emirate of Dubai. These processes of instant and spectacular urbanization have been possible in part through the labour of hundreds of thousands of construction workers from countries such as India, Pakistan, Nepal, China, Bangladesh, and Sri Lanka (Table 5.1).

These workers enter the UAE on one- or two-year temporary work visas and tend to work six days a week, often for 10–12 hours a day (Buckley, 2012). Workers typically must rely on their employer for almost everything. This includes housing accommodation, often in employer-provided, dormitory-style labour camps; transportation, typically provided through company buses, which are workers' only source of transport in the city; and residency visas, which under the UAE's *kafala*, or residency sponsorship system, tie these workers to a single employer who not only employs them but who also possesses the almost unilateral power to revoke their permission to work and live in the country. Having a workforce whose mobility, economic security, and residency are almost entirely controlled by its employer/sponsor often

Table 5.1 Gulf-focused real estate and infrastructure development funds, 2007–2009

Fund Name	Backers/Partners	Fund Locale	Target Investors	Initial Worth (USD$ mil.)	Investment Locale(s)
MENA Infrastructure Fund	Dubai International Capital and HSBC	Dubai	GCC*	$500	Middle East and North Africa
Real Estate Fund	Dubai Investments	Dubai	GCC	$750	Dubai and Ajman
Global Real Estate Ijarah Fund	Global Investment House	Kuwait	GCC	$300	Middle East and North Africa
Second Gulf Real Estate Fund	Global Investment House	Kuwait	Global	$500	Gulf region
BM-CSAM Private Equity Fund LP and BM Private Equity Fund II LP	Bank Muscat and National Bank of Oman	Oman	-	-	Gulf region
GCC Real Estate Fund	Global Investment House and the Central Bank of Bahrain	Bahrain	-	$100	Gulf region
Akar GCC Real Estate Fund	Al Bilad Bank (Saudi Arabia)	Saudi Arabia	GCC	-	Gulf region
Dar Istithmar Lel Aqarat	Isthithmar and AIG Global real estate	Dubai	GCC		Middle East, North Africa, India, Pakistan, Turkey

*Refers to national citizens of Gulf Cooperation Council countries.
Michelle Buckley.

translates into very high levels of productivity on the construction site.

At the same time, these jobs can offer a valuable livelihood for those lucky enough to get a fair contract with a decent employer. However, for many, even good construction jobs are frequently undermined by the large sums of money that local recruitment agents in their hometowns often charge to secure a job in the Gulf. Young and middle-aged men from South Asia seeking a job on one of UAE's construction sites typically have to pay a large sum of money—often equivalent to one or more years of their overseas salary—to local labour recruiters or other intermediaries in order to secure employment, work permits, and transportation to the region. Poorer migrants who go to work in construction tend to pay more for the chance to get a job in the Gulf. For example, those who do not have enough money to pay outright for a work visa from a recruitment agent in their hometown often have no choice but to take out a high-interest loan from local moneylenders. As a result, most lower-waged migrants who enter the UAE's construction markets do so by taking on a major economic burden before they even leave their home countries. Subsequently, once a worker is in the UAE, the labour-market outcomes for him can depend on a multitude of intersecting factors. If workers get paid the wage that was initially promised to them, if the value of the UAE's currency remains strong relative to that of their home country currency while they are abroad, if they do not get injured on the job, and if they do not have other unanticipated fees illegally extracted from them by their employer, then their chances of paying down their migration debt quickly and starting to save are higher than they might be for workers who were not so lucky. Considering the complexity of risks involved in migrating to construction labour markets like Dubai's, it is not surprising that while some manage to pay off the debts from these fees fairly quickly, others, particularly poorer and less-skilled migrants, can spend years paying off migration debts and manage to save very little by the end of their contract (Buckley, 2013).

While the high costs of migration can place workers in the position of having to work for years to pay off heavy debts, workers often confront another set of problems when they get to the UAE and start working. Because construction often operates with many subcontractors and complicated payment systems, from the project management company on down, construction has in the past been an industry where workers have frequently not been paid on time; sometimes workers can be without a paycheque for months at a time. Some employers also keep workers' wages for several months to make sure they do not leave to go work illegally for another employer. All of these practices are illegal, and in recent years the UAE government has taken positive steps, such as requiring all workers to have a bank account so that officials can ensure workers get paid properly and on time. Additionally, despite the fact that unions and collective bargaining are also illegal under UAE federal law, since the mid-2000s construction workers have been informally organizing to collectively demand better wages and living conditions (Buckley, 2012).

Thus, from the withholding of workers' wages by employers in the UAE to workers' unauthorized labour strikes, the "formal" processes of instant, state-led urbanization unfolding in Dubai is itself transected by many informal relationships. While some of these are absolutely essential to the city's rapid urban development agenda, others pose a significant challenge to it.

Incremental Urbanization and Informal Urbanism

In contrast to the large, centrally planned and executed megaprojects being built in Dubai, the main form of urbanization has been, and continues to be, incremental in character. For example, in North America prior to the Second World War, urbanization was not on a large scale but a process of accretion without a prevailing master plan (Echanove and Srivastava, 2010). Master planning has only become a more pervasive way of managing urban growth and change in the twenty-first century. Despite this recent shared history, incremental urbanization in the global South is often not understood in the same way as gradual urban growth in the global North. Rather, it is often depicted as an anomaly and a hallmark of poverty, conjuring images of slum squalor and inadequate infrastructures. Accordingly, there has been a tendency to view incremental forms of urbanization associated with informal urbanism as the sole purview of the urban poor in the global South. In reality, though, housing procurement in the formal sector in both the global North and the global South is incremental. As explained by Wakely and Riley (2011: 1):

> Only a minute segment of any society . . . has the resources to purchase outright or construct their dwellings as a one-off event. Upper- and middle-income households with regular incomes and collateral have access to long-term credit—housing loans and mortgages—that may take between 15 and 30 years to redeem. Households with low or irregular incomes and no access to formally recognized collateral construct minimal basic dwellings, which they extend and improve as resources become available.

The "minimal basic dwellings" constructed by urban dwellers are most pervasive in the global South and are usually referred to as informal housing/settlements. Variously known as *favelas* (Brazil), *barriadas* (Peru), *villas miserias* (Argentina), shanty towns, and squatter settlements, informal settlements vary in size, age, and character within and between cities, but they share similar characteristics such as precarious land tenure, inadequate access to basic services, and scrap building materials. Informal settlements defy the imposition of order by operating outside of the purview, control, and authorization of the state and refusing to comply with official rules and regulations, especially those concerning zoning and construction laws. The UN-Habitat Programme (2003) offers a widely accepted definition of informal settlements as residential areas where a group of housing units have been constructed on land to which the occupants have no legal claim or which they occupy illegally, or as unplanned settlements and areas where housing is not in compliance with current planning and building regulations.

Even though incremental development is not unique to the global South, most of the urbanization taking place in the global South is incremental in character. A contributing factor to the rapid growth of cities in the global South is **rural–urban migration**, both voluntary and involuntary. Voluntary migration often occurs when migrants are in search of better job opportunities and economic security in the city, particularly in the informal sector, although some are able to secure employment in manufacturing and service industries (Levien, 2011). Involuntary migration, on the other hand, often results from rural land dispossession, displacement caused by corporate **land grabbing** or migration owing to war, conflict, or climate change. Land grabbing is often facilitated by the state as part of a neoliberal development-policy goal to increase local/global investment through the commodification and privatization of rural land. In India, for example, acquired land is frequently converted to Special Economic Zones (SEZs) for transnational exporting companies. These companies engage in industrial and commercial infrastructure production and pay low taxes and tariffs. While the companies and the state profit largely from rural land acquisition, rural inhabitants can become displaced with no alternative sources of livelihood (Mishra, 2011). Their most viable alternative is to migrate to urban areas to seek other livelihood options.

Once in the city, rural migrants face the challenge of securing housing and a living. Over the last 50 years, the rise in property prices has displaced the urban poor; they have been unable to compete with the middle class and the wealthy for limited legal real estate. Thus, informal land and housing markets often become the most viable option for low-income individuals. Currently, a large percentage of urban dwellers live in precarious settlements: 61.7 per cent of the urban population in Africa live in informal settlements, while in Asia 30 per cent live in informal settlements and 24 per cent in the Latin America and Caribbean region (UN-Habitat, 2014).

The growth of informal settlements is not only attributed to rural–urban migration but also to the neoliberal urban redevelopment agenda that has become pervasive in many cities of the global South. In this sense, spectacular and globally marketed forms of urbanism, discussed earlier in the chapter, are not separate from these forms of informality; indeed, they are often directly linked to the production of informal urbanism in contradictory and troubling ways. Urban megaprojects and other ambitious urban "renewal" and revitalization schemes have been used to gain planning exceptions, and this entails the circumvention of established protocols governing development planning, such as protocols for public consultation, municipal plans and zoning, and land-use bylaws. Large-scale formal development projects can thus legitimize the displacement of poorer residents whose lives and livelihoods do not "fit" in with plans to re-engineer urban space. The process of re-engineering urban space often entails the framing of informal urban settlements as anachronistic, illegal, and the antithesis to modernity. As a consequence, in many cities of the global South, informal settlements are being demolished and their inhabitants evicted; unwillingness to comply has in some cases led to state-sanctioned violence. Upon successful demolition, these reclaimed lands are valued as prime real estate and are often used for business development, such as the building of malls (with international retail), resorts, hotels, or expensive housing units. Those who are displaced are forced to find other housing solutions, and these are often more insecure than the places from which they have been evicted. For example, evictions that have taken place in Chennai, India, to pave the way for new commercial and apartment buildings have left the majority of informal housing dwellers homeless or have led to their relocation in questionable resettlement areas that are more crowded, are environmentally unsafe, and/or have limited access to urban infrastructures and facilities (Rajagopalan, 2013).

Given all this, there is a growing gap between a vision for a cosmopolitan world-class city and the lived realities of the urban majority. Amin (2013) argues that instead of having the urban poor relegated to the realm of the periphery (legally, socially, and geographically), it would be worthwhile to view the city as a shared common space in which the poor and the rich have equal entitlement. In this sense, the poor would be able to exercise their right to the city in their everyday life. The concept of the **right to the city** prioritizes the rights of urban dwellers over those of capital and the state and also entails their right to participate in decisions that pertain to the production of urban space and the right to appropriate (access, occupy, use, and produce) urban space without punishment (Harvey, 2012; Purcell, 2002). A key part of efforts to articulate pro-poor agendas for urban growth and development has focused on challenging pejorative portrayals of informality and framing them in new ways.

Re-framing Informality

Challenges to formal redevelopment projects and accompanying negative narratives about informal spaces of poverty have emerged through the concept of **subaltern urbanism** (Roy, 2011a). Subaltern urbanism often emphasizes the entrepreneurialism and political agency of the urban poor while highlighting their lived experiences and survival and subversion tactics. Accordingly, informality is a form of grassroots revolution that is a deliberate opposition to state bureaucracy. This viewpoint promotes the concept of **heroic informality**, whereby informality is celebrated as resistance, entrepreneurial and spontaneous, with an absence of bourgeois calculations (Varley, 2013). This concept also enables informality to be extolled as an anti-colonial practice that resists the legacies of hegemonic European planning, and as one that is ecologically virtuous through the use of recycled materials for construction (Arabindoo, 2011).

The notion of heroic informality further highlights the integral role that informal urban dwellers play in the economic and social development of the city. Without the labour and the entrepreneurialism of the urban poor, the city's ability to thrive would decline. As a case in point, Dharavi, in Mumbai, India—one of the largest informal settlements in Asia—is a million-dollar economy that provides food to Mumbai and exports crafts and manufactured goods (Echanove and Srivastava, 2009). The urban poor are also portrayed as heroic because they come up with resourceful responses to inadequate shelter and deficient services and infrastructure despite municipal neglect and indifference to their plight. To illustrate, Pyati and Kamal (2012) highlight the important intervention the creation of community libraries has made in Bangalore's informal settlements. The libraries not only meet educational needs for residents in these settlements, but also provide spaces for community forums and foster a sense of place and belonging.

Although the conceptualization of informality as heroic is important, it also has the proclivity to undermine harsh realities, lived experiences, and desires for change. Moreover, by over-romanticizing the "make-do" characteristics of informality, heroic informality does little to challenge structural roots of poverty and inequality and further contributes to municipal inaction. In this vein, Bayat (2000: 545) has pointed out that informality is not always indicative of deliberate collective political action but is a "quiet encroachment of the ordinary people on the propertied and powerful in order to survive and improve their lives." This encroachment enables morally justified illegality wherein inhabitants of informal settlements engage in a form of **pirate urbanism** in which existing legal, high-end media and technological infrastructures that are otherwise exclusionary and inaccessible are siphoned to enable access for everyday use and entrepreneurial purposes (Simone, 2006). For example, illegal settlements often do not have access to legal electricity connections, and thus they facilitate their own access by hooking wires to "formal" connection points to siphon off electricity.

Although quiet encroachment often involves advances that tend to be individual and gradual, when informality is criminalized or threatened, it can at times be defended in a collective and overt manner (Bayat, 2000) such as in **occupancy urbanism** whereby surplus real estate is appropriated illegally by urban residents and government efforts to entrench master planning is defied and thwarted. Occupancy urbanism is manifested in many ways. One example is through the use of urban populism whereby low-income urban dwellers take advantage of their large population numbers to secure tenuous access to land and municipal services in exchange for political and electoral loyalties (Vasudevan, 2015). Another example entails the engagement of poor urban citizens in social movements that challenge their displacement and exclusionary master planning; such activities include the reoccupation of evicted sites and the appropriation of land that has been slated for urban redevelopment (Roy, 2012).

While it is worthwhile to humanize the informal and to see the heroic and agency, it is also important to contest the proclivity with which illegality and informality are associated with the urban poor (Roy, 2011a). Entrepreneurial governments are vested in constructing "informality" through the legal mechanisms they put in place and in choosing which land claims and building plans to designate as illegitimate and who to criminalize (Varley, 2013). Wealthy elites also engage in informal production of urban space, but their class power enables the glossing over of their illegality because they have the finances to produce space that is aesthetic, "world class," and conducive to global capitalist expansion. For example, in promoting mass tourism in Thailand, luxury resorts encroach on protected lands and ignore land-use planning regulations designed to inhibit ecological harm (Pleumarom, 2008). There are also pervasive practices of displacing farmers through the land grabbing of fertile agricultural land—land that is often slated for building industrial factories, shopping malls, and housing complexes. Thus, the illegalities of wealthy urbanites can sometimes be lauded as examples of urban development success, while the illegalities of the poor are decried as failures of urban development (Roy, 2011a).

Informal Urbanization and Informal Urbanism: Constructing Oke-Foko, Ibadan

Ibadan, Nigeria, is the fifth-largest city in West Africa with a population of about 3 million people; it has, moreover, an average annual population increase of 131,000 (UN-Habitat, 2014). It has 11 autonomous local governments areas, 5 of which are considered urban, while 6 are considered semi-urban (Figure 5.5). Ibadan is currently undergoing urban renewal projects in efforts to make it a world-class city. A major challenge identified by the city's urban planners is that of Ibadan's "unsightly" areas. Among these is the district of Oke-Foko, an indigenous precolonial settlement located in the inner core of the city. In 2009, it had a population of 51,871 and a projected annual growth of 2.8 per cent (Adewale and Amole, 2015). The houses in Oke-Foko are in close proximity to one another, and some are in dilapidated and crowded conditions (Figure 5.6). Most houses have little or no access to a potable water supply or sanitation facilities. Yet, in a recent study it was found that the perceptions that urban residents in Ibadan have about the challenges and potential solutions to the lack of municipal services in Foko depend in part on their gender and class (Adeniyi Ogunyankin, 2014).

Oke-Foko is referred to as a "slum" in daily parlance in Ibadan. This reference is often loaded with negative connotations and is used to capture unsavoury sanitary conditions and poor housing conditions (UN-Habitat, 2003). However, the word "slum" is in quotation marks to signal that it is also a politicized term. The term is often used from an outsider's perspective and is a value judgment on those living in areas designated as "slum." The outsider's perspective often discounts the insider's perspective and masks the experience and everyday reality of those living there (Sanjek, 1990). For example, with regard to sanitation, there is a prevailing assumption among some middle-class women in Ibadan that low-income women, particularly those who live in "slum" areas, are unsanitary and need to be taught proper sanitation methods and cleanliness. One middle-class woman characterized a low-income area by the "dirtiness all over the place. Now [since] there was rain, there's flooding. There was flooding because of blocked

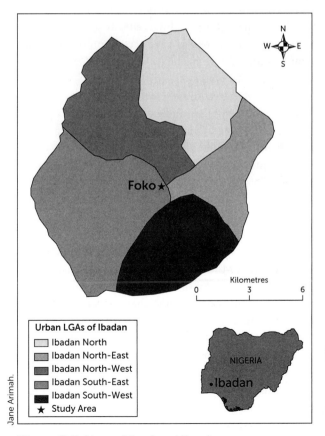

Figure 5.5 Map of Ibadan, Nigeria

Figure 5.6 A house in Foko, Ibadan

drainage. If the people don't throw garbage in the drain, there won't be blockages" (Interview, 2011, July 18). In contrast, low-income women often reiterated that there should be better waste management available to them: "[The government] should give us a place to dump our waste so our neighbourhood will not be rough. So we can be neat" (Interview, 2011, July 16). Instead of displaying a heightened concern about how unsanitary these women are, municipal leaders should focus more on alleviating the daily burdens women experience as a result of living where they do. However, the government's approach has been to engage in urban renewal planning to make informal settlement areas like Foko look more orderly and modern (Figure 5.7). This often involves demolition and displacement.

From the perspectives of urban planning officers, "development control is very difficult" in Oke-Foko (Interview, 2011, March 16). In reference to demolition exercises, one senior urban planner mentioned that "it is by force that we are doing it. This morning if not because of the police protection, the people tend to attack us during this exercise . . . even the government cannot force them to leave" (Interview, 2011, March 16). Because Oke-Foko is in the precolonial core of the city, urban planners stipulate that it is difficult to engage in urban renewal projects because the community members they consulted with are attached to their ancestral homes and have thus refused government offers to engage in renewal. However, community consultation is often unrepresentative of the community and commonly only includes male voices, as women are often excluded from consultation and planning processes.

Unlike the men, the low-income women in Oke-Foko appear to have little or no sense of attachment to where they live. Because men are rarely engaged in tasks that make the place unbearable for the women (e.g., collecting water and cooking outside), they can easily gloss over the problems that highlight the impracticalities of the space they live in. The women in Foko consistently expressed dissatisfaction with where they live mainly because the lack of infrastructure increases their daily reproductive work. In this particular area, the houses where the women are living do not have toilets, kitchens, or running water. Moreover, there is no waste disposal system; women have to walk between 15 minutes to one hour to collect water and to dispose of waste (including feces). These women want major changes to occur in their area, such as reconstruction of crumbling buildings and provision of basic infrastructures. For example, the women strongly expressed that it is the government's duty to intervene, as it is well within their rights as citizens to have access to proper sanitation facilities. As such, the lack of the basics of toilets, bathroom, and water is something many women in Oke-Foko want to address.

The women also indicated the importance of having their voices heard in urban planning and development decisions. They argued that having more women in decision-making would improve Ibadan's development because women, they assert, have better knowledge of the city and of water and sanitation issues. Female Oke-Foko residents similarly noted that if they were included in urban planning and decision-making processes, they would invest in accessible infrastructure for the masses.

While the lack of urban services and informal settlement together pose specific challenges to the lives of lower-income women in Ibadan, urban informality is not simply a condition that characterizes cities or the lives of urban residents in a city like Ibadan; it is a defining condition across the globe (Roy and AlSayyad, 2004). As we explore in the text box below, we discover that various relations of informality also transect urban life in cities across Canada.

Figure 5.7 Houses in Ibadan, Nigeria, that contravene planning regulations by being so close to the street and each other

Urban Spectacle and Informality in Canadian Cities

Most cities across Canada, both big and small, are characterized by a variety of unauthorized, quasi-legal forms of settlement and occupancy urbanism. The temporary and informal appropriation of private and public spaces by homeless populations—from public parks to private doorways—is an integral facet of Canadian urban landscapes. Additionally, recent investigations into the proliferation of basement apartments in Vancouver and Toronto have concluded that both cities have very large numbers of illegal rental apartment units. While Ontario only legalized basement apartments in 2012 in response to acute housing shortages, tens of thousands of residents across the Greater Toronto Area (GTA) are still living in "illegal" basement units, either because no building permit was registered to construct them, because they do not meet strict municipal building codes for fire safety and other standards, or because some GTA municipalities prohibit these kinds of "secondary suites" entirely (Carville, 2014; CMHC, 2016).

Canadian cities are also key sites of deliberately non-conformist acts of informal settlement. Toronto's "Tent City," a community of hundreds of people living in makeshift shelters underneath the Gardiner Expressway in 1998, was forcibly removed by the municipal government in 2002 (CBC News, 2002). Being far more than simply the growth of temporary shelters, Tent City became a historic site of protest and dialogue over the lack of affordable housing in Toronto. Meanwhile, the long legacy of squatted buildings—occupied by anti-poverty and affordable housing advocates in Peterborough, Montreal, Halifax, Vancouver, and Toronto—is an example of occupancy urbanism taking place across Canada and reflects a set of practices often meant to challenge and resist unfair and unjust housing policies and growing income inequality in Canada's urban centres. Examples such as these illustrate the ways that incremental urbanization and informal urbanisms transect urban life in Canadian cities—yet they are rarely framed as such.

Meanwhile, outside Canada's borders, many globally mobile Canadian professionals—whether as part of immense engineering multinationals, boutique architecture firms, or transnational consultancy companies—are closely involved in spectacular urban development projects across the globe. The Canadian architecture firm NORR, for example, was involved in building the Burj Khalifa in Dubai. Closer to home, however, spectacle and urban form have manifested most recently through the government's hosting of the Pan Am and Parapan Am Games in Toronto in 2015. Requiring almost CAD$100 million in state financing, these investments have been heavily trained on improvements in the built environment, including the construction of a special transit depot for the Games, the development and refurbishment of sporting venues across the city, and the commissioning of ostentatious sculptures—such as the rainbow-coloured TORONTO sculpture that was erected at Toronto City Hall in 2015 for the Games—that are now set to become permanent fixtures in the urban landscape. The objective of these investments of course goes well beyond infrastructure needed for the Games themselves; this was summed up succinctly in local media coverage of the Games in which it was announced that Toronto had finally come of age, "crack[ing] Lonely Planet's list of top cities in the world for 2015, not just for hosting the Pan Am Games, but also for its buzzing nightlife and vibrant dining scene" (CTV News, 2015). Popular discourse like this illustrates how spectacular sporting **urban mega-events** like these are also global place-making strategies designed to influence cities' status within the global economy and attract investment. It also illustrates how these events are powerful tools to legitimize the reorientation of public resources (in this instance, transit infrastructure, taxpayer dollars, and recreational venues) towards activities aimed at generating revenue from tourism, real estate investment, and domestic and transnational commerce.

Conclusion

At a time when it is common to hear that we are living in an "urban age" in which supposedly more people are moving to cities than ever before in human history, the popular media has been rife with narratives about spectacular, informal, incremental, and instant forms of urban growth and change. In order to dig below the surface of these narratives, it is imperative to scrutinize the social, economic, and political relationships that underpin such processes on the ground. In doing so, this chapter has unpacked some of the forms of "instant," "informal," "spectacular," and "incremental" urban growth and change as distinct, but also often interconnected, processes. While formal urbanization and incremental/informal urbanization are often billed as competing strategies of urban development, they are not always geographically distant—in fact, they are often intimately connected to each other. As we have seen above, instant forms of urbanization and the marketing of spectacular forms of urbanism through signature urban development projects can often spawn new forms of informality as employers capitalize on the needs of vulnerable mobile populations, or as displaced residents must carve out new informal places to live and ways of making a living in the city.

Finally, what can be learned by comparing and contrasting the very different forms of urbanization in Dubai and Ibadan? The first and most general insight is that urbanization is a geographically, socially, and economically *uneven* process that takes many different forms, enlists many different kinds of social relations, and often engenders a diversity of politics on the ground. In Dubai, a key facet of the politics of urbanization involved confrontational labour struggles wrought by migrant workers, while in Ibadan, it took the form of lower-income women contesting patriarchal state planning activities to improve the quality of their domestic life. More broadly, there is a deeply gendered politics to the forms of work that go into producing urban built environments. Incremental forms of building and domestic improvement, like those in Ibadan, are unwaged and overwhelmingly performed by women. Such forms of incremental city-building are often considered a form of domestic labour and are thus frequently undervalued and often under-studied in geography. In contrast, waged work in construction in Dubai is almost entirely performed by men (indeed, the gender dimensions of this paid work are further illustrated by the fact that migrant workers in Dubai are often hired on the basis of masculine notions of toughness, piousness, insensitivity to the desert climate, and physical strength). Processes of urbanization and its articulation with different forms of urbanism are thus *always* political in some way because they enlist and shape the social and institutional relations that govern gender, sexuality, citizenship, class, race, and ethnicity.

Perhaps most importantly, this chapter has emphasized that while terms like "instant urbanization" or "informal urbanisms" relate to real processes taking place in urban areas, they are also imperfect and problematic concepts whose use always needs to be investigated critically. This is because the sheer complexity of urban change currently afoot often exceeds the capacity of these terms to describe them adequately, while at other times these labels can be applied to urban locales in one-sided or stereotypical ways. These kinds of processes, moreover, apply to Canadian urban realities; in this sense, we must not only cast a critical eye on how these labels are applied to particular places in the global South, but also ask why they are not applied more often to urban life in the global North.

Key Points

- This chapter distinguishes between different forms of urbanization and urbanisms, focusing in particular on instant and incremental urbanization and spectacular and informal urbanisms.
- Examples of large and spectacular urban development projects demonstrate how urban built environments can be used to make symbolic claims to geopolitical or transnational economic power while providing crucial productive, speculative, and rent-based opportunities for generating profit.
- Low-income urban dwellers often rely upon informal settlements for survival owing to poverty and the unaffordability of "legal" housing.

- Dubai's experiments with urbanization are illustrative of the ways that some states are using speculative and spectacular real estate assets to create new opportunities for profit-making and place-branding through the built environment.
- Low-income urban dwellers in Ibadan are interested in participating in urban planning decision-making processes; women, in particular, want their everyday experiences to inform urban planning.
- Canadian urban landscapes are important sites of *both* urban informality and spectacular urbanization.
- In cities across Canada, urbanization is transected by processes taking place *outside of, at the limits of,* and *in opposition to* the law.

Activities and Questions for Review

1. Work in groups, and on a map of the world draw flows of migrant labour and money that can help to explain global processes of "instant" urbanization.
2. In small groups, discuss the merits of moving beyond the dualistic understanding of informality as "failure" or "heroic."
3. What examples of spectacular urbanism have you seen in Canadian cities? Share your examples with the class and identify together what characteristics these examples have in common.

Acknowledgments

The research on Dubai was generously supported by the Social Sciences and Humanities Research Council of Canada. The research on Ibadan was carried out with the aid of a grant from the International Development Research Centre in Ottawa, Canada.

References

Adeniyi Ogunyankin, G.O.A. (2014). *It's my right to fix the city: Women, class, and the postcolonial, politics of neoliberal urbanism in Ibadan, Nigeria.* PhD diss., York University, Toronto.

Adewale, B., and Amole, B. (2015). Housing needs in Ibadan core area: A case study of Oke-Foko. In *Responsive built environment: The proceedings of environmental design and management international conference.* Ile-Ife, Nigeria: Obafemi Awolowo University, Faculty of Environmental Design and Management, pp. 282–291.

Amin, A. (2013). Telescopic urbanism and the poor. *City, 17*(4): 476–492.

Arabindoo, P. (2011). Rhetoric of the "slum": Rethinking urban poverty. *City, 15*(6): 636–646.

Bayat, A. (2000). From "dangerous classes" to "quiet rebels" politics of the urban subaltern in the global south. *International Sociology, 15*(3): 533–557.

Buckley, M. (2011). *Building the global gulf city: Tracing transnational geographies of capital and labour in Dubai, UAE.* DPhil diss., University of Oxford, UK.

Buckley, M. (2012). From Kerala to Dubai and back again: Construction migrants and the global economic crisis. *Geoforum, 43*(2): 250–259.

Buckley, M. (2013). Locating neoliberalism in Dubai: Migrant workers and class struggle in the autocratic city. *Antipode, 45*(2): 256–274.

Buckley, M., and Hanieh, A. (2014). Diversification by urbanization: Tracing the property-finance nexus in Dubai and the Gulf. *International Journal of Urban and Regional Research, 38*(1): 155–175.

Carville, O. (2014). Is your basement apartment safe? Bylaw delays across the GTA are leaving tenants at risk. *Toronto Star.* Retrieved from http://www.thestar.com/news/gta/2014/10/02/is_your_basement_apartment_safe_bylaw_delays_across_the_gta_are_leaving_tenants_at_risk.html

CBC News. (2002). Homeless evicted from Toronto's "tent city." *CBC News: Canada.* Retrieved from http://www.cbc.ca/news/canada/homeless-evicted-from-toronto-s-tent-city-1.339489

Chan, K.W. (2010). The global financial crisis and migrant workers in China: "There is No Future as a Labourer; Returning to the Village has No Meaning." *International Journal of Urban and Regional Research, 34*(3): 659–677.

CMHC (Canada Mortgage and Housing Corporation) (2016). Permitting secondary suites. Retrieved from http://www.cmhc-schl.gc.ca/en/inpr/afhoce/afhoce/afhostcast/afhoid/pore/pesesu/pesesu_001.cfm

CTV News (2015). Pan Am Games put Toronto in international spotlight. *CTV News* (July 9). Retrieved from http://www.ctvnews.ca/lifestyle/pan-am-games-put-toronto-in-international-spotlight-1.2460877

Echanove, M., and Srivastava, R. (2009). Taking the slum out of "Slumdog." *New York Times*, February 21. Retrieved from http://www.nytimes.com/2009/02/21/opinion/21srivastava.html

Echanove, M., and Srivastava, R. (2010). The village inside. In S. Goldsmith and L. Elizabeth (Eds.), *What we see: Advancing the observations of Jane Jacobs*. Oakland, CA: New Village Press, pp. 135–148.

Government of Dubai (2013). Dubai in Figures: 2013. Dubai Statistics Centre. Dubai, United Arab Emirates. Retrieved from https://www.dsc.gov.ae/en-us/Publications/Pages/publication-details.aspx?PublicationId=6

Harvey, D. (2012). *Rebel cities: From the right to the city to the urban revolution* (2nd ed.). New York: Verso.

Hessler, P. (2007, June). Instant cities. *National Geographic Magazine*.

Human Rights Watch (2010). *From the tiger to the crocodile: Abuse of migrant workers in Thailand*. New York.

International Labour Organization (ILO) (2011). *Baseline study to assess gender disparities in construction sector jobs*. ILO Country Office for Pakistan. Islamabad.

Levien, M. (2011). Special economic zones and accumulation by dispossession in India. *Journal of Agrarian Change*, 11(4): 454–483.

Mishra, D.K. (2011, April 6–8). Behind dispossession: State, land grabbing and agrarian change in rural Orissa. *International Conference on Global Land Grabbing*.

Molavi, A. (2007). Sudden city. *National Geographic*, 211(1): 94, 96, 98, 100, 102–108, 110, 112–113.

Monk, D.B., Graham, S., and Campbell, D. (2007). Introduction to urbicide: The killing of cities? *Theory & Event*, 10(2).

Ong, A. (2011). Hyperbuilding: Spectacle, speculation, and the hyperspace of sovereignty. In A. Roy and A. Ong (Eds.), *Worlding cities*. West Sussex, UK: Wiley-Blackwell, pp. 205–226.

Pleumarom, A. (2008). Mekong tourism—model or mockery? In *IFIs and tourism: Perspectives and debates*. Bangalore: EQUATIONS, pp. 33–50.

Purcell, M. (2002). Excavating Lefebvre: The right to the city and its urban politics of the inhabitant. *GeoJournal*, 58(2–3): 99–108.

Pyati, A.K., and Kamal, A.M. (2012). Rethinking community and public space from the margins: A study of community libraries in Bangalore's slums. *Area*, 44(3): 336–343.

Rajagopalan, K. (2013). The human cost of slum clearance. World Policy Institute. Retrieved from http://www.worldpolicy.org/blog/2012/09/18/human-cost-slum-clearance

Rankin, K.N. (2011). Assemblage and the politics of thick description. *City*, 15(5): 563–569.

Richardson, H.W., and Nam, C.W. (Eds.) (2014). *Shrinking cities: A global perspective*. New York: Routledge.

Roy, A. (2011a). Slumdog cities: Rethinking subaltern urbanism. *International Journal of Urban and Regional Research*, 35(2): 223–238.

Roy, A. (2011b). Urbanisms, worlding practices and the theory of planning. *Planning Theory*, 10(1): 6–15.

Ro, A. (2012). Urban informality: The production of space and the practice of planning. In R. Weber and R. Crane (Eds.), *Oxford Handbook of Urban Planning*. Oxford University Press, pp. 691–705.

Roy, A., and AlSayyad, N. (Eds.) (2004). *Urban informality: Transnational perspectives from the Middle East, Latin America, and South Asia*. Lanham, MD: Lexington Books.

Sanjek, R. (1990). Urban anthropology in the 1980s: A world view. *Annual Review of Anthropology*, 19(1): 151–186.

Shatkin, G. (2011). Planning privatopolis: Representation and contestation in the development of urban integrated megaprojects. In A. Roy and A. Ong (Eds.), *Worlding cities*. West Sussex, UK: Wiley-Blackwell, pp. 77–97.

Shatkin, G. (2014). Contesting the Indian city: Global visions and the politics of the local. *International Journal of Urban and Regional Research*, 38(1): 1–13.

Simone, A. (2006). Pirate towns: Reworking social and symbolic infrastructures in Johannesburg and Douala. *Urban Studies*, 43(2), 357–370.

United Nations, Department of Economic and Social Affairs, Population Division (2014). *World urbanization prospects: The 2014 revision, highlights*. New York: United Nations.

UN-Habitat (2003) *The challenge of slums: Global Report on Human Settlements 2003*. New York: Earthscan Publications.

UN-Habitat (2013). *State of the world's cities 2012-2013*. New York: Earthscan Publications.

UN-Habitat (2014). *Voices from the slum*. Kenya: United Nations Human Settlements Programme.

Vaid, K.N. (1999). Contract labour in the construction industry in India. In D.P.A. Naidu (Ed.), *Contract labour in South Asia*. Geneva: ILO, Bureau for Workers' Activities.

van den Berg, M. (2015). Imagineering the city. In R. Paddison and T. Hutton (Eds.), *Cities and economic change: Restructuring and dislocation in the global metropolis*. Los Angeles: Sage, pp. 162–175.

Varley, A. (2013). Postcolonialising informality? *Environment and Planning D: Society and Space*, 31(1): 4–22.

Vasudevan, A. (2015). The makeshift city: Towards a global geography of squatting. *Progress in Human Geography*, 39(3): 338–359.

Wakely, P., and Riley, E. (2011): Cities without slums: The case for incremental housing. *Cities Alliance Policy Research and Working Papers*, 1: 1–55.

Zwingle, E. (2002). Where's everybody going? *National Geographic*, 22(5): 72–99.

II Urban Policy and Planning

6

Urban Policy and Governance

Austerity Urbanism

Betsy Donald and Mia Gray

Introduction

This chapter introduces the concept of **austerity** and uses it as a lens to examine the changing character of urban policy and **governance**, especially in relation to the North American city. Austerity, as a policy and a **discourse**, has been around since at least the 1930s, but it has resurfaced as a government policy since the 2007–2008 financial crisis and is now widespread and pervasive (Bailey et al., 2015; Kitson et al., 2011). In economics, austerity refers to government policies that seek to reduce budget deficits and may include a combination of spending cuts, reducing or freezing labour costs, tax increases, privatization, and reconfiguring public services and the welfare state (Whitfield, 2013). Austerity is often presented as a policy alternative to more **Keynesian-inspired policy** attempts to reduce deficits by stimulating demand through increased government spending.

Some cities have been under austerity for decades, especially in the global South under **structural adjustment programs** imposed by the **International Monetary Fund.** Throughout the 1980s and 1990s, the IMF required countries to embrace structural adjustments in their national economies in exchange for IMF loans. In response, many countries in the global South imposed austerity, privatization, and deregulation in order to encourage the market and minimize the role of the state (Chang, 2003). This played itself out unevenly across space; cities are often dynamic centres of national economies and changes in the national relationship to the global economy as a result of structural adjustment have had transformative effects on urban economies. Differences in cities' industrial mix, history, and political and policy frameworks across the global South inevitably resulted in very different localized responses to structural adjustment programs and austerity (Harris and Fabricius, 1996).

Until recently, this type of austerity had been confined mostly to cities in the global South; however, austerity has now become a common policy response in the United States and throughout European states, such as Greece, Ireland, Spain, and Portugal. Again, although academics and policy-makers often refer to this as a national issue, we argue that it is often an urban phenomenon. In the United States, **austerity urbanism** has been used to describe austerity measures that have been imposed on cities, especially since the global financial crisis of 2007–2008 (Peck, 2012). Urban authorities in several cities in the United States have been required to cut back on essential services, lay off public-sector workers, control spending, and reduce debt (Jonas et al., 2015). They are doing this because they have either been required to do so by higher levels of government or have run out of money. The geographical spread of austerity policies and programs varies across the country, with some regions, cities, and neighbourhoods hit harder than others. Several **"Sun Belt" cities** in California and Nevada were hit particularly hard with the subprime mortgage housing market collapse (Bardhan and Walker, 2011; Davidson and Ward, 2014; Davidson and Kutz, 2015). Other **"Rust Belt" cities** and neighbourhoods in places like Buffalo, Detroit, and Baltimore were also affected—particularly those that had a long history of deindustrialization, weak economic growth, and racial discrimination.

This chapter examines how austerity is playing out at the urban level in the North American context. It explains why austerity has become such a pervasive idea in the governing of many North American (especially American) cities, paying attention to the links between the 2007–2008 global financial crisis and a broader state crisis with implications for cities. Renowned urban geographer Jamie Peck (2012) has coined the term "austerity urbanism" to capture this recent phenomenon, as well as two related ideas: "crisis" and "distress." The chapter draws on two American case studies—the recent municipal bankruptcies of Detroit, Michigan, and Stockton, California—to illustrate these concepts. While both examples could be considered extreme cases of austerity urbanism, they nevertheless illustrate the varied ways in which austerity plays out at the local level. Both examples also underscore the complex interplay between broad global forces, particular state structures, and local and historically contingent factors.

Finally, this chapter examines the relevance of the issues addressed to the Canadian urban context. While American-style bankruptcy is unlikely to happen in Canadian municipalities for a variety of institutional reasons, this chapter nevertheless shows how the *narratives* of austerity have migrated north to Canada and are having an impact on who gets what in the politics and planning of redistribution in Canadian cities. The common perception is that Canada has emerged relatively unscathed from the financial crisis. However, this chapter argues that Canadian austerity is often hidden. Canada's collective narrative projects the country as a "kinder, gentler society" than its southern neighbour, and this self-image extends down to the city level, as many of our cities are known around the world as providing high levels of services and quality of life (Banting and Myles, 2013; Donald and Gray, 2015).

Austerity Urbanism, Municipal Crisis, and Distress

In recent years, the concept of "austerity urbanism" has emerged to capture the idea that cities and their governments have been living under constrained public-sector budgets for some time—hence the use of the term "urbanism," which indicates austerity as a *character* of urban life and city planning. As Jamie Peck (2012: 626) explains, "austerity budgeting in the public sector, selectively targeting the social state, is a long-established trait of **neoliberal governance**, but it has been enforced with renewed systemic intensity in the period since the Wall Street crash of 2007–2008." He, like other urban geographers and political economists, explains the link between the Wall Street crash and state cutbacks with consequences for city budgets. This link is summarized below:

> When the credit crisis hit in 2007, risk was heavily concentrated among leveraged financial institutions at the heart of the global financial system in New York City and London. The first casualty was the financial institutions themselves, which were forced to contract employment leading to the loss of thousands of jobs in the financial sector. Nevertheless, the largest impact was quickly registered through the wider crisis-induced recession. The depth of the recession and prolonged nature of the recovery was exacerbated by governments' struggle to manage their indebtedness (itself in part the result of bailing out the banking sectors and recapitalizing credit markets). To manage their indebtedness, many governments turned to austerity policies to address high levels of public indebtedness absorbed during the heady days of the subprime lending spree. (Donald et al., 2014: 5)

As noted above, austerity is not a new idea, but is often evoked as a policy frame to financial crisis (Blyth, 2013; Clavel, 1980; Peck, 2012). Social scientists have been using the concept of **crisis** for decades as a way of interpreting various periods of profound social and economic change in the urban experience under capitalism (Fainstein, 2010; Hackworth, 2007; Harvey, 1989a; Logan and Molotch, 2007; Macleod and Goodwin, 1999; Mitchell, 2003; Peck and Tickell, 2002; Wyly et al., 2009). Throughout this history, capitalism as a system has been viewed as one which is inherently crisis-prone and certain cities have become emblematic symbols of growth (and decline) throughout this history—Liverpool, as a key industrial/deindustrializing city during British industrial growth/decline; Detroit, as the iconic city of American Fordist industrial might and then decay; and Las Vegas, as a symbol of the rise (and fall) of the post-industrial/subprime mortgage city. On the one hand, while some writers interpret the current financial crisis as something new, others see crisis as inherent in the cyclical nature of advanced capitalism, documenting a series of them over the last 100 years. Today questions arise as to how municipalities are now coping with new rounds of austerity. What kinds of austerity policies are being imposed from other levels of government? How are municipalities coping with these requirements? What kind of distress are municipalities now in?

Distress is an important concept in urban public finance and can be used to determine the fiscal and overall health and well-being of a municipality. Traditionally, public-finance experts measure distress in terms of the local fiscal health of a municipality and use indicators such as financial management, fiscal stress, credit-rating reports, and intergovernmental transfers to systematically measure the health of a municipalities' finances and its ability to pay for essential services (Bird and Slack, 1993; Levine et al., 2012; Slack and Chattopadhyah, 2013). Another literature, however, explores distress in terms of the financial, social, and economic situation of the people living in the city. Indicators here include population, income, poverty rates, employment rates, and other characteristics of well-being, such as **social inclusion** and citizen engagement (Gray et al., 2012). This chapter argues that these latter characteristics of **civic life** form an important component of a city's well-being. Recent examples of austerity on the national and the urban level have seen the displacement of elected officials with appointed financial managers, leading in some cases to the "abandonment of the democratic process" (Donald et al., 2014: 6). In the case of Greece and Italy, for example, the elected heads of state (the prime ministers) were replaced in 2012 by unelected "technocrats" who were charged with pushing through politically unpopular reductions in state spending. These state-led programs were encouraged by the political and financial elites of the European Union and the European Central Bank. Similarly, in cities like Detroit, Michigan, and Stockton, California, unelected "administrators" assumed control of municipal finances in 2011–2013 and pushed through austerity measures that reduced the size of the public sector and slashed public pensions.

Who and What Is to Blame?

There is widespread recognition that the 2007–2008 financial crisis (also referred to in the literature as the Great Recession) has exacerbated the ability of municipalities to deliver services, but there is less agreement as to the root causes in some of the more crisis-ridden places. In the United States, California municipalities, such as Stockton and Vallejo, were seen as ground zero for the subprime mortgage housing market collapse (Bardhan and Walker, 2011). The large and rapidly growing population, the booming construction industry, the particular demographic makeup, and the lack of regulation of mortgage-related businesses, made the state of California prime territory for mortgage predators. The **subprime mortgage crisis** was initially a global banking crisis that coincided with the global economic recession of 2007-2008. Essentially, this banking crisis was triggered by a corrupt mortgage-lending system that passed off risk to others, which led to the erosion of lending standards. Low-income, racialized, and high-risk borrowers became targets of highly leveraged financial institutions making risky loans on unfavourable terms. When the credit crisis hit, risk was highly concentrated among several highly levered financial institutions, but then it was quickly passed down to high-risk borrowers who saw the rapid decline in their home prices leading to mortgage delinquencies and home foreclosures. At the height of the recession in 2008, some cities (e.g., Vallejo, California) witnessed a surge in foreclosure rates on homes, leading to a rapid decline in property-tax revenue for that municipality (Davidson and Kutz, 2015).

Depending on the approach taken, scholars have offered various explanations on *who* and *what* is to blame for this fiscal situation. The first explanation suggests that the cities and localities themselves are to blame for decades of irresponsible profligate spending. Urban writers in this camp cite various examples of ill-conceived building projects, corruption, local mismanagement, an expanding public-sector, and "gold-plated" public pensions (Rozansky, 2012). Harvard economist Edward Glaeser (2013: 1) calls Detroit's decision to file for bankruptcy "the necessary step to bring fiscal sanity to an awful mess created by 50 years of private-sector decline and public-sector irresponsibility." Other scholars disagree with this analysis and offer up a second explanation: urban scholars in this camp argue that municipalities are victims of broader structural processes such as the neoliberal fiscal environment and entrepreneurial governance models that hamper their ability to make effective public-finance decisions (Harvey, 1989b; Peck, 2012). Peck (2014) contends that under neoliberal forms of governance, austerity is a particularly urban phenomenon because it has been used to rewrite the social contract between citizens and the state and local governments under the cover of budget constraints. This is significant, he notes, because US public services in their postwar Keynesian form—or at least what remains of them—are now essentially delivered at the state and local level. A third perspective contains elements of both perspectives, with scholars in this camp generally placing greater emphasis on the root of today's urban fiscal crisis as hidden in longer histories of persistent racial and class inequality, discrimination, and deindustrialization. For example, Sugrue reminds us that we cannot make sense of Detroit's contemporary political economy without understanding its long history of racial tension and segregation (Sugrue, 1996; see also Clapham et al., 2012; Wyly et al., 2009).

Responses and Implications for Urban Theory and Policy

Municipal responses to the global financial crisis and subsequent austerity measures have varied: they include everything from bankruptcies (especially in the American context), which can include the takeover of a municipality by an appointed financial manager; to budget cutbacks; to pension slashing; to increased privatization and/or risky-development projects. We have also seen an increased emphasis on "responsible localism," "performance budgeting" (United States), and promoting the volunteer sector through "Big Society" (United Kingdom) (Warner and Clifton, 2014; Lobao and Adua, 2011). There have also been "push-backs" in the form of protests and the emergence of new social movements, both on the left and right. One example on the right side of the political spectrum is the **Tea Party** movement, which emerged in 2009 after Barack Obama's federal election. The Tea Party is known for its conservative policies and is active in the Republican Party, where it pushes for less government intervention and lower taxes. An example on the left side of the political spectrum is the emergence in 2011 of the **Occupy Wall Street** grassroots movement, which first emerged in Zuccotti Park in New York City. The movement quickly grew to be a worldwide social movement against social and economic inequality. It was spurred on when the US government prioritized bailing out the banking sector immediately after the 2007–2008 financial crisis at the expense of homeowners and more disadvantaged people affected by the subprime mortgage crisis (Mayer, 2013).

In many ways, these "push-back" responses confirm the second and third perspectives outlined above, of a prolonged and extended period of **neoliberal urbanism** that began in the 1970s with increased privatization of state assets, and of an increased downloading and off-loading of services from the state to individuals, families, and communities. But today's realities and responses are also linked to longer hidden histories of racial and class segregation, discrimination, and deindustrialization in the making of the modern city. From a policy perspective, there seems to be a growing concern that many of the austerity responses are not working and will only further harm already-depressed places—not only in terms of their economy's ability to rebound, but also in terms of wider social harmony and inclusion as the poorest and most vulnerable in society take the brunt of the budget cuts. While these are still early days, there is a movement to push back against central state austerity measures, with alternative policy models on the horizon (Warner and Clifton, 2014).

Stockton, California: From Boom to Bust in the Sun Belt

Stockton is a city of 300,000 people, one of California's 20 largest cities, which lies roughly 130 kilometres (80 miles) east of San Francisco (Figure 6.1). Stockton's economy expanded rapidly in the 2000s as the increased availability of credit allowed the city to grow rapidly, particularly its property markets. In 2007, the city of Stockton was riding the wave of the housing-induced financial boom. Property tax revenues had more than doubled over the previous seven years, and sales tax receipts were up by 65 per cent (Farmer, 2014). The city started an ambitious series of speculative urban regeneration projects—what many scholars call **entrepreneurial urbanism**—issuing municipal bonds to fund a marina, a downtown arena, a minor league ballpark, a new city hall, and a large parking garage. When the housing market crashed and the financial crisis hit, Stockton's finances were in serious trouble (Figure 6.2). Falling

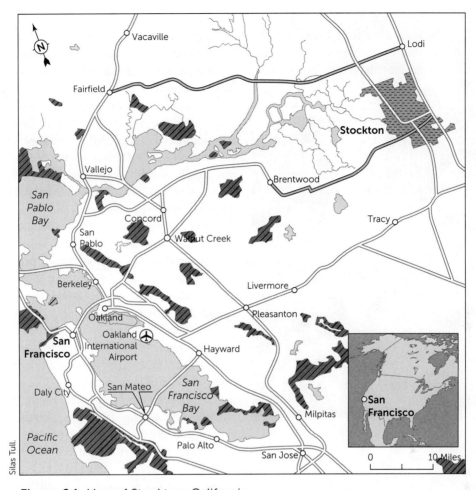

Figure 6.1 Map of Stockton, California

housing prices led to falling property tax revenue, while increased unemployment contributed to lower sales tax revenues. The city's falling revenue streams were countered with growing expenses, as more city residents called upon unemployment and other benefits.

By June 2012, Stockton had already been in an officially declared state of fiscal emergency for two years. Urban analysts highlight Stockton's over-reliance on revenue fueled by new growth. Michael Coleman, a fiscal consultant to the League of California Cities, a public-sector urban advocacy group, argues that once things started to go wrong, city officials gambled on risky debt-financing schemes, which ultimately were extremely bad bets for the city. This is evident in the city's skyrocketing debt repayment—the city paid 600 per cent more in debt repayment in 2012 than it had five years earlier (Farmer, 2014).

In an effort to avoid bankruptcy, the city focused on reducing services and cutting the number of public-sector workers. By June 2012, public services had been cut, including the police and fire fighters, and nearly half of non-safety city employees were laid off. Public-sector workers were the focus of many of the cuts: they were required to take unpaid furlough days, the employee health-care plan was phased out, wages were reduced by as much as 23 per cent, and public-sector union contracts were renegotiated. In all, the city cut over $90 million in annual expenses. Even with these drastic measures, Stockton still did not have the funds to keep running. On June 28, 2012, the city of Stockton filed for Chapter 9

Figure 6.2 Stockton, California, foreclosure
Inman News, 2008.

Figure 6.3 The city of Stockton goes into bankruptcy
REUTERS/Kevin Bartram.

bankruptcy, effectively meaning that the city could apply through the courts to reorganize its debt with all its creditors (Holeywell, 2012) (Figure 6.3).

How did Stockton get to this point? Like many California cities, Stockton was badly affected by the Great Recession. By the time it declared bankruptcy, property prices had fallen by 70 per cent, unemployment had reached a high of 22 per cent, and household income was 20 per cent below the state average (Farmer, 2014). However, Stockton, like other Californian cities, has additional budgetary restrictions, embedded in Proposition 13, that leaves it little room for fiscal manoeuvre. Proposition 13 was an anti-tax initiative introduced in 1978 to cap local property taxes in California. The law required a two-thirds majority vote (by legislature or citizens) to approve any increase in tax rates. This encourages Californian cities to pursue more entrepreneurial, or speculative, urban projects in an attempt to promote future revenue streams, such as Stockton's redevelopment bonds (Davidson and Ward, 2014; Anderson, 2012).

Post-bankruptcy life in Stockton included the withdrawal of many more public services. The city closed libraries and public parks, and the city's police force lost over a quarter of its personnel (Rudolf, 2012). However, it is significant that despite enormous pressure from creditors (including legal representation), Stockton chose to not cut its employees' pensions. City Manager Bob Deis wrote an open letter to Stockton's creditors in the *Wall Street Journal* in which he justifies the decision not to cut pensions. Deis (2012) ends the letter in a defiant tone: "[I]instead of challenging this bankruptcy from corporate offices, our creditors should stay in Stockton for a hot summer week with their windows wide open." The city officially exited bankruptcy on February 25, 2015.

Stockton is only one of many Sun Belt cities to officially declare bankruptcy following the financial crisis of 2007–2008. Other cities, such as San Bernadino and Manmouth Lakes in California, followed patterns very similar to Stockton's, although each city has a different political and economic history. The Stockton case shows that the financial crisis, while an uneven process, hit booming regions of the United States as well as older industrial areas. The case also highlights the ways in which events beyond the control of the city interact with local decision-making to produce conditions ripe for urban austerity. Austerity cuts were severe and affected the basic provision of public services in

Stockton. The final settlement was also significant because unlike other cities (such as Detroit, see below) the final settlement excluded the restructuring of pension provisions to city employees. The Stockton case was the second-largest municipal bankruptcy in the United States; the largest was the city of Detroit. Although the details and economic structure of the two cases were different, there were many similarities that bear careful scrutiny.

Detroit's Rise and Fall: From Fordist Success to Bankruptcy

In 2013, Michigan's Republican governor, Rick Snyder, declared a "financial emergency" in Detroit and the city was put under emergency measures and eventually filed for Chapter 9 bankruptcy. The city was in serious financial trouble with unsecured debts and liabilities of over $18 billion, the largest municipal bankruptcy in US history. (Holeywell, 2013). The emergency measures overturned the elected government in Detroit and replaced it with an emergency manager appointed by the governor of Michigan. The democratic deficit this represented was a particularly salient issue in a city with a troubled history around issues of race, where much of the population already felt disenfranchised. This represented an "abandonment," or at least a temporary withdrawal, of the exercise of democratic rights as an element of **urban citizenship**. "Urban citizenship" is a term that highlights the importance of cities in promoting *practices* of citizenship, particularly the engagement and participation of diverse groups in political, social, and economic life. The temporary withdrawal of democratic rights reminds us that austerity is not simply a budget-cutting exercise; it can also fundamentally restructure the political system and mechanisms of local accountability.

Detroit's rise was inextricably linked with the rise of the auto industry (Figure 6.4). The city was the symbol of the golden age of Fordism and the unofficial capital of the United States labour movement. The city's booming auto industry, along with a strong union movement, spread affluence to semi-skilled autoworkers in the late 1940s and early 1950s. In 1950, Detroit was a teeming metropolis of 1.8 million people; by 1960, it had the highest per capita income in the United States (Bureau of Labor Statistics, numerous years). African Americans also found well-paid manufacturing jobs and a foothold in the industry. However, deindustrialization hit Detroit hard—in 1980, 29 per cent of the city's jobs were in manufacturing, but that figure was reduced to only 15 per cent by 2007 (McDonald, 2014). Also, the issue of race was and is ever present—with pronounced segregation, disinvestment, and **white flight** (and middle-class black flight) into the suburbs, which continue to this day. The spatialization of race and class divisions between the majority black city of Detroit and its white suburbs has reinforced the county and state's predisposition to withhold financial and political support for Detroit (Clement and Kanai, 2015) (Figure 6.5).

Figure 6.4 View from Woodward and Garfield in midtown Detroit

The financial crisis of 2007–2008 had enormous repercussions for Detroit. Total revenue dropped sharply. In 2007, it was $1.4 billion, but by 2014 this had dropped to $972 million (Holeywell, 2013). While much of the auto industry had long ago decoupled from Detroit and had experienced decades of difficulties, the bankruptcy of General Motors and Chrysler Group in 2009 was still an important symbolic watershed. The auto industry was still a potent symbol for the city. Other problems preceded the financial crash: Detroit owed hundreds of millions of dollars to banks and bondholders because of earlier, ill-conceived, risky investments and bond issues. Additionally, Detroit had gambled on interest-rate swaps, which forced it to make big payments to the banks as interest rates fell.

Instead of being able to invest in basic services such as police, fire, and blight removal, the city was diverting

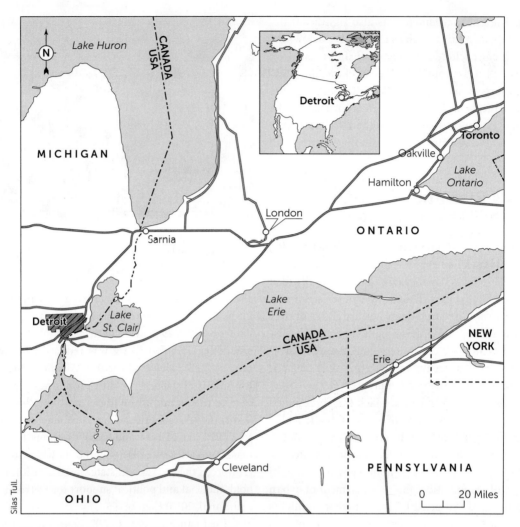

Figure 6.5 Location of Detroit, Michigan

nearly 5 per cent of its budget just to pay its debt (Holeywell, 2013). By 2012, many basic services were lacking in Detroit: most police stations were only open for eight hours a day, 40 per cent of street lights did not work, only one-third of the ambulances were running, and two-thirds of the parks were permanently closed. A corrupt local government reinforced the problems (Yaccino, 2013). Those residents economically able continued to leave the city, and by 2010, roughly only 700,000 residents remained (Tompor, 2015).

Detroit's bankruptcy created new political coalitions, and there were fierce negotiations over where austerity cuts would fall (Figure 6.6). Striking firefighters challenged the cuts, activists protested against widespread water shutoffs, and pensioners overwhelmed the public electronic filing system with protest letters. Despite this, the public sector was hard hit, as the city's workforce experienced pay cuts of 20 per cent and a severe reduction in numbers, while (unlike in Stockton) public-sector unions and pensioners saw reductions in their pensions (Dolan, 2015). Some city assets were privatized, such as the water and sewage system, while others remained in the public sphere. Interestingly, the successful political coalitions came together around the **cultural economy**. The Detroit Institute of Art, for example, was only saved by the $816 million

Detroit emerged from bankruptcy in late 2014, but at an enormous cost to the city, its residents, and its employees. Detroit is an extreme case, but it highlights problems that have afflicted many cities. As Anthony Minghine, associate executive director and chief operating officer of the Michigan Municipal League, another public-sector urban advocacy group, puts it, "The unfortunate part of Detroit and the mismanagement through the years, the corruption and everything else, is that it [the corruption] became a good sort of excuse for why this happened to them. But in reality, it's happening to everybody. It just got to them faster" (Kaffer, 2015: 1).

The scale and scope of Detroit's bankruptcy and the austerity measures that followed were extreme. The case shows the vulnerable position of some former industrial cities that now bear the scars of deindustrialization and previous rounds of economic restructuring. Detroit's municipal finances were already under pressure when the financial crisis of 2007–2008 started and deindustrialization, demographic change, and racialized politics took their toll. Where the austerity cuts would fall was fiercely contested, but despite this, the basic provision of public services was severely affected, pension benefits were cut, and city assets were sold. Just as importantly, the Detroit case highlights the precarious nature of urban citizenship, which can come under extreme pressure under conditions of austerity (see Harper, 2015).

Of course, urban austerity has not been confined to the United States. Canadian cities also experienced austerity measures after the global economic recession of 2007–2008, although often in ways that we describe as hidden. The following text box explores why and how Canadian cities experienced austerity, the ways in which this resembled the trends in the United States, and how it differed.

Figure 6.6 Poster advertising "Day against Austerity (Detroit)"

"grand bargain"—a coalition of national and local foundations, the United Auto Workers, the State of Michigan, and the Detroit Institute of Art, which together financed a new fund to save the city from selling art from the museum to pay off Detroit's creditors (Kaffer, 2015).

Austerity, Bankruptcy, and Canadian Cities?

Most urban experts agree that the possibility of an American-style bankruptcy of a Canadian municipality is unlikely. The rules in Canada are different, with much more regulatory and fiscal oversight than in American cities and towns. Municipalities are creatures of provincial governments, which means

continued

that the provinces and territories determine municipal powers and sources of revenues. Municipalities are not permitted to run operating deficits, and there are restrictions on the amount of money they can borrow from capital markets. Provinces are responsible for most of the major social expenditures, and there is an implicit shared state responsibility for municipal debt. Provinces receive large, unconditional transfers from the federal government, and for some provinces and municipalities transfers from the federal government are more important sources of revenue than their own taxes. If a municipality gets into trouble with its finances, the province will intervene in a variety of ways, including by rearranging municipal boundaries, taking over functions, or, in extreme cases, taking over the municipality's finances. In Ontario in the early 1930s, for example, more than 40 municipalities and school boards defaulted on their obligations, leading the province to set up more stringent controls over municipal finances. According to public-finance experts, most Canadian municipalities score well on traditional fiscal health measures like credit-rating reports and financial management (Shafroth, 2014).

While a full-blown American-style bankruptcy is unlikely in Canada, many urban experts are still concerned about the ability of Canadian municipalities to pay for local services and address all the needs of the people living in the city. Like their American counterparts, Canadian cities depend on property taxes for much of their revenue, so a shrinking tax base can cause a vicious cycle of declining services and declining revenues. This is a particular concern in slow-growing towns and cities, such as Thunder Bay, Ontario, or eastern Canadian cities like Halifax, Nova Scotia, and Saint John, New Brunswick. Many of these municipalities struggle to keep aging infrastructure up-to-date. Even fast-growing municipalities (such as Toronto) or ones with stable growth (like Kingston) with a deeper property tax base struggle to pay for infrastructure, especially public transit, and some social services, such as social housing, daycare subsidies, and family-support services. This is because for the past two decades more-senior levels of government have been downloading and offloading many services to municipalities but without providing the corresponding revenue to help municipalities pay for those services.

Some critical urban scholars interpret this downloading trend within the wider context of neoliberal development and the demise of the Fordist-Keynesian system (Harvey, 2005; Peck, 2014). Neil (2004) labels this "state-rescaling," which reflects the tendency of functions of the central state to be rescaled territorially, with sub-national governments assuming greater roles in both economic growth and redistribution. This means that municipalities in Canada (and elsewhere) must either become **entrepreneurial cities** in order to promote economic development and thereby increase their tax base, or engage in austerity policies such as cutting services to balance their budgets or hold the line on property tax increases. A recent example facing the city of Kingston is the province's decision to cut back on the child-care subsidy budget. The subsidy goes to families that need assistance paying for daycare so that the parents can either hold a minimum-wage-paying job or return to school for more training. Now the city has to decide if it is going to pay for the subsidy through property tax revenues or reduce the service. One of the dilemmas facing the city is that if it decides not to pick up the subsidies, there is a risk that many of those in the program will give up their studies or their minimum-wage jobs and end up on social assistance. This will place additional demands on the city because Ontario Works (the province's social assistance program), unlike daycare subsidies, is a mandatory program for municipalities (Schliesmann, 2013; see also Donald and Hall, 2016).

Crucially, in some cases, local politicians adopt the austerity narrative as a political device to shift spending priorities, even when their municipalities are not in debt. Once seen as a form of politics constrained to the American context, many anti-government ideas have migrated into Canadian municipal politics. For example, research on the overall spending and municipal revenue patterns in Ontario municipalities between 2009 and 2012 shows that while overall spending and municipal

Table 6.1 Ontario municipalities operating expenses by function, as percentage of total operating expenses 2009–2012

	2009	2010	2011	2012
General government	4.2	4.6	5.3	4.6
Protection services	16.5	16.7	17.3	17.5
Transportation services	22.2	22.2	22.0	22.8
Environmental services	14.2	14.1	14.3	14.9
Health services	4.9	5.0	5.2	5.1
Social and family services	19.2	18.7	16.9	16.8
Social housing	6.3	6.4	6.0	5.9
Recreation and cultural services	9.6	9.3	10.1	9.6
Planning and development	2.2	2.2	2.1	2.1
Other	0.7	0.8	0.8	0.8
TOTAL	100.0	100.0	100.0	100.0

Ministry of Municipal Affairs and Housing, Toronto.

revenue patterns appear stable in this period, the priority areas of spending have changed, as has the makeup of revenue sources (Table 6.1). Priorities appear to be moving away from social redistribution functions towards greater emphasis on protection services. These strategies have particular implications for certain groups and classes in society and raise questions about the role of **hidden austerity** in people's right to access services in their city. While many factors contribute to increasing inequality in Canadian cities, the evidence is convincing that inequality has increased at the same time that social redistribution has been fading (Banting and Myles, 2013). David Hulchanski's multi-year study of six Canadian cities (Halifax, Montreal, Toronto, Winnipeg, Calgary and Vancouver) is providing factual material on the growing income gap between the rich and the poor and the progressively smaller middle-income group, which has social implications at the neighbourhood and city level (for up-to-date statistics and reports, see http://neighbourhoodchange.ca; see also Hulchanski, 2010; Murdie et al., 2014).

Conclusion

This chapter has explored the concept of austerity and applied it to the urban level with a particular emphasis on how austerity policies and discourses can affect the planning and priority spending of urban governments. Austerity is not a new concept but has taken on particular significance in the global context as municipalities across North America deal with constrained public-sector budgets and increasing service demands from their constituents. Municipal governments have responded in a variety of ways to a more austere way of planning or what Peck (2012) has called "austerity urbanism": everything from increased privatization, to public-sector job and pension cutbacks, to municipal bankruptcies. The two case studies of Detroit, Michigan, and Stockton, California, are examples of extreme forms of austerity in the North American context, with both these cities declaring bankruptcy in a last-ditch effort to address major fiscal distress. How they came to these decisions reflects the complex interplay between global forces, state-level peculiarities, and local and historically contingent factors. In both cases, austerity was not only a top-down imposed process of state restructuring but also a bottom-up one, for many local decisions and actions over time facilitated the austerity response (see also the study on Vallejo by Davidson and Kutz, 2015). Although municipal bankruptcies are unlikely to happen to a Canadian municipality for a variety of institutional and regulatory reasons, we nevertheless show how discourses of austerity have migrated north and have shaped the spending priorities of municipal politicians who, in some cases, adopt austerity discourses even when their urban finances are in relatively good shape. The emerging evidence is that spending priorities are shifting away from redistributive functions such as child-care, affordable housing, and social services. The young, the poor, racialized communities, and the elderly are the groups most affected by these changing priorities in spending.

Austerity is much more than a neutral policy to reduce budget deficits. It is intended to function as an alternative to other, more progressive policy approaches like guaranteed basic income programs that seek to stimulate demand. Austerity, as a policy and a discourse, is now widespread and pervasive with long-term implications for cities and the people who live in them. It is a phenomenon that has played out differently across cities and states in North America—from the withdrawal of basic public services in places such as Detroit and Stockton that had a particular impact on the racialized poor in certain neighbourhoods to the use of austerity narratives to justify changing budget priorities in Ontario that again had an asymmetrical impact on social groups. Understanding this variation requires sensitivity to the ways in which local institutions, political structures, history, and the extent of budgetary crisis mediate larger economic and political changes such as fiscal crisis, state downloading, and rescaling. We need more research to understand this variation in local austerity (see Davidson and Kutz, 2015, for example). We still need to explore why and how austerity takes such different forms, how this affects often the poorest and most vulnerable urban residents, and the ways it functions to restructure notions of accountability and civic democracy at the local level. The case studies of Detroit and Stockton show how austerity is not just a top-down process of state restructuring but is also contingent on particular local decisions taken in time and space. The cases also highlight how citizens have found ways to resist and challenge austerity urbanism. With recent events in Greece and around the world, we also have emerging examples of push-backs against austerity-imposed policies. The extent to which these push-backs may lead to viable alternatives to austerity and neoliberal forms of urbanization remains to be seen in the years ahead.

Key Points

- Austerity as a policy and discourse existed since at least the 1930s, but the contemporary use of austerity is particularly local in nature.
- Although austerity is occurring in many cities of the global North, how it is realized varies.
- Austerity is the outcome of a complex interplay between global forces and local, contingent factors.
- The case of austerity in Stockton, California, illustrates the profound collapse of the city as a result of the housing-induced boom and global financial markets.
- The case of austerity in Detroit, Michigan, was the outcome of both long-standing deindustrialization, suburbanization, and the systemic racialization of city politics and the specificity of the short-term financial crisis.
- The policy and the narrative of austerity have transferred north to Canadian municipalities, and while municipal bankruptcy is unlikely in Canadian cities, we do see "hidden austerity" or the shifting of urban budget priorities away from redistribution.
- The shifting of urban budget priorities away from redistribution is happening at the same time that we are witnessing a growing income inequality within our cities.

Activities and Questions for Review

1. Work in pairs to develop a list of some of the ways in which austerity intensifies inequality.
2. How does austerity play out differently in Stockton and Detroit? Illustrate these differences with symbols in a picture that represents each city.
3. In groups, debate the extent to which austerity is a necessary urban policy response to the global financial crisis. What could cities do differently in response to the neoliberal discourse of austerity?

Acknowledgments

We would like to thank the Cambridge Political Economy Society Trust (Centre for Business Research, Cambridge Judge Business School, University of Cambridge) and the British Academy for their funding of our research project, "Regimes of Austerity: Economic Change and the Politics of Contraction" (2015–2018). This chapter benefited from the initial stages of this research.

References

Anderson, M.W. (2012). Democratic dissolution: Radical experimentation in state takeovers of local governments. *Fordham Urban Law Journal*, 39(3): 577–623.

Bailey, N., Bramley, G., and Hastings, A. (2015). Introduction: Local responses to "Austerity." *Local Government Studies*, doi:10.1080/03003930.2015.1036988

Banting, K., and Myles, J. (Eds.) (2013). *Inequality and the fading of redistributive politics*. Vancouver: University of British Columbia Press.

Bardhan, A., and Walker, R. (2011). California shrugged: Fountainhead of the Great Recession. *Cambridge Journal of Regions Economy and Society*, 4(3): 303–322.

Bird, R., and Slack, E. (1993). *Urban public finance in Canada*. Toronto: Butterworth.

Blyth, M. (2013). *Austerity: The history of a dangerous idea*. Oxford: Oxford University Press.

Brenner, N. (2004). Urban governance and the production of new state spaces in Western Europe, 1960–2000. *Review of International Political Economy*, 11(3): 447–488.

Bureau of Labor Statistics, United States. Numerous years. State and Area Employment, Hours, and Earnings.

Chang, H.-J. (2003). The market, the state and institutions in economic development. In H.-J. Chang (Ed.), *Rethinking development economics*. London: Anthem Press.

Clapham, D.F., Clark, W.A., and Gibb, K. (Eds.) (2012). *The Sage handbook of housing studies*. London: Sage.

Clavel, P. (1980). *Urban and regional planning in an age of austerity*. New York: Pergamon Press.

Clement, D., and Kanai, M. (2015). The Detroit future city: How pervasive neoliberal urbanism exacerbates racialized spatial injustice. *American Behavioral Scientist*, 59(3): 369–385.

Davidson, M., and Kutz, W. (2015). Grassroots austerity: Municipal bankruptcy from below in Vallejo, California. *Environment and Planning A*, 47, 1440–1459.

Davidson, M., and Ward, K. (2014). "Picking up the Pieces": Austerity urbanism, California, and fiscal crisis. *Cambridge Journal of Regions, Economy and Society*, 7: 81–97.

Deis, B. (2012). A message from the city that went bankrupt. *Wall Street Journal*, September 28. Retrieved from http://www.wsj.com/articles/SB10000872396390443995604578002200588578188

Dolan, M. (2015). In Detroit bankruptcy, art was key to the deal: "Grand bargain" involving city-owned collection paved way for exit from court. *Wall Street Journal*, November. Retrieved from http://www.wsj.com/articles/in-detroit-bankruptcy-art-was-key-to-the-deal-1415384308

Donald, B., Glasmeier, A., Gray, M., and Lobao, L. (2014). Austerity in the city: Economic crisis and urban service decline? *Cambridge Journal of Regions, Economy and Society*, 7(1): 3–15.

Donald, B., and Gray, M. (2015). Canada's hidden austerity in contemporary municipal governance. Working Paper, Department of Geography, Queen's University and Cambridge University.

Donald, B., and Hall, H. (2016). The social dynamics of economic performance in the public-sector city: Kingston, Ontario. In D.A. Wolfe and M. Gertler (Eds.), *Growing urban economies: Innovation, creativity, and governance in Canadian city-regions*. Toronto: University of Toronto Press, pp. 311–333.

Fainstein, S.S. (2010). *The just city*. Cornell, NY: Cornell University Press.

Farmer, L. (2014). Exiting municipal bankruptcy only a step in road to recovery. *Governing the states and localities*. Retrieved from http://www.governing.com/topics/finance/gov-stockton-municipal-bankruptcy-exit.html

Glaeser, E. (2013). In Detroit, bad policies bear bitter fruit. *Boston Globe*, July 23. Retrieved from http://www.bostonglobe.com/opinion/2013/07/23/detroit-bad-policies-bear-bitter-fruit/PHY6nyuDHfvpr3qreIPahK/story.html

Gray, M., Lobao, L. and Martin, R. (2012). Making space for wellbeing. *Cambridge Journal of Regions, Economy and Society*, 5: 3–13.

Hackworth, J.R. (2007). *The neoliberal city: Governance, ideology, and development in American urbanism*. Cornell, NY: Cornell University Press.

Harper, H. (2015). Local financial crisis and the democratic process: A case study of Michigan's emergency manager law. Working Paper, University of California–San Diego, Department of Sociology.

Harris, N., and Fabricius, I. (1996). *Cities and structural adjustment*. London: Routledge.

Harvey, D. (1989a). *The urban experience*. Baltimore: Johns Hopkins University Press.

Harvey, D. (1989b). From managerialism to entrepreneurialism: The transformation in urban governance in late capitalism. *Geografiska Annaler. Series B. Human Geography*, 71(1): 3–17.

Harvey, D. (2005). *A brief history of neoliberalism*. Oxford: Oxford University Press.

Holeywell, R. (2012). Stockton, Calif., is largest city to file for bankruptcy. *Governing the states and localities*. Retrieved from http://www.governing.com/news/local/gov-stockton-california-bankruptcy.html

Holeywell, R. (2013). Detroit files for Chapter 9 bankruptcy. *Governing the states and localities*. Retrieved from http://www.governing.com/blogs/view/Detroit-Bankruptcy.html

Hulchanski, J.D. (2010). *The three cities within Toronto: Income polarization among Toronto's neighbourhoods, 1970–2005*. Toronto: Cities Centre, University of Toronto.

Jonas, A.E.G., McCann, E., and Thomas, M. (2015). *Urban geography: A critical introduction*. Wiley Blackwell.

Kaffer, N. (2015). Detroit faces the same challenges after bankruptcy. *Detroit Free Press*. Retrieved from http://www.freep.com/story/opinion/columnists/nancy-kaffer/2015/01/04/detroit-revenue-bankruptcy/21215969/

Kitson, M., Martin, R., and Tyler, P. (2011). The geographies of austerity. *Cambridge Journal of Regions, Economy and Society*, 4(3): 289–302.

Levine, H., Scorsone, E.A., and Justice, J.B. (2012). *Handbook of local government fiscal health*. Burlington, MA: Jones & Bartlett Publishers.

Lobao, L.M., and Adua, L. (2011). State rescaling and local governments' austerity policies across the USA, 2001–2008. *Cambridge Journal of Regions, Economy and Society*, 4(3): 419–435.

Logan, J.R., and Molotch, H.L. (2007). *Urban fortunes: The political economy of place*. Berkeley: University of California Press (20th anniversary edition).

McDonald, J. (2014). What happened to and in Detroit. *Urban Studies*, 51(16): 3309–3329.

Macleod, G., and Goodwin, M. (1999). Space, scale and state strategy: Rethinking urban and regional governance. *Progress in Human Geography*, 23(4): 503–527.

Mayer, M. (2013). First world urban activism: Beyond austerity urbanism and creative city politics. *City*, 17(1): 5–19.

Mitchell, D. (2003). *The right to the city: Social justice and the fight for public space*. New York: Guilford Press.

Murdie, R., Maaranen, R., and Logan, J. (2014). Eight Canadian metropolitan areas: Spatial patterns of neighbourhood change. Retrieved from http://neighbourhoodchange.ca/documents/2015/04/murdie-etal-2014-spatial-patterns-nhood-change-1981-2006-rp234.pdf

Peck, J. (2012). Austerity urbanism: American cities under extreme economy. *City*, 16(6): 626–655.

Peck, J. (2014). Pushing austerity: State failure, municipal bankruptcy and the crises of fiscal federalism in the USA. *Cambridge Journal of Regions, Economy and Society*, 7(1): 17–44.

Peck J., and Tickell, A. (2002). Neoliberalizing space. *Antipode*, 34(3): 380–404.

Rozansky, J. (2012). San Bernardino's route to bankruptcy. *City Journal*, July 18. Retrieved from http://www.city-journal.org/2012/cjc0718jr.html

Rudolf, J. (2012). Stockton bankruptcy may force "mass exodus" of police during crime wave. *Huffington Post*, August 23. Retrieved from http://www.huffingtonpost.com/2012/08/23/stockton-bankruptcy-police-crime_n_1826100.html

Schliesmann, P. (2013). City braces for childcare cuts. *thewhig.com*, January 21. Retrieved from http://wwwthewhig.com/2013/01/21/city-braces-for-child-care-cut

Shafroth, F. (2014). Why cities can't go bankrupt in Canada or Germany: There's a lot America can learn from these two countries about how to avert municipal bankruptcies. Retrieved from http://www.governing.com/columns/public-money/gov-municipal-debt-traps-nein.html

Slack, E., and Chattopadhyah, R. (2013). *Governance and finance of metropolitan areas in federal systems*. Oxford: Oxford University Press.

Sugrue, T. (1996). *The origins of the urban crisis: Race and inequality in postwar Detroit*. Princeton, NJ: Princeton University Press.

Tompor, S. (2015). Detroit retirees to see pension cuts. *Detroit Free Press*, March 2. Retrieved from http://www.freep.com/story/money/personal-finance/susan-tompor/2015/02/27/detroit-orr-pension-checks-cuts/24144513/

Warner, M.E., and Clifton, J. (2014). Marketisation, public services and the city: The potential for Polanyian counter movements. *Cambridge Journal of Regions, Economy and Society*, 7: 45–61.

Whitfield, D. (2013). *Unmasking austerity*. Paper published through the Australian Workplace Innovation and Social Workplace Centre, University of Adelaide report. Retrieved from http://www.cpsu.asn.au/upload/Research/WISeR_unmasking-austerity%20FINAL.pdf

Wyly, E., Moos, M., Hammel, D., and Kabahizi, E. (2009). Cartographies of race and class: Mapping the class-monopoly rents of American subprime mortgage capital. *International Journal of Urban and Regional Research*, 33(2): 332–354.

Yaccino, Steven. (2013). Kwame M. Kilpatrick, former Detroit mayor, sentenced to 28 years in corruption case. *New York Times*, October 10.

7. Land Use and Creativity in Post-industrial Cities

Alison L. Bain and Rachael Baker

Introduction

The restructuring of the global economy has had significant, although differential, impacts on cities around the world. The movement of traditional manufacturing away from North American, European, and British cities into regions of the global South, where real estate, wages, and conditions of labour cost significantly less, reflects a geographic shift in industrial production but also a change in the relationship between labour, capital, and place. Such a reconfiguration of the global supply chain perpetuates North American and European economic dominance and the expansion of unfair and unsafe working conditions in industrializing cities of the global South. In this exploitative global labour market, cities compete with one another to attract foreign investment. Urban competition and urban growth have thus become the buzzwords of the twenty-first century for city leaders, particularly for those in the global North grappling with the loss of an industrial economic base and the decline in blue-collar and middle-income jobs.

Western urban scholarship has emphasized how former industrial cities in the global North have become hollowed out by the globalization of capital and are in need of post-industrial regeneration (Curran, 2010). Motivated by the decline of their industrial economies and urban landscapes, many urban policy-makers in older industrial cities in North America, mainland Europe, and Britain share the common goal of kick-starting a new era of post-industrial growth (Zukin and Braslow, 2011). Urban policy in the global North has thus shifted away from industrial growth, which drove urban development in the late nineteenth and early twentieth centuries, to concentrate instead on marketing **quality of life** and leisure opportunities for residents, investors, and tourists. A key urban policy objective is to hide "the shadow" by replacing negative perceptions of the city as a place of disinvestment, decay, crime, and poverty with images of growth, vitality, and prosperity (Short, 1999). A renewed city image can influence where businesses locate, which in turn can impact where inward investment is directed and what new infrastructure of everyday life is built for local residents. For over a decade, mobile and influential international creative city-planning and policy consultancy discourses have advocated that urban managers develop culture-led urban revitalization strategies to positively re-image cities and to reclaim the central city for higher-end, consumptive-driven land uses.

In light of the creative and cultural turn in neoliberal urban policy, this chapter demonstrates that the development of land—a key process in the construction and growth of cities—involves contestations that can augment racialized and classed social exclusions. The ways in which urban land is put into use are inevitably the result of complex negotiations between different urban actors with varying political agendas and everyday struggles. This chapter, however, emphasizes the power of the urban planning profession and urban policy discourses in influencing, shaping, and controlling property development. The argument is made that a privileging of the **creative class** and **creative-city theory** in urban planning and policy distorts land-use investment priorities and augments socio-spatial disparities. In what follows, this chapter discusses the challenges facing **post-industrial cities** in the global North, explores how urban policy is mobile, and reflects upon the impacts of creative city policies in the cities of Detroit, USA, and Leipzig, Germany. These are two cities dealing in different ways with urban decline and urban shrinkage through the strategic development and management of land for urban agriculture and cultural production. The chapter concludes by examining how the management of decline is also a significant urban policy issue in Canada, particularly for small- and mid-sized cities, with scholars advocating for a policy shift away from growth to a focus instead on sustainability, livability, and quality of life.

Understanding the Urban Post-industrial Challenge

The term "post-industrial" first surfaced in political economic theory in the 1960s and 1970s and was featured in Daniel Bell's *The Coming of Post-industrial Society* (1973). Scholarly analysis of the post-industrial era coincided with the reorganization of North American industrial manufacturing at the onset of what Bell predicted would be the start of a knowledge-based economy. Its application to cities aligned the transforming urban sphere with the everyday experiences of working-class and marginalized communities, marking the beginning of the more frequent use of the term "post-industrial" within urban theory.

In the twenty-first century, some urban scholars have taken a more aestheticized approach to industrial decline. Alice Mah, in her book *Industrial Ruination, Community, and Place* (2012), for example, characterizes post-industrial cities through an aesthetic process of aging and general decrepitude that takes place over time. The "ruination" of abandoned lots, brownfield sites, vacated industrial buildings, chained factory gates, and rusting former instruments of manufacturing are united for Mah (2012: 3) by a common aesthetic that reveals the visual results of disinvestment. Given that mass unemployment, factory closures, and longer-term decline of industrial sites happen over a period of years and often decades, ruination helps to frame industrial decline as a process as much as a material and economic marker on the landscapes of post-industrial cities. Mah's understanding of deindustrialization as operating as part of a longer continuum stands in contrast to scholars who describe it as a series of economic and political processes that have come and gone (Hall, 1997; Montgomery, 2011).

The US Rust Belt and the Canadian Golden Horseshoe regions have traditionally been dominant centres of academic inquiry into processes of urban deindustrialization in the post-industrial era. Although an international border separates these two regions, they tend to be couched under the single banner of the North American Rust Belt (Cowie and Heathcott, 2003). Together these regions mark a footprint around Lake Ontario, Lake Erie, and Lake Michigan, connecting what were once North America's largest industrial manufacturing cities. Such a regional perspective is useful for appreciating the local political economic impacts of industrial decline. As well, urban geographers often simply differentiate industrial and post-industrial cities by the presence or lack of traditional manufacturing sectors within a city or region.

One of the struggles faced by post-industrial cities is the task of overcoming the social and economic challenges that accompany industrial disinvestment. Such place-based maladies uphold reputations of urban decline in post-industrial cities and equate local realities of poor health outcomes, poverty, structural neglect, and vacancy as natural consequences of industrial disinvestment. Part of the post-industrial landscape is the presence of property vacancy, abandonment, disrepair, and blight over time. Popular representations of post-industrial cities as sites of decay or urban wilderness construct these cities as in need of development and revitalization. A common urban policy response to this decline and to the exodus of large, traditional manufacturing employers is to identify new economic drivers that can both provide a new employment base and help to redefine a city's image. Knowledge-based technological and creative industries have become sought after as replacements. In the United States, the cities of Baltimore, Buffalo, Chicago, Detroit, and Pittsburgh; and in Canada, the cities of Hamilton and Kitchener, have all sought to attract medical facilities and educational institutions, along with cultural entrepreneurs, to central city neighbourhoods—what is often termed the "meds and eds" strategy (Newman and Safransky, 2014). Such dramatic urban-rebranding initiatives and their accompanying infrastructural investment priorities and target demographics can be disruptive to local and long-standing notions and representations of place that are deeply seeded within urban culture and everyday experiences. The rebranding of post-industrial cities for the sake of salvaging their economic base can interrupt and potentially misrepresent local narratives of belonging, risking the creation of factionalism between creative and professional workers and those who identify deeply with a city's working-class culture as well as with the working-class along racialized lines.

Creative City Policy Mobilities: Favouring the Creative Class

Cities are made through specific policy practices and representations. Learning from the policy practices of other cities is a key part of urbanism. Much inter-city learning occurs through urban **policy mobilities** initiated by urban policy-makers who function as transfer agents. Influenced by political ideology, institutional arrangements, and path dependencies, urban policy-makers work to identify and circulate successful urban policy models and to adapt them to local conditions (McCann, 2011a). Urban policies "are not inherently mobile, but rather are bundled, packaged, (mis)represented, projected, shared, communicated, disregarded or aspired to through complex power-laden social and political processes" (Barber, 2014: 1185). Urban policy knowledge, best practices, and models are a product of dynamic exchanges within and between the cities where they are elaborated. They form the discursive "connective tissue" of cities, influencing public discourse and setting the agenda of possibilities for urban change (McCann, 2011b). The production of policy knowledge involves "a politics of persuasion and coalition-building in and through which long-term and effective consensus is established over the definition of key problems and specific rationalities and technologies through which problems will be addressed" (Temenos and McCann, 2012: 1394). Access to urban power allows individuals, groups, organizations, and institutions to variously define what constitutes an urban problem and to intervene to develop solutions. And while power can be a productive force if shared relationally, often access to it is unequal, even in democratic models of urban politics. People do not all have the same ability to have their interests represented, their voices heard, or their needs met, particularly when a dominant policy discourse defines how different urban interests and actors interact.

An urban policy discourse that has dominated land-use development in post-industrial cities, setting the terms of debate and investment priorities for the last decade, is Richard Florida's (2002, 2004) creative-class thesis, which urges cities to deploy **cultural capital** as an entrepreneurial asset (Bridge, 2006).

The concept of the creative class is intensely controversial, as are Florida's suppositions, research methods, and policy recommendations. Florida's argument (2002) is that rooted in an analysis of American cities, a creative class drives urban economic growth and innovation. This class has two parts: a "super-creative core" of professionals "whose economic function is to create new ideas, new technology and/or creative content" (Florida, 2002: 8) in the fields of fine arts, media, entertainment, science, engineering, education, and research; and "creative professionals" in knowledge-based industries like business, finance, law, and health care. The earning potential and quality-of-life needs of these two creative-class groups are very different. Bankers, lawyers, engineers, and doctors have higher salaries and often also different urban amenity and social-service requirements than lower-income, more precariously employed professional cultural workers like visual artists, writers, dancers, actors, and musicians, who frequently work on contract, lack health benefits and pensions, and struggle to find affordable work space (Bain and McLean, 2013). Corporate elites may be more concerned with access to luxurious and secure accommodation, private schools and recreation facilities, and high-end restaurants, stores, and entertainment than with the provision of basic public services like public schools, community centres, parks, and libraries. These very different lived experiences of cities and expectations about urban infrastructure and service investment priorities have opened creative-class theory to significant critique for its conceptual "fuzziness" (Markusen, 2006); aggregation of unrelated occupational groups with very different work-based cultures, skills, and incomes (Krätke, 2010); minimization of the significance of social reproductive labour (Parker, 2008); racialized exclusions (Catungal and Leslie, 2009); privileging of middle-class rather than working-class cultural capital (McLean, 2014); reproduction of homonormativities (Oswin, 2014); politically expedient policy malleability (Peck, 2005); and policy aggravations between the use of culture for social inclusion and its use for economic growth (Krivy, 2013). Also unaddressed by Florida, but important to consider, is that grassroots community organizations often employ collaborative rather than competitive creative practices to express local struggle and address local needs in ways that can build community resilience. Urban agriculture, as a land-use practice, for example, works through community organizations to address local food insecurity needs while also transforming representations of stigmatized post-industrial landscapes as dirty and contaminated into productive community resources. In transformating land and neighbourhood place-identities through the growing and selling of food, urban farmers engage in creative practices, but this occupational group is not directly named as members of Florida's creative class.

The global appeal of Florida's creative-class urban policy approach is that human rather than financial capital is framed as the key to post-industrial growth through public-private partnerships. Thus, the rise of a knowledge-intensive and culture-based urban economy presents as an appealing, relatively low-cost strategy for cash-strapped cities to attract and to retain creative residents. Cities with little money to invest can celebrate surface transformations to the public realm brought about, for example, by the building of bicycle lanes and the commissioning of public art. In place of factories they can brand arts districts and encourage gentrification through the incursion of hipster bars, cafés, and farmers markets into low-income neighbourhoods and public spaces. Such cultural amenities are intended to foster "an atmosphere of social tolerance for men and women who express their 'differences' from traditional, mainstream ways of life" (Zukin and Braslow, 2011: 131). But commodified tolerance offers neither overt acceptance nor direct support for marginalized social groups (e.g., gays and lesbians, transgender individuals, refugees, immigrants, racialized people, senior citizens, or people with disabilities). Cultural producers are not agents of social and cultural diversity. Instead, public officials and real estate developers have deployed the symbolic and cultural capital (formal educational qualifications and "creative" skills) of an artistic mode of production and associated preferences for certain lifestyles and urban spaces to extend and to consolidate "a normalized neoliberal urban rule" (Novy and Colomb, 2013: 1821).

The celebration of urban culture as a powerful force shaping cityscapes is apparent in the cultural planning and cultural policy literature with its terms that highlight culture's significance and concentration in particular places: creative milieu, cultural quarters, cultural districts, and arts districts. These terms all reinforce an idea popular with cultural economists—that creativity is grounded in particular geographic, social, and economic contexts and not in others (Mizzau and Montanari, 2008). As Stolarick and Florida (2006: 1802) assert, "the general creative milieu of a place with a prominent presence of artists, musicians, and other creative people increases overall creativity and innovation by providing stimulus and inspiration" and allows "important interactions and spillovers" to occur. In post-industrial cities, urban cultural policy is dominated by this Floridian idealization of the value of creative clusters as sites of capital accumulation, cultural innovation, and social inclusion (Landry, 2000). The cluster concept is, however, little more than an overused land-use regulation tool to transform cultural production into an economically successful sector (Van Heur, 2009).

To initiate urban revitalization, city leaders in the global North frequently seek to brand neighbourhoods with a concentration of cultural activity as cultural quarters and cultural districts (Evans, 2009; Mommaas, 2004). Cultural quarter is the terminology popular in Britain, while cultural district and arts district are the terms most commonly used within North America (Brooks and Kushner 2001). In this chapter we use the term arts district. Branding a neighbourhood as an arts district usually means that it is of mixed use, with differently sized cultural venues that host programming for artistic production and consumption that sustain the local cultural economy (Montgomery, 2008). Arts districts have become a standard urban revitalization and economic development strategy to enhance networking and innovation opportunities for members of the creative class.

In urban studies scholarship, a distinction is made between *formal* (resulting from state-led cultural planning and policy and public investment) and *informal* arts districts (emerging organically, spontaneously, and dynamically through the bottom-up practices of local stakeholders in the absence of a civic plan) (Chapple et al., 2010). In practice, however, what are designated as "formal" or "informal" or, to use Vivant's (2009) notion, as "in" and "off" cultural production systems are often economically interdependent and a by-product of the agendas of different social actors. Arts districts often develop where the seeds of a cultural presence already exist (Roodhouse, 2006)—suggesting that an initial informal presence is formally capitalized upon. Financial and political support to develop local arts and culture scenes is commonly asserted to be an essential public policy directive (Mizzau and Montanari, 2008: 667), but this is not always the case. In the face of municipal budgetary cutbacks, many informal arts districts, especially grassroots cultural initiatives with no economic imperative, receive little civic recognition. It is important to emphasize that top-down cultural planning can also be constraining because it usually privileges economic, marketing, and hard infrastructure priorities over bottom-up community-building and the quality and extent of informal interaction (Power and Scott, 2004).

When culture and creativity are exploited to manufacture new place-identities for inner-city industrial and dilapidated areas, the goal of such rebranding is to increase the economic value of the land. Urban areas with distinctive natural and built heritage assets and high visibility within the cityscape are targeted for reinvestment. Areas of the city once dominated by large-scale, Fordist industrial production have been disassembled and adaptively reused as sites of consumption. As Stevens and Ambler (2010: 529) assert, such a transition "is not tidy, complete or permanently fixed" because "the post-industrial urban landscape is fragmented with 'gaps' and marginal spaces that are ill-defined, under-developed and less regulated." The gaps in this "terrain vague" can create openings for alternative claims to urban space with less profit-driven, controlled, and privatized orientations, such as artistic interventions, community gardens, collective kitchens, and urban forests (Mariani and Barron, 2014).

In cities, land is a finite resource and a basic urban commodity that needs to be shared among different social groups and used for different functional purposes. Land is governed by property rights that dictate who can use land and how. While land is bounded and its boundaries and categories of use are policed and reinforced through urban planning and policy, land and land use can also be reconfigured. In a creative city framework, land has been revalorized through human ingenuity (as is encapsulated in the concept of creativity—imaginative creative processes that produce something new). Civic leaders of post-industrial urban economies have become preoccupied with the multiplier effects of culture and creativity—"both as a productive force in its own right and as a critical component of the 'soft infrastructure' necessary to compete for mobile capital investment, jobs, people, and tourist spending" (Novy and Colomb, 2013: 1822). The case studies that follow from Detroit and Leipzig reveal that access to affordable land has helped to sustain grassroots, low-cost creative and cultural activities. The danger, however, is that such innovative, temporary uses of land merely get "absorbed into the 'software' of neoliberal urban renewal" (Holm and Kuhn, 2011: 655) as creative distractions from the more costly infrastructural investments needed in jobs, education, and housing in multi-ethnic, working-class, "disadvantaged" neighbourhoods (Blokland and Nast, 2014).

Detroit: A City of Ruin, Creativity, and Land Excess

Detroit is one of North America's most photographed post-industrial cities (Millington, 2013). Its high rates of residential and industrial property abandonment have made the city a destination for urban explorers and photojournalists documenting the ruinous remains of post-industrial America. In photographs, the Motor City is commonly depicted as an empty, abandoned, and apocalyptic landscape. To manage the city's excess land and housing stock—its vacant, abandoned, and tax-foreclosed properties—the city established the Detroit Land Bank Authority in 2008 (http://www.buildingdetroit.org/). The **land bank** operates an ongoing public auction of "banked" properties and carries out demolition and land clearance when residential, industrial, and commercial sites fall beyond repair. The city's excess housing stock is then managed through sale by the county at public auction, where speculative buyers, long-time residents, and the city's new creative class have been able to purchase housing at rates substantially below market value.

Detroit's approximately 90,000 vacant lots are the result of decades of industrial decline, white- and middle-class flight to the suburbs, and, most recently, the aftermath of the subprime mortgage crisis, which produced a 25 per cent decline in the city's population in the first decade of the twenty-first century. Detroit's population has shrunk by over a million people in the last half century, and approximately 25 per cent of this loss occurred between 2000 and 2010. In 2015, Detroit had a population of 688,701, 83 per cent of which is African American. Several dramatic declines in population have exacerbated racial and class inequalities festering in Detroit since the 1950s and have resulted in dramatic disparities in population density, municipal service delivery, and housing security throughout the city's 59 neighbourhoods.

The city of Detroit and the state of Michigan have been presented with a number of proposals outlining potential redevelopment strategies for managing and redeveloping the city's expanses of vacant terrain. The proposal that has received the most attention in the city's post-bankruptcy period is the Detroit Future City (DFC, 2012) "right-sizing" plan that proposes to re-concentrate city services and residents in neighbourhoods that have maintained relatively high-population densities. The plan is celebrated by urban planners and policy-makers for its projected repurposing of 104 square kilometres (40 square miles) of land in the city. Detroit Future City proposes redeveloping land into "distinct and regionally competitive neighborhoods" comprising "the assets that make a city attractive" (DFC, 2012: 205). This right-sizing strategy has been actively worked on by city administrators since 2010 when then-mayor David Bing introduced a demolition plan that would carry out 10,000 household and structural demolitions during the course of his four-year term in office. Removing blighted or under-utilized properties from private ownership has been a practice of the city since the infamous 1967 racial uprisings that caused extensive fire damage to over 2000 structures. Attempts to centralize the city's population and consolidate the operation and maintenance of civic infrastructure into "core" neighbourhoods come at a cost that unevenly impacts city residents. While the mass demolition of vacant and dilapidated homes and buildings addresses city leaders' desires to eliminate blight, it simultaneously erases long-standing structural markers of neighbourhoods and city culture in addition to depending on the forcible eviction of residents who reside in underserviced areas of the city.

At the forefront of redevelopment goals is the integration of green infrastructure into the city's resources and the creation of walkable and "live and make" neighbourhoods. Green and walkable neighbourhoods constitute a redevelopment strategy that transforms formerly industrial areas into spaces of economic, social, and creative growth through mixed-use developments of residential, cultural, and commercial space. Such areas are intended to attract a creative workforce who will contribute entrepreneurial, artistic, and creative capital to the reinvention of these formerly industrial neighbourhoods.

Carried out in the name of urban renewal, redevelopment in Detroit is packaged as a project that will pull the city out of its post-industrial slump by replacing blighted neighbourhoods with new green (e.g., urban forests, park expansion, urban farms, and greenways running parallel to the city's extensive highway and road system) and blue (e.g., retention ponds and river resurfacing) infrastructure; all of this is done in accordance with the Fifth Amendment of the US Constitution, which conditions that private property cannot be taken for public use without just compensation and that all proceedings of eminent domain must contribute to public use or to the greater good (Kotlyarevskaya, 2005). All "right-sizing" efforts will inevitably depend on some citizens' homes and neighbourhoods being sacrificed for the supposed greater good of urban renewal.

In the spring of 2015, Wayne County, where Detroit is located, seized the titles of approximately 21,000 occupied homes in the city, impacting approximately 57,000 residents across Detroit. Many of these foreclosed homes and properties were then sold through auction. By 2011, 25 per cent of households in the city had entered some degree of foreclosure brought about by the 2007 US mortgage crisis. The impact of property dispossession disproportionately affected African American households—a clear marker of the structural racism that have existed in the US housing market and the lack of consideration for racial equality given to planning processes more generally. Take, for example, an elderly homeowner who was evicted from her family home of over 60 years after the property was sold at auction in May 2015 for $20,000 because she had not been able to keep up with property tax payments (Sheehan, 2015). A recent amendment to state law prohibited this resident—and all homeowners who fall delinquent on property tax payments—from participating in the auction, excluding them from repurchasing their foreclosed homes. At auction, many homes are purchased as income properties or fixer-upper projects by people who are looking to be part of the city's rebirth. In May 2015, 279-square-metre (3000-square-foot) houses with up to six bedrooms were available in the Detroit Land Bank auction listings, with bids from $250 to $1000. A former executive of a multinational company recently moved to Detroit from Singapore and purchased 150 homes, some for as low as $500. His plan is to refurbish and rent these properties as a way to "save neighbourhoods" (Ager, 2015). A lack of financial capital and social mobility does, however, prevent many Detroiters from purchasing even low-cost residential properties or from buying back homes that were lost in the foreclosure crisis. In Detroit, structural racism plays a large part in who has the ability to own land or to have guaranteed housing security.

For the last decade, a common interim land use for vacated industrial and residential lots in Detroit has been urban agriculture. Urban farmers in Detroit, many of whom are newcomers, focus their efforts on addressing the local food needs of growers, consumers, and marginalized communities (Tracey, 2011; DeLind, 2002). The Detroit Garden Resource Program estimates that between 1500 and 2000 food-producing farms and gardens currently operate in the city (MSNBC, 2014) (Figure 7.1). Such extensive use of urban land for agriculture, a relatively footprint-heavy activity, has been possible through informal claims to the neighbouring lots of absentee homeowners and through parklands left to pasture by the city.

The conversion of so much urban land to agricultural use is possible because the city's vast, vacant acreage is discursively framed as "empty" space (Safransky, 2014). Although the Michigan Right to Farm Act of 1981 prohibits agricultural operations within the city, Detroit city administrators have predominantly supported urban agricultural initiatives (Kalish, 2011). Such municipal support stems from the belief that farming deters "undesirable activities" (e.g., survival sex, drug-dealing, violence, organized crime, car break-ins, arson, dumping, and squatting) and helps to stabilize neighbourhoods.

Efforts to revitalize or, at the very least, maintain Detroit through greening and creative projects are visible throughout the city and are not solely driven by city- and state-led redevelopment initiatives (Figure 7.2). Detroit's growing creative class of visual, theatrical, musical, and

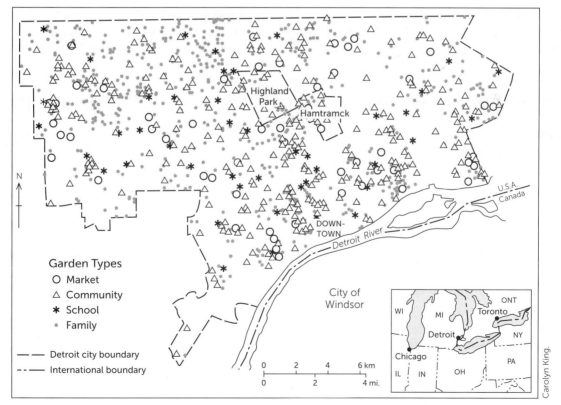

Figure 7.1 Map of market, community, school, and family gardens in Detroit, Michigan

Figure 7.2 Brother Nature Organics farm located in Detroit's North Corktown neighbourhood, where an owner works among rows of salad greens

Figure 7.3 Farnsworth Street community garden on Detroit's east side is used all year

urban agricultural workers is responsible for a number of grassroots projects in the city that address local needs and create shared resources. The urban farms found throughout the city put otherwise underutilized or **brownfield development** lots back into productive use while addressing local concerns over access to affordable and fresh foods (Figure 7.3). Art installations throughout the city illustrate messages of Black Power, community resilience, and post-apocalyptic spectacle (the Heidelberg project, www.heidelberg.org/; Ice House, http://icehousedetroit.blogspot.com/; Detroit Beautification Project, https://vimeo.com/44699407; and Illuminated Mural, http://detroitfunk.com/illuminated-mural/). Creative and green land uses are informal, institutional, and occasionally city or state funded. As the city's redevelopment plan rolls out, informal, creative, and community uses of land have been, and continue to be, challenged by new policy frameworks that demand greater control over neighbourhood planning. Although the strategic Detroit Future City framework features plans to increase green and creative land uses and clusters in the city, some urban farmers who grow on squatted land have faced demolition of their farms without notice or forced eviction to make room for projected future development. Property ownership and regulation adherence appear to be necessary for green and creative project survival.

This Detroit case study has shown that land is a valuable urban asset. It has illustrated the power of art and agriculture to reclaim blighted urban land and to put it back into productive use. The labour of the creative class's **artistic precariat** (Bain and McLean, 2013) often does not generate a **living wage**, yet it is being siphoned off to fuel the engine of urban neoliberal development. Who benefits from this investment of labour and capital in land over the long term? A real concern is that the aesthetic orientation of such culture-led regeneration projects will, over time, drive up the cost of living and local property values, making areas less affordable for low-income residents and small businesses. A similar concern has produced tensions in Leipzig over land-use redevelopment trajectories.

Leipzig: From Shrinking City to Creative City

Leipzig is the largest city in the Free State of Saxony in eastern Germany. It currently has a population of nearly 550,000; however, it experienced a phase of long-term shrinkage that extended from the 1930s, when it had a population of 713,000, until the late 1990s, when it had a population of 437,000 (Florentin, 2010; Lang, 2012; Liebmann and Kuder, 2012; Rink et al., 2012). Pre–Second World War Leipzig experienced dynamic industrialization and urbanization. It benefited from its location at the intersection of two traditional European routes of commerce: Via Regia (a royal highway in the Middle Ages that ran from Santiago de Compostela, Spain, east through Leipzig and terminated in Moscow, Russia) and Via Imperii (an imperial road in the Roman Empire that extended from the Adriatic Sea through Leipzig to the Baltic Sea). Leipzig has had a long history as a centre of trade, culture, publishing, and industry.

In postwar East Germany under socialism, the city was protected from market forces and competition, but it lost its status as an intellectual, cultural, and financial capital. Leipzig became a second-tier city used by the German Democratic Republic (GDR) socialist regime as an international platform to display its industrial and manufactured goods at trade fairs (Bontje, 2005). A site of heavy-machinery production and coal mining, Leipzig saw its industrial base decay owing to disinvestment. It became a heavily polluted city with a steadily declining—though relatively homogeneous and locally rooted—population and deteriorating, inner-city, prewar housing stock (Burdack et al., 2009).

In the decade after East and West Germany reunified, Leipzig continued to suffer sustained demographic shrinkage in the wake of economic restructuring, deindustrialization, and a loss of 90,000 manufacturing jobs, which resulted in high rates of unemployment, outmigration, housing vacancies, and ongoing inner-city decline. An urban revitalization process was initiated in the 1990s that sought to address urban shrinkage as a "housing problem" (Bernt et al., 2014; Bernt, 2009) with the motto "regeneration in stock" (Haase et al., 2012). Urban regeneration policy and investment priority were given "to the inner city that had been for decades abandoned, socially and symbolically devalued and emaciated" (Haase et al., 2012: 1182) (Figure 7.4). Central city dwellings have since been refurbished for middle-class families, and small cultural enterprises that could facilitate "soft gentrification" in East Leipzig

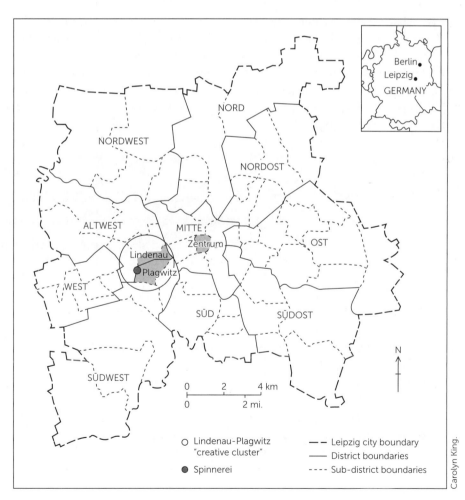

Figure 7.4 Map of the Lindenau-Plagwitz "creative cluster" in Leipzig, Germany

have been funded. In addition, urban roundtable discussions have been held to foster participatory planning and context-driven micro-solutions; a guardian house program (*Wächterhaus*) has been established to prevent unused buildings from deteriorating and to provide affordable space for cultural production; and an urban re-imaging process is underway (Rall and Haase, 2011) (Figure 7.5). Scholars and urban policy-makers optimistically frame Leipzig's "greatest potential" as "its high degree of spatial opportunities" and "relaxed density" presented in vacant apartment buildings, suburban housing, offices, industrial facilities, and brownfield redevelopment sites; it affords "freedom, spaces and places for creativity at discount prices without forcing people to instant economic success" (Lange et al., 2008: 1).

Leipzig's urban renewal and reurbanization have been driven by access to national transfer payments and European Union public subsidies that have permitted swift urban decision-making and reinvestment in infrastructure (Kabisch et al., 2010). A new airport has been built; national connections have been strengthened through train and highway networks; and a new media industry cluster developed (Bathelt, 2002). The city, ironically referred to by locals as "hype-zig" because of all of the media attention it has received as the "new Berlin," has become an important migration destination for young people—students, apprentices, and early-stage professionals—seeking vocational and higher education, cultural vitality, and affordable, high-quality housing. Some urban planning and policy attention has also been directed towards social cohesion—dealing with the poverty, access to amenities, ethnic segregation, and socio-spatial inequalities deriving from demographic change (Cortese et al., 2014).

Central to Leipzig's **urban imagineering** is the marketing of the city's cultural and subcultural spaces, landscapes, and heritage to young people, cultural workers, and tourists (Coles, 2003; van Leeuwen and Nijkamp, 2011). City officials have capitalized on the international renown Leipzig achieved within the art world for the creative incubation function performed by the Academy of Visual Arts and the associated New Leipzig School movement of modern art painters who trained and taught at the academy and whose work commands

Figure 7.5 A *Wächterhaus* guardian house program project on Karl-Heine Straße in the Lindenau and Plagwitz neighbourhoods

Alison Bain.

high prices at international art fairs, exhibitions, and auctions (Jakobi, 2014). Neo Rauch, perhaps the most famous of the contemporary New Leipzig School artists, is based at the Spinnerei, a 125-year-old former cotton-spinning mill (the largest in continental Europe in 1907) and now commercially successful arts district located in the Lindenau and Plagwitz neighbourhoods in West Leipzig (see Figure 8.4). With the slogan "from cotton to culture," redevelopment of the obsolete cotton-yarn factory complex began in 1992 and gathered momentum over the next decade. The post-industrial architectural setting of the Spinnerei (www.spinnerei.de) hosts 100 studios for emerging and established artists, 14 galleries, and numerous public and private arts institutions (Figure 7.6). Nearby, industrial warehouses have been converted into studio and exhibition space. Creative-city discourses have been hard at work in Leipzig, positioning the Spinnerei at the heart of the Lindenau-Plagwitz creative cluster and generating EU funds to develop and sustain its cultural infrastructure. The heavily marketed Spinnerei arts district has been placed central stage in Leipzig's reurbanization strategy, concentrating much of the city's cultural infrastructure in an arts district.

Such concentrating of reinvestment in the cultural sector is not without its detractors. Tension has mounted in nearby residential neighbourhoods where anarchist and communist activists advocate keeping neighbourhoods ugly as a way to resist gentrification. Graffiti crews are active in some central-city neighbourhoods, targeting new residential developments with paint bombs, graffiti, and window smashing. The selective "hate-on" for middle-class residential and commercial redevelopment is written into Leipzig's urban landscape (Figure 7.7). With spray cans, paint, and squatted buildings, some politically left-leaning local residents are working aggressively to defend urban space from gentrification and commercialization, challenging the pacification of neighbourhoods through culture and advocating for a free and accessible public realm that is socially and culturally mixed. Graffiti and the tactic of squatting are expressions of an insurgent politics that seeks to disrupt and declassify, making unauthorized claims to land (Swyngedouw,

Figure 7.6 The Spinnerei, a 125-year-old former cotton-spinning mill and now a commercially successful arts district in West Leipzig

Alison Bain.

Figure 7.7 Paint-bombing of a new middle-class residential development in Leipzig's central city

2011). They are critiques of the urban status quo and an exertion of pressure from below to demand urban change.

This Leipzig case study reveals the transformative dynamism of culture to address urban decline when strategically positioned within creative-city planning and policy discourses. Such a culture-led policy agenda has significant financial support from multiple levels of government—municipal, German, and European Union. It is widespread economic and political support such as this that has allowed the top-down vision of a creative rather than a shrinking city to be implemented so swiftly. The imposition of this creative-city vision for Leipzig has been challenged from the bottom up by politically active residents who oppose the valorization of urban land for its exchange value rather than its use value. Through their actions, these grassroots activists raise fundamental questions about whether property ownership should be the dominant voice in decisions about what to do with urban land. Such grassroots activism has long been active in Canada as well. Following in the activist footsteps of Jane Jacobs, particularly her mobilization of a citizen-led campaign to successfully oppose the Spadina Expressway, there is a tradition of opposition to large-scale land-use development projects. In small- and mid-sized Canadian cities, however, public policy debate on urban development trajectories, as the following text box examines, is increasingly about managing a slowing of urban growth.

Managing Post-industrial Urban Decline in the Canadian Urban System

The 2011 Canadian census reveals an uneven inter-urban geography of urban growth and decline. Population growth is concentrated in Canada's major urban centres, while the remaining 60 per cent of the urban system, particularly small and mid-sized cities, is growing slowly (40 per cent) or is declining (18 per cent). In Canada, 85 of 147 census agglomerations (CAs) and census metropolitan areas (CMAs) experienced slow growth or decline between 2001 and 2011 (Donald and Hall, 2015). While questions of what to do with derelict brownfield, waterfront, and industrial lands have preoccupied Canadian urban planners and policy-makers for decades, in the twenty-first century the management of decline has become a significant urban policy issue. In the face of declining fertility rates, loss of manufacturing and resource-sector employment, and a shift away from state redistribution policies intended to address spatial inequalities, urban areas across the country are grappling with new patterns of uneven growth.

If demographic growth can no longer be relied upon to create jobs, reinforce the tax base, underwrite the cost of infrastructure development and service delivery, and solve social problems, then a new policy framework needs to be found. Drawing on the experiences of Sudbury, Ontario, Donald and Hall (2015: 267) emphasize "smart decline" and "qualitative development" as viable policy paths forward. Such strategic decline-oriented planning techniques focus on who or what will remain, consider how land and buildings can be recycled and infrastructure adapted, and focus on "the existing built form, [by] promoting redevelopment, infilling, and conservation." The challenge, they argue, is to shift the policy mindset away from quantifying urban development in terms of expanded population growth, services, and infrastructure in the city and to focus instead on planning for sustainability, livability, and quality of life.

Urban decline is an issue of particular concern in Canada's industrial heartland in the Golden Horseshoe region of southwestern Ontario. The American Rust Belt and the Canadian Golden Horseshoe regions represent the former and now struggling urban manufacturing hubs of North America. The Golden Horseshoe, Canada's half of the Great Lakes heartland, experienced a different trajectory of industrial transition following the recession of the 1980s than its American counterparts. Golden Horseshoe industrial cities experienced urban growth throughout the 1970s, which in the Canadian context was funded federally, provincially, and municipally—unlike in the United States, however, where urban growth was based on a "homerule" approach of greater economic and political autonomy from the state (Garber and Imbroscio, 1996). By the mid-1980s, the impacts of the recession on neighbouring American cities and the decline of manufacturing in Quebec framed southern Ontario as an exception during the broader industrial downturn. However, an American spillover narrative of deindustrialization influenced the Canadian federal government to form the Task Force on Housing and Urban Development in 1965, described by High (2003: 37) as the "diagnosis of urban decline" arriving before the "disease." Canadian cities, in southern Ontario in particular (e.g., Hamilton, Kitchener, London, Oshawa, St Catharines, and Windsor), braced for the storm before it arrived. Whereas industry-related urban decline in the United States was considered an inevitable part of doing business, Canadian manufacturing workers and companies largely blamed American exceptionalism and dominance for the downturn—a nationalist standpoint that diffused Canadian fear of susceptibility to the global market (High, 2003). Despite the different political climates, Canadian and American post-industrial cities share a number of similarities, from concerns over brownfield remediation and the drive for high-density land use, to concerns over attracting creative-class and professional industries to boost the local economies and reinvigorate the urban core.

In Canada in the wake of the 2008–2009 financial recession, large and mid-sized former industrial cities outside of Toronto in southern Ontario have all had to grapple with the significant loss of manufacturing jobs, rates of unemployment above the national average, an aging population, and the out-migration of young people (Radwanski, 2014). These are all cities with a legacy of reputable universities and colleges, cultural institutions, middle-class residential neighbourhoods, heritage properties, industrial warehouses, affordable real estate, and a low cost of living. These are cities that are struggling to reinvent themselves by "attracting and keeping well-educated, entrepreneurial citizens committed to community-building and capable of creating wealth and quality of life around them" (Radwanski, 2015, n.p.)—an urban creative class. The new cultural economy, with its emphasis on the amenity, spectacle, leisure, and incubating functions of the central city, is being positioned, as it has been in so many post-industrial cities in the global North, as the new industry (Shearmur and Hutton, 2011).

Conclusion

Through the case studies of Detroit and Leipzig, this chapter has revealed cities to be intentional artifacts produced out of a legacy of collective planning and policy decisions about how to manage urban growth, development, and decline. Urban planning, an increasingly bureaucratic regulatory activity, is the profession with the societal legitimacy to engage in large-scale socio-spatial interventions. Such planned interventions into the urban fabric involve complex balancing and problem-solving acts that are made all the more challenging in post-industrial cities. While urban planners may not hold the reigns of political or financial power, their visions for the future of cities are often comprehensive and persuasive, and there is a real danger that they speak to urban development as a predominantly middle-class project. The Detroit Future City plan and the creative city vision for Leipzig, as manifest in the Spinnerei arts district, are two differently scaled and in-process examples of property-led, amenity-centred **place-branding** initiatives to promote profit-driven urban development for the well-educated and culturally curious middle-class. A valid concern raised by these two examples is that the creative "buzz" of urban agricultural initiatives and collective spaces of artistic life will be co-opted and instrumentalized by city administrators to the neglect of the everyday need of lower-income, local populations for living-wage jobs, good schools and daycare, and affordable housing. The urban planning and policy decisions that political elites have made, and continue to make, certainly have a consequential geography that can create lasting structures of unevenly distributed advantage and disadvantage (Soja, 2010). While such decisions can negatively affect the everyday lives of urban residents who fall outside of the creative class, their effects can also be changed through forms of social and political action.

In post-industrial cities like Detroit and Leipzig, conflict is emerging from the cultural turn in contemporary urban development—that is, from urban planning and policy practices that are characterized as entrepreneurial, neoliberal, and oriented towards the creative class. Although some cultural producers are becoming increasingly politicized, there is certainly greater scope for members of the creative class to more extensively challenge growth-oriented entrepreneurial policy agendas that appropriate culture and creativity in potentially socially divisive ways. There is also opportunity to deepen urban cultural scholarly, planning, and policy understandings about how mutually supportive relationships between art, culture, and local communities can be forged that encourage more socially just approaches to the redevelopment of land in post-industrial cities. The urban landscape is a repository of meaning; it is a palimpsest that carries meaning forward from the past but also opens up manifold interpretive possibilities into the present and the future for how land can function and how it can be used in socially inclusive ways.

Key Points

- Cities are intentional artifacts produced out of a legacy of mobile collective-planning and policy decisions about how to develop land in order to effectively manage urban growth, development, and decline.
- Contemporary urban policy in the global North exploits creativity and culture as urban redevelopment tools to market quality of life and leisure opportunities for middle-class residents, investors, and tourists.
- Land is a finite urban resource and a basic urban commodity that needs to be governed, shared among, and used by, different groups of people for different functional purposes.
- Case studies from Detroit and Leipzig show that agriculture and art are powerful practices for reclaiming blighted urban land and giving it new use and exchange value.
- In Detroit and Leipzig, culture-led regeneration projects have driven up the cost of living and local property values, making inner-city areas less affordable for low-income residents, communities of colour, and small businesses.

- Patterns of uneven growth characterize Canada's urban system, demanding a shift in policy mindset, particularly in small and mid-sized post-industrial cities, away from managing growth to managing decline by planning for sustainability, livability, and quality of life.

Activities and Questions for Review

1. Create a list of post-industrial cities in Europe and North America that you have visited or know about. Using insights from this chapter, identify some of the ways in which the creative and cultural turn in urban planning and policy may have influenced the way land is being developed in the cities on your list.
2. Speed-read the current headlines for the main section of a national or city newspaper in either online or paper format. In your review, what current print media examples did you find of urban planning and policy, and what issues did they raise?
3. Introduce yourself to a classmate whom you do not already know and talk about different examples of urban agriculture that you have each seen. How do your examples illustrate some of the ways in which urban agriculture is used as a tool of urban reinvention, both by communities and in redevelopment strategies?
4. Do a Google image search for "arts and culture" for a city of your choice. What do these images reveal about how arts and culture have been instrumentalized by civic leaders?

Acknowledgments

Research for the case studies in this chapter was made possible through an Ontario Graduate Scholarship, an International Collaboration Grant, and a Minor Research Grant from York University. We extend our thanks to the many farmers in Detroit who continue to welcome curious researchers.

References

Ager, S. (2015). Taking back Detroit. *National Geographic*, May 1: 56–83.

Bain, A.L., and McLean, H. (2013). The artistic precariat. *Cambridge Journal of Regions, Economy, and Society*, 6(1): 93–111.

Bathelt, H. (2002). The re-emergence of a media industry cluster in Leipzig. *European Planning Studies*, 10(5): 583–611.

Barber, L. (2014). (Re)making heritage policy in Hong Kong: A relational politics of global knowledge and local innovation. *Urban Studies*, 51(6): 1179–1195.

Bell, D. (1973). *The coming of post-industrial society.* New York: Basic Books.

Bernt, M. (2009). Renaissance through demolition in Leipzig. In L. Porter and K. Shaw (Eds.), *Whose urban renaissance? An international comparison of urban regeneration strategies.* New York: Routledge, pp. 75–83.

Bernt, M., Haase, A., Großmann, K., Cocks, M., Couch, C., Cortese, C., and Krzysztofik, R. (2014). How does(n't) urban shrinkage get onto the agenda? Experiences from Leipzig, Liverpool, Genoa, and Bytom. *International Journal of Urban and Regional Research*, 38(5): 1749–1766.

Blokland, T., and Nast, J. (2014). From public familiarity to comfort zone: The relevance of absent ties for belonging in Berlin's mixed neighbourhoods. *International Journal of Urban and Regional Research*, 38(4): 1142–1159.

Bontje, M. (2005). Facing the challenge of shrinking cities in East Germany: The case of Leipzig. *Geojournal*, 61: 13–21.

Bridge, G. (2006). It's not just a question of taste: Gentrification, the neighbourhood, and cultural capital. *Environment and Planning D*, 38: 1965–1978.

Brooks, A., and Kushner, R. (2001). Cultural districts and urban development. *International Journal of Arts Management*, 3(2): 4–15.

Burdack, J., Lange, B., and Ehrlich, K. (2009). *Creative Leipzig? The views of high-skilled employees, managers, and transnational migrants.* ACRE Report 8.6. Amsterdam: University of Amsterdam.

Catungal, J.P., and Leslie, D. (2009). Contesting the creative city: Race, nation, and multiculturalism. *Geoforum*, 40(5): 701–704.

Chapple, K., Jackson, S., and Martin, A.J. (2010). Concentrating creativity: The planning of formal and informal arts districts. *City, Culture, and Society*, 1(4): 225–234.

Coles, T. (2003). Urban tourism, place promotion, and economic restructuring: The case of post-socialist Leipzig. *Tourism Geographies*, 5(2): 190–219.

Cortese, C., Haase, A., Großmann, K., and Ticha, I. (2014). Governing social cohesion in shrinking cities: The cases of Ostrava, Genoa, and Leipzig. *European Planning Studies*, 22(10): 2050–2066.

Cowie, J., and Heathcott, J. (2003). *Beyond the ruins: The meanings of deindustrialization*. Ithaca, NY: Cornell University Press.

Curran, W. (2010). In defense of old industrial spaces: Manufacturing, creativity, and innovation in Williamsburg, Brooklyn. *International Journal of Urban and Regional Research, 34*(4): 871–885.

DeLind, L.B. (2002). Place, work, and civic agriculture: Common fields for cultivation. *Agriculture and Human Values, 19*(3): 217–224.

Detroit Future City (DFC) (2012). *2012 Detroit Strategic Framework Plan*. Detroit, MI: Inland Press.

Donald, B., and Hall, H. (2015). Slow growth and decline in Canadian cities. In P. Filion, M. Moos, T. Vinodrai, and R. Walker (Eds.), *Canadian Cities in Transition: Perspectives for an Urban Age* (5th ed.). Don Mills, ON: Oxford University Press, pp. 258–273.

Ehlenz, M.M., Birch, E.L., and Agness, B. (2014). *The power of "eds and meds" urban universities investing in neighbourhood revitalization and innovation districts*. University of Pennsylvania: Penn Institute for Urban Research.

Evans, G. (2009). Creative cities, creative spaces, and urban policy. *Urban Studies, 46*(5–6): 1003–1040.

Florentin, D. (2010). The "perforated city": Leipzig's model of urban shrinkage management. *Berkeley Planning Journal, 23*: 83–101.

Florida, R. (2002). *The rise of the creative class: And how it is transforming work, leisure, community, and everyday life*. New York: Basic Books.

Florida, R. (2004). *The flight of the creative class*. New York: Harper Business.

Garber, J., and Imbroscio, D. (1996). "The myth of the North American city" reconsidered: Local constitutional regimes in Canada and the United States. *Urban Affairs Review, 31*(5): 595–624.

Haase, A., Herfert, G., Kabisch, S., and Steinführer, A. (2012). Re-urbanizing Leipzig (Germany): Context, conditions, and residential actors (2000–2007). *European Planning Studies, 20*(7): 1173–1196.

Hall, P. (1997). Modeling the post-industrial city. *Time and Space: Geographic Perspectives on the Future, 29*(4–5): 311–322.

High, S. (2003). *Industrial sunset: The making of North America's rustbelt, 1969–1984*. Toronto: University of Toronto Press.

Holm, A.J., and Kuhn, A. (2011). Squatting and urban renewal: The interaction of squatter movements and strategies of urban restructuring in Berlin. *International Journal of Urban and Regional Research, 35*(3): 644–658.

Jakobi, S. (2014). *Leipzig's visual artists as actors of urban change: Articulating the intersection between place and attachment, professional development and urban pioneering*. Unpublished MA thesis, Department of Culture, Media and Creative Industries, King's College London.

Kabisch, N., Haase, D., and Haase, A. (2010). Evolving reurbanisation? Spatio-temporal dynamics as exemplified by the East German city of Leipzig. *Urban Studies, 47*(5): 967–990.

Kalish, J. (2011). "The gift of Detroit": Tilling urban terrain. National Public Radio. Retrieved from http://www.npr.org/2011/10/02/140903516/the-gift-of-detroit-tilling-urban-terrain

Kotlyarevskaya, O.V. (2005). "Public use" requirement in eminent domain cases based on slum clearance, elimination of urban blight, and economic development. *Connecticut Public Interest Law Journal, 5*(2): 197–231.

Krätke, S. (2010). "Creative cities" and the rise of the dealer class: A critique of Richard Florida's approach to urban theory. *International Journal of Urban and Regional Research, 34*(4): 835–853.

Krivy, M. (2013). Don't plan! The use of the notion of "culture" in transforming obsolete industrial space. *International Journal of Urban and Regional Research, 37*(5): 1724–1746.

Landry, C. (2000). *The creative city*. London: Earthscan.

Lang, T. (2012). Shrinkage, metropolitanization, and peripheralization in East Germany. *European Planning Studies, 20*(10): 1747–1754.

Lange, B., Burdack, J., Thalmann, R., Manz, K., Nadler, R., and Dziuba, C. (2008). *Creative Leipzig: Understanding the attractiveness of the metropolitan region for creative knowledge workers*. ACRE Report 5.6. Amsterdam: University of Amsterdam.

Liebmann, H., and Kuder, T. (2012). Pathways and strategies of urban regeneration—Deindustrializing cities in Eastern Germany. *European Planning Studies, 20*(7): 1155–1172.

McCann, E. (2011a). Veritable inventions: Cities, policies, assemblage. *Area, 43*: 143–174.

McCann, E. (2011b). Urban policy mobilities and global circuits of knowledge: Toward a research agenda. *Annals of the Association of American Geographers, 101*(1): 107–130.

McLean, H. (2014). Digging into the creative city: A feminist critique. *Antipode, 46*(3): 669–690.

Mah, A. (2012). *Industrial ruination, community, and place: Landscapes and legacies of urban decline*. Toronto: University of Toronto Press.

Mariani, M., and Barron, P. (Eds.). (2014). *Terrain vague: Interstices at the edge of the pale*. New York: Routledge.

Markusen, A. (2006). Urban development and the politics of a creative class: Evidence from a study of artists. *Environment and Planning A, 38*(10): 1921–1940.

Markusen, A. (2014). Creative cities: A 10-year research agenda. *Journal of Urban Affairs, 36*: 567–589.

Millington, N. (2013). Post-industrial imaginaries: Nature, representation, and ruin in Detroit, Michigan. *International Journal of Urban and Regional Research, 37*(1): 279–296.

Mizzau, L., and Montanari, F. (2008). Cultural districts and the challenge of authenticity: The case of Piedmont, Italy. *Journal of Economic Geography, 8*(5): 651–673.

Mommaas, H. (2004). Cultural clusters and the post-industrial city: Towards the remapping of urban cultural policy. *Urban Studies, 41*(3): 507–532.

Montgomery, A.F. (2011). The sight of loss. *Antipode, 43*(5): 1828–1850.

Montgomery, J. (2008). *The newest wealth of cities: City dynamics and the fifth wave*. Aldershot, UK: Ashgate.

MSNBC (2014). Detroit's secret weapon against food insecurity. Retrieved from http://www.msnbc.com/msnbc/detroit-gardening-weapon-against-food-insecurity

Newman, A., and Safransky, S. (2014). Remapping the Motor City and the politics of austerity. *Anthropology Now, 6*(3): 17–28.

Novy, J., and Colomb, C. (2013). Struggling for the right to the (creative) city in Berlin and Hamburg: New urban social movements, new "spaces of hope"? *International Journal of Urban and Regional Research, 37*(5): 1816–1838.

Oswin, N. (2014). Queering the city: Sexual citizenship in creative city Singapore. In M. Davidson and D. Martin (Eds.), *Urban politics: Critical approaches*. London: SAGE, 139–155.

Parker, B. (2008). Beyond the class act: Gender and race in the "creative city" discourse. In J. DeSena (Ed.), Gender in an urban world. *Research in Urban Sociology, 9*: 201–232. Bingley, UK: Emerald.

Peck, J. (2005). Struggling with the creative class. *International Journal of Urban and Regional Research, 24*: 740–770.

Power, D., and Scott, A. (Eds.) (2004). *Cultural industries and the production of culture*. London and New York: Routledge.

Radwanski, A. (2014). After the gold rush: The long, slow decline of the nation's industrial heartland. *Globe and Mail*, May 30. Retrieved from http://www.theglobeandmail.com/news/politics/after-the-gold-rush/article18923563/

Radwanski, A. (2015). Rust Belt revival: Lessons from southwest Ontario from America's industrial heartland. *Globe and Mail*, January 16. Retrieved from http://www.theglobeandmail.com/news/national/rust-belt-revival-lessons-for-southwestern-ontario-from-americas-industrial-heartland/article22489159/

Rall, E.L., and Haase, D. (2011). Creative intervention in a dynamic city: A sustainability assessment of an interim use strategy for brownfields in Leipzig, Germany. *Landscape and Urban Planning, 100*: 189–201.

Rink, D., Haase, A., Grossmann, K., Couch, C., and Cocks, M. (2012). From long-term shrinkage to re-growth: The urban development trajectories of Liverpool and Leipzig. *Built Environment, 38*(2): 162–178.

Roodhouse, S. (2006). *Cultural qu4rters: Principles and practices*. Bristol, UK: Intellect.

Safransky, S. (2014). Greening the urban frontier: Race, property and resettlement in Detroit. *Geoforum, 56*: 237–248.

Shearmur, R., and Hutton, T. (2011). Canada's changing city-regions: The expanding metropolis. In L. Bourne, T. Hutton, R. Shearmur, and J. Simmons (Eds.), *Canadian urban regions: Trajectories of growth and change*. Don Mills, ON: Oxford University Press, pp. 99–124.

Sheehan, P. (2015). Revitalization by gentrification. *Jacobin*. Retrieved from https://www.jacobinmag.com/2015/05/detroit-foreclosure-redlining-evictions/

Short, J.R. (1999). Urban imagineers: Boosterism and the representation of cities. In A. Jonas and D. Wilson (Eds.), *The urban growth machine: Critical perspectives two decades later*. Albany, NY: State University of New York, pp. 37–45.

Soja, E. (2010). *Seeking spatial justice*. Minneapolis: University of Minnesota Press.

Stevens, Q., and Ambler, M. (2010). Europe's city beaches as post-fordist place-making. *Journal of Urban Design, 15*(4): 515–537.

Stolarick, K., and Florida, R. (2006). Creativity, connections, and innovation: A study of linkages in the Montreal region. *Environment and Planning A: Society and Space, 38*: 1799–1817.

Swyngedouw, E. (2011). Interrogating post-democratization: Reclaiming egalitarian political space. *Political Geography, 30*(7): 370–380.

Temenos, C., and McCann, E. (2012). The local politics of policy mobility: Learning, persuasion, and the production of a municipal sustainability fix. *Environment and Planning A, 44*(6): 1389–1406.

Tracey, D. (2011). *Urban agriculture: Ideas and designs for the new food revolution*. Gabriola Island, BC: New Society Publishers.

United States Census Bureau (2012). Annual household income. [Data file] Retrieved from *http://quickfacts.census.gov/qfd/states/26/2622000.html*

United States Department of Labour (2015). Detroit unemployment rate. [Data file] Retrieved from http://www.bls.gov/lau/lacilg10.htm

Van Heur, B. (2009). The clustering of creative networks: Between myth and reality. *Urban Studies, 46*: 1531–1552.

van Leeuwen, E., and Nijkamp, P. (2011). The importance of e-services in cultural tourism: An application to Amsterdam, Leipzig, and Genoa. *International Journal of Sustainable Development, 14*(3/4): 262–289.

Vivant, E. (2009). How underground culture is changing Paris. *Urban Research and Practice, 2*(1): 36–52.

Zukin, S., and Braslow, L. (2011). The lifecycle of New York's creative districts: Reflections on the unanticipated consequences of unplanned cultural zones. *City, Culture, and Society, 2*: 131–140.

8

Socialist and Post-socialist Cities in the Twenty-First Century

Lisa B.W. Drummond and Douglas Young

Introduction

Urban life in the twenty-first century has been shaped in quite significant ways by the thoughts, practices, ideologies, and social organization of the twentieth century. This chapter focuses on the urban legacies of twentieth-century socialism and explores their impact on urban policy, spatial form, and everyday life in the twenty-first century. With the neoliberal model of **city-building** appearing to be in crisis mode, new approaches to urban issues are urgently needed.

Neoliberalism has been the dominant paradigm of urban governance since the 1980s in many urban regions around the world; its core beliefs of free markets and free individuals have led to a realignment of the relationships among state (government and its agencies), market (business enterprises), and civil society (citizens). Governments have retreated from direct intervention in many aspects of urban life, and where they are still active, their regulatory frameworks have been loosened. Examples of these changes include cuts to spending on social programs (e.g., welfare and social housing), the privatization of urban infrastructure (e.g., transportation systems and water and sewage treatment), a reduction in personal income tax and corporate tax rates, and a loosening of urban planning and environmental protection regulations. In many cities, these changes—implemented over the past 30 years or so—have generated a crisis situation of intense socio-spatial polarization, heightened air and water pollution, and increased traffic congestion. This chapter considers the possibility that a critical reassessment of the legacies of socialist models could provide valuable lessons for urban policy-makers and citizens alike as they confront the present-day crisis of neoliberal governance. Such a reassessment is of particular relevance to the urban policy and planning debates currently underway in many Canadian cities where the question of neighbourhood revitalization can be conceptually reframed as a confrontation with the urban legacies of the twentieth century.

While in the non-socialist world **socialism** was, and is, often simplistically equated with communism (taken to mean a society where there was no private ownership of property, all economic activity was state-owned, and there was only one political party), actually existing socialism has taken many forms, ranging from social democracy to repressive state socialism. The different socialisms display great variation both in the degree of personal and political freedom enjoyed by citizens and in the extent of state control of economic activity. Their differences also reflect the pre-socialist conditions of each society that experienced a transition to socialism. In other words, each socialist nation transitioned from and under different conditions, which then influenced the form and experience of its particular socialism: some were urban and industrial (as was East Germany), while others were non-urban and agricultural (as was Vietnam); some were democracies, while others were not; some were European and imperial powers (Russia, for example), while others were colonies and part of far-flung empires (Mozambique, for example). The common thread of the "social" that runs through them all is a commitment to the modernist idea of universal progress (i.e., a better life for all members of society), a desire to create a socially de-differentiated society (in which social-class differences disappear or at least are muted), and a heightened degree of state intervention in processes of city-building and urban governance.

In subsequent sections this chapter discusses the literature on **socialist** and **post-socialist cities** and identifies two core questions that have been debated by urban scholars: was or is there a distinctly socialist city; and what was or is socialist about socialist urbanism? Two case studies are used, from Hanoi, Vietnam, and Stockholm, Sweden, to illustrate different visions of socialist urbanism and different contemporary fates for socialist urban projects. The chapter concludes by reflecting upon the legacies of twentieth-century urban ideas in Canada in light of the broader trend to socio-spatial polarization and the widespread adoption in Canada of neoliberal modes of urban governance.

Incarnations of Socialist Cities

Socialist cities across the spectrum of socialism exist around the globe. The most intense period of socialist city-building occurred in the mid-twentieth century, but started as early as the First World War (1914–1918) in what became the Soviet Union, and continues today. After the Second World War (1939–1945) and the consequent decolonization of most European empires, communist or socialist parties, which had led nationalist and anti-colonial movements, came to power in many newly independent countries. In Asia, the countries of China, Vietnam, and North Korea, for example, engaged in socialist urban projects throughout the 1960s, 1970s, and 1980s. Many new African countries arising out of decolonization were similarly socialist or communist post-independence, such as Ethiopia, Benin, Congo-Brazzaville, Mozambique, and Algeria, among others. In Latin America, although decolonization for the most part happened much earlier, socialist or communist rulers came to power mainly in the mid-twentieth century, in Cuba, Nicaragua, and Guyana, for example. In Europe, although the Soviet Union had existed and expanded since 1922, the Eastern Bloc, as it became known, of communist states formed at the end of the Second World War was composed of territories allied with or captured by the Soviets during the war; these included, for example, the German Democratic Republic (East Germany), Poland, and Hungary. In a list of the socialist countries of Europe, Sweden must be included as an exemplar of social democracy, as we discuss below in one of our two case studies in this chapter. In social democracies, while economic activity is largely capitalist in terms of ownership and organization, the state plays a leading role in establishing social welfare and in determining overall urban policy. Urban planning processes are organized such that they are controlled by the state, and changes to built and natural urban environments are tightly regulated, unlike, for example, the urban planning process in Canada, where regulations can be quite flexible and property owners have many powers as-of-right related to land that they own. While there still exists a number of communist and socialist states, such as China, Vietnam, and Cuba, the majority have become "post-socialist," many of them in Europe very quickly after the fall of the Soviet Union in 1991 (Andrusz et al., 1996). Even in those still ruled by avowedly communist parties, very few remain isolationist (North Korea is perhaps the only remaining example), while others have opened up in various ways and to varying degrees to global flows, of people, ideas, and, of course, capital.

All urban places can be considered representations of the economic and social systems within which they are situated, whether those systems are socialist or capitalist. This means that the physical design of urban space and the processes whereby urban space is shaped and reshaped, the role of various actors in creating and governing those spaces, the ways of living that unfold in them (some perhaps organically, some perhaps by dictate), and the type of economic activity that is present can all be taken as a reflection of particular forms of social relations, economic systems, and cultural values. In some cities it is possible to read several layers of historical change in terms of social and economic systems and ideas about urban planning. A particularly striking example of this is Berlin, a city which has experienced several waves of dramatic change throughout the past 100 years, including being divided into a capitalist Berlin and a socialist Berlin.

In the early twentieth century, Berlin was the capital of the Wilhelmine Empire, then of the social democratic Weimar Republic and from 1933–1945 of the Nazi Reich. It was an occupied city immediately following the Second World War, then a divided city for 40 years with the eastern half being the capital of the socialist German

Democratic Republic or GDR (known as East Germany in the West) and the western half an island of capitalist urbanism within the GDR. Unification of West and East Germany in 1990 meant that East Berlin was no longer a capital city, though in 1999 it became the capital of unified Germany. Each of those periods of dramatic change generated physical transformations of parts of the city and changes in the social relations and cultural values embodied in its residents. With so many layers visible today, Berlin can be considered a **palimpsest**—a text in which old layers of meaning can be read through several newer layers. Berlin is perhaps unique in that so many layers are visible. While living in a city-as-palimpsest can provide everyday life with a rich sense of urban history, at the same time living with the legacies of a previous era's urbanism can present challenges for citizens and for policy-makers.

Building Socialist Cities

In the literature on socialist and post-socialist cities (see, for example, French and Hamilton, 1979; Pickvance, 2002), two key areas of debate stand out. First, was there or is there a distinctly socialist city? Second, what precisely was socialist about socialist urbanism? In what follows these two questions are answered.

The first question requires determining whether there are or were socialist urban characteristics that mark socialist cities as distinctly different from cities developed in capitalist societies. It must also be determined whether those socialist characteristics are or were universal or, instead, varied across space and time. For example, are European and Asian socialist urbanisms as distinct as they are generally assumed to be? Under socialism, the city was intended to reflect socialism's desired transformation of social relations and social life and to be itself an instrument of that transformation. This chapter maintains that it would be impossible to claim that there was one distinct and universal model of "the socialist city" just as it would be impossible to identify a universal capitalist city without variations across space and time. There is a tendency, in the literature on socialist urbanism, to look primarily at the cities of Eastern Europe and the Soviet Union (e.g., Hirt, 2008; Horak, 2007; May, 2003). From that perspective, it is tempting to insist on a particular set of characteristics as being universal to socialist cities. These universals would include

- a desire to overcome the problems of the pre-modern city that were attributed to the failings of capitalist urbanization (cities that were dirty, crowded, and dark);
- grand public spaces for parades and gatherings and for demonstrations of the masses;
- monumentalism in architecture and dominance of the city centre in urban design;
- efficiency in urban functioning (e.g., traffic that moved smoothly and safely, people residing close to work and having their basic needs met within a short distance of home); and
- a **de-differentiated city**, with housing, amenities/facilities, and workplaces evenly distributed without inequalities between neighbourhoods.

Beyond these features, which are often taken to be universal, socialist urbanism took varied forms according to the place, the time, and the type of socialism that drove the construction of particular cities in particular countries. Further to the left on the socialist spectrum, urbanism would emphasize collective life and the centrality of the state (as in, for example, the Soviet Union, Vietnam, and China); further to the right, urbanism might be less de-differentiated and less monumental (as in Sweden). In the middle, cities reflected the various key characteristics to different degrees and in different mixes (e.g., in the GDR and Algeria). Indeed, many of the concerns of socialist urban policy-makers were also shared by urbanists in the capitalist world; for example, the desire to overcome the problems of the pre-modern city and to achieve efficiency in a city's functions was characteristic of all twentieth-century **modernist urbanism**, and not just socialist urbanism. And, looking at architectural debates about the city over the course of the twentieth century, it is clear that ideas flowed back and forth between socialist and non-socialist countries and within the socialist bloc.

This leads to a second key question. If there is no universal model of socialist urbanism, what, then, is or was socialist about socialist urbanism? Was it the design, the planning, the methods used to construct the city, the mode of use of the city in everyday life, or the mode of urban governance?

The Power of Architecture, Design, and Urban Planning

As noted above, there was a fair degree of synchronicity between socialists and modernists in the early decades of the twentieth century, and many individual modernist architects in Western Europe in the 1920s (a decade of particularly intense debate about and experimentation with the shape of urban space) were socialist in their political beliefs. Rather like socialism confronting the perceived failures of the city of industrial capitalism, "[m]odernism confronts contemporary life rather than seeking to escape: it tries to redirect, improve or at least enliven present-day realities" (Wright, 2008: 7). Modernism's confrontational impulse is driven by the desire to advance the social project of universal progress.

Many of the architectural and planning concepts that became widespread throughout socialist and non-socialist countries in the twentieth century originated in the congresses held by CIAM, the Congrès Internationaux d'Architecture Moderne (the International Congresses of Modern Architecture) (Mumford, 2000). In existence from 1928 until 1959, CIAM was one of the most influential organizations in the realm of architecture and urban planning in the world, with members drawn mostly from the field of architecture and mostly from Europe (among the non-Europeans were a few Canadians). One of CIAM's leading figures, the Swiss-born, Paris-based architect Le Corbusier, was an early admirer of Soviet ideas about architecture and urbanism, especially the more comprehensive approach taken by Soviet architects to designing for everyday urban life. This he contrasted to CIAM's tendency to focus on discrete elements of design (Mumford, 2000: 44). Other members of CIAM took up key positions as architects, planners, and university professors in many countries. For example, a Swedish member of CIAM, Sven Markelius, became director of the Stockholm Planning Commission in 1944, a position that allowed him to plan the shape of postwar Stockholm, discussed in detail below (Mumford, 2000; Pass, 1969).

From time to time and from place to place over the course of the twentieth century, socialist policy-makers officially adopted certain architectural styles in an attempt to visually differentiate their buildings from those in capitalist societies. For example, upon the establishment of the GDR in 1949, planners and government officials searched for a national tradition in architecture, rejecting modernism as Western and capitalist (Tscheschner, 2000: 260). Official buildings as well as apartment blocks (dubbed "workers' palaces") (The Wende Museum, www.wendemuseum.org/collections/stalinallee) were constructed in a neoclassical style with rich architectural detail, very different from the simple and stark forms of the International Style modern architecture then gaining popularity in capitalist countries. But from the mid-1950s on, in European socialist countries, including the GDR, there was a turn away from neoclassicism and an embrace of modernism as the appropriate architectural style of socialist urbanism (in part, because neoclassical buildings proved very expensive to build). At the larger scale of urban design, while many Western urban regions dispersed into low-density suburbia, in Central and Eastern Europe and the Soviet Union that kind of suburbanization was rejected. Instead, new peripheral developments in those countries took the form of large districts of apartment blocks.

These ideas moved between Europe and Asia through the urban specialists and transfers of aid that circulated in socialist networks (Logan, 1995; Schwenkel, 2012). Urban planners and architects were sent from the USSR, for example, to advise on and draft a plan for the complete redesign of Hanoi (never substantially implemented), and Vietnamese students were sent to study architecture and planning in the USSR (Logan, 1995). Socialist countries also supported Vietnam's postwar efforts to rebuild in part through financing and building specific infrastructure. Schwenkel (2012) details a GDR initiative to build housing blocks in Vinh, a city that had been almost completely flattened by bombing at the end of the US war in Vietnam. German architects visited the site and designed the blocks in accordance with the socialist design principles in force in the GDR; thus, this is one of the few housing complexes in Vietnam to feature mainly self-contained apartments. Similarly, Russia financed and either built or advised on the design of factories, bridges, collective housing blocks, and all manner of urban infrastructure and architecture, including, in Hanoi, an enormous and impressive Viet-Xo Friendship Palace for cultural events on the site of the former French colonial exhibition hall (Logan, 2000). Sweden was also an important supporter, sending hundreds of Swedish advisers to design and oversee the building and running of factories, such as the Bai Bang Paper

Mill built with Swedish funding. Through the bodies of advisers traveling to work on urban projects and of students sent to study urban sciences in other socialist countries, as well as through the urban projects themselves, ideas about what cities should be—how they should be organized and how they should look—thus flowed around the socialist bloc.

On a theoretical plane, there were intense debates in European socialist countries from the 1920s through to the 1950s about the appropriate form socialist urbanization should take. One of the debates was between urbanists—those who favoured concentrated and relatively dense developments—and dis-urbanists, who favoured relatively low-density settlements stretched along transportation routes) (Mumford, 2000: 45). Around 1950, the decision was taken in the Soviet Union and the socialist countries of Central and Eastern Europe to adopt the urbanist approach. In the GDR, this planning approach was codified in a document called "The 16 Principles of Urban Development" (Tscheschner, 2000). In actual fact, however, the first priority of European socialist countries established in the late 1940s (countries such as the GDR and Poland) was rebuilding the nation's industrial capacity, which had been severely damaged in the Second World War, and urban issues were largely overlooked. Eventually urban living conditions (housing in particular) deteriorated so badly that the state was forced to address them. In the GDR, for example, urban housing had not been improved since the 1930s. In the early 1970s, the state launched a massive program of new housing construction in GDR cities that lasted until the country was absorbed by West Germany in 1990. Some scholars argue that, in the case of the GDR, the failure to address urban living conditions for such a long time contributed to the growth of an urban-based opposition movement that aided the ultimate downfall of the socialist regime (Strom, 1996). The housing that was built at the edge of Berlin and of most other cities in socialist Central and Eastern Europe and the Soviet Union was in large districts planned according to the principles of a document developed by CIAM and Le Corbusier in the 1930s and 1940s called the Athens Charter (Jeanneret-Gris, 1973). This charter stipulated that a "good city" would be one separated into different functional zones (e.g., dwelling, work, recreation, transport) and that residents should live in apartment blocks surrounded by abundant greenspace. In the GDR, the largest single district developed—Marzahn in East Berlin—comprised 60,000 apartment units and housed 150,000 people (Young, 2005).

In contrast to the adoption of a more or less standardized vision of socialist urbanization in the Soviet Union and Central and Eastern Europe, socialist countries in Asia displayed a wide range of attitudes towards urbanization in general. Communist parties in Asia (notably in China, Vietnam, and, most radically and tragically, Cambodia) have tended to mistrust urban places and urbanites. This mistrust has ranged from ambivalence (in Vietnam and Laos) to suspicion and anti-urbanism (China in the 1960s and 1970s, when educated urban youth were sent to the countryside to "learn from the peasants"), to avowedly anti-urban purges (Cambodia under Pol Pot, 1975–1979), when cities were emptied and many urbanites were starved to death or executed in the "Killing Fields" (Kiernan, 2014; Vickery, 1984). In Vietnam, ambivalent mistrust led to the demographic flooding of the cities, particularly the capital, Hanoi, with cadre (Communist Party members) from the countryside. These newly urbanized peasant cadre had unimpeachable communist credentials and were therefore trusted and rewarded with positions in the new regime's bureaucracy and its new industrial enterprises, unlike many of the urbanites who were seen to have supported the French colonial system and were considered to be living "bourgeois" lifestyles.

Exploiting Pre-fabricated Construction Methods and Volunteer Labour

It was not just the particular design of space that distinguished socialist urbanism, but also, in many cases, the methods used to physically create it. As was the case in East Germany and Vietnam, many socialist countries embraced industrialized construction techniques, a process whereby buildings are assembled entirely from prefabricated components. This technique required the construction of factories where the components (concrete panels for walls and floors) would be produced and a distribution system that would move the panels to construction sites. Industrialized production offered the possibility of furthering the development of a country's economy while at the same time producing large volumes of much-needed housing. Prefabricated construction was adopted as the only way of building throughout socialist Central and Eastern Europe and the Soviet Union, beginning in the 1960s. In those countries, 70 million apartments that were to be home to 170 million people were built from prefabricated concrete panels (Csagoly, 1999). The downside of the turn to the industrialization of construction, at least in the GDR, was that the traditional building industry disappeared. It should be noted that in several Western European non-socialist countries—Britain, for example—architects and builders experimented with industrialized construction in the 1960s but on a much smaller scale than in socialist countries. In most non-socialist countries, the results were considered unsatisfactory and the period of experimentation was short-lived.

In many socialist cities, so-called volunteer or social labour provided an important resource for implementing urban projects (Rolandsen, 2008: 109). Hanoi residents in the 1960s and 1970s, for example, were assigned to teams and required to work several days per month on urban infrastructure projects, including repairing the system of dikes that protect the city from river flooding and even building a city park that remains the largest green space in the central city (Wells-Dang, 2012). Such labour was also invoked to construct the collective residential estates that were needed to house the rapidly expanding urban population (discussed in more detail below).

Governing Everyday Life

The socialist urbanism ideals that drove socialist urban development in the period up to the 1990s envisaged a city that provided full support to all residents, that would in turn allow urbanites to participate fully in urban life, working efficiently and being engaged in wholesome self- and society-improving recreational activities. Since all citizens of working age were expected to be employed in productive work, socialist urban systems provided support to the reproductive activities usually relegated to the household. As much as possible, for reasons of both economy and ideals, such activities were collectivized. Thus daycares and canteens were included in the designs of new housing estates and factories. In some cities, factories operated 24 hours a day, and so daycare and transit operated around the clock. In some socialist cities, housing was built with private sleeping quarters and shared bathing and kitchen facilities, as in the Vietnamese collective housing estates discussed in the case study below. In European socialist countries, households generally lived in self-contained apartments but with collectivized services available within the building or the estate. Most socialist urban governments have had to discontinue such extensive support because of lack of funds. As Vietnam and China, for example, have turned to **market socialism**, previously subsidized services (e.g., education and medical care) have increasingly become user-pay systems, and collective housing is increasingly under threat (Andreff, 1993; Beresford, 2008). Other forms of state involvement in urban everyday life continue, however, such as social mobilization campaigns, which seek to inculcate specific behaviours in the population. In Vietnam, the Cultured Family campaign (and the Cultured Neighbourhood/District) aims to produce law-abiding citizens who comply with various state social policies such as those on family size (Drummond, 2004).

The capacity and authority of socialist states to assemble land, plan, finance, and build urban projects, and (in the case of residential development) assign dwelling units to occupants distinguishes them from most capitalist states. They have the capacity because of their direct ownership and control of substantial amounts of resources and the authority because of the political organization of the state (in state socialist and communist societies, there is only one political party and democratic debate of urban issues is not possible). In capitalist societies the state may have a desire to engage in large-scale city-building, but it generally lacks the capacity and the authority to do so—it does not directly control all of the necessary resources and its more democratic politics constrains its authority to act. In state socialist societies, large sectors of the economy were, and are, centrally planned and controlled by state agencies, and a significant amount of land (in the case of the Soviet Union, all land) was, and is, publicly owned. Even in social democratic Sweden, local governments have engaged in extensive **land-banking**, purchasing large tracts of land beyond the developed edge of cities decades in advance of their actual development. Land-banking lowered the eventual cost of property development and also gave the local planning authorities more control over the planning and development process. There is a striking quality to socialist urbanism given that it is meant to represent a different way of living, different in terms of how everyday life is organized and in terms of social relations. The creation of socialist urban space was centrally planned and directed, characteristics that can be felt when experiencing socialist, formerly socialist, or post-socialist cities today. This aspect of breaking with past urbanisms i.e., capitalist urbanisms, meant that often (though not always) socialist urbanism embraced modernism in terms of urban design and architecture, following modernism's emphatic break with tradition. In the case of cities in Central and Eastern Europe and the Soviet Union, the vast scale of development and redevelopment after the Second World War and the attempts there to totally organize everyday life through the provision of a wide range of services, such as 24-hour daycare, provide a material representation of what Judit Bodnár (2001: 186) considers "arguably the most thorough, large-scale **modernization** project of our time, state socialism." Capitalist city-building, on the other hand, can be seen to be less centrally planned, sometimes almost chaotic and anarchic in nature. While property capital is certainly eager in its desire to shape and reshape urban space in capitalist societies, and indeed we see examples of "the perennial gale of creative destruction" (Schumpeter 1994: 84) and many very bold insertions into urban space, the processes and results of urban change in capitalist cities can appear haphazard in comparison to those of socialist cities.

The next section explores the nature of socialist and post-socialist cities in more detail by considering the examples of two very different cities. These two case study cities have been chosen, one in Asia and one in Europe, as representing different visions of socialist urbanism, different methods implemented to realize those visions, and different trajectories from their inception in the twentieth century through to the present. The first case study city, Hanoi, continues to be a socialist city today (albeit in a country, Vietnam, that has embraced market socialism), and the second case study city, social democratic Stockholm, remains an icon of enlightened twentieth-century city-building in a neoliberalizing Sweden. In Hanoi, the twentieth-century development of collective housing estates and their twenty-first-century fate are considered, a trajectory that offers insight into the changing context of the city and the contest between market forces and socialist planning in urban development. In Stockholm, the case study is the satellite suburb of Vällingby, widely praised when developed in the 1950s as a representation of a distinctively **Swedish model** in social policy and urban planning and design and still acknowledged as a special achievement in the history of Swedish urbanism. The two case studies address the following questions: How is their socialist and modernist past discursively framed—as a problem or a possibility? In what way can current urban politics in the case study cities be considered confrontations with twentieth-century socialist legacies?

Collective Housing in Hanoi, Vietnam

Hanoi's urban projects of the 1950s to 1980s represent an urbanism from the far left of the socialist spectrum (Figure 8.1). After a long, gruelling, and destructive anti-colonialist movement, full independence from France and removal of the French from Vietnamese soil were finally achieved, in 1954, in what became North Vietnam (the Democratic Republic of Vietnam). The Hanoi that was "liberated" in October 1954 was a city of about 400,000 people, and a city that had suffered much damage. Hanoi had limited infrastructure to begin with, and what it did have was in poor condition; it also had very little industry, as colonial regimes tend to favour resource extraction over industrial development. Over the next two decades, the city again found itself at war, this time against the United States and the US-backed regime in South Vietnam, a conflict that caused extensive damage to the existing urban infrastructure and necessitated several rounds of evacuation of the city, sometimes leaving only "indispensable" bureaucrats and workers, with over half of the population ordered to the countryside to escape intense periods of bombing raids, such as in the periods 1965–1968 and 1972–1973 (Thrift and Forbes, 2007: 146).

In this unpromising context, the communist government confronted urgent demands for housing as well as urban infrastructure repairs and extension. All were priorities, and housing became an ongoing crisis: housing was already a scarce commodity under the French, some existing stock was damaged by war, and new stock was needed to house the inflow of rural migrants who came to work in the new industries. Under state socialism, all land became state owned, which facilitated the urban renovations undertaken, and many urban projects were completed with the use of at least some social labour by Hanoi residents, as discussed above. The urban solution adopted to address the city's housing demands was that developed and urged upon the Vietnamese by the Soviet Union: the *mikrorayon* model of housing estates set up to be self-contained neighbourhoods (Drummond et al.,

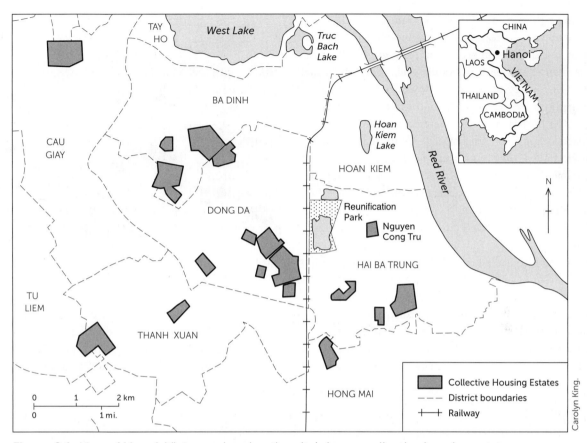

Figure 8.1 Map of Hanoi, Vietnam showing the city's larger collective housing estates.

2013; Schenk, 2005; Young et al., 2011). These *Khu Tap The* (KTT), as they were called in Vietnamese, embodied in their design a social ideal of collective life and a concomitant shift from emphasis on the private sphere and the family to an emphasis on the public sphere and the individual's relationship to the state (Humphrey, 2005). Many KTT were built in Hanoi, often with the design, material, or financial assistance of other socialist countries. Some are small, built to house workers at a specific factory or other worksite; some are vast, housing residents working in different enterprises or government agencies. Some of the earlier KTT, dating from the 1950s and 1960s, were constructed with amenities such as canteens, daycares, recreational facilities, and green spaces in the public areas between buildings. They were, upon completion, considered the most modern and desirable housing in Hanoi (Logan, 2000). Later, KTT tended to be constructed with fewer of the amenities, but most featured a design whereby households were allocated separate living and sleeping spaces while cooking and washing facilities were shared. Most units had only one or two rooms and were allocated according to household size and importance, with higher-ranking cadre getting larger quarters and sharing their cooking and bathing facilities among fewer households. This design was economical in a time of shortages but also spoke to the ideal of collective life in larger, state-organized communities over private life within families.

Living with the legacy of this social ideal and the specific conditions of this built form in Hanoi has been difficult. The structures themselves have been poorly maintained owing to a lack of funds, and they were often, for the same reason, built with low-quality materials (Logan, 1995). Once built, many degraded quickly, and today are in poor to unsafe condition. From the 1980s, as the Communist Party relaxed collectivization in all spheres, residents of the KTT began to divide up their collective facilities into small, often awkward, household units. Under new land title laws in the 1990s, residents of the KTT were able to purchase their dwelling units. Green spaces were colonized by those residents fortunate enough to have been allocated a ground-floor flat, and households built outwards wherever possible, even at higher storeys, by adding or extending balconies and sometimes by occupying public spaces such as roofs. Services such as canteens were discontinued and their spaces occupied by residents, either sufficiently brazen or sufficiently senior, for various purposes from storage to commerce or small-scale production, even to living space (Figure 8.2). The public narrative of the KTT turned from desirability to dereliction.

In contemporary Hanoi, a city of over 6.5 million that is communist-run but market socialist, the KTT seem to be considered a failed experiment by planners, officials, and the media. Thanks to this perception, the KTT, whose distinct residential housing forms were and continue to be home to a large proportion of the city's population, do not make it into the state's selective commemoration of the past. The state constructs and maintains a foundational mythology around the socialist republic that prominently features images, sites, and stories of Ho Chi Minh, the first president of an independent Vietnam, and of the Communist Party's foundation and anti-colonial activities (Tai, 2001) (Figure 8.3). None

Figure 8.2 Resident-adapted but derelict *Khu Tap The* (collective housing block) in Hanoi, Vietnam

Figure 8.3 Hanoi residents light incense at the mausoleum of Vietnam's first president and founder of the Communist Party, Ho Chi Minh, at the Lunar New Year 2015
Lisa Drummond.

of the KTT, even the most iconic Nguyen Cong Tru KTT, which was the first built in Hanoi and endowed with the most amenities, are slated for preservation; some have already been demolished, and many are awaiting demolition and reconstruction as condominiums by private developers (Drummond et al., 2013; Young et al., 2011).

While some features of the socialist urban everyday persist, from the ubiquitous neighbourhood loudspeakers broadcasting local news every morning to national social mobilization campaigns such as the Cultured Family, the KTT, haphazardly refashioned by individual households and in serious disrepair, seem to signal only their failure to achieve a full socialist transformation of daily life. As a consequence, these built ideals are quickly vanishing, to little public or state lament.

Vällingby: A Satellite District of Stockholm, Sweden

The site of our second case study, Vällingby, was developed as a satellite of Stockholm, the capital of Sweden, in the 1950s (Figure 8.4). A "social contract" between labour, capital, and the state was agreed in Sweden in the 1930s (Mattsson and Wallenstein, 2010). The compromise that

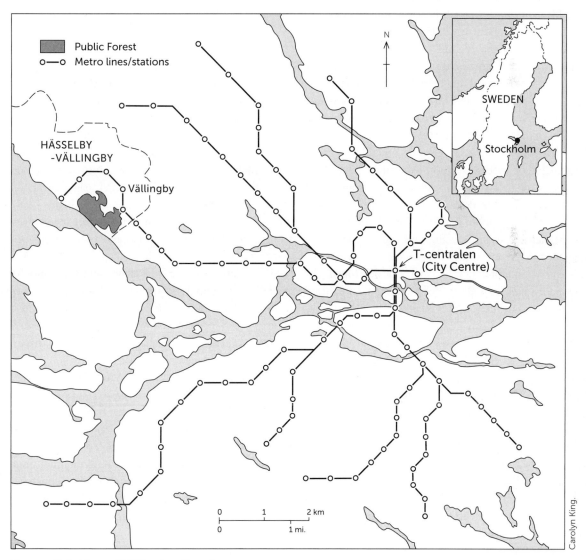

Figure 8.4 Map of Vällingby, Sweden

was achieved (in what is often referred to as the "**Swedish model**") between often warring opponents was that labour would organize into trade unions, they would be well paid by their employers in exchange for no labour strife, and a generous array of social services (e.g., health, education, housing, child care, and pensions) would be provided to all citizens on a socialized basis by the state. This coupling of capitalist Fordist economic activity and a Keynesian social welfare state was meant to generate a productive economy, a high standard of living for citizens, and social peace. David Harvey describes the Swedish welfare state as having "been organized around the ideals of a redistributive socialism with progressive taxation and a reduction of income inequality and poverty achieved in part through the provision of elaborate welfare services" (Harvey, 2005: 112). Sweden, in the decades following the Second World War, was widely considered to have among the most highly developed conditions of social welfare in the world. In Harvey's words, "[p]robably nowhere in the Western world was the power of capital more democratically threatened in the 1970s than in Sweden" (Harvey, 2005: 112).

In the context of the "large-scale Social Democratic restructuring of Swedish society" (Mumford, 2000: 166), urban planning in Stockholm attempted to encompass all aspects of everyday life, not just physical space. "The General Plan" for Stockholm, developed under the direction of CIAM member Sven Markelius and completed in 1952, anticipated a rapid increase in population and the need for a substantial amount of new housing to be developed. The General Plan called for new development to be organized into suburban satellite towns of 50,000 to 100,000 people. They were to be relatively dense (row housing and apartments, with only 10–15 per cent of dwellings in the form of single family dwellings), and each new satellite district was developed on a subway line linking it to the historic centre of Stockholm. The new districts were developed as ABC towns, ABC standing for *Arbete, Bostad, Centrum* (Work, Housing, Centre). In other words, they were to be relatively self-sufficient with jobs, dwellings, shopping, and other services and not just dormitory suburbs.

Vällingby was the first complete ABC town developed and is located 12 kilometres (7.5 miles) from central Stockholm (Figure 8.5). The development site was almost entirely publicly owned, having been purchased by the city of Stockholm between 1927 and 1931 in anticipation of future development. The total site area

Figure 8.5 Vällingby in 1960
Public Domain/Cogg bildbyrå/Sven Markelius, arkitekt. Arkitektur Förlag.

was 900 hectares (2250 acres), including 180 hectares (450 acres) of forest that was to remain in its natural state and was next to an arm of the sea (Popenoe, 1977). The plan for Vällingby was prepared from 1949 to 1951 (Pass, 1969). A new subway line, linking the site to central Stockholm, was built before any residents moved into their new homes. The subway opened in 1950, the first neighbourhood was ready for its residents in 1951, and Vällingby Centrum opened in 1954. The Centrum (with shops, a medical centre, churches, a library, a post office, and a community centre) sits on top of the subway station (Popenoe, 1977). Adjacent to the Centrum are slender apartment buildings (point towers) of 10 and 11 storeys each. These are quite small buildings by Canadian standards and contain only four apartments per floor. Moving away from the centre, there is a decline in building height and density first to three- and four-storey apartment blocks, then to row houses, and finally to single detached housing. A total of 18,850 dwellings were built.

Vällingby is reflective of what has come to be called Swedish **modernism**; in Sweden "modernism was 'softened'" (Mumford, 2000: 165). Urban design characteristics of this approach were that it was neither monumental nor heroic, it favoured brick and wood as construction materials rather than concrete, site planning was picturesque and varied rather than rational and Cartesian, high- and low-rise buildings were mixed, and apartment buildings took the form of point towers rather than long slab blocks (Mumford, 2000: 166–167).

The population of Vällingby district peaked at about 60,000 in 1960, with 24,000 in the core area. The core area now houses about half that (Nyström and Lundström, 2006), reflecting more the overall drop in the birthrate in Sweden and in household size in the Stockholm region, as well as the aging of Vällingby's inhabitants, than any sudden or drastic unpopularity of the district. As well, many of the apartments in the original buildings are small by present-day standards and not considered suitable by contemporary families. Some new housing has been built recently, with other residential projects in the planning stage, and Vällingby Centrum (Figure 8.6) has seen a recent large addition to its shopping facilities. The general reputation of Vällingby within the Stockholm metropolitan region is that it is a reasonable place to live, if somewhat boring and far from the centre (a 25-minute subway ride being considered far in Stockholm terms). At the same time, it is understood that it is a unique place and represents something important in local urban planning and design history. That uniqueness is formalized with architectural heritage preservation for the central part of the district. Alterations to the original urban vision are treated very carefully: a recent proposal for a new 17-storey point tower was subject to intense public discussion because it deviated from the standard building height of the original buildings. No such public debate or input has been invoked in Hanoi's plans to alter or even demolish its socialist housing estates.

Ultimately, Vällingby appears (to both Swedes and many outsiders) a quite livable place: it offers a variety of well-maintained housing, good shopping, good public transit, abundant and attractive open space, and proximity to water. The compromise of social democracy matched by the compromise of the Swedish version of modernism has worked well for its past and current residents. The legacies of its twentieth-century origins are perceived positively, unlike the case of Hanoi.

Having considered the socialist and post-socialist urbanisms of the two case study cities, Hanoi and Stockholm, as well as their twenty-first-century fates, the subsequent text box focuses on some of the challenges Canadians face today in living with the legacies of twentieth-century urban ideas.

Figure 8.6 Vällingby in 2015

What Relevance Does a Consideration of Socialist and Post-socialist Cities Have for Canadian Urban Areas?

This chapter argues that the relevance of socialist and post-socialist cities lies in understanding urban space and urban ways of living as reflections of a particular economic system, a particular set of social relations, and cultural values. The urban spaces created in Canadian cities in the twentieth century can thus be understood as reflections of Canadian society and economy during that period. Today, residents in many Canadian cities are confronted with the question of how livable the previous century's urbanisms are. In material terms, do they meet the needs of urban residents and workers; in cultural terms, do they meet their desires?

How, then, does the "good city" today compare to visions of the "good city" from the previous century? In answering this question, there is scope to consider the role that different **urban actors** play in the processes of city-building—of bringing to material reality visions of how people could or, some would say, should live in urban regions. Underpinning the wide variety of socialist urbanisms of the twentieth century was a conviction that all urban residents deserved equal access to housing, transport, jobs, and social welfare. There is evidence in many contemporary Canadian cities of increasing sociospatial segregation and polarization of the sort that twentieth-century socialist urbanisms attempted to overcome. Is this sort of segregation and polarization inevitable and/or acceptable? What can be done and who is to challenge it? To understand how this shift towards **social polarization** occurred, one must consider both the urban policies that shaped urban space in Canada in the last century and the urban policies in force today. In this regard, it is striking to note that Canada does not have a federal urban policy, unlike many other countries. Certainly the federal government affects urban life through its jurisdiction over immigration, finance, and transportation, but there is no coherent set of coordinated policies specifically targeting the country's urban regions, where so many Canadians live. This is in sharp contrast to the case of countries that had, or continue to have, socialist regimes in which the nation-states had, or have, a very clear urban policy.

In the twentieth century, Canadian urban policy-makers eliminated whole urban districts because they were considered to be unlivable, usually because they housed high-density populations of low-income households. An example of such interventions was the erasure of the African-Canadian district in Halifax called Africville (Nelson, 2009). What often replaced targeted areas were large public-housing projects, as in the case of the then working-class Toronto neighbourhood Cabbagetown, which was redeveloped in the 1940s and 1950s as Canada's largest public-housing project, Regent Park (James, 2010). In other parts of the urban region, extensive suburban districts were created in the decades following the Second World War. Today, such urban interventions are criticized. On the one hand, the large housing projects are often now home to racialized households living in poverty, the buildings are often materially in need of repair, and their locations, today not well served by public transit, are seen to isolate communities away from employment opportunities. On the other hand, the vast expanses of car-facilitating suburbs are today seen to carry enormous environmental costs, and often health costs as well.

The critique of the Regent Park approach to urban interventions did lead to attempts in some Canadian cities to build mixed-income neighbourhoods—the St Lawrence neighbourhood in Toronto being an example of that approach (Lewinberg, 2000). However, the cancellation by the federal government (and by some provincial governments) of funding for non-profit housing construction in the 1990s has made the development of mixed-income neighbourhoods extremely difficult, if not impossible. In the twenty-first century, neoliberalizing tendencies in national and municipal policies and the retreat of all levels of government from urban investment have emphasized the role of the market in determining urban form, rather than state direction. "The market"

> now decides which areas of the city are worthy of investment, and that worth is determined by the potential for profit. Today's cities, particularly those that figure prominently, or wish to, in the global network of cities, are often sharply polarized between urban "glamour zones" and zones of urban deprivation (Sassen, 2000). Investment is massively oriented towards waterfront condominiums and glass-and-steel office towers. The complexes of towers that house low-income and struggling households are far less attractive to global capital.
>
> The urban "ideals" that shape contemporary Canadian cities are, in many ways, the antithesis of socialist urbanism and its de-differentiated city. In her book *Dreamworld and Catastrophe: The Passing of Mass Utopia in East and West*, Susan Buck-Morss (2000) describes how many twentieth-century "dreamworlds" (visions of utopian urbanisms, both capitalist and socialist) resulted in urban "catastrophes." While the tendency might be to degrade if not erase the legacies of those urban visions, she argues instead that they have the potential to inspire much needed utopian thinking in the twenty-first century. Realizing that potential requires working through the ruins of capitalist and socialist "dreamworlds" (Buck-Morss, 2000), retrieving the dreams in those ruins, and building on them. In other words, contemporary urbanism is not irretrievably neoliberal.

Conclusion

A consideration of socialist urbanisms permits critical reflection upon who shapes urban space and for what purpose. All socialist urbanisms have involved a heightened degree of state intervention in the processes of creating and governing urban space ranging from democratic to authoritarian. In attempting to overcome what were considered the failings of capitalist urbanization of the nineteenth century, socialist states wrested power (to varying degrees) from capital to remediate those failings and shape urban space according to socialist urban ideals.

The two case studies in this chapter from Hanoi and Vällingby provide valuable insight into the capacity and the willingness of the state to shape new spaces and new ways of living. In the case of Hanoi, the new socialist state of the early 1950s was eager to produce new space and ways of living. The state was all-powerful to do so but had limited resources. Today, the state continues to hold considerable power over urban space but is less willing to directly shape and produce urban projects itself, preferring instead that the market do so. Much of the urban space developed from the 1950s to the 1980s is devalued today, and the experiment in directly influencing everyday life through urban form largely abandoned. In Vällingby in the 1950s, the state was willing to play a lead role in shaping space. While the hegemony of social democracy began to decline in the 1980s and today Sweden has become neoliberal (certainly in housing provision), the state still plays an active and interventionist role, granting preservation status to much of Vällingby and providing substantial power to local authority planners.

As this chapter has shown, socialist urbanism espouses a specific idea of the "good city," one that functions efficiently, serves all residents equally, and celebrates the urban, emphasizing the core through monumental architecture or providing expansive spaces for citizen gatherings, or both. Socialist urbanism emerged as a response to, and proposed a solution to, what were seen as the failures of earlier urbanisms that had produced cities that did not function smoothly or did not serve all residents well. But each socialist city, as every city, also reflects its path-dependent production, the urban fabric it inherited, the economic or political constraints of its governors and planners, and the cultural values or preferences of its people. All, however, are built manifestations of an ideal of urban life, a utopian vision of human–urban interaction, a belief in a better urban future. As we noted above, what were seen as solutions to urban problems in the twentieth century are in many cases now seen as problems themselves, as in turn our economic, political, and cultural environments have changed. As noted in the text box above, this applies to Canadian urban areas as well. It remains to be asked: What today is our ideal of the "good city" and of a "good" urban life? What can we learn from socialist experiments in the urban to inform our own attempts to produce better cities?

Key Points

- Socialism exists on a spectrum, and actually existing socialisms have taken many forms, with socialist cities reflecting this variety.
- A number of key characteristics of socialist urbanism can be identified—the design, the planning, the methods used to construct the city, the mode of use of the city in everyday life, and the mode of urban governance; each socialist city embodies some of these in different ways and to different degrees.
- Socialism and modernism addressed shared concerns about the pre-modern city, and their resolutions are often similar, though differently achieved and with different goals for twentieth-century urban society.
- Hanoi is representative of a communist socialism where the state implemented urban planning and development and sought to provide housing through large-scale estates whose design emphasized shared facilities and encouraged a collective approach to daily life; under "market socialism" these estates, already deteriorating and their shared facilities divided up for individual use, are not considered worthy of preservation as part of the city's urban heritage.
- Stockholm is representative of a particular model of twentieth-century socialist urbanism, one in which local government took the lead role in planning the city's postwar expansion in the form of satellite districts like Vällingby.
- The fate of socialist urban projects undertaken in the mid-twentieth century varies from preservation, where the projects are considered to have been successful in achieving their urban goals, to demolition, where these projects are considered to embody failure to achieve a full socialist transformation.
- As Canadians, we need to think carefully about the fate of our socialist and modernist urban experiments, the relevance of the urban social goals they represented, and the urban society we currently experience.

Activities and Questions for Review

1. Work in pairs to list the universal characteristics that almost all socialist cities have in some combination.
2. From your list above, identify what is distinctive about socialist modes of governance. How does socialist governance facilitate the undertaking of large-scale urban projects?
3. In a small group, compare and contrast the fates of Sweden's and Vietnam's socialist housing models. Why are these fates so different?
4. Create a diagram of the different features of a "good city" as understood within both socialist urbanism and neoliberal urbanism. Which version of a "good city" do you find most appealing and why?

Acknowledgments

This chapter is based on research funded by the Social Sciences and Humanities Research Council of Canada (Standard Research Grant 410-2010-2617).

References

Andreff, W. (1993). The double transition from underdevelopment and from socialism in Vietnam. *Journal of Contemporary Asia*, 23(4): 515–531.

Andrusz, G., Harloe, M., and Szelenyi, I. (Eds.) (1996). *Cities after socialism: Urban and regional change and conflict in post-socialist societies*. Oxford: Blackwell.

Beresford, M. (2008). Doi Moi in review: The challenges of building market socialism in Vietnam. *Journal of Contemporary Asia*, 38(2): 221–243.

Bodnár, J. (2001). *Fin de millénaire Budapest: Metamorphoses of urban life*. Minneapolis: University of Minnesota Press.

Buck-Morss, S. (2000). *Dreamworld and catastrophe: The passing of mass utopia in East and West*. Cambridge, MA: MIT Press.

Csagoly, P. (1999). Post-com pre-fab. *Cental European Review*. Retrieved from http://www.ce-review.org/99/25/best25_csagoly.html

Drummond, L.B.W. (2004). The "modern" Vietnamese woman: Socialization and fashion magazines. In L.B.W. Drummond and H. Rydstrom (Eds.), *Gender practices in contemporary Vietnam*. Singapore: Singapore University Press.

Drummond, L.B.W., Kip, M., and Young, D. (2013). The afterlife of Soviet urbanism: Contemporary attitudes to their socialist landscapes in Hanoi and Berlin. RC21 Conference 2013. Sociology of Urban and Regional Development, International Sociological Association. Berlin, August 31, 2013.

French, R., and Hamilton, I. (Eds.) (1979). *The socialist city: Spatial structure and urban policy*. Chichester and New York: Wiley.

Harvey, D. (2005). *A brief history of neoliberalism*. New York: Oxford University Press.

Hirt, S. (2008). Landscapes of post-modernity: Changes in the built fabric of Belgrade and Sofia since the end of socialism. *Urban Geography*, 29(8): 785–809.

Horak, M. (2007). *Governing the post-communist city: Institutions and democratic development in Prague*. Toronto: University of Toronto Press.

Humphrey, C. (2005). Ideology in infrastructure: Architecture and Soviet imagination. *Journal of the Royal Anthropological Institute*, 11(1): 39–58.

James, R. (2010). From "slum clearance" to "revitalization": Planning, expertise and moral regulation in Toronto's Regent Park. *Planning Perspectives*, 25(1): 69–86.

Jeanneret-Gris, C.E. (1973). *The Athens Charter*. New York: Grossman Publishers.

Kiernan, B. (2014). *The Pol Pot regime: Race, power, and genocide in Cambodia under the Khmer Rouge, 1975–79*. New Haven, CT: Yale University Press.

Lewinberg, F. (2000). The St. Lawrence neighbourhood: A lesson for the future. In N. Byrtus, M. Fram, and M. McClelland (Eds.), *East/West: A guide to where people live in downtown Toronto*. Toronto: Coach House Books.

Logan, W.S. (1995). Russians on the Red River: The Soviet impact on Hanoi's townscape, 1955–90. *Europe Asia Studies*, 47(3): 443–468.

Logan, W.S. (2000). *Hanoi: Biography of a city*. Sydney: University of New South Wales Press.

Mattsson, H., and Wallenstein, S. (2010). Introduction. In H. Mattsson and S. Wallenstein (Eds.), *Swedish modernism: Architecture, consumption and the welfare state*. London: black dog publishing

May, R. (2003). Planned city Stalinstadt: A manifesto of the early German Democratic Republic. *Planning Perspectives*, 18: 47–78.

Mumford, E. (2000). *The CIAM discourse on urbanism, 1928–1960*. Cambridge, MA: MIT Press.

Nelson, J. (2009). *Razing Africville: A geography of racism*. Toronto: University of Toronto Press.

Nyström, L., and Lanström, M.J. (2006). Sweden: The death and life of great neighbourhood centres. *Built Environment*, 32(1): 32–52.

Pass, D. (1969). *Vällingby and Farst—from idea to reality: The suburban development process in a large suburban city*. Stockholm, Sweden: Staens Institute for Byggnadsforskning.

Pickvance, C. (2002). State socialism, post-socialism and their urban patterns: Theorizing the Central and Eastern European experience. In J. Eade and C. Mele (Eds.), *Understanding the city: Contemporary and future perspectives*. Oxford: Blackwell.

Popenoe, D. (1977). *The suburban environment: Sweden and the United States*. Chicago: University of Chicago Press.

Rolandsen, U.M.H. (2008). A collective of their own: Young volunteers at the fringes of the party realm. *European Journal of East Asian Studies*, 7(1): 101–129.

Sassen, S. (2000). New frontiers facing urban sociology at the Millennium. *British Journal of Sociology*, 51(1): 143–159.

Schenk, H. (2005). Between the imperfect past and the conditional future. *Directors of Urban Change in Asia*, 48.

Schumpeter, J. ([1942] 1994). *Capitalism, socialism and democracy*. London: Routledge.

Schwenkel, C. (2012). Civilizing the city: Socialist ruins and urban renewal in central Vietnam. *positions*, 20(2): 437–470.

Strom, E. (1996). *Planning the global city: The politics of development in the new Berlin*. Unpublished diss., Rutgers University, New Brunswick, NJ.

Tai, H.T.H. (2001). *The country of memory: Remaking the past in late socialist Vietnam* (Vol. 3). Los Angeles: University of California Press.

Thrift, N., and D. Forbes. ([1986] 2007). *The price of war*. New York: Routledge.

Tscheschner, D. (2000). Sixteen principles of urban design and the Athens Charter? In T. Scheer et al. (Eds.), *City of architecture/Architecture of the city*. Berlin: Nicolai.

Vickery, M. (1984). *Cambodia: 1975–1982*. Sydney/Hempstead: Allen & Unwin.

Wells-Dang, A. (2012). *Civil society networks in China and Vietnam*. London: Palgrave Macmillan.

Wright, G. (2008). *USA: Modern architectures in history*. London: Reaktion Books.

Young, D. (2005). Still divided? Considering the future of Berlin-Marzahn. *Berichte zur Deutschen Landeskunde*, 79(4): 437–456.

Young, D., Drummond, L.B., and Kip, M. (2011). Still haunted by ghosts? Comparing socialisms and post-Socialisms in Hanoi, Berlin, and Stockholm. Presented at *Cities after Transition* conference, Bucharest, Romania.

9

Urban Planning, Indigenous Peoples, and Settler States

Ryan Walker and Sarem Nejad

Introduction

Most **Indigenous peoples** in Australia, New Zealand, and Canada live in urban areas. In Australia and New Zealand, the figures are 76 per cent and 85 per cent respectively (Kukutai, 2013; Taylor, 2013). In Canada, roughly 56 per cent of the Aboriginal population live in urban centres (Indigenous and Northern Affairs Canada, 2014). In colonized territories such as Western settler states, planning has been used as an instrument of control over the use and exchange of land, economic expansion, and social engineering to minimize the influence and expression of non-dominant cultures and knowledge systems (Matunga, 2013). Yet a majority of discussion about Indigenous peoples in urban planning and policy circles, as well as by urban scholars, still suffers from a lack of attention to what a "good" urban life means for Indigenous peoples and what **indigeneity** signifies for places we call urban. We explore these conceptual and practical challenges in this chapter.

The term "Indigenous peoples" is used to encompass descendants of the original inhabitants of settler countries around the world, including Aboriginal peoples in Canada and Australia. "Aboriginal" is the legal term used in Canada's Constitution Act of 1982 to refer to First Nations, Métis, and Inuit peoples. It is also the term used most often in Australia to refer to Indigenous peoples there. The term "Māori" refers to the Indigenous peoples of New Zealand. According to the United Nations Declaration on the Rights of Indigenous Peoples (2007), there are over 370 million Indigenous people in roughly 90 countries across Africa, the Pacific, Asia, the Americas, and Europe. Most live in contexts where they must coexist territorially and politically with settler societies in what are often referred to as **settler states** that have been superimposed, often by force and violence, onto Indigenous lands and community structures. Indigenous peoples are "among the most impoverished, marginalized and frequently victimized people in the world" (United Nations, 2007: 3).

The persistence of racism reproduces disparities between Indigenous and non-Indigenous peoples in settler states. The ideology of "**democratic racism**" (Henry and Tator, 2009) in Western (settler) states like Canada casts its discursive influence throughout our most central societal institutions, such as the media, the legal system, education, and social services. Democratic racism is linked to the notion of a "**whitestream**" society (Denis, 1997), where whiteness is perceived as neutral and objective, and societal institutions—descended mostly from white European society—are perceived to treat everyone "equally." Racist discourse can include claims that Indigenous peoples are unable to control or administer services and infrastructure on behalf of society as a whole. Patricia Wood's (2003) examination of the relationship between the city of Calgary and the Tsuu T'ina nation over construction of a road on First Nation land to serve the needs of the city provides a study of how racist discourse can easily be dispensed by mainstream societal institutions to support aims of dispossession and cast negative stereotypes of Indigenous peoples as unable to administer public goods.

This chapter provides an overview of some key features of the academic literature on Indigenous peoples living in cities, arriving at a point of departure for case studies from Sydney, Australia, and Auckland, New

Zealand. Sydney and Auckland are the largest cities in each country and have been among the most prominent and influential sites for the development of Indigenous movements for the reclamation of Indigenous political and territorial rights in those countries. These case studies examine the historic and contemporary contours of what coexistence between Indigenous and settler peoples looks like and consider where the fault lines lie. Drawing upon the Canadian experience, discussion then turns to how the sense of civic identity might be deepened by enhancing indigeneity in urban spaces. The chapter concludes by emphasizing the urgent need to create conditions for an Indigenous-inclusive citizenship and to address the erasure of Indigenous materialities and imaginaries from urban space in settler states like Australia, New Zealand, and Canada. While the focus is on these three English-speaking settler states, those interested in looking at other settler contexts may wish to explore the urban Sámi experience in Scandinavian countries (e.g., Nyseth and Pedersen, 2014) or the Ainu in Japan (Nakamura, 2015).

Placing Urban Indigeneity

Cities in settler societies like Australia, New Zealand, and Canada are located on the traditional territories of Indigenous peoples, and in contrast to the belief that these territories used to be empty, unused, and undermanaged, they have been places for settlement, trade, civilization, and culture for centuries prior to the arrival of European settlers (Matunga, 2013; Miller, 2011; Morgan, 2006; Jojola, 2013). Connected to this prior occupancy is the continuing inherent right of self-determination (United Nations, 2007). This is as true in urban areas, where the majority of Indigenous peoples in these countries live, as it is in rural and remote contexts. Yet the debate about how self-determination—an Indigenous group right—factors into a common citizenship that all are supposed to share within a nation-state is vigorous and ongoing.

Roger Maaka and Augie Fleras (2005) have conceptualized this as a dichotomous debate of *universal* versus **Indigenous-*inclusive*** citizenship. Universal citizenship centres on the common identity all share within a nation-state framework, the entitlements and responsibilities this confers, and the shared practices it reproduces. Although persistently supported by a majority within settler publics, it fails to account for the fact that Indigenous societies were determining their own affairs prior to the arrival of settler newcomers and never ceded their collective rights to continue doing so in perpetuity.

The alternative view is the notion that common national citizenship can include Indigenous self-determination as a fundamental aspect of state–society relations. Creating, enhancing, or protecting spaces for self-determination, with meaningful consequences in urban planning, would advance Indigenous-*inclusive* citizenship (Walker and Barcham, 2010). Métis author Chris Andersen (2013) makes the fundamental point that self-determination should include the notion that **Indigenous "density"** exists and is complex, and it should not simply be a matter of trying to track discrete instances of Indigenous difference from the cultural mainstream. Density hinges on the separate standing of Indigenous political communities owing to prior occupancy, international and national covenants, treaties, and constitutions. Indigenous self-determination based in a recognition of density is a crucial correction to a focus upon difference "because it offers a form of conceptual autonomy that allows for the creative position-takings forced upon us by our resistance to and incorporation of colonial rationalities and intervention strategies, and likewise, does not require us to continually demonstrate difference from whitestream normativity as a basis for collective authenticity" (Andersen, 2013: 268).

If Indigenous-inclusive citizenship is challenging in general terms at the national level, it is even more so at the urban scale where mainstream society promotes a particular view of where the legitimate territorial basis for *authentic* indigeneity lies. There have for many decades been cohesive and identifiable urban Indigenous communities in cities all across Australia, New Zealand, and Canada (Walker and Barcham, 2010; Morgan, 2006). Kay Anderson (2000) examines distinctions made in Australian public consciousness between "savagery" and "civility," and how metropolitan urban life has been correlated with (white) "civilized" culture and is perceived as incompatible with (black) Indigenous "savage or natural" culture. Indigenous culture is cast publicly as belonging somewhere "out there" in discrete rural and remote Australian communities. As far as it is perceived to exist at all within settler society, Indigenous culture has its own bounded spaces—discrete rural, remote, and reserve communities—within which to reproduce itself and maintain its vitality. And yet authentic indigeneity is not discarded upon entering city limits, though the contentious process of othering does generate a public perception of spatial transgression. Indeed, Indigenous urban communities have grown and become stronger in spite of the challenges to their existence and legitimacy. One example lies in Sydney's inner city. Indigenous households from all over Australia, including from within suburban Sydney, have moved to live in proximity to a large urban Indigenous community in the inner city, centred on "the Block" owned by the Aboriginal Housing Company and the numerous urban institutions that grew out of it (e.g., health, legal, housing, cultural, arts) (Morgan, 2006).

Contemplating the place of contemporary and historic indigeneity in our cities is an important exercise for developing our knowledge as urban scholars. Equally important is examining how we might *plan for* urban coexistence among Indigenous and non-Indigenous peoples. It is this planning challenge for settler states like Australia, New Zealand, and Canada that is explored in the next section.

Planning for Urban Coexistence

Planning and policy-making have long been colonial means of facilitating sovereignty in and over territories in settler states, while cities have been the headquarters for producing and reproducing colonial relations (Porter, 2010; Shaw, 2007). In Western planning thought, the relationship between people and land is forged through individual private property ownership and exchange. On epistemological and ideological grounds, this sometimes conflicts with the ways in which Indigenous communities carry out group tenure and stewardship over land, which emanate from systems of knowledge and world views, as well as community title (Henderson, 2000; Lane and Hibbard, 2005). In the absence of a strong focus on the inclusion of non-dominant cultures in planning processes—Indigenous cultures in particular—the social, political, and spatial segregation of Indigenous communities and their disengagement from many local government processes, including planning, occur (Walker and Belanger, 2013).

Based on her research in Australian and Canadian contexts, Libby Porter (2013) asserts that collaborative planning between Indigenous communities and the mainstream planning system has advanced further in non-urban contexts, such as in relation to natural resource management and environmental planning, than it has in urban areas. She uses the concept of "coexistence" to discuss how diverse and seemingly adverse values, aspirations, rights, and political communities can potentially be accommodated in urban planning systems alongside one another:

> Coexistence might most easily be defined as "sharing space" in more just, equitable, and sustainable ways (see Howitt and Lunkapis, 2010) where Indigenous rights can be seen to coexist alongside other rights, both of which are constituted as two or more possible cultural expressions of claims in space and place. The existence of one does not render the existence of the other obsolete.... Coexistence requires some kind of "rough equality" (Fraser, 1995) and a recognition of the other as an equally viable and legitimate party to conflicts about place. (Porter, 2013: 291)

Critiquing urban planning approaches in Melbourne, Australia, Porter (2013) explains that the incorporation of indigeneity in urban planning has been mostly based on a social welfare model centred on the provision of services by Indigenous organizations to address problems of poverty, homelessness, unemployment, and addiction, among others. Heritage protection is another area in which Australian cities create some space for considering Indigenous peoples, though they provide a limited and anachronistic portrayal of indigeneity in the city, "frozen in pre-colonial time, and entirely unrelated to property rights, governance, and law" (Porter, 2013: 288). Normalizing coexistence in cities will involve not only the recognition of inherent Indigenous self-determination rights, but also the deconstruction or unlearning of the colonial culture of planning (Porter, 2010). This unlearning will require critiquing the structure of planning processes and renegotiating its values, knowledge systems, and power relations between settler and Indigenous communities.

Ted Jojola (2013) and Hirini Matunga (2013) have done much to articulate the principles of **Indigenous planning**. It is a paradigm being reclaimed as a parallel tradition to settler planning institutions in response to the "violence of colonialism" (Matunga, 2013: 3; see Holmes, Hunt, and Piedalue, 2014, for a discussion of the continuing violence of colonialism against Indigenous peoples in Western settler states). Indigenous planning has been practised by Indigenous communities for hundreds, if not thousands, of years prior to the superimposition of colonial planning systems upon Indigenous lands and institutional structures. It has a strong tradition of resistance and a staunch commitment to political change in the context of coexisting alongside settler planning systems and the established power relations that give them voice while others (e.g., Indigenous systems) are silenced (see Listerborn, 2007, and Mitlin, 2013, for discussions of how women and the urban poor in the global South endure silencing by the professional planning system as well). Matunga (2013) positions Indigenous planning as a theory and practice of internalized self-definition and externalized advocacy. For Indigenous peoples, critical questions include these:

> Whose future? Who decides what this future should or could look like? Who is doing the analysis and making the decisions? Who has the authority, the control, the final decision-making power? Whose values, ethics, concepts, and knowledge? Whose methods and approaches? What frameworks, institutions, and organizations are being used to guide the planning processes that most affect Indigenous peoples? Where are Indigenous peoples positioned in the construction of that future? (Matunga, 2013: 4)

Processes of Indigenous planning connect people (e.g., kinship networks, tribe, nation), place (e.g., land, environment, resources), knowledge (e.g., traditional, contemporary), and values and world view (e.g., attitudes, beliefs, ethics, principles) with decisions (e.g., processes, institutions) and practices (e.g., applications, approaches) to enhance the well-being of the community (Matunga, 2013). Decision-making processes and practices vary by community but often have common aspects, such as striving for consensus, using traditional values to evaluate options and expected outcomes, and recognizing the wisdom of Elders. All of these aspects require time and attention to purposeful elocution to safeguard trust and cohesion within the community (Jojola, 2013).

The eminent Māori architect Rau Hoskins explains that an Indigenous cultural landscape is whatever Indigenous peoples see, hear, and feel that resonates with their Indigenous world view (Hoskins, 2014). In countries colonized by settler governments, however, the *materiality* (i.e., presence, structure, and physical quality) and *memory* (i.e., recall of experience or existence) of Indigenous communities have to a great extent been erased (Matunga, 2013). The material evidence of Indigenous peoples was removed, taking away the memory of Indigenous places, sites, resources, and settlements and in their place erecting colonial buildings, churches, parks, city survey and street patterns, all modeled after the "mother country" to create a new materiality and memory familiar to the settler public (Matunga, 2013).

The two case studies that follow examine tensions between the perceived place of indigeneity at the urban scale and how Indigenous-inclusive citizenship in the city is being articulated against the trenchant and pervasive discourses of universalism and whitestream cultural neutrality. The remainder of the chapter focuses on two key questions. First, how can a transformative coexistence between Indigenous and non-Indigenous peoples be mediated through the planning system in cities? Second, how are Indigenous communities reclaiming their materiality and memory through Indigenous planning in the largest cities of Australia (Sydney) and New Zealand (Auckland) and in cities across Canada?

Indigenous Resistance and Planning for the Future: The Aboriginal Housing Company, Sydney, Australia

The Aboriginal Housing Company (AHC), in the Sydney inner suburb of Redfern, started delivering housing in 1973 on what is known as "the Block" (a block of land owned by the AHC, a short distance from downtown Sydney) (Figure 9.1). With respect only to recent history, the 1920s and 1930s were a time when Aboriginal migrants from around the state of New South Wales (NSW) were gaining freedom to move from segregated rural reserves and assimilation policies were gaining currency. They worked on the railway network, making their way into inner Sydney and the Eveleigh Railway workshops, and subsequently, when the Depression hit, stayed on in Redfern with family for mutual support (Anderson, 1993; Pitts, 2001). Redfern had therefore been a gathering point for Aboriginal people for many years before the

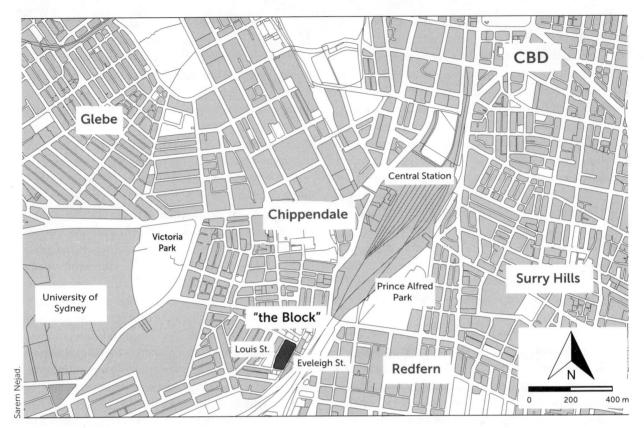

Figure 9.1 Map of "the Block" in Sydney, Australia

AHC was created. In the 1950s, there was a rural recession in NSW and Aboriginal migration increased into Redfern, where Aboriginal people could find inexpensive housing, public transport, and employment in the Eveleigh Railway Yards (Anderson, 1993). Over time a diverse Aboriginal community in Redfern developed, with people coming from traditional territories all over Australia. Kinship networks led to more migration and continued growth of the community.

In the 1960s and early 1970s, the Redfern community became a launching point for a mobilizing Aboriginal rights movement, in tandem with the civil rights and Black Power movements in the United States (Foley, 2001; note that "black" is an identity label often used in relation to Aboriginal people in Australia, owing partly to this early affinity with the civil rights and Black Power movements in the USA). In 1972, hundreds of Aboriginal demonstrators congregated at Redfern Park to launch a national land rights march that served to attract nationwide media coverage (Anderson, 1993).

The Redfern Aboriginal community, with the assistance of non-Aboriginal community members, including students from the nearby University of Sydney, developed institutions such as the Aboriginal Legal Service and Aboriginal Medical Service, both of which grew into national programs, and a number of cultural institutions, including the Black Theatre. The Black Theatre was established in 1972, an important institution in the creative resistance accompanying the emerging Indigenous land rights movement and urban Indigenous identity in Redfern, Sydney. Early work included politically charged street performances that tackled issues like police racism against Indigenous activists (Pollock, 2008). Full theatre plays—no less political—were also staged, as well as acting, dancing, and writing workshops, before the Black Theatre closed in 1977.

The availability of adequate and affordable housing was becoming increasingly limited by the early 1970s as the Aboriginal community grew to over 10,000. Discrimination by landlords against Aboriginal people increased, and the old inner suburban housing stock was being replaced with higher-cost market housing, either new or renovated. Housing in this area was largely terrace units, 80–100-plus years old (Pitts, 2001). In 1972, a young Aboriginal man was chased by police into an abandoned home on Louis Street (what would later be part of the AHC Block) where he had been squatting. In this house and neighbouring abandoned units were a number of Aboriginal squatters who police then arrested and charged for trespassing (Anderson, 1993; Pitts, 2001). When they were sentenced to jail time, priests from nearby St Vincent's presbytery offered alternative lodgings instead of jail. They opened a Catholic school hall behind the presbytery to homeless Aboriginal people as a form of emergency housing (Wain, 1976). Some of the people who became involved in a supporting role at this makeshift emergency housing centre formed an Aboriginal Housing Committee. Bob Bellear (a law student with Aboriginal ancestry) and his wife Kaye (a nurse) became centrally involved in the committee and Bob became its spokesperson. The housing committee called for the creation of an Aboriginal housing co-operative, clustered together at the Block to further strengthen the sense of Aboriginal community.

A cluster of 41 houses—the same terrace houses the Aboriginal squatters occupied when they were arrested—were owned by a developer who had ambitions of rehabilitating them into middle-income dwellings. Through a process of negotiation with a trade union sympathetic to the pursuits envisioned by the Aboriginal Housing Committee, the developer agreed to allow homeless Aboriginal people to squat in two of the unoccupied units (Wain, 1976). The housing committee used its base in these two homes to demand that Prime Minister Gough Whitlam's Commonwealth government provide Redfern's Aboriginal people with better housing (Anderson, 1993). Whitlam was elected late in 1972 on an ambitious platform focusing on social welfare, urban affairs, and Aboriginal rights. When the Aboriginal Housing Committee submitted a formal request to the minister in the newly created Department of Aboriginal Affairs, it was well received as a flagship case for implementing self-determination in an urban community. The committee had requested funding to purchase an entire block where the terrace houses were located and to build and renovate homes there for the Aboriginal community; this plan was to be carried out by a new Aboriginal Housing Company (AHC).

There was considerable protest from other local residents, as well as from the South Sydney council, and apprehension that a black (Aboriginal) ghetto was being created. A number of approaches were used to try to deter the proposal from going ahead, including prohibiting the extended-family housing desired by the AHC (Anderson, 1993). The AHC recruited architecture professor Col James (from the University of Sydney) to assist with designing the envisioned Aboriginal housing co-operative (Anderson, 1993; Wain, 1976). Plans were also made to establish community services, such as a preschool run by Aboriginal mothers, a cultural centre, and a workshop for trades training.

Notwithstanding the local opposition from some area residents and the local council, the commonwealth government still agreed to fund the Aboriginal housing proposal. The AHC was thus incorporated in July 1973 to deploy the initial grant under the Aboriginal Housing Association Grant Program from the commonwealth government to purchase and continue restoring the terrace houses on the Block (Department of Aboriginal Affairs, 1973; Pitts, 2001). The AHC now owned a city block of land in inner Sydney as well as the houses on it, and it was the landlord responsible for choosing tenants and managing the properties.

The 1980s were a time of decline on the Block with social tensions, a rise in crime and in drug use, and a lack of social and recreational activities (Pitts, 2001). It was at this time that the media's portrayal of the "Redfern Problem" began to entrench itself in the public consciousness (Pitts, 2001). In response to escalating social problems, some of the derelict houses on the Block were demolished by the AHC in the 1980s to make space for a park. As the 1990s unfolded, heroin use became an increasing problem, dealing an enormous blow to the strength of the Redfern Aboriginal community.

To reverse the trend of deteriorating housing and community on the Block, the AHC began a community social planning process in 1999, drawing on the expertise of a community planner, Angela Pitts, from the University of Sydney (following in the path that Col James had cut decades earlier). The national and international award-winning Community Social Plan (Pitts, 2001) that was produced to guide the regeneration of the Block—known as the Pemulwuy Redevelopment Project—has been the mainstay of work undertaken by the AHC over the past decade. Pemulwuy was the name of a famous

Aboriginal warrior from the Eora tribe (whose clans and traditional territory are based in the Sydney area, including the Block). The Pemulwuy Project is centred on a mixed-use redevelopment of the Block, including 62 affordable housing units for low- to middle-income Aboriginal families, commercial and retail space, a public civic space, art gallery, student housing (the Block is within walking distance of the University of Sydney), and sporting and child-care facilities.

The greatest challenge to implementing the Pemulwuy plan was the adoption of the Redfern-Waterloo Authority's (RWA) Built Environment Plan in August 2006. The RWA was created in response to the deteriorating circumstances in the Redfern area, which came to a head in February 2004 when the Redfern riots became national and international news. A 17-year-old man from the Aboriginal community died trying to get away from local police on his bicycle when on his way home from his girlfriend's house on Valentine's Day. The ensuing riots in response to this incident brought to the fore years of frustration within the Aboriginal community about government neglect of social welfare issues, rights, and continuing bad relations with police in the area. Parliament introduced the RWA in order to immediately improve the area and foreclose possible future unrest. Bypassing the usual planning approaches centred on democratic process and participation, the RWA set about to create and implement a 10-year built environment plan focusing on infrastructure, job creation, and human services, with no focus on situated indigeneity.

The Pemulwuy plan, begun several years earlier, had an uneasy relationship with the new RWA plan. The minister responsible for the RWA indicated that he would not support the Pemulwuy plan, arguing instead that Aboriginal housing would be dispersed into other areas of the city (Turnbull, 2005). The RWA plan, unlike the locally produced Pemulwuy plan (Government of New South Wales, 2004; Welsh, 2005), would have disempowered the AHC and moved the Aboriginal community away from this central base of Aboriginal identity and political empowerment in the centre of Australia's largest city. Unwilling to have its plans sidelined, the AHC created the Pemulwuy Vision Taskforce, made up of professionals and academics, to assess the feasibility of its plan and lodged its application with the New South Wales Department of Planning in 2007. In 2009, the Pemulwuy Project Concept Plan was approved by the New South Wales government, and the tenacity of Aboriginal self-determination over time held out in this urban context once again.

The Pemulwuy plan is still at the centre of the AHC's aspirations at the Block, although particularly over the past couple of years there has been vigorous debate about the staging of development between some Aboriginal community members, led by Elder Jenny Munro (one of the founders of the AHC) and the AHC itself. Unable to secure government funding to develop the 62 units of affordable housing for Aboriginal families at the outset, the AHC planned to develop the commercial space first, hoping to then leverage resources to create the affordable housing subsequently. Community members, led by Munro, argued that if the affordable housing was not developed first or at least concurrently, the Block would turn its back on its history as the heart of Aboriginal identity in Sydney, with Aboriginal families displaced by commercial development. An Aboriginal tent occupation was set up on the Block for roughly 15 months to stall implementation of the Pemulwuy Project's commercial components, and the AHC ultimately had to file a court challenge to evict the protesters and begin development.

The Block remains a contested site for Indigenous planning in the centre of Australia's largest city, most recently involving conflict between different Aboriginal visions for the site (Figure 9.2). In August 2015, the AHC received a favourable decision from the Supreme Court, allowing development of the long-awaited Pemulwuy Project to proceed. The tent occupation began to disperse immediately after the court decision, though with some

Figure 9.2 "the Block," Aboriginal Housing Company (Redfern), Sydney

good news as well. The federal Aboriginal Affairs minister announced that the government would provide a financial grant and assist the AHC to secure additional financing to develop the 62 units of affordable housing before or concurrently with the commercial development (*Sydney Morning Herald*, 2015). It appears that the AHC and Aboriginal community members may be close to implementing the plan's social and commercial components as envisioned over a decade ago, through a combination of their community planning efforts and the continuing use of resistance and protest to steer a course for plan implementation.

While the AHC became the first experiment of the newly elected Whitlam government in advancing its Aboriginal self-determination and urban affairs agendas (Anderson, 1993; Department of Aboriginal Affairs, 1973), the establishment of the AHC is attributable to Aboriginal community activism and resistance to Australia's narrow view of universal citizenship that excluded the density of Indigenous political and cultural communities. The Block in Redfern is significant to the Aboriginal peoples in urban and rural/remote areas and is often referred to as the "Black Heart" of Australia for its political, spiritual, and cultural significance (Government of New South Wales, 2004).

The AHC case study shows how a block of land in the centre of Australia's largest city was the site for the settler state and Indigenous peoples to contest the place of urban indigeneity and where Indigenous peoples from around the country have struggled for decades to create an Indigenous-inclusive society (Walker and Barcham, 2010). The AHC's Pemulwuy plan ran up against the institutionally racist state-imposed Redfern-Waterloo Authority's Built Environment Plan for the same area, challenging the notion that Indigenous and settler claims to urban space could coexist with authority and legitimacy. Instead, the state government of NSW sought to "render the existence of the other obsolete" (Porter, 2013: 291). The Aboriginal community's persistent work over decades to reclaim their materiality and memory (Matunga, 2013) in Sydney—including the most recent competing visions among Aboriginal community members and the AHC over the balance of social and commercial development—shows no sign of ending in the near future. Robust Indigenous planning requires discussion and debate among Indigenous community members, according to the principles and processes set forth by the community (Jojola, 2013; Matunga, 2013). It will be important to learn over time how the leadership of local community members and Elders intertwines with the work of the AHC to produce and reproduce Indigenous space in the "Black Heart" of Australia. Attention is now directed to a second case study on urban planning, Indigenous peoples, and the contested nature of indigeneity, this time in New Zealand's largest city.

Reclaiming Indigenous Planning: Ngāti Whātua Ōrākei, Auckland, New Zealand

Ngāti Whātua Ōrākei is a *hapū* (i.e., a number of extended-family groups related by genealogical descent, referred to also as a sub-tribe) of the Ngāti Whātua *iwi* (referred to also as a tribe), whose traditional territory includes Auckland, New Zealand's largest city. The Ngāti Whātua Ōrākei *hapū* derives from three *hapū* traditionally based in the area: Te Taoū, Ngāoho, and Te Uringutu (Ngāti Whātua Ōrākei Trust, 2012). For over 260 years Ngāti Whātua have exercised customary authority over lands in the Auckland area, leading back to the chief Tuperiri's consolidation of land and resources. Shortly after signing the Treaty of Waitangi in 1840 thousands of hectares of Ngāti Whātua land were made available by chiefs to Governor Hobson for development of a European settlement, in exchange for education, medicine, and trade opportunities (Ngāti Whātua Ōrākei Trust, 2012).

Within 10 years of signing the Treaty, control over much of the traditional territory was lost (almost 33,590 hectares [83,000 acres]) and the chief, grandson of Tuperiri, approached the Native Land Court to confirm the remaining 283-hectare (700-acre) Ōrākei Block of land under Ngāti Whātua's communal ownership, ensuring it would not be broken up under individual titles. In 1868 the Land Court affirmed the *hapū*'s communal ownership authority and that indeed the 283 hectare (700 acres) of land was inalienable to any person. Thirty years later, in 1898, however, the Land Court divided most of the Ōrākei Block into individual titles (Ngāti Whātua Ōrākei Trust, 2012). Indigenous communal ownership of the site had thus disintegrated. This approach in colonial settler states like New Zealand, Australia, and Canada has been a foundation of the colonial culture of planning that overwrote Indigenous group land tenure and stewardship guided by Indigenous world views and kinship networks (Jojola, 2013; Matunga, 2013).

In 1951, Ngāti Whātua Ōrākei were evicted from the last remaining 5 hectares (13 acres) of land held on the Ōrākei Block on Okahu Bay (just outside the Auckland Central Business District) and relocated a short distance away in 35 state (public) housing units. The buildings, homes, and *marae* (the cultural centre of Māori communities) on Okahu Bay were torn down and burned. The relocation was justified under the Public Works Act with the rationale that the Okahu Bay settlement was unsanitary and prone to flooding. The flooding has been attributed to settler modifications of the landscape, such as the installation of an embankment and sewerage pipe and the construction of Tamaki Drive along the edge of the bay in 1929.

For many years Ngāti Whātua Ōrākei were without a meeting house or *marae*. The design of the state houses at Ōrākei in 1951 was not conducive to communal living, with three-bedroom houses divided by fences and private sections. After being state-housing tenants on their traditional land for over 25 years, and after years of protest, dating back to the turn of the century, over the alienation of communally owned land, a branch of the *hapū* occupied Bastion Point at Ōrākei for over 500 days, when the government announced plans in 1976 to subdivide an area of 24 hectares (60 acres) there (Ngāti Whātua Ōrākei Trust, 2012). This resulted in a large mobilization of police and military to evict the occupiers for trespassing on Crown land (even though it was actually Ngāti Whātua Ōrākei ancestral land); 222 people were arrested, most of whom were Ngāti Whātua. At the same time, Elders were working to have a 4-hectare (10-acre) block returned that had been taken under the Public Works Act and was not being used for its intended purpose. The end result was that a settlement was reached whereby the 35 state houses (and the land they were built on) were sold to Ngāti Whātua Ōrākei for what was considered a nominal price for land so close to Auckland's central business district.

In 1986, the leadership of Ngāti Whātua Ōrākei lodged a claim at the Waitangi Tribunal on the basis of having wrongly lost its 283-hectare (700-acre) Ōrākei Block, which ought to have remained in communal *hapū* ownership. In 1987, the Waitangi Tribunal produced its recommendations and the government responded with the Ōrākei Act 1991, compensating the *hapū* with financial resources to assist with housing and other development and returning to collective *hapū* ownership about 32 hectares (80 acres) of the original 283 (700) lost. Most of the land was to be left as open space, a park reserve for the use of Māori and all citizens of Auckland—co-managed by the *hapū* and city council—and parts of the land were developed, or will be developed, by Ngāti Whātua Ōrākei Trust (Figure 9.3).

In later negotiations with the Housing New Zealand Corporation in the mid-1990s, roughly 100 additional state houses on the Ōrākei Block were transferred to the trust. In the early 1990s, about 8 hectares (20 acres) of the settlement land were subdivided to build market housing for owner occupation on collectively owned *hapū* land. Descendants of the common ancestor Tuperiri could apply for a licence to occupy a parcel of the land and build a house. It was a means for bringing families back to the traditional Ōrākei lands. On another parcel of the settlement land, the trust developed the Eastcliffe Retirement Village, open to residents of any background, not only Ngāti Whātua. This has been a source of income for the trust, growing its asset base from the original settlement. More culturally specific to the needs of the *hapū*, a complex of *kaumātua* (Elder) flats was built with trust funds close to the *marae* and meeting house. These different types of housing and open space development are a few examples of how the Ngāti Whātua Ōrākei Trust has been stewarding development on their ancestral land, now held once again in community ownership, to provide an inviting home for *hapū* families and secondarily for Auckland citizens in general (e.g., Eastcliffe Retirement Village) (for a fuller examination, see Ngāti Whātua Ōrākei Trust, 2012; and Boffa Miskell, 2013) (Figure 9.4).

In December 2005, a Papakāinga (Māori settlement for habitation) planning process was begun to determine how the community would like to develop over the following decades, applying customary values and principles, and creating a place for *hapū* families to come back to and feel at home in the city. The Papakāinga planning process has addressed housing, services, restaurants, offices, governance, and other sectors in the context of a large site plan for the Ōrākei land that is oriented around areas of cultural importance and according to design principles important to the *hapū*. Design principles pertain to the physical architecture and orientation of built form in relation to landmarks, heritage, and community interaction, creating a sense of Indigenous place.

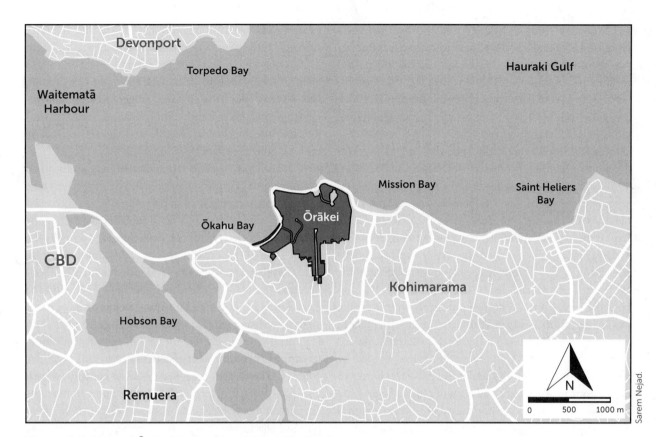

Figure 9.3 Map of Ōrākei in Auckland, New Zealand

They also pertain to social considerations, such as designing in such as way that children have the right of way in public spaces of the community (Boffa Miskell, 2013). The Papakāinga planning process has also addressed the development of culturally appropriate infrastructure, such as the prospects for environmentally friendly energy sources, native flora, and waste and storm water disposal methods that are environmentally sustainable, the relationship between land-development decisions and stewardship over water, and the adherence to Māori cultural practices in general (Ngāti Whātua Ōrākei Trust, 2012; Boffa Miskell, 2013; for discussions of Māori cultural practices applied in planning, see Awatere et al., 2013; and Matunga, 2013). Ngāti Whātua Ōrākei and Auckland City Council recently created a Joint Management Agreement for administering the official planning regulatory functions pertaining to lands owned by the *hapū* at Ōrākei. This partnership aims to ensure that the overall city plan for Auckland and the specific plans of Ngāti Whātua Ōrākei are taken into account together, and that the leadership of both the settler and Māori communities in that territory exercise their authority in mutually respectful ways.

Figure 9.4 Development on Ngāti Whātua Ōrākei Trust land, Auckland

This case study of Ngāti Whātua Ōrākei in Auckland, New Zealand, exhibits the movement over time between colonial dispossession of Māori lands, to a partial reclamation of those lands starting in the 1990s, and ultimately to a set of contemporary circumstances in which the ancestral *hapū* is re-territorializing its traditional lands in Auckland, forging a relationship of coexistence (Porter, 2013) with the settler state through its own Indigenous planning processes (Jojola, 2013; Matunga, 2013) and a joint management agreement. With the passage of time we can see **Indigenous density** and complexity (Andersen, 2013) developing in New Zealand's largest metropolis. All of this is occurring against a backdrop of significant change in the settler and Indigenous communities there. Notably, the recommendations of the Waitangi Tribunal that led to the Ōrākei Act 1991, discussed above, have now been significantly supplemented by the important Treaty of Waitangi Claim settlement in favour of Ngāti Whātua Ōrākei that Parliament passed into legislation in 2012. The claim was lodged in 1993 by Ngāti Whātua Ōrākei. The settlement covers a large span of territory in Auckland, and includes an apology from the Crown, an agreed historical account, and cultural and commercial redress (New Zealand Herald, 2012). The threads of Indigenous materiality and memory (Matunga, 2013) that were pulled apart over a century ago in Auckland are being woven together again by the Indigenous peoples of that land. Tools for co-producing (Walker and Belanger, 2013) the urban planning framework and creating space for parallel planning traditions (Matunga, 2013) now appear to be taking form.

The key to recapturing self-determination in this urban context, following a period of land loss and forced removal, was the transfer of financial and land assets back to the *hapū*. Without these, self-determination on the site—for example, the housing developments in the 1990s and the Papakāinga planning process over the past decade that set the stage for continued residential, commercial, open space, environmental, and cultural development for years to come—would not have been possible. It was not only the reacquisition of lands, however, but also the reclamation of the communal land tenure system by the Ngāti Whātua Ōrākei Trust that was important. The disaggregation of communally held land in 1898 into individually titled plots by the settler government's Land Court, followed by the selling of these plots over time on the open market, was as destructive to Indigenous materiality and memory in Auckland as the appropriation of the land outright by the non-Indigenous settler government would have been. The imposition of a Western land-title system was destructive of the state of Indigenous-inclusive citizenship and coexistence that existed between a self-determining Indigenous people and the settler society moving onto their land.

The cases of Ngāti Whātua Ōrākei in Auckland and the Aboriginal Housing Company in Sydney have some interesting characteristics in common. Indigenous resistance, on a recurring and continuous basis over many years, was necessary in both cases in order to press the settler state to respect Indigenous rights. Both also have land that they can use as a foundation for further development, as a place to stand as Indigenous peoples in the city—land that has high cultural, spiritual, and historic value as well as high real estate value given its proximity to central cities in each country. Indigenous traditional territories encompass entire cities and regions in settler countries, lands that have been used to generate wealth and quality of life disproportionately for non-Indigenous citizens. When Indigenous peoples re-territorialize their place in the urban sphere, they create new potential to generate the material and socio-cultural conditions that can enhance urban quality of life for their communities. Yet examples of this occurring are still rare in Western settler states. What is far more common is that settler societies apply a notion of universal citizenship that is imagined as culturally neutral. When gaps inevitably occur between the socio-economic circumstances of those privileged by cultural dominance and the socio-economic circumstances of those whose cultures are marginalized in the universal mainstream, the social welfare state is often seen as the only solution and social programs are targeted at those most "in need." The return of land, the sharing of resource wealth, and the coexistence of Indigenous and non-Indigenous governance frameworks as tools for self-determination are too rarely used alongside welfare state measures. Without these combined changes, Indigenous-inclusive citizenship in our Western cities will remain very much incomplete. The chapter turns next to a discussion of urban planning and Indigenous peoples in the Canadian context and of how it relates to and supplements what has been learned from the Australian and New Zealand experiences.

Indigenous Urbanism and the Re-territorialization of Canadian Cities

The second half of the twentieth century was a time of vigorous debate about the urbanization of Aboriginal peoples (i.e., their movement from reserves and rural communities to cities and towns). The early decades of the twenty-first century are now seeing attention turn to **Indigenous urbanism**. With Indigenous urbanism, the focus is shifting towards the study of how Aboriginal peoples participate in the creation and re-territorialization of urban space and enjoyment of a good urban life (Newhouse and FitzMaurice, 2012; Fawcett, Walker, and Greene, 2015).

Aboriginal peoples living in Canadian cities or moving regionally between cities, towns, reserves, and rural areas are often within their own traditional, national, and treaty territories (Peters, 2015), or they recognize that they are in the traditional territories of other Aboriginal peoples. Non-Aboriginal people are often unaware of the Indigenous cultural landscapes they inhabit and share. With over half of Aboriginal people in Canada residing in urban areas, and with a population growth rate that is the highest in the country, it is important to consider the implications for the planning of a comparatively young and fast-growing urban population group (Peters, 2015). Planning as a scholarly and professional discipline is concerned with future-seeking. In other words, planners seek to understand the current state of urban affairs, to learn of the future aspirations of urban communities, and to implement policies, programs, and developments that intentionally bend current trends towards the sought-after future(s). What are Canadian cities doing to take into account the rising numbers of Aboriginal people, their cultural resurgence, and to de-colonize urban space?

Just as the Aboriginal Housing Company in Sydney and Ngāti Whātua Ōrākei in Auckland use land in the city as a basis upon which to pursue broader community goals for their respective Indigenous communities, so too are land and resources important to Aboriginal people in Canadian cities. **Urban reserves** are one interesting example of how First Nations are re-territorializing urban space. An urban reserve might be a parcel of downtown land with an office building on it (Figure 9.5 shows Saskatoon's newest urban reserve, a downtown land parcel where the Yellow Quill First Nation is developing an office building). It could also contain housing developments, an educational institution, or a gas station. Anything that you might find in a neighbourhood permitted by a city's land-use plan could conceivably occur on an urban reserve. The first urban reserve in a large Canadian city was a commercial complex in an industrial area of Saskatoon, created in 1988 by the Muskeg Lake Cree Nation (Sully et al., 2008) (Figure 9.6).

Figure 9.5 Yellow Quill First Nation urban reserve, Saskatoon

Figure 9.6 Muskeg Lake Cree Nation urban reserve, Saskatoon

continued

Urban reserves are satellites of principal reserves located elsewhere (e.g., the principal reserve of Muskeg Lake Cree Nation is about 100 kilometres (62 miles) north of Saskatoon) and are acquired on a willing buyer–willing seller basis in the real estate market, typically with financial resources derived from land-claim settlements. Once the First Nation acquires the urban land parcel, it applies to Indigenous and Northern Affairs Canada to acquire reserve status on the land. This is done under the federal government's Additions to Reserves Policy. Urban reserves are becoming common in Saskatchewan and Manitoba, catalyzed in large part by the financial resources allocated to many First Nations through the Saskatchewan and Manitoba Treaty Land Entitlement (TLE) Framework Agreements, which were established in the 1990s and have been used to acquire urban land holdings. These agreements were struck between the federal, provincial, and many First Nations governments to settle specific land claims that would fulfil outstanding debts of land owed to First Nations by the Crown. Recall the comparable process of claims filed by Ngāti Whātua Ōrākei at the Waitangi Tribunal in New Zealand in the late 1980s and early 1990s over debts of land discussed in the previous section.

Urban reserves are still rare in other parts of Canada, but will likely accelerate over time as success stories continue to spread from the Prairies across the rest of the country and First Nations develop the financial resources and community aspiration to acquire new (urban) land. There are currently 44 urban reserves in Saskatchewan, with an additional 49 planned (Belanger, 2014) (Figures 9.5 and 9.6). Prior to the federal government approving a conversion to reserve status, an agreement dealing with four key issues must typically be reached between the First Nation and the municipality: (1) compensation paid in lieu of property and school taxes once the land is converted to reserve; (2) the type and financing of municipal services to be extended to the new reserve; (3) bylaw development and compatibility with surrounding municipal land uses; and (4) a joint consultative process and dispute-resolution mechanism (Sully et al., 2008).

Land and resources deployed for Indigenous community self-determination are evident in urban social planning and policy as well. The Lu'ma Native Housing Society (lu'ma means "new beginnings" in Coast Salish) has operated hundreds of social and affordable housing units for Aboriginal households in Vancouver since the 1980s. One interesting recent development, opened in 2013 by Lu'ma, is the David Pranteau Aboriginal Children's Village. This 24-unit development, housing foster children and families, also features space for programming (e.g., wood carving), counseling and support, and some commercial units at street level. It was designed by Nisga'a architect, Patrick Stewart, for Lu'ma. A driving social concern behind this project was that foster children tend to have high residential mobility, moving from home to home, which negatively affects their childhood development. The Aboriginal Children's Village flips this problem on its head. Foster children are allowed to remain, making this their home. It is the adults who move on if they no longer serve as foster parents. The children stay and continue to be supported. The Aboriginal Children's Village serves as an example of how Indigenous communities in Canadian cities are creating Indigenous-inclusive spaces, both in a material and a socio-cultural sense. In this case, it is not just the design of the building and grounds, but also the programming within that represents the production of Indigenous urban space.

Finally, in Ottawa, Canada's capital, it is possible to see how indigeneity is being designed into the materiality and memory of one of the city's signature public spaces, Lansdowne Park. Over the past few years the city of Ottawa has worked with the Algonquins of Ontario to reflect Algonquin history, culture, and art in the city's revitalization of the park. The Lansdowne Park Revitalization Project includes, for example, an Algonquin food vessel basket-weave pattern for the paving of Aberdeen Square (Figure 9.7), which houses the Ottawa Farmers' Market. The park landscaping plan also includes trees of cultural significance to Algonquins,

Figure 9.7 Aberdeen Square Algonquin basket-weave pattern, Lansdowne Park, Ottawa

an ethno-botanical garden, and a teaching circle (Figure 9.8). Other Algonquin-themed programming, public art, and wayfinding may also be incorporated as the revitalization project continues. The revitalization of Lansdowne Park is an example of how indigeneity might become part of the shared spaces of the city, public spaces that welcome all citizens but are designed to be inclusive of Indigenous peoples' sense of place; coexistence in urban areas requires not only separate spaces for indigeneity, but prominence in shared spaces as well. For this to be achieved, working relationships between Indigenous (e.g., Algonquins of Ontario) and non-Indigenous (e.g., City of Ottawa) parties

continued

Figure 9.8 Algonquin teaching circle, Lansdowne Park, Ottawa

require careful attention so that the objective-setting, the planning, and the fine details of development implementation are all carried out in a way that satisfies everyone. It is important to recall that spaces owned by Indigenous communities in a city and the shared public spaces owned by the city are within Indigenous traditional territories. On the one hand, the case studies from Sydney and Auckland, as well as the discussion of urban reserves in Canada, focused mostly on land owned and developed by Indigenous peoples and their organizations. On the other hand, the discussion of the relationship between the city of Ottawa and the Algonquins of Ontario in Lansdowne Park draws attention to re-territorializing traditional Indigenous lands in the shared public spaces of the city.

As Canadian society delves earnestly into the work of implementing means of **reconciliation** between settler and Indigenous peoples following the 2015 report of the Truth and Reconciliation Commission, urban planning and policy will be important areas of concern. So much of what occurs in Indigenous affairs is projected at a nation-to-nation level, often in federal and international spheres. The power of many small acts of reconciliation at the urban scale should not be underestimated where

Indigenous urbanism is being cultivated and where settlers from all over the world live side by side with the original occupants of the land, learning ways of "living together differently without drifting apart" (Maaka and Fleras, 2005: 300). Canada is certainly not the first country to undertake a truth and reconciliation process. In the mid-1990s, South Africa instituted the first process of this kind as part of its transition out of **apartheid**. Australia held a similar process later in the 1990s. Truth and reconciliation commissions can help national communities bear witness to the deplorable human rights abuses associated with the forced removal of Indigenous children to church-run residential schools or to non-Indigenous foster homes, among other acts of racism and colonial violence. They can also mark a strong beginning in a process of healing and redistribution between Indigenous and non-Indigenous peoples. In order to make a lasting impact that will be felt across generations, however, the recommendations of the commission must be diligently followed up with specific actions that improve quality of life and reduce the social distance between Indigenous and non-Indigenous peoples. Urban policy and planning will play a critical role in translating broad goals of reconciliation into the tangible lived experience and production of de-colonized spaces in cities, where most Indigenous and non-Indigenous people live.

Conclusion

It is urgent that cities in settler countries like Australia, New Zealand, and Canada create the conditions for an Indigenous-inclusive citizenship, respectful coexistence in space and place, and the reclamation of Indigenous materiality and memory at the urban scale (Porter, 2013; Matunga, 2013; Walker and Barcham, 2010). Put another way, Indigenous urbanism requires the recognition of Indigenous density (Andersen, 2013) in the cultural, social, economic, and environmental circumstances that lead to the participation in and enjoyment of an urban life (Newhouse and FitzMaurice, 2012). Density requires an appreciation for the separate place from which Indigenous political communities derive their legitimacy and authenticity in what would be an Indigenous-inclusive common citizenship, not an endless search for the ways that Indigenous urbanism differ from the cultural mainstream. Whether Indigenous planning produces outcomes that seem to be the same, different or hybrid approaches traceable to the influence of (and resistance to) settler rationalities are not the key concern. Instead the focus should be on recognizing that the complexity of indigeneity in cities, in all its organic hybridity, is dynamic and ongoing. What is most pressing is the need to stabilize the extent to which our settler institutions enable sufficient discursive and procedural space to operationalize self-determination on the basis of the Indigenous political separateness that runs parallel to settler communities. Producing planning and policy frameworks that create places of intersection for these parallel traditions is one of the most exciting areas of work for Indigenous and non-Indigenous urbanists.

Case studies of the Aboriginal Housing Company in Sydney, Australia, and the Ngāti Whātua Ōrākei Trust in Auckland, New Zealand, have shown the contested nature of indigeneity in the largest cities of these settler countries. Examples from the Canadian context, from Saskatoon, Vancouver, and Ottawa, give an impression of a country moving towards progressive engagement at the urban scale between settler and Indigenous peoples. Progress is slow, however, and when it occurs at all, it is against a backdrop of continuing racism (Lashta, Berdahl, and Walker, 2016) and the devastating intergenerational legacy of residential schools (in Canada and Australia) and other socio-cultural experiments upon the Indigenous peoples using the colonial instruments of planning (e.g., dispossession of land and land title regimes in all three countries). Future research and thoughtful practice is required in this sub-field of urban policy and planning. It is certainly one of the most pressing and interesting challenges facing those who would aim to de-colonize the twenty-first century city.

Key Points

- Most Indigenous peoples in Western settler countries like Australia, New Zealand, and Canada live in urban areas.
- Cities in Western settler states are situated in the traditional territories of Indigenous peoples, where the Indigenous right of self-determination continues to exist.
- Indigenous planning is a tradition parallel to the Western planning system, and it is being reclaimed after having been marginalized for generations by the Western settler state.
- The Aboriginal Housing Company in Sydney, Australia, teaches us about the power of resistance over time to create Indigenous spaces in the city.
- Ngāti Whātua Ōrākei in Auckland, New Zealand, shows us the evolution from colonial dispossession to the reclamation of Indigenous planning in the city.
- Examples from Saskatoon, Vancouver, and Ottawa demonstrate how Indigenous peoples are re-territorializing Canadian cities, sometimes in partnership with municipal administrations.
- As Canada implements actions following from the Truth and Reconciliation Commission, there will be a critical role for urban policy and planning.

Activities and Questions for Review

1. Identify and debate in a small group some of the ways that the planning system can foster a transformative coexistence between Indigenous and non-Indigenous peoples in cities.
2. Drawing on examples from the chapter, discuss with a classmate how Indigenous communities are reclaiming their materiality and memory in Sydney, Australia, and Auckland, New Zealand.
3. Imagine that you are a part of a private consultancy team with five of your classmates; you have been hired by a provincial government to evaluate what Canadian cities are doing to take into account the rising numbers of Aboriginal people in cities, their cultural resurgence, and to de-colonize urban space. What major criticisms would you have and what recommendations would you provide?

Acknowledgments

The empirical work for the case studies in Sydney and Auckland was funded by the Canada Mortgage and Housing Corporation (CMHC) and carried out by Ryan Walker (the first author) and Manuhuia Barcham. We acknowledge the National Aboriginal Housing Association for articulating the need for international comparative research, which prompted CMHC to fund it. A significant portion of the work included in this chapter was funded by an Insight Grant from the Social Sciences and Humanities Research Council of Canada held by the first author. We are grateful to the many people who participated in interviews and conversations over the past decade in Australia, New Zealand, and Canada.

References

Andersen, C. (2013). Urban Aboriginal planning: Towards a transformative statistical praxis. In R. Walker, T. Jojola, and D. Natcher (Eds.), *Reclaiming Indigenous planning*, 260–282. Montreal: McGill-Queen's University Press.

Anderson, K. (1993). Place narratives and the origins of inner Sydney's Aboriginal settlement, 1972–73. *Journal of Historical Geography, 19*: 314–335.

Anderson, K. (2000). "The beast within": Race, humanity, and animality. *Environment and Planning D, 18*: 301–320.

Awatere, S., Harmsworth, G., Rolleston, S., and Pauling, C. (2013). Kaitiakitanga o ngā ngahere pōhatu— "Kaitiakitanga of urban settlements." In R. Walker, T. Jojola, and D. Natcher (Eds.), *Reclaiming Indigenous planning*. Montreal: McGill-Queen's University Press, pp. 236–259.

Belanger, Y. (2014). *Ways of knowing: An introduction to native studies in Canada* (2nd ed.). Toronto: Nelson Education.

Boffa Miskell. (2013). *Ōrākei Papakāinga Masterplan*, for Ngāti Whātua Ōrākei. Auckland, NZ: Boffa Miskell Ltd.

Denis, C. (1997). *We are not you: First Nations and Canadian modernity*. Toronto: Broadview Press.

Department of Aboriginal Affairs (1973). Redfern Housing Project. *Newsletter, 4*: 2–3.

Fawcett, R.B., Walker, R., and Greene, J. (2015). Indigenizing city planning processes in Saskatoon, Canada. *Canadian Journal of Urban Research, 24*(2): 158–175.

Foley, G. (2001). Black power in Redfern 1968–1972. *The Koori history website*. Retrieved from www.kooriweb.org/foley/essays/essay_1.html

Government of New South Wales (2004). *Inquiry into issues relating to Redfern and Waterloo: Final report*. Sydney: Standing Committee on Social Issues.

Henderson, J.Y. (2000). Postcolonial ghost dancing: Diagnosing European colonialism. In M. Battiste (Ed.), *Reclaiming Indigenous voice and vision*. Vancouver: University of British Columbia Press, pp. 57–76.

Henry, F., and Tator, C. (2009). *The colour of democracy: Racism in Canadian society* (4th ed.). Toronto: Nelson.

Holmes, C., Hunt, S., and Piedalue, A. (2014). Violence, colonialism, and space: Towards a decolonizing dialogue. *acme: An International E-Journal for Critical Geographies, 14*(2): 539–570.

Hoskins, R. (2014). Our faces in our places—The development of Māori cultural landscapes in Aotearoa–New Zealand. Presentation at Coming to a Common Place: Indigenous Peoples and Urban Design, University of Manitoba, Winnipeg, November 18.

Howitt, R., and Lunkapis, G. (2010). Coexistence: Planning and the challenge of Indigenous rights. In J. Hillier and P. Healey (Eds.), *The Ashgate research companion to planning theory: Conceptual challenges for spatial planning*. Farnham, UK: Ashgate, pp. 109–133.

Indigenous and Northern Affairs Canada (2014). *Urban Aboriginal peoples*. Ottawa: Minister of Indigenous and Northern Affairs Canada. Retrieved from www.aadnc-aandc.gc.ca/eng/1100100014265/1369225120949

Jojola, T. 2013. Indigenous planning: Towards a seven generations model. In R. Walker, T. Jojola, and D. Natcher (Eds.), *Reclaiming Indigenous planning*. Montreal: McGill-Queen's University Press, pp. 457–472.

Kukutai, T. (2013). The structure of urban Māori identities. In E. Peters and C. Andersen (Eds.), *Indigenous in the city: Contemporary identities and cultural innovation*. Vancouver: University of British Columbia Press, pp. 311–333.

Lane, M.B., and Hibbard, M. (2005). Doing it for themselves: Transformative planning by Indigenous peoples. *Journal of Planning Education and Research, 25*(2): 172–184.

Lashta, E., Berdahl, L., and Walker, R. (2016). Interpersonal contact and attitudes towards Indigenous peoples in Canada's prairie cities. *Ethnic and Racial Studies, 39*(7): 1242–1260.

Listerborn, C. (2007). Who speaks? And who listens? The relationship between planners and women's participation in local planning in a multi-cultural urban environment. *GeoJournal, 70*: 61–74.

Maaka, R., and Fleras, A. (2005). *The politics of indigeneity: Challenging the state in Canada and Aotearoa New Zealand*. Dunedin, NZ: University of Otago Press.

Matunga, H. (2013). Theorizing Indigenous planning. In R. Walker, T. Jojola, and D. Natcher (Eds.), *Reclaiming Indigenous planning*. Montreal: McGill-Queen's University Press, pp. 3–32.

Miller, J. (2011). The Papaschase band: Building awareness and community in the city of Edmonton. In H. Howard and C. Proulx (Eds.), *Aboriginal peoples in Canadian cities: Transformations and continuities*. Waterloo, ON: Wilfrid Laurier Press, pp. 53–68.

Mitin, D. (2013). A class act: Professional support to people's organizations in towns and cities of the global South. *Environment & Urbanization, 25*(2): 483–499.

Morgan, G. (2006). *Unsettled places: Aboriginal people and urbanisation in New South Wales*. Kent Town, AU: Wakefield Press.

Nakamura, N. (2015). Being Indigenous in a non-Indigenous environment: Identity politics of the Dogai Ainu and new Indigenous policies of Japan. *Environment and Planning A, 47*: 660–675.

Newhouse, D., FitzMaurice, K. (2012). Introduction. In D. Newhouse, K. FitzMaurice, T. McGuire-Adams, and D. Jette (Eds.), *Well-being in the urban Aboriginal community: Fostering Biimaadiziwin, a national research conference on urban Aboriginal peoples*. Toronto: Thompson Educational Publishing.

New Zealand Herald (2012). Arduous treaty claim to end today, November 15. Retrieved from www.nzherald.co.nz/nz/news/article.cfm?c_id=1&objectid=10847563

Ngāti Whātua Ōrākei Trust (2012). *Ngāti Whātua Ōrākei Iwi Management Plan*. Auckland: Ngāti Whātua Ōrākei Trust.

Nyseth, T., and Pedersen, P. (2014). Urban Sámi identities in Scandinavia: Hybridities, ambivalences and cultural innovation. *Acta Borealia, 31*(2): 131–151.

Peters, E. (2015). Aboriginal people in Canadian cities. In P. Filion, M. Moos, T. Vinodrai, and R. Walker (Eds.), *Canadian cities in transition: Perspectives for an urban age* (5th ed.). Toronto: Oxford University Press.

Pitts, A. (2001). *Community social plan: Pemulwuy reconstruction project*. Sydney, AU: Aboriginal Housing Company.

Pollock, Z. (2008). National black theatre. *Dictionary of Sydney*. Retrieved from www.dictionaryofsydney.org/entry/national_black_theatre

Porter, L. (2010). *Unlearning the colonial cultures of planning*. Burlington, VT: Ashgate.

Porter, L. (2013). Coexistence in cities: The challenge of Indigenous urban planning in the twenty-first century. In R. Walker, T. Jojola, and D. Natcher (Eds.), *Reclaiming Indigenous planning*. Montreal: McGill-Queen's University Press, pp. 283–310.

Shaw, W. (2007). *Cities of whiteness*. Malden, MA: Blackwell.

Sully, L., Kellett, L., Garcea, J., and Walker, R. (2008). First Nations urban reserves in Saskatoon: Partnerships for positive development. *Plan Canada, 48*(1): 39–42.

Sydney Morning Herald (2015). Tent embassy prepares to leave the Block as Indigenous housing "guaranteed," August 27. Retrieved from www.smh.com.au/nsw/tent-embassy-prepares-to-leave-the-block-as-indigenous-housing-guaranteed-20150826-gj87yd.html

Taylor, J. (2013). Indigenous urbanization in Australia: Patterns and processes of ethnogenesis. In E. Peters and C. Andersen (Eds.), *Indigenous in the city: Contemporary identities and cultural innovation*. Vancouver: University of British Columbia Press, pp. 311–333.

Turnbull, G. (2005). Actions speak louder than words: Redfern-Waterloo's recent experience with consultation. *Indigenous Law Bulletin*, 6: 21–24.

United Nations. (2007). *Declaration of the Rights of Indigenous Peoples*. United Nations General Assembly. New York: United Nations.

Wain, G. (1976). Redfern Aboriginal Housing Project. *Polis*, 3: 3–6.

Walker, R., and Barcham, M. (2010). Indigenous-inclusive citizenship: The city and social housing in Canada, New Zealand and Australia. *Environment and Planning A*, *42*: 314–331.

Walker, R., and Belanger, Y. (2013). Aboriginality and planning in Canada's large Prairie cities. In R. Walker, T. Jojola, and D. Natcher (Eds.), *Reclaiming Indigenous planning*. Montreal: McGill-Queen's University Press, 193–216.

Welsh, R. (2005). The future of Redfern should be based on need not greed. Address to ANTAR Forum, Redfern Town Hall, June.

Wood, P.K. (2003). A road runs through it: Aboriginal citizenship at the edge of urban development. *Citizenship Studies*, 7: 463–479.

10 Urban Policy and Planning for Climate Change

Daniel Aldana Cohen

Introduction

The planet is urbanizing and warming at once (Brenner, 2013; Steffen et al., 2009). The former is unstoppable. The latter is a danger to the survival of human civilization as we know it. Even if catastrophic **climate change** is forestalled, some extreme weather is unavoidable and will continue to pose grave threats to city dwellers worldwide. There is also consensus among a wide range of scholars and institutes that if cities are reshaped and transformed in the right ways, they can reduce the **greenhouse gas** (GHG) emissions that ultimately cause climate change and extreme weather; since cities are where people live, they are also crucial sites for defending people against the impacts of climate change (Bulkeley and Betsill, 2013; Glaeser, 2011; Gore and Robinson, 2009; Rode and Burdett, 2011; Ross, 2011). The question is, what is the "right way" to plan for climate change? One strand of debate on urban **climate policy mitigation** focuses on the technical challenges that cities face: their contribution to GHG emissions and their vulnerability to extreme weather. A second strand interprets these in more socially nuanced terms: the relationship between wealth and trade patterns and GHG emissions and the different ways that different communities are vulnerable to extreme weather. A third strand combines the first two in order to assess which kinds of climate policy are most politically feasible and, ultimately, socially just (Romero Lankao, 2007, 2012).

The stakes are high. Urban life is responsible for approximately half of the world carbon dioxide emissions (Seto et al., 2014). Meanwhile, it is estimated that if just 724 of the world's largest cities conducted more pro-density planning, they could collectively reduce GHG emissions, between a 2012 baseline and 2030, by 14.4 gigatonnes of carbon—that is over half of the carbon stored in currently recoverable oil in Alberta's tar sands. Of course, not all cities are alike. Richer cities are responsible for more carbon than poor cities, and large cities for more carbon than small ones. Richer cities are better able to withstand the impacts of extreme weather. But as explored below, there are also important within-city differences that some carbon accounting methods spotlight, in particular the importance of high incomes and consumption-rich lifestyles as drivers of high carbon emissions. And even within wealthy cities, large numbers of vulnerable people may by intensely exposed to environmental harms. While the relationships that cities have to climate change differ enormously according to many factors, questions of politics, equity, and responsibility are ubiquitous.

Thus, in this chapter, urban policy and planning for climate change are explored in a way that reflects the three strands listed above, but the third—the question of political feasibility—is viewed as central for ultimately assessing how to move forward. First, the core issues involving cities and climate are reviewed: how do urban activities cause climate change, directly and indirectly, through the emissions of GHGs, which trap heat in the atmosphere? Second, a variety of threats that climate change poses to cities, and their social implications, are surveyed, followed by a review of the changing nature of urban climate politics. Two global case studies that speak to the social and political dimensions of urban climate policy are explored: São Paulo's efforts to reduce its emissions and New York City's experience of Hurricane

Sandy in 2012. The São Paulo study exemplifies the difficulty of implementing low-carbon policy even when there is considerable political will to do so at the very top of the governance structure. The New York City study illustrates that even in the wealthiest, most powerful cities, there remain huge disparities in vulnerability to climate change, which translates into very different ideas about what must be done. The challenges facing Canadian cities are discussed in terms of two challenges: reducing emissions while addressing equity concerns and preparing for increased heat as the planet warms. The chapter concludes by reinforcing the idea that it is impossible to reduce an urban region's carbon footprint, or to adequately protect residents from extreme weather, without the formulation and implementation of climate policies that substantively address social and economic equity.

Cities and Climate Change

Any discussion of urban climate policy requires a solid grasp of two sets of dynamics. First, what role do urban activities play in causing the GHG emissions that trap heat in the atmosphere (the most important GHGs being carbon dioxide, methane, and nitrous oxide)? Second, what are the impacts of climate change on cities and their residents? Once these are understood, a better assessment can be made of the evolution of urban climate politics in the late twentieth and early twenty-first centuries.

Cities' Contribution to Climate Change: Direct and Indirect Emissions

There are two leading ways of understanding how activities in cities relate to the GHG emissions that cause climate change (Stockholm Environment Institute–U.S., 2012; Yetano Roche et al., 2013). The most common is **production-based carbon accounting**; this is how cities around the world now count and report their emissions. The idea is to draw a line around an area (e.g., a city's jurisdiction) and count all the carbon emitted by activities in that area, like driving a car or burning heating oil (these are also termed *scope 1 emissions*). Typically, one then adds to that total the GHG emissions associated with the power plants that power the city, and those resulting from the eventual disposal of the city's solid waste (or *scope 2 emissions*). Airports have usually been excluded from this count. One reason is that getting precise data from airport managers has been challenging; a second is that since airports would drive up cities' emission counts, there would be little enthusiasm for forcing the issue or for making estimates based on available data.

The great advantage of this accounting strategy is that it focuses policy-makers' attention on things that city governments can in many cases directly shape—like transit, land use, and building codes—in order to cut emissions: a policy field called "**mitigation**" (Bulkeley, 2013). The main analytic upshot of this method of counting, for scholars and policy-makers, has been the idea that to reduce cities' (and suburbs') GHG emissions, those cities should be densified and public transit, cycling, and walking should be encouraged over car use (Glaeser and Kahn, 2010; Hillman, 1996; Owen, 2009). **Densification** is helpful because it can reduce travel distances, make non-car travel more practical, and, by concentrating more people into fewer, larger buildings (in which to live and work), those buildings are liable to use energy more efficiently. For example, if one lives in an apartment, the heat that escapes from one unit can help to heat the one next door.

The problem with production accounting is that cities are not snow globes; and in post-industrial cities, especially those in countries in the global North, many of the key, greenhouse-gas-intensive activities that make everyday life possible occur elsewhere, from the food that city dwellers eat to the clothes and computers that we buy and use (Heinonen et al., 2013; Peters et al., 2011; Seto et al., 2014). Moreover, people with different levels of wealth, living in areas of similar density, may be indirectly responsible for emitting greatly varying amounts of GHG (Heinonen, Kyrö, and Junnila, 2011). Another issue is that with the globalization of production and the growing international division of labour, it is problematic to delineate social and ecological impacts only within the borders of any particular city.

An alternative, **consumption-based carbon accounting** method captures urban consumers' total GHG footprint (individuals, government, and businesses). It works by calculating the full life cycle of GHG emissions for each good and service—like the energy required to produce, transport, and power a computer—and assign that full footprint to the final consumer. The footprint concept builds on the notion of an **ecological footprint**, an older metric that measures the amount of land required to provide the resources and the waste absorption space to sustain a particular person or area's consumption. The consumption method, especially when calculated in the more narrow and precise terms of carbon, closely approximates the actually existing relationship between the things that people do in cities and the globalization of production and consumption that underlie how and what people do in cities. As a result, consumption-based carbon accounting provides a clearer picture of how socio-economic inequalities are expressed in terms of the overall environmental impact of various incomes and lifestyles.

Indeed, these methods paint a different picture of dense, post-industrial cities whose consumption-based emissions profiles are much higher than those produced with the production method. Consumption counts of Seattle, San Francisco, and London have found *per capita* emissions two to three times higher than the production counts find (British Standards Institution, 2014; Stockholm Environment Institute–U.S., 2012). Other studies, of European and American cities, have found that the benefits of density can be almost cancelled out by individual consumption when lifestyle and class are controlled for (Heinonen et al., 2013; Minx et al., 2013; Ummel, 2014). And there is evidence, based on a slightly different model, that the densest US cities tend to be surrounded by the most energy-inefficient, sprawling, GHG-intensive suburbs (Jones and Kammen, 2014).

Consumption studies are crucial because prosperous cities, like prosperous countries, have in many ways "off-shored" GHG-intensive activities, from farming to manufacturing, to other countries. How can urban policies address this? At one level, consumption studies show that transportation is much less GHG-intensive than making or using a good or service; a study of Seattle found that for everything the city consumed, transportation accounted, on average, for just 10 per cent of the GHG footprint (Stockholm Environment Institute–U.S., 2012). Buying local is usually not a meaningful solution. Critical sustainability scholars are increasingly arguing that the priority should be to organize cities in a way that de-emphasizes excessive material consumption (of items like extra T-shirts and red meat), while facilitating care work and low-carbon leisure. For instance, cultural consumption tends to be less carbon-intensive than the consumption of many goods. On average, $10 spent on a movie ticket is associated with less than half as many carbon emissions as $10 spent on clothing (Berners-Lee, 2011; Stockholm Environment Institute–U.S., 2012). As the urbanist Mike Davis (2010: 43) writes, "the cornerstone of the low-carbon city, far more than any green design or technology, is the priority given to public affluence over private wealth" (see also Cohen, 2014; Jackson, 2009; Schor, 2011). In other words, when one thinks of public basketball courts, public squares, public cultural institutions, public parks, or public beaches all buzzing with activity, these are key features of an urban world whose built environment and social relations are oriented towards collective social interactions in shared spaces; the affluence of the community is invested in people being brought together rather than in isolated experiences of wasteful consumption. Davis's key insight here is that thinking about cities and climate change together prompts us to imagine the imperative of social transformations that are more wide-ranging than simply substituting solar panels for coal-fired power plants. The whole culture of urban living is at stake in addressing climate change.

Cities' Vulnerability to Climate Change: Water and Heat

But, of course, cities are not just sites for reducing pollution; they are also where people live. Moreover, no matter how quickly emissions are cut, climate change that is already guaranteed will have profound and lasting impacts. Some carbon dioxide molecules will stay in the atmosphere for hundreds of years. Others will linger for millennia, much longer, for example, than Stonehenge (the prehistoric remains of a ring of standing stones built some 5000 years ago in England) has existed (Archer, 2011). Even if we rapidly reduce GHG emissions and avoid runaway climate change, extreme weather will increase for decades to come, and sea levels will rise for even longer. It is estimated that the *already inevitable* global average sea-level rise, in 2000 years, will be 4.8 metres (16 feet)—well over Hurricane Sandy's 4.2-metre (14-foot) peak storm surge in Manhattan's Battery Park (Strauss, 2013). (That storm caused nearly USD$70 billion in damages.) It is true that not all increases in dangerous weather are inevitable. For instance, in the coming decades, under a low GHG emissions scenario, only 555 US cities located on the coast would be threatened by sea-level rise; higher emissions would threaten 900 more (Strauss, 2013). But cities must nonetheless prepare for extensive changes to their environment, a field of climate policy called "**adaptation**."

The expectation of ever-rising sea levels has prompted a global debate about "**resilience**," a concept very often used with reference to cities and that designates places' ability to bounce back from disasters (Pizzo, 2015; Rosenzweig and Solecki, 2014). Some view resilience as a vague term that at its worst celebrates the ruggedness of the poor (when we should really be improving their situation) and that ultimately comes from counter-insurgency military thinking. Others see it as a useful concept for thinking through ways that cities can "climate-proof" themselves from sea level. For instance, resilience advocates tend to criticize the idea of building sea walls and other types of fortification, suggesting instead that cities should use soft barriers and "green infrastructure," like regenerated oyster beds and restored wetlands, to slow storm surges and absorb heavy rainfall. The idea is to let the water in in a controlled fashion, whose capacity can be developed over time, rather than build rigid barriers that will eventually become obsolete. Although "resilience" remains a controversial term, there appears to be a growing consensus around the need for more flexible approaches to rising seas. The American foundation the Rockefeller Institute has funded a "100 Resilient Cities" initiative to spread this perspective around the world. Dutch engineers and water managers have developed decades of expertise in managing water and have travelled the world to support this perspective (Funk, 2014), although some political ecologists think that even the Dutch are too old-fashioned, and urge us to be even bolder in embracing perpetual flooding (Mathur and da Cunha, 2009). The flipside is a tentative and controversial conversation about "managed retreat" in the face of climate change. Some communities will simply have to up and move—ideally of their own volition (Koslov, 2014).

This raises a second issue faced by cities: rainfall. The debate about flash floods increasing owing to increasingly heavy storms is similar to the debate about rising seas. Let us look instead at the problem of drought. Climate change will likely compound an already growing tendency towards water scarcity around the world (Schewe et al., 2014). How will cities cope with increasingly scarce water? Cities that can afford to do so have been exploring water recycling—known pejoratively as toilet-to-tap systems—and desalination plants (expensive factories that would convert ocean water into drinkable water) (Fishman, 2011). Cities like Las Vegas and Perth have also imposed strict limitations on outdoor water use. But rationing *indoor* water use remains controversial in rich and poor cities, as residents who have achieved a 24-hour service are extremely unwilling to give it up (see also Cohen, 2016).

Moreover, in the cities of the global South, where many people lack access to regular drinking water and, even more, to basic sanitation services, city governments have had to struggle to develop and implement fair and effective water governance, to prepare to further curb wasteful consumption by the affluent, to treat all sewage and industrial effluent, and to use local water more efficiently (Bakker, 2010). A crucial challenge is that water politics inevitably implicate a wide range of socio-ecological struggles, including, most profoundly, those related to housing and land use (Rademacher, 2009). In cities with splintered infrastructures (Graham and Marvin, 2001) where, for instance, water and sanitation services only reach certain (generally affluent) urban spaces, and where the governance structures that would ensure universal service are lacking or fractured, it is challenging to isolate and address one dimension of inequality (and hence vulnerability to climate change) at a time.

Perhaps the most subtle and deadly everyday threat posed by climate change is heat waves (Stone Jr, 2012). Heat waves have killed more Americans than any other natural disaster in the past 100 years (Klinenberg, 2002). Because of the **urban heat island** effect, where concrete and asphalt absorb and radiate heat, in-city temperatures are increasing more than twice as fast as average global surface temperatures. These are liable to be especially devastating in the generally hotter cities of the global South, especially large cities with little greenery and extensive slums whose buildings lack proper ventilation. Moreover, in many cases, a combination of poverty and spotty electricity grids means that air conditioning is less widely available. Measures like painting roofs white, planting trees and gardens, building green roofs, and building green walls can all reduce the rising heat. But these solutions are expensive and require maintenance; even well-funded efforts to plant trees, like the Million Tree Challenge in New York City, have struggled to reach their goal of engaging volunteers in thinking about climate change and maintaining trees once planted (Fisher, Svendsen, and Connolly, 2015). The challenges of even launching such a wide-ranging program are of course even greater in cities with much tighter budgets. Meanwhile, scholars have found that in North America, trees tend to be clustered in affluent neighbourhoods—yet another case of the urban poor being disproportionately vulnerable to climate change (Landry and Chakraborty, 2009).

Urban Climate Politics: From the C40 to the People's Climate March

Since the Rio Earth Conference of 1992, there has been growing attention to urban climate policy via policy networks like the UN-associated International Council for Local Environmental Initiatives (ICLEI), now renamed Local Governments for Sustainability. In 2005, the then London mayor Ken Livingstone founded what would become the C40 Cities Climate Leadership Group, a growing network of large and usually prosperous cities that revitalized the urban climate policy discussion thanks to the C40 members' wealth and disproportionate exposure to media and academic study (Bulkeley and Betsill, 2013).

In order to capture the way that urban climate policy has involved both inter-city policy networks, global policy think tanks, and city-based policy-makers' relationship to other levels of government in their own country (mainly regional and global), Harriet Bulkeley and Michelle Betsill (2003) have developed an enormously influential **multi-level governance climate model** to explain and study urban climate policy-making. The model is designed precisely to help one understand how policy-makers think about and act on climate change. And scholars using the model have focused on the capacity of networks like ICLEI and the C40 to spread a common set of measurement tools and policy visions (like densification) around the world.

But if 10 years ago cities seemed to be on the cusp of leapfrogging national governments in climate policy-making thanks to "win-win" pro-density policies that would improve cities' quality of life, economic dynamism, and GHG emissions (e.g., Kahn, 2006), today's scholars confront a troubling gap between rhetoric and action (Bulkeley, 2011; Ryan, 2015). Many cities have had some successes. London's **congestion charge**, for example, which levies a fee on cars and trucks that enter the downtown core, reduced emissions by 16 per cent in the affected area while raising well over £100 million (CAD$200 million) per year in increased funding for public transit (C40 Cities, 2011; Richards, 2005). Yet it is difficult to compare cities' progress with a great deal of rigour. GHG accounting and reporting methods differ still, although there are now efforts by groups like the World Resources Institute to develop common standards. Overall, Bulkeley (2011) has written that cities' climate policy progress is "depressingly similar" to that of countries, as their efforts to implement far-reaching emission reductions have stagnated just as severely, despite their early promise.

Some scholars have argued that while prosperous cities in the global North must find a way to cut their GHG emissions faster, cities in the global South should focus on climate policies that improve the lives of the poor by reducing their vulnerability to ecological crises, typically through adaptation measures (Romero Lankao, 2007). Still, pro-poor policies can achieve mitigation. For example, increasing public transit and reducing car travel, or increasing the efficiency of car engines, will reduce high levels of **air pollution** and therefore reduce lung disease while increasing the mobility options of the poor (Romero Lankao, 2007).

With both mitigation and adaptation, however, it has been a persistent challenge to translate often long-term and abstract climate policy agendas into the kind of viscerally urgent agenda that ordinary people, journalists, non-environmentalist social movements, and politicians can rally behind. There is only mixed evidence that extreme weather increases people's support for climate policies (Brulle, Carmichael, and Jenkins, 2012). The politics of densification are fraught because a global wave of gentrification has unleashed widespread resistance to urban redevelopment policies; many feel that "green" urban policies are just an excuse to raise property values and cater to the lifestyle aspirations of wealthy consumers and young professionals (Checker, 2011; Cohen, forthcoming; Curran and Hamilton, 2012). Efforts to relocate poor people living in vulnerable areas in cities in the global South have stumbled because communities have resisted displacement, which they view as only superficially concerned with the environment and instead mainly an effort to create cities for and of an affluent middle class (Chatterjee, 2003; Rademacher, 2009).

The irony is that because poorer people consume less and typically advocate a model of the city with more public transit and denser affordable housing with jobs and services nearby, they are in a sense hidden protagonists for low-carbon urbanism (Cohen, forthcoming). Yet despite this potential, and often because of other battles over redevelopment and land use in sensitive areas, environmentalists have typically ignored the potential of poor people's social movements to make a significant, positive contribution to urban climate politics.

There are reasons, however, to believe that these divisions may be overcome. The People's Climate March in New York on September 21, 2014, co-organized by climate activists and community-based environmental justice organizations, saw approximately 400,000 demonstrators march through Manhattan. The march was led by groups of poor, racialized "front-line" communities—namely, groups hailing from parts of the United States (and elsewhere) that are most vulnerable to climate change (in large part, communities located along coastlines or in urban areas with little tree cover, from Staten Island's north shore to Englewood in Chicago). The growing partnership between such community-based groups and many traditional environmentalist organizations has the potential to transform urban climate politics, building a grassroots movement that is distinct and autonomous from the "multi-level governance" networks of government policymakers and global policy think tanks. This new movement frames its objectives in terms of **climate justice**.

According to this framing, those whose wealth (contemporary and historic) has done the most to cause climate change also tend to be those least vulnerable to climate change's impacts. Achieving climate justice would mean that the prosperous contribute the most both to slashing GHG emissions and to helping vulnerable communities adapt to extreme weather. What this means in practice, however, is unclear, especially in large cities. The politics of climate change initially developed in diplomatic circles, with large consensus around the idea that one should distinguish countries in the global North from those in the global South (Ciplet, Roberts, and Khan, 2015). But not only are those clean divisions fracturing (for instance, because of China's vertiginous economic growth and increased emissions), but within cities of both the global North and the global South, stark social, economic, and ethno-racial divisions also have an impact on local efforts to address climate change. Two case studies illustrate these complexities, one focused on emission-reduction efforts in São Paulo, the other on the response to extreme weather in New York City.

São Paulo's Efforts to Cut Emissions

The municipality of São Paulo is Brazil's largest city, with a population of 11 million, nested in a greater metropolitan area of over 20 million inhabitants (Figure 10.1). The urban region is South America's economic capital, home to thriving financial and service sectors and still tightly connected to a regional geography of high-value manufacturing. Throughout the twentieth century, the city has grown explosively, and from mid-century onward city leaders encouraged a growth pattern based on low-density residential sprawl and car transportation. As a result, the city's roads have grown extraordinarily congested, with commuters facing some of the world's longest travel times. A Crédit Suisse report found that the city's average commute of 43 minutes was the second longest in a list of the world's major cities, behind only Shanghai (Bevins, 2015). An additional result is the breathing of highly polluted air by city residents. Meanwhile, there is an increasing shortage of affordable housing, with one-third of the municipality's residents housed in informal or substandard settlements, and there is an acute lack of housing options in the city's central areas where the highest proportion of jobs is found, many in the low-paying and often informal sector (see Cohen, forthcoming). Since the turn of the millennium, easing congestion and improving affordable housing access have been some of the population of São Paulo's top demands.

São Paulo's efforts to reduce its GHG emissions reflect both global trends and these local challenges (Cohen, forthcoming). Under the centre-left Workers' Party mayor Marta Suplicy, from 2001 to 2004, the city government persuaded ICLEI–Latin America to relocate to São Paulo. The city commissioned a high-quality audit of its GHG emissions and sent a technical team to London to learn from efforts being made there. Suplicy's government also made path-breaking improvements to the city's bus system to ease the commuting life of the poor (who lived in the periphery but worked in more central areas) by adding dozens of kilometres of dedicated bus lanes. But no one in government discussed the low-carbon benefits of this bus policy. In 2004 the centre-right won the mayoral elections, and in 2006 another centre-right politician took office, Gilberto Kassab. Climate policy was stagnating.

Then Kassab visited New York City in 2007 for a C40 summit. There, New York City officials insisted that pro-density low-carbon policy could attract attention and investment, spur real estate development, and improve quality of life (these claims were perhaps a bit exaggerated). Kassab immediately instructed his green secretariat to draft a climate law along these lines. In 2009, with Kassab's support, São Paulo passed a climate law with unanimous

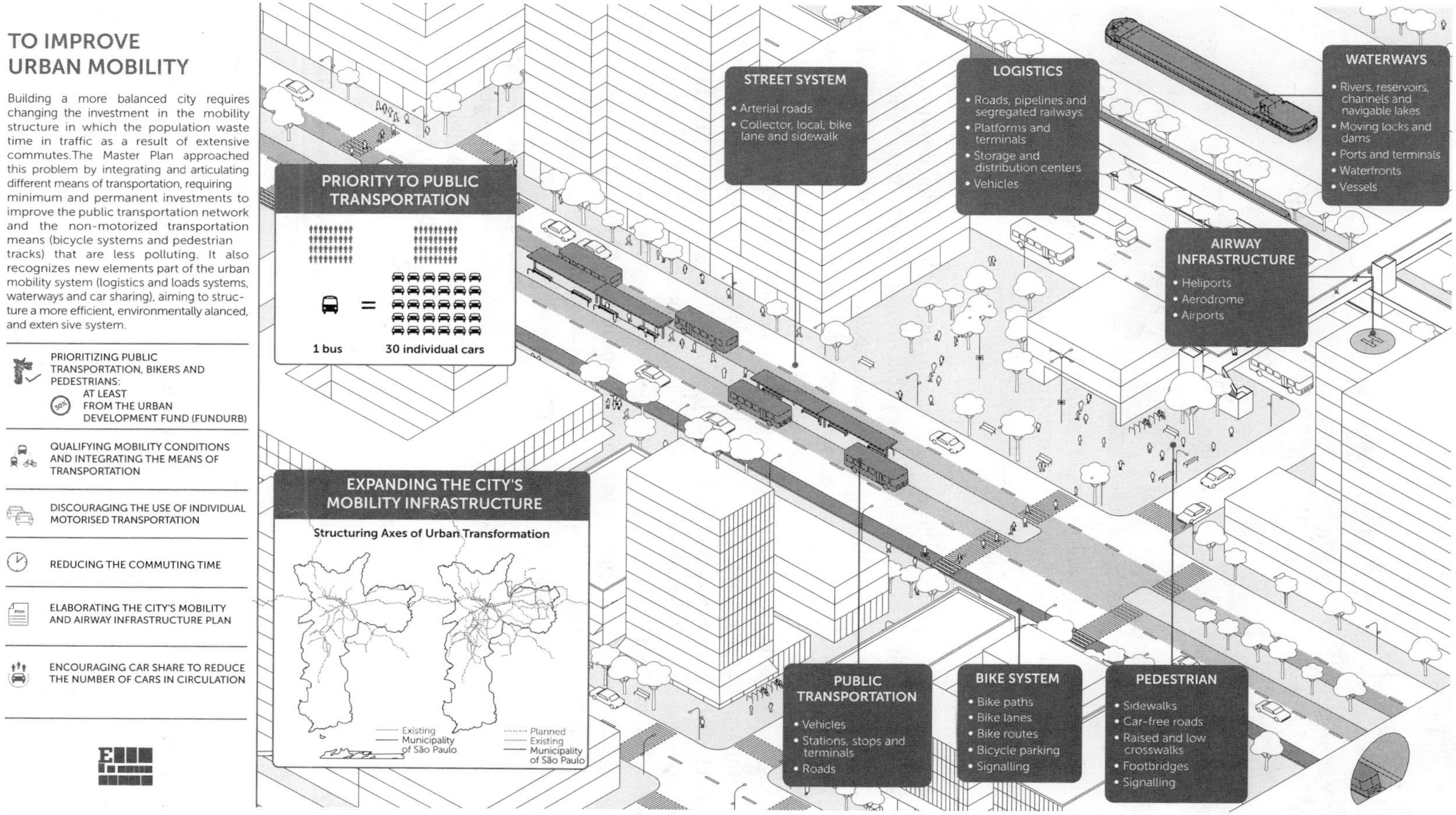
TO IMPROVE URBAN MOBILITY
Building a more balanced city requires changing the investment in the mobility structure in which the population waste time in traffic as a result of extensive commutes.The Master Plan approached this problem by integrating and articulating different means of transportation, requiring minimum and permanent investments to improve the public transportation network and the non-motorized transportation means (bicycle systems and pedestrian tracks) that are less polluting. It also recognizes new elements part of the urban mobility system (logistics and loads systems, waterways and car sharing), aiming to structure a more efficient, environmentally alanced, and exten sive system.
PRIORITIZING PUBLIC TRANSPORTATION, BIKERS AND PEDESTRIANS: AT LEAST 30% FROM THE URBAN DEVELOPMENT FUND (FUNDURB)
QUALIFYING MOBILITY CONDITIONS AND INTEGRATING THE MEANS OF TRANSPORTATION
DISCOURAGING THE USE OF INDIVIDUAL MOTORISED TRANSPORTATION
REDUCING THE COMMUTING TIME
ELABORATING THE CITY'S MOBILITY AND AIRWAY INFRASTRUCTURE PLAN
ENCOURAGING CAR SHARE TO REDUCE THE NUMBER OF CARS IN CIRCULATION
PRIORITY TO PUBLIC TRANSPORTATION
1 bus = 30 individual cars
EXPANDING THE CITY'S MOBILITY INFRASTRUCTURE
Structuring Axes of Urban Transformation
Existing
Municipality of São Paulo
Planned
Existing
Municipality of São Paulo
STREET SYSTEM
• Arterial roads
• Collector, local, bike lane and sidewalk
LOGISTICS
• Roads, pipelines and segregated railways
• Platforms and terminals
• Storage and distribution centers
• Vehicles
WATERWAYS
• Rivers, reservoirs, channels and navigable lakes
• Moving locks and dams
• Ports and terminals
• Waterfronts
• Vessels
AIRWAY INFRASTRUCTURE
• Heliports
• Aerodrome
• Airports
PUBLIC TRANSPORTATION
• Vehicles
• Stations, stops and terminals
• Roads
BIKE SYSTEM
• Bike paths
• Bike lanes
• Bike routes
• Bicycle parking
• Signalling
PEDESTRIAN
• Sidewalks
• Car-free roads
• Raised and low crosswalks
• Footbridges
• Signalling

Figure 10.1 Infographic of planned changes to São Paulo's transportation system, Brazil

The English-language section of São Paulo's Secretariat of Municipal Development, on its website Urban Management [Gestão Urbana], http://gestaourbana.prefeitura.sp.gov.br/master-plan/

Figure 10.2 Housing movements marching in central São Paulo to demand more affordable housing in the city core

Figure 10.3 Central São Paulo, as seen from an apartment building, with City Hall the black building on the left

support in city council, pledging to cut emissions by 30 per cent by 2012. International institutes praised São Paulo's leadership, and in 2011 it hosted the biannual C40 summit. But the city's emissions did not go down.

Two aspects of this story are telling. First, three-quarters of the city's emissions were due to automobile traffic. But instead of continuing the prior mayor's dedicated bus lane efforts, which slashed congestion, Kassab's administration focused overwhelmingly on recreational bike lanes in an effort to appeal to new middle-class cultural preferences. Second, the city decided to densify and revitalize the downtown. But the project it pursued, called Nova Luz, would have expelled thousands of low-income residents and small shopkeepers from the designated area. These people formed a coalition and resisted fiercely. In fact, housing movements in the area had long been occupying buildings that stood empty, abandoned, yet were surrounded by subway stops and quality public services. Such movements demanded more affordable housing in the city core, where many poor people worked (Figure 10.2). The realization of this vision would likely reduce GHG emissions, but the city's environmental policy-makers ignored it. Instead, they saw the housing movements as short-sighted obstacles to their own modernizing vision (Figure 10.3). Ultimately, the housing movements and small shopkeepers prevailed and Nova Luz was cancelled.

The next mayor, in office from 2013 to 2016, again represented the centre-left Workers' Party. He hardly ever spoke about climate change, but through a new master plan, he passed policies that would see an increase in affordable housing downtown (and alongside new public-transit corridors). He built hundreds of kilometres of dedicated bus and bicycle lanes. Policy-makers expect these measures, which are rarely framed in climate policy terms, to nonetheless reduce emissions significantly. The new planning priorities of the Workers' Party, if implemented in the coming years, will benefit the climate while improving quality of life, and they will do so by prioritizing "public affluence" (Davis, 2010). The São Paulo case suggests that there is a real difference between the intent and the effectiveness of urban policies with a climate dimension, and that difference depends on how central the needs of the poor and the working class are to the policies in question. The different ways that social classes relate to climate politics is just as relevant when it comes to safeguarding city residents from the kinds of extreme weather that climate change will increasingly unleash in urban regions.

New York City after Superstorm Sandy

On October 29, 2012, Hurricane Sandy slammed into the northeastern seaboard of the United States, flooded New York City's subway system and 90,000 of its homes and businesses, and killed 43 people (Figure 10.4). Two million others temporarily lost electricity, many for days. It is never possible to attribute any extreme weather event to climate change alone, but Sandy's damage was clearly compounded by climate-linked rising sea levels

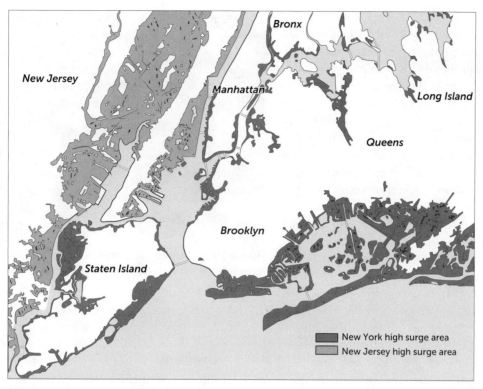

Figure 10.4 Map of flooded areas in New York, USA
US Center for Disease Control and Prevention.

and likely reflects the stronger storms that scientists expect to result from warmer air, which holds more moisture (Sobel, 2014). In the wake of the hurricane, the magazine *Bloomberg Businessweek* ran a cover headline synthesizing the mainstream media attitude: "It's global warming, stupid" (Barrett, 2012). The question many have asked since is how extreme weather linked to climate change affects the climate politics of a place. In New York City, a group of researchers formed the Superstorm Research Lab to find out, interviewing 81 New Yorkers affected by the storm. The interviewees were divided into four groups: senior government officials involved in disaster management and recovery; leaders of non-governmental organizations (NGOs) and community groups involved in disaster recovery; volunteer responders, most involved with the spontaneous activist network of 60,000 people, Occupy Sandy; and people directly affected by the storm. The Lab found that, following the storm, two very distinct narratives emerged, a development that the Lab called "A Tale of Two Sandys" (Cohen and Liboiron, 2014; Superstorm Research Lab, 2013). The first Sandy was the story told by government elites and some of the largest and most influential NGOs, like the Red Cross. According to this narrative, the storm represented a vicious interruption of the status quo, and the top priority was to repair physical things, from bridges to houses. In this story, the storm was an acute, short-term emergency. Long-term recovery planning, which the city also engaged in, was considered to be needed, and it focused on physical infrastructures—albeit here, too, many gaps have remained. Major rebuilding projects that were allocated hundreds of millions of dollars in federal aid are still in the design phase, and the municipal program to rebuild over 20,000 homes languished. In mid-2014, over one and a half years after the storm, not one of over 20,000 homeowners who "applied for help rebuilding their homes in the city had seen work begin" (Buettner and Chen, 2014).

The second Sandy was the story of community groups and grassroots activists. For them, Sandy exacerbated a long-term, ongoing social and economic crisis. Recovery was needed, but there was no sense in returning to the status quo. For example, a survey of public housing in affected areas of the city found that after the storm 45 per cent of housing units were damaged by mould.

But *before* the storm, 35 per cent of units were *already* damaged by mould (Alliance for a Just Rebuilding et al., 2014: 2). Moreover, neighbourhoods that were flooded were more black, poor, and elderly than other New York City neighbourhoods, and public-transit users across the city were intensely inconvenienced by widespread subway closures (Faber, 2015). According to many academics and community groups who were badly affected by the storm, especially those who were already engaged, or became engaged, in the environmental justice movement, poor and racialized communities should always expect to suffer the most from extreme weather (see also Kersten et al., 2012). Plus, since displaced tenants are not systematically tracked, in a city where nearly three-quarters of residents are tenants, there are no reliable numbers on exactly how many people were displaced, for how long, or what their precise demographic characteristics were. Because such communities are disproportionately vulnerable, they are called "front-line" communities. And in formulating climate policies to safeguard people from the extreme weather that it is too late to prevent, this narrative argues that front-line communities should be granted a seat at the planning table. They should, in fact, provide leadership since they are experiencing the effects of climate change more severely than anyone else (see Figure 10.5). And climate policies should include long-term efforts to enhance everyone's economic and social security, not just to rebuild roads and bridges and set up stronger storm defences.

This case illustrates the ways that climate change–linked extreme disasters may be understood in different ways by different people, depending on their position in a city's structures of inequality and power. Climate-linked extreme weather is not an exceptional phenomenon; rather it interacts with cities' long-standing social and economic insecurities. We cannot prepare for increasing extreme weather without engaging with the academic and everyday knowledge about inequality and urban life that geographers, sociologists, and community groups have been building up for generations. As the following discussion of Canadian cities illustrates, climate policy interacts with long-standing equity challenges in cities large and small, and all across Canada.

Figure 10.5 The remains from a home in Staten Island, New York, destroyed by Hurricane Sandy, June 2013

Canadian Cities and Climate Change: Suburbs, Transit, Equity

Canadian cities have in many cases actively confronted the imperative to reduce their GHG emissions, prepare for increased extreme weather, and address long-standing inequalities and the increasing pressure of rising home prices. Because Canada's cities are so geographically and morphologically diverse, only a few generalizations hold across contexts.

The first of these concerns emission reductions. As we have seen, intelligent densification is crucial to reducing emissions. Research suggests, for instance, that Toronto's downtown neighbourhoods are its most carbon efficient (VandeWeghe and Kennedy, 2008). Of course, as we have seen, to the extent that density is linked to high levels of income and profligate consumption, the ecological benefits of density are muted. Moreover, Toronto, like Calgary, Vancouver, and other Canadian cities, is surrounded by sprawling suburbs in which people travel by car (inner suburbs are another story and beyond the scope of this brief discussion). Calgary, for example, was the first city government to develop ecological footprint reduction targets (undertaking an ecological footprint study in 2005). Vancouver is an especially interesting case of a city that has participated in a broad range of award-winning, livable-density programs, in part to slash carbon emissions (Jones, 2012). In 2006 the city of Vancouver launched an EcoDensity initiative and in 2008 adopted an EcoDensity Charter for the city that requires environmental sustainability to be the primary goal in all urban planning decisions, with housing affordability and livability kept in mind. The eight pillars of Vancouver's sustainability vision are these: a completely walkable community; a low-impact transportation system; green building; flexible open space; green infrastructure; a healthy food system; community facilities and programs; and economic development. And yet Vancouver's pro-density policies have had the unexpected consequence of contributing to surging home prices that are beyond the reach of middle- and working-class families, who have tended to move to the city's sprawling suburbs, which of course are not affected by Vancouver's ecologically minded planning regulations (Peck, Siemiatycki, and Wyly, 2014).

Suburbs are also urban environments, and slashing emissions in these spaces must be a top priority (Cohen, 2014). One strategy might be to compel—legally—new developments to be much denser; this is the approach pursued by Ontario's Places to Grow Act. High-rises are built around commercial centres and connected to transit hubs, a strategy known as transit-oriented development (but it should be noted that at least in public the plan was not framed in terms of climate change or carbon). Beyond the dense nodes issue, legislation can be adjusted to make it easier for suburban households in traditional subdevelopments to divide up and rent parts of their homes, increasing population density without having to build whole new structures (see also Charmes and Keil, 2015; Hayden, 1980).

This increased density will make public transit in sprawling suburbs more efficient and cost-effective. Although heavy and light rail are effective in moving large numbers of commuters into city cores, these are impractical for moving smaller numbers around suburban communities. For shorter trips, cycling lanes and comfortable, reliable bus service should be priorities. In between rail and normal bus service is bus rapid transit (BRT), in which buses have dedicated lanes and riders prepay to board at stops. Toronto has built a BRT line on its northern Highway 7 corridor; Halifax has also built a BRT line. But changes to transit and housing infrastructure alone are not enough. Public services such as daycares and arts centres (like the Rose Theatre in Brampton) are also crucial to fostering a sense of connection, community, and shared public space in suburban communities (Cohen, 2014). Without establishing such "public affluence" in suburbs, it will be difficult to sustain political support for the large public investments needed for high-quality densification. Residents need to feel that such densification improves their quality of life.

The second priority must be to prepare for harsher and more frequent extreme weather. The most pervasive threat is heat (Stone Jr, 2012). Most of Canada's large cities have continental climates, and peak summer heat is already oppressive. It will only get hotter. One aspect of preparing for heat waves is social. Public-health plans are needed so that the most vulnerable and isolated people can be reached; elderly people without air conditioning may need to be temporarily moved to public cooling centres (Klinenberg, 2002). Another aspect is infrastructural. All forms of greenery, by virtue of evapotranspiration, cool the air around them. Trees, green roofs, and green walls are all helpful. However, it is important for public policy to ensure that these are not just perks for wealthy neighbourhoods, but top priorities for lower-income areas where the need for such investments is greatest.

It is worth noting that the mayor of Toronto from 2003 to 2010, David Miller, was a chair of the C40, the global low-carbon urban policy network. Miller has been considered a visionary climate policy-maker around the world. His administration planned for major public-transit improvements through a light-rail-oriented plan called Transit

City. He also undertook initiatives to increase the energy efficiency of low-income buildings in a relatively affordable fashion. Miller was, by any measure, an urban climate policy leader. The problem was, he never sold the whole city's population on these plans. The next mayor was the controversial Rob Ford, who campaigned against Miller's legacy; Ford accused Miller of waging a "war against the car." Once elected, Ford canceled Transit City and slowed the construction of bicycle lanes. The city's newest mayor, John Tory, has proposed a plan called SmartTrack, which reprises Transit City's theme of a rail grid but is smaller in scope and would work by increasing traffic on existing commuter rail lines. At the time of writing, it was unclear whether this more capital-intensive plan would be implemented in Tory's first term, and if so, what the nitty-gritty details would be.

Overall, if far-sighted climate policies are going to survive and thrive, they need to be formulated and communicated in such a way as to win widespread community support. It is imperative that such policies not be seen, as happened in Toronto under Ford, to simply reflect the lifestyle preferences of "downtown elites"—a sometimes unfair characterization but occasionally a politically effective one nonetheless. Low-carbon policies need to make substantive contributions to social and economic equity. And they must then be framed in those terms. Detailing the kind of participatory governance strategies that this could entail is beyond the scope of this chapter. The essential takeaway is that without a widespread community buy-in, such long-term planning is likely to be brushed aside in favour of shorter-term concerns. Life is hard. It is not enough to simply implement "best practices" gleaned from expert proposals. Creative and sustained political engagement with a range of communities will be required to ensure that such policies stick.

Conclusion

As desirable as it may be, the integration of urban policy and planning in general with climate-focused policy and planning will be uneven, regardless of the fact that the core issues of the urban form—housing (and buildings), transit, and land use—are also the core issues of urban climate politics. The key questions will become: (1) how effective, in a technical sense, are climate policies in reducing GHG emissions and addressing climate vulnerability?, and (2) how are these policies interacting with existing social and economic inequalities—are these being exacerbated by climate policies or diminished? As this chapter has shown, climate policies that exacerbate inequalities (in a real or perceived way) may lead to backlash, slowing progress in a serious and damaging way.

As a general rule, we know more about why cities should pass far-reaching climate legislation than we do about what happens when they try. We know more about the technical parts of the story than about the human ones. In the words of the urban political ecologist Matthew Gandy (2014: 215), "We already know how a zero-carbon city might intersect with fields such as architecture, ecology, or engineering, but the socio-political dimensions are much more ill-defined."

Because the socio-political dimensions of urban issues are as spatial as they are social, human geographers are uniquely positioned to conduct research on the "socio-politics" involved in urban climate policy and, ideally, to help shape better policies for the years to come. The newest frontiers of research involve climate policy and the suburbs; debates on **green gentrification** and how to implement climate policies that do not exclude the most vulnerable; and the general intersection of climate policy and a widespread desire by city dwellers to have a better quality of life. In each of these fields, the issues of social inequality merge with technical challenges around reducing GHG emissions and mounting a defence against extreme weather.

As this chapter has documented, for both technical and socio-political reasons, it is impossible to reduce an urban region's carbon footprint or to defend its residents against extreme weather without formulating and implementing climate policies that make substantive contributions to social and economic equity. Indeed, there is an increasing convergence between the technical tensions and challenges of climate policy in particular and broader, long-standing tensions in social and economic policies in cities. The result is an increasingly wide range of perspectives and projects around cities and climate change, as seen in the evolution from the narrow debates of the early 2000s to the much more significant engagement of civil society exemplified by the 2014 People's Climate March. Finally, as the case studies of São Paulo and New York City have shown, along with a survey of Canadian urban climate policy, social and economic inequalities continue to play a crucial role in shaping climate policy.

In sum, the conjuncture of issues and challenges is enormously complex. No simple "solution" can solve all of these problems. But a combination of critical and practical thinking is desperately needed and should prove enormously helpful. This is a project that should interest and engage any student who is excited by the thought of shaping our urban future.

Key Points

- Cities' contribution to climate change can be calculated in two ways: production-based accounting and consumption-based accounting.
- Cities' vulnerability to climate change results principally from sea-level rise, flash floods, drought, and heat waves.
- Urban climate politics began as a largely expert-driven policy process but are increasingly being redefined by grassroots and community engagement.
- São Paulo's unanimously passed climate law foundered because the main efforts to implement it ignored the transit and housing needs of the poor.
- New York City's experience of Hurricane Sandy shows that elites and community groups understand climate-linked extreme weather in very different ways, with only the second emphasizing the importance of long-standing social and economic inequalities.
- Canadian cities must find creative and politically resonant strategies to densify their suburbs in order to reduce carbon emissions.
- Canadian cities must take care to prioritize the needs of the poor and vulnerable as they prepare for increasing heat waves.

Activities and Questions for Review

1. What is innovative about consumption-based accounting of GHG emissions and how does it change popular, practitioner, and scholarly understandings of cities and climate change?
2. As a class, debate whether it is fair to hold individuals and businesses responsible for the full carbon footprint of all the goods and services that they consume.
3. In pairs, discuss how climate change has affected your life. What adaptations, if any, have you made individually or collectively to address climate change?
4. What are the reasons why low-carbon policies agreed upon by experts may fail to be implemented? Pretend that you are a member of a civil society group outside of government, and in collaboration with 10 of your classmates identify strategies that you could use to pressure politicians to implement more aggressive low-carbon policies.

Acknowledgments

I thank this book's editors for their keen editorial insights and their patience. For my own research, which I cite in the case studies of New York and São Paulo, I acknowledge funding support from a Social Sciences and Humanities Research Council (SSHRC) doctoral award, the Hertog Global Strategy Initiative at Columbia University, and New York University's Institute for Public Knowledge. I also thank the Superstorm Research Lab, the mutual aid research collective with which I conducted research on New York's climate politics after Hurricane Sandy. I thank Liz Koslov and Max Liboiron, my colleagues in the Lab, for sharing their photographs for this chapter. I thank the Center for the Study of the

Metropolis, in São Paulo, for hosting me for several months of my fieldwork there. Finally, and most important, I thank all of my informants and interviewees, who over the course of my research took time out of their busy lives and their important work to share their knowledge and experiences with me.

References

Alliance for a Just Rebuilding et al. (2014). *Weathering the storm: Rebuilding a more resilient New York City Housing Authority post-Sandy.* New York.

Archer, D. (2011). *The long thaw: How humans are changing the next 100,000 years of Earth's climate.* Princeton, NJ: Princeton University Press.

Bakker, K. (2010). *Privatizing water: Governance failure and the world's urban water crisis.* Ithica, NY: Cornell University Press.

Barrett, P.M. (2012). It's global warming, stupid. *Bloomberg Businessweek*, November 1.

Berners-Lee, M. (2011). *How bad are bananas? The carbon footprint of everything.* Vancouver, BC: Greystone.

Bevins, V. (2015). The São Paulo commute: Walk, bus, train, train, train, bus, walk. Repeat. *Los Angeles Times*, March 13.

Brenner, N. (2013). Theses on urbanization. *Public Culture*, 25(1): 85–114.

British Standards Institution (2014). *Application of PAS 2070—London, United Kingdom. An assessment of greenhouse gas emissions of a city.* London: BSI Standards Limited.

Brulle, R.J., Carmichael, J., and Jenkins, J.C. (2012). Shifting public opinion on climate change: An empirical assessment of factors influencing concern over climate change in the U.S., 2002–2010. *Climatic Change*, 114: 169–188.

Buettner, R., and Chen, D.W. (2014). Hurricane Sandy recovery program in New York City was mired by its design. *New York Times*, September 14. Retrieved from http://www.nytimes.com/2014/09/05/nyregion/after-hurricane-sandy-a-rebuilding-program-is-hindered-by-its-own-construction.html

Bulkeley, H. (2011). Cities and subnational governments. In J.D. Dryzek, RB. Norgaard, and D. Scholsberg (Eds.), *The Oxford handbook of climate change and society.* Oxford: Oxford University Press, pp. 464–478.

Bulkeley, H. (2013). *Cities and climate change.* London: Routledge.

Bulkeley, H., and Betsill, M.M. (2003). *Cities and climate change: Urban sustainability and global environmental governance.* New York and London: Routledge.

Bulkeley, H., and Betsill, M.M. (2013). Revisiting the urban politics of climate change. *Environmental Politics*, 22(1): 136–154.

C40 Cities. (2011). Case study: London's Congestion charge cuts CO2 emissions by 16%. Retrieved from http://www.c40.org/case_studies/londons-congestion-charge-cuts-co2-emissions-by-16

Charmes, E., and Keil, R. (2015). The politics of post-suburban densification in Canada and France. *International Journal of Urban and Regional Research*, 39(3): 581–602.

Chatterjee, P. (2003). Are Indian cities becoming bourgeois at last? *Body City: Siting Contemporary Culture in India*, 2(11–12): 170–185.

Checker, M. (2011). Wiped out by the "greenwave": Environmental gentrification and the paradoxical politics of urban sustainability. *City & Society*, 23: 210–229.

Ciplet, D., Roberts, J.T., and Khan, M.R. (2015). *Power in a warming world.* Cambridge, MA: MIT Press.

Cohen, D.A. (2014). Seize the Hamptons. *Jacobin*, Fall issue: 151–159.

Cohen, D.A. (2016). The rationed city: The politics of water, housing, and land use in drought-parched São Paulo. *Public Culture*, 28(2): 261–289.

Cohen, D.A. (forthcoming). The other low-carbon protagonists: Poor people's movements and climate politics in São Paulo. In M. Greenberg and P. Lews (Eds.), *The city is the factory.* Ithica, NY: Cornell University Press.

Cohen, D.A., and Liboiron, M. (2014). New York's two Sandys. *Metropolics*, October 30.

Curran, W., and Hamilton, T. (2012). Just green enough: Contesting environmental gentrification in Greenpoint, Brooklyn. *Local environment: The International Journal of Justice and Sustainability*, 17(9): 37–41.

Davis, M. (2010). "Who will build the ark?" *New Left Review*, no. 61: 29–46.

Faber, J.W. (2015). Superstorm Sandy and the demographics of flood risk in New York City. *Human Ecology*, 43(3): 363–378.

Fisher, D., Svendsen, E., and Connolly, J. (2015). *Urban environmental stewardship and civic engagement: How planting trees strengthens the roots of democracy.* New York: Rutgers.

Fishman, C. (2011). *The big thirst: The secret life and turbulent future of water.* New York: Free Press.

Funk, M. (2014). *Windfall: The booming business of global warming.* New York: Penguin.

Gandy, M. (2014). *The fabric of space: Water, modernity, and the urban imagination.* Cambridge, MA: MIT Press.

Glaeser, E.L. (2011). *Triumph of the city.* New York: Penguin.

Glaeser, E.L., and Kahn, M.E. (2010). The greenness of cities: Carbon dioxide emissions and urban development. *Journal of Urban Economics*, 67(3): 404–418.

Gore, C.D., and Robinson, P. (2009). Local government response to climate change: Our last, best hope? In H. Selin and S.D VanDeveer (Eds.), *Changing climates in North American politics.* Cambridge, MA: MIT Press, pp. 152–173.

Graham, S., and Marvin, S. (2001). *Spintering urbanism: Networked infrastructures, technological mobilities and the urban condition.* New York: Routledge.

Hayden, D. (1980). What would a non-sexist city be like? Speculations on housing, urban design, and human work. *Signs*, 5(4): 170–187.

Heinonen, J., Jalas, M., Juntunen, J.K., Ala-Mantila, S., and Junnila, S. (2013). Situated lifestyles: The impacts of urban density, housing type and motorization on the greenhouse gas emissions of the middle-income consumers in Finland. *Environmental Research Letters*, 8(3): 35–50.

Heinonen, J., Kyrö, R., and Junnila, S. (2011). Dense downtown living more carbon intense due to higher consumption:

A case study of Helsinki. *Environmental Research Letters*, 6(3): 1–9.

Hillman, M. (1996). In favour of the compact city. In M. Jenkins, E. Burton, and K. Williams (Eds.), *The compact city: A sustainable urban form?* London: E & FN Spon, pp. 30–36.

Jackson, T. (2009). *Prosperity without growth: Economics for a finite planet.* New York: Routledge.

Jones, C., and Kammen, D.M. (2014). Spatial distribution of U.S. household carbon footprints reveals suburbanization undermines greenhouse gas benefits of urban population density. *Environmental Science and Technology*, 48(2): 895–902.

Jones, S. (2012). A tale of two cities: Climate change policies in Vancouver and Melbourne—Barometers of cooperative federalism? *International Journal of Urban and Regional Research*, 36(6): 1242–1267.

Kahn, M.E. (2006). *Green cities.* Washington, DC: Brookings Institution.

Kersten, E., Morello-Frosch, R., Pastor, M., and Ramos, M. (2012). *Facing the climate gap: How environmental justice communities are leading the way to a more sustainable and equitable California.* University of Southern California. Program for Environmental and Regional Equity, Los Angeles..

Klinenberg, E. (2002). *Heat wave: A social autopsy of disaster in Chicago.* Chicago: University of Chicago Press.

Koslov, L. (2014). Fighting for retreat after Sandy: The ocean breeze buyout tent on Staten Island. *Metropolitics*, April 24.

Landry, S.M., and Chakraborty, J. (2009). Street trees and equity: Evaluating the spatial distribution of an urban amenity. *Environment and Planning A*, 41(11): 2651–2670.

Mathur, A., and da Cunha, D. (2009). *Soak: Mumbai in an estuary.* New Delhi India: Rupa.

Minx, J., Baiocchi, G., Wiedmann, T., Barrett, J., Creutzig, F., Feng, K., Förster, M., Weisz, H., and Hubacek, K. (2013). Carbon footprints of cities and other human settlements in the UK. *Environmental Research Letters*, 8(3): 1–10.

Owen, D. (2009). *Green metropolis: Why living smaller, living closer, and driving less are the keys to sustainability.* New York: Riverhead.

Peck, J., Siemiatycki, E., and Wyly, E. (2014). Vancouver's suburban involution. *City*, 18(4–5): 386–415.

Peters, G.P., Minx, J.C., Weber, C.L., and Edenhofer, O. (2011). Growth in emission transfers via international trade from 1990 to 2008. *Proceedings of the National Academy of Sciences of the United States of America*, 108(21): 8903–8908.

Pizzo, B. (2015). Problematizing resilience: Implications for planning theory and practice. *Cities*, 43: 133–140.

Rademacher, A. (2009). When is housing an environmental problem? Reforming informality in Kathmandu. *Current Anthropology*, 50(4): 513–533.

Richards, M. (2005). *Congestion charging in London: The policy and the politics.* London: Palgrave Macmillan.

Rode, P., and Burdett, R. (2011). Cities: Investing in energy and resource efficency. In United Nations Environmental Program (Ed.), *Towards a green economy: Pathways to sustainable development and poverty eradication*, pp. 449–489. Retrieved from http://web.unep.org/greeneconomy/sites/unep.org.greeneconomy/files/publications/ger/ger_final_dec_2011/Green%20EconomyReport_Final_Dec2011.pdf. See p. 2.

Romero Lankao, P. (2007). Are we missing the point? Particularities of urbanization, sustainability and carbon emissions in Latin American Cities. *Environment and Urbanization*, 19(1):159–175.

Romero Lankao, P. (2012). Governing carbon and climate in the cities: An overview of policy and planning challenges and options. *European Planning Studies*, 20(1): 7–26.

Rosenzweig, C., and Solecki, W. (2014). Hurricane Sandy and adaptation pathways in New York: Lessons from a first-responder city. *Global Environmental Change*, 28: 395–408.

Ross, A. (2011). *Bird on fire: Lessons from the world's least sustainable city.* New York and London: Oxford University Press.

Ryan, D. (2015). From commitment to action: A literature review on climate policy implementation at city level. *Climatic Change*, 131(4): 519–531.

Schewe, J., et al. (2014). Multimodel assessment of water scarcity under climate change. *Proceedings of the National Academy of Sciences of the United States of America*, 111(9): 3245–3250.

Schor, J.B. (2011). *True wealth: How and why millions of Americans are creating a time-rich,ecologically light, small-scale, high-satisfaction economy.* New York: Penguin Books.

Seto, K.C., et al. (2014). Human settlements, infrastructure and spatial planning. In *Climate change 2014: Mitigation of climate change.* Contribution of Working Group III to the Fifth Assessment Report of the Intergovernmental Panel on Climate Change. Cambridge: Cambridge University Press.

Sobel, A. (2014). *Storm surge: Hurricane Sandy, our changing climate, and extreme weather of the past and future.* New York: Harper Wave.

Steffen, W., et al. (2009). Planetary boundaries: Exploring the safe operating space for humanity. Ecology and Society, 14(2): 32.

Stockholm Environment Institute–U.S. (2012). *Greenhouse gas emissions in King County: An updated geographic-plus inventory, a consumption-based inventory, and an ongoing tracking framework.* Seattle.

Stone Jr, B. (2012). *The city and the coming climate: Climate change in the places we live.* Cambridge and New York: Cambridge University Press.

Strauss, B.H. (2013). Rapid accumulation of committed sea-level rise from global warming. *Proceedings of the National Academy of Sciences of the United States of America*, 110(34): 13699–13700.

Superstorm Research Lab (2013). *A tale of two Sandys.* White Paper. New York.

Ummel, K. (2014). *Who pollutes? A household-level database of America's greenhouse gas footprint.* Washington, DC: Center for Global Development.

VandeWeghe, J.R., and Kennedy, C. (2008). A spatial analysis of residential greenhouse gas emissions in the Toronto Census Metropolitan Area. *Journal of Industrial Ecology*, 11(2): 133–144.

Yetano Roche, M., Lechtenbömer, S., Fischedick, M., Gröne, M.-C., Xia, C., and Dienst, C. (2013). Concepts and methodologies for measuring the sustainability of cities. *Annual Review of Environment and Resources*, 39(1): 519–547.

III Urban Forms

11

Gentrification, Gated Communities, and Social Mixing

Nicholas Lynch and Yolande Pottie-Sherman

Introduction

The material form of every city is inherently connected to desires and fears. Our neighbourhoods, our streets, our parks, indeed all of our built structures shape and are shaped by particular visions of society, some projecting utopian ideals, others embodying reactions to threats, real or perceived. Of course, whose desires and fears take shape in the city is rarely equitable. In the pursuit of urban growth and development, a project largely dominated by urban elites, a city's "livability" is never guaranteed for all. While many urban elites seek to sustain or to acquire global city status, the dramatic restructuring of local economies—with accompanying employment shifts, changes in social service delivery, and evolving property development—often results in rising concerns about poverty, crime, housing affordability, and social polarization. Urban **inequality**, segregation, and exclusion represent some of the outcomes of materializing the desires and fears of a select few. But while some divisions in our cities, both physical and social, are deeply embedded, many forms of urban development have intensified this process.

This chapter explores three specific but interconnected urban forms that represent spaces of division in contemporary cities: **gentrification**, **gated communities**, and **social mixing**. The first of these, gentrification, is a widely used but variously understood concept that traces the social and material upgrading of urban neighbourhoods. Evolving from the renovation of working-class houses and the displacement of lower-income residents in North America and Europe, gentrification now includes large-scale redevelopment projects in globalizing cities throughout Latin America, Africa, and Asia. It is thus described as a global urban phenomenon (Smith, 2002), and geographers are exploring its new forms, from new-build condominium towers to loft spaces rebuilt in the shells of abandoned industrial buildings and empty schools.

The second context concerns gated communities, a diverse form of housing that restricts public access to streets and homes often through the use of gates and security services. From Lagos to Calgary, the construction of gates, walls, and high-tech security features intended to keep certain people out are now commonplace. As these communities have received much attention in the media, research on them has blossomed in the last several decades and has begun to uncover how new forms of privatized housing affect security and segregation in the contemporary city. But lingering questions remain: do these radical spaces share similar functions and forms across all urban geographies? Are the driving forces that shape the demand for this "bunker mentality" (Atkinson and Blandy, 2007) the same in the global South as they are in the global North?

The third and last context explores the case of mixed communities—a resoundingly popular but contested policy that seeks to bridge some of the gaps created by exclusionary urban forms like gentrification and gated communities. Over the last three decades, social mixing has become a mantra for urban policy-makers, in both the global North and global South, who are grappling with issues of urbanization and housing inequality.

This chapter begins with a contextual section that unpacks these three urban forms and explores their role in shaping the social life of urban residents. Next, it highlights two related case studies from London and New York City—cities that sit atop the global urban hierarchy and represent important spaces of urban development and neighbourhood change. The first case study traces a trend of gentrification in London, that of the **adaptive reuse** of former public buildings for upscale housing. Here, London's expansive housing market, inspired by its global status, is pushing the boundaries of gentrification to new post-institutional spaces. Schools and churches, for instance, are now prime targets in the making of luxury housing for a select few. The second case study explores the rise of "**poor doors**" in new-build housing in New York City. In this case, some developers are circumventing a public-sector initiative to encourage social mixing and social housing in new housing construction (called an "inclusionary housing program") by building in two doors—one for low-income tenants and one for upper-income owners. The rise of "poor doors" highlights the persistent challenge in bridging socio-economic divides in the city. Considering the Canadian context, the next section highlights how gentrification, gated communities, and social mixing have evolved and had an impact on cities across Canada—from its largest urban areas to its emerging metropolitan regions. The chapter concludes by recognizing that just as patterns of gentrification, gating, and mixing are going global, so too has resistance.

Geographies of Gentrification: New Forms of Gentrification in the Contemporary City

> I'm for democracy and letting everybody live but you gotta have some respect. You can't just come in when people have a culture that's been laid down for generations and you come in and now gotta change because you're here? Get the <expletive> outta here. Can't do that!
>
> —Spike Lee (in Michael and Bramley, 2014)

In 2013, Spike Lee, a Hollywood producer-director, shocked a live audience at a Black History Month event with an acerbic off-the-cuff speech concerning the dramatic changes in his hometown of Brooklyn, New York. Besides citing "hipsters" and new wealthy residents as the vanguard of a "disrespectful" change in his neighbourhood, Lee argued that these new arrivals are also guilty of a Christopher Columbus syndrome—the propensity to act as if they had not only discovered a new neighbourhood but had settled it too (Michael and Bramley, 2014). Perhaps not surprisingly, the "Spike Lee rant," as some in the news-media call it, made international headlines and sparked worldwide debate about the process of neighbourhood change that urban geographers formally call gentrification. It would seem that Spike Lee struck a nerve.

It is important to note that gentrification is not a new phenomenon and neither is it restricted to neighbourhoods in the United States. Indeed, it was in 1964 that a British sociologist named Ruth Glass published a report that reshaped the study of neighbourhood change. With a focus on London, UK, Glass was the first to detail the "upgrading" taking place in neighbourhoods in the borough of Islington (Figure 11.1). In particular, Glass (1964) noted that older **working-class** quarters were invaded by an expanding group of **middle-class** bohemian Londoners—housing consumers who had the financial means to buy and refurbish what were relatively cheaper but centrally located properties. Importantly, Glass highlighted that this creeping process also caused the displacement of long-standing blue-collar communities, the result of which still rings true today as more affluent housing consumers consistently seek to extract capital (social and economic) in devalued inner-city spaces and thus often price out less affluent residents.

Subsequent articulations of Glass's concept have attempted to model the process in coherent stages. In neighbourhoods like Kitsilano in Vancouver, Kreuzberg in Berlin (Dangschat, 1988), or the Marais Quartier in Paris (Carpenter and Lees, 1995), early stages saw "pioneering" gentrifiers, primarily artists and smaller countercultural subgroups of the middle class, investing their own labour in the design and renovation of older housing stock, commonly referred to as "**sweat equity**." In general, the "mundane" aesthetics of working-class neighbourhoods were eventually incorporated into an edgier urban cachet. In the classic **stage model of gentrification**, an influx of wealthier groups of the middle class—largely "liberal" media and creative types as well as younger professionals—join existing residents. Typically, neighbourhood land values rise through new rounds of renovation and in response to the increasing interests of developers and real estate speculators. In time, rising demand for cheaper neighbourhoods helps to diffuse patterns of gentrification and upscaling into nearby areas, intensifying the displacement of lower-income communities. In later stages, financially secure but risk-averse

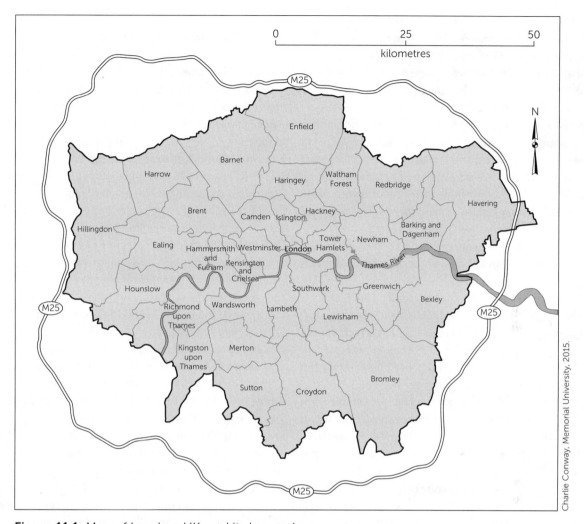

Figure 11.1 Map of London, UK, and its boroughs

groups of urban professionals, along with savvy property developers and large retailers, routinely make forays into established neighbourhoods. In response to soaring property values, rental housing stock is often broken up or lost to the private market while local commercial properties are renovated to cater to more affluent clients.

For many, the routine spatialization of gentrification understood through the paradigm of the stage model goes a long way in explaining the changes found in neighbourhoods in many urban regions. But how predictable, tidy, or rule-based is this phenomenon? The brief answer is that gentrification is neither a completely nor a fully mapped process. Contrary to these popular narratives, gentrification is remarkably uneven and fragmented (Ley and Dobson, 2008). In fact, in a relatively short time since Ruth Glass's observations, the character of neighbourhood change and gentrification has both differentiated across urban contexts but also circulated from the global North to the global South (Smith, 2002). Indeed, the origins, motives, and forms of gentrification are still deeply contested, and numerous studies of the process use various, sometimes conflicting, definitions and methods in their research (Lees et al., 2008).

Beyond the pioneer or classic gentrification, contemporary researchers now study the process in numerous contexts and in a variety of spaces, investigating what David Ley (1996) calls "a geography of gentrification." For instance, scholars now routinely explore the relationships between gentrification and **globalization**. In this case, Neil Smith (2002) has argued that gentrification is a "global urban strategy" that is connected to new socio-economic policies (i.e., **urban neoliberalism**) and state restructuring that privileges capitalist production over **social reproduction**. In short, cities around the world increasingly favour urban revitalization that caters to more affluent residents over the housing and consumption needs of less wealthy citizens. Beyond the global North, cities like Cape Town, South Africa (Visser and Kotze, 2008), and Shanghai, China (He, 2007), for instance, have received increasing attention as state governments, along with urban developers, actively facilitate neighbourhood upgrading, inner-city revitalization, and new rounds of capital investment. Large cities of the global South, like their northern neighbours, display troubling trends of housing-affordability issues coupled with rising income inequality and the displacement of lower-income residents to poorly serviced peripheral areas.

In addition to its changing global context, gentrification is proceeding in a diversity of socio-economic spaces such as gated communities, suburban neighbourhoods, and even rural places (Álvarez-Rivadulla, 2007). The last of these, rural gentrification, highlights the geographical reach of neighbourhood change. Focused largely on the British context, geographers have brought attention to the fact that numerous rural areas outside of large cities have become destinations for wealthy "counterurbanists" (Phillips, 2010) who engage in **counterurbanization** in search of a pastoral lifestyle. Indeed, a surge of urban middle-class migrants to rural England and Scotland, for example, have raised rural house prices, encouraged higher-order retailing, and, for many local residents, adversely transformed the culture of rural life (Stockdale, 2010). Outside of the global North, recent research also points to an expanding process of rural gentrification in key regions of China. Qian et al. (2013), for instance, describe an artist-led rural gentrification in the rapidly transforming village of Xiaozhou, located in China's southern metropolis of Guangzhou. In this case, "grassroots" artists in search of "slower-paced lifestyles" and "romantic rural living" have sparked the commodification and gentrification of local neighbourhoods and profoundly reshaped local housing geographies in the region.

Besides the changing contexts and spaces of gentrification, considerable urban geographical research is also exploring the new material forms of gentrification. While classic gentrification was once largely relegated to inner-city working-class housing, urban geographers have noted novel morphologies in the contemporary process. For instance, recent debates have circulated around the notions of "post-industrial" (Zukin, 1982) and new-build gentrification (Davidson and Lees, 2010). In terms of the former,

Sharon Zukin (1982) traced the conversion of manufacturing buildings to living spaces, what she called "living **lofts**." Focused on the warehousing district south of Houston Street in Lower Manhattan (SoHo), Zukin highlighted the processes involved in transforming this post-industrial area to a thriving artists district and later to an upscale residential market. From the outset, artists were the innovators of the emerging loft trend, as many in their ranks targeted a growing number of relatively cheap but uniquely large industrial properties for their potential as live-work spaces. From what seemed like a Manhattan oddity, loft conversions have since spread globally to other deindustrializing and gentrifying cities, like Montreal (Podmore, 1998), Beijing (Wu, 2004), and Johannesburg (Frenzel, 2014).

Loft-style housing thus represents an important contributor to a growing list of upscale real estate that is transforming the socio-economic landscapes of inner cities around the world. As Zukin (1982: 256) has argued, lofts are part and parcel of the long-term transformations of a city's political economy, contributing to the deindustrialization and gentrification of the urban core. Indeed, lofts not only offer upscale living spaces close to the city centre and within reach of popular urban amenities (e.g., waterfronts, entertainment, food, and culture), but they are also part of a process that commodifies urban space for a select few by pricing out other potential homeowners/renters and changing the social and political environment of local neighbourhoods. In much the same way, new-build **condominium** towers have become a central aspect of contemporary gentrification. In the last several decades the rise of residential towers, from Hong Kong to New York, represents the maturation and mutation of the gentrification process (Davidson and Lees, 2010). Many argue that the proliferation of towers, especially in downtown spaces, has not only reinvented the inner city for middle- and upper-class homeowners, but has also transformed the wider socio-economic landscape of downtowns by encouraging higher-order retailers (e.g., boutiques, cafés, art and design stores) and services (e.g., niche grocery stores, dog daycares, fitness studios) that cater to more affluent residents (Rosen and Walks, 2013).

Gentrification and its study have continued to expand in response to the increasing socio-spatial and economic diversity of the contemporary city. Importantly, however, gentrification is merely one phenomenon to receive the attention of contemporary urban geographers. Another key trend involves gated communities.

Beyond the Gates: Enclaving and Gating Communities in the City

It is now over four years since the shooting death of Trayvon Martin, a 17-year-old African-American high school student. While walking to his father's house in a gated community ("The Retreat at Twin Lakes") in Sanford, Florida, Martin was confronted and later fatally shot by the community's neighbourhood-watch coordinator, George Zimmerman, a 28-year-old mixed-race Hispanic man. While this tragic event caused tremendous public outcry, some in response to rising tensions from racial profiling and questions surrounding the morality of Florida's Stand Your Ground statute, there were commentators who argued that the poorly planned, exclusionary built environment of the gated community was a factor in Martin's death (Goodyear, 2013). In particular, critics of gated communities argue that some design features intended to restrict and monitor public access (e.g., gates, walls, fences, **closed-circuit television** [CCTV] cameras, and security guards) paradoxically compromise safety rather than increase it. As Rich Benjamin (2012), a cultural critic, puts it: gated communities churn a vicious cycle by attracting like-minded residents who seek shelter from outsiders and whose physical seclusion then worsens paranoid group-think against outsiders. In the Martin–Zimmerman case, Martin was regarded as a trespasser, an unwanted outsider who had seemingly transgressed both the tightly defined physical boundaries and the "private" socio-political space of the community.

While the Sanford event has drawn renewed attention to gated communities, it is important to note that in many countries enclavism and fortification have a long history. For some nations, such as China and England, walls and gates are intimately linked to political-historical legacies like imperialism and feudalism. In the present, these features now offer a new appeal. As Rosen and Grant (2011: 779) explain, their remaining walls have become mythologized symbols of a glorious past, used to fuel tourism. In other contexts, gated communities represent entirely different purposes. There is, however, little consensus on what actually counts in defining this urban form. Like the case of gentrification, the global development of gated communities is characterized by a remarkable diversity in physical form and motivation. For instance, even a working definition offered by geographers Atkinson and Blandy (2007: viii)—walled or fenced housing developments to which access is restricted, characterized by legal agreements that tie residents to a common code of conduct and (usually) collective responsibility for management—cannot fully capture the range of morphological differences across different geographical contexts. We now see a continuum of gating whereby a variety of design features, from the *soft* (e.g., symbolic booms and facade half-walls in suburban developments) to the *hard* (e.g., guard towers in elite housing estates), mark the landscape to achieve a form of private securitized living. But the continuum of form is also linked to a continuum of motivation. While some studies explain the proliferation of gated communities as a reaction to rising perceptions and lived experiences of crime, other research considers symbolic aspects like investment potential, lifestyle, and status as motivations for enclavism (Low, 2003).

The first explanation states that gated communities are a socio-cultural response by affluent residents seeking both to evade crime and to correct the systemic failure of government to provide satisfactory security (Blakely and Snyder, 1997). Consistent demand for defensible neighbourhoods in places like Mexico (Giglia, 2008) and Brazil (*condominos fechadoes* or closed housing estates) (Coy, 2006), for instance, is largely attributed to long histories of population segregation, urban poverty, and, mounting drug, gang, gender, and race-related violence. Yet even in relatively secure nations, the persistence of hard gating has proceeded apace, in part as a physical means to exclude unwanted, often racially defined others, as with the Trayvon Martin case.

Besides highlighting the fear of others, the second explanation claims that the function of gated communities is also linked to a demand in housing developments that sell lifestyle or trade on status, luxury, and exclusivity. In the case of the former, master-planned retirement villages like SunCity, Arizona, are modestly guarded developments built around central facilities offering medical care, community services, entertainment, and golf courses. In the case of the latter, prestige communities like SunCity, Ghana (based on the Arizona template), typically offer residents more elaborate security features as a means to protect property, but are also designed with high-quality construction and aesthetics, what Setha Low (2003) calls the construction of niceness, to act as a product of social differentiation. In other words, a home in a luxury gated community represents a "positional good," a commodity that affirms one's position and class in society.

Regardless of motivation, the popularity of gated communities is as strong as ever. Indeed, gated communities are especially fashionable in nations with rapid urban population growth and changing immigration profiles (Atkinson and Blandy, 2007). Yet, from the global North to the global South, neighbourhood fortification is routinely marketed as a panacea to a variety of urban problems, some real and some not. Still, there are voices that try to balance this discourse. Recently, the UN-Habitat Executive Director, Joan Clos, called on nations to move away from segregated housing, arguing that "[t]he ideal city is not one with gated communities, security cameras, a futuristic scene from *Blade Runner*, dark and dramatic, with profound unhappiness" (Provost, 2014, n.p.). Instead, Clos called for more democratic and "open" cities through renewed approaches to the design and planning of "public spaces" for the use of all urban citizens, and for a concerted effort by governments and aid donors to consider urban policies that increase the provision of services that empower the disenfranchised and poor rather than the rich (Provost, 2014).

New Urban Forms for New Urban Environments: Mixed Communities

Gentrification and gating are two processes by which exclusion is mapped onto the built environment of cities. These processes homogenize urban landscapes through direct or indirect **displacement** (in the case of gentrification) or through outright physical exclusion (in the case of gated communities). In this case, we look beyond gentrification and gated communities to consider emerging geographic research on mixed communities.

Although social-mix policy has roots in nineteenth-century Britain, it experienced a revival in the 1990s, particularly in Britain and the United States (Lees et al., 2012). Policy-makers throughout the global North have embraced mixed communities as the antidote to **socio-spatial segregation**. Social mixing refers to policies that seek to increase neighbourhood diversity by mixing occupants across socio-economic lines. These policies take a variety of forms, ranging from quotas that reserve housing units for low-income residents, to vouchers aimed at dispersing poor residents, to redevelopment strategies that encourage the middle-classes to move into stigmatized or racially segregated areas. An implicit assumption of such strategies is that disadvantaged neighbourhoods trap poor residents, blocking outward and upward mobility (Wilson, 1987). Social-mix advocates therefore see place diversity as a keystone of social equity (Adams et al., 2013). Once targeted primarily at dispersing spatial concentrations of the urban poor, social mixing has evolved to target other groups, including racial minorities and immigrants (Fincher et al., 2014).

For geographers, the central debate surrounding mixed communities concerns whether or not their associated policies actually produce inclusivity (Rose et al., 2013). On its surface, social mix appears to be benign and policy-makers have adopted it widely as a mainstay of urban policy (Lees et al., 2012: 9). But there are many reasons to be critical of these discourses. Bridge et al. (2012) argue that social mixing amounts to gentrification by stealth because it uses the language of inclusivity to promote middle-class encroachment in disadvantaged urban areas. In Dublin, for example, the city council used social mix as a justification for dismantling social housing (Lawton and Punch, 2014: 866). The long-standing focus of social-mix proponents on marginalized groups (rather than on equally pervasive concentrations of the white, upper-class group) reflects an inherent class and racial bias.

As Lees et al. (2008: 199) note, the irony of social-mix policies is that they usually (intentionally or unintentionally) spur gentrification, eventually displacing lower-income groups and producing greater homogeneity. However well intentioned, mixed-community projects have a tend to succumb to housing market pressure (Adams et al., 2013). Low-income residents who are included in mixed-income housing may experience exclusion *in situ*—for example, through a diminished capacity to influence decision-making (Lees, 2014). Critics have recently called attention to "poor doors" in mixed-housing in New York City and London, where tenants who pay lower rates are forced to use separate entrances (Navarro, 2015). Other scholars are skeptical of the actual potential for co-mingling in mixed communities given the tendency of people to self-segregate even at the micro-level (Butler and Robson, 2003). Furthermore, other scholars argue that interactions are likely to be characterized by tension rather than by peaceful co-existence if and when mixing is achieved (Matejskova and Leitner, 2011).

New Geographies of Gentrification? Post-Institutional Adaptive Re-use and Loft Living in London, UK.

In February 2012, London and New York City squared off in a good-humoured, public-relations-inspired debate to prove which city is the best in the world (Florida, 2012). In what was dubbed the "Battle of the Giants," the mayor of London, Boris Johnson, narrowly edged (in a fabricated points system) New York's deputy mayors with his city's long list of statistics and accomplishments. Sitting atop the global urban hierarchy (with New York and Tokyo), the Greater London Metro Area boasts a population of over 14 million, is arguably the world's leading financial centre, and has a remarkably expansive advanced service industry comprising finance, insurance, and real estate (FIRE) sectors. In recent years, London has also become a leading cosmopolitan city, with increasing numbers of professional and economic migrants and the construction of globally inspired office and housing redevelopments like the remodeled Canary Wharf or the recently completed Shard building. In short, in recent decades dramatic transformations in land use along with changing social, cultural, and economic values have made London a unique world city, rivaled by few as a distinct command-and-control centre in an expanding global-urban network.

But besides this winning list, the intensification of London's seemingly incessant pace of development has resulted in long-standing patterns of social upscaling and lower-income displacement (Butler and Robson, 2003) along with new geographies of gentrification. As in other global cities, London's loft sector, for one, has become an important element in an expanding terrain of unaffordable housing, especially in the city's inner boroughs. Although industrial lofts (warehouses and manufacturing buildings) remain the most visible form of loft adaptive reuse, a variety of redundant institutional properties are now considered hot commodities in London's housing market. These properties range from formerly publicly managed facilities like schools and hospitals to religious institutions and their various spaces of worship (Lynch, 2016).

The closure of public facilities is often connected to political-economic shifts, some of which are linked to the emergence of new politics that foster urban entrepreneurial and neoliberal strategies at the expense of social services. Such agendas have led, in some instances, to the resale of properties as a means to meet public-sector austerity measures. Public schools, for one, are assets that school boards routinely sell in response to demographic shifts and budget cuts (Basu, 2007). Buying and converting old schools into lofts is an increasingly attractive option for housing developers, especially in competitive real estate markets in metropolitan regions like London where there are ample properties and where demand for unique properties is high.

The redevelopment of a public school into "The Village Lofts" in the Battersea neighbourhood (borough of Wandsworth), for instance, is but one of many loft conversions in the city. Not only situated in a gated community, the Village Lofts are part of a wider gentrification process along the length of the metropolitan Thames where outmoded buildings have been transformed into some of the most desirable residential space in London (Davidson and Lees, 2010: 404). Selling close to £1 million (approximately CAD$2 million) in the real estate market, these unique lofts highlight the value of a central location in the city and underline the role of rich architectural and historic fabric as a means to (re) make urban neighbourhoods for elite housing consumers.

Church conversions are another case worthy of note. In this instance, numerous local congregations are fighting to keep historic worship spaces alive amid shrinking congregation numbers, increasing property maintenance fees, and competition for urban space. For many congregations these rising costs are simply too much to bear, and as with closed schools, the choice to sell property in the local real estate market represents a viable option. In recent years, numerous churches throughout London have been declared redundant and sold to loft developers. For instance, developments like the "Notting Hill Lofts" in Notting Hill (borough of Kensington and Chelsea), the "Sanctuary" in Clapham Junction (borough of Wandsworth), and the "Honour Oak Lofts" in East Dulwich (in the borough of Southwark) (Figure 11.2) point to the ongoing role of redundant urban churches in the upscale housing market. Each of these projects also reflects a different context of gentrification: from *mature* gentrification, to *recent* gentrification, to gentrification-*in-process*. In particular, the completion of the Notting Hill Lofts in 2009 came at a later stage in what is an area of relatively mature gentrification. Once considered "the capital's trendiest district" (Bradshaw, 1999), this West London loft conversion has affirmed the elite status of the local housing stock (a 2015 listing has a unit on sale for over

Figure 11.2 Luxury church conversions in London's gentrified and gentrifying neighbourhoods (from right to left: Notting Hill Lofts, the Sanctuary, Honour Oak Lofts)

£3.5 million, approximately CAD$7 million) and added to the prestige of the neighbourhood by encouraging capital investment and high-quality development, especially relating to its local tourist and consumption spaces.

In comparison, the Sanctuary in Clapham Junction represents a new project in an area of recent gentrification. Close to the Village Lofts described above, the Sanctuary is a four-unit luxury conversion of an historic Anglican church on the local high street. According to the property agents in charge, each property features impressive suites and furnishings that would please the most discerning of buyers. Beyond this branding message, the Sanctuary's builders are quick to link the project to the established character of the local neighbourhood, increasingly defined by its key amenities: proximity to the city centre and mass transit, as well as access to the River Thames and a vibrant high street.

The last example, the Honour Oak Lofts in East Dulwich, represents a project in an emerging space of gentrification. Located in South London, the neighbourhood has undergone tremendous but gradual change in the last several years—from a student destination driven by lower rental prices and proximity to university campuses, to an area targeted by young, middle-class professionals who are buying first homes. Furthermore, a procession of new pubs, restaurants, delis, and boutiques has moved in, making the neighbourhood an up-and-coming area. East Dulwich also sits adjacent to the rapidly gentrifying Brixton neighbourhood, a contentious space and flashpoint for recent "Reclaim Brixton" events—an anti-gentrification and social protest movement that is fighting austerity measures and neoliberal policies (Hill, 2015) (Figures 11.3 and 11.4).

Figure 11.3 Graffiti along a new housing development in Brixton, London

Figure 11.4 Reclaiming Brixton: Community members and local activists attend Another Lambeth Is Possible, an open conference to discuss the housing crisis and community solutions

In sum, rather than being solely tied to the decline of industrialization, a story well told, these cases of post-institutional adaptive reuse, from schools to churches, demonstrate how contemporary lofts are spaces where transforming cultural values and changing economic circumstances encourage a re-valorization of the built environment. But for whom are these spaces built, and who do they exclude? How might these new developments exacerbate local housing pressures? What other futures could these once public buildings have had? While further research is required to answer these lingering questions, it is clear that gentrification has evolved to take root in new spaces in the contemporary city. The case of church lofts points directly to an emergent geography of gentrification, a process that is both reflecting and producing new values and new approaches to urban living. And like the luxury condominium towers under construction in New York City, these new values have the potential to exclude, marginalize, and displace.

Poor Doors and Inclusionary Zoning in New York City, United States

On April 29, 2015, a crowd gathered at 250 South Street to protest the Extell Development Company's latest luxury condominium project on New York's Lower East Side—one of five boroughs that make New York the largest city in the United States (Figure 11.5). With approximately 8.5 million residents and a thriving and diverse economy, the "Big Apple" is consistently ranked among

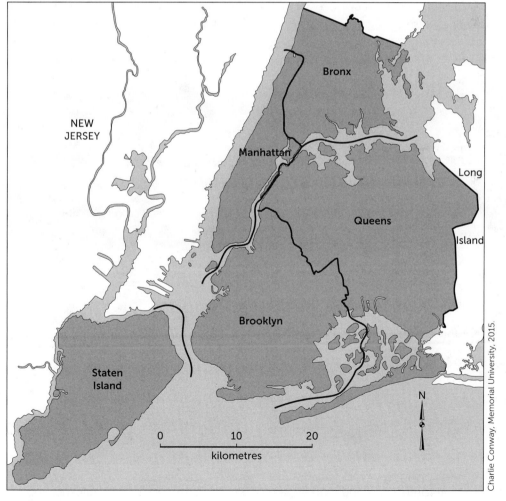

Figure 11.5 Map of New York City, USA, and its boroughs

the top global cities. Attempting to cash in on the rising demand for residential space in the city's urban core, Extell has begun construction on a slew of new developments that include what critics call "poor doors." The new tower at 250 South Street has received particular scorn from protesters, since it has been designed with separate entrances for low-income tenants.

One Riverside Park, for example, is a 33-storey luxury condominium tower at the corner of Riverside Boulevard and 62nd Street. Tenants who buy units at the market price (some as high as USD$25 million) will enter from a well-lit, hotel-style entrance on Riverside Boulevard, managed by a 24-hour concierge. They will also have access to an extravagant range of conveniences: a movie theatre, a bowling alley, and a fitness facility that boasts 3716 square metres (40,000 square feet) of amenities, including an indoor pool, a climbing wall, a basketball court, and a golf simulator (Extell, n.d.). Tenants who have been selected by lottery for one of 55 affordable housing units will enter through a separate entrance on 62nd Street (Figure 11.6). These units will range in price from USD$833 (per month) for a studio to USD$1082 (per month) for a two-bedroom apartment and will be open to applicants with yearly incomes between USD$30,240 and $50,340. They will be clustered together and physically separated from the market-rate tenants (Navarro, 2015).

Extell took advantage of New York City's Inclusionary Housing Program (IHP), which offers a **floor area ratio** (**FAR**) bonus in designated districts to developers who include a certain percentage of housing units below the market rate. The IHP incentivizes affordable housing in areas of the city undergoing substantial new residential development (Department of City Planning, n.d.). In high-density districts, primarily in Manhattan, developers can expand their FAR if they reserve units for low-income tenants whose incomes are below 80 per cent of the area median income and/or moderate-income tenants whose incomes are below 125 per cent of the area median income (Zukin et al., 2009: 63). For every square foot reserved for affordable

Figure 11.6 Poor doors in New York City (left: rear view of One Riverside Park showing the affordable housing units in the foreground and market-rate condominium towers in the background; right: One Riverside Park's "affordable housing" poor door at 470 West 62nd Street)

housing in high-density districts, developers can expand their FAR between 0.1 and 0.3 square metres (1.25 and 3.5 square feet) (Department of City Planning, n.d.). Although this aspect of the IHP was established in 1987, since 2009 a stipulation of the zoning law allowed affordable and market-rate sections of the development to be treated as two separate buildings (Quinn, 2015); hence, the poor-door phenomenon is a recent development in New York City.

Separate entrance arrangements are not unique to the Big Apple. They also have a long history in London, UK, where the demand for luxury real estate is high and the need for affordable housing is acute. In both London and New York City, the poor-door phenomenon is symptomatic of wider housing crises that have accompanied the emergence of these cities at the top of the global urban hierarchy since the 1980s. In New York City, two periods of economic boom since 1980—one from 1980 to 1989 and another from 1996 to 2006—spurred major increases in home prices. During the latter period, for example, the average price of a home in New York City rose by 173 per cent (Furman Centre, 2014). In 2013, the **median monthly gross rent** was USD$1244, or about USD$300 more than the US average per renting household. This figure masks much larger increases in *asking rents*, however, because it includes approximately 1 million rent-stabilized apartments in New York City as well as a small percentage of rent-controlled apartments. In 2013, the **median asking rent** (listed in rental advertisements on Street Easy) was USD$2900 per month; and in eight neighbourhoods in Manhattan it was above USD$3000 (Furman Centre, 2014).

While rents have skyrocketed, median household incomes have risen only a few percentage points. The mismatch places a rising **rent burden** on households. The rent burden is a measure of housing affordability based on the proportion of household income devoted to rent. Generally, households that spend more than 30 per cent of their income on rent are considered to be rent burdened. Those that spend over 50 per cent of income on rent are considered to be severely rent burdened. In 2013, 54 per cent of households in New York City were rent burdened, compared to 28.5 per cent in 1970 (Marom and Carmon, 2015). Lower-income households are disproportionately affected by the rent burden. In 2013, 80 per cent of low-income households in New York City paid more than 30 per cent of their income to rent. New York City's poor doors are a by-product of an inclusionary zoning policy that ensures these lower-income households have access to new private housing developments that would otherwise be far out of reach.

Separate-entrance developments like One Riverside Park have generated substantial controversy. Critics view the separation of entrances and amenities as a form of income-based segregation, allowing the wealthy to maintain their distance from the lower income communities. In response, New York City's Mayor Bill de Blasio, whose platform included affordable housing as a key component, inserted a provision in the real estate property tax law, requiring shared entrances for market-rate and affordable units. This new provision does not apply to One Riverside Park, which despite the public outcry received more than 88,000 applications for its 55 affordable units (Navarro, 2015). The overwhelming interest in this development indicates the extent of the housing crisis in New York City (Figure 11.7).

Figure 11.7 Protesters during "Turn the Tide" march following Hurricane Sandy—seeking affordable housing in New York City

The poor-door phenomenon provides two major lessons for students of urbanization. First, cities can approach affordable housing in different ways. New York City's inclusionary zoning policy is an example of a market-based strategy. It allows real estate developers to build larger developments in exchange for a guarantee that some units will be priced at below-market rates. In other words, the policy incentivizes affordability. Poor doors illustrate one of the pitfalls of market-based approaches. Inclusionary zoning is intended to improve access to affordable housing, but while it has contributed to an increase in the total number of units available at below-market rates, it has paradoxically produced a rather stark new form of income-based segregation.

Second, these projects need to be considered in the context of long-standing narratives about social mixing that subscribe to a utopian vision of cross-class interaction. Many poor-door critics see separate entrances as antithetical to social mixing, preferring mixed-income developments that encourage interaction between income groups to developments where residents are ushered in through different doors. But this critique assumes that lower-income tenants *want* to mix with upper-income tenants when they may actually fear disempowerment through mixed-housing arrangements. The concept of poor doors has moved north to Canada's largest cities, where several new condominium developments are being designed with separate entrances for residents of city-subsidized rental units. The phenomena of gentrification, gated communities, and social mix are clearly also alive and well in Canadian cities.

Gentrification, Gated Communities, and Social Mix in Canadian Cities

Over the last three decades, the patterns and processes of social upgrading, residential displacement, and unaffordable housing have spread from the country's largest cities to many of its smaller but emerging metropolitan areas. By the mid-1980s, a limited roster of neighbourhoods in cities like Montreal (Plateau and Vieux Montreal), Toronto (Yorkville, Cabbagetown/Don Vale), and Vancouver (Kitsilano, Fairview Slopes) had firmly set the scene of change. Some of the earliest forms of gentrification followed the classic stage model, as they tended to progress from the recolonization patterns of local artists and some of the less affluent but culturally savvy professionalizing middle class. In other instances, upgrading and displacement were found in historic neighbourhoods with interesting housing stock and appealing architecture, as well as in areas close to central amenities like parks, sports facilities, and cultural institutions (Pottie-Sherman, 2013).

Toronto, for instance, is a well-studied city when it comes to issues of gentrification and neighbourhood change. For decades, geographers have described in various detail the social and physical transformations that have occurred across the city and region: massive ongoing waterfront redevelopment schemes (Hoyle et al., 1988); local community upgrading (Slater, 2004); the construction of luxury and niche lofts (Bain, 2006); and the production of high-rise enclavism (Rosen and Walks, 2013). Leslie Kern (2008), for instance, uses a feminist perspective to explore a gendered social geography of new-build gentrification in Toronto. In this case, an increasing number of affluent and professional women are seeking condominium ownership as a means to build financial autonomy and realize new live-work-play balance in the city centre. Recent work has also uncovered troubling trends in many of Toronto's neighbourhoods. Hulchanski (2010), in particular, shows how Toronto is polarized into three distinct cities based on income change in the population: an expansion of lower-income neighbourhoods in the city's postwar suburbs; a consolidation of upper-income neighbourhoods, mostly in the inner city; and shrinking middle-class neighbourhoods in what were Toronto's older inner suburbs. Many of these dramatic transitions are part of wider processes of change, including expansive growth of suburban regions, dis-investments in the older inner suburbs, and substantial reinvestments in inner-city areas, often in the form of gentrification, as a means to cater to the growing population of urban professionals.

It is increasingly clear, however, that beyond Canada's largest cities gentrification is playing a key role in reshaping smaller but emerging metropolitan regions. The inner cities of Victoria, Winnipeg, and Kitchener-Waterloo, for instance, are in a process of rapid redevelopment because younger upwardly mobile professionals are seeking access to centralized character housing. Following closely are new retail establishments and public institutions (e.g., satellite university campuses) that are revitalizing and remaking inner-city space.

Much like gentrification, gated communities are a relatively recent phenomenon in Canada. Some of the earliest privatized communities emerged in the 1960s and were considered experiments in recreation development (Townshend, 2006). Private and semi-private golf course communities, such as Lake Bonavista in Calgary, offered residents separated and distinguished spaces that not only facilitated sport but also created niche spaces for upper-class consumption and lifestyle.

Apart from the private recreational community, the more contemporary master-planned or subdivided gated community, the model of which was imported from the United States, did not appear until the 1990s (Walks, 2010). By some estimates, the number of these communities in Canada surpassed 300 by 2004, with a significant concentration in British Columbia (Grant and Mittlesteadt, 2004). In Kelowna, BC, for instance, a number of gated communities, such as Canyon Ridge, offer housing options that cater to the increasing demands of retirees and seasonal residents who are seeking private, often waterfront, living. It is important to note, however, that very few of the developments throughout Canada exhibit the fortress-style compounds characteristic of many other countries. As Grant and Mittlesteadt (2004) show, rare are the communities that offer hard security options like guards or full-length walls. Rather, the demand for gates, in various forms, is predicated on privacy, on community identity, and on securing high-quality amenities. Thus, the form and function of the Canadian gated community are less about creating fortress-like bunkers and more about reflecting a demand for spaces or landscapes that communicate social status and facilitate lifestyle choices. In fact, in the heart of Canada's growing metropolitan areas, condominium towers are emerging as key spaces of exclusion, described by some scholars as "vertical gated communities" (Ghosh, 2014). In Toronto and Vancouver, the leading edges of the Canadian "condo-boom," high-rise towers have gained popularity because they offer ways of densifying the city, reducing automobile dependence, and catalyzing mixed land use (Rosen and Walks, 2013). But such spaces, especially in the inner city, also represent a distinct form of new middle-class gentrification and privatization, made possible by high entry costs (i.e., expectations of ownership), monitored building entrances, and comprehensive surveillance systems.

In this way, the form and function of the gated community is ever evolving. As Alan Walks (2010: 10) notes, the popularity of new gated communities in Canada, whether in a vertical or conventional form, is also potentially linked to rising perceptions of crime (even though crime rates are dropping) and high immigration rates in **gateway cities** like Toronto, Vancouver, and Calgary. Linked to the global economy, these cities are the foci for increasing flows of business migrants, foreign workers, and refugees, a context that is encouraging new demands for separated housing away from these "others" with high gates, armed guards, and video surveillance (Walks, 2010).

Emerging evidence from Canadian cities underscores the paradox of social-mix discourse. Despite the rationale of diversity used to promote gentrification, Walks and Maaranen (2008) found that income and ethnic diversity declined in the gentrifying neighbourhoods of Toronto, Vancouver, and Montreal between 1971 and 2001. Yet, across Canada, the idea of social mix has found new champions.

This discourse is particularly evident in Toronto, where the Toronto Community Housing Corporation (TCHC) has undertaken two large-scale, mixed-income public housing developments in the last decade—the Don Mount Court Project and

continued

the Regent Park Revitalization Plan—with a third on the way in the Lawrence Heights neighbourhood. The Don Mount Court Project on Dundas West was an early example of such redevelopment projects in Canada (August 2014; for other early projects, see Hulchanski, 1984). A public housing project constructed in the late 1960s, Don Mount (now named Rivertowne) came to be associated with concentrated poverty, crime, and social isolation. From 2002 to 2010, it was the focus of a $102-million regeneration scheme. The project's 232 original units of public housing have been replaced with 187 mixed-income townhouse condominiums linked by design features intended to help engender interconnections within the community (Kearns Mancini Architects, 2013).

Although these developments have been lauded as exemplary cases of mixed-community planning, recent research by August (2014) demonstrates, at Rivertowne, the consequences of such developments for low-income tenants. She observed a series of meetings and events that were intended to promote interaction between public housing and market-rate tenants. Not only were these meetings characterized by conflict, but they also had serious consequences for low-income tenants. As August (2014) explains, during these meetings, middle-class residents often used their political influence and social capital to the detriment of tenant safety and quality of life.

In Montreal, some community organizations have embraced mixed communities (Rose et al., 2013). Community activists in Hochelaga—an area of the city that experienced deindustrialization, population loss, and disinvestment—supported controlled gentrification of the neighbourhood in order to encourage middle- and upper-income investment. The Lavo Project of the 1990s involved the conversion of a brownfield (former industrial) site into a residential development and was supported widely by the community as an opportunity to rebalance the neighborhood's income diversity, perceived to have been disrupted by the movement of former middle-income residents to the suburbs (Rose et al., 2013). This case illustrates the pervasiveness of the social-mix rationale, which may detract from Montreal's long history of social housing as a product of struggles—followed by negotiations—between governmental and community stakeholders.

While Canadian cities give us much to celebrate, the emergence of gentrification, the creation of gated communities, and the politics of social mixing represent important issues of concern. Geographers, in particular, have consistently warned us that trends like these are leading to tremendous, potentially detrimental change in the quality of life for many Canadians. As the spectres of inequality, segregation, and exclusion rise throughout Canada's urban system, it is increasingly important that students of urbanization remain informed of the dynamic forces that continue to shape the material, social, and economic livability of Canada's diverse urban communities.

Conclusion

This chapter has explored three urban forms that highlight the social, economic, and political complexity in contemporary city making. In particular, the focus on gentrification, gated communities, and mixed communities has stressed that the morphology of any city and its neighbourhoods both shape and are shaped by particular visions of society. Gentrification and gated communities, for instance, reflect patterns and processes of exclusion in the city, while the social-mixing agenda underlines the complex politics associated with an attempt by the state to produce the opposite, inclusivity.

In the case of gentrification, rising urban property values, widening income inequalities, and rapid sociodemographic change have seriously limited the ability of different, often lower-income, groups to live in the city centre. Importantly, inner-city upgrading by urban elites has gone mainstream, taking root in previously overlooked urban contexts, like schools and churches, or intensifying in areas with long-standing patterns of gentrification. So, too, as cities inevitably change, some residents are taking to partitioning and fortifying their private spaces. In such cases, gated communities are built in and around metropolitan areas as a means to ensure security, communicate status, and realize (private) lifestyles. Bricks and mortar are fashioned into gates, hard defensible surfaces that are the material manifestation of fear and represent the demand, by some, for conspicuous distance and exclusion from others.

In exploring social mixing, this chapter has also pointed to the enduring complexity of interventions in the urban forms and the social landscapes of the city. This is not to say that interventions should be abandoned or alternative ways of building our cities should not be sought, but rather that our intentions in reshaping our city can have unforeseen consequences.

As we continue to trace the social and material evolution of the city, it is clear that more attention needs to be paid to how, why, and, perhaps most importantly, for whom are urban spaces being (re)defined. Academics and urban planners point out, for instance, the need for more research concerning how processes of globalization (i.e., flows of people, ideas, cultures) touch down in local areas to impact upon the form and function of gentrification, gated communities, and neighbourhood-based policies. While this agenda is particularly salient in world cities of the global North, like London and New York City, it is important not to neglect cities that have long been viewed as peripheral. Cities of the global South are being rapidly remade and have adapted to accommodate the dramatic impacts of global neoliberalism, urbanization, and surging populations seeking work in the city. In megacities like Mumbai, Shanghai, and Johannesburg, for instance, numerous urban quarters are being built and retrofitted with defensible spaces and surveillance technologies, while others are rebranded to entice revitalization and gentrification. How then are the dreams and fears of urban leaders in these places shaping, directing, or intervening in the making of exclusive city spaces?

But beyond such questions, it is necessary to recognize that while patterns of gentrification, gating, and mixing are going global, resistance to such forms is globalizing as well. Indeed, with various results, local communities are mobilizing against exclusion, inequality, and segregation. How might resistance movements across urban contexts and at more local levels affect the morphology of the city? During public demonstrations in the spring of 2015, protesters in the Brixton neighbourhood of London, England, circulated their list of solutions to gentrification, many of which reached beyond the local scale: build more houses for working people; put caps on rent; stop evictions; prioritize repair over regeneration; and build a livable city (Hill, 2015). Envisioning a livable city, a city of social as well as racial and economic equality, is for many a central agenda for today's urbanizing world, and it will take efforts from both experts and members of local communities to realize this goal.

Key Points

- Gentrification is a process characterized by the upgrading of working-class housing, the construction of new-build towers, and the conversion of loft spaces in city centres primarily by middle- and upper-income groups.
- Gated communities refer to planned neighbourhoods that use design features (walls, gates, fences, guards) as a way to control public access in response to fear of crime and/or as a symbolic means to build lifestyle, and reflect one's status in society.
- Social mixing is an urban policy agenda that seeks to increase neighbourhood and tenure diversity by mixing tenants across socio-economic lines.
- Loft development and adaptive reuse in London, England, represent an emergent form of gentrification in which post-industrial and post-institutional buildings are incorporated into the social upgrading of local neighbourhoods.
- The poor-doors phenomenon represents one outcome of market-based inclusionary housing approaches that are intended to address housing-affordability crises by offering incentives to developers if they include units at below market rates.
- Gentrification, gated communities, and social mix have evolved and impacted cities across Canada in a variety of ways: gentrification has spread from Canada's largest urban centres to a number of its emerging urban regions; various forms of gated communities are being built in both suburban and urban contexts; and social-mixing policies are on the rise as "solutions" to many of Canada's urban problems.

Activities and Questions for Review

1. What elements of the stage model of gentrification have you seen play out in an urban landscape, and how?
2. Visit a gated community, either in person or online. What are some key features that you can see and what purpose do they serve?
3. As a class, debate whether social mixing is a "cure all" solution to socio-economic divisions in the city.
4. In small groups, mark on a map of the largest city in your province or territory areas where gentrification, gated communities, and social mix can be found. What does your map tell you about patterns of residential development and social exclusion in this city?

References

Adams, D., Tiesdell, S., and White, J.T. (2013). Smart parcelization and place diversity: Reconciling real estate and urban design priorities. *Journal of Urban Design, 18*: 459–477.

Álvarez-Rivadulla, M.J. (2007). Golden ghettos: Gated communities and class residential segregation in Montevideo, Uruguay. *Environment and Planning A, 39*(1): 47–63.

Atkinson, R., and Blandy, S. (Eds.) (2007). *Gated communities: International perspectives*. Routledge: London.

August, M. (2014). Negotiating social mix in Toronto's first public housing redevelopment: Power, space and social control in Don Mount Court. *International Journal of Urban and Regional Research, 38*(4): 1160–1180.

Bain, A.L. (2006). Resisting the creation of forgotten places: Artistic production in Toronto neighbourhoods. *Canadian Geographer/Le Géographe canadien, 50*(4): 417–431.

Basu, R. (2007). Negotiating acts of citizenship in an era of neoliberal reform: The game of school closures. *International Journal of Urban and Regional Research, 31*: 109–127.

Benjamin, R. (2012). The gated community mentality. *New York Times*, 29.

Blakely, E.J., and Snyder, M.G. (Eds.) (1997). *Fortress America: Gated communities in the United States*. Brookings Institution Press.

Bradshaw, P. (1999). It's not just a handy carnival location you know. *The Guardian*, April 27. Retrieved from http://www.theguardian.com/theguardian/1999/apr/27/features11.g22

Bridge, G., Butler, T., and Lees, L. (Eds.) (2012). *Mixed communities: Gentrification by stealth?* Chicago: Policy Press.

Butler, T., and Robson, G. (2003). *London calling: The middle classes and the re-making of inner London*. Oxford: Berg.

Carpenter, J., and Lees, L. (1995). Gentrification in New York, London and Paris: An international comparison. *International Journal of Urban and Regional Research, 19*(2): 286–303.

Coy, M. (2006). Gated communities in Latin American megacities: Case studies in Brazil and Argentina. *Environment and Planning B: Planning and Design, 29*(3): 355–370.

Dangschat, J. (1988). Gentrification: Der Wandel innenstadtnaher Wohnviertel. *Soziologische Stadtforschung*, pp. 272–292.

Davidson, M., and Lees, L. (2010). New-build gentrification: Its histories, trajectories, and critical geographies. *Population, Space and Place, 16*: 395–411.

Department of City Planning (n.d.). Zoning tools: Inclusionary zoning. City of New York. Retrieved from http://www.nyc.gov/html/dcp/html/zone/zh_inclu_housing.shtml

Extell (n.d.). One Riverside Park. Retrieved from http://oneriversidepark.com

Fincher, R., Iveson, K., Leitner, H., and Preston, V. (2014). Planning in the multicultural city: Celebrating diversity or reinforcing difference? *Progress in Planning, 92*: 1–55.

Florida, R. (2012). New York vs. London, in debate form. *Atlantic City lab*. Retrieved from http://www.citylab.com/politics/2012/02/new-york-vs-london-debate/1228

Frenzel, F. (2014). Slum tourism and urban regeneration: Touring inner Johannesburg. *Urban Forum, 25*(4): 431–447.

Furman Centre. (2014). *State of New York City's housing and neighbourhoods in 2014*. New York: NYU Furman Centre.

Gayle, D. (2015). Brixton at standstill as crowds register frustration at gentrification, *The Guardian*, April 26. Retrieved from http://www.theguardian.com/uk-news/2015/apr/26/brixton-at-standstill-as-crowds-register-frustration-at-gentrification

Ghosh, S. (2014). Everyday lives in vertical neighbourhoods: Exploring Bangladeshi residential spaces in Toronto's inner suburbs. *International Journal of Urban and Regional Research, 38*(6): 2008–2024.

Giglia, A. (2008). Gated communities in Mexico City. *Home Cultures, 5*(1): 65–84.

Glass, R. (1964). *London: Aspects of change*. London: MacGibbon and Kee.

Goodyear, S. (2013). The threat of gated communities. *CityLab*, July 15. Retrieved from http://www.citylab.com/crime/2013/07/threat-gated-communities/6198

Grant, J., and Mittelsteadt, L. (2004). Types of gated communities. *Environment and Planning B: Planning and Design, 31*(6): 913–930.

He, S. (2007). State-sponsored gentrification under market transition: The case of Shanghai. *Urban Affairs Review, 43*(2): 171–198.

Hill, D. (2015). Brixton's anti-gentrification protest: Identifying the problems is one thing, fixing them is another. *The Guardian*. Retrieved from http://www.theguardian.com/cities/davehillblog/2015/apr/28/brixton-anti-gentrification-protest-reclaim-foxtons-estate-agent

Hoyle, B., Pinder, D., and Husain, M. (1988). *Revitalising the waterfront: International dimensions of dockland redevelopment*. Toronto: Belhaven Press.

Hulchanski, D. (1984). *St. Lawrence and False Creek: A review of the planning and development of two new inner city neighbourhoods*. Vancouver: School of Community and Regional Planning, University of British Columbia (Planning Papers 10).

Hulchanski, D. (2010). *The three cities within Toronto: Income polarization among Toronto's neighbourhoods, 1970–2005.* Toronto: Cities Centre, University of Toronto.

Kearns Mancini Architects Inc. (2013). *Don Mount Court, Rivertowne.* Retrieved from http://kmai.com/projects/don-mount-court-rivertowne

Kern, L. (2008). Gender, condominium development and the neoliberalization of urban living. *Urban Geography, 28*(7): 657–681.

Lawton, P., and Punch, M. (2014). Urban governance and the "European city": Ideals and realities in Dublin, Ireland. *International Journal of Urban and Regional Research, 38*: 864–885.

Lees, L. (2014). The urban injustices of new Labour's "New Urban Renewal": The case of the Aylesbury Estate in London. *Antipode, 46*(4), 921–947.

Lees, L., Butler, T., and Bridge, T. (2012). Introduction: Gentrification, social mix/ing and mixed communities. In G. Bridge, T. Butler, and L. Lees (Eds.), *Mixed communities: Gentrification by stealth?* Chicago: Policy Press, pp.1–16.

Lees, L., Slater, T., and Wyly, E. (Eds.) (2008). *Gentrification.* London and New York: Routledge/Taylor & Francis Group.

Ley, D. (1996). *The new middle class and the remaking of the central city.* Toronto: Oxford University Press.

Ley, D., and Dobson, C. (2008). Are there limits to gentrification? The contexts of impeded gentrification in Vancouver. *Urban Studies, 45*(12): 2471–2498.

Low, S.M. (2003). *Behind the gates: Life, security, and the pursuit of happiness in Fortress America* (Vol. 35). New York: Routledge.

Lynch, N. (2016). Domesticating the church: The reuse of urban churches as loft living in the post-secular city. *Social & Cultural Geography*, doi: 10.1080/14649365.2016.1139167

Marom, N., and Carmon, N. (2015). Affordable housing plans in London and New York: Between marketplace and social mix. *Housing Studies, 30*(7): 993–1015.

Matejskova, T., and Leitner, H. (2011). Urban encounters with difference: The contact hypothesis and immigration integration projects in eastern Berlin. *Social and Cultural Geography, 12*: 717–741.

Michael, C., and Bramely, E.V. (2014). Spike Lee's gentrification rant—transcript: "Fort Greene Park is like the Westminster dog show," *The Guardian*, February 26. Retrieved from http://www.theguardian.com/cities/2014/feb/26/spike-lee-gentrification-rant-transcript

Navarro, M. (2015). 88,000 applicants and counting for 55 units in "Poor Door" building. *New York Times*, April 20. Retrieved from http://www.nytimes.com/2015/04/21/nyregion/poor-door-building-draws-88000-applicants-for-55-rental-units.html?_r=0

Phillips, M. (2010). Counterurbanisation and rural gentrification: An exploration of the terms. *Population, Space and Place, 16*(6): 539–558.

Podmore, J. (1998). (Re)reading the loft living habitus in Montreal's inner city. *International Journal of Urban and Regional Research, 22*: 283–302.

Pottie-Sherman, Y. (2013). Vancouver's Chinatown night market: Gentrification and the perception of Chinatown as a form of revitalization. *Built Environment, 39*(2): 172–189.

Provost, C. (2014). Gated communities fuel Blade Runner dystopia and profound unhappiness. *The Guardian*, May 2. Retrieved from http://www.theguardian.com/global-development/2014/may/02/gated-communities-blade-runner-dystopia-unhappiness-un-joan-clos

Qian, J., He, S., and Liu, L. (2013). Aestheticisation, rent-seeking, and rural gentrification amidst China's rapid urbanisation: The case of Xiaozhou village, Guangzhou. *Journal of Rural Studies, 32*: 331–345.

Quinn, S. (2015). The poor door debate is a waste of our collective breath. *Globe and Mail*, May 8. Retrieved from http://www.theglobeandmail.com/news/british-columbia/the-poor-door-debate-is-a-waste-of-our-collective-breath/article24351968

Rose, D., Germain, A., Bacqué, M.H., Bridge, G., Fijalkow, Y., and Slater, T. (2013). "Social mix" and neighbourhood revitalization in a transatlantic perspective: Comparing local policy discourses and expectations in Paris (France), Bristol (UK) and Montréal (Canada). *International Journal of Urban and Regional Research, 37*: 430–450.

Rosen, G., and Grant, J. (2011). Reproducing difference: Gated communities in Canada and Israel. *International Journal of Urban and Regional Research, 35*(4): 778–793.

Rosen, G., and Walks, A. (2013). Rising cities: Condominium development and the private transformation of the metropolis. *Geoforum, 49*: 160–172.

Slater, T. (2004). North American gentrification? Revanchist and emancipatory perspectives explored. *Environment and Planning A, 36*: 1191–1213.

Smith, N. (2002). New globalism, new urbanism: Gentrification as global urban strategy. *Antipode, 34*(3): 427–450.

Stockdale, A. (2010). The diverse geographies of rural gentrification in Scotland. *Journal of Rural Studies, 26*(1): 31–40.

Townshend, I.J. (2006). From public neighbourhoods to multi-tier private neighbourhoods: The evolving ecology of neighbourhood privatization in Calgary. *GeoJournal, 66*(1–2): 103–120.

Visser, G., and Kotze, N. (2008). The state and new-build gentrification in central Cape Town, South Africa. *Urban Studies, 45*(12): 2565–2593.

Walks, R.A. (2010). Electoral behaviour behind the gates: Partisanship and political participation among Canadian gated community residents. *Area, 42*(1): 7–24.

Walks, R.A., and Maaranen, R. (2008). The timing, patterning, and forms of gentrification and neighbourhood upgrading in Montreal, Toronto, and Vancouver, 1961 to 2001. Citiescentre.utoronto.ca, May (211): 1–110.

Wilson, W.J. (1987). *The truly disadvantaged*. Chicago: University of Chicago Press.

Wu, F. (2004). Transplanting cityscapes: The use of imagined globalization in housing commodification in Beijing. *Area, 36*(3): 227–234.

Zukin, S. (1982). *Loft living: Culture and capital in urban change.* New Brunswick, NJ: Rutgers University Press.

Zukin, S., Trujillo, V., Frase, P., Jackson, D., Recuber, T., and Walker, A. (2009). New retail capital and neighborhood change: Boutiques and gentrification in New York City. *City & Community, 8*: 47–64.

12

Unequal and Volatile Urban Housing Markets

Alan Walks and Dylan Simone

Introduction

The residential **housing stock** makes up a significant proportion of the total fixed capital of any given city. Housing represents the largest single expenditure that households face, and comprises the majority of wealth accumulated by home-owning households. **Housing markets** are highly localized and at the same time increasingly global in reach as residential **land titles** and **mortgages** issued against housing collateral are turned into financial commodities traded on global securities markets. In most globally connected cities, a number of households live transnational lives (sometimes living and working in multiple countries). Housing is thus one of the most important factors in understanding contemporary urbanization processes, including social and spatial inequalities in wealth **accumulation** among different social groups and neighbourhoods.

This chapter begins with a discussion of the basic structure of residential **housing systems** and markets in countries in the global North. It then analyzes changes in market structures and housing policies, including the globalization of housing finance and its relationship to the 2008 global financial crisis (GFC). The effects of these changes are then examined in relation to the volatility experienced in housing production and prices, and to inequities related to access to housing and mortgage lending, particularly in the United States, which has the most globalized housing and mortgage markets, but also through case studies of the Irish and Spanish housing bubbles in relation to the Dublin and Barcelona metropolitan regions. The chapter then discusses the implications for Canadian urban housing markets, comparing pre- and post-crisis conditions, as well as showcasing a study of Montreal. The chapter concludes with the reminder that the effects of housing bubbles and the global financial crisis vary widely among and within national contexts, as do their social implications and policy responses.

The Structure of Housing Systems and Markets

Housing markets are one component of the overall housing system in any given nation, built through social and political negotiation, conflict, and policy shifts over many years. Most housing systems involve a series of market, non-market, and quasi-market forms of housing. In non-market forms, occupants are typically selected on the basis of need rather than on ability to pay, and occupants pay for housing on the basis of a proportion of their income instead of on what the market will bear. Such forms include housing built and maintained by the state (often called public housing), rental housing provided by non-profit agencies (which is often but not necessarily subsidized by the state), and limited-equity co-operatives and rental co-housing initiatives in which residents share in governance and maintenance decisions. The proportion of the occupied housing stock in non-market forms ranges from over 40 per cent in Hong Kong, to around 16 per cent in France and the United Kingdom, to less than 5 per cent in certain countries in the global North such as Canada (4.2 per cent: CMHC, 2014), Australia (3.9 per cent: ABS, 2011), Spain (roughly 2 per cent: INE, 2011), and, notably, the United States (1.0 per cent: calculated from HUD, 2007; and US Census

Bureau, 2008). Even the latter nations, however, reveal significant variation across metropolitan areas. In New York City, for instance, public housing makes up 5.4 per cent of the housing stock—more than five times the US national average (as calculated from NYCHA, 2014; and US Census Bureau, 2013). In many cities, the proportion of the housing stock in non-market forms has fallen considerably since the late 1980s as a result of both neoliberal policies (discussed below) and rapid growth of the private-market stock since that time.

Market forms of housing involve different kinds of tenures, including freehold ownership, **condominium** (strata) tenure, rent-to-own arrangements, and market rental (leasehold) tenures. In the global North, the proportion of the population that own their housing ranges from a low of 44 per cent in Switzerland to over 90 per cent in some Eastern European countries (see the EuroStat housing statistics database at http://appsso.eurostat.ec.europa.eu/nui/show.do). Within nations, the rates of home ownership and the kinds of tenure rights associated with this differ significantly between places. Quasi-market forms of housing, meanwhile, can involve a range of options, including state-subsidized "affordable ownership" programs, as well as Singapore's 99-year tradeable leases on housing units. In the global South, meanwhile, a very large proportion of the housing stock is in market or quasi-market forms, with the vast majority of the latter involving semi-legal or illegal squatter settlements in which residents have limited tenure or land rights.

Under market-based forms, housing is allocated according to willingness and ability to pay. In competitive markets, the differential prices or rents attained are typically due to variance in quality, maintenance, size, level of amenity, and location, within the context of the prevailing correspondence (or "equilibrium") between the supply of units and the level of effective demand. The latter is itself a function of disposable income, degree of optimism, access to credit, and the availability of alternatives. Together, such features influence the amount different users of land and housing are willing and able to bid, in terms of rents or purchase prices, for occupation. Such factors are the bases for the "hedonistic" (preference-satisfying) models of price and rent determination. The latter fit ideologically within the larger "consumer sovereignty" perspective regarding housing allocation and urban location, which assumes that cities have developed "naturally" according to common demand preferences that the market expresses through price channels. This can be contrasted with political economy approaches and other critical perspectives that see access to housing and urban development, as well as the evolution of the "market" itself, as anything but natural, and instead as an expression of differential power among unequal groups (based on class, race, gender, and culture), including the power of landowners and developers.

Under the mainstream consumer sovereignty perspective, prevailing urban forms are said to be the expression of consumer preferences in relation to the trade-off between accessibility and space. This is the core feature of the **bid-rent models** of urban land use, in which each different sector (and each sub-sector within the larger residential sector) bids different amounts per unit of land based on accessibility to employment, typically in relation to distance from the **central business district** (CBD). Offices, retail, and high-density residential sub-sectors (apartments and condominiums) use land more intensively and thus are willing and able to out-bid other users for space in the central city, whereas uses with flatter curves—industrial, low-density (single-family), residential, and, particularly, agricultural uses—provide the winning bids for space near the fringes (Miron, 2000). When the accessibility benefits of being near transportation routes are added to the models, a variegated land-value surface is produced, with a strong peak at the centre coupled with peaks of diminishing intensity at the intersections of key transportation routes as distance from the CBD increases (Berry, 1963).

There is a long history of racial, ethnic, gender, and class discrimination in the labour and housing sectors of many cities and nations that challenges mainstream economic approaches. The work of Massey and Denton (1993) and Wacquant (2008) shows how racial discrimination in the United States compelled African-Americans to live in particular urban districts and, furthermore, how the legacies of racial settlement patterns, coupled with state retrenchment and neglect enabled the reproduction of racialized housing markets even after civil rights legislation eroded formalized racial ordinances and covenants. One form of racial discrimination involves **redlining**, in which lenders refuse to provide mortgages in neighbourhoods seen as high risk. Redlining was initiated in the 1930s by the US Federal Housing Administration (FHA), which used four different colours to represent neighbourhood trajectories that became the basis for judging their level of credit worthiness (see Aalbers, 2011; Massey and Denton, 1993). **Reverse redlining**, also referred to as "yellowlining" and "greenlining," is a process through which lenders identify and target certain neighbourhoods for aggressive loan marketing. Meanwhile, a number of scholars have argued that the politics of **zoning** have increased population densities in certain (particularly poorer) locales and led to much lower average population concentrations in wealthier areas within US cities (Fischel, 1999). Mainstream economists characterize these issues as mere "market distortions," resulting in allocations that are different from what one might expect given "purely" economic preferences and variables. However, given that the urban by its very definition involves the bringing together of strangers, producing a terrain of ever-shifting political coalitions and conflicts (Simmel, 2002; Isin, 2002), we might consider the social and cultural politics that derive from the city inherent to the structure of urban housing markets. In turn, we might reject any theoretical approach, such as neoclassical economics, that seeks to define such factors as outside the realm of the market. Markets are inherently social, political, and cultural phenomena.

Evolution of Housing and Mortgage Markets

Land markets have always been a feature of cities in the global North. While formal developer-builders had long been active in the construction of tenements and other apartment buildings in the large cities of Europe and North America, a significant portion of the low-rise residential landscape built at the suburban fringe, particularly in North America and Australasia, was self-built at least until the First World War (Harris, 1996). The first models for non-market housing included the utopian communities built in the United Kingdom and elsewhere in Europe in the mid-1800s and the Boundary Estate housing project built in London in the 1890s (see Harloe, 1995). Yet, for the most part, it was not until the 1920s in Germany and the United Kingdom, and the 1940s and 1950s in North America and other European nations, that public (state-provided) housing projects were planned on any scale. And it was only after the Second World War that non-market forms of housing became sufficiently common to begin to challenge private-market provision (Harloe, 1995).

Across the global North, until the mid-1930s states had little direct influence, either through state-run insurance programs or through regulation, on lending practices in private-mortgage markets. Banks were largely not in the business of mortgage lending, and if they were, they would often require at least 50 per cent equity in a down-payment, with short amortization terms (for Canada, see Poapst, 1993). After the Second World War, many countries in the global North sought to develop mortgage markets according to social and political objectives. National regulatory bodies were set up to monitor, regulate, and in some cases directly insure mortgages issued by lenders. The United States in 1938 had the FHA begin insuring mortgages issued by approved lenders and established the Federal National Mortgage Association ("Fannie Mae") to purchase these mortgages and set mortgage-lending terms. This system effectively dictated the kinds of mortgage products lenders could issue if they wanted them to be insured (with most such insurance protecting the lenders, not the borrowers), and encouraged banks to become more involved in mortgage lending. Mortgages issued according to these terms, which set limits on loan-to-value ratios, amortization terms, loan amounts, and property purchase prices, are called "conforming" mortgages.

Some countries followed the US by setting up organizations with similar mandates and functions. The state-run Japan Housing Loan Corporation (JHLC), for instance, was set up in 1950 with the objective of funding new housing construction, to act as a mortgage lender of last resort, and to compel builders to meet higher construction standards set by the state (Seko, 1994). The subdivision of urban land and the development of mortgage credit proceeded more slowly in the global South. Some nations, such as Mexico and Brazil, followed with similar organizations, although their impacts were typically weaker, with the housing they helped build often swamped by the rapid spread of unplanned squatter settlements.

From the early postwar period until the early 1980s, mortgage-lending systems in many countries ensured the smooth flow of funding for new housing construction in both the market and non-market sectors. However, while purchasers deemed most credit-worthy were able to access mortgages and move into new housing typically located in new suburbs, neighbourhoods with older housing were often subject to redlining and denied access to mortgage credit by commercial lenders and government mortgage institutions. Race and other social characteristics played a role in limiting access to credit. In the United States, African-American neighbourhoods were much more likely to be redlined than other neighbourhoods (Aalbers, 2011; Massey and Denton, 1993). In general, because mortgage credit was significantly more difficult to access for older housing, demand for housing in the inner cores of cities around the globe declined in relative, and often in absolute, terms, leading to declining prices relative to the suburbs. Groups who could access other forms of funding, particularly landlords, but also ethnic groups who could access funds through informal family and community channels, often ended up owning this older housing (Murdie, 1986, 1991). One result was the shift of older inner-city housing towards rental tenure, as well as the spatial concentration of particular social, racial, and ethnic/immigrant groups within inner-city districts through the early postwar period.

High and rising interest rates in the 1970s and 1980s made the direct funding of public and social housing more costly for governments, which then looked to private-sector solutions. In the United States, housing and mortgage corporations were restructured and began innovating in the trade and securitization of **mortgage-backed securities** (MBS). MBS involve the pooling of many mortgages into a single security. Under the US system, state-sponsored organizations sell bonds and stocks and then use the proceeds to purchase conforming FHA-insured mortgages packaged into MBS by banks, which frees up their balance sheets and allows them to continue lending to new borrowers. This system uses the financial power of the state to promote credit creation on behalf of private-sector lenders, but on the terms dictated by the state, and can be effective in doing so as long as the underlying loans continue to be paid and the market remains buoyant. The success of this state-sponsored "public label" system encouraged the major Wall Street investment banks to set up a parallel private system, using special-purpose vehicles (SPV) to raise funds for purchasing and securitizing MBS. Under this "private label" system, the state no longer has control over lending terms. In Europe, a similar system developed by which private-sector "covered bonds" rose in prominence as the main financial securities funneling credit to mortgage markets.

From the late 1980s through the 2000s, housing systems across the global North underwent transformation. Eyeing the US model, many nations sought to encourage the securitization of mortgages as a way to encourage private lenders to issue mortgages in the context of declining public commitments to non-market housing (Walks and Clifford, 2015). In the United Kingdom, the Thatcher government privatized the most desirable public-housing units (known as council housing) through its "right-to-buy" scheme, discouraged the building of any new social housing, deregulated the building societies (which were until then the main providers of mortgage credit), and opened up the mortgage market to non-traditional lenders (Ford, 1994). Through the **Hope VI** program, many social-housing communities in the United States were demolished and rebuilt as "mixed" tenure communities, but with fewer units overall and with the "mix" articulated in the difference between the very separate living quarters of the market-ownership units and the socially subsidized units (often on completely separate streets or districts) (Goez, 2013; Smit, 2013). As nations across the globe adopted variations of these policies to differing degrees, non-market housing sectors increasingly lost out to private-market initiatives; and large sections of cities—often poor areas sitting on highly accessible land near the core—were redeveloped with housing intended for the middle classes and elites (August, 2014; Lees, and Wyly, 2008). With the **gentrification** of many inner cities and the displacement of lower-income households into less-accessible neighbourhoods containing less-desirable market housing and "residualized" (under-maintained and often ignored) non-market housing, cities became increasingly segregated by class, as well as by race and immigrant status (Johnston, Poulsen, and Forrest, 2007; Massey and Denton, 1993; Walks, 2011).

Nonetheless, such shifts promised lower government budget deficits and the promise of private-sector investment into housing. Many nations brought in their own securitization schemes and encouraged competition in the mortgage insurance business through the granting of new licences to private-sector mortgage insurers. Many nations in the global South, such as Mexico and Brazil, eyeing the success of securitization schemes in the global North, founded new institutions with the express purpose of promoting the expansion of mortgage lending (de Mendonca and Barcelos, 2015; Soederberg, 2014).

Securitization was also tried as a means of attracting private-sector funding into non-market (social) housing sectors, but those securities had little market demand, and so the vast majority of new investment found its way into the private-mortgage market, further skewing public policies away from non-market solutions (Walks and Clifford, 2015). The securitization of housing assets can thus be seen as an important component of the neo-liberalization of housing policy. It was also a factor in the production of greater volatility, and crisis, in **housing markets**.

Financialization, Volatility, and Crisis

Recent shifts in housing finance are part of the wider financialization of the economy in which "profits accrue primarily through financial channels rather than through trade and commodity production" (Krippner, 2005: 174). Securitization of MBS allows for spatially fixed houses to become globally tradable financial assets, thus producing financial liquidity out of the geographical fixity of urban housing markets (Gotham, 2009). The rise of private-label MBS and their subsequent use in collateralized debt/mortgage obligations (CDOS/CMOS), in which bunches of MBS are pooled and then "tranched" (divided into different securities) based on the level of perceived risk (with the different tranches sold to different investors with different risk appetites), helped promote the "financialization of home" (Aalbers, 2008).

The growing popularity of mortgage securitization practices across the globe (both public and private label) led to a rapid increase in the amount of financial securities offered on the market and, in turn, to the amount of credit available for the purchase of housing and land. This had consequent effects in both driving up housing and land prices and spurring new housing and office production. The overall result was to encourage investment in housing and urban redevelopment rather than in traditional production sectors—a global example of what Harvey (1978) calls **capital switching**. The effects of this massive switch of capital investment into land and housing markets (Harvey's "secondary circuit") have been uneven over time and across space, as evidenced by the volatility in house price fluctuations in different national housing markets (Figure 12.1). Whereas the housing markets in nations such as Canada witnessed relatively smooth incremental gains in relative prices over time, other markets boomed before the financial crisis and then crashed once it hit. The degree of boom and bust

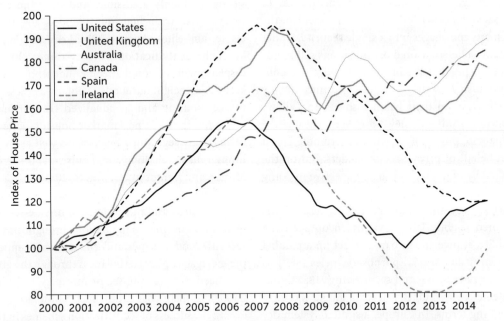

Figure 12.1 House prices in real terms, selected countries, 2000–2014 (100 = Jan. 1996, except for Calgary, Toronto, Hamilton, and Ottawa 100 = Jan. 1999)

Alan Walks and Dylan Simone / The Economist.

varies across contexts, and only some of these markets have "recovered."

Such trajectories are not happenstance, but relate to the very different ways that credit and investment flow, as well as economic growth, are articulated spatially and socially within and between nations. The forms and directions taken by these flows are themselves influenced by the kinds of social, institutional, and regulatory practices that shape mortgage lending and other forms of investment in each place. In the United States, the profusion of mortgage credit, coupled with the fact that, with securitization, lenders no longer needed to hold the mortgages on their balance sheets but instead could sell them to intermediaries for packaging into MBS, meant that lenders had an incentive in raising the volumes of their lending yet little interest in ensuring that borrowers could pay back the loans. This augmented existing tendencies towards predatory forms of lending (Immergluck, 2009; Wyly et al., 2009). The vast profits involved in mortgage lending made the stakes high, and the George W. Bush administration sided with lenders in "pre-empting" key state-level, anti-predatory-lending legislation (Immergluck, 2009, 2011). This added fuel to the fire and helped spur the origination of new kinds of **"subprime" mortgage** products with predatory terms (Immergluck, 2009).

Subprime mortgages are officially those offered to borrowers with "less-than-prime" credit scores. They were justified by the theory of risk-based pricing, which promotes the extension of credit to borrowers with greater risk of default, but at higher interest rates to compensate for the additional risk, in the name of "market completion" (Ashton, 2009; Wyly et al., 2009). Squires (2009: 21) outlines key characteristics among subprime mortgages in the United States: (1) high interest rates and fees; (2) initial "teaser" interest rates that reset upward after a few years; (3) high pre-payment penalties; (4) loans issued on the value of the collateralized property instead of the borrower's ability to pay; (5) "loan flipping" characterized by frequent refinancing; (6) "balloon" payments in which the term ends with a substantial principal payment, compelling the borrower to refinance beforehand; and (7) negative amortization. In the decade leading up to the American housing crisis, many such characteristics were included in subprime mortgage contracts in order to get borrowers to constantly refinance, generating substantial profits for subprime lenders and brokers.

The US financial system began unraveling in earnest once the (originally triple A-rated) CDOs stopped performing and large investment banks on Wall Street began making bets against them (Engel and McCoy, 2011; Engelen et al., 2011). Notably, it was only those scholars and analysts working from heterodox economic perspectives—particularly from those based on the work of Marx, Minsky, the post-Keynesian school, and the Austrian school—who predicted in advance the crisis that ensued, which is why the common refrain from the "orthodox" mainstream (including heads of the US Federal Reserve) is that "no one saw this coming" (Bezemer, 2009). The American experience has proven the neoclassical school teachings to be (perhaps unwittingly) complicit in producing the economic crisis and in encouraging predatory lending, if not also completely useless as economic theory (Keen, 2011; Smith, 2010).

The American **housing bubble**, and the deep housing market correction and "Great Recession" that subsequently engulfed the country once it burst, led to the bankruptcy of some of the largest American lenders and to disparate effects among American cities. One important effect on the housing market has been a rapid increase in foreclosures that could not be absorbed by the market, even when the properties were offered at steep discounts on auction, leaving US banks with record numbers of "REO" properties (properties that could not be sold at foreclosure and are officially recorded as "real estate owned" by the lenders) (Immergluck, 2009). **Foreclosure** typically involves a series of steps that varies within and between nations. In some American states (and in some Canadian provinces), lenders can initiate a "short sale" (or a "power-of-sale" in Canada) to force the sale of a property without having to foreclose or take ownership of it. The rapid rise of defaults and foreclosures led to an unprecedented response from the US federal government and the US Federal Reserve. In an attempt to prevent the collapse of the US economy, they resorted to massive bailouts of the financial sector and automobile companies, historic stimulus/spending packages, and new programs and policies intended to encourage the banks to lend and households to borrow and spend (Engelen et al., 2011; Immergluck, 2011; Walks, 2010).

Foreclosures and mortgage defaults in the United States have augmented, rather than softened, neighbourhood-based inequalities by class and race. Across many American cities, lower-income households,

African-Americans, and non-white Hispanics were more likely to have been targeted for subprime loans, revealing the class and racial biases of predatory finance in the lead-up to the crisis (Hernandez, 2009; Wyly et al., 2009). As a result, the rate of foreclosure has generally been more severe since the crisis in the urban neighbourhoods where these groups are most concentrated (Immergluck, 2009; Rugh and Massey, 2010).

Immergluck (2011) uncovers significant unevenness in the degree to which American cities have suffered from declining house prices and rising numbers of foreclosures, with just over one-third (37 per cent) of all US metropolitan statistical areas (MSAs), comprising half of all mortgaged properties, seeing a rapid rise in foreclosures during the crisis (Immergluck, 2011: 138). These high-foreclosure MSAS display two distinct patterns. In slow-growing and deindustrializing cities that had already experienced high foreclosure rates in the run-up to the crisis, new foreclosures were disproportionately concentrated in central-city neighbourhoods. Many of these neighbourhoods contain housing built before the Second World War, and for much of the postwar period they disproportionately housed the poor and racialized communities, particularly African-Americans. Cities that fit this pattern include Detroit, Cleveland, Atlanta, Minneapolis, and Denver (Immergluck, 2011). The pattern of foreclosures is very different in many "hot-market" cities that were growing in advance of the crisis, where foreclosures had initially been rare. In these metropolitan areas, the foreclosures have been disproportionately concentrated in the newer suburbs located at the urban fringe (Immergluck, 2011). In the suburbs, Hispanic households have been significantly more likely to experience foreclosure (Niedt and Martin, 2013). Cities in this category include San Francisco, Los Angeles, Las Vegas, and Miami. Although the United States formally exited the recession in late 2009, house prices only began rising again in late 2012, and by the end of 2014 were still 23 per cent lower than when the crisis hit (Figure 12.1). The subprime/foreclosure crisis augmented class and racial inequality in US cities. Many lower-income households were effectively stripped of their equity and left deeper in debt, while wealthier households and higher-income neighbourhoods were far less likely to suffer from house-price declines, foreclosures, or equity loss (Grinstein-Weiss et al., 2015; Schafran and Wegmann, 2012).

While the United States enjoys disproportionate media and scholarly scrutiny, its institutions and the ways its cities have developed are somewhat atypical in a global context, and one cannot assume that the effects of financialization and crisis in American urban housing markets are similar to those in other nations. Even in nations that experienced equally deep housing-market downturns as the United States did, the effects on urban development and the articulation of urban social inequalities vary considerably. This can be seen in relation to two European nations that experienced unprecedented housing bubbles and subsequent busts, Ireland and Spain, and their key cities, Dublin and Barcelona.

The Impact of Ireland's Housing Bubble on Dublin

From the mid-1990s until the global financial crisis of 2008, Ireland sustained one of the largest housing bubbles in recent history. During this period, Ireland was often referred to as the "Celtic Tiger" (Allen, 2009). Through the 1990s Ireland witnessed a neoliberal turn focused on deregulation, the prioritization of free-market principles, and the privatization of public services (Kitchin et al., 2012). Ireland's Celtic Tiger era can be divided into two distinct periods. First, from the early to the late 1990s, increases in **foreign direct investment** (FDI) in export-oriented manufacturing and services sectors fueled a growing population, declining unemployment, and a rising standard of living (Honohan, 2010). Financial deregulation, coupled with a low corporate tax rate, enticed many transnational companies to invest in Ireland. As the country was part of the European Union (EU), this provided an entry point for firms wanting to do business in the EU (Kitchin et al., 2012; Mercille, 2014). Much of this new investment migrated to Dublin. Dublin is the capital and largest city in the Republic of Ireland, situated on the northeast coast. The population in 2011 was recorded to be over 525,000 people in the city proper and 1,270,000 people metro-wide, representing approximately 208,000 and 465,000 private households, respectively (Republic of Ireland Central Statistics Office, 2011).

By the early 2000s, exports had declined, while lower interest rates—partly resulting from Ireland's joining the Eurozone and the European Economic and Monetary

Union—fueled a construction boom (Mercille, 2014). Tax structures were also favourable towards the construction sector, and Irish banks could not keep up with all of the demand for credit, so they began borrowing from foreign institutions, which heavily favoured the residential property market (Mercille, 2014). New housing prices rose by 382 per cent across the country between 1991 and 2007, and by even more (429 per cent) in Dublin; and the number of new units built grew twice as fast as the population (71.9 per cent versus 30.1 per cent) (Honohan, 2010; Kitchin et al., 2012).

The bursting of the housing bubble came quickly with the global financial crisis in 2008. While many workers lost their jobs (many of whom then decided to migrate to other countries for work, alleviating some of the labour-market impacts) and a number of people lost their homes, the largest financial losses were concentrated among speculators who had purchased multiple properties for capital gains or to let out (erroneously betting that demand would continue rising) and among developers whose new housing estates were not yet complete at the time of the global financial crisis. Many of the latter "**ghost estates**," including numerous Dublin condo buildings and suburban subdivisions, remained frozen in time afterwards, with piles of dirt and digging machines sitting in exactly the same places as on day the crisis hit (Figure 12.2) (Kitchin et al., 2014). Since the crisis, Dublin has experienced emigration owing to job loss, climate-related issues such as flooding, and demographic shifts that have led to changes in the socio-spatial composition of the city (see Jeffers, 2013).

Irish governments, banks, labour unions, and private firms all profited during the boom, but once this was over the responsibility for the bailouts fell onto the public. The national government guaranteed bank deposits and liabilities, and the state moved to protect wealthy developers and speculators from losses, including instituting a plan to have social-housing providers rent some of the ghost estates from pinched developers (Jeffers, 2013). Moreover, in its bailout plan, Ireland (as well as Spain) pursued a different strategy from that of many other nations, creating a public "bad bank"—in the Irish case the National Asset Management Agency (NAMA)—to purchase toxic loans from other banks, with the idea that this would help keep the private banks solvent. The results were mixed, however, and Ireland was forced to request an €85-billion joint bailout from the International Monetary Fund (IMF), the European Union, and the European Central Bank (ECB) (the "troika") at the end of 2010 (Kitchin et al., 2012; Mercille, 2014; see also Bresnihan and Byrne, 2015). Ireland was left with a massive inventory of new houses and condo units built for populations that never materialized. Many of these unfinished ghost estates are found in suburban and exurban estates surrounding Dublin (Figure 12.3),

Figure 12.2 The unfinished "Red Arches" estate in Baldoyle, Dublin Region, Ireland, 2013

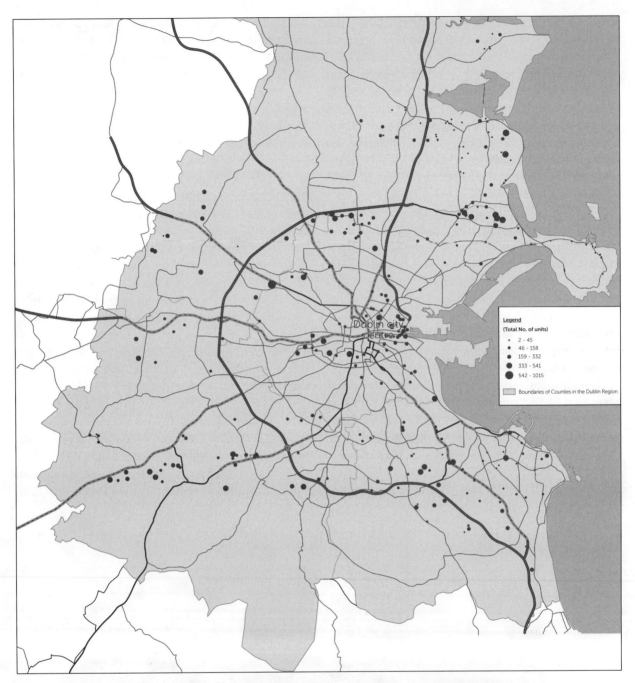

Figure 12.3 Map of unfinished (ghost) estates in the Dublin Region, Ireland, 2010

Aggregated by the author from maps at the county level published by the National Housing Development Survey, 2010, Department of Environment, Community and Local Government, Republic of Ireland, http://www.environ.ie/en/DevelopmentHousing/Housing/UnfinishedHousingDevelopments/NationalHousingDevelopmentSurvey2010/

as well as in smaller cities and clusters of condominiums near the CBD. While the state encourages developers to finish those units that were not complete at the time of the global financial crisis, a number of these estates now house low-income households, squatters, and desperate populations (Kitchin et al., 2014).

The Spanish Housing Bubble and Its Aftermath in Barcelona

Spain also experienced a spectacular housing bubble that burst after the eruption of the global financial crisis. In Spain, the number of annual housing starts almost tripled between 1996 and 2006 (from 288,034 to 760,179), and over this time the country built new units at twice the rate of the rest of the EU (Burriel, 2011; Mercille, 2014). Cumulatively, over 5.5 million housing units were started in this 10-year period, which represents more than all the housing starts of France, Italy, and Germany combined (Burriel, 2011). In addition to building many new units in the large cities (e.g., Barcelona and Madrid), a large amount of new housing was built in coastal cities for foreign buyers or tourism workers. The population of Barcelona, Spain's second-largest metropolis, grew by 17 per cent between 1998 and 2009, representing over 730,000 new residents (Bayona-Carrasco and Gil-Alonso, 2012). Rapid growth, coupled with the housing bust, transformed the social and spatial composition of cities such as Barcelona and Madrid and encouraged various attempts to manage growth in the name of promoting sustainability (Diaz-Pacheco and Garcia-Palomares, 2014). In Barcelona, de-concentration and suburbanization, coupled with gentrification, had by 2008 produced a neighbourhood landscape marked by class and immigration status differences (Figure 12.4).

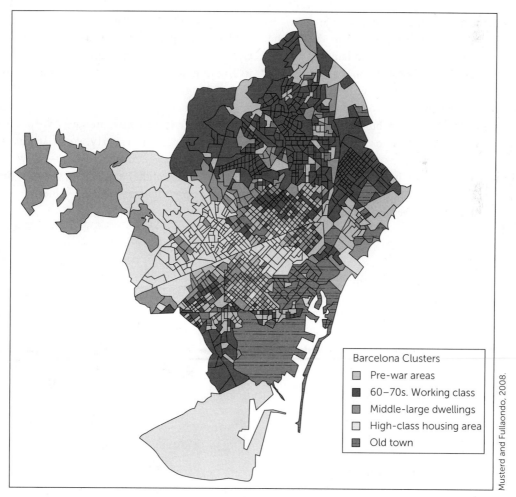

Figure 12.4 Map of neighbourhood-based housing clusters in Barcelona, Spain

A number of factors contributed to the trajectory of Spain's housing bubble (Burriel, 2011). Economically, historically unprecedented interest rates between 1997 and 2006 fueled easy access to credit (Conefrey and Gerald, 2010). In Spain, not only is there a strong cultural preference for ownership over renting, but as in many other nations in the global North, many aspire to own second properties. Furthermore, with Spain entering the European monetary union, investors from other nations could now easily purchase property in Spain. Nearly one-third of all housing demand during 1997–2006 came from foreign investors purchasing second properties, much of this in coastal cities or in the larger cosmopolitan cities such as Barcelona (Burriel, 2011). Immigrants as a percentage of the total population in the Barcelona metropolitan area increased from 2 per cent in the year 2000 to over 15 per cent in 2008, more than in the Spain writ-large (13 per cent) (Martori and Apparicio, 2011). The supply of land, furthermore, was stimulated by a 1996 law (later struck down on constitutional grounds, but implemented nonetheless by most municipalities) that classified much of the country as-of-right as "suitable for development"; this led to over-building, particularly in areas frequented by foreigners or tourists. For a while, many actors—households, firms, and governments—came to believe in the possibility of unlimited growth, and so they did not plan for the possibility that the economy might go into reverse (Burriel, 2011; Conefrey and Gerald, 2010).

The pain of the bursting housing bubble has been deep and shared widely among those in the working and middle classes, including many immigrants. Spain is somewhat distinct in not allowing debtors to expunge their mortgage debt through bankruptcy (Dowsett, 2013; De Weerdt and Garcia, 2015). Those who have lost their jobs and have their houses foreclosed and repossessed still have to pay their mortgage debts back to the banks, which are often allowed to continue charging high-interest penalties and other fees afterwards, sometimes through liens placed on debtors' incomes (Daley, 2010; Dowsett, 2013). While this has helped keep the banks afloat (along with a "troika" bailout worth €40 billion), afflicted debtors typically have to sell their businesses, cars, and furniture to do so, and take any job they can find, while many immigrants to Spain who are caught up in this trap can never save enough to return to their country of origin (Dowsett, 2013). This has squeezed many of Spain's youth out of the labour market. Since the global financial crisis, Spain's youth have suffered unemployment rates of over 50 per cent, producing a "lost generation" (Burgen, 2013) and giving rise to collective social movements such as the Platform of Mortgage Victims (PAH). PAH started in Barcelona and has spread to other cities, aiming to reform mortgage and bankruptcy laws while improving accessibility to housing (De Weerdt and Garcia, 2015). While the impact of Spain's regressive bankruptcy laws might not be visible in the heart of Barcelona (Figure 12.5), bankruptcies and foreclosures are very evident in the early postwar working-class neighbourhoods shown in Figure 12.4.

Many foreign investors have also suffered huge losses, and the Spanish government has responded with a special plan (based on similar schemes in Malta, Cyprus, and some Caribbean nations) to resuscitate its housing market, or at least dissuade foreign owners from selling; it hoped to accomplish this by granting Spanish passports ("golden visas") and eventually full Spanish citizenship to any foreigner who purchased and maintained Spanish property worth at least €500,000 (see Berdonces, 2013, for details). It is not yet clear how many investors might be taking up the offer or whether it is having any effect on housing values.

Figure 12.5 The heart of Barcelona: Catalonia Square

Canadian Housing Markets before and after the Global Financial Crisis, with a Focus on Montreal

The Canadian housing market is often seen to have avoided the financial crisis (Lynch, 2010) because of its oligopolistic banking structure and state regulatory functions (Bordo et al., 2014). While the claim made by mainstream economists is that "no financial institutions here needed bailout funds" (Londerville, 2010: 6), in fact, Canadian banks received substantial emergency funds just like their counterparts in the United States, the United Kingdom, and elsewhere (Macdonald, 2012; Walks, 2014), and at relatively similar levels (Walks, 2014). However, it is true that Canadian banks weathered the storm better than banks in other countries did. Furthermore, the Canadian housing market did not decline or correct to nearly the same degree as those in other countries, such as the United States or Britain, and real house prices continued to rise in a relatively smooth fashion even after the global financial crisis. The explanation for this can be found in the way that state institutions regulate the private mortgage market and in the way the state responded to the crisis.

A key feature of the Canadian system is the role of the state as the main funder of both non-market forms of housing and mortgage insurance for conforming private residential mortgages (Walks and Clifford, 2015). The Central (later "Canada") Mortgage and Housing Corporation (CMHC) was established in 1946 with the dual aims of building non-market (public and social) housing and insuring mortgages issued by lenders for the purchase of housing. The latter insurance protects the lenders (not the borrowers) in the case of default caused by job loss on behalf of the borrowers, thus making the origination of conforming mortgages virtually without risk for the private lenders. Mortgage insurance has been offered since 1954 by CMHC for mortgages that conform to particular amortization terms (until 2006, 25 years or less) and down-payment requirements (originally 20 per cent, reduced to 5 per cent in the 1990s). While housing-market downturns have always accompanied recessions, the role of the CMHC traditionally kept house prices relatively stable and allowed for the smooth functioning of the private-mortgage market.

The securitization of mortgages into MBS, modeled on the American public label system, was introduced in the late 1980s, although it was not until the Canada Mortgage Bonds (CMB) program in 2001—allowing MBS to be purchased outright from banks, enabling them to continue lending unconstrained by reserve limits—that the practice of mortgage securitization really took off (Walks, 2014; Walks and Clifford, 2015). This was followed by relaxation of the criteria for conformance with CMHC mortgage insurance standards, including, for a while, 40-year amortizations and zero-down (no equity) mortgages (these terms were incrementally scaled back since the global financial crisis; see Walks, 2012). During the global financial crisis, the federal government initiated a new program—the Insured Mortgage Purchase Program (IMPP)—through which the CMHC was told to purchase MBS from the banks on behalf of Canadians, so that the banks would keep on issuing mortgages. The IMPP worked so well that banks ramped up their lending, and house prices not only recovered from their temporary drop but went on to surpass new records each year since the end of the global financial crisis. Because all of the mortgages insured by CMHC and 90 per cent of the value of mortgages insured by the private-sector mortgage insurers (Genworth and Canada Guaranty) are guaranteed by the federal government, Mohindra (2010) has called Canada's mortgage insurance system a "high taxpayer vulnerability model."

The easy access to mortgage credit created by this system has also meant rapid rises in house prices in many Canadian cities over the time period (Figure 12.6). While the largest metropolitan areas, particularly Toronto and Vancouver, have

continued

the highest housing costs in Canada, it is actually a number of smaller metropolitan areas that have witnessed the most rapid increases in prices. House prices in Winnipeg and Quebec City, for example, increased at faster rates than in Toronto or Vancouver (Figure 12.6). The former are not cities known for high rates of immigration, rapid population growth, or low rates of unemployment, the traditional drivers of demand. Instead, the rapid rise of house prices in smaller cities like these is due to the fact that looser credit, spurred by the securitization of mortgages, allowed banks to take more risks with their lending. The increasing cost of housing, coupled with the fact that very little non-market housing has been built in most Canadian cities since the early 1990s (Walks and Clifford, 2015), has pushed households into an owner-occupied sector requiring ever-increasing amounts of debt to access, driving the debt burden among Canadian households to record levels. Household debt has risen from a household average of less than 90 per cent of disposable income in 1990 to over 163 per cent in 2014 (Cansim Table 3780123, 3780125, 3800001). There is a clear spatial relationship between house prices and debt levels in Canada's cities. Indeed, once various socio-demographic and metropolitan-level variables are controlled for, with every $10,000 increment in average house price, the level of household debt as a proportion of disposable income rises by 2.94 per cent at the CMA level and by an additional 1.14 percentage points at the neighbourhood level across Canadian metropolitan areas (Walks, 2013: 177–178).

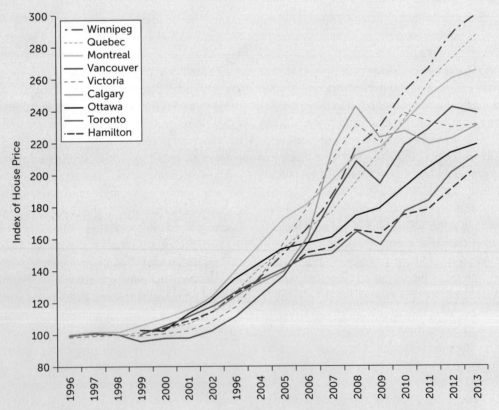

Figure 12.6 House Value Index, selected Canadian metropolitan areas, 1996–2013. (100 = Jan. 1996, except for Calgary, Toronto, Hamilton, and Ottawa 100 = Jan. 1999)

Calculated from Teranet 11-City House Price Index, http://www.housepriceindex.ca.

Since the GFC, both the household and state sectors became dangerously leveraged. Such a scenario has constrained government capacities to intervene and led to increasing class and generational differences in access to wealth and housing affordability. Hurst (2011) found that Canadian households in the lowest-income band are over four times more likely to pay 40 per cent or more of their disposable income on debt service than are middle-income households, and over ten times more likely to have such payments than are high-income households. Rising indebtedness also makes urban areas, and many neighbourhoods within them, more vulnerable to economic downturns. Poorer neighbourhoods have significantly higher debt-to-income ratios than richer neighbourhoods, while neighbourhoods containing more immigrants and certain visible minorities also reveal higher debt loads (Walks, 2013: 177–178). This adds to the already-significant vulnerabilities faced by the poor, by young families, and by immigrants and visible minorities in Canada's cities, as high local debt levels can lead to lower levels of local spending and employment, bankruptcies, declining property values, foreclosures, and lower property tax revenue for local municipalities that often translates into declining public-service levels.

Montreal is the CMA with the lowest level of homeownership (55 per cent) among Canadian metropolitan areas (compared to 69 per cent for Canada as a whole). However, Montreal reveals a relatively typical geographic distribution of household debt, much of which is in the form of mortgage debt (Figure 12.7). In Montreal, the lowest total

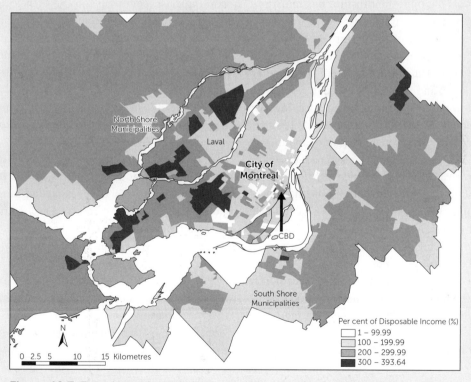

Figure 12.7 Total household debt as a proportion of disposable income, by census tract, Montreal CMA 2012

Simone and Walks, from custom data ordered from Environics Analytics.

continued

debt values (as a proportion of household disposable income) are found in census tracts (neighbourhoods) near the CBD, where many new condo units have been built (Rosen and Walks, 2013), and in the eastern part of Montreal. The highest debt burdens extend westward in a "V" shape, covering parts of Laval and the West Island suburbs. These are the areas of Montreal that contain more anglophones and allophones (speakers of languages other than French or English, many of whom are immigrants). The municipality of Saint-Laurent—an industrial suburb with a mixed population—has some of Montreal Island's most highly indebted neighbourhoods, with average household debt levels over 300 per cent of annual disposable income.

While the literature suggests that rising indebtedness portends a regressive urban socio-spatial restructuring within Canadian metropolitan areas, Canadian cities will not necessarily replicate the patterns of inequality and segregation experienced by their US urban counterparts. Outcomes depend on the kinds of political choices that are made, both those predating the onset of crisis and those negotiated in its aftermath, at multiple scales of **governance**, including the national, provincial, and municipal levels.

Conclusion

Housing markets are important institutions undergirding social, economic, cultural, and political changes in cities. One key aspect of neoliberal urbanism has been the use of securitized lending to stimulate private-sector mortgage issuance and the development of secondary mortgage markets, and with them a rise in home ownership. While this has produced benefits for investors and for households that wish to access credit, it has also led to an explosion of household debt and, due to bank bailouts and other "stimulative" responses (including state purchases of MBS), government debt. If governments respond to their growing liabilities with a politics of austerity and cutbacks in wages and/or government services, and if poor households face increasingly constricted choices owing to high debt levels, it becomes more likely that growing indebtedness will be translated into rising urban unemployment, house-price corrections, and rising household and neighbourhood inequality. The bubbling and collapsing of housing markets in countries such as the United States, Ireland, and Spain have led to a restructuring of their urban economies—which had become increasingly geared towards urban land development and which then found themselves with fewer jobs once the bubble burst—as well as the physical and social geographies of their cities. In cities that were growing fast in the run-up to the global financial crisis, it is more often the suburban rings that have experienced the worst effects of the bursting housing bubble, while in slow-growing deindustrializing cities, the worst effects have often been felt within inner-city working-class communities. Nevertheless, the effects of housing bubbles and the global financial crisis have varied widely among and within nations, as have their social implications. This is evident in the different trajectories found for the American, Irish, Spanish, and Canadian case studies.

Since 2000, many of the financial innovations arising in the global North were imported into cities in the global South, including the practice of mortgage securitization and the extension of loans to lower-income populations. In addition to this, many private-sector micro-credit lenders have expanded throughout the global South, offering credit to very low income households at high interest rates (Soederberg, 2014). As a result, debt levels in many nations and cities in the global South are catching up with those in the global North, along with similar attendant problems, from rising foreclosures to farmer suicides. While the global financial crisis did not (yet) have the same effects in the global South (with many such nations experiencing economic booms between 2008 and 2014, not busts), many cities in the global South saw the value of housing rise rapidly. The housing boom meant that affordable rental housing began to disappear from accessible locations within many cities, driving up rents and displacing poorer households. In some cities, particularly those in China but also in Saudi Arabia and other countries, whole new residential districts and even new cities have been built, but with few actual residents as of the time of writing (Moser, Swain, and Alkhabbaz, 2015). It remains too soon to say how the financialization of housing will affect cities in the global South, except to note that these effects will be uneven and in some cities auger a very dramatic restructuring of the urban fabric.

Key Points

- Housing markets are one component of the overall housing system in any given nation in which they have been built through social and political negotiation, conflict, and policy shifts over many years.
- In the postwar era, many countries in the global North developed housing systems according to social and political objectives, but since the late 1980s they have favoured private-market provision over non-market forms of housing.
- Securitization of MBS allows for spatially fixed houses to become globally tradable financial assets, thus producing financial liquidity out of the geographical fixity of urban housing markets.
- Rapid house price increases in a given city or nation are often fueled by increasing ease of access to mortgage credit; the increasing adoption of mortgage securitization practices across the globe led to an increase in the amount of credit available for the purchase of housing and land, and contributed to the rise of housing bubbles.
- The way that housing markets and mortgage markets operate in the United States has augmented, rather than ameliorated, inequalities related to class and race.
- The bursting of Ireland's housing bubble left many unfinished housing estates in Dublin, particularly in the newer suburbs, contributing to changes in the socio-spatial composition of the city.
- The pain of the bursting housing bubble in Barcelona is deep and shared widely among those in the working and middle classes, including many immigrants. Spain's distinctiveness in relation to its bankruptcy laws has led to new social and political movements seeking to change legislation around housing-related debt.
- A key feature of the Canadian housing system for the majority of the postwar period is the important role played by the state.
- There is a strong reflexive relationship between rising housing costs and mortgage debts. The vast majority of the increase in household debt levels in Canadian cities is due to the larger mortgages required to purchase housing on the private market, as well as to the ease by which households have been able to access mortgages.

Activities and Questions for Review

1. Explain in your own words to a classmate what redlining is. What role does redlining play in shaping where people can live?
2. What relationship exists, if any, between mortgage-backed securities (MBS) and the global financial crisis? List the potential positive and negative contributions of MBS and other financial innovations to socio-spatial (in)equality in cities today.
3. What can foreclosures tell us about the racial, class, and gendered differences in cities? In a small group, compare and contrast the experience of foreclosures in two different cities.
4. Using the information from the Canadian text box, discuss as a class why house prices in Canada's private housing market have shown more resilience than the housing markets of other countries in the global North.

Acknowledgments

The research that went into this chapter was supported by the Social Sciences and Humanities Research Council of Canada (SSHRC). The authors would like to thank SSHRC for this support. Special thanks as well to Cian O'Callaghan for use of his photograph in Figure 12.2.

References

Aalbers, M.B. (2008). The financialization of home and the mortgage market crisis. *Competition and Change, 12*(2): 148–166.

Aalbers, M.B. (2011). *Place, exclusion and mortgage markets*. Malden, MA: Wiley-Blackwell.

ABS (2011). *Housing occupancy and costs, 2011–12* (Table 4130: Dendogram of selected household characteristics). Canberra: Australian Bureau of Statistics. Retrieved from http://www.abs.gov.au/ausstats/abs@.nsf/Latestproducts/4130.0Main%20Features22011-12?opendocument&tabname=Summary&prodno=4130.0&issue=2011-12&num=&view=

Allen, K. (2009). *Ireland's economic crash: A radical agenda for change*. Dublin, Ire.: Liffey Press.

Ashton, P. (2009). An appetite for yield: The anatomy of the subprime mortgage crisis. *Environment and Planning A, 41*(6): 1420–1441.

August, M. (2014). Negotiating social mix in Toronto's first public housing redevelopment: Power, space and social control in Don Mount Court. *International Journal of Urban and Regional Research, 38*(4): 1120–1140.

Bayona-Carrasco, J., and Gil-Alonso, F. (2012). Suburbanisation and international immigration: The case of the Barcelona Metropolitan Region (1998–2009). *Tijdschrift voor Economische end Sociale Geografie, 103*(3): 312–329.

Berdonces, A. (2013). Property investment and Spanish residency law explained. Spanish Property Insight, July 2013 edition. Retrieved from http://www.spanishpropertyinsight.com/residency/property-investment-spanish-residency-law-explained

Berry, B.J.L. (1963). *Commercial structure and commercial blight*. Chicago: University of Chicago, Department of Geography, Research Paper No. 85.

Bezemer, D.J. (2009). "No one saw this coming": Understanding financial crisis through accounting models. University of Munich: MPRA Economics Paper, June 16. Retrieved from http://mpra.ub.uni-muenchen.de/15892/1/MPRA_paper_15892.pdf

Bordo, M.D., Redish, A., and Rockoff, H. (2014). Why didn't Canada have a banking crisis in 2008 (or in 1930, or 1907 . . .)? *Economic History Review*, doi: 10.1111/1468-0289.665.

Bresnihan, P., and Byrne, M. (2015). Escape into the city: Everyday practices of communing and the production of urban space in Dublin. *Antipode, 47*(1): 36–54.

Burgen, S. (2013). Spain youth unemployment reaches record 56.1%. *The Guardian*, August 30. Retrieved from http://www.theguardian.com/business/2013/aug/30/spain-youth-unemployment-record-high

Burriel, E.L. (2011). Subversion of land-use plans and the housing bubble in Spain. *Urban Research and Practice, 4*(3): 232–249.

City of Toronto (2006). *Rental housing supply and demand indicators*. Toronto: City of Toronto, City Planning, Policy and Research Division. Retrieved from https://www1.toronto.ca/city_of_toronto/social_development_finance_administration/files/pdf/housing_rental.pdf

CMHC (various years). *Canadian housing observer*. Ottawa: Canada Mortgage and Housing Corporation (annual report).

Conefrey, T., and Gerald, J.F. (2010). Managing housing bubbles in regional economies under EMU: Ireland and Spain. *National Institute Economic Review, 211*(1): R27–R44.

Daley, S. (2010). In Spain, homes are taken but debt stays. *New York Times*, October 27. Retrieved from http://www.nytimes.com/2010/10/28/world/europe/28spain.html?_r=0

de Mendonca, H.F., and Barcelos, V.I. (2015). Securitization and credit risk: Empirical evidence from an emerging economy. *North American Journal of Economics and Finance, 32*: 12–28.

De Weerdt, J., and Garcia, M. (2015). Housing crisis: The Platform of Mortgage Victims (PAH) movement in Barcelona and innovations in governance. *Journal of Housing and the Built Environment*, doi: 10.1007/s10901-015-9465-2.

Dias-Pacheco, J., and Garcia-Palomares, J.C. (2014). Urban sprawl in the Mediterranean urban regions in Europe and the crisis effect on the urban land development: Madrid as study case. *Urban Studies Research*, Article ID 807381: 1–13.

Dowsett, S. (2013). Insight: In Spain, banks buck calls for mortgage law reform. *Reuters*, February 26. Retrieved from http://www.reuters.com/article/2013/02/26/us-spain-mortgage-reform-idUSBRE91P09Q20130226

Engel, K.C., and McCoy, P.A. (2011). *The subprime virus: Reckless credit, regulatory failure, and next steps*. New York: Oxford University Press.

Engelen, E., Erturk, I., Froud, J., Johal, S., Leaver, A., Moran, M., Nilsson, A., and Williams, K. (2011). *After the great complacence: Financial crisis and the politics of reform*. Oxford: Oxford University Press.

EuroStat (multiple years) Housing Statistics (database). Retrieved from http://appsso.eurostat.ec.europa.eu/nui/show.do

Fischel, W. (1999). Does the American way of zoning cause the suburbs of metropolitan areas to be too spread out? In A. Altschuler et al. (Eds.), *Governance and opportunity in metropolitan America*. Washington, DC: National Academy Press, pp. 151–191.

Ford, J. (1994). *Problematic home ownership*. Loughborough, UK: Joseph Rowntree Foundation.

Goetz, E. (2013). The audacity of Hope VI: Discourse and the dismantling of public housing. *Cities, 35* (SI): 342–348.

Gotham, K.F. (2009). Creating liquidity out of spatial fixity: The secondary circuit of capital and the subprime mortgage crisis. *International Journal of Urban and Regional Research, 33*(2): 355–371.

Grinstein-Weiss, M., Key, C., and Carrillo, S. (2015). Homeownership, the great recession, and wealth: Evidence from the survey of consumer finances. *Housing Policy Debate*, doi: 10.1080/10511482.2014.971042

Harloe, M. (1995). *The people's home? Social rented housing in Europe and America*. Oxford, UK: Blackwell.

Harris, R. (1996). *Unplanned suburbs: Toronto's American tragedy*. Baltimore: Johns Hopkins University Press.

Harvey, D. (1978). The urban process under capitalism. *International Journal of Urban and Regional Research, 2*: 101–131.

Hernandez, J. (2009). Redlining revisited: Mortgage lending patterns in Sacramento 1930–2004. *International Journal of Urban and Regional Research*, 33(2): 291–313.

Honohan, P. (2010). *The Irish banking crisis: Regulatory and financial stability policy 2003–2008*. Dublin: Irish Central Bank. Retrieved from http://www.bankinginquiry.gov.ie

HUD (US Department of Housing and Urban Development). (2007). *Performance and accountability report*. Washington, DC: HUD.

Hurst, M. (2011). *Debt and family type in Canada*. Ottawa: Statistics Canada. Article 11430 of Cat. no. 11-008-X, April 21, 2011. Retrieved from http://www.statcan.gc.ca/pub/11-008-x/2011001/article/11430-eng.pdf

Immergluck, D. (2009). *Foreclosed: High-risk lending, deregulation, and the undermining of America's mortgage market*. Ithaca, NY: Cornell University Press.

Immergluck, D. (2011). The local wreckage of global capital: The subprime crisis, federal policy, and high-foreclosure neighbourhoods in the US. *International Journal of Urban and Regional Research*, 35(1): 130–146.

INE (Instituto Nacional Estadistica). (2011). *Housing census: Dwelling by type*. Madrid: INE.

Isin, E. (2002). *Being political: Genealogies of citizenship*. Minneapolis: University of Minneapolis Press.

Jeffers, J.M. (2013). Double exposures and decision making: Adaptation policy and planning in Ireland's coastal cities during a boom-bust cycle. *Environment and Planning A*, 45: 1436–1454.

Johnston, R., Poulsen, M., and Forrest, J. (2007). Ethnic and residential segregation in US metropolitan areas, 1980–2000: The dimensions of segregation revisited. *Urban Affairs Review*, 44(4): 479–504.

Keen, S. (2011). *Debunking economics: The naked emperor dethroned*. London: Zed Books.

Kitchin, R., O'Callaghan, C., Boyle, M., Gleeson, J., and Keaveney, K. (2012). Placing neoliberalism: The rise and fall of Ireland's Celtic Tiger. *Environment and Planning A*, 44: 1302–1326.

Kitchin, R., O'Callaghan, C., and Gleeson, J. (2014). The new ruins of Ireland? Unfinished estates in the post-Celtic Tiger era. *International Journal of Urban and Regional Research*, 38(3): 1069–1080.

Krippner, G. (2005). The financialization of the American economy. *Socio-Economic Review*, 3(3): 173–208.

Lees, L., Slater, T. and Wyly, E. (2008). *Gentrification*. London: Routledge.

Londerville, J. (2010). *Mortgage insurance in Canada: Basically sound but room for improvement*. Ottawa: McDonald-Laurier Institute for Public Policy. Published online November 2010 at http://www.macdonaldlaurier.ca/files/pdf/MortgageInsurance.pdf

Lynch, K. (2010). Avoiding the financial crisis: Lessons from Canada. *Policy Options*, 31(5): 12–15.

Macdonald, D. (2012). *The big banks big secret*. Toronto: Canadian Centre for Policy Alternatives.

Martori, J.C., and Apparicio, P. (2011). Changes in the spatial patterns of the immigrant population of a southern European metropolis: The case of the Barcelona metropolitan area, 2001–2008. *Tijdschrift voor economische en sociale geografie*, 102(5): 562–581.

Massey, D.S., and Denton, N.A. (1993). *American apartheid: Segregation and the making of the underclass*. Cambridge, MA: Harvard University Press.

Mercille, J. (2014). The role of the media in sustaining Ireland's housing bubble. *New Political Economy*, 19(2): 282–301.

Miron, J. (2000). Cities as real estate. In T.E. Bunting and P. Filion (Eds.), *Canadian cities in transition: The twenty-first century* (2nd ed.). New York: Oxford University Press, pp. 154–172.

Mohindra, N. (2010). *Mortgage finance reform: Protecting taxpayers from liability*. Vancouver: Fraser Institute research paper. Published online at http://www.fraserinstitute.org/publicationdisplay.aspx?id=15934

Moser, S., Swain, M., and Alkhabbaz, M.H. (2015). King Abdullah economic city: Engineering Saudi Arabia's post-oil future. *Cities*, 45: 71–80.

Murdie, R. (1986). Residential mortgage lending in Metropolitan Toronto: A case study of the resale market. *Canadian Geographer*, 30(2): 98–110.

Murdie, R. (1991). Local strategies in resale home financing in the Toronto housing market. *Urban Studies*, 28: 465–483.

Musterd, S., and Fullaondo, A. (2008). Ethnic segregation and the housing market in two cities in Northern and Southern Europe: The cases of Amsterdam and Barcelona. *ACE: Architecture, City and Environment*, 3(8): 93–115.

National Housing Development Survey. (2010). Maps of unfinished estates at the county level. Republic of Ireland: Department of Environment, Community and Local Government. Retrieved from http://www.environ.ie/en/DevelopmentHousing/Housing/UnfinishedHousingDevelopments/NationalHousingDevelopmentSurvey2010

Niedt, C., and Martin, W. (2013). Who are the foreclosed? A statistical portrait of America in crisis. *Housing Policy Debate*, 23(1): 159–176.

NYCHA (New York City Housing Authority). (2014). Facts about nycha. New York City: NYCHA. Retrieved from http://www1.nyc.gov/assets/nycha/downloads/pdf/factsheet.pdf

Poapst, J.V. (1993). Financing of post-war housing. In J.R. Miron (Ed.), *House, home and community: Progress in housing Canadians, 1945–1986*. Ottawa: Canada Mortgage and Housing Corporation, pp. 94–109

Republic of Ireland Central Statistics Office (CSO). (2011). Population statistics, 2011. Retrieved from http://www.cso.ie/en/statistics/population/populationofeachprovincecountyandcity2011 and http://www.cso.ie/px/pxeirestat/Statire/SelectVarVal/Define.asp?Maintable=CDD10&Planguage=0

Rosen, G., and Walks, A. (2013). Rising cities: Condominium development and the private transformation of the metropolis. *Geoforum*, 49: 160–172.

Rugh, J.S., and Massey, D.S. (2010). Racial segregation and the American foreclosure crisis. *American Sociological Review*, 75: 629–651.

Schafran, A., and Wegmann, J. (2012). Restructuring, race, and real estate: Changing home values and the new California metropolis, 1989–2010. *Urban Geography*, 33(5): 630–654.

Seko, M. (1994). Housing finance in Japan. *Housing markets in the US and Japan*. National Bureau of Economic Research (NBER)/University of Chicago Press, pp. 49–64. Reprinted and retrieved from http://www.nber.org/chapters/c8821.pdf

Simmel, G. ([1903] 2002). The metropolis and mental life. Republished in G. Bridge and S. Watson (Eds.), *The Blackwell city reader*. Oxford: Blackwell, pp. 11–19.

Smith, J. (2013). The end of public housing as we knew it. *Urban Research and Practice*, 6(3): 276–296.

Smith, Y. (2010). *Econned: How unenlightened self interest undermined democracy and corrupted capitalism*. New York: Palgrave Macmillan.

Soederberg, S. (2014). *Debtfare states and the poverty industry*. London: Routledge.

Squires, G.D. (2009). Inequality and access to financial services. In J. Niemi, I. Ramsay, and W.C. Whitford (Eds.), *Consumer credit, debt, and bankruptcy: Comparative and international perspectives*. Oxford: Hart Publishing, pp. 11–31.

The Economist (2015). Global house prices: Location, location, location—house prices in real terms (Graphic Detail). *The Economist*, April 16. Retrieved from http://www.economist.com/blogs/dailychart/2011/11/global-house-prices?zid=305&ah=417bd5664dc76da5d98af4f7a640fd8a

US Census Bureau (2008). *Statistical abstract of the United States*. Washington, DC: US Census Bureau.

US Census Bureau (2013). State and county quick facts: New York City. Washington, DC: US Census Bureau. Retrieved from http://quickfacts.census.gov/qfd/states/36/3651000.html

Wacquant, L. (2008). *Urban outcasts: A comparative sociology of advanced marginality*. Cambridge, UK: Polity.

Walks, A. (2010). Bailing out the wealthy: Responses to the financial crisis, Ponzi neoliberalism, and the city. *Human Geography*, 3(3): 54–84.

Walks, A. (2011). Economic restructuring and trajectories of socio-spatial polarization in the twenty-first century Canadian city. In L.S. Bourne, T. Hutton, R. Shearmur, and J. Simmons (Eds.), *Canadian urban regions: Trajectories of growth and change*. Toronto: Oxford University Press, pp. 125–159.

Walks, A. (2012). *Canada's new federal mortgage regulations: Warranted and fair?* Toronto: University of Toronto Cities Centre Research Bulletin #46. Retrieved from http://neighbourhoodchange.ca/wp-content/uploads/2012/12/Walks-2012-New-Mortgage-Regulations-Canada-CC-RB-461.pdf

Walks, A. (2013) Mapping the urban debtscape: The geography of household debt in Canadian cities. *Urban Geography*, 34(2): 153–187.

Walks, A. (2014). Canada's housing bubble story: Mortgage securitization, the state, and the global financial crisis. *International Journal of Urban and Regional Research*, 38(1): 256–284.

Walks, A., and Clifford, B. (2015). The political economy of mortgage securitization and the neoliberalization of housing policy in Canada. *Environment and Planning A*, 47(8): 1624–1642.

Wyly, E., Moos, M., Kabahizi, E., and Hammel, D. (2009). Cartographies of race and class: Mapping the class-monopoly rents of American subprime mortgage capital. *International Journal of Urban and Regional Research*, 33(2): 343–364.

13 Urban Public Spaces, Virtual Spaces, and Protest

Ebru Ustundag and Gökbörü S. Tanyildiz

Introduction

Urban public spaces around the world have been a vital part of everyday life for citizens. Classical urban theorists since the early twentieth century, such as Georg Simmel, Louis Wirth, Lewis Mumford, and Jane Jacobs, have underlined the significance of strangers' encounters in public spaces for **civic life** and participatory democracy. Parks, plazas, squares, and **streets** have historically provided a place of collectivity, belonging, and identity, where citizens can express their solidarity as well as their resistance to various socio-political discourses. Historically, urban public spaces have also been sites of struggle for diverse social groups, where they redefined **citizenship**, including expanding existing rights and articulations of new ones. These struggles, which take place through and within the materiality of urban public spaces, are significant for the visibility, appropriation, and **public address** opportunities they provide for marginalized and underrepresented social groups (Iveson, 2007, 2009). These are struggles not only for participation in and appropriation of urban spaces, but also for the claim of being public and hence of being political (Iveson, 2007; Staeheli and Mitchell, 2008).

There has been, moreover, a range of tactics and strategies used by various social groups to challenge normative definitions of **public space**. New forms of digital technology have also catalyzed users to engage in new forms of activism, protest, dissent, and resistance; widespread usage of digital technologies has resulted in the emergence of not only new political **subjectivities** and new forms of direct action, but also in new **socio-spatialities**. Over the last decade, the advancements in digital technologies accompanied by the high visibility of online **social networking sites** have significantly altered forms as well as practices of communication. Social networking sites have become important spaces for social groups to raise social and political awareness, and also to coordinate direct actions in and through urban public spaces. Hashtags like #occupy, #arabspring, #IdleNoMore, #ferguson, and #BlackLivesMatter have also been utilized to build digital and material alliances as well as solidarities among various social groups. These new forms of dissent and insurgence successfully challenge inherited conceptions of the public, of urban public space, of citizenship, and of liberal democracies.

This chapter examines multidimensional understandings of the power dynamics embedded in contemporary urbanization in relation to urban public spaces, virtual spaces, and new forms of citizenship. Following an overview of the emergence and contemporary understandings of public space, it establishes how different practices of citizenship, such as **urban citizenship** and **digital citizenship**, have emerged. In order to explain the dynamic relationship between citizenship, public space, and new digital technologies of the contemporary moment, two case studies of Istanbul and New York City are provided to demonstrate how marginalized social groups mobilize physical and virtual resources in order to claim new rights and expand upon existing ones. These case studies further address how new digital technologies are deployed by urban social movements and contribute to the making of public spaces. Finally, by way of two examples, the significance of digital urban citizenship to democratic participation and making rights claims in the Canadian urban context is examined.

The Evolution of Public Space, Urban Citizenship, Digital Citizenship, and the Decline of Public Space

Several dualities (e.g., nature versus culture, orient versus occident, north versus south, men versus women) have come to contextualize how geographers have defined the genealogies of power relations at various scales. One of these key dualisms is public versus private. Despite the pitfalls of binary thinking, its legacy is strong. The term "public" carries associations (e.g., "public parks," "public swimming pools," "public libraries," "public opinion," "public media," and the "public sector") that differ from those attached to the term "private" (e.g., "private spaces," "private interests," and the "private sector." The notion of public is associated with openness, commonality, and accessibility; these are predominantly idealized and romanticized interpretations of the public that celebrate access, enjoyment, and participation. The notion of private, on the other hand, is associated with ownership and equated with restrictions in access. Not all spaces are clearly delineated, however, and "in-between spaces"— what are often referred to as privatized public spaces (e.g., malls and restaurants)—tend to confuse categorization; discursively understood as public spaces, they fit into definitions of private spaces with strict regulations about access and usage. This section explores these understandings, outlining both the emergence and decline of urban public space and its relationship to contemporary debates about urban citizenship and digital urban citizenship.

The Emergence of Urban Public Space

While the boundaries and meanings of public and private spaces are fluid (Staeheli and Mitchell, 2008), it is possible to trace back to ancient Greece the socio-economic conditions that resulted in what is commonly understood as a clear distinction between public and private in contemporary societies in the global North, where public space, signifying the power and dominance of politics, was only accessible to men and property owners (Bowlby, 1999). In ancient Greece, the political and commercial life of the city was closely tied to the *agora*, the central amphitheatre, where major political issues were discussed in a structured way (Bridge, 2009). The agora was thus the centre of communal life, functioning as both a market and a meeting place. It represented a democratic ideal of the **public sphere**—despite the exclusionary nature of this space with respect to women, children, foreigners, and slaves—as an arena of political deliberation and participation in which citizens (those who were male and property owning) could vote on issues of government and justice.

The agora in ancient Greece raises some key themes about public space that are still relevant today: public space can have multiple functions; it is a democratic space where citizens interact and discuss urban issues; it is also used for commercial purposes; it is an informal meeting place and community space; and it has restricted access, with some people having more rights than others. However, the geopolitical division of Greece into approximately a thousand city-states contradicted the democratic form of ancient Greece that emerged in the agora. These different city-states were in endless local wars with one another over resources, mercantile advantage, and agricultural land. These wars resulted in an increasing militarization that gradually eroded the democratic character of city-states. In 338 BCE at the Battle of Chaeronea, the Macedonian army of Alexander the Great put an end to both the independence of Greek city-states and Greek warfare (Faulkner, 2013). With the destruction of city-state democracies under the imperial system, the notion of the agora became historically defunct. However, the rule of democracy, in its fundamental relation to public space, was to be re-established in Western Europe with a burgeoning bourgeoisie in a new political regime.

In Western Europe, before the **Industrial Revolution**, socio-economic relations were community based and the social reproduction of everyday life was not founded on a clear distinction between working and living spaces (Domosh and Seager, 2001). People mostly lived and worked in one location. For example, farmers lived and worked on their farms, while artisans had their workshops located on the lower floor of the building in which they lived. The invention of the steam engine in 1698 and its impact on the **mode of production** drastically changed socio-economic relations and introduced a clear-cut separation between living and work places, for both the working class and middle class (Domosh and Seager, 2001). Rapid industrialization in the nineteenth century, although initially rural-based, was soon to force working-class women and men to leave their rural communities and move to urban areas to work, where the concentration of the labour force generated savings for industrial capital (Federici, 2004). The emergence of factories as workplaces in urban areas drastically changed the social and economic fabric of citizens' everyday life practices.

Waged work was now defined as an activity that was performed outside the home. Capitalism and **patriarchy** significantly regulated the social **division of labour** as well as the social reproduction of everyday life. While men eventually came to be the only ones who worked in factories and dominated public spaces, working-class women's primary roles were increasingly confined to household duties in the private space of the home, including care for children and the elderly, although their informal roles as "managers" of their communities allowed them the freedom of residential streets (Domosh and Seager, 2001), and some women never had the capacity to leave paid work, be it (predominantly) in service, agriculture, or sex work. Middle-class women were more clearly denied a role in public life and were often relegated to suburban life away from the dynamism of the city. In other words, the separation between waged and unwaged work systematically reinforced the binary of the home (private) and the world (public) along gendered lines and, therefore, ascribed a masculine character to urban public space. However, while certain public spaces were designed for masculine consumption (e.g., taverns, pubs, and coffee houses), the emergence of large-scale department stores and of shopping districts in early nineteenth-century Western Europe provided middle-class women with a legitimate reason to be on the streets and also with a gateway to mass consumer culture, and working-class women with another form of employment. However, while serving to legitimate middle-class women's mobility, these activities also served as barriers to women's entry into other spheres of urban public life, particularly politics (Domosh and Seager, 2001; Strange, 1995).

By the late nineteenth century, industrial capitalism had introduced urban planning as a profession, to ensure and control the efficient and effective mobilization of goods, people, capital, and information within urban public spaces. It was also to alter who was seen as worthy of participating in urban life (Wilson, 1991; Joyce, 2003). Accompanied by the increase in purchasing power of the middle classes and the availability of mass consumption, urban planning became an important tool to regulate strangers' encounters in everyday life. Urban planning has also played an important role in regulating a gendered industrial city. While masculine urban public spaces were designed as spaces of work, politics, and leisurely activities, feminized private spaces were governed by the Victorian understanding of femininity and domesticity (Domosh and Seager, 2001). This gendered spatial arrangement was facilitated by the urban planning practices of land use, zoning, and transportation. Urban planning also facilitated class-based residential areas, via zoning practices. While the working class lived close to environmentally polluted industrial zones, middle and upper classes enjoyed residential areas, with access to greenery and other social amenities. In Ontario, Hamilton stands out as a classic example of how zoning, in an industrial city, created a class-based division of neighbourhoods.

This separation of the locations of paid and unpaid work, the emergence of mass consumption, and the evolution of urban planning processes have all had significant impacts on liberal democratic understandings and representations of public and private spaces, as well as of who had the right to participate and be present in these spaces. The prioritization of the public sphere as the main domain of politics, privileging (white, heterosexual, able-bodied) men, has been the dominant form of power dynamics in urban public spaces. In countries of the global North, the privileging of certain bodies in public places has resulted in understandings of inclusion and exclusion, of those who fit and of those who do not fit into dominant discourses of urban public spaces. These understandings are explored in the following sections.

Contemporary Understandings of Public Space

Globalization and urbanization, along with their accompanying processes of migration, have resulted in a proliferation of social relations in urban public spaces over the last three decades. In this context of global urbanization, various public spaces (e.g., parks, squares, and marketplaces) allow strangers to encounter one another and to experience urban life (Amin, 2006; Massey, 2005; Valentine, 2008; Young, 1990). Scholars like Sandercock (1997), Sennett (1997), and Zukin (1995) have underlined the vital connection between urban public space and civic culture in relation to vibrant and livable cities. The "thrown togetherness" (Massey, 2005: 181) of people in their classed, gendered, racialized, and sexualized diversity produces urban public spaces "as a site of difference" (Young, 1990: 240). Contemporary urban claims and struggles that make and remake urban citizenship take place in these sites of difference. Contemporary literatures on public space, the public sphere, and **counter-publics** are pivotal to understanding the socio-political character of the everyday encounters of citizens.

In order to explicate how shared common understanding of the "lifeworld," which is a pre-reflective perception of world as it is experienced, is produced in this atmosphere of difference, the German philosopher Jürgen Habermas suggested a paradigm shift from the analysis of modes of production that condition the very processes of globalization and urbanization to the analysis of communicative action of face-to-face relations between various social groups across their differences (Habermas, 1987). With this shift, Habermas contributed to bourgeois liberal conceptualizations of the public sphere in which the consensus of plural political subjects is achieved through constant deliberation over time, as opposed to a more radical Marxist approach according to which the contradictory interests of different social classes cannot be reconciled through mutual understanding, but by a revolutionary transformation of existing social structures.

Based on normative ideals of democracy, the Habermasian public sphere has played a vital role in understandings of the formation of public opinion. For Habermas, "the idea of the public sphere is that of a body of 'private persons' assembled to discuss matters of 'public concern' or 'common interest'" (Fraser, 1990: 58). Habermas's conceptualization is based on the idea that the public sphere, as a place of social intercourse, plays an important role in the transference of the interests of bourgeois society to the structures of the state via various forms of representation, including free speech and the right to assembly. Habermas argues that private interests and social categories of difference can be bracketed (set aside), to allow participants to become equals through rational argument and debate in an "inclusive" manner (Fraser, 1990; Cassegard, 2014). The positive aspects of this process were that it opened up the possibility of collective participation and universal interest through rational-critical debates, rather than through social status. However, the Habermasian notion that this public space is open and accessible to all has been critiqued.

Various scholars, especially feminists and including geographers (Staeheli and Mitchell, 2008), have underlined the limits and exclusionary nature of this particular understanding of the public sphere in relation to the bracketing of difference (Fraser, 1990; Ryan, 1997). The feminist scholar Nancy Fraser (1990) highlights the exclusionary nature of the Habermasian public sphere, where bracketing works to the advantage of dominant groups. In her critique, Fraser (1997: 81) conceptualizes counter-publics as "parallel discursive arenas where members of subordinated social groups invent and circulate counter-discourses, which in turn permit them to formulate oppositional interpretations of their identities, interests and needs." The creation of counter-publics offers spaces of action as well as collective resistance for marginalized social groups, including contemporary social movements like the Occupy movement, the **Arab Spring** uprising, and the Quebec student movement. Fraser reminds people, however, that not all counter-publics are committed to social justice, equality, or greater democratic participation. It is also important to emphasize that taking part in counter-publics is risky because of the potential of being labeled as deviant, disorderly, and criminal and because of the potential emotional and mental costs of such engagement.

Michael Warner (2002) was critical of Fraser's work in his book *Publics and Counterpublics*. He suggests that rational critical debate should not be the exclusive or only standard for identifying something as public or counter-public. He emphasizes, like Iveson (2007), that publics are always under construction—they are made, sustained, and altered through public address; the act of addressing a public is what brings that public into existence. Warner's ideas about publics and counter-publics are a little different than those of Habermas and Fraser. Warner (2002, 114) treats publics as having a poetic function—what he calls "poetic world making" or "the performative dimension of public discourse." By this he means that through the circulation of texts, counter-publics can make and maintain alternative worlds for themselves. Counter-public discourse, then, is not only about debating a point. Counter-publics are about bringing new social worlds into existence. The feminist political scientist Iris Marion Young (1990) also challenges Fraser's understanding of the public sphere and counter-publics, arguing that there is a need for a "heterogeneous public" within which challenges and contestations can take place.

Many critical geographers today understand public space not as a site in which consensus and freedom is secured, but as a space that is in a constant process of formation and deformation by ongoing contestation over competing social meanings, values, and recognition (Mitchell, 2003; McCann, 1999). While the Habermasian ideal public sphere is claimed to be universal and "spatially undifferentiated" (Smith and Low, 2006: 5), feminist critiques, including Fraser's and Young's, provide a framework for "spatializing public sphere theory" (Smith and Low, 2006: 5). As demonstrated in this section, conceptualizations of the public sphere and public space are vital for understanding the socio-spatialization of practices of democracy and the struggles of citizenship. The growing literatures on urban citizenship and the right to the city in geography demonstrate how various struggles that take place through and within the materiality of urban public spaces are significant to understanding the socio-economic struggles of various marginalized, excluded, and discriminated communities.

Urban Citizenship

It is now well documented that urban public spaces have been important places in the making of urban citizenship (Isin, 2000; Mitchell, 2003; Smith and Low, 2006). Formal understandings of citizenship denote the position or status of being a citizen, with the addition of defining who is a **stranger** or an outsider. This definition traditionally refers to membership status of a nation-state, defined by political boundaries, such as being a Canadian citizen or a Bangladeshi citizen. Citizenship in this liberal democratic context, is defined as a status and describes the legal relationship between a political subject and the **polity**. Within this liberal democratic framing, rights are also accompanied by political and social obligations, including paying taxes and contributing to the social and moral order of the polity. In addition to signifying membership status, citizenship can also be understood as a practice through which citizens exercise their civil (e.g., freedom of speech, thought, and faith), political (e.g., voting and assemblage), and social rights (e.g., security, health, and economic welfare).

The political theorist Hannah Arendt uses the phrase the "right to have rights" to denote that there is a tension between citizenship as status and citizenship as practice. In terms of the latter and analogous to George Orwell's (2008: 112) famous quote "All animals are equal but some more than others," there are discrepancies in ability of particular groups to exercise their rights or articulate new ones. Struggles of inclusion by various social groups, including women, people of colour, queers, the disabled, and others, have been a vital area of inquiry for urban scholars. Within this framework of citizenship, as practice it is possible to link genealogies of citizenship to group struggles either to exercise their rights, articulate new rights, or struggle to expand existing rights (Isin and Wood, 1999). These include struggles for redistribution (e.g., Indigenous land claims) and recognition (e.g., children's rights and disability rights) (Young, 1990).

While status framings of citizenship have been traditionally attached to the nation-state, scholars, especially geographers, have been contesting this formal definition via the practices of citizenship approach that emerge at various other scales, especially urban and cyber spaces. As argued by Sassen (1996) and others (Isin, 2000), urban public spaces have always been strategic sites for struggles and contestation over distribution and recognition. The appropriation and use of urban public space have been essential for social groups to make their demands visible and public (Mitchell, 2003). Introduced by the French Marxist Henri Lefebvre (1968), the concept of the **right to the city** has been important to urban theorists seeking to develop understandings of urban citizenship (Holston, 1998; Holston and Appadurai, 1999; Isin and Wood, 1999). The right to the city includes the citizen's right to access, occupy, and use urban space, as well as to the possibility of creating new spaces. For Lefebvre (1968: 174), "[T]he right to the city manifests itself as a superior form of rights: right to freedom, to individualization in socialization, to habit and to inhabit. By empowering dwellers to participate in the use and production of urban space, right to the city enables control over the production of urban spaces."

Purcell (2003) explains that Lefebvre's right to the city includes the right to participate in everyday life for urban dwellers. It is a call to redistribute power relations and shift control away from capital and the state to empower citizens. In the twenty-first century, increasing attention has also been given to cyber spaces and urban citizenship.

Digital Citizenship

Three hundred hours of YouTube video are uploaded every minute (www.youtube.com/yt/press/statistics.html). Around 40 per cent of the world's population has an Internet connection today that provides rapid and extensive connection among various social groups (www.internetlivestats.com/internet-users). Speed, low cost, accessibility, and the extensive geographical reach of social networking sites and other new digital technologies have significantly changed our relationships with ourselves and with others. For example, throughout our everyday life practices and mobilities, we frequently encounter people who use their personal digital devices, such as cellphones and portable music players, to facilitate their own private spaces within public spaces. These new technologies "create mobile bubbles of private and parochial interaction within the public realm" (Hampton et al., 2010: 705), shifting how we use, represent, and imagine material and virtual spaces.

Both critical digital studies and citizenship studies examine the modalities and effects of these new socio-technical arrangements (Isin and Ruppert, 2015). While it is agreed that the use of the Internet through advanced technological devices has radically changed the ways in which people participate in politics and that many contemporary political struggles take place in online forums across geographical boundaries, the character of this change and of its subjects remain debatable.

Critical digital studies tend to argue that this change articulates a break from the ways that politics are typically carried out. According to this field of inquiry, which explores "the dominant codes of technology, politics, and culture in the digital era" (Kroker and Kroker, 2013: 14), with new socio-technical arrangements, the Internet constitutes a semi-autonomous sphere of politics. Accordingly, more traditional political actors, like citizens, are replaced with new political subjects of the Internet such as hackers and cyber-activists. On the other hand, the field of citizenship studies tends to assume that the change produced by new digital technologies has not ruptured the ways that politics is typically carried out. New technologies have only introduced new means through which the relationship between nation-states and their citizens is administered. Thus, citizenship studies often focus on analyzing emerging forms of e-government and e-citizenship (Isin and Ruppert, 2015).

Benefiting from the insights of both fields, the geographer Engin Isin and the sociologist Evelyn Ruppert have sought to recast the relationship between new digital technologies and their effects on politics and political subjects. However, unlike these two approaches, their approach does not invest primarily in either the discontinuity or the continuity of politics. Rather, Isin and Ruppert examine the ways in which politics and political subjects "shape and are shaped by socio-technical arrangements" (2015: 10). Thus, for instance, instead of abandoning the notion of public space and the citizen's place in it in relation to enacting politics, they are able to alter traditional understandings of public space by relating it to the political activities that take place in virtual space. According to them, virtual space is not separate from physical space and is an essential space of politics. For Isin and Ruppert (2015: 7), the Internet "has radically changed the meaning and function of being citizens." They call this new modality of politics, which is performed in and through new socio-technical arrangements, digital citizenship, and these new subjects, digital citizens.

The importance of the notions of digital citizenship and digital citizens can be easily seen at play in **urban social movements** such as the recent Occupy, Black Lives Matter, and Arab Spring social movements. In these movements, marginalized communities use the Internet and new digital technologies as main information and communication channels to disseminate their dissenting views and to challenge dominant political powers. However, while it is important to understand how mobile technologies enable new forms of organizing and practices of democracy for social change, it is also important to underline how mass usage of digital technologies and incorporation of mobile technologies also subject citizens to new forms of control and **surveillance**.

The Decline of Urban Public Space

As a result of neoliberal and neoconservative urbanism, over the last 40 years there has been increasing regulation, privatization, and diminishment of urban public spaces in cities across the global North and South (Keil, 2009). Urban renewal projects, accompanied by gentrification processes, have displaced vulnerable social groups (e.g., refugee and immigrant families as well as homeless queer youth) from the public spaces they use (Smith and Low, 2006; Smith, 1992; Sorkin, 1992). Various strategies, including changes in urban design (such as removing and redesigning benches to prevent loitering or sleeping) and the incorporation of closed-circuit television systems (CCTV), systematically and intentionally discriminate against those who do not fit into the marketable image of the urban life.

Since the early 1980s, urban public spaces across the globe have been largely shaped by discourses of **revanchist urbanism** (Mitchell, 1997; Smith, 1996, 1998; Swanson, 2007). Neil Smith (1996) introduced the concept of the revanchist city to describe the specific social conditions of neoliberal urbanism as it unfolded in 1990s New York City. Revanchist urbanism describes intentional and aggressive displacements of marginalized populations, especially from the regenerated urban core, via various urban and legal practices. In addition to criminalization of the urban poor, architectural design (e.g., changing bench structures to prevent sleeping and using sprinklers to discourage sleeping in parks) has been used as an additional displacement method. Financial and political pressures to appropriate prime urban land have led to greater emphasis in maintaining private interests and private property through the introduction of surveillance technologies and private security guards. The introduction of **zero-tolerance policies** (e.g., banning street panhandling and loitering), aggressive practices of urban policing (e.g., carding and racial profiling), systematic displacement of vulnerable populations (e.g., incarceration based on petty crimes), and eviction of major social service providers from downtown cores have been successful not only in reducing the amount of urban public space, but also in reshaping the physical and social landscapes of urban public spaces. Accompanied by rapid processes of gentrification, these discourses not only systematically restrain citizens' right to the city, but also create zones of privilege. The two case studies in this chapter—Istanbul's Gezi Park protests and New York's Morris Justice Project—clearly demonstrate how social groups successfully act, organize, and resist these aggressive discourses.

The Gezi Park Protests in Istanbul

Previously known as Byzantium and Constantinople, Istanbul has been one of the most significant urban centres in history. It served as the capital of many empires for sixteen centuries, from the time of the Byzantine Empire to the Ottoman Empire (Çelik, 1993). Today, with its 14.5 million inhabitants, Istanbul is the most populated city in Turkey, as well as one of the most populous cities in the world (World Population Review, 2015). Because of its social, historical, economic, and cultural centrality, Istanbul has been a category of desire for **civil society** and the state alike. Therefore, the ownership and use of the spaces of Istanbul have become contentious subjects among its citizens, the bourgeoisie, and the state. The most recent and significant example was the conflict over an urban development plan through which the increasingly authoritarian state wanted to achieve cultural hegemony and contribute to the capital accumulation of the bourgeoisie at the expense of its citizens.

Tearing down the walls and uprooting the trees of Gezi Park, one of the few green spaces left in Taksim Square (Figure 13.1), contractors working for Istanbul's metropolitan municipality began demolishing the park in the evening of May 27, 2013. The demolition of Gezi Park was part of an urban transformation plan enforced by Turkey's autocratic Adalet ve Kalkinma Partisi (AKP) (Justice and Development Party) government. According to this plan, Gezi Park was to be replaced with a modern "historic" complex that would consist of a replica of a nineteenth-century Ottoman artillery barracks, a shopping mall, and a mosque (Figure 13.2). Demanding "transparency and the opportunity for input in the redesign of Taksim Square" and reminding the government and city officials that "urban development plans should take into consideration those residing in nearby neighborhoods as well as those who work in and visit the area" (Taksim Platformu, 2014: n.p.), members of a neighbourhood civil society organization and a group of environmentalist youth stopped the demolition through a peaceful sit-in performance.

Despite constant police brutality and the absence of mainstream media coverage, news of the protests spread through social media. Twitter and Facebook were the main channels of communication through which protestors developed, shared, and coordinated

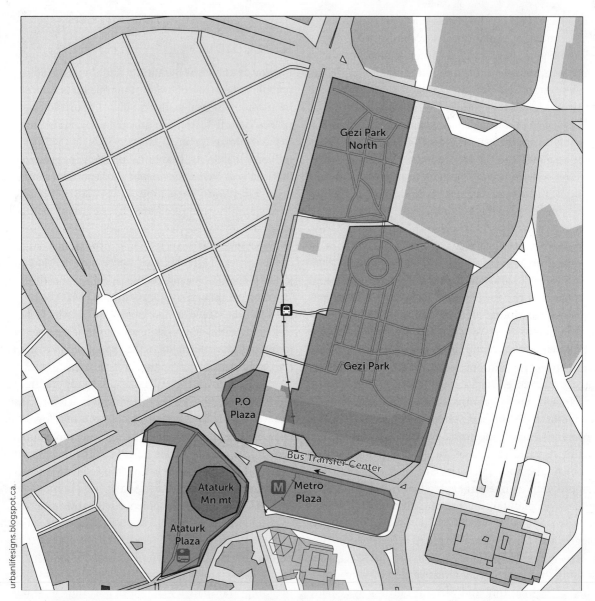

Figure 13.1 Map of Gezi Park in Istanbul, Turkey, in 2013

the strategies and tactics of the Gezi Park protests. Yusuf Sarfati (2015: 38) notes that "the hashtag #direngezipark (#resistgezipark) was used in 950,000 tweets during the uprising's first day, 1.8 million times in the uprising's first three days, and more than 4 million times in the first eleven days." Consequently, the protests snowballed into countrywide demonstrations protesting the authoritarian and anti-democratic rule of the AKP government, and thus Turkey experienced the largest popular protests of its 90-year-history (Oncu, 2014). Between May 28 and the first week of September 2013, "5,532 protests were organized across [80 out of 81] provinces of Turkey. Approximately 3,600,000 people attended the protests, and 5,513 were detained by the police, 189 were arrested. 4,329 were wounded, and five demonstrators were killed" (International Federation for Human Rights, 2014: 6) (Figure 13.3). As result, Gezi Park became a crucial socio-political reference point for contemporary Turkey.

Turkey's Gezi Park protests are an important example of how public space is reappropriated, produced, shared,

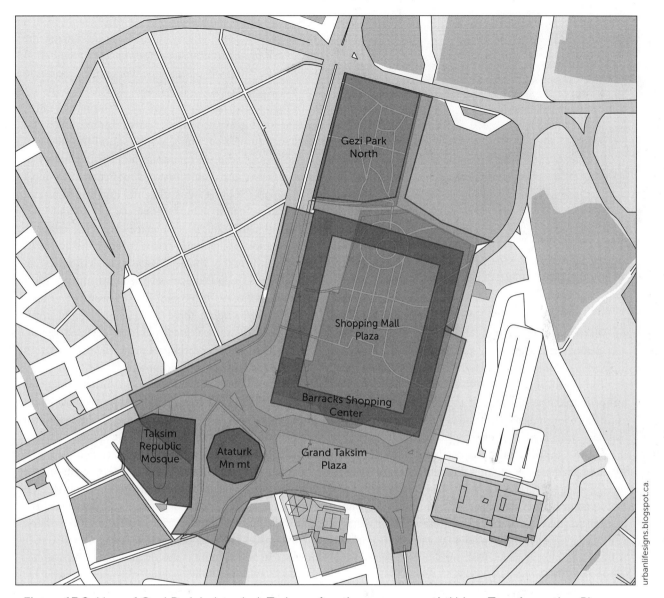

Figure 13.2 Map of Gezi Park in Istanbul, Turkey, after the government's Urban Transformation Plan

and maintained by those who participate in social movements; this example also offers us an opportunity to rethink public space in a context of global urbanization. In particular, the Gezi Park protests demonstrate how public spaces are sites of contention and critique, how urban social movements use public spaces as places that accommodate political differences, and how public and virtual spaces are produced in and through one another.

First, as Don Mitchell argues (1995: 115), "public space is the product of competing ideas about what constitutes that space—order and control or free, and perhaps dangerous, interaction—and who constitutes the "public." In the case of Gezi Park, the government officials argued that they themselves embodied the public interest because the public had elected them. Therefore, demolishing and commodifying the park was a legitimate action, and those who opposed the urban transformation plan could not be seen as a part of the legitimate public, with the Prime Minister Recep Tayyip Erdogan referring to the protestors as "drunkards," "looters,"

Figure 13.3 A protestor being showered with tear gas by the police in Gezi Park in 2014

and "extremists" in order to degrade them. Against this anti-democratic argument, protestors reappropriated the park by occupying it and living there. For the 20 days that they lived in the park before their removal, protestors produced a public space that rendered their concerns visible to the broader public and the government, which had been ignoring these concerns for over a decade.

Second, unlike a bourgeois masculinist public space, which brackets and eliminates differences (Fraser, 1990), the public space produced by the Gezi Park protestors highlighted the specificities of different political groups. Feminists, queers, anarchists, environmentalists, socialists, Marxists, and anti-capitalist Muslims did not collapse into each other. On the contrary, they shared the same space through a constant deliberative process. For instance, the protestors initially referred to the government as "faggots" and "son of a bitch." After feminist protestors explained that LGBT folks and sex workers were also active participants of and contributors to the protests, the slogans of the protests were divested from their sexist, heterosexist, and normative implications.

Third, the Gezi Park protests were "grounded in a material place . . . [and had] a particular preoccupation with public space" (Ors, 2014: 492). However, the protests were able to produce associational public spaces beyond its natal place through the use of social media, performance, and graffiti. Mainly produced through the use of social media, the virtual space of the protests contributed to the production of public spaces across the country.

Following Seyla Benhabib (1993), it can be argued that virtual space itself became a public space during the Gezi Park protests because protestors acted together, in concert (despite their disagreements), on the Internet to coordinate social and political consciousness of Turkey. Political performances and their wide distribution in the virtual space generated public spaces through a "playful construction of a community that bound together disparate groups who want, among other things, a say in how their world [is] administered" (Snyder, 2014: n.p.). Virtual space is not a transcendental universe of disembodied non-sociality. On the contrary, as Isin and Ruppert insist (2015: 28), it "is a space of [social] relations between and among bodies acting through the Internet."

Since the end of the Gezi Park protests in the fall of 2013, Recep Tayyip Erdogan was elected as the president of the Republic of Turkey and the AKP won two general elections. Drawing on these events, some might argue that despite the oppositional public and virtual spaces that the Gezi Park protests produced, the authoritarian rulers of Turkey remained in power. However, as Tanyildiz argues, "the specter of Gezi is haunting Turkey. Its memory is vivid and it is still the main reference point in the political arena" (forthcoming). Indeed, Halklarin Demokratik Partisi (the Peoples' Democratic Party), with its openly pro-Kurdish, socialist, feminist, and environmentalist politics and its popularity among the oppressed, exemplifies the ongoing influence of the Gezi Park protests. The future of Turkey will depend on the political actions of its peoples in and through public and virtual spaces. In Istanbul, as in New York City, state power disciplines public spaces and particular bodies within them. The following case study examines the racialized dimensions of public space in New York City.

The Morris Justice Project in New York City

"Race" and processes of racialization are deeply engraved in the very formation of settler colonial nation-states (Bannerji, 2000). For many academics and activists, the United States is considered to be a prime example of Himani Bannerji's proposition. A relational tripartite regime of domination composed of "the settler colonization of Indigenous Americans, the transatlantic slave trade, and all [US] colonized . . . labour" (Morgensen, 2011: 65) lies at the heart of the state formation of the

United States. Even though some advancement has been achieved by long-standing social movements, this regime of social injustice continues and predominates. For instance, American-led imperial wars and capitalist economic impositions still force displacement and migration of the citizens of societies in the global South, and ongoing internal colonization minoritizes and racializes an array of people living on the Indigenous lands of the United States (Tuck and Yang, 2012).

One recent example that raises questions about public space, virtual space, and protests—about how the state apparatuses in the United States use "race" and racialization processes to administer its citizens' lives in urban centres—is the police's treatment of people of colour, in particular African-American men. According to Mapping Police Violence, a research collaborative that collects data on police killings in the United States, "at least 335 Black people have been killed by police in the U.S. in 2015" (Mapping Police Violence, 2015; see also www.theguardian.com/us-news/ng-interactive/2015/jun/01/the-counted-police-killings-us-database). In July alone, "31 Black people [were] killed by police—one Black person every 24 hours" (Mapping Police Violence, 2015: n.p.). Against this brutality, there have been nationwide protests in American cities. The most powerful recent protests include the 2015 Baltimore protest, which was organized to demand accountability for the death in police custody of 25-year-old African-American Freddie Gray; the 2014–2015 Ferguson protests, which began after the fatal shooting of the 18-year-old African-American youth Michael Brown by a police officer; and the #BlackLivesMatter movement that began after the acquittal in 2012 of the murderer of 17-year-old African-American high school student Trayvon Martin (Black Lives Matter, 2012).

Within this context of police violence, the Morris Justice Project (MJP) emerged in New York City's South Bronx. Born out of the neighbourhood mothers' care for their sons' lives and named after one of the main avenues of the Bronx, the MJP comprises a collaborative research team of neighbourhood residents and members of the Public Science Project, the City University of New York Graduate Center, John Jay College, and Pace University Law Center. Practising participatory action research, the team conducted focus groups and comprehensive surveys in the 40-block neighbourhood, which is 96 per cent black/Latino(a), to unearth the residents' experiences of policing (Figure 13.4). According to an MJP report, 75 per cent of community members in the survey said they had been stopped by police; a total of 60 per cent reported being asked to move by the police when standing outside their own building, standing in a group, standing on a street corner, standing outside a friend's or a family member's building, in the park, inside their building, in the subway, at bus station/stop, or at school; and 63 per cent reported feeling targeted by the police (Morris Justice Project, 2011).

This report suggests that the urban public spaces of New York City are constantly negotiated and structured by, among other factors, "race." The state, as the sole claimant of *"the monopoly of legitimate use of physical force* within a given territory" (Weber, 1994: 310, emphasis in original), excludes the residents of this community of colour from the public space in the area in which they live and work by enforcing a discriminatory and abusive policing regime through one of its apparatuses, the New York Police Department (NYPD). Stemming from colonial histories and reproduced in contemporary power relations, racial stereotypes are used to determine who is allowed to appear in urban public spaces.

In other words, the MJP reveals that "race" is one of the main signifiers in the production and reorganization of New York City's urban public spaces. Through an oppressive regime of policing, the state produces neighbourhoods of colour as "ghettos" in which crime runs amok. Leaving existing "socio-economic spatial segregation" (Zieleniec, 2007: 168) unexamined, this production deepens the structural racial inequalities that constitute the urban fabric of New York City. Furthermore, it constructs residents of these racialized neighbourhoods as suspects who require constant "order-maintenance" and receive "zero-tolerance." Consequently, "selling loosies [a single cigarette], dancing on the trains, riding bicycles on the sidewalk, having a beer outside, standing on the corner with a group, smoking marijuana, being in the park after-hours" (Morris Justice Project, 2011: n.p.) become serious crimes and indicators of even greater crimes to come. Thus, access to public spaces for people of colour living in these neighbourhoods becomes police-controlled and limited.

In order to break this racist and abusive policing regime, the MJP initiated a series of events, including advocating for new city legislation, reaching out to those who do not live in the neighbourhood, meeting with elected officials, reporting back to the South Bronx community, defending their rights in court, and speaking back to the NYPD (Morris Justice

We are the **Morris Justice Project**, a group of researchers from our neighborhood, CUNY, and Pace University Law School. We are concerned about policing in our neighborhood. Please help us by taking our 4-part survey. It asks important questions about your experiences with police and will take only about 15 minutes. Once we have the results we will be holding community meetings to talk and act on what we found. Thanks and stay tuned!

Part I: Neighborhood Experiences with Police

Directions: We are interested in the experiences of people who live, work, or go to school within the outlined part of the map.

1. I _____ in the neighborhood (check all that apply):
    - O Live
    - O Work
    - O Go to school

2. **CIRCLE the nearest street corner** where you live, work, or go to school. If more than one applies to you, try to label them as clearly as you can.
(Remember, this survey is anonymous, so **do not** give us your exact home address!)

3. How many years have you have lived, worked, or gone to school in the neighborhood? _____.

4. How old were you the first time you were stopped by police? _____.
   Never stopped? Check this circle O

5. Please circle the number of times you've been stopped in **the last year**?

   0 – 1 – 2 – 3 – 4 – 5 – 6 – 7 – 8 – 9 – 10 – 11 +

   if more than 10, exactly how many? _____.

6. Using the map, please **PLACE an "X"** on the locations where police stopped you **in the last year**. And **PLACE** a **CIRCLE** where you live, work, to go to school. ⟶

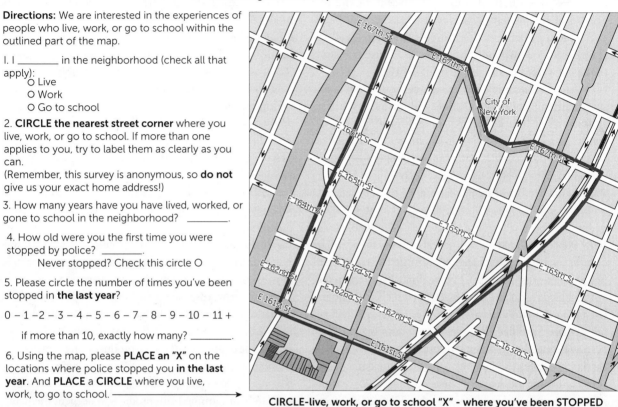

CIRCLE-live, work, or go to school "X" - where you've been STOPPED

Figure 13.4 Map of the Bronx neighbourhood in New York City, USA, used in the Morris Justice "critical mapping project" to investigate residents' experiences with police
morrisjustice.org, 2013.

Project, 2011). Through these events, an atmosphere of dialogue, debate, and discussion, which is crucial for the existence of public spaces, was engendered. The MJP also initiated creative performances, producing the neighborhood as a new public space. For instance, the MJP organized a "sidewalk science" project in which people took over their neighbourhood sidewalks, voicing their concerns by chalking messages on them, and an "envisioning alternatives" project encouraged reimagining the concept of community safety from the viewpoint of the residents (Morris Justice Project, 2011).

The MJP challenges the prevalent dichotomous thinking about public and virtual space. It demonstrates that in the twenty-first century, where states and their apparatuses have intensified surveillance and oppression, it has become more difficult to appear in public spaces without the simultaneous presence of well-organized virtual spaces. In this respect, the MJP illustrates that it is untenable to hold that the material (public space) and immaterial (virtual space) binary exists as two independent realms of reality. Instead, by making its findings, actions, and documents available in virtual spaces, such as on their web page, Tumblr and Vimeo, and by not abandoning direct action in public space, the MJP suggests that public space and virtual space are produced in and through one another in an interdependent manner, even though they might require distinct tactics and strategies in accordance with their specific circumstances.

As the two case studies have demonstrated, citizens take social problems that are particularly crushing for them to the streets and squares of their respective countries. In these public spaces, they voice their concerns about existing social relations through embodied interactions that are conditioned by the technologies of the day, such as the use of social media. Among other acts of citizenship, these practices of urban protest produce the political culture and history of a given country. In crossing the border from the United States to Canada, people deploy practices of urban citizenship through intersections of public space and social media to contest racialized and gendered relations of those most disenfranchised. Protest movements that have garnered a global presence have first been mobilized in Canadian urban public spaces.

Canadian Urban Public Spaces and Protest

In the post-industrial era, Canadian metropolitan centres have experienced declines and transformations in their public spaces, as well as the introduction of new regulations to govern urban citizenship. Aggressive neoliberal and neoconservative urbanization strategies to increase the marketability of the city have resulted in downtown cores becoming spaces of gentrification and exclusiveness (Hulchanski, 2010). As a response, several social groups have adopted strategies to fight against spatial exclusion, displacement, and gentrification.

Historically, public-space activism in major Canadian cities has covered a range of social, political, and environmental issues. Over the last decade, in Toronto, for example, several campaigns have been mounted to claim public space and to fight against the corporatization and commercialization of urban public spaces, including guerrilla gardening, downtown de-fencing, advocacy against mobile billboards, and advocacy against ad-infused trash receptacles (Pask, 2010). Vancouver has seen advocacy and community-capacity building, including street art, public-space films, non-compliant billboards, community gardening, community composting, guerrilla way-finding projects, and community policing (Pask, 2010). These campaigns, among many others initiated by the Toronto Public Space Committee (www.publicspace.ca) and the Vancouver Public Space Network (http://vancouverpublicspace.ca), have been successful in bringing community members together, as collectives, while mobilizing social change via capacity building. Like their counterparts across the globe, these campaigns have successfully incorporated social media.

Canadian social movements like #SlutWalk (www.facebook.com/SlutWalkToronto) and #IdleNoMore (www.idlenomore.ca), have quickly become worldwide protest movements via widespread usage of digital technologies (Church, 2012). On January 24, 2011, while giving a talk on health and safety to a group of students at Osgoode Hall Law School in Toronto, police constable Michael Sanguinetti made the following statement: "I've been told I'm not supposed to say this—however, women should avoid dressing like sluts in order not to be victimized." After his remarks, there was an outrage in social media. On April 3, 2011, 3000 people took part in the first "SlutWalk" in Toronto, wherein participants claimed their right to the city's public spaces and protested women's appearance being used as an excuse for any type of assault. In a short period of time, SlutWalk has become an important intersectional feminist

continued

movement, with marches happening in cities across Canada:

> Toronto (www.facebook.com/SlutWalkToronto)
> Hamilton (www.facebook.com/slutwalkham)
> London (www.facebook.com/LondonsSolidaritySlutWalk)
> Montreal (www.facebook.com/SlutWalk-Montreal-147489268653143)
> Vancouver (https://twitter.com/slutwalkvan)
> Edmonton (https://edmontonslutwalk.wordpress.com)
> Winnipeg (https://slutwalkwinnipeg.wordpress.com)
> St John's (www.facebook.com/SJSlutwalk)

The year 2015 marked the fifth-year anniversary of SlutWalk, with increased participation from various social justice organizations across Canadian cities. The success of SlutWalk's international feminist activism has, Kaitlynn Mendes (2015) argues, been largely due to the formation and work of counter-publics via Facebook, Twitter, and blogs.

Idle No More is another urban social movement that emerged in Canada. In November 2012, Nina Wilson, Sheelah McLean, Sylvia McAdam, and Jessica Gordon, four women from Saskatchewan, started a Facebook page to discuss ideas for action. Gordon, who is from Pasqua 4 Treaty Territory, decided to name the page "Idle No More." Their first major activity was a teach-in held in Saskatoon. The Idle No More social movement opposes neocolonial socio-economic and environmental policies and practices and honours Indigenous sovereignty to protect land and water. The movement underlines the significance of the interconnectedness of race, gender, sexuality, class, and other identity constructions in understanding the complexities of power, domination, and oppression. In a short time, via Facebook and Twitter, various alliances and solidarities have been built around the movement and across various urban centres. On December 10, 2012, the day Idle No More launched online, 2800 participants issued over 14,600 tweets (https://warriorpublications.wordpress.com/2013/12/11/one-year-of-idle-no-more-update-and-analysis). Social media enabled Idle No More to gain rapid worldwide support from various social justice advocates, environmentalists, and other groups (Wotherspoon and Hansen, 2013). Based on their analysis of 500,000 tweets, Dahlberg-Grundberg and Lindgren (2014) argue that the #IdleNoMore tag has successfully made ties to other transnational movements (e.g., Occupy).

In an interview with Leanne Betamasamosake Simpson, Eugene Boulanger (2014: 318) explains that in the Idle No More movement "[s]ocial media provides us a space from which we can broadcast our own narratives—from a grassroots level. Rather than waiting and working with the confines of corporate broadcast media, clear narratives were communicated between diverse groups across the world in mere seconds."

In three years, via rallies, protests, sit-ins, and dances, Idle No More became an ongoing transnational movement for justice and Indigenous self-determination. In Canada, the movement has been successful in undertaking issues around First Nations by launching a new crowd-funding project (One House, Many Nations, www.indiegogo.com/projects/one-house-many-nations#) and pushing for a national inquiry into missing and murdered Indigenous women, which the federal Liberal government initiated in 2016.

Canadian social movements like SlutWalk and Idle No More have successfully utilized social media to connect various struggles, bodies, and places in a novel way. While most of the Canadian urban literature is focused on understanding how neoliberal and neoconservative discourses are unfolding in the three largest cities—Toronto, Vancouver, and Montreal—it is important to think about how these processes unfold in smaller towns and cities. While scholars underline new forms of activism in suburbs (Parlette and Cowen, 2010), small and mid-sized urban centres in Canada are under significant pressures resulting from de-industrialization but remain under-researched. And yet, despite their scholarly invisibility, several social movements are active within these smaller centres, such as Food Not Bombs, Critical Mass, Occupy, Palestinian Solidarity networks, and Refugee Welcome. In Niagara Falls and St Catharines,

> for example, advocates of Marineland Animal Defense (http://marinelandanimaldefense.tumblr.com) have successfully utilized hashtags #endcaptivity, #marineland, and #emptythetanks, not only to organize ongoing protests in Niagara Falls, but to raise global awareness regarding intersectionalities of non-human animal and labour abuses. Across Canada, social media have played a significant role in identifying whose voices are being heard, as well as whose voices are being silenced, and in bringing people together to facilitate social change and the building of alliances and solidarities across various sites.

Conclusion

Around the world, contemporary urban landscapes have been shaped by the violent and aggressive discourses of late capitalism, accompanied by neoliberal urbanism. With increased commercialization and privatization, critical urban analyses of urban dwellers' right to participate and right to appropriate are essential to our understandings of contemporary power struggles. This chapter has thus provided theoretical and conceptual frameworks on urban public spaces, digital spaces, and protests that are supported by examples from contemporary social movements as they unfold, in and through, various contested urban and digital spaces. Assemblages of bodies and crowds coming together via the materiality of urban public space denote not only the emergence of new urban political subjectivities but also the possibility of the end of public space as we know it. Critical understandings of the new subjectivities and socio-spatialities that are emerging as a result of urban protests are significant for social activists as they reconsider various tactics and strategies, but they also give us new ways of theorizing public space and counter-publics.

Marginalized, excluded, and discriminated social groups have been successful in utilizing social networking sites and digital spaces to provide connection, visibility, public address, and real-time information sharing. By introducing new ways of organizing and communicating, these new struggles and claims have not only challenged our previous conceptualizations of subjectivities and socio-spatialities, but have also made us rethink new forms and practices of politics, citizenship, and democracy. From Instagram followers to cyber-anarchists, advancement in digital technologies and social media have also significantly changed our relationship to ourselves and others. In Istanbul, when access to Twitter, YouTube, and Facebook was blocked by the government in its attempt to disrupt widespread social uprising, the streets of the neighbourhood of Taksim were graffitied with alternative ways to go online. From the streets of Istanbul, to New York City, to Canadian cities both large and small, the emergence of new forms of dissent and insurgency have been challenging the normative understandings of politics, citizenship, and urban public space. As demonstrated in this chapter, social media have provided expression, solidarity, connectivity, accessibility, and openness to many people, yet it has also been a space of governance, surveillance, criminalization, and control.

Key Points

- Historically, urban public spaces like squares, markets, and plazas, have been an important part of democratic practices, as they define the power dynamics of inclusion and exclusion in the civic and political formation of citizenship.
- Systems of capitalism and patriarchy, mass consumption, and urban planning have played important roles in the emergence of public and private spaces.
- In urban public space, strangers become the public and (re)produce publicness through everyday encounters.

- Digital citizenship is a new political modality of the contemporary advanced technological era in which online participation and rights claims are considered to be essential to democratic participation, thus expanding the notion of public space highlighting the imbrication of material and virtual space.
- Commercialization, privatization, and securitization of urban public space have been the major reasons for the socio-spatial and legal exclusion of vulnerable groups, as well as the major reasons for the decline in urban public space.
- The Gezi Park protests in Turkey developed in opposition to a neoliberal urban transformation plan that attempted to turn a public park into private property. Protestors with different political claims challenged what public space means by re-appropriating and reproducing the park as a place in which they could critique the increasingly authoritarian character of the government.
- The Morris Justice Project in the South Bronx emerged in the context of racialized power dynamics and police violence, contributing to the reorganization of New York City's urban public spaces. Local residents sought to stop racist, discriminatory, and abusive regimes of policing by producing their neighbourhood as a new public space through creative and virtual performances. In Canada, various social movements (e.g., Idle No More and SlutWalk) use social media to demonstrate new forms of dissent and insurgency that challenge understandings of the public, political, and urban public spaces.

Activities and Questions for Review

1. Based on your own experience as an undergraduate student, debate the publicness of space on your university campus. Pay particular attention to how different student groups and your university administration negotiate access to these spaces.
2. Use your cellphone to have a text-based discussion with a classmate about how you use social media in your everyday lives and how your usages illustrate the interdependence of public and virtual spaces. How would this discussion have been different, if at all, had you had a face-to-face conversation instead?
3. In pairs, reflect critically upon the intersections of your own positionality and relative privilege (e.g., racialization, gender, class, sexuality, and age) and consider how these could affect upon your ability to make rights claims in public space.
4. As a class, debate the extent to which urban social movements and activism can survive without urban public space.

Acknowledgments

We would like to express our gratitude to Linda Peake and Alison Bain for their invitation to write this chapter and for their generous editorial support throughout its various stages.

References

Amin, A. (2006). The good city. *Urban Studies, 43*(5–6): 1009–1023.

Bannerji, H. (2000). *The dark side of the nation: Essays on multiculturalism, nationalism and gender.* Toronto: Canadian Scholars' Press.

Benhabib, S. (1993). Feminist theory and Hannah Arendt's concept of public space. *History of the Human Sciences, 6*(2): 97–114.

Black Lives Matter (2012). *About us.* Retrieved from http://blacklivesmatter.com/about

Boulanger, E. (2014). Seeing strength, beauty and resilience in ourselves. In Kino-nda-niimi Collective (Ed.), *Winter we danced: Voices from the past, the future, and Idle No More movement.* Manitoba: ARB Books, pp. 316–320.

Bowlby, S. (1999). Public/Private division. In L. McDowell and J. Sharp (Eds.), *A feminist glossary of human geography.* New York: Arnold, pp. 222–224.

Bridge, G. (2009). Reason in the city: Communicative action, media and urban politics. *Urban and Regional Research, 33*(1): 237–240.

Cassegard, C. (2014). Contestation and bracketing: The relation between public space and public sphere. *Environment and Planning D, 32*(4): 689–703.

Çelik, Z. (1993). *The remaking of Istanbul: Portrait of an Ottoman city in the nineteenth century.* Berkeley and Los Angeles: University of California Press.

Church, E. (2012). SlutWalk sparks worldwide protest movement. Retrieved from http://www.theglobeandmail.com/news/toronto/slutwalk-sparks-worldwide-protest-movement/article583076

Dahlberg-Grundberg, M., and Lindgren, S. (2014). Translocal frame extensions in a networked protest: Situating the #IdleNoMore Hashtag *ic-Revisra Cientifica de Informacion y Communicacion, 11*: 49–77.

Domosh, M., and Seager, J. (2001). *Putting women in place: Feminist geographers.* New York and London: Guilford Press.

Faulkner, N. (2013). *A Marxist history of the world: From neanderthals to neoliberals.* London: Pluto Press.

Federici, S. (2004). *Caliban and the witch: Women, the body and primitive accumulation.* New York: Ak Press.

Fraser, N. (1990). Rethinking the public sphere: A contribution to the critique of actually existing democracy. *Social Text*, no. 25/26: 56–80.

Fraser, N. (1997). *Justice interruptus: Critical reflections on the 'postsocialist' condition.* London: Routledge.

Habermas, J. (1987). *The philosophical discourse of modernity.* Cambridge, MA: MIT Press.

Hampton, K.N, Livio, O., and Goulet, L.S. (2010). The social life of wireless urban spaces: Internet use, social networks, and the public realm. *Journal of Communication, 60*(4): 701–722.

Herriot, L. (2015). Slutwalk: Contextualizing the movement. *Women Studies International Forum, 53*: 22–30.

Holston, J. (1998). Spaces of insurgent citizenship. In L. Sandercock (Ed.), *Making the visible invisible: A multicultural planning history.* Berkeley: University of California Press.

Holston, J., and Appadurai, A. (1999). Cities and citizenship. *Public Culture, 8*(2): 187.

Hulchanski, D. (2010). *Three cities within Toronto: Income polarization among Toronto's neighbourhoods.* Centre for Urban Communities Research Bulletin.

International Federation for Human Rights (2014). *Turkey: Gezi, one year on.* Retrieved from http://www.fidh.org/IMG/pdf/turkey_avril_2014_uk_web.pdf

Isin, E. (2000). Introduction: Democracy, citizenship and the city. In E.F. Isin (Ed.), *Democracy: Citizenship and the global city.* London: Routledge.

Isin, E., and Ruppert, E. (2015). *Being digital citizens.* London and New York: Rowman & Littlefield International.

Isin, E., and Wood, P. (1999). *Citizenship and identity.* London: SAGE.

Iveson, K. (2007). *Publics and the city.* London: Blackwell.

Iveson, K. (2009). The city versus the media? Mapping the mobile geographies of public address. *International Journal of Urban and Regional Research, 33*(1): 241–245.

Joyce, P. (2003). *Rule of freedom: Liberalism and the modern city.* London: Verso.

Keil, R. (2009). The urban politics of roll-with-it neoliberalization. *City, 13*(2–3): 231–246.

Kroker, A., and Kroker, M. (2008). *Critical digital studies: A reader.* Toronto: University of Toronto Press.

Lefevbre, H. (1968). *La droit à la ville.* Paris: Anthropos.

McCann, E.J. (1999). Race, protest and public space: Contextualizing Lefebvre in the US city. *Antipode, 31*(2): 163–184.

Mapping Police Violence (2015). Reports. Retrieved from http://mappingpoliceviolence.org/reports

Massey, D. (2005). *For space.* London: SAGE.

Mendes, K. (2015). *Slutwalk: Feminism, activism and media.* London: Palgrave Macmillan.

Mitchell, D. (1995). The end of public space: People's Park, definitions of the public and democracy. *Annals of the Association of American Geographers, 85*(1): 108–133.

Mitchell, D. (1997). The annihilation of space by law: The roots and implications of anti-homeless laws in the United States. *Antipode, 29*: 303–335.

Mitchell, D. (2003). *The right to the city: Social justice and fight for the public space.* New York: Guilford Press.

Morgensen, S.L. (2011). The biopolitics of settler colonialism: Right here, right now. *Settler Colonial Studies, 1*(1): 52–76.

Morris Justice Project, The. (2011). Action. Retrieved from http://morrisjustice.org/action

Oncu, A. (2014). Turkish capitalist modernity and the Gezi revolt. *Journal of Historical Sociology, 27*(2): 151–176.

Ors, I.R. (2014). Genie in the bottle: Gezi Park, Taksim Square, and the realignment of democracy and space in Turkey. *Philosophy and Social Criticism, 40*(4–5): 489–498.

Orwell, G. (2008). *Animal farm.* London: Penguin Books.

Parlette, V., and Cowen, D. (2010). Dead malls: Suburban activism, local spaces, global logistics. *International Journal of Urban and Regional Research, 35*(4): 794–811.

Pask, A. (2010). Public space activism, Toronto and Vancouver: Using the banner of public space to build capacity and activate change. In J. Hou (Ed.), *Insurgent public space: Guerrilla urbanism and the remaking of contemporary cities.* London: Routledge.

Pinch, S. (2014). Slutwalk three years later. Where has the movement taken us? Accessed December 11, 2015, at http://rabble.ca/news/2014/07/slutwalk-three-years-later-where-has-movement-taken-us

Purcell, M. (2003). Citizenship and the right to the global city: Re-imagining the capitalist world order. *International Journal of Urban and Regional Research, 27*(3): 564–590.

Ryan, M. (1997). *Civic wars: Democracy and public life in the American city during the nineteenth century.* Los Angeles: University of California Press.

Sandercock, L. (1997). *Towards Cosmopolis: Planning for multicultural cities.* London: Wiley.

Sarfati, Y. (2015). Dynamics of mobilization during Gezi Park protests in Turkey. In I. Epstein (Ed.), *The whole world is texting: Youth protest in the information age.* Rotterdam, Boston, and Taipei: Sense Publishers, pp. 25–45.

Sassen, S. (1996). *Losing control? Sovereignty in an age of globalization.* New York: Columbia University Press.

Sennett, R. (1997). *The fall of public man.* New York: N.W. Norton.

Smith, N. (1992). New city, new frontier: Lower East Side as Wild, Wild West. In M. Sorkin (Eds.), *Variations on a theme park: The new American city and the end of public space.* London: Routledge.

Smith, N. (1996). *The new urban frontier: Gentrification and the revanchist city.* New York: Routledge.

Smith, N. (1998). Giuliani time. *Social Text*, *57*: 1–20.

Smith, N., and Low, S. (2006). Introduction: The imperative public space. In S. Low and N. Smith (Eds.), *Politics of Public Space*. London: Routledge.

Sorkin, M. (1992). See you in Disneyland. In M. Sorkin (Ed.), *Variations on a theme park: The new American city and the end of public space*. London: Routledge.

Snyder, S. (2014). Gezi Park and the transformative power of art. *Roar Magazine*, January 8. Retrieved from http://roarmag.org/2014/01/nietzsche-gezi-power-art

Staeheli, L., and Mitchell, D. (2008). *The people's property? Power, politics and the public*. New York: Routledge.

Strange, C. (1995). *Toronto's girl problem: The perils and pleasures of the city*. Toronto: University of Toronto Press.

Swanson, K. (2007). Revanchist urbanism heads south: The regulation of Indigenous beggars and street vendors in Ecuador. *Antipode*, *39*(4): 708–728.

Taksim Platformu (2014). What do we want? Retrieved from http://www.taksimplatformu.com/english.php

Tanyildiz, G.S. (forthcoming). The Gezi protests: The making of the next left generation in Turkey. In R. Hadj-Moussa and M. Ayyash (Eds.), *Generations and protests in the Middle East and Mediterranean*. Leiden and Boston: Brill.

Tuck, E., and Yang, W. (2012). Decolonization is not a metaphor. *Decolonization: Indigeneity, Education & Society*, *1*(1): 1–40.

Valentine, G. (1998). Living with difference: Reflections on geographies of encounter. *Progress in Human Geography*, *32*(3): 323–337.

Warner, M. (2002). *Publics and counterpublics*. Cambridge, MA: MIT Press.

Weber, M. (1994). *Political writings*. Cambridge: Cambridge University Press.

Wilson, E. (1991). *Sphinx in the city*. London: Palgrave.

World Population Review (2015). World Cities—Istanbul Population. Retrieved from http://worldpopulationreview.com/world-cities/istanbul-population

Wotherspoon, T., and Hansen, J. (2013). The "Idle No More" movement: Paradoxes of First Nations inclusion in the Canadian context. *Social Inclusion*, *1*(1): 21–36.

Young, I.M. (1990). *Justice and the politics of difference*. Princeton, NJ: Princeton University Press.

Zieleniec, A. (2007). *Space and social theory*. London: SAGE.

Zukin, S. (1995). *Culture of cities*. London: Wiley.

Urban Geopolitics
War, Militarization, and "The Camp"

Nicole Laliberté and Dima Saad

Introduction

What are the limits of urban studies? Are (inter)national military strategies, political treaties, and transnational social movements appropriate topics to consider as we study cities? As Stephen Graham (2004: 171) asserts in his studies of the new **military urbanism**, the "urban" "tended to remain hermetically separated from the 'strategic' [as] . . . the overwhelmingly 'local' concerns of modern urban social science were kept rigidly apart from (inter)national ones." A disciplinary divide was created between what was considered appropriate for urban studies—the local, civil, and domestic—and the political processes that were occurring at multiple scales—the (inter)national, the military, and the strategic. These latter areas were left to other fields such as international relations, military studies, and geopolitics. This divide is now recognized as a false one and this chapter argues that **geopolitics** is indeed an understudied yet integral aspect of urbanization. Accordingly, it shows how challenging this division between the local and the (inter)national through an **urban geopolitics** illuminates complex political processes that not only contribute to contemporary urban forms but also challenge our very definitions of "the urban."

If we bring a geopolitical lens to the study of urbanization, the connections between multiply scaled processes are made visible; that is, an urban geopolitical approach can investigate processes of securitization to draw connections between international military interventions and national security via the management of urban places. With this approach we can also analyze the social movement networks that shape multiple local processes of urbanization through their transnational connections and simultaneously acknowledge and challenge state-centric power structures. Moreover, an urban geopolitical analysis, when focused on Canada, can reveal the project of **settler colonialism** which has separated settlers from Indigenous populations through the racially coded spatial distinctions between "reserve" and "city."

This chapter focuses urban geopolitical analysis on the issues of militarization and forced displacement, not only because they highlight the importance of the multi-scalar analysis inherent to an urban geopolitics, but also because they allow questions to be raised about popular conceptions of urbanity; they force a questioning of our own definitions of what an urban space is and how it is implicated in geopolitical negotiations. In other words, the chapter examines how **war** and militarization contribute to the creation of unique urban forms as well as how these urban forms contribute to the practices of war and militarization. The chapter begins with a brief discussion of geopolitics in relation to critical theories of place and scale. It moves to definitions of war and militarization in order to make connections between the politics of war and militarization and contemporary urban spaces. Then, it examines the politics of urbanization in the context of spaces that are formed as a result of forced displacement. In line with these arguments, case studies of the refugee camps Awere in Uganda and al-Baqa'a camp in Jordan demonstrate the importance of geopolitical contexts with regard to the specific forms of urbanism that constitute these camps. The final section addresses the relevance of militarization and displacement in an analysis of Canadian urban geopolitics. It discusses the ways in which the formation of the Canadian national identity relies on the systematic policing of "othered" bodies in urban(izing) spaces. Urban geopolitics, therefore, views the urban as more than just a site of geopolitical action; rather, it posits that urbanization itself is a geopolitical process.

Building an Urban Geopolitics of Militarization

This section discusses concepts that are related to geopolitics and militarization and that offer a foundation for the subsequent discussion of what an urban geopolitical analysis of militarization looks like. The discussion, in turn, provides the groundwork for the specific focus of this chapter, the urban geopolitics of forced displacement due to war and conflict.

Understanding Geopolitics

Any discussion of urban geopolitics needs to begin by clarifying the term geopolitics. While its exact definition varies by discipline, it is defined here as the multiscalar intersections of territoriality, spatiality, violence, and resistance via political processes. Within the field of geography, there have been three distinct approaches to studying geopolitics. The earliest tradition is that of classical geopolitics, a "geography as destiny" approach in which the world is explained through an analysis of the geographical relations between nations and states (Flint, 2012). Geopolitics, following this framework, can refer to the field of geography's contribution to statecraft and the consolidation of state authority (Mackinder, 1904). **Critical geopolitics**, which took hold in the 1990s, challenges the conservatism and state-centric focus of classical geopolitics to expose its biases and political agendas (Dalby, 1991; O'Tuathail, 1996; Flint, 2012). Although critical geopolitics is generally unsympathetic towards state-building projects, it still maintains a focus on traditionally defined geopolitical actors. Like critical geopolitics, feminist geopolitical analyses decentre the state and make connections between power relations at multiple scales. Drawing heavily upon contemporary work on the social and political construction of scale (e.g., Cox, 1998; Marston, 2000; Herod and Wright, 2002), feminist geopolitical scholars critique the heteronormative assumptions of critical geopolitics (Peake, 2013) while arguing for the investigation of the everyday and seemingly "apolitical" spaces as sites of significant geopolitical analysis (Kuus, Sharp, and Dodds, 2013; Dowler and Sharp, 2001; Staeheli and Kofman, 2004). These spaces are, as Dowler and Sharp (2001: 169) argue, a "lens through which the everyday experiences of the disenfranchised can be made more visible." This chapter draws predominantly on these everyday spaces and on the work of critical and feminist geopolitical scholars.

Critical geographic theories of **place** and **scale** allow the connections between spaces and processes that were traditionally separated into the categories of "urban," "national," or "global" to be understood as mutually constitutive. Historically, geographers promoted the idea of place as representative of the "local" scale, as a static entity, an inward-looking enclosure with a fixed meaning and singular identity (Massey, 1994). Feminist and critical geographers have critiqued such a perception of place for its tendency to obfuscate the construction and maintenance of hegemonic power relations at multiple scales that can make it difficult to imagine the means by which to change them (Massey, 1994; Cox, 1998; Herod and Wright, 2002). Instead, they argue for a relational understanding of scale, including that of the **body**, in which the local is not only connected to the global, but implicated in its production (Massey, 1994; McDowell, 1999; Marston, 2000). It is from this theoretical position that feminist geopoliticians have studied geopolitical processes through embodied experiences.

By connecting people's everyday lives to geopolitical processes via the urban, feminist geopolitical analysis offers insight into what an urban geopolitics can look like. For example, Anna Secor (2001) calls for feminist counter-geopolitics, or alternate spatializations of lived urban politics, by examining how women in Istanbul, Turkey, negotiate, contest, and (re)shape urban spaces through formal and informal political activities. Sara Smith (2012) brings geopolitics into the intimate spaces of the home in the Ladakh region of India by showing how decisions of marriage and child-rearing are tied to nationalist claims on territory. Cindi Katz (2006) examines how military and government officials strategically manipulated everyday urban spaces in post-9/11 New York City to produce feelings of fear and insecurity within the population through the use of terror alert warnings and the physical presence of military personnel to justify the increased securitization of the city. Romola Sanyal (2014) argues that urban forms and processes are inseparable from claims of citizenship and political agency for many contemporary refugees and displaced persons. While all of these are active avenues for geopolitical analysis, it is Sanyal's analysis that serves as the guidepost for the arguments presented here; this chapter follows the connections that Sanyal draws between the geopolitics of urbanization and the processes of militarization and forced displacement. For one to understand these links, however, key terms related to militarization need to first be clarified.

War and Militarization

Historically, the term "war" has conjured up images of military conflict between armies, often associated with territorial conflicts between or within states. This image is consistent with a classical geopolitical analysis of war as a process of statecraft (Mackinder, 1904; Owen, 2000). While such a perspective positions war as part of a political project, it does not question the category of war. In feminist theories of war, however, the very process of labeling particular forms of **violence** as war is understood to be a political project (Cooke, 1996; Sjoberg, 2006). Through such interrogations of the category of war, critical scholars highlight the nationalist and state-based agendas that separate state-sanctioned violence from other forms of violence, such as domestic violence (Borer, 2009; Chan, 2011). They effectively trouble the war/peace binary by showing how the violence of war is not separate from the violence of **everyday life** (Sjoberg, 2006; Enloe, 2007). By demonstrating how gender regimes, processes of "othering," militarization, and economic systems of inequality blur the war/peace divide, critical scholars have brought attention to the multiple forms of insecurity simultaneously produced and obscured by traditional narratives of war. Hence, the term "war" is used in this chapter with the understanding that it is a political marker deployed to characterize certain types of violence as abnormal and separate from normal "peaceful" social relations.

In close relation to the concept of war are the processes of militarization and militarism. Enloe (2000: 281, emphasis in original) defines **militarization** as "the step-by-step process by which something becomes *controlled by*, *dependent on*, or *derives its value from* the military as an institution or militaristic criteria." An analysis of militarization therefore takes us beyond the actions of militaries and into complex social relations produced through and in relation to militaristic institutions and ideas. As a complementary concept, **militarism** is a militarized system and set of ideas that a society adopts to sustain peace and/or prepare for war (Dowler, 2012). In what follows, the discussion demonstrates how an urban geopolitical analysis can highlight processes of militarization and militarism as they shape, destroy, and (re)construct urban landscapes.

Urban Geopolitics of War and Militarization

Contemporary warfare no longer follows the typical twentieth-century image of a battlefield flanked by the armies of two opposing states, for there is rarely a spatial division between the front line and the home front. Warfare is increasingly taking place in cities—in doorways, manufacturing districts, subway tunnels, high-rises, and busy intersections rather than in open fields, forests, and distant mountains. The urban home front has become the front line as cities increasingly become strategic targets. For those who wish either to cripple their enemy or to claim the enemy's territory, cities hold many strategic targets, including transportation hubs, energy grids, communication networks, and manufacturing infrastructures. Cities can also function as strategic targets owing to their symbolic value, seen, for example, in the bombing of the World Trade Center in New York City. While the bombing was clearly targeted at weakening the economic systems of the United States and international markets, it is the symbolic power of the strike that has had the strongest and longest-lasting impact.

In his analysis of the effects of war on cities, geographer Kenneth Hewitt (1983: 258) writes, "There is . . . a direct reciprocity between war and cities. The latter are the more thorough-going constructions of collective life, containing the definitive human places. War is the most thorough-going or consciously prosecuted occasion of collective violence that destroys places." While we acknowledge the ability of violence to destroy places—what Hewitt calls "**place annihilation**"—we should also explore how war and violence can *construct* places. Accordingly, this chapter examines the ways in which unique urban forms are created through war in order to reveal how urban systems are integrated into and constitutive of war and processes of militarization.

In areas where there is (or has been) active urban warfare, it is possible to see an immediate challenge to the idea of the urban as "local." As Appadurai (1996: 152, original emphasis) explains,

> In the conditions of ethnic unrest and urban warfare that characterize cities such as Belfast and Los Angeles, Ahmedabad and Sarajevo, Mogadishu and Johannesburg, urban war zones are becoming armed camps, driven wholly by *implosive* forces that fold into neighborhoods the most violent and problematic repercussions of wider regional, national, and global processes.

Appadurai references the critical theories of place and scale mentioned earlier in this chapter. Places—cities in this context—do not exist in isolation at the local scale. They are (re)produced and destroyed through multi-scalar processes that involve the interaction of the regional, national, and global with the so-called local to create the particularities of a given place. An urban geopolitical approach can highlight the intricacies of these processes, partially achieved through integrating discussions of militarization and militarism into understandings of urban warfare.

Armed combat is only one aspect of the larger systems of militarization that shape contemporary cities. When common debates about "security" enter into urban governance, planning, and everyday life, urban spaces become increasingly militaristic. As mentioned above, Katz (2006) points to the presence of uniformed military at intersections and to signs in subway stations depicting threat and insecurity as constituting the normalization of militarized security in city life. Indeed, the language and tactics designed to provide urban security are not developed in a vacuum, but rather are part of (inter)national military projects. As Stephen Graham (2011: xvii) states in his study of the new military urbanism,

> [t]his synergy, between foreign and homeland security operations, is [a] key feature of the new military urbanism . . . the resurgence of explicitly colonial strategies and techniques amongst nation-states such as the US, UK and Israel in the contemporary "post-colonial" period involved not just the deployment of the techniques of the new military urbanism in foreign war-zones but their diffusion and imitation through the securitization of Western urban life.

Graham continues to describe the ways in which military interventions designed by imperial governments for people "over there" are turned back upon their own populations. For example, the construction of security zones around financial centres in London and New York directly draw upon strategies to protect overseas bases and green zones; and Israeli **drones** designed to control the Palestinian population are now used by police forces in North America, Europe, and East Asia (Graham, 2011: xviii). While this transference of military strategy is not a novel concept, especially for controlling minority and "undesirable" populations, in the contemporary age these strategies are now also being employed to control the majority population through the politics of fear.

Through an urban geopolitics that is attentive to processes of securitization and militarization, it is possible to study the role of military industries in shaping economic and political urban processes. It is also feasible to examine urban developments around military bases in order to understand how the needs and desires of a militarized workforce shape surrounding areas. It is equally productive to research how military agendas pervade national politics and determine not only how resources are distributed across different cities and across different parts within those same cities, but also how they can lead to the development of new urban settlements.

Following Graham (2004, 197), this chapter asserts that a specifically urban geopolitics of militarization opens up the following lines of inquiry: analyzing the political mechanisms that "cloak" the geopolitical and strategic spatialities of the urban; exposing the "hidden" militarized histories and spatialities that connect urban planning and state terror; studying the characteristics of city spaces and the infrastructure that make them a prime target for militarized violence as well as the effects of this violence on people's everyday lives; analyzing the connections between the geopolitics of war and empire and the political economies of production, consumption, migration, the media, and resistance; and uncovering militarized research, also known as "shadow" urban research.

As part of the larger project of employing urban geopolitical analysis to understanding militarization, this chapter focuses on the urban geopolitics of forced displacement. Examining the processes by which large populations are forcibly removed from one place and relocated to "camps" offers insight into how militarization constitutes urbanization as well as how such processes of urbanization simultaneously support and challenge systems of militarization. Displacement camps, therefore, offer complex sites of geopolitical negotiations through which varied forms of urbanism are constituted.

Urban Geopolitics of Forced Displacement

In scholarly discussions of forced displacement engendered by conflict, a distinction is commonly made between **refugees** and **internally displaced persons** (IDPs). At the most basic level, this distinction is one of political geography. A refugee, according to the 1951 Refugee Convention, is an individual who has fled their country of citizenship for fear of persecution (UN, 1951). If this same person were to flee their home but not leave their country, not cross a national border, then they would be classified under the category of internally displaced persons. One might consider this merely a question of semantics if it were not coupled with higher levels of commitment by the United Nations High Commission on Refugees (UNHCR) and other international actors to provide support and assistance to refugees and the governments of the countries hosting them than to IDPs. While there are many policies and programs in place to deal with the humanitarian crises of mass displacement for both refugees and IDPs, most of these policies are notably focused on short-term needs such as food, water, and emergency health care rather than on long-term concerns such as housing or urban(izing) infrastructure. This standard of relief planning is made evident in the UN's *Handbook for Emergencies* (2007), which is 569 pages long and includes a wide range of protocols for managing logistics from vehicle log sheets to nutritional standards but only contains 19 pages related to shelter.

According to the UNHCR's latest *Statistical Yearbook*, by the end of 2014, there were 59.5 million individuals forcibly displaced as a result of persecution, conflict, generalized violence, or human rights violations (UNHCR, 2015). This represents an increase of 8.3 million from the previous year, which is the highest annual increase on record. The UNHCR (2015: 8) puts this number in context as follows:

> Some 19.5 million persons were refugees: 14.4 million under UNHCR's mandate and 5.1 million Palestinian refugees registered by UNRWA. The global figure included 38.2 million internally displaced persons and 1.8 million asylum-seekers. The total was the highest recorded level in the post–World War II era. If these 59.5 million persons were a nation, they would make up the 24th largest in the world.

The report goes on to state that about a third of the refugee population were living in planned/managed refugee camps, while the rest were living in private accommodations, usually in urban areas. The authors do not give equivalent information for IDPs other than to say that most IDPs return to rural areas after displacement (UNHCR, 2015).

A core aspect of what makes refugee and IDP camps distinct forms of urbanism is their assumed temporality. The assumption that camps are designed to address an immediate and short-term need calls for informal (and perhaps non-existent) urban plans and governance (Cuny, 1977). This is, in part, a political choice motivated by the desire to maintain positive relations between displaced populations and "host" communities: "There's an unspoken assumption in host countries," said Jeff Crisp, the head of Policy Development and Evaluation for the United Nations Refugee Commission, "if we make life difficult or make it uncomfortable, people will go home sooner" (quoted in Lewis, 2008: n.p.). In other words, the intended transiency of camps is meant to structurally convey that the displacement is temporary. This is despite the fact that people living in camps are staying put for increasingly longer periods. According to the UNHCR (2006), a refugee in 1993 could expect to live in a camp for 9 years; by 2004, this had increased to 17 years. The politics of the narrative of "temporary" displacement becomes painfully transparent with these numbers. Whose security and whose needs are prioritized by developing informal urban settlements meant to keep people feeling "uncomfortable" for decades?

This narrative of temporality is part of what Sanyal (2014: 563) argues is the active informalization of camps: "The ambivalent and often contradictory approaches to 'helping' refugees [has] led to the informalization of their spaces and livelihoods. Struggles over 'rights' to remain, property and identity, [have] had an impact upon urban politics, and the development of camps and colonies."

It is these spaces of in-betweenness—of being neither a bona fide resident of a location nor a temporary guest—that Yiftachel (2009: 250) describes as "**gray spaces**" or "developments, enclaves, populations and transactions positioned between the 'lightness' of legality/approval/safety and the 'darkness' of eviction/destruction/death." Gray spaces are discursively constructed as existing at the margins of society, but in a context of forced displacement lasting for generations. Gray spaces can be the everyday spaces at the centre of life for millions of people. In these cases, to imagine that these spaces are somehow temporary or tangential to "normal" urban processes is to perpetuate the geopolitical narrative that war, violence, and oppression are temporary disruptions in an inherently peaceful world. As we noted in the above discussion of the term "war," the constructed temporality of displacement can be a political project to erase the violence(s) of people's everyday lives through narratives of "normal life" as life supposedly without violence. The trick is in shaping the narrative of *which* violence(s) are seen as problematic and *which* are accepted as part of the status quo.

When discussing the urbanization of refuge, Sanyal (2014: 561) asks the following rhetorical question, which she effectively answers through her analysis of the varied forms of urbanization tied to the politics of displacement: "'Camps' themselves can come to resemble slums of thousands of people, especially when they have existed for many years. Are these just forms of 'emergency urbanism' used to house displaced populations, or are they new and emerging geographies that require new theorizations of urbanity?"

While examining the idea of "**emergency urbanism**" and related forms of informal urbanism associated with displacement, Sanyal (2014) asks us to question the line between the informal urbanity of slums and informal settlements and those of refugee camps. What happens when we shift our perception of "the urban" from the economic hubs of global cities to the margins? Is informal urbanism the new norm? Whose stories become prioritized? Whose politics are considered "global"?

According to the UNHCR, refugee camps are considered a separate category from both rural areas and cities. As far back as 2005, its *Statistical Yearbook* stated: "UNHCR protection and assistance programmes are generally implemented at the field level. A key question in every project is the settlement pattern of the assisted population: are refugees living in camps, in urban areas or in rural areas among the local population?" (UNHCR, 2007: 55).

Given how flexible the criteria are for defining urban settlements, how much they vary across nations, and how they are used to perpetuate different definitions in different places, it begs the question of why refugee camps and IDP camps are explicitly not described as urban places. The vast majority of them fulfil most, if not all, of the criteria for defining a place as urban. As anthropologist Michel Agier (2002: 232) prompts, can and should the refugee camp be considered an urban settlement not only in the sense of these criteria of urbanization but also in the sense of urbanism—that is, of camps also being spaces of urban sociability and political agency? If not, why not, and how does the camp manage to remain a "stunted city-to-be-made" (Agier, 2002: 237) if it embodies most preconditions for urbanization and urbanism? It is here that the question of politics resurfaces: What is the political context in which the spatial arrangement of the camp exists, and how does this context shape the interpretation of these densely populated but supposedly "non-urban" spaces?

To answer these questions, this chapter turns to two case studies that allow us to see how context matters in understanding the geopolitics of militarized urbanization. The first turns to an IDP camp in northern Uganda during the 20-plus-year war between the Lord's Resistance Army and the Ugandan government. This study focuses on showing how the history of this place is important in understanding the politics of displacement and the spatial organization of humanitarian efforts within a militarized context. But the examination of the socio-spatial relationships between the state, the displaced population, and international humanitarian organizations is only a partial story. More actors—including other rebel armies, the United States and its War on Terror, and transnational peace initiatives—are implicated in this story. A full geopolitical analysis would examine their linkages and interconnections. The second case study switches our attention to al-Baqa'a Refugee Camp in Jordan. Again, long histories of militarization and claims to land have affected the socio-spatial politics of everyday life. Here, the temporal politics of explicitly refraining from calling a refugee camp "urban" is examined with particular attention given to how this refusal of the urban label is used to shape the meaning of "the camp" for differently situated actors at various scales. As in the first case study, the focus is on a few actors in order to highlight the intersection of both top-down and bottom-up politics in the everyday practices of the camps.

"Protected Villages" without Protection: Awere IDP Camp, Uganda

At its height, Awere Camp in Northern Uganda was home to approximately 21,000 internally displaced persons from the Acholi ethnic group (Government of Uganda, 2005) (Figure 14.1). Awere was one of dozens of such IDP camps scattered across Northern Uganda that provided a temporary home to the 1.8 million people displaced by the war between the Lord's Resistance Army (LRA) and the Ugandan Government that raged from the mid-1980s to the mid-2000s. Just like the war, the so-called temporary displacement went on for much longer than expected, with some people living in IDP camps for 10–15 years. These IDP camps, which at times housed up to 60,000 people in very tight quarters, were never discussed by the government or humanitarian organizations as urban spaces, even when their populations dwarfed those of most regional towns and cities. To understand the politics behind this assumed non-urbanity, it is necessary to understand the politics behind the construction and maintenance of the camps.

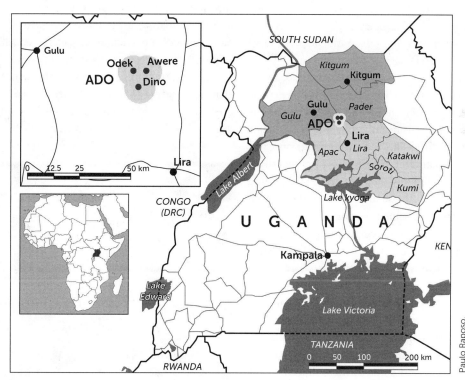

Figure 14.1 Map of Awere IDP Camp, Uganda. The shading in the northern districts reflects the intensity of war activity in the region with darker grey shading indicating more activity.

One starting point for this story is the colonial period (1896–1962) in the region that would become Uganda and in which the British implemented a divide-and-rule colonial policy. The British created a divide between what they considered the "civilized" ethnic groups in central and southern Uganda and the ethnic groups in the north, such as the Acholi, which the British considered "primitive" and warlike (Atkinson, 1994). This colonial north/south divide was reinforced in the postcolonial period by Yoweri Museveni and his National Resistance Army (NRA) during their rebellion against President Milton Obote's regime in the mid-1980s (Branch, 2011; Dolan, 2011). Museveni strategically invoked the north/south narrative as a way to blame Obote (who was from the north) for the violence of the rebellion. This narrative also helped Museveni bring together formerly divergent groups from southern Uganda. In so doing, he reified the image of northerners, particularly the Acholi, as primitive warriors who threatened the modern nation-state (Laliberté, 2013). Upon taking power, Museveni and his National Resistance Movement (NRM) party maintained this north/south divide as central to their national politics. They excluded the Acholi from national power and suppressed independent local leadership (Branch, 2011).

Following Museveni's rise to power in 1986, many rebellions formed in the north of the country to depose him and the NRM. By 1987, however, as other rebellions ended through defeat or negotiation, the only major rebel group left was the LRA led by Joseph Kony and comprised predominantly of people of the Acholi ethnic group. What had been a low-intensity conflict escalated in 1994 following the collapse of peace negotiations. In 1996, the Ugandan government began to forcefully relocate people in the war-torn region from their homes to what the government called "protected villages." While the government maintained that this forced displacement protected civilians, the lack of security or basic services in the so-called protected villages made it clear that the primary security concern was for the government. The forced displacement served to secure the state by physically separating Acholi civilians, who were assumed to be collaborators due to their shared ethnic identity, from Acholi rebels (Mwenda, 2010).

This was the context in which the IDP camps were created in northern Uganda. People living in the area tell stories of the army coming to their homes and telling them they had 24, 48, or (if they were lucky) 72 hours to abandon their homes and go to the protected villages. Anyone found outside of the designated areas after that time would be considered a rebel and shot on the spot. Sometimes people were actively chased out of their homes by bombings and by soldiers setting fire to the grass thatch roofs. (Acholi Religious Leaders Peace Initiative, 2001). When they arrived at the camps, there was nothing there, no food, no housing, and no security. People had to sleep in the grass, hoping that neither the rebels nor the army would see them. Their insecurity increased in these so-called protected villages, since they no longer lived spread out in isolated homesteads but were forced to stay together in a known location (Figure 14.2). This was particularly problematic for children, who were the targets of abduction and forced recruitment by the LRA.

Adam Branch (2009) argues that the only reason people stayed in the camps—the only reason the forced internment of the Acholi population was able to proceed—was because international humanitarian organizations arrived early on in the creation of the camps and used them as a site from which to distribute relief aid, food aid in particular. Once created, they were sustained by the government's strategic instrumentalization of humanitarian resources. Branch argues that rather than responding to a humanitarian crisis, these camps created a humanitarian crisis by forcibly interning almost 2 million people without adequate food, shelter,

Figure 14.2 An IDP camp in Northern Uganda where the inhabitants had to build their own homes

health care, or water. Excess mortality rates in the camps (the number of people who died not from natural causes or from armed violence but from the conditions of the camp) were estimated at 1,000 people per week during the height of the displacement (IRIN, 2005). This crisis justified more interventions, which further formalized the urban structure of the camps. As humanitarian organizations set up food distribution points in the camps, they created emergency health clinics, and they facilitated a governance structure to promote the efficient functioning of the camps. The humanitarian organizations, in other words, facilitated the urbanization process in order to create an efficient system for distributing relief aid.

These camps developed as a form of emergency urbanization. They were not designed to support long-term housing or sustainable livelihoods. As is standard with humanitarian interventions, food, water, and health care were prioritized over housing or urban infrastructure. This was a purposeful form of forced informal urbanization designed to perpetuate a narrative of protection and support while maintaining conditions of insecurity and dependency. The story, however, does not end there. The people living in the camps were neither passive victims nor clients of the LRA, the government, or the humanitarian organizations. For example, in Awere camp, a group of women took advantage of their forced proximity and the international organizations' policies of gender mainstreaming to create ADO, a women's human rights organization. Together with colleagues in Dino and Odek, two neighbouring IDP camps, this group sought to fight the violence of life in the camps, from child neglect to domestic violence to military abuses of power (Laliberté, 2014) (Figure 14.3). It was the conditions of the war—the spatial, social, and political conditions—that created both the need and the opportunity for this group of individuals to come together for collective action.

In summary, Awere IDP camp was created as a result of military strategy and maintained through humanitarian interventions. However, the types of political agency realized in and through these congested spaces of emergency urbanism were not confined to the geopolitical agendas of national and international agencies. The people living in these spaces also shaped their meaning and possibilities. In the next section, this chapter examines how these same processes of militarization, displacement, and humanitarianism play out in a

Figure 14.3 ADO members in 2010

different context and thereby construct a different form of urbanism and urban geopolitics.

The Geographies of Permanent Temporality: Al-Baqa'a Refugee Camp, Jordan

The Arab-Israeli war of 1948 engendered the dispossession of about 800,000 Palestinians, who comprised 80 per cent of the population inhabiting the territory upon which the state of Israel was established (Pappé, 2007). More than 500 Palestinian villages, towns, and cities were depopulated and destroyed (Sa'di, 2008). Houses, homes, libraries and personal collections, fields, farms, land plots, and other material assets of the newly exiled Palestinian inhabitants were seized and distributed to Jewish settlers (Farah, 2003). This intense series of events, which Palestinians refer to as *al-Nakba* (the Catastrophe), frayed the networks and the institutions that had organized Palestinian society.

Faced with few alternatives, a significant number of the displaced Palestinians sought refuge in the neighbouring states of Syria, Lebanon, and Jordan (Morris, 1991), although some were able to stay in (the still existing) refugee camps within Palestine. While the majority who fled to neighbouring states were convinced that they would eventually be able to return to Palestine (Khalidi, 1992), the international community made extensive arrangements to indicate otherwise: in 1949, the United Nations established the United Nations Relief and Works Agency (UNRWA), a separate organization designed to exclusively facilitate the "resettlement" and "repatriation" of the displaced Palestinians in Arab states

(Bocco, 2009: 248). When a second Arab-Israeli war displaced an additional 325,000–350,000 Palestinians in 1967, UNRWA erected new emergency camps to further accommodate the growing refugee populations (Farah, 2003). In spite of, and perhaps due to, these recurring political misfortunes, the desire to return to their homeland became central to the Palestinian (trans)national narrative of dispossession and struggle against overwhelming odds (Khalidi, 1992: 29).

The geographies of these longings can be clearly observed in al-Baqa'a, Jordan's largest Palestinian refugee camp (Alnsour and Meaton, 2014) (Figure 14.4). In a busy intersection at the edge of the camp, a mural depicting Al Aqsa mosque, which often symbolizes the Palestinian claim to Jerusalem, reads, "We are returning" (Figure 14.5). As such, al-Baqa'a's 104,000 (registered) inhabitants (UNRWA, 2016), in the process of continually "re-territorializing" their histories, inscribe the space within the camp as Palestinian (Farah, 2003: 187). They name shops, retail stores, other local businesses, community centres, and even their children after Palestinian villages and cities (Davis, 2010). Additionally, most housing units, which gradually replaced tents after the establishment of the camp in 1967, are crowded, half-finished shelters that are in need of walls, steps, and ceilings (Alnsour and Meaton, 2014). With "permanent-looking" housing viewed as a sign of forfeiting the political right to return, most units are "hesitant structures" that mirror a long state of temporality (Farah, 2003: 188). Indeed refusing to normalize their condition of exile (Petti, 2013), many of al-Baqa'a's

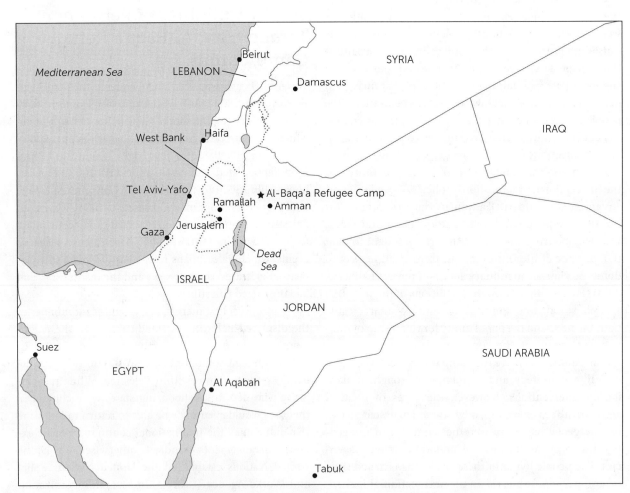

Figure 14.4 Map showing location of al-Baqa'a Refugee Camp, Jordan

Figure 14.5 In one of al-Baqa'a's main intersections, street art portrays symbols of Palestinian claims to the homeland and reads, "We are returning."

Figure 14.6 A depiction of al-Baqa'a's unfinished infrastructure. The lack of drainage is evident after rainfall, the rods of incomplete buildings are visible, and the effects of insufficient garbage collection are tangible.

inhabitants choose to remain in the camp, even if the opportunity to leave presents itself (Al-Qutub, 1989).

Over time and with the prolonged exceptionality of Palestinian exile, the infrastructure of al-Baqa'a contracted the issues that are usually associated with urban centres: overcrowding, water and electricity shortages, sewerage malfunctions, irregular garbage collection, and uneven public service provision (Cuny, 1977) (Figure 14.6). Today, these issues are especially compounded, not only because an unplanned, unsupervised, "haphazard" urbanization occurred in al-Baqa'a (Alnsour and Meaton, 2014: 69), but also because the space is classified as a "camp" under the administration of the resource-poor UNRWA rather than as a municipality under the administration of the Jordanian government. But why is this the case? If al-Baqa'a can "ecologically" and "demographically" be classified as urban (Al-Qutub, 1989: 106), then why is the space that is contained within its boundaries perceived as a "camp"—a "non-urban" conglomeration—rather than as an urban settlement? What would be at stake if al-Baqa'a was classified as urban rather than as a refugee camp?

Multiple actors construct imagined and physical borders around al-Baqa'a in order to reinscribe its spatial identity as a refugee camp rather than as an urban settlement. As shown above, the inhabitants maintain the identity of al-Baqa'a as a refugee camp with the specific goal of underscoring its existence as a spatial symbol of Palestine (Farah, 2003). Administratively, UNRWA also classifies al-Baqa'a as a refugee camp, albeit with the intent of reproducing an image of the inhabitants as victims—as "refugee[s] in need"—thereby justifying expenditures and mobilizations for funds (Farah, 2003: 186). Following a similar logic, the Jordanian state is keen on representing al-Baqa'a as a "*Jordanian* place for *refuge*" (Farah, 2003: 186, emphasis in original) in order to portray the camp as a clearly defined spatial manifestation of the "financial and material burden" that Jordan has undertaken in the context of Palestinian refugee crises. Had the Jordanian state classified al-Baqa'a as an urban centre instead of as a refugee camp, foreign delegations, especially those representing donors, would not have been able to visibly recognize the contained presence of a Palestinian refugee population. In the cases of both UNRWA and the Jordanian state, discourses that justify humanitarian aid promote the displaced as "voiceless victim[s]" (Sanyal, 2012: 634) who are deemed unworthy of an urban designation—the quintessential democratic space (Petti, 2013).

Evidently, the "extraterritorial" and extra-temporal condition of al-Baqa'a in Jordan brings forth a new form of urbanity as well as a new set of complex social relations (Petti, 2013; Sanyal, 2012) that scholarly considerations of refugee camps need to consider. Here, additional questions need to be posed: which actors are worthy of constituting accepted forms of urbanity? What can we learn from the subversive practice of being inside, yet also outside, the landscapes of power upon which nation-states are built (Petti, 2013)? How can we expand definitions of the political subject and thus of urbanity? What role do foreign nations, namely those who represent sources of humanitarian aid, play in the designation of refugee camps as decidedly non-urban? In the text box that follows, a multi-scalar discussion of Canada's role in these urban geopolitics of forced displacement is offered.

Canadian Urban Geopolitics: Mythologies of National Security

Owing to its geographic location, Canada is unlikely to experience large waves of displaced persons walking over its borders as happens in other countries that neighbour multiple areas of armed conflict. This does not mean, however, that Canada is not implicated in the urban geopolitical questions that are tied to forced displacement. This section employs three different analytical lenses to interpret Canadian urban geopolitics: Canadian international interventions, Canadian policing of refugee and "othered" bodies at home, and Canada as a white settler society.

According to the Canadian national mythology, Canada is a peacekeeper—a peaceful and supportive nation. This belief exists in stark contrast to the way its southern neighbour, the United States, is perceived—as a military aggressor that perpetuates violence and generates forced displacement around the world (Mackey, 2000). Thus, when the Canadian Conservative government deployed military forces to the Kandahar province in Afghanistan in 2006, it was widely seen as a historical rupture in Canadian foreign policy. David Jefferess (2009) argues, however, that the perception of this event as a rupture is a fallacy. Despite the national mythos of Canada as peacekeeper, there is a long history of Canadian aggression around the world, with Canada's active military role in the 1991 US-led war against Iraq, the torture and murder committed by Canadian "peacekeepers" in Somalia, and Canadian participation in the NATO bombardment of Serbia and Kosovo as just a few recent examples of Canadian international military aggression (Jefferess, 2009). Similarly, the image of Canada as peacekeeper masks its role as a weapons exporter (Wezeman and Wezeman, 2015). It also masks the histories of Canadian-based mining corporations that have been implicated in conflict in many parts of the world (Gordon and Webber, 2008). Such information suggests that Canada has not always been an innocent bystander in armed conflict around the world. For good and for bad, the consequences of such military and corporate interventions—including the ways in which they shape processes of urbanization around the world—must be acknowledged if a proper analysis of Canadian geopolitical actions were to be conducted.

Shifting focus from displaced persons camps "out there" to the policing of "othered" bodies in Canada shows how geopolitical negotiations are part of urban policy and design. Canada has a long and complex history of racialized immigration policies designed to promote a particular idea of the nation—from the Chinese Immigration Act (1923) to the Continuous Passage Act (1908). When the Canadian Charter of Rights and Freedoms was adopted in 1982, all explicitly racist immigration policies were eliminated and Canada sought to reimagine itself as a nation that is open and welcoming to immigrants. However, policies still exist to deskill incoming professionals and to limit the rights of temporary workers, rendering access to full participation in the Canadian nation unequal. Furthermore, recent research shows how calls for increased national security have led to the uneven policing of different bodies. Deborah Cowen (2007) reveals how embodied practices of national security in Canadian ports create uneven geographies of citizenship within which workers can lose their job if there is "reasonable suspicion" of terrorist affiliation. It is not difficult to imagine how these security clearances may favour certain bodies and certain histories over others. Controlling who has access to different urban spaces—in this case, transportation hubs—is part of defining who is perceived as part of the nation and who is perceived as a threat to it.

With regard to **asylum-seekers**, Canadian geopolitics are particularly complicated. While ostensibly keeping its doors open to all asylum-seekers, the Canadian government instituted the category of "Designated Countries of Origin" (DCO) in its asylum application process in 2012. According to this policy, individuals from certain countries have limited rights in applying for asylum. The government claims that people from many places try to claim asylum in Canada, and given the time and resources it takes to deal with these "false" claims, DCOs have been identified that include countries that do not normally produce refugees, but supposedly do respect human rights and offer state protection (www.cic.gc.ca/english/refugees/reform-safe.asp). Thus, the Canadian government decides *a priori* which countries are and are not "refugee producers." The political relationships between Canada and other countries, therefore, inform individuals' access to rights of citizenship and the engagement of political urban spaces. Tensions between politics occurring at the site of the body and federal policy are often manifest in urban places. For example, in 2010, facing persistent lobbying from Toronto activists attempting to protect women who had experienced violence, the Toronto office of the Canadian Border Service Agency (CBSA) agreed not to enter women's shelters to apprehend undocumented migrants and women who had been denied asylum (Abji, 2013). The federal CBSA overruled this through a policy mandate that gave CBSA officers across the country the right to enter women's shelters. The struggle over governmental jurisdiction and the rights of victims of violence continues to shape the urban landscape of Toronto and other urban centres.

Finally, any discussion of Canadian urban geopolitics in relation to forced displacement is not complete without an acknowledgment of the systematic patterns of racialized exclusion that rest at the heart of Canadian urbanization. Indigenous reserves, after all, were created to spatially confine First Nations, Métis, and Inuit peoples away from urban areas (Razack, 2002)—namely, from the spaces that were emerging as markers of whiteness and, by association, of "progress." Despite these efforts, the out-migration of Indigenous peoples from reserves to cities has been increasing since the 1950s. While only 6.7 per cent of Indigenous peoples lived in urban centres in 1951, census figures show that this had increased to 53.2 per cent in 2006 (Peters and Andersen, 2013). Despite these trends, the legacies of colonial exclusion in urban centres persist. Following the Indian Act's (1876) regulation of identity and residence for Indigenous peoples, the Canadian federal government has hitherto assumed responsibility for the provision of social services *solely* for those who live on reserves (Peters, 2006). Thus, many urban Indigenous institutions seek to mitigate such impediments to an Indigenous presence in cities by providing "status-blind" services and by building a sense of community among culturally diverse populations. Peters (2006) posits that Winnipeg is among the urban centres with the most well-developed set of urban Indigenous institutions, listing around 30 agencies that are interlinked, self-managed, and that provide a variety of services—from child care, to community development, to employment training, to language education. Besides their essential commitment to service provision, such institutions are also integral in the process of shaping an urban Indigenous collective identity and establishing the roots of urban self-government. While emphasizing certain elements of traditional practices, they also manage to adopt an all-inclusive approach to forging networks of support in the context of systematic marginalization (Andersen, 2013).

Evidently, Canadian urban geopolitics have their basis in a colonial nation-building project. Although the meaning and politics of urban spaces do not remain static and therefore can also be used to challenge the settler narrative, xenophobia, and militarization, it is still necessary to address the historical use of urban spaces in exclusionary practices of nation-building and in the legacies of colonial projects today.

Conclusion

Just as the local cannot be separated from the global, so too the urban cannot be separated from the (inter)national. Critical urban geopolitics as a field of study explicitly engages with the multiple connections between political processes that are variously scaled as local, national, or global and that all manifest at the site of the urban. This approach challenges us to examine various forms of urbanization, including forms not labeled as such for political reasons. This chapter has demonstrated the importance of such an analysis by focusing on the geopolitics of militarized urbanization through a discussion of displaced persons camps in Uganda and Jordan. In an examination of the specific political landscapes upon which camps are established, it becomes clear that urbanization also occurs in and through the uneven geographies of militarization, humanitarian aid, and resource distribution around the world.

This chapter, moreover, indicates that it is important to remember that urban geopolitics is not an approach to studying "those cities" "over there," but rather a focus on the political processes that create (dis)connections between the imaginaries of "here" in Canada and "over there." As Stephen Graham (2004: 188) asserts, "global geopolitical tensions, and attempts to bolster 'homeland security,' have telescoped into policies shaping immigration controls, social policies addressing asylum seekers, and local policies toward multicultural and diasporic communities in cities." And, as the text box asserts, Canadian foreign policies have implications both at home and abroad. While this chapter's case studies reveal how processes of racialized, gendered, and classed urbanization—even if it is not labeled as such—emerge (partially) as a consequence of foreign policy, it also argues that those very same processes formed the social organization of urban centres in Canada through the deliberate non-integration of First Nations, Métis, and Inuit communities. Based on the settler-colonial geographic imaginary that was fulfilled through the Indian Act, these communities were displaced from their indigenous lands onto reserves, underdeveloped and isolated lands that were historically established at a distance from urban places. The persistence of such spatialized inequities becomes evident upon the examination of current Indigenous out-migration to urban centres, where systemic disadvantages relating to service provision, unemployment, poverty, and gendered violence characterize the issues facing many urban—"non-status"—Indigenous communities. Seeing these geopolitical connections and how they manifest in everyday urban spaces provides the opportunity to imagine avenues for affecting change through them. With regard to critical and feminist urban geopolitical analysis, this focus on social change is not a diversion from but rather a core motivation of the work.

Key Points

- Critical and feminist geopolitics refer to subfields of human geography that investigate the entangled power dynamics that inform the processes shaping world politics, while urban geopolitics is a move to understanding the urban as part of these multiscalar processes.
- The language and practices of "securitization" are key to understanding the ways in which militarization and militarism are implicated in urban geopolitical projects.
- The forced displacement of people into "camps," whether as refugees or internally displaced persons, is a political project.
- The politics of not calling camps "urban" often draws upon justifications for humanitarian interventions, efforts to control and contain a displaced population, and maintaining the right to return to a homeland.
- In Northern Uganda, the construction and maintenance of displacement camps functioned as a forced internment of a minority ethnic group by the government, facilitated by international humanitarian organizations and furthering the government's geopolitical goals.
- The Palestinian desire to return to their homeland shapes the geography of al-Baqa'a Refugee Camp in Jordan, traceable not only through the ways in which the inhabitants reinscribe the space as Palestinian, but also through the infrastructure of the camp that mirrors a long state of temporality.
- Examining Canadian urban geopolitics in terms of militarization and displacement raises questions about Canadian international interventions, Canadian policing of refugee and "othered" bodies at home, and Canada's history and present as a white settler society.

Activities and Questions for Review

1. In a small group, identify some of the ways in which geopolitics influence processes of urbanization.
2. Create a Venn diagram that illustrates the overlapping and distinct features of refugee camps and cities.
3. Use the two case studies in this chapter to investigate the role of temporality in shaping the geographies of refugee camps, IDP camps, and urban centres. How does the assumed transience of displacement affect the infrastructure of these spaces respectively?
4. How are militarization and displacement relevant to Canadian urban geopolitics both domestically and abroad? Discuss this with reference to the material in the Canadian text box.

Acknowledgments

The authors would like to thank the members of ADO for sharing their experiences of displacement. We would also like to acknowledge those who financially supported the research for the case study on Northern Uganda, including the National Science Foundation, the Society of Woman Geographers, and Pennsylvania State University's Africana Research Center, Department of Geography, and Department of Women's Studies.

References

Abji, S. (2013). Post-nationalism re-considered: A case study of the "No One Is Illegal" movement in Canada. *Citizenship Studies*, 17(3–4): 322–338.

Acholi Religious Leaders Peace Initiative (2001). *Let my people go: The forgotten plight of the people in the displaced camps in Acholi*. Gulu, Uganda: Acholi Religious Leaders Peace Initiative.

Agier, M. (2002). Between war and city. *Ethnography*, 3(3): 317–341.

Alnsour, J., and Meaton, J. (2014). Housing conditions in Palestinian refugee camps, Jordan. *Cities*, 36: 65–73.

Al-Qutub, I.Y. (1989). Refugee camp cities in the Middle East: A challenge for urban development policies. *International Sociology*, 4(1): 91–108.

Andersen, C. (2013). Urban aboriginality as a distinctive identity, in twelve parts. In E. Peters and C. Andersen (Eds.), *Indigenous in the city: Contemporary identities and cultural innovation*. Vancouver: University of British Columbia Press, pp. 46–68.

Appadurai, A. (1996). *Modernity at large: Cultural dimensions of globalization*. Minneapolis: University of Minnesota Press.

Atkinson, R. (1994). *The roots of ethnicity: The origins of the Acholi of Uganda before 1800*. Philadelphia: University of Pennsylvania Press.

Bocco, R. (2009). UNRWA and the Palestinian refugees: A history within history. *Refugee Survey Quarterly*, 28(2–3): 229–252.

Borer, T.A. (2009). Gendered war and gendered peace: Truth commissions and postconflict gender violence: Lessons from South Africa. *Violence against Women*, 15(10): 1169–1193.

Branch, A. (2009). Humanitarianism, violence, and the camp in Northern Uganda. *Civil Wars*, 11(4): 477–501.

Branch, A. (2011). *Displacing human rights: War and intervention in Northern Uganda*. Oxford: Oxford University Press.

Chan, S.-H. (2011). Beyond war and men: Reconceptualizing peace in relation to the everyday and women. *Signs*, 36(3): 521–532.

Cooke, M. (1996). *Women and the war story*. Berkeley: University of California Press.

Cowen, D. (2007). Struggling with "security": National security and labour in the ports. *Just Labour*, 10: 30–44.

Cox, K. (1998). Spaces of dependence, spaces of engagement and the politics of scale, or looking for local politics. *Political Geography*, 17(1): 1–23.

Cuny, F.C. (1977). Refugee camps and camp planning: The state of the art. *Disasters*, 1(2): 125–143.

Dalby, S. (1991). Critical geopolitics: Discourse, difference and dissent. *Environment and Planning D: Society and Space*, 9(3): 261–283.

Davis, R. (2010). *Palestinian village histories: Geographies of the displaced*. Stanford, CA: Stanford University Press.

De Montclos, M.-A. P., and Kagwanja, P.M. (2000). Refugee camps or cities? The socio-economic dynamics of the Dadaab and Kakuma Camps in Northern Kenya. *Journal of Refugee Studies*, 13(2): 205–222.

Dolan, C. (2011). *Social torture: The case of Northern Uganda, 1986–2006*. New York: Berghahn Books.

Dowler, L. (2012). Gender, militarization and sovereignty. *Geography Compass*, 6(8): 490–499.

Dowler, L., and Sharp, J. (2001). A feminist geopolitics? *Space and Polity*, 5(3): 165–176.

Enloe, C. (2000). *Maneuvers: The international politics of militarizing women's lives*. Berkeley: University of California Press.

Enloe, C. (2007). *Globalization and militarism: Feminists make the link*. Lanham, MD: Rowman & Littlefield.

Farah, R. (2003). Palestinian refugee camps: Reinscribing and contesting memory and space. In C. Strange and A. Bashford (Eds.), *Isolation: Places and practices of exclusion*. New York: Routledge, pp. 181–195.

Flint, C. (2012). *Introduction to Geopolitics*. New York: Routledge.

Gordon, T., and Webber, J.R. (2008). Imperialism and resistance: Canadian mining companies in Latin America. *Third World Quarterly*, 29(1): 63–87.

Government of Uganda (2005). *Health and mortality survey among internally displaced persons in Gulu, Kitgum and Pader districts, Northern Uganda*. Kampala, Ministry of Health.

Graham, S. (2004). Postmortem city. *City*, 8(2): 165–196.

Graham, S. (2011). *Cities under siege: The new military urbanism*. New York: Verso Books.

Herod, A., and Wright, M.W. (Eds.) (2002). *Geographies of power: Placing scale*. Oxford: Blackwell.

Hewitt, K. (1983). Place annihilation: Area bombing and the fate of urban places. *Annals of the Association of American Geographers*, 73(2): 257–284.

IRIN (2005). Uganda: 1,000 displaced die every week in war-torn North—Report. IRINnews. Retrieved from http://www.irinnews.org/report/56063/uganda-1-000-displaced-die-every-week-in-war-torn-north-report

Jefferess, D. (2009). Responsibility, nostalgia, and the mythology of Canada as a peacekeeper. *University of Toronto Quarterly*, 78(2): 709–727.

Katz, C. (2006). Banal terrorism: Spatial fetishism and everyday insecurity. In D. Gregory and A. Pred (Eds.), *Violent geographies: Fear, terror and political violence*. London: Routledge, pp. 349–362.

Khalidi, R. (1992). Observations on the right of return. *Journal of Palestine Studies*, 21(2): 29–40.

Kuus, M., Sharp, J., and Dodds, K. (2013). *The Ashgate research companion to critical geopolitics*. Burlington, VT: Ashgate.

Laliberté, N. (2013). In pursuit of a monster: Militarization and (in)security in Northern Uganda. *Geopolitics*, 18(4): 875–894.

Laliberté, N. (2014). Building peaceful geographies. In N. Megoran, P. Williams, and F. McConnell (Eds.), *Geographies of peace*. London: I.B. Tauris.

Lewis, J. (2008). The exigent city. *New York Times*, June 8. Retrieved from http://www.nytimes.com/2008/06/08/magazine/08wwln-urbanism-t.html

McDowell, L. (1999). *Gender, identity and place: Understanding feminist geographies*. Minneapolis: University of Minnesota Press.

Mackey, E. (2000). "Death by landscape": Race, nature, and gender in Canadian nationalist mythology. *Canadian Woman Studies*, 20(2): 125–130.

Mackinder, H.J. (1904). The geographical pivot of history. *Geographical Journal*, 23(4): 421–437.

Marston, S. (2000). The social construction of scale. *Progress in Human Geography*, 24(2): 219–242.

Massey, D. (1994). *Space, place, and gender*. Minneapolis: University of Minnesota Press.

Morris, B. (1991). *The birth of the Palestinian refugee problem, 1947–1949*. New York: Cambridge University Press.

Mwenda, A. (2010). Uganda's politics of foreign aid and violent conflict. In T. Allen and K. Vlassenroot (Eds.), *The Lord's Resistance Army: Myth and reality*. London: Zed Books, pp. 45–58.

O'Tuathail, G. (1996). *Critical geopolitics: The politics of writing global space*. Minneapolis: University of Minnesota Press.

Owen, J.M. (2000). *Liberal peace, liberal war: American politics and international security*. Ithaca, NY: Cornell University Press.

Pappé, I. (2007). *The ethnic cleansing of Palestine*. Oxford: Oneworld.

Peake, L. (2013). Heteronormativity. In K. Dodds, M. Kuus, and J. Sharpe (Eds.), *Companion to critical geopolitics*. London: Ashgate, pp. 89–108.

Peters, E. (2006). "[W]e do not lose our treaty rights outside the . . . reserve": Challenging the scales of social service provision for First Nations women in Canadian Cities. *GeoJournal*, 65(4): 315–327.

Peters, E., and Andersen, C. (Eds.) (2013). *Indigenous in the city: Contemporary identities and cultural innovation*. Vancouver: University of British Columbia Press.

Petti, A. (2013). Beyond the state: The refugee camp as a site of political invention. *Jadaliyya*. Retrieved from http://www.jadaliyya.com/pages/pedagogy/commentary/rlcicuzxblvxeuilxl

Razack, S. (Ed.) (2002). *Race, space, and the law: Unmapping a white settler society*. Toronto: Between the Lines.

Sa'di, A.H. (2008). Remembering Al-Nakba in a time of amnesia: On silence, dislocation and time. *Interventions*, 10(3): 381–399.

Sanyal, R. (2012). Refugees and the city: An urban discussion. *Geography Compass*, 6(11): 633–644.

Sanyal, R. (2014). Urbanizing refuge: Interrogating spaces of displacement. *International Journal of Urban and Regional Research*, 38(2): 558–572.

Secor, A. (2001). Toward a feminist counter-geopolitics: Gender, space and Islamist politics in Istanbul. *Space and Polity*, 5(3): 191–211.

Sjoberg, L. (2006). *Gender, justice, and the wars in Iraq, a feminist reformulation of just war theory*. Lanham, MD: Lexington Books.

Smith, S. (2012). Intimate geopolitics: Religion, marriage, and reproductive bodies in Leh, Ladakh. *Annals of the Association of American Geographers*, 102(6): 1511–1528.

Staeheli, L., and E. Kofman (2004) Mapping gender, making politics: Toward feminist political geographies. In L. Staeheli, E. Kofman, and L. Peake (Eds.), *Mapping women, making politics: Feminist perspectives on political geography*. New York: Routledge, pp. 1–14.

UN (1951). *Convention relating to the status of refugees*. Geneva, Switzerland.

UNHCR (2007). *Statistical yearbook for 2005*. Geneva, Switzerland: UNHCR.

UNHCR (2006). *The state of the world's refugees 2006: Human displacement in the new millennium*. Geneva, Switzerland: UNHCR.

UNHCR (2007). *Handbook for emergencies*. Geneva, Switzerland: UNHCR.

UNHCR (2015). *Statistical yearbook for 2014*. Geneva, Switzerland: UNHCR.

UNRWA (2016). Where we work. UNRWA. Accessed January 15, http://www.unrwa.org/where-we-work/jordan/camp-profiles

Wezeman, P., and Wezeman, S. (2015). Trends in international arms transfers, 2015. *Stockholm International Peace Research Institute (sipri) Fact Sheet*. Stockholm, Sweden.

Yiftachel, O. (2009). Critical theory and "gray space": Mobilization of the colonized. *City*, 13(2–3): 246–263.

IV Urban Lives

15 Placing the Transnational Urban Migrant

Harjant S. Gill and Margaret Walton-Roberts

Introduction

Across the global North, a robust system of higher education (colleges and universities) has historically offered a reliable path to economic opportunity and **class mobility**. While citizens of countries in North America and Europe remain the primary beneficiaries of the educational opportunities available, globalization and neoliberal economic policies have made it easier for others in the global South to travel to North America and Europe to take advantage of the same opportunities, as long as they can afford to do so. The effects of globalization on higher education, however, are far-reaching and often unpredictable, and it is to the urban scale that we turn to understand this. As Holston and Appadurai (1996: 189) note, in an "era of mass migration [and] globalization of the economy... cities represent the localization of global forces as much as they do the dense articulation of national resources, persons, and projects." In particular, a study of urban educational infrastructure can illuminate how transnational migration is facilitated and, in the process, knits together physically distant urban regions across continental divides.

While the term "globalization" is often used to characterize processes and phenomena that are "unhinged" from the control of specific territories and histories, the concept of transnationalism signals how processes and practices (such as human mobility and migration) are deeply structured by specific places and their histories and, more importantly, by the *interaction* between places. Following Grewal and Kaplan's (1994: 13) use of the term "**transnationalism**," this chapter also "problematize[s] a purely location politics of global-local or centre-periphery in favour of... the lines cutting across them." Using a transnational approach to understand phenomena such as migration allows for the mapping of roles played by different actors in cross-boundary movements, without national location or origin limiting the scope of analysis (Perez, 2004). A transnational framework for understanding migration critically examines how social groups (be they migrants, bureaucrats, or social institutions) at various spatial scales (urban, regional, and national) are engaged in adapting to, resisting, initiating, or exploiting various globalization processes to their own, and their communities' benefit.

India is currently one of the major source regions for transnational migrants and a country undergoing profound urbanization (Brosius, 2010; Kapur, 2010). Over the past decade, a dynamic and entrepreneurial educational infrastructure has emerged across Indian cities to facilitate the international movement of students. The linguistic and cultural diversity of India, however, can foster a strong sense of regionalism that can eclipse nationalism. Many Indian citizens see themselves as regional rather than national subjects—identifying themselves as Punjabi or Gujarati before seeing themselves as Indian. Regional identities influence global **migration** decisions in that aspiring student migrants from provincial villages rarely move to megacities like Delhi and Mumbai in their quest to travel abroad other than to pass through their international airports. Instead, the preparation for international educational migration takes place in small **provincial cities** and in regional capitals, where potential migrants access migration-related services. These provincial cities have transformed themselves to meet the growing needs of educational migrants. Attention to provincial cities within migration studies and urban geography also marks the shift away from a traditional fixation on the binary origin-destination model, and "provincializes" a historically Western-dominated framework of migration, mobility and citizenship (Vora, 2013: 5).

This chapter focuses on one such provincial city and regional capital, Chandigarh—a small city and a union territory located 260 kilometres north of New Delhi; it is a city where prospective student migrants access migration-related services and acquire the educational, vocational, and linguistic training needed to qualify for a study-abroad visa. Chandigarh is an exceptional site in migrant journeys because it has represented, from its very conception, a microcosm where regional, national, and global ways of being converge. For Punjabi emigrants the city acts as an in-between space—neither the point of origin nor the destination, yet necessary nonetheless in the overall journey to a more prosperous future characterized by transnational mobility, material consumption, and middle-class status. Prospective migrants see their stay in Chandigarh as a rehearsal for their upcoming lives in the global North in countries like Australia, Canada, the United Kingdom, and the United States.

The globalization of higher education and increased study-abroad opportunities are reshaping both Indian and Canadian cities. This chapter examines international educational migration through the city of Chandigarh in northern India in order to assess how the city's urban form, function, and infrastructure service this migratory desire. Consideration is given to how transnational processes related to migration involve intimate choices, indicating ethnic, gender, and religious identities of the migrant. It is the migration journeys of students that connect multiple urban regions across national borders through an educational/migration economy and culture. The chapter concludes with an examination of the impact that international students can have on urban centres in Canada, exploring the influence of universities and colleges as loci of changing socio-spatial diversity and innovation in Canadian cities.

Student Travel, Mobility, and Transnationalism

Migration, as Perez (2004: 7) notes, "whether voluntary or involuntary, is fundamentally about power relations—between countries, economies, and individuals—and it raises important questions about the nature and scope of power hierarchies, including those of race, class, gender, sexuality, and nation." The differences between the haves and have-nots are no longer simply characterized through collective wealth and purchasing power. Being able to travel, migrate, or move across regional and national boundaries in order to seek a better future elsewhere is an important form of **social capital** that is increasingly valued in a globally interconnected, yet economically uneven, world (Schiller et al., 1995; Basch et al., 1994; Urry, 2007; Dicken 2011). Travel is also not simply a journey from one destination to another. As Clifford (1997: 35) notes, travel is an ongoing process that "denotes a range of material, spatial practices that produce knowledges, stories, traditions, comportments, musics, books, diaries, and other cultural expressions." Where one is born and where we are allowed to claim citizenship determines where we can travel, migrate, study, work, and live (Benhabib and Resnick, 2009; Howard Hassmann and Walton-Roberts, 2015).

For college students across North America who have American and Canadian passports, spending a semester or two in a foreign country with unfamiliar language and customs is the equivalent of an extended vacation infused with opportunities for the kind of engaged learning rarely possible back home in a college classroom. American and Canadian students often study abroad to expand their cultural repertoire and résumés, and to become attractive job candidates for multinational industries. For students from Punjab, however, study abroad is regarded as a once-in-a-lifetime chance to leave India and thereby escape a society struggling with limited opportunities, excessive bureaucracy, corruption, and poor environmental conditions. Most Punjabi students apply for admission in Canadian and American educational institutions with the intention of finding employment upon the completion of their studies and potentially claiming permanent residence and citizenship in their host countries (Hawthorne, 2010). In exchange for these opportunities, universities benefit by often charging higher tuition fees to foreign students for the same education without providing much in the way of financial aid. Colleges and universities across North America and Europe have exponentially increased their foreign student enrolment, realizing the source of revenue it offers. The revenue generated means education itself becomes a significant economic "export" for countries in the global North such as Canada, Australia, the UK, and the USA. For example, over 450,000 Indian students now spend in excess of USD$13 billion per annum to gain education abroad (Bhandari and Blumenthal, 2011).

The receiving countries benefit from the revenue, but also from the potential of employing highly skilled graduates who enhance their national economic development. Multinational corporations also benefit from these transactions by having unprecedented access to a newly educated foreign workforce seduced by a promise of vocational opportunities that might lead to permanent residency in exchange for subpar salaries, reduced benefits, and inadequate labour conditions. This is part of a broader phenomenon commonly referred to as **brain drain** (where one country gains skilled people at the expense of the other country), but more recently it has been conceptualized as a model of **brain circulation** in which highly skilled foreign graduates promote global development to the benefit of both sending and receiving nations (Saxenian, 2006). The attractiveness of an overseas education and the employment and citizenship options it can lead to generate immense marketing opportunities in India. International migration also creates the opportunity for financial remittances, where migrants send the money they have earned home to their family members and communities. In much of Punjab and other North Indian states, advertisements for education abroad and money transfer services create a streetscape that is testament to the global imaginary that increasingly structures young people's daily life (Figure 15.1). This phase of international connectivity is not entirely new in this part of India.

Punjab's Migration Culture

One way to understand these migration pathways is to consider why some places in one country are connected to other places in another country. In the case of Indian Punjab, the region has developed historical migration links with the UK and Canada because of a century or

Figure 15.1 Billboard advertising study-abroad opportunities dominates the provincial Punjabi landscape

more of colonial connections. Colonial networks are also largely responsible for the large Indian diasporic communities present in the UK, Canada, Australia, and South Africa. The emergence of these settlement patterns for communities from Punjab can be partially understood by tracing the role of British colonialism in the integration of the Punjab region into a global economy through improved transportation, Western education, and agricultural intensification (Talbot and Thandi, 2004). From the late 1800s onwards, increased agricultural production in this region of India led to a rise in income, but also in land taxes, so that by the turn of the twentieth century agriculturalist families began to invest in sending family members abroad so they could send their earnings home in order to protect landholdings. The British colonial authorities also enlisted members of the Sikh religious community (mostly from Punjab) into the army, and this also contributed to the ability of the Punjabi Sikh community to travel abroad (Fox, 1985; Sharma, 1997; Bhatti and Qalb-i-Abid, 2007). These early processes of connection between India and sites abroad began a transnational process of Punjabi community formation. Some of these communities did very well economically, and when they returned to their home villages in India, they demonstrated their wealth by hosting celebrations, building large ostentatious homes, and engaging in other forms of conspicuous consumption (Walton-Roberts, 2004). The resulting transformations in the landscape encouraged others to follow the same route to secure opportunities abroad in order to improve their families' well-being and status. This has resulted in a social context where many Punjabi families see migration promising them "class-based progress, material comforts, and utopian aspiration of modernity" (Mooney, 2011: 32). Within increasingly competitive Indian urban environments that boast a rapidly growing middle class, access to land wealth and education alone no longer ensures secure employment and social mobility (Brosius, 2010; Mooney 2011; Nisbett, 2009). "India represents a set of limitations that middle class (Punjabis) seek in ever-increasing numbers to escape by emigrating," notes Mooney (2011: 159), and migration is the "accelerated ticket to the urban middle class." Mooney (2011: 170) goes on to explain that over the past decade, migration and vocation abroad have surpassed even a career in Indian government or the Indian armed forces in terms of the prestige it bestows on individuals and their families: "Migration is now the singular stuff of Punjabi dreams of family progress," she concludes. Accessing migration opportunities, however, has become more difficult, since skilled rather than unskilled labour is in greater demand and potential migrants must have the language and vocational skills necessary to be part of the transnational workforce. India is also focused on increasing the educational capacity of its young population, so educating young people for continued opportunities abroad is one approach to handling India's large youth population.

Everyday conversations among Punjabi youth are filled with local colloquialisms and witticisms referencing transnational migration. Commonly used expressions such as *Kabootar baazi* (to play the game of pigeon) and *Khamb lag gaye* (to grow wings) suggest the degree to which transnational migration is part of the daily linguistic discourse as well as central to the narrative of being successful (see *Sent Away Boys* by Gill, 2016). Punjabis regularly refer to cities and suburbs with large diasporic populations (e.g., Surrey and Mississauga in Canada or Southall and Coventry in England) as just an "extension of Punjab" or a "second Punjab." However, unlike the labour migrations in the 1970s and the 1980s, which were often opportunistic and undertaken haphazardly by the men in Punjabi families, today's migration is an event that is actively sought and strategically planned by the entire family (Chopra, 2010; Mooney, 2011). As Chopra (2010: 113) notes, "Families strategize to enable emigration, taking loans to finance migrant journeys, for example, activating networks to acquire travel papers, and, most of all, activating transnational family networks."

For Punjabi parents seeking to send their sons and daughters abroad, there are two commonly accessed avenues for migration: through transnational kinship networks, which entails arranged marriages or family reunification programs, or through student or work visas that would hopefully lead to permanent residency and ultimately to citizenship in the host country. Decisions about how and in what to invest in terms of migration opportunities are made with scrutiny and care. These decisions also determine how much schooling each sibling receives to enhance his or her migration prospects (Chopra, 2005). Mooney (2011: 171) refers to this strategic phenomenon as "migration-aimed-education." Chopra (2010) notes that Punjabi parents often invest heavily in the eldest daughter's education and dowry to ensure that she would make an attractive bride for a Non-Resident Indian (NRI) husband, a person of Indian origin now living permanently in another country. Upon migrating, sisters are often expected to sponsor their brothers and fathers, thus providing a way for the men in the family to emigrate as well. In families with multiple sons, the eldest son is often held back in schooling and taught farming with the expectation that he would look after his family's property and land while the younger sons (and daughters less so) are sent to educational institutions in provincial urban centres like Chandigarh to gain the technical skills that would enhance their opportunities to emigrate and to satisfy various English-language testing requirements, including the International English Language Testing System (IELTS). These tests are required for most international student visas and have become a central feature of the migration economy that accompanies international student flows. Urban centres that provide educational migration services aim to assist people to pass the IELTS and other language testing systems (Figures 15.2 and 15.3) and engage in the process to secure opportunities to travel, study, and work abroad. Upon migrating and settling abroad, family members are expected to remit money home, helping retain and improve their family's fortune and property (Chopra, 2010; Mooney, 2011), as well as to invest in the region's economy (Walton-Roberts, 2004).

Enabled through transnational kinship networks and sustained by the technologies of globalization, migration from Punjab and other parts of India is no longer a one-way journey out of the country but a circular process in which visiting and remitting money home

Figure 15.2 IELTS institutes and migration-related service industry established in Chandigarh's modernist city centre, Sector 17

is just as important a feature in the narrative of being successful as moving abroad in the first place. Chopra (2010: 113) explains: "[M]igration is therefore a process that fosters and nourishes kinship links, while creating new ones through marriage, work, even childcare." Circular migration, and the flow of people and capital that accompanies the process, provides a "flexible survival strategy" in a global economy, enhancing migrants' socio-economic status (Perez, 2004: 13). Migration and mobility for young Punjabis build the narrative of success for the family, and for the individual concerned, it represents collective success (or failure).

Given that the ability to study abroad requires considerable financial investment and the stakes are often high, who is able to gain access to these opportunities? What are the criteria and processes involved in qualifying for a student visa? How do families make specific decisions about sending their children to study abroad? How are their journeys financed? Answers to these questions are just as varied as the experiences of the people who are on the move. The following case study seeks to answer these questions in relation to the provincial city of Chandigarh.

A Modern City in Global Times: Chandigarh

Spread over 114 square kilometres (44 square miles) in the footsteps of Shivalik Hill where the first Himalayan range begins, Chandigarh is an autonomously controlled Union Territory that serves as the joint administrative capital

Figure 15.3 Mastering English-language skills is regarded as the essential step in successful student migration

for the two neighbouring states, Punjab and Haryana (Figure 15.4). Compared to other Indian cities, Chandigarh is a relatively young city, developed in the 1950s after the 1947 partition of British India into the nations of India and Pakistan. As a result of the Partition, pre-independence Punjab lost its capital city of Lahore. Chandigarh became independent India's first modern city; in some ways it can be seen as consolation for the trauma and loss of the Partition. Chandigarh was envisioned as a city of the future, one that would turn the country in a new direction. To achieve such a bold desire, India's first prime minister, Jawaharlal Nehru, commissioned French-Swiss architect Charles-Edouard Jeanneret, better known as Le Corbusier, along with his team of French architects and urban planners, to design the city (Prakash, 2002). From the time of its conception, Chandigarh has embodied the aspirations of the modernist movement in architecture that influenced much of mid-twentieth century urbanism in Europe and across North and South America.

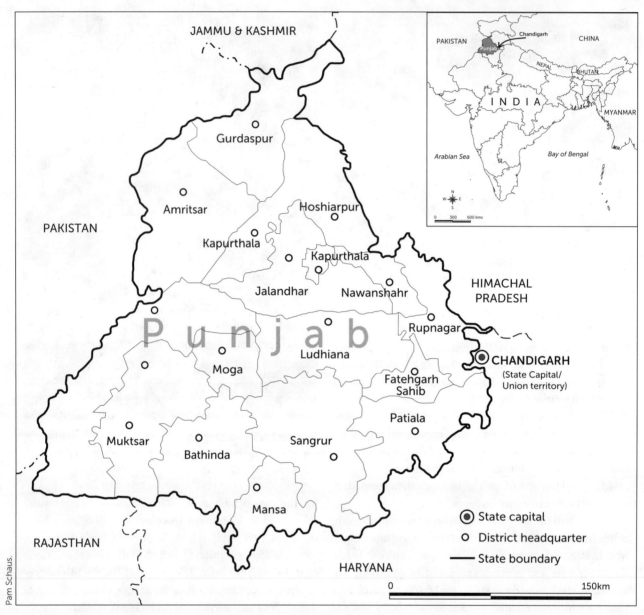

Figure 15.4 Map of Punjab District, India locating Chandigarh

Present-day Chandigarh remains buried under the weight of its modernist legacy. Signs of neoliberal investment and economic development fueled by globalization are most visible in satellite towns surrounding the city. Over the last 20 years, sparsely populated villages and farmlands surrounding Chandigarh have been replaced with gated residential communities, headquarters for multinational technology companies, and exclusive shopping malls and multiplex cinemas promoting material consumption and the lifestyle associated with the growing urban middle class in India (Figure 15.5). Meanwhile, Chandigarh's landscape remains limited by Le Corbusier's myopic vision of what a modern city can and should look like. Based on Le Corbusier's master plan, the city is divided into 55 rectangular sectors or enclaves laid out on a rectangular grid. Chandigarh's

Figure 15.5 New office development, Ram Darbar Colony, Chandigarh

government and planning commission has carefully controlled the appearance of any new commercial development, uniformly applying the sensibilities of the modernist tradition.

As Prakash notes, "one of the significant and overwhelming characteristics of Chandigarh is that it is not a visibly Indian city" (2002: 33). He describes Chandigarh's architecture as "stark and clean," lacking "profuse ornamentation," with neatly planned streets that are largely devoid of the disorder and chaos often found in other Indian cities (2002: 33). He goes on to assert that many Indians consider Chandigarh to be "un-Indian" largely because it is "self-consciously constructed as a 'modern' city" (2002: 33). The erasure of local heritage from Chandigarh's planning echoes Holston's (1989: 9–10) critique of modernist cities founded on the principles of de-historicization and de-contextualization in which only the "imagined future is posited as the critical ground in terms of which to evaluate the present" and in which "the teleological view of history dispenses with a consideration of intervening actors and intentions, of their diverse sources and conflicts . . . thus, modernism's relation to history is strangely disembodied."

Chandigarh is part of a rapidly globalizing India and its emerging knowledge society. This reflects what Nisbett (2009: 7) describes as "changes in human society emerging through the growing use of information, and of information and communication technologies in mediating society, economy, and global culture in the last quarter of the twentieth and the beginning of the twenty-first century." Originally planned for 500,000 inhabitants (Kalia, 1999), Chandigarh has a population that, according to the 2011 Indian Census, has nearly doubled to 960,787. Over the past two decades, Chandigarh's surrounding areas have developed into several different satellite towns to accommodate the growing population and influx of migrants to the region from all over the country. Commenting on Chandigarh's population growth, Walton-Roberts and Pratt (2005: 177) note that it "epitomizes India's socio-economic duality, simultaneously manifest in business sectors filled with computer training centres, private banks, immigration consultants and other services directed at increasingly globally mobile consumers; and instant ramshackle shelters at the edges of the city for those who flock from the neighbouring poverty-stricken states." While Chandigarh's architectural aesthetic remains shackled to its past, its economy and demographic makeup has witnessed a stark transformation precipitated by the globalization of higher education and transnational student migration (Gopinath, 2014).

Socio-spatialities of Transnational Educational Migration in Chandigarh

Economic hardship coupled with agricultural change has resulted in farmers moving out of the village, and this has meant that their children are increasingly attending urban schools and colleges (Jodhka, 2006). Year after year, aspiring students from the surrounding Punjab and Haryana regions migrate to Chandigarh to access the educational opportunities and immigration-related services in the city. Chandigarh figures prominently in the overall experiences of these transnational student migrants precisely because of its modernist architecture and legacy. Its seemingly foreign landscape with its emphasis on form and functionalism, the middle-class lifestyle of its residents, and the emphasis on English medium education all add to the city's allure of being an "un-Indian city"; indeed, outsiders refer to it as "almost like abroad." Punjabi families perceive Chandigarh as an educational hub within the region to which they can send their sons (and to a lesser extent their daughters) to gain the education and skills needed for geographical, economic, and social mobility. With the abundance of migration-related services that have become available in the last 10 years with the growth of the "immigration industry," Chandigarh serves as a stepping stone for transnational migrants with global aspirations.

Encapsulated within Chandigarh's growing immigration industry are various small immigration brokers, privately owned language-training and test-preparation centres, and large-scale global resettlement agencies that promise to serve as a one-stop shop for all of the needs of prospective migrants seeking to study and relocate abroad. With limited governmental oversight, these private institutes deploy lofty rhetorical claims juxtaposed against idyllic cityscapes from around the world to capitalize on the migrants' dreams of living abroad. Their offices range in size and resources from single-room dwellings in neglected office buildings to multi-storey storefronts with gleaming marble-covered lobbies, attended by a staff of professionally dressed "agents" whose goal is to sell every service imaginable to prospective student migrants, including language training, help with university application forms, temporary job placements, scheduling airport pickups, and making living arrangements upon their arrival at their destination. The owners of these agencies and the consultants they employ are often transnational migrants themselves who have access to the most current information about immigration policy changes. If visa applications are denied, some immigration consultants even partner up with legal advisors to challenge the consulate's decisions—effectively transnationally brokering a path to migration (Walton-Roberts, 2011a).

The city's immigration industry dominates the ephemeral spaces where public culture is displayed. Billboards, posters, and signage advertising immigration services and IELTS training are ubiquitous throughout the city. Phrases like "reach your potential by studying abroad" accompanying images of planes taking off dominate the city's shopping plazas and markets. Whereas the more prominent institutes and training centres are concentrated in the city centre, the smaller and less-expensive establishments are scattered around the peripheries of the city and throughout the satellite towns where commercial real estate is more affordable. Unlike the locally owned businesses in Chandigarh, which are often named after family ancestors or Hindu gods and goddesses, most of these agencies and institutes have anglicized names that draw on transnational educational discourses (e.g., the British Council, Brain Tree, Competition Cell, Mr. Visa, and Western Overseas).

Members of Punjab's landowning **caste**, the *Jat* Sikhs, comprise the largest demographic of educational migrants from this region (Mooney, 2011). Unlike working-class families across the region, many of whom can barely afford the application fees needed to apply for a student visa, landed *Jat* families have been able to finance migrations through land sales and remittances from family members already settled abroad. On average, families often spend upwards of between CAD$2500 and $3000 on application fees, accessing migration services, and the costs associated with living in Chandigarh before visa applications are submitted. In addition, most foreign consulates also require applicants to provide proof of CAD$20,000–$25,000 in liquid assets as a condition for being granted a student visa, sums often beyond the reach of non-landed families. Even after obtaining a visa, the costs related to necessary medical evaluations, travel, and relocation could easily add up to yet another CAD$2000–$3000. As transnational migration increasingly becomes one of the only viable paths to class mobility for families across North India, access to this path remains restricted based on caste and class hierarchies.

The choice to seek immigration-related services in Chandigarh, despite the high cost of doing so, is driven partly by Punjabi families imbuing brokers and agencies based in Chandigarh with a greater sense of credibility and legitimacy. Fearing immigration-related scams, cases of which are rampant across the region and dominate the local news media, Punjabis are increasingly cautious about how and where to invest their resources. Because of the autonomy Chandigarh's administration enjoys and an overall stricter adherence to rules and regulations across its administrative branches, migrants and their families often perceive service providers based in the city as more legitimate than those based in other urban areas of Punjab, which fall under the jurisdiction of the notoriously corrupt Punjab government and police. Higher fees, tuition rates, and the cost of living and studying in Chandigarh are carefully weighed against the perception of the certainty of transnational migration and the long-term benefits that might result from such investments.

Chandigarh's role in facilitating transnational migration is solidified by the colonial imprint on the Indian education system in urban India, which continues to privilege English as the language of global India. Acquisition of English language skills is a primary qualifier for education- or vocation-based migration. While middle-class Indians growing up in Chandigarh possess near-native fluency, having studied in English medium schools, for those arriving from surrounding Punjab and Haryana regions who grew up with Hindi or Punjabi medium education, English-language skills are often the greatest hurdle between them and their dreams of transnational migration. Qualifying for a student visa requires prospective migrants to first clear IELTS or Test of English as a Foreign Language (TOEFL) exams that test for English-language proficiency. Most prospective migrants clear these tests upon repeated attempts interspersed among language training at coaching centres across Chandigarh.

While the fee for taking the IELTS exam averages around Rs 10,500 (around CAD$200), coaching centres charge anywhere from Rs 15,000 to Rs 25,000 (CAD$300–$500) for courses that last from three to six months. Many student migrants who struggle to clear their IELTS exams go on to struggle academically upon going abroad. Instances of academic probation and expulsion followed by the loss of the student visa are not uncommon. Those who lose their student visas upon relocating abroad rarely return to India unless forcefully deported; instead most crossover to join the pool of undocumented migrants who survive on the whims of the unregulated underground economies of cities like Toronto, New York, London, and Sydney.

Chandigarh's service economy has also evolved to accommodate the needs of the temporary student residents; it includes a whole series of privately owned and often informal businesses. For the city's long-term residents, the migrants offer new sources of revenue. Walking through the residential neighbourhoods in the city centre (sector 17), one encounters numerous multi-storey houses where one or more floors have been converted into accommodation for paying guests, Internet cafés, or student hostels. Paying-guest accommodations (simply referred to as PGs) are sublets or sparsely furnished rooms, managed by private homeowners and shared among groups of young men and women, mostly from Punjab and Haryana, all at various stages of preparing for their upcoming journeys. Flyers advertising *tiffin* specials (home-cooked food delivered in lunch boxes) are plastered on lampposts throughout the residential sectors, offering migrants a cheaper alternative to eating out in restaurants and eliminating the need to cook their own food at home. Even local restaurants and barbershops cater to migrants' needs at discounted "student" prices. Young men and women use their network of friends and classmates or word-of-mouth recommendations to seek out these services.

Social Life for Transnational Migrants in Chandigarh

For most Punjabi migrants from rural areas, the time they spend in Chandigarh represents their initial taste of freedom from the daily familial restraints and an opportunity for them to live on their own. While the migrant demographic includes young men and women from across the region, the experiences of the two groups differ radically. Migration is often a deeply gendered process (Benhabib and Resnik, 2009; Agustín, 2007), and this is true of the Punjabi migration experience. While changes in gender relations in India are evident, "[t]he continuing importance of arranged marriage . . . and dowry . . . in modern India bolsters the argument that globalization may have a long way to go before gender scripts woven into the Indian stratification system begin to unravel" (Desai and Andrist, 2010: 683). In a conventionally patriarchal society where notions of honour and shame are paramount in safeguarding familial status and arranged marriages continue to be the norm, unmarried Punjabi women, their bodies and their movements, are carefully monitored by family members (Jejeebhoy and Sathar, 2001; Walton-Roberts, 2015).

Access to migration-related opportunities are thus far more restricted for Punjabi women, who have traditionally moved as brides of Non Resident Indian (NRI) grooms and less often as students or vocational migrants (Mooney, 2006). Families that do allow their daughters to pursue student migration enforce restrictions on where they can or cannot go and what they can or cannot study. Their mobility is often determined by the availability of extended family members, family friends, or other relatives willing to police (or chaperone) their actions. Young women are limited in the type of mobility they can engage in because families are concerned about safety, but also by how their behaviour might be interpreted by the community—it might reflect poorly on the families' honour and social status. Punjabi women's mobility is often policed by both the men and women in their families. This process can also be seen at the state level in terms of how the Indian government itself controls the migration of unskilled females (e.g., unmarried women under 30 years of age are restricted in their ability to migrate overseas). While these controls are presented as protective measures for the safety of women, they actually serve to increase and, in some cases, create new gendered vulnerabilities (Kodoth and Varghese, 2012). Unmarried Punjabi men, on the other hand, enjoy free rein, being able to move and travel when they want and wherever they desire (Gill, 2012). Nevertheless, the family desire for social mobility can override the caution many families might hold with regard to female mobility. As opportunities for international migration increasingly align with educational qualifications, some families recognize that it is their daughters who can best achieve the family's desires for social and spatial mobility. However, the manner in which young female migrants inhabit urban spaces like Chandigarh tends to differ from the experiences of their male counterparts.

Middle-class Punjabi men take for granted the privilege of mobility and the ability to migrate first to Chandigarh, and later abroad. Away from the watchful gaze of their families, young Punjabi men often characterize their life in Chandigarh as being filled with *moj-maasti* (exuberance and fun), shaped through new experiences, including their first romantic and sexual encounters. When not attending classes or training sessions, young Punjabi men spend their time walking around in Chandigarh's city centre or nearby malls, shopping at global brand-name showrooms, learning how to look like a transnational migrant. Alongside their formal training, Chandigarh's urban and cosmopolitan sensibilities offer informal lessons on how to adapt their appearance and mannerisms to their image of a global traveler. This period is akin to a rehearsal for their upcoming lives. In addition to buying new wardrobes, many Sikh migrants also forgo their unshorn hair and turbans upon arriving in Chandigarh (Gill, 2016). Along with the offices of immigration brokers and training institutes, the city centre is also lined with several barbershops where young Sikh migrants get their first haircuts. The reasons for cutting their unshorn hair differ widely among Sikh men in India (See *Roots of Love* by Gill, 2011). They range from wanting to avoid the daily inconvenience of having to care for unshorn hair to the desire to look modern and hip, emulating the appearance of their favourite film stars. Most prospective Sikh migrants, however, are also intuitively aware that their turbans might invite discrimination and associations with religious extremism upon their migrating abroad (Kalra, 2005; Puar, 2008). The fear of being racially profiled is often the underlying motivation for Sikh migrants to forgo their unshorn hair and turbans. Chandigarh is thus one stop in a process of transformation for migrants who anticipate where and how they will experience social exclusion and inclusion (see Gill, 2011). This experience continues if they successfully reach international locations to carry on their studies, and in Canada international students are increasingly seen as important actors in changing the social and spatial fabric of the cities to which they have come to settle.

International Students and Urban Change in Canada

The interest in international student migration in India is highly significant for Canada. International students have become an increasingly important dimension of Canada's educational and immigration policy landscape through the development of coherent pathways from educational to working visa status. Buoyed by federal and provincial government internationalization initiatives, annual international student migration to Canada increased from 25,000 to 30,000 in the early 1980s to nearly 100,000 students in recent years (Williams et al., 2015). More than ever, students from traditional and non-traditional source countries are enrolling in a variety of programs in a growing number of post-secondary institutions (PSIs). International students represent a highly coveted population group, since they are typically young, educated, and highly skilled, and they also inject needed funds and experience into the post-secondary sector during a period of declining enrolments and restructured public expenditure. The growth of the international student population in Canada also has implications for the nature of Canadian urban change and regional development. One of the more sustained debates in the last 10 years regarding urban development has been in reference to Richard Florida's work. His research has attracted tremendous attention from various sectors, including municipal governments, many of which are rushing to rethink urban regional development from within Florida's framework of the three T's: technology, talent, and tolerance. Florida interprets technology and talent as flows that need to be attracted to a place, and thus, for him, place matters in ways that go beyond issues of local tax rates and physical infrastructure. Tolerance, the third factor that makes some places better than others in Florida's (2002) model, is demonstrated by diversity or "low barriers to entry"; "the places most likely to mobilize the creative talents of their people are those that don't just tolerate differences but are *proactively inclusive*" (Florida, 2005: 39, emphasis in original). While talent, technology, and tolerance may offer an appealing alliteration, Florida really means talent, technology, and proactively inclusive communities.

Florida's tolerance factor is represented through three indices: the presence of gays and lesbians (the gay index), bohemians/artists (the bohemian index), and immigrants and racial diversity (the mosaic index). The correlation of these indices to regional economic success has been the subject of much critical speculation and analysis (Peck, 2005; Storper, 2013). One outcome of this increased attention to innovation and creative communities has been a more detailed consideration of how universities contribute to the three T's. The university's role in technology and talent has been explored in some detail (Wolfe, 2005; Florida, 1999), but the place of tolerance and the university is less explored. However, the experiences of foreign students and scholars are increasingly an important focus of attention (Walton-Roberts, 2011b). A more detailed exploration of how the university might contribute to building diverse and tolerant communities has not been widely explored, but the general consensus (in North America at least) is that universities can contribute positively to building social cohesion (but see Ratcliffe [2006] for a different perspective from the United Kingdom). As Gertler and Vinodrai (2006: 21) have argued, universities "contribute to the creation of open and tolerant places, which in turn helps to create the necessary conditions for attracting and retaining talent, thereby making the entire process mutually reinforcing. In other words the university has a much *wider role* to play in the community that reaches well beyond industry collaboration, technology transfer, and commercialization."

International students directly influence the educational sector through the injection of human, social, and financial capital, but beyond that, it is important to assess the implications of an international student presence in terms of urban and regional development. A recently completed report on international students in Ontario indicates that this province is the primary and, over time, an increasingly important destination for international students bound for Canada (Williams et al., 2015). Ontario-bound international students show a greater and growing tendency to study in the college sector, with over 50 per cent of new entrants in 2012 registered for college- rather than university-based programs. Over the last 10 years, gender parity in international student entries has shifted to a more male bias (60 per cent male in 2012 compared to 52 per cent in 2000). India and China are the dominant sources of international student entries. This marks a major shift from source regions in 2000, which were more likely to be European. International students tend to be concentrated in areas with major colleges and universities. For example, the most common Ontario locations of study for international students during 2000–2012 were overwhelmingly Toronto (54 per cent or 116,294 of the total), followed by Ottawa (10 per cent or 21,131), and Middlesex County, Waterloo

continued

Region, and Hamilton (5 per cent each). Indian and Chinese students, who formed nearly 60 per cent of entries in 2012, tend to be clustered in and around Toronto, with notable populations in Ottawa, Waterloo, London, and Windsor.

Educational institutions not only become part of the global economy of knowledge and generate employment and tax income for local communities, but they can also act as sites for increasing cultural diversity and multicultural enrichment (Walton-Roberts, 2011b). The distribution of international students to urban regions with colleges or universities indicates their significance as the locus of increasing economic and cultural value and diversity within their local communities. The future of economic innovation and change is often cited as emerging from education and skills development, and universities and colleges are increasingly cited as necessary partners for communities that want to achieve sustainable growth. Issues of cultural diversity and creativity are seen to be promoted through educational institutions that are open to diverse populations and play a transformative leadership role within their communities. In smaller towns and cities, universities and colleges are often the first places to host diverse populations through their engagement with international students, faculty, and staff. Aligned with these transformations in the diverse social population, the built spaces of PSIs create opportunities for physical change in the urban structure; the development of open public educational spaces can transform urban regions that previously suffered from economic decline. Public education institutions, for example, play a particularly important role in generating new arts and recreational spaces that are oriented to the broader community and can promote interactions across generational, cultural, and community difference.

Conclusion

In an era of transnationalism, immigration into urban areas is a global phenomenon that, albeit unevenly, has affected large numbers of the world's cities and urban regions. This chapter has shown how cities and urban regions can be transformed through processes related to transnational educational migration and globalization. Rather than a typical global city, it is a provincial city, Chandigarh, that has been the focus here, with its growing international role as an educational hub for transnational migrants. Chandigarh was designed and built as a self-consciously modern city, produced to mark India's postcolonial independence, and is seen by Indians as one of the most modern, clean, and well-organized cities in India. The urban landscape of Chandigarh represents over half a century of transformation; through the confidence of postcolonial independence to the vagaries of globalization and Indian financial liberalization, the city has emerged as a hub to aid the mobility of migrants through the formal and informal services they need to continue the journey. The modernization and social mobility spurred on in this context offer both threat and promise—hold on to what you have but also compete to secure more.

Young people arrive in Chandigarh on their migration journeys in order to access educational and migration-related opportunities and services. The families of these young migrants are often intimately involved in the decision-making processes that result in their temporary settlement in the city. City services reflect the demands and needs of these migrants, from the creation of a paying-guest accommodation sector to the proliferation of Internet service providers, coffee shops, and hairdressers. A migration-education economy has formed in the city, providing the services migrants need to access work and study-abroad opportunities; language and vocational training centres and immigration brokers, who assist in the process of securing the necessary visas, abound in the city. From there onwards, migrant journeys continue on to mid-sized cities in Canada and the United Kingdom (among other countries), where study-based migration subsequently leads to more long-term work and permanent migration.

In the destination locations, migrant journeys also play a role in reshaping cities. While many students locate in the largest cities (such as Toronto in Canada or London in England), migrant routes also embrace mid-sized cities, where international student migrants adjust to their new surroundings while contributing to social and economic change in the urban fabric of these communities. In Canada the growth of international student flows from India have focused on the largest cities in the largest provinces, but also smaller mid-sized cities such as Waterloo and Hamilton. In these destinations, international students contribute to transforming the social and spatial nature of urban communities by enhancing diversity and potentially strengthening the talent, tolerance, and technology aspects of community. Places of education such as colleges and universities thus become central actors in the transformation of their surrounding communities. Researchers argue that this cultural transformation is representative of the kind of innovative and creative change that is needed in the global economy. The reality is likely more complex, however, as international students do not always become the fully engaged citizens such urban regional-development theories envision. For international students, the failure to complete programs of study and the financial hardships they face may truncate their ability to achieve the goal of work experience, permanent residence, or citizenship. This leads to their exclusion from the rights of full membership in the city, forcing many of them into the shadows of undocumented and marginal status.

These provincial towns and cities are important staging posts in a migration process that contributes to economic change in an age where high-skilled, knowledge-rich economic development is highly sought after. The process of student migration links multiple urban regions together in a chain of immense economic value. Students move through these systems and carry with them financial, social, and cultural resources, their experiences varied across and intimately linked to differences of class, gender, religion, and region. The international student journey becomes the thread that integrates seemingly distant and distinct urban regions into the fabric of modern global knowledge economies.

Key Points

- Transnational migration is an indication of how social groups at different spatial scales adapt, resist, initiate, and/or exploit various globalization processes to their own, and their communities', benefit.
- In the Indian context of Punjab, where a regional culture of migration is well developed, transnational migration is a form of educational development and social mobility for young people.
- Far from being rendered inconsequential by globalization, provincial cities that act as infrastructural hubs in transnational migration have an opportunity to refashion their landscapes and economies.
- In Chandigarh, educational migration is embedded in the material landscape of the city and its infrastructure, including immigration brokers, language-training and test-preparation centres, global resettlement agencies, Internet cafés, and student hostels.
- Social life for transnational migrants in Chandigarh is deeply gendered and classed.
- In Canadian cities, international students contribute significantly to the urban economies where they settle; they spend money on accommodation, products, and entertainment and may add to the skilled labour pool.

Activities and Questions for Review

1. In pairs, list the positive and negative impacts that transnational flows of individuals have on urban and regional communities that send their brightest and most affluent students abroad.
2. How has the education sector of Chandigarh's economy changed the urban landscape of this provincial Indian city?
3. Are new immigrants to your community socially included or excluded in the urban environment? Form two teams to debate this issue, using examples such as the presence of amenities, services, and spaces of daily interaction (e.g., homes, schools, public transit, shopping malls, restaurants, places of worship, public parks, community centres, and recreational spaces).
4. As a student, how have you changed the urban area where you study and how has it changed you? Keep a journal for one week so you can note down the nature of these interactions and their outcomes.

Acknowledgments

This chapter has been developed based on the authors' research over a number of years, funded by various agencies, including the Social Sciences and Humanities Research Council of Canada, the Shastri Indo-Canadian Institute, the Wenner-Gren Foundation, and the Woodrow Wilson National Fellowship Foundation.

References

Agustín, L.M. (2007) *Sex at the margins: Migration, labour markets and the rescue industry*. New York: Zed Books.

Agrawal, V. (2002). Privatizing the public: A strategy for transit oriented development in Chandigarh. In J. Takhar (Ed.), *Celebrating Chandigarh*. Ahmedabad: Mapin Publishing, pp. 283–293.

Basch, L., Schiller, N.G., and Blanc, C.S. (Eds.) (1994). *Nations unbound: Transnational projects, postcolonial predicaments, and deterritorialized nation-states*. New York: Routledge.

Benhabib, S., and Resnik, J. (Eds) (2009). *Migrations and mobilities: Citizenship, borders and gender*. New York: NYU Press.

Bhandari, R., and Blumenthal, P. (Eds.) (2011). *International students and global mobility in higher education: National trends and new directions*. New York: Palgrave Macmillan.

Bhatti, F.M., and Qalb-i-Abid, S. (2007). *East Indian immigration into Canada 1905–1973*. Lahore: Pakistan Study Centre.

Brosius, C. (2010). *India's middle class: New forms of urban leisure, consumption and prosperity*. New Delhi: Routledge.

Chopra, R. (2005). Sisters and brothers: Schooling, family and migration. In R. Chopra and P. Jeffery (Eds.), *Educational regimes in contemporary India*. New Delhi: SAGE, pp. 299–315.

Chopra, R. (2010). *Militant and migrant: The politics and social history of Punjab*. London: Routledge.

Clifford, J. (1997). *Routes: Travel and translation in the late twentieth century*. Cambridge, MA: Harvard University Press.

Desai, S., and Andrist, L. (2010). Gender scripts and age at marriage in India. *Demography*, 47(3): 667–687.

Dicken, P. (2011). *Global shift (sixth edition): Mapping the changing contours of the world economy*. New York: Guilford Press.

Florida, R. (1999). The role of the university: Leveraging talent, not technology. *Issues in Science and Technology*, 15(4): 67–73.

Florida, R. (2002). The economic geography of talent. *Annals of the Association of American Geographers*, 92(4): 743–755.

Florida, R. (2005). The rise of the creative class. *Regional Science and Urban Economics*, 35(5): 593–596.

Fox, R.G. (1985). *Lions of the Punjab: Culture in the making*. Los Angeles: University of California Press.

Gertler, M.S., and Vinodrai, T. (2006). *Creativity, culture and innovation in the knowledge-based economy: Opportunities and challenges for Ontario*. Report prepared for the Ontario Ministry of Research and Innovation October 1. Retrieved from http://www.utoronto.ca/onris/research_review/WorkingPapers/WorkingDOCS/Working06/Vinodrai06_Creativity.pdf

Gill, H. S. (2016). What the Sikh turban means to masculinity in these transnational times. *In Plainspeak* September 15. Retrieved from http://www.tarshi.net/inplainspeak/voices-sikh-turban-masculinity-transnational/

Gill, H.S. (2012). Masculinity, mobility and transformation in Punjabi cinema: From *Putt Jattan De* (Sons of Jat Farmers) to *Munde UK De* (Boys of UK). *South Asian Popular Culture*, 10(2): 109–122.

Gopinath, D. (2014). Characterizing Indian students pursuing global higher education: A conceptual framework of pathways to internationalization. *Journal of Studies in International Education*, 19(3): 283–305.

Grewal, I., and Kaplan, C. (Eds.) (1994). *Scattered hegemonies: Postmodernity and transnational feminist practices*. Minneapolis: University of Minnesota Press.

Hawthorne, L. (2010). Demography, migration and demand for international students. In C. Findlay and W. Tierneyeds (Eds.), *Globalisation and tertiary education in the Asia-Pacific: The changing nature of a dynamic market*. Singapore: World Scientific Publishing, pp. 93–120.

Holston, J. (1989). *The modernist city: An anthropological critique of Brasilia*. Chicago: Chicago University Press.

Holston, J., and Appadurai, A. (1996). Cities and citizenship. *Public Culture*, 8(2): 187–204.

Howard Hassmann, R., and Walton-Roberts, M. (Eds.) (2015). *The human right to citizenship: A slippery concept*. Philadelphia: University of Pennsylvania Press.

Jejeebhoy, S.J., and Sathar, Z.A. (2001). Women's autonomy in India and Pakistan: The influence of religion and region. *Population and Development Review*, 27(4): 687–712.

Jodhka, S.S. (2006). Beyond "crises": Rethinking contemporary Punjab agriculture. *Economic and Political Weekly*, 41(16): 1530–1537.

Kalia, R. (1999). *Chandigarh: The making of an Indian city*. New Delhi: Oxford University Press.

Kalra, V. (2005). Locating the Sikh *pagh*. *Sikh Formations*, 1(1): 75–92.

Kapur, D. (2010). *Diaspora, development and democracy: The domestic impact of international migration from India*. Princeton, NJ: Princeton University Press.

Kodoth, P., and Varghese, V.J. (2012). Protecting women or endangering the emigration process. *Economic and Political Weekly*, 47(43): 56–66.

Mooney, N. (2006). Aspiration, reunification and gender transformation in Jat Sikh marriages from India to Canada. *Global Networks*, 6(4): 389–403.

Mooney, N. (2011). *Rural nostalgias and transnational dreams: Identity and modernity among Jat Sikhs*. Toronto: University of Toronto Press.

Nisbett, N. (2009). *Growing up in the knowledge society: Living the it dream in Bangalore*. New Delhi: Routledge.

Peck, J. (2005). Struggling with the creative class. *International Journal of Urban and Regional Research*, 29(4): 740–770.

Perez, G. (2004). *The near northwest side story: Migration, displacement and Puerto Rican families*. Berkeley: University of California Press.

Prakash, V. (2002). *Chandigarh's Le Corbusier: The struggle for modernity in postcolonial India*. Seattle: University of Washington Press.

Puar, J. (2008). "The turban is not a hat": Queer diaspora and the practices of profiling. *Sikh Formations*, 4(1): 47–91.

Ratcliffe, P. (2006). Higher education, "race" and the inclusive society. In W. Allen, M. Bonous-Hammarth, and R. Teranishi (Eds.), *Higher education in a global society: Achieving diversity, equity and excellence*. Amsterdam, London: Elsevier, pp. 131–148.

Saxenian, A. (2007). *The new Argonauts: Regional advantage in a global economy*. Cambridge, MA: Harvard University Press.

Schiller, N.G., Basch, L., and Blanc, C.S. (1995). From immigrant to transmigrant: Theorizing transnational migration. *Anthropological Quarterly*, 68(1): 48–63.

Sharma, K.A. (1997). *The ongoing journey: Indian migration to Canada*. New Delhi: Creative Books.

Storper, M. (2013). *Keys to the city: How economics, institutions, social interaction, and politics shape development*. Princeton, NJ: Princeton University Press.

Talbot, I., and Thandi, S. (Eds.) (2004). *People on the move: Punjabi colonial, and post-colonial migration*. Oxford: Oxford University Press.

Urry, J. (2007). *Mobilities*. Cambridge: Polity Press.

Vora, N. (2013). *Impossible citizens: Dubai's Indian diaspora*. Durham, NC: Duke University Press.

Walton-Roberts, M. (2004). Returning, remitting, reshaping: Non-resident Indians and the transformation of society and space in Punjab, India. In P. Jackson, P. Crang and C. Dwyer (Eds.), *Transnational spaces*. London: Routledge, pp. 78–103.

Walton-Roberts, M. (2011a). Multiculturalism already unbound. In M. Chazan, L. Helps, A. Stanley, and S. Thakkar (Eds.), *Home and native land: Unsettling multiculturalism in Canada*. Toronto: Between the Lines Press, pp. 102–122.

Walton-Roberts, M. (2011b). Immigration, the university, and the welcoming second tier city. *Journal of International Migration and Integration*, 12(4): 453–473.

Walton-Roberts, M. (2015). Femininity, mobility and family fears: Indian international student migration and transnational parental control. Journal of Cultural Geography, 32(1): 68–82.

Walton-Roberts, M., and Pratt, G. (2005). Mobile modernities: A South Asian family negotiates immigration, gender and class in Canada. *Gender Place and Culture*, 12(2): 173–196.

Williams, K., Williams, G., Arbuckle, A., Walton-Roberts, M., and Hennebry, J. (2015). *International students in Ontario's post-secondary education system 2000–2012*. Research Report for Higher Education Quality Council Ontario.

Wolfe, D. (2005). The role of universities in regional development and cluster formation. In G.A. Jones, P.L. McCarney, and M. Skolnik (Eds.), *Creating knowledge, strengthening nations: The changing role of higher education*. Toronto: University of Toronto Press, pp. 167–194.

Filmography

Gill, Harjant S. (dir.) (2011). *Roots of love: On Sikh hair and turban* (26 mins.).

Gill, Harjant S. (dir.) (2016). *Sent away boys* (43 mins.).

16

The Urban Poor
The Urban Majority and Everyday Life

Sabin Ninglekhu and Katharine Rankin

Introduction

By 2014, 54 per cent of world's population resided in urban areas. By 2050, that figure will have risen to 66 per cent, meaning that the urban population will have increased by 2.5 billion people—90 per cent of whom will be concentrated in Asia and Africa. Moreover, a third of today's urban population lives in slums (UN, 2014: 5), and poverty in absolute terms has also increased at a much faster rate in urban areas than in rural areas; global **poverty**, it is clear, has become an urban phenomenon. With changing patterns of urbanization, the nature of poverty itself is changing all over the world, with increases, for example, in *working* poverty or the number of poor people who earn a regular income and work for a living but live in households that fall below the poverty line (Stapleton and Kay, 2015).

Given these trends, a great deal of academic scholarship has been devoted to defining and measuring poverty, to understanding how the nature and the spatial distribution of poverty has changed with its urbanization, and to probing the links between poverty and inequality. Poverty can be understood in both absolute terms (e.g., measures that reveal inability to afford a minimum level of nutrition and basic needs) or in relative terms (e.g., measures that reveal the disadvantaged position of poor people relative to society as a whole) (Davis, 2006). The problem with such efforts to measure poverty is that they cannot portray the voices and experiences of the urban poor themselves, and thus the latter are rarely taken into consideration in the development of urban theory and the planning of cities. In today's predominantly neoliberal world, the poor are typically stereotyped as uneducated and inept, if not lazy and burdensome to the system. Even the classic critical accounts that address the systemic causes of poverty and **inequality** tend to circumscribe the political agency of the urban poor. One promising development for urban studies along these lines comes from critical ethnographers working in the global South; their orientation to community-based research increasingly focuses on perspectives of the urban poor (Appadurai, 2002; Benjamin, 2008; Yiftachel, 2009). These analytical traditions can be combined with some of the literatures in urban planning and urban theory that have effectively broached normative questions about "what is to be done" and what is and should be a "**right to the city**" (Harvey, 2008; Marcuse, 2009; Purcell, 2003). In fostering dialogue across these traditions—urban theory, urban planning, subaltern studies, and postcolonial studies—this chapter asks: What do we know of the places the urban poor occupy, the social and economic practices in which they engage, and their forms of social belonging? Particular attention is given to the resourcefulness of this majority group in adapting to ever-changing social and economic conditions.

Our consideration of "the urban majority and everyday life" is thus located at the intersection of two key strands of urban scholarship: first, scholarly engagements with the idea of "the right to the city" coming out of a Marxist approach to industrialized cities of the global North (Purcell, 2002; Harvey, 2008; Fainstein, 2009; Marcuse et al., 2009); and second, a "critical global South urbanism" that draws on the contributions of postcolonial studies (Appadurai, 2000, 2002;

Chatterjee, 2004; Simone, 2005, 2008, 2010, 2014, 2015). Braiding these two strands together offers some useful concepts for understanding contemporary urban struggles—namely, **everyday life, urban revolution,** and **inhabitance**—concepts that we derive primarily from the opus of French Marxist philosopher Henri Lefebvre, but which we interrogate and develop further with recourse to postcolonial studies. The chapter illustrates these concepts and puts them to work in the analysis of a case study dealing with the claims of the landless poor for a right to inhabit the city in Kathmandu, the capital of Nepal. The chapter focuses on the experience of a squatter settlement that has historically faced threats of displacement resulting from municipal infrastructural projects. The case takes the neighbourhood as its scale of analysis for the purpose of probing the relationship between everyday life and the right of the urban poor to inhabit the city. It is also presented in relation to poverty in Canadian cities to draw out, empirically, links between poverty and its analytics in the global North and global South.

The chapter begins with a discussion of the literatures on the right to the city and postcolonial urbanism. This section argues for joining the insights from postcolonial urbanism on the messiness and uncertainty of people's attempts to thrive in the city with the broader political thrust of a Marxist approach oriented towards radical transformation. The second section engages the concepts generated in the previous section to present the case study of a **squatter settlement** in Kathmandu, Nepal. In so doing, it illustrates the ways in which development interventions aimed at city rebuilding inevitably contain within them the threat of **dispossession** and **displacement** for the urban poor. It also reveals how the poor navigate the complex socio-political and bureaucratic terrains of the cities they inhabit to make claims about **citizenship** and belonging. Drawing on the concepts highlighted in the case study, the chapter illustrates the transnational migration flows that connect Canada's largest metropolis to Kathmandu, the common struggles of displacement and dispossession, and a fresh perspective on the issue of property that comparison allows. The chapter concludes by developing a key line of inquiry: under what conditions can the urban poor be resourceful in claiming their right to the city, and under what conditions can these claims contribute to a meaningful process of social transformation?

Right to the City and Postcolonial Urbanism

We begin by shedding some light on the idea of the "right to the city" as a form of critical urban scholarship and a mode of political activism. To do so, one must acknowledge that this idea has assumed several different incarnations. It has been taken up as a philosophical concept and a political theory (Lefebvre, 1996), and it has been adopted as a slogan for various social movements, and thus as a class project with specific goals and objectives (Slater, 2009). But it has also been advanced in a more liberal "human-rights" framing by international humanitarian organizations, such as the UN-Habitat and UNESCO (Brown, 2013). Our discussion focuses on the first two aspects because they seek to question and challenge the dominant system of capitalist relations and the forms of political liberalism that help stabilize that system—rather than accepting that system as a given and necessary way of organizing the world.

One of the key concepts that is developed in the Lefebvrian right-to-the-city tradition is that of everyday life. This concept is crucial for foregrounding how poverty is experienced. At the same time, we detect a tendency to anticipate, and indeed advocate for, that everyday experiences with poverty and associated injustices point towards a utopian socialist future. In this anticipated future, social relations ultimately assume a more egalitarian and equitable form. This form may come via major social mobilizations or via more gradual processes resulting from everyday modes of social relations amongst the poor, marginalized, and dispossessed. Some contemporary work in postcolonial studies, particularly that of anthropologist AbdouMaliq Simone (2004, 2005, 2008, 2014, 2015), offers a slightly different perspective rooted in an ethnographic approach to everyday life; **ethnography** centres on mundane everyday experiences that reveal how the poor use the city in a flexible way to move their lives forward. This approach takes a great deal of care in attributing agency to the urban poor, but resists the tendency to attribute a socially transformative trajectory to their practices. This section argues that a postcolonial perspective on everyday life offers promising directions for understanding the experience of, and imagining possibilities for, the urban majority.

Right to the City: Core Principles

The "right to the city" (RTTC) as a philosophical concept and political theory is rooted in French philosopher Henri Lefebvre's (1968) seminal work, *Writings on Cities*. This formulation was inspired by student uprisings in Paris that same year. Lefebvre's formulation of the right to the city transcends liberal democratic civic rights in that it is the everyday experience of living in the city—rather than formal citizenship or legal residency status—that forms the basis of rights claims (Purcell, 2014: 142). Substantively, Lefebvre (translated in Kofman and Lebas, 1996: 174) argues, "The right to the city manifests itself as a superior form of rights: right to freedom, to individualization in socialization, to habit and to inhabit." For Lefebvre, the onus is on **urban inhabitants** to lead the charge, but we may ponder their propensity to do so and what might be the implications of this interpretation of rights for the practice of planning, which has a strong orientation towards equity and social justice.

The seminal works of urban theorists David Harvey and Peter Marcuse follow Lefebvre's right to the city framework in their engagement with existing urban struggles; that is, while retaining Lefebvre's commitment to the urban as a philosophical project, they also refine an interpretation of the right to the city in relation to contemporary struggles in the city waged for and by the culturally oppressed and socio-economically alienated (Marcuse, 2009). Such struggles make it clear that we are not here advocating "all rights for all people," because some already have more rights than they need or deserve, including the right to oppress and exploit others; this right is protected in a liberal-democratic capitalist system as much as in more authoritarian and feudal contexts, such as what has been prevalent in Nepal until recently. Defending the right to the city thus requires limiting rights for some and extending rights for others (Marcuse, 2009, 2012). The crucial question that Marcuse forces into the right-to-the-city scholarship is this: How is it possible to strategically unite politics across divergent interests, and especially between those who are materially deprived through the relations of production and those who are culturally oppressed through the inability to freely express their various identities. And, extrapolating from Marcuse's insights, we might ask: what role might there be for planners in anticipating and forging common ground across this difference?

David Harvey's contribution to the discussion explicitly grapples with the temptation to interpret the right to the city according to a liberal human rights framework. He contends that the latter is inadequate for the Lefebvrian project of reclaiming urban life because it leaves intact unexamined capitalist and imperialist social relations. The right to the city, on the contrary, must be centrally concerned with the relationship between urbanization and capitalism. Capital absorption through urban development entails the capture of valuable land and resources from low-income communities, a process that Harvey (2008: 34) refers to as **accumulation by dispossession**. This formulation draws upon Lefebvre's emphasis on space as being central to the demand for the right to the city. Urban struggles, Harvey argues, commonly transpire over and around the issues of access to space and the production of space, and it is around these issues that, in practice, rights are made and unmade. To resist and reverse this trend of accumulation by dispossession, Harvey (2008: 40) calls for a revolution that "has to be urban in the broadest sense of the term, or nothing at all." An urban revolution is required to assert and claim the right to the city, as elaborated further below.

At the centre of Lefebvre's conception of right are two ideas—inhabitance and urban life. A Lefebvrian right to the city calls for the right of inhabitants to urban life, regardless of their legal status as property owners or citizens. Thus, the right to the city is a social and cultural right to access, and produce, amenities and spaces in the city. In this sense it must be sharply distinguished from a "human rights" conception wherein rights are confined to and defined by the logic of property ownership and citizenship within a sovereign entity under the ambit of liberal democracy. The idea of inhabitance—"the right to appropriate urban space" and "the right to participate centrally in the production of urban space" (Purcell, 2003: 577)—further concretizes the right to the city. "If inhabitants hold a central role in the decisions that produce urban space," argues geographer Mark Purcell (2003: 579), "then property ownership can no longer confer a dominant voice in decisions about what to do with urban land." In short, the right to the city is a right for the inhabitants of the city and must be yoked to inhabitance, as opposed to property ownership, as an everyday practice and politics of lived reality. How, more specifically, does everyday life fit in to this formulation, and what is its significance for urban revolution?

Everyday Life and Urban Revolution

Everyday life is invoked most fundamentally as a philosophical category in Henri Lefebvre's work. As such, it allows us to read the ways in which both capitalism and socialism are latent in people's everyday experiences in the city. Capitalism is latent in the ways that people participate individualistically in consumer cultures that support commodity production; socialism, for Lefebvre, is latent in the ways that people seek community with one another (although, as Marcuse along with numerous ethnographies of communal violence remind us, the quest for community can take decidedly regressive forms that run counter to the principles of socialism). Within this formulation, the role of critical urban scholarship is to document both latencies, and especially to scrutinize signs of socialistic modes of relating in mundane everyday activity in diverse social settings, including neighbourhood community life, workplaces, religious institutions, and universities. Lefebvre's work falls short of either prescribing or illustrating how this task could be achieved in practice, and for that more methodological consideration and the substantive insights it yields, we will turn, in the subsequent subsection, to the scholarship on critical global South urbanism and postcolonial studies. The significance of the philosophical contribution, however, requires some further elaboration for one to assess its potent political implications.

Following in the Lefebvrian tradition, everyday life may be understood as a critique of *actually lived* everyday life, which in modern industrialized societies largely takes place through cultures of (commodity) consumption. Marxist theorist Guy Debord (1962: 23) comments on the critique of the lived everyday:

> Many technologies do, in fact, more or less markedly alter certain aspects of everyday life . . . telephones, television, music on long-playing records, mass air travel, etc. These developments arise anarchically, by chance, without anyone having foreseen their interrelations or consequences. But there is no denying that, on the whole, this introduction of technology into everyday life ultimately takes place within the framework of modern bureaucratized capitalism and tends to reduce people's independence and creativity.

It may be concluded that conditions for transformative politics could best be found in spaces of everyday life that have not been encompassed by cultures of consumption. What kind of everyday life would this be? And more importantly, whose everyday life would it likely be? These are not questions that Lefebvre explicitly asks but which can perhaps clarify the philosophical argument and political stakes. To do so, we can recognize that when everyday life is filled with struggles, there is likely less capacity (and time) for self-oriented, mundane acts of commodity consumption. This contention comes out of the pragmatics of **precarity**. If one does not have adequate shelter or health care or access to basic facilities, one does not likely have much capital to participate in cultures of commodity consumption. And one is perhaps more likely to collaborate and co-operate with others in relations of mutual help and community. We can think of this condition as a sort of socialistic mode of everyday life born out of necessity to accumulate "wealth in people" as opposed to merely wealth in things.

The idealized scenario of poor people being more likely to collaborate in relations of mutual help than those who have more capacity to invest in commodity consumption is not meant to suggest that the poor do not engage or aspire to engage in such consumption habits. Of course they do. Nor should it impute some kind of romanticized solidarity among the poor. Of course slum societies are rife with conflict and power differentials just like any other. Rather, this depiction of the *relative* propensity of the poor to secure their livelihoods in a relational way, through investing in relations with others, is meant to pose the possibility that everyday livelihoods of the poor might offer some raw materials for a kind of gradual and permanent "urban revolution" (Lefebvre, 2003)—a transformative force that would extend the right to the city to the socio-economically alienated and the culturally oppressed. Perhaps there are moments in the everyday life of squatters in Kathmandu, for example, that signal an alternative politics of getting things done in a way that leads to incremental gains towards social transformation. Perhaps the informal entrepreneurs, street vendors, small-time traders, hawkers, and mechanics in the different districts and neighbourhoods of Dakar, Jakarta, Douala, Phnom Penh, or Abidjan that Simone (2004, 2005, 2008, 2010, 2015) similarly consults to build a picture of *"The City Yet to Come"* provide cues about the resourcefulness and collectivist orientations that are available for confronting poverty and inequality, should planners learn how to recognize them. The case study of Kathmandu, Nepal, and the text box about Toronto, Canada, in this chapter broach this possibility in a relational way. They are intended to stand alongside accounts of more formal, organized modes of socialistic expression, such as the **Occupy Wall Street** movement, that manifest widespread consciousness about the injustices and inequities produced by speculative capitalism.

"Urban revolution" as Lefebvre conceived it, might be distinguished from the more common interpretation of revolution as a spectacular, explosive, and usually violent event or movement in which resistance takes on a collective, overt, and dramatic form. It is particularly apt for interpreting the politics of the urban majority in the current conjuncture. As much as the global financial crisis raised consciousness about the "99 per cent" (the political slogan coined by the Occupy Wall Street movement to evoke the concentration of wealth among the top 1 per cent of income earners compared to the other 99 per cent of the American population), it is fair to say that we still live in the "TINA" moment; Margaret Thatcher's assertion, while British prime minister, that "There Is No Alternative" (to capitalism) seems to have assumed hegemonic proportions and entered into popular common sense. Even within Nepal, which for much of the last decade was regarded as a modern-day vanguard of people-centred leftist politics, the Maoist movement that was once propelled by mass support to political power has now been declared by reputed scholars and journalists to have ended.

In planning theory, too, those who seek to promote the **just city** have elected to confine their horizons to the existing capitalist order and the possibilities for redistributing wealth and opportunity within it. An emerging "just cities" literature advocates the pragmatics of the socially just capitalist city over the unattainable utopia of the socialist city (Fainstein 2009). In other words, disaffection with "big R" epochal revolution prevails. Under these conditions, perhaps "urban revolution"—the proposition that the seeds of a socialistic mode of relations are incipient in the everyday life of the oppressed and the alienated—does offer some solace and some possibilities (Lefebvre, 2003). Perhaps the oppressed and the alienated may furnish the modes of sociality and mutual dependency that could inspire a program for socializing the economy and everyday life with the support of a wider constituency. And perhaps this formulation of an urban revolution rooted in cultivating the socialist self and geared towards securing a generalized right to the city *does* indeed present an alternative.

Contributions from Postcolonial Scholarship

Lefebvrian scholars are explicit in their conviction that a philosophical understanding of everyday life "is as urban and political as it is philosophical" (Goonewardena, 2009: 216). But however much the concept is developed in relation to lived experience and urban struggles, an orientation towards conveying fine-grained accounts of everyday life among the alienated and oppressed is necessary to inform the activities of scholars with a bent towards normative questions of "what is to be done." No ethnographic content is furnished for Lefebvrian everyday life, and no political program. And here is precisely where postcolonial urbanism can help—by bringing analytical tools and political commitment to understanding urban experience from the standpoint of the marginal. Ethnographic renderings can contribute representations of the immense resourcefulness contained within everyday life as well as of its potential to convey an aspirational politics of the poor. The review of the literature in this chapter centres on the contributions of anthropologist Simone, but also pays tribute to a burgeoning literature on global South urbanism that has furnished the fields of urban studies and development studies with some useful concepts that will also be briefly considered here, namely, "**deep democracy**" (Appadurai, 2002) and "**political society**" (Chatterjee, 2004).

Political society, in Partha Chatterjee's (2004) framing, is a popular mode of politics of the poor. It is a space carved out by **subalterns**—the systematically unprivileged and underrepresented—for advancing their interests and securing their livelihoods outside the ambit of liberal democracy; in contrast to civil society, political society operates at the margins of formally regulated markets and is constituted through practices that may violate the law and subvert civic norms of good citizenship. Arjun Appadurai's (2002) concept of deep democracy interjects an explicitly positive valuation of subaltern practice: the contingencies of informality require sharing, co-operating, self-censorship, and mutual regulation, which are more deeply democratic than the conventions of political liberalism to which civil society must conform. These politics, in Appadurai's (2002: 40) words, entail "us[ing] the knowledge of the poor to leverage expert knowledge, redeem[ing] humiliation through a politics of recognition, and enabl[ing] the deepening of democracy among the poor themselves."

These important contributions signal the self-organizing capacities of those lacking the formal rights that accompany citizenship, formal employment, property ownership, and other modes of official recognition. Deep democracy and political society may indeed suffice analytically to capture the informal solidarities forged by subaltern groups in the ongoing effort to secure basic services and the right to human flourishing, which we can think of as "the right . . . to the full development of [one's] innate intellectual, physical and spiritual potentials in the context of wider communities" (Friedman, 2000: 466). However, such categories also circumscribe the politics of the poor with a certainty and prescription that in practice amounts to only part of a more ambiguous picture. Simone, in several ethnographic accounts of African and Southeast Asian cities (2004, 2005, 2008, 2010, 2015), attends to the unplanned and unexpected intersections and encounters of everyday life, as well as to the unanticipated and politically ambiguous outcomes. Simone (2004: 242) alludes to "thousands of daily stories where urban actors find their lives suddenly transformed. Immersed in daily struggles to make a living to survive the economic chaos and political manipulations around them, some unexpected directions are opened up; some are survived and some are not."

Such stories interweave to imbue the urban life of the poor with uncertainty. Taking note of such uncertainty allows us to look into the slum and acknowledge multiple logics infusing everyday life. A thousand stories circulate; some conjoin to build a coherent narrative, others do not. Simone (2015: 17) argues, that "[t]he articulations and divides are full of complexities and deceptions: histories of apparent resourcefulness have often raised more problems than they have addressed. . . . What often look like substantial assets of social capital, democratic practice, and social collaboration can be highly murky manoeuvres of opportunism and trickery." Foregrounding more spontaneous and unorganized alliances, as well as the uncertainty of their outcomes, does not diminish the imperative of the right to the city as a political aspiration. In fact, Simone contends that the mode of representation is itself emancipatory insofar as it refuses to engage encompassing, prescriptive categories or to assert binaries or draw boundaries around types of political practice. Lefebvre (1987: 9) himself also privileges the mundanity of everyday life, but he vests it more directly with emancipatory potential—"Why wouldn't the concept of everydayness reveal the extraordinary in the ordinary?" he asks, the "extraordinary" being the potential for a radical opening contained within the unpredictable mundane. The difference with Simone is that the right to the city itself is somewhat less certain both procedurally and substantively.

As such, if the right to the city is to be seen as a politics of possibility, the politics is such that "collective actions coalesce and mutate in light of new urban developments" (Simone, 2008: 197). The content of political aspirations is itself mutable. Simone's contribution is to make visible the messiness and uncertainty contained within everyday life and the right to the city as a collective project. In such a formulation, the right to the city may be more accurately understood as "the right to be messy and inconsistent, or to look disordered . . . [the right to] thrive in unanticipated ways" (Simone, 2010: 331).

Now, Simone himself acknowledges that the challenge for a radical transformative project lies in the dearth of "an overarching institutional logic or public discourse capable of tying its heterogeneous residents together in some conviction of common belonging or reference" (Simone, 2005: 323). Conjoining Lefebvre's philosophical contributions with the postcolonial urbanism of Simone and others suggests a role for ethnography in documenting the messiness and uncertainty of people's attempts to thrive in the city; at the same time, we can think about a role for planners in identifying the conditions of possibility for successful attempts to thrive, so that these might become the basis for an overarching vision of the right to the city.

Everyday Life in a Kathmandu Squatter Community: From Resourcefulness to Urban Revolution?

This section offers a case study of Acharya Tole, a small squatter settlement in Kathmandu, Nepal, whose inhabitants' struggle against eviction illustrates well the possibilities and challenges for claiming a right to the city (Figure 16.1). The case study traces how a state-led development initiative encounters the livelihood of the poor as a site for struggle over staking claims for the right to the city. The case study is part of a broader research project on urban poverty, citizenship, and right to the city that grew out of collaborative research on the politics of neighbourhood associations (Ninglekhu and Rankin, 2008) and formed the basis for Ninglekhu's (2016) doctoral dissertation. The dissertation's central aim is to consider the conditions that enable the urban poor to make claims for the right to the city. The data for the case study were collected during the summer of 2015 through qualitative interviews with both residents and resident-activists of Acharya Tole and with representatives of the local ward office and the Kathmandu Valley Development Authority (KVDA), a municipal government body that oversees infrastructure and land use planning in the ctiy.

Sukumbasi is the Nepali term for squatter communities that lack a recognized legal or legitimate claim to property ownership or tenancy status. The majority of Nepal's *sukumbasi* are migrants from other parts of the country outside Kathmandu. Like many cities in the global South that are experiencing rapid population growth owing to rural–urban migration, Kathmandu has many squatter settlements—64 as of 2011 (Lumanti, 2011) (Figure 16.2). Some are readily visible as one moves about the city's major thoroughfares, such as those lining the banks of the Bagmati River and its tributaries that cut through the centre of the city (Figure 16.3). Some are hidden in pockets of the densely built-up inner city, lying in the shadows of major institutions and structures, such as hotels, gated condominiums, and private hospitals. Still others can be found on the urban periphery, scattered among the unplanned housing developments that have eaten up the country's most arable agricultural soil. Some, like the 12-household settlement that constitutes the case study in this section, are very small in size but have been in existence for nearly three decades; others are less than a decade old but contain upwards of 200 households.

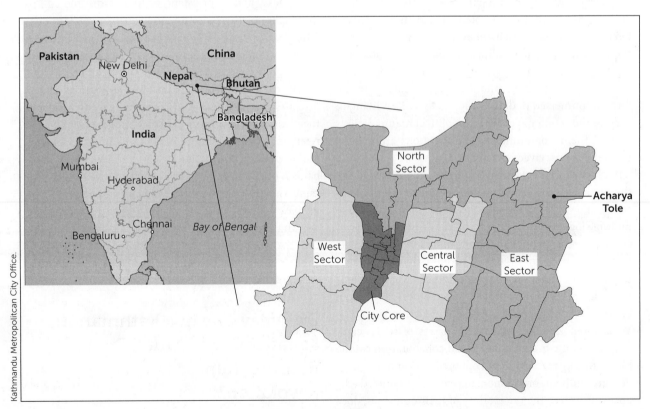

Figure 16.1 Map of Kathmandu Metropolitan City, Nepal

Figure 16.2 A slum settlement in Sinamangal, western end of Kathmandu

Figure 16.3 A slum settlement on the banks of the Bagmati River in Kathmandu

Deep Democracy?

Acharya Tole is located in downtown Kathmandu, sandwiched between two neighbourhoods of middle-class, single-family homes and apartment buildings—Gaurighat to the east and Boudhadhwar to the west. There are a total of 300 households in these three neighbourhoods, including the 12 *sukumbasi* households in Acharya Tole. There are many such *sukumbasi* settlements scattered amid residential neighbourhoods in Kathmandu, all of which have long histories of residents collaborating to manage the contingencies of everyday life in places that have largely fallen off the grid of urban services. In Acharya Tole, water, sewage, and financial services are all examples of basic services that residents have had to organize collectively for themselves—in a manner reminiscent of Appadurai's claims about forms of deep democracy in the slums of Mumbai, India.

This case study tells a story of how the legacy of socialized modes of everyday life in Acharya Tole came to underwrite a somewhat haphazard but nonetheless effective resistance to encroachments on *sukumbasi*'s right to the city. The resistance in turn raises key issues regarding municipal bureaucratic practice from an equity and justice perspective. In response to a city-wide road expansion project, a committee comprising the residents of Gaurigat, Boudhawar and Acharya Tole was formed in the summer of 2014. It was the first of its kind in the city—a collective that transcended the class cleavage between the property-owning middle-class and the property-less residents. The committee (referred to as the residents' association) formed around a collective demand for compensation when residents' land was to be expropriated for the road widening. In making this demand, the residents' association made no distinction between property owners and the propertyless. The backstory behind the formation of the association and how it articulated a collective stance is critical for understanding the connection between everyday life and urban revolution in the Lefebvrian sense.

Spectres of Violence

In December 2013, residents of Acharya Tole returned home from their daily work to find a notice hanging by their doorknobs. It was from KVDA. The residents were notified that the front part of their homes would be dismantled to make space for the widening of the road's right-of-way by 3 metres (10 feet). The residents' first response was panic, even though public debate over the ongoing citywide road expansion project had been raging and should have served as a forewarning. The project had been expropriating and removing homes, sidewalks, green islands, and other structures to accommodate an 11-metre-wide (36-foot) standard for all arterial roads in Kathmandu.

The notice recalled spectral images of the recent past, when the municipal government and local property owners had coalesced to dispossess the sukumbasi of their homes. In the early 2000s, local property owners and municipal bureaucrats had joined forces to build parkland and a road through the sukumbasi settlement. *Sukumbasi* were told to leave, either by vacating their houses or by taking their houses away with them. At first they merely pleaded against this action, arguing that "our homes do not have wheels to be taken with us," as one resident put it. When faced with the appearance of a municipal bulldozer that commenced extending the road, however, their efforts assumed a more organized and embodied form. They began by covering the streets of the neighbourhood with rocks, with potholes interspersed among them, to obstruct traffic and complicate the bulldozing process. "We told them that the street is ours," another resident explained. "We told them, the only way you can go past it is by either riding over us, or through imaginary flyovers." It quickly became apparent that pleas and physical obstructions would not suffice to stop the planned developments; this is when the *sukumbasi* began to mobilize in more tactical ways. Already pathways of communication and modes of organizing had been established through the collective action needed to provision the community with water, sewage, and financial services. By informally mobilizing long-standing relations with low-level municipal bureaucrats, the *sukumbasi* obtained the names of the homeowners who had petitioned the municipal government for their eviction. The names were announced and circulated within Acharya Tole. Subsequently, when these homeowner neighbours walked through the streets of Acharya Tole, residents would stare at them and spit on the street. This politics of shaming seemed to work, as the homeowner constituency asked the government to halt the road and park project and it did. At that time, in 2000, the level of aggression was subtly pitched to resist an incursion on the right of inhabitance, while also avoiding more spectacular forms of violence.

In 2013, things had changed. *Sukumbasi* households weren't the only ones targeted by the state, and neither were the neighbouring property owners advocating the project. Still, there was reason for the *sukumbasi* to be especially afraid. Most *sukumbasi* in Kathmandu lack formal citizenship status because regulations and practices mandate that property ownership is a precondition for obtaining citizenship. The Citizenship Act was revised in 2006 to officially separate citizenship from property ownership, but bias against the landless continues to infuse the bureaucracy. In practice, landless applicants are typically denied even when they have furnished all the required documentation. Bureaucrats steeped in long-standing discriminatory political cultures continue in practice to demand *lalpurja*—landownership certificates—and they are free to do so in the absence of regulatory oversight within a dysfunctional state bureaucracy. The ongoing exclusion from citizenship rights via landlessness also ensures that *sukumbasi* lack many municipal rights that propertied citizens in the city can access. One of these is the right to compensation in the case of expropriation of a private home for public works such as the road expansion project in question.

After six months of lobbying together, residents of Gaurighat, Boudhadhwar, and Acharya Tole were all able to claim compensation for the expropriation of their homes. The significance of this crucial victory derives from the fact that until that moment, the state had always acted with impunity to remove *sukumbasi* from public lands in the name of public purpose, whether road widening, environmental rehabilitation, or the control of nuisance. In this case, that impunity was checked by an unlikely coalition of property owners and squatters to whom the state was made accountable via the public demands of the association. For the residents of Acharya Tole, obtaining compensation for the expropriation marked a radical intervention in the sense that they were building from small acts of solidarity and community to generate incremental gains towards a longer, gradual process of extending the right to the city. Nepal's *sukumbasi* openly celebrate the victory in Acharya Tole as an incremental gain towards achieving their central demand for land rights and all the amenities and services that accrue to property owners (Figure 16.4).

The Politics of Everyday Life

What might have been the conditions of possibility for this unlikely cross-class alliance that led up to the victory over compensation (see Tanaka 2009)? This crucial question requires a serious engagement with the politics of everyday life. Ninglekhu (2016) shows that one condition of possibility was certainly the prior experience with activism—organizing co-operatives for access to services—born out of a socialistic mode of everyday life, itself necessitated by conditions of poverty and

KVDA, the government body mandated with road expansion works. The regular meetings were made possible through informal connections afforded by ties forged by members of the residents' association. It was too early to know whether this particular alliance would endure, or whether it would just reflect a fleeting and merely ad hoc alignment of interests. Regardless of the long-term relationship between landowners and squatters, the unanticipated link with propertied neighbours at this conjuncture served a critical function in the much longer arc of the sukumbasi's struggle for the right to the city in Kathmandu. It would afford the movement another basis for claiming a right to the city on terms that the state would understand. Though squatters still lacked landownership documents, they now had compensation receipts from the municipality. The state had given legal recognition to their residences.

Figure 16.4 Residents of Acharya Tole rebuilding their homes after road expansion

marginality. The account here suggests additionally that new kinds of rights (not just for inhabitance but also for recognition as householders) were made possible through unanticipated networks of solidarity with the very neighbours who had formerly posed their right to the city in opposition to *sukumbasi*'s right to the city. Perhaps those neighbours had learned their lesson. Perhaps they feared another shaming or a more overtly violent response if they sought to protect their own property in the face of the 2013 road-widening initiative, while ignoring the fate of their *sukumbasi* neighbours. Perhaps the *sukumbasi* had earned their neighbour's respect as fellow inhabitants through their successes with local co-operatives or through the passionate demonstration of their history of belonging and their right to the city. As one member of the residents' association put it, as if endorsing the calls of critical urban scholars for the right to the city, "[t]hose who have already inhabited should not be uprooted."

Whatever the reason, the residents' association constituted a forum where *sukumbasi* demands were given legitimacy and an institutional form from which to undertake more direct advocacy with the municipal state (compared to such tactics as shaming or obstructing bulldozers with rocks and potholes). The association's members included an ex-officer of the Nepal police, the principal of a local elementary school, and a respected entrepreneur (quoted earlier). The association used formal channels to organize meetings with the head of

Inhabitance

To elaborate this point, it is important to emphasize that another condition of possibility must surely have been the fact that, unlike most *sukumbasi* settlements in the city, Acharya Tole had a cadastral blueprint that provided evidence of long-time *sukumbasi* inhabitance. The residents of Acharya Tole had obtained the blueprint from the District Land Office in 1990. This blueprint maps only those land uses that have a "legal" function in the city. It is an official document that identifies all 12 houses in Acharya Tole; this is an unusual circumstance, because *sukumbasi* reside on public land that is usually characterized as "vacant" or "unused" in government documents and maps. The blueprint provided a key basis upon which *sukumbasi* of Acharya Tole might claim their right to compensation for expropriated land (Figure 16.5).

This politics of claiming compensation allows us to think of the right to the city as an incremental and open-ended process, achieved through "murky manoeuvres of opportunism and trickery" (Simone, 2015: 15). Such an understanding of everyday life in the city opens space for thinking about the right to the city as a project of mutable aspirational politics. One of the Acharya Tole resident-activists intimated that "[t]he road extension, as it turns out now, was a blessing in disguise. We will never feel completely safe because we do not have the home or landownership documents yet. But we now

Figure 16.5 Acharya Tole in the aftermath of road expansion

have compensation receipts from the government. The receipts will add further credibility to our demands [for the right to shelter and title to squatted lands]." Urban revolution is thus filled with "mutable aspirations" in "varying degrees of realization" in which, as Simone (2005: 323) put it, "the pursuit of aspirations itself largely depends on what kinds of connections residents can put together between diverse infrastructures, spaces, populations, and economic activities of the city."

Like Acharya Tole, squatter and other marginalized communities all over the world, including Canada, are at constant risk of displacement. The forces of displacement are the local and national states that clamour for the status of world-class city, as well as propertied residents and investors who wish to partake in this status and the material rewards it would bring (Ghertner, 2015). Development driven by global competition entails "cleaning up" these communities, removing nuisance, greening abandoned lands, and improving infrastructure, particularly transportation infrastructure that would improve the flow of people and commerce. At the same time, the Acharya Tole account offers an opportunity to defamiliarize ourselves with the system of private property that prevails in Canada and much of the world, and to imagine other forms of tenure that might make it possible for the urban majority to claim a right to the city.

Urbanity: Parallel Formations in Canadian Cities

As elsewhere in the world, Canada's population is becoming increasingly urbanized, and as much as 90 per cent of Canada's total population growth in recent years has been in its urban areas. Between 2001 and 2006, the population of Canada's six largest cities (Toronto, Montreal, Vancouver, Ottawa-Gatineau, Calgary, and Edmonton) grew eight times faster than Canada's rural areas, and in 2006, these six cities alone accounted for 45 per cent of Canada's total population (Statistics Canada, 2008). With the increase in urbanization, there has been a dramatic rise in homelessness across Canada's major cities. Nearly a quarter of a million Canadians experience homelessness in a given year, including those who are unsheltered, in emergency shelters, and or provisionally sheltered, moving from place to place, "couch surfing," or residing in institutional settings such as hospitals or prisons; many more are at risk of homelessness at any time. Indeed, a 2013 Ipso Reid poll suggests that "as many as 1.3 million Canadians have experienced homelessness or extremely insecure housing at some point during the past five years" (Gaetz et al., 2013: 5).

Key determinants of homelessness in Canada include a lack of government investments in affordable housing; the declining income and spending power of nearly half the population in the last decade; and family and domestic violence. The homeless population comprises all groups within the population, but families with children are the fastest-growing demographic, many fleeing dangerous situations; and Aboriginal peoples are overrepresented across the country, as are other marginalized and racialized groups (Gaetz et al., 2013: 5, 26, 42). As a testament to the problem of homelessness, there are numerous temporary encampments and mini-squats in and around Canada's major cities in addition to some larger and more permanent settlements that have come and gone, such as Toronto's Tent City. This latter site grew along the harbour from the late 1990s until 2002, when 100 or more inhabitants were forcibly removed by police after being "evicted" by Home Depot, which owned the land where the settlement was located (Laird, 2007: 48–49). More recently, temporary squatter settlements linked to the Occupy movement, which protested economic inequality and an exploitative global financial system, popped up in cities across Canada but were quickly and forcibly shut down by police. Before they were shut down, however, there were movements within them towards community activism, cross-class solidarity, and community democratic governing—similar in some ways to those found in Acharya Tole (e.g., Crowe, 2007; Sitrin, 2012).

Even so, in contrast to Nepal and many other countries in the global South, in Canada, public provision of health and education under a decent welfare system provide some measure of support across the socio-economic spectrum. Policies aimed at promoting multiculturalism seek to create a tolerant environment in which groups of different ethno-cultural backgrounds can inhabit the city together, with at least some appreciation of difference, if not mutual respect. Land, however, is treated as a commodity, and the rights of property owners take precedence over those of other inhabitants in the city. While access to health and education are rights, housing is not. Displacement pressure derives from the pressures on housing affordability associated with gentrification, the rising cost of living, and precarious labour markets.

Canadian cities are major destinations for immigrants from the global South who move to Canada seeking the opportunity for education, income, and political stability. Not surprisingly, new immigrants from the global South tend to face poverty conditions—arriving with few assets, with educational or professional credentials that have no standing in the Canadian context, with the responsibility to send money to their families in their countries of origin, and with having to manage the high cost of living. In Toronto, a key destination for immigrants to Canada, they settle primarily in the urban periphery, in inner suburban neighbourhoods where poverty is increasingly concentrated. Toronto's downtown has become the near-exclusive purview of the middle and upper classes (Hulchanski, 2010). At the same time, growing evidence suggests that the inner suburbs have become a new frontier for Toronto's strategy of promoting the knowledge economy: as the downtown increasingly prices out the so-called creative classes considered critical to fueling innovation, the solution is to redevelop the still-affordable inner suburbs. Thus the city of Toronto, in partnership with private-sector developers, has initiated a strategy of facilitating the development of creative economy hubs, or planned enclaves for arts-oriented amenities and services, in association with a major investment in transportation infrastructure that could better link the suburbs with the downtown. The community of Weston, for example, has suffered a long economic decline associated first with deindustrialization and subsequently with patterns of socio-spatial inequality that have concentrated poor new immigrants on the city's periphery; today, rapid development is imminent, catalyzed by the development of transit infrastructure and arts-oriented residential and commercial development (Rankin and McLean, 2014).

Cities as seemingly disparate as Toronto and Kathmandu show similar patterns of development, expropriation, dispossession, and displacement. Just as the road-widening projects of Kathmandu were intended to improve mobility and flow, two major transportation projects in Toronto aim to ease congestion—the Union Pearson Express linking the downtown to Pearson International Airport, and the

continued

Eglinton Crosstown light-rail transit (LRT) linking the eastern and western inner suburban zones. These are the projects that have created opportunities for new condominium and commercial development in Weston and several other inner suburban neighbourhoods housing high concentrations of new immigrants and low-income tenants. The so-called mobility hubs are celebrated as a way to revitalize neighbourhoods that have been vilified as disinvested, crime ridden, and economically stagnant. Seen from the perspective of the city's new immigrant populations that predominate in these neighbourhoods, these developments spell certain gentrification of some of the few areas of the city that have remained affordable (Rankin and McLean, 2014).

In this light, the experiences of *sukumbasi* in Kathmandu are relevant for understanding socio-spatial inequality in cities like Toronto. Although the context may be different, the threat of displacement profoundly shapes urban experience for large and growing populations living in conditions of economic precarity. Centring the experience of socio-economically alienated groups helps to foreground the continual threat to those without property and without citizenship or permanent residence status. Such an understanding leads to questions about conditions under which resourcefulness and socialistic modes of everyday life might generate claims-making in Canadian contexts that could catalyze a challenge to market-driven development and broach the vexing question of how to revitalize a low-income area without displacing its current residents. How could the Acharya Tole case, together with a philosophical and methodological orientation to everyday life, alert us to the possibilities for unplanned, spontaneous cross-class alliances that might find a common stake in forestalling the predictable gentrification trajectory. An approach rooted in everyday life, with cues from both Marxist and postcolonial traditions, attends as well to the more mundane instances of encounter and association that may or may not generate a common stake.

A second insight that derives directly from the case of Acharya Tole relates to property tenure. In Canada, struggles over the right to the city are commonly waged around the displacements wrought by increases in land value and rent as neighbourhoods gentrify. These struggles and the literature about them (e.g., Walks, 2008, 2013) are premised on an understanding that private property predominates and indeed has no alternative. Nepal's *sukumbasi*, by contrast, seek to secure title to homes that they have built on common land—usually vacant public lands. Their purpose is to secure the right to the city by accessing rights of citizenship that have customarily been linked to property ownership. But this is premised on the presumption that common lands exist as an alternative form of land tenure that could sustain relatively equitable forms of social reproduction (Ghertner, 2015). Thus, the *sukumbasi* struggles in the global South serve as important reminders that the private-property systems prevailing in the global North represent one end of a wider spectrum of land tenure. Non-private forms of tenure have historically sustained populations lacking in significant private assets; they also remind us that a private-property model of tenure is just that—a model created through human intention and action rather than a naturally occurring phenomenon—and it can be challenged and modified just like any other mode of market regulation.

Conclusion

This chapter has broached the right to urban life for the urban majority, in both philosophical and political terms. It advocates an interpretation of rights that extends beyond political liberalism to encompass a deeper interpretation entailing the right to inhabitance or to appropriate urban spaces for shelter, work, and play. We are specifically concerned with the rights of the economically marginalized and culturally oppressed who constitute the urban majority; this is because their right to the city may be directly undermined by the ways in which those enjoying dominant cultural status and economic security seek to defend their right to the city on exploitative terms. We are concerned with the matter of *whose* rights for another reason as well: because everyday life among the urban majority assumes a certain sociality deriving from precarity, a spontaneous socialized response to the requirements for urban services that are denied by the state can become the basis for unexpected connections and alliances, which may create the grounds for transformative change processes.

Urban revolution thus demarks not so much the spectacle of violence (although embodied tactics may be required) but rather a gradual social transformation that arises when the sociality of everyday life forges a constituency and a demand to extend the right to the city to the urban majority. The Kathmandu case study shows a deliberately unspectacular instance of urban revolution. It is unspectacular because it involves a small number of households and because no spectacular protest or resistance was staged, and yet it reveals how the murky manoeuvres born out of everyday life can catalyze social transformation. In this instance, winning the right to compensation for expropriated land for squatters lacking formal ownership papers was both an incremental and a monumental step towards the possibility of expanding the right to the city. Ethnographic approaches that can trace local political histories were necessary to identify the unanticipated and messy processes through which property owners and landless came to see their fates entangled and their interests vis-à-vis the state aligned. The cross-class alliance, and particularly the opportunities it afforded *sukumbasi* to navigate within the neighbourhood as well as to access state institutions, gave more force to the sukumbasi demands than the embodied tactics of protest that had been attempted earlier.

However, much urban revolution may be the product of murky manoeuvres and mutable aspirations—not planned in a linear way, but instead born of an unpredictable intersection of people, institutions and infrastructures. It is thus useful to consider the conditions that make urban revolution possible. Taking stock of the conditions of possibility for urban revolution can perhaps be best achieved through comparative ethnographic research on the everyday life of the oppressed and the alienated. Under what conditions could everyday life generate urban revolution? Under what circumstances is the radical potential of everyday life subverted to state projects or market deepening? Such a comparative project could go a long way towards assessing a role for planners with long-standing commitments to an inclusive urbanism in engaging everyday life of the poor as a site for imagining transformative politics and urban revolution.

Key Points

- The urban poor are resourceful in securing basic services and a right to human flourishing; urban policy and planning should support rather than undermine this resourcefulness, and the first step is to develop approaches to understanding poverty that are rooted in the experience of the urban poor.
- Ethnographic methods can reveal not only the resourcefulness of the urban poor, but also the messiness and uncertainty of their attempts to thrive in the city, which can result in unexpected opportunities to unite politics across divergent interests.
- The imperative of a right to the city for the urban majority is best articulated through a combination of postcolonial and Marxist urbanism. The former helps to emphasize the significance of inhabitance and everyday life; the latter underscores the ambiguity and uncertainty that imbues everyday life and thus makes the radical quest for a right to the city a mutable and unpredictable project.
- Urban revolution need not entail organized, spectacular movements of resistance; it may be achieved through a gradual and incremental process of change rooted in modes of sociality and inhabitance among the urban majority.
- Squatters in Kathmandu, Nepal, took significant steps towards securing a right to the city when they secured their right to compensation for their homes expropriated during a municipal road-widening project; they did so by forging cross-class alliances with property-owning neighbours, extracting crucial information from low-level officials, and building on modes of co-operation developed over time through efforts to gain access to basic services.

- Canadian cities exhibit similar patterns of displacement and dispossession of socio-economically marginalized groups, as was described in the Kathmandu case study. The case study helps to emphasize the threat experienced by those groups, as well as the wider range of conceivable land tenure forms than is commonly acknowledged in North American contexts.

Activities and Questions for Review

1. Explain to a classmate in your own words what the concept of right to the city means.
2. If urban revolution is understood as a gradual process of transformative change, what examples from your own everyday life could illustrate urban revolutionary potential?
3. Drawing on the case study of Acharya Tole, identify and characterize specific practices used by the squatters to initiate urban change.
4. Working in a small group, discuss how the case study of squatter experiences in Kathmandu, Nepal, informs your understanding of poverty in Canada.

Acknowledgments

We gratefully acknowledge funding support for this research from the Asian Institute and Centre for South Asian Studies at the University of Toronto. Thanks go to Kanishka Goonewardena and Prasad Khanolkar for helpful conversations that contributed to refining our argument, as well as to friends at the Society for Preservation of Shelter and Habitat in Nepal, long-time collaborators in advocating for the rights of the sukumbasi inhabitants of Kathmandu.

References

Appadurai, A. (2000). Spectral housing and urban cleansing: Notes on millennial Mumbai. *Public Culture, 12*(3): 627–651.

Appadurai, A. (2002). Deep democracy: Urban governmentality and the horizon of politics. *Public Culture, 14*(1): 21–47.

Benjamin, S. (2008). Occupancy urbanism: Radicalizing politics and economy beyond policy and programs. *International Journal of Urban and Regional Research, 32*(3): 719–729.

Brown, A. (2013). The right to the city: Road to Rio 2010. *International Journal of Urban and Regional Research, 37*(3): 957–971.

Chatterjee, P. (2004). *The politics of the governed, reflections on popular politics on most of the world*. New York: Columbia University Press.

Crowe, C. (2007). A home of their own in Tent City. *Toronto Star*, April 8. Retrieved from https://www.thestar.com/news/2007/04/08/a_home_of_their_own_in_tent_city.html

Davis, M. (2006). *Planet of slums*. New York: Verso.

Debord, G. (1962). Perspective for conscious changes in everyday life. Situationist International Online. Retrieved from http://www.cddc.vt.edu/sionline/si/everyday.html

Fainstein, S. (2009). Planning and the just city. In P. Marcuse, J. Connolly, J. Novy, I. Olivo, C. Potter, and J. Steil (Eds.), *Searching for the just city: Debates in urban theory and practice*. London and New York: Routledge, pp. 19–39.

Friedmann, J. (2000). The good city: In defense of utopia thinking. *International Journal of Urban and Regional Research, 24*(2): 460–472.

Gaetz, S., Donaldson, J., Richter, T, and Gulliver, T. (2013). *The state of homelessness in Canada, 2013*. Toronto: Canadian Homelessness Research Network Press. Retrieved from http://www.wellesleyinstitute.com/wp-content/uploads/2013/06/SOHC2103.pdf

Ghertner, D. A. (2015). *Rule by aesthetics: World-class city making in Delhi*. Oxford: Oxford University Press.

Goonewardena, K. (2009). Urban studies, critical theory, radical politics: Eight theses for Peter Marcuse. *City, 13*(2): 208–218.

Harvey, D. (2008). The right to the city. *New Left Review, 53*: 23–40.

Hulchanski, J.D. (2010). *The three cities within Toronto: Income polarization among Toronto's neighbourhoods, 1970–2005*. Cities Centre, University of Toronto.

Laird, G. (2007). Homelessness in a growth economy: Canada's 21st Century Paradox. Canadian Women's Health Network website, http://www.cwhn.ca/en/node/40566

Lefebvre, H. (1984 [1968]). *Everyday life in the modern world*. Trans. S. Rabinovitch. New Brunswick, NJ: Transaction.

Lefebvre, H. (1987). The everyday and everydayness. *Everyday Life, 83*: 7–11.

Lefebvre, H. (1991 [1947/1958]). *Critique of everyday life* (Vol. 1). Trans. J. Moore. London: Verso.

Lefebvre, H. (1996). The right to the city. In E. Kofman and E. Lebas (Eds.), *Writing on cities*. London: Blackwell, pp. 63–184.

Lefebvre, H. (2003). *The urban revolution*. Minneapolis: University of Minnesota Press.

Lumanti. (2011). *A situation analysis of urban poor communities in Kathmandu and Lalitpur*. Lumanti Support Group for Shelter, Lalitpur.

Marcuse, P. (2009). From critical urban theory to the right to the city. *City*, *13*(2–3): 185–197.

Marcuse, P. (2012). Whose right(s) to what city? In N. Brenner, P. Marcuse, and M. Mayer (Eds.), *Cities for people, not for profit: Critical urban theory and the right to the city*. Abingdon/New York: Routledge.

Marcuse, P., Connolly J., Novy J, Olivo I., Potter, C., and Steil, J. (2009). *Searching for the just city: Debates in urban theory and practice*. London and New York: Routledge.

Ninglekhu, S. (2016). *Mutable aspirations and uncertain futures: Everyday life and the politics of urgency in Kathmandu, Nepal*. Unpublished doctoral thesis. University of Toronto, Canada.

Ninglekhu, S., and Rankin, K. (2008). Neighborhood associations as civic space in Kathmandu: Progressive and regressive possibilities. In A. Danière and M. Douglass (Eds.), *The politics of civic space in Asia: Building urban communities*. New York: Routledge, pp. 151–174.

Purcell, M. (2002). Excavating Lefebvre: The right to the city and its urban politics of the inhabitant. *Geojournal*, *58*(2–3): 99–108.

Purcell, M. (2003). Citizenship and the right to the global city: Reimagining the capitalist world order. *International Journal of Urban and Regional Research*, *27*(3): 564–590.

Purcell, M. (2014). Possible worlds: Henri Lefebvre and the right to the city. *Journal of Urban Affairs*, *36*(1): 141–154.

Rankin, K.N., and McLean, H. (2014). Governing the commercial streets of the city: New terrains of disinvestment and gentrification in Toronto's inner suburbs. *Antipode*, *47*(1): 216–239.

Roy, A., and Ong, A. (Eds.) (2011). *Worlding cities: Asian experiments and the art of being global*. Malden, MA, and Oxford, UK: Wiley-Blackwell.

Simone, A. (2004). *For the city yet to come: Changing African life in four cities*. Durham, NC: Duke University Press.

Simone, A. (2005) The right to the city. *Interventions*, *7*(3): 321–325.

Simone, A. (2008). The politics of the possible: Making urban life in Phnom Penh. *Singapore Journal of Tropical Geography*, *29*(2): 186–204.

Simone, A. (2010). *City life from Jakarta to Dakar: Movements at the crossroads*. New York and London: Routledge.

Simone, A. (2014). Cities of uncertainty, Jakarta, the urban majority, and inventive political technologies. *Theory, Culture and Society*, *30*: 243–263.

Simone, A. (2015). The urban poor and their ambivalent exceptionalities: Some notes from Jakarta. *Cultural Anthropology*, *56*(S11): 15–23.

Sitrin, M. (2012). Horizontalism and the Occupy movements. *Dissent*, Spring. Retrieved from https://www.dissentmagazine.org/article/horizontalism-and-the-occupy-movements

Slater, T. (2009). Missing Marcuse: On gentrification and displacement. *City*, *12*(2–3): 293–311.

Stapleton, J., and Kay, J. (2012). *The working poor in the Toronto region: Mapping working poverty in Canada's richest city*. Metcalf Foundation.

Statistics Canada. (2008). Retrieved from http://www.statcan.gc.ca/pub/11-008-x/2007004/10313-eng.htm

Tanaka, M. (2009). From confrontation to collaboration: A decade in the work of the squatters' movement in Nepal. *Environment and Urbanization*, *21*(1): 143–159.

UN (2014). World Urbanization Project. Retrieved from http://esa.un.org/unpd/wup/highlights/wup2014-highlights.pdf

Walks, A. (2008). *The timing, patterning, and forms of gentrification and neighbourhood upgrading in Montreal, Toronto, and Vancouver, 1961 to 2001*. Research Paper 211. Cities Centre, University of Toronto.

Walks, A. (2013). *Income inequality and polarization in Canada's Cities: An examination and new form of measurement*. Research Paper 227. Cities Centre, University of Toronto.

Yiftachel, O. (2009). Critical theory and "gray space": Mobilization of the colonized. *City*, *13*(203): 247–263.

17 Women in Cities

Linda Peake and Geraldine Pratt

Introduction

In 1989, Suzanne Mackenzie, a Canadian urban feminist geographer responsible for developing much of the urban theory that considered women and the gendered relations between women and men as essential elements for urban theory as opposed to optional add-ons, published an influential book chapter entitled "Women in the City." Urbanization, she noted, is a deeply gendered process, meaning that women and men experience urban environments differently and cities are misunderstood when **gender** relations are ignored. Over 25 years on, this chapter pluralizes Mackenzie's title to draw attention to the variety of urban contexts, as much in the global South as the global North, that demand gender analyses. Moreover, another pluralization requires consideration, one that has become more obvious in the last quarter century, and this is the proliferation of gender identifications and disidentifications, including living between genders as transgendered. There is obviously a need to revisit knowledge production about the lives of women in cities, and not least because of the combination of deepening inequalities after the 2007–2008 global fiscal crisis (Dorling, 2014) and lives on the move, particularly migrations moving into the largest cities in the global South. In particular, there is a need to better understand the lives of those women who form the largest group in cities, the working poor, who face increasing levels of **inequality**.

This chapter argues that the experience of inequality is best understood through the insecurities of daily urban life and that gendered urban insecurities can be analyzed as having three dimensions. *Economic insecurity* is the condition of not having a stable income or other resources with which to secure an adequate standard of living. This insecurity not only narrows economic opportunities and life-chances for women, but also slows down economic growth (Woetzel et al., 2015). Urban economic inequalities intersect with issues of environmental sustainability, increasing *environmental insecurity,* which is the inability to cope with environmental risks or changes as well as reduced access to resources, including food, water, and energy. It is the poorest (and among these, women and children) who, despite having the smallest **ecological footprint** (Satterthwaite, 2008), are most at risk and vulnerable to urban environmental changes through, for example, climate change, pollution, and other toxic hazards. *Social insecurity* refers to the socio-spatial limitations on a woman's place and rights within the city and their physical and emotional consequences. Inequality makes it harder for cities to balance diversity and tolerance, which results in increased socio-spatial polarization and increased policing and security measures and contributes to the production of precarious social relations and segregated environments of fear, tension, and conflict, often borne out by **violence** against women. Inequality matters, therefore, for women's urban social lives, for personal safety and security, and for its effects on an urban sociality in which diverse groups can peacefully coexist on safe streets and in city spaces.

This chapter is concerned with these issues of gendered equity, belonging, and justice that lie at the heart of addressing how women in urban places live out their lives (Peake and Rieker, 2013). Preceded by a brief discussion of the dominant analytical framings that have been employed in the study of women's urban lives, two of the longest-standing analyses of women in cities are explored: issues of gendered fear, safety, violence, and rights to the city and the intertwining strands of women's engagement in activities of waged work (production) and unwaged work (**social reproduction**). These issues are further explored through two case studies that address the similarities of the daily challenges facing women in very differently sized cities in the global South, namely Manila, Philippines, and Georgetown, Guyana. Further parallels are then drawn between these studies and similar challenges facing women across Canadian cities in relation to waged work, daycare, and gendered violence. The conclusion suggests directions for future analyses of "women in cities."

Analytical Framings of Studies of Women in Cities

While the ways in which the urban lives of women have been portrayed and interpreted have varied across regions, disciplines, and over time, three distinct themes through which women's lives have been investigated have emerged. One central mode of understanding women's place in the city is through the lens of waged and unwaged work. Questions about which sectors of the economy have been open to women and about women's concentration in low-paid, service-sector, and informal-sector jobs, in addition to their dual role in time-consuming activities of social reproduction, have been addressed in relation to **patriarchy** and capitalism and most recently neoliberalism. A second theme is the pernicious grip of notions of "good" versus "bad" women. Gender relations, sexuality, and morality are regulated (or policed) through physical and emotional violence, and have been used to dictate the places in the city that are considered appropriate for women to be (or not to be); such logics sustain urban landscapes of inclusion and exclusion for women (Tonkiss, 2005). Views on the extent to which urbanization has increased women's autonomy through economic, social, and sexual opportunities have been counteracted by those that emphasize the constraints still imposed on women's behaviour and the risks and violence they can face in both public and private urban spaces. A third and more recent theme is that of women's bodily experiences of the urban (Kwan, 2002). Although initially focusing on emotions of fear versus those of pleasure, this field is expanding to cover a range of women's affective and sensory experiences of the urban. These binary distinctions of women and the urban—production versus reproduction, good versus bad, and fearful versus emancipated—were first formulated in the late nineteenth century but have remained dominant in urban studies in the twentieth century and beyond. It was not until the latter decades of the twentieth century that the feminist questioning of these hierarchical and essentialized divides started to coalesce, with some of the earliest feminist publications written by Canadian urban scholars such as Damaris Rose, Gerda Wekerle, and the aforementioned Suzanne Mackenzie (see Mackenzie and Rose, 1983; Wekerle, 1984).

Although a substantive multidisciplinary literature on women in cities started to accumulate in academic, planning, and policy circles in the 1960s, it was to take different paths in countries of the global North and South. The initial interest in studies in the global South was in women who migrated to cities, albeit understood primarily within a masculinist analytical framework that viewed women's agency within the narrow confines of primarily functioning to serve the domestic and sexual needs of men. Feminist approaches started to emerge in the late 1970s, addressing women in cities in relation to development issues, such as the **new international division of labour** and the impact of **structural adjustment programs** (Peake, 1998) on women's engagement in production, reproduction, and community engagement (see Table 17.1). In the global North, work on women in cities was first taken up by feminist scholars in the early 1970s, as a result of **second-wave feminism**. Feminist urban geography research had a broad focus with an initial interest in theoretical critiques of Marxist conceptions of production, social reproduction, and urban form (Breughel, 1973; Burnett, 1973) but also in empirical studies, such as Jacqui Tivers's (1977) study of the child-care needs of single mothers in the United Kingdom. This literature has expanded vastly over the last five decades to address a large number of themes (see Table 17.2). Doubtless, there are other examples not yet documented or not yet circulating in the Western feminist discourses that underlie the tables accompanying this chapter. In the remainder of this section, attention is directed to two of the most commonly addressed of these realms of investigation—gendered fear, safety, violence, and rights to the city and women's participation in processes of reproduction and production.

Table 17.1 Feminist research on gender in the urban global South

Research Foci	1970s	1980s	1990s	2000s+
Women and urbanization processes, especially rural to urban migration	X	X	X	X
Women and work: formal and informal sectors; women as active in production and reproduction (in the 1970s with a focus on basic needs policy and in the 1980s with a focus on the implications of the new international division of labour); feminist critiques of women's engagement in sex work.	X	X	X	X
Land, housing, and human settlements		X	X	X
Urban poverty; women's triple role in production, reproduction, and community management				
Household structures; household survival strategies; distribution of resources within the household		X	X	X
Urban politics, especially squatter movements; urban service provision such as water, child care, and transportation		X	X	X
Domestic spaces; the home			X	X
Urban crisis survival strategies; impact of structural adjustment programs; urban citizenship			X	X
Urban violence, domestic and economic			X	X
Specific groups of workers, especially transnational migrant workers and sex workers			X	X
Urban environments: degradation of urban environments; women farmers in urban areas; urban feminist ecology			X	X
Urban planning and women			X	X
Gendered nature of urban space			X	X
Urban health; women's reproductive health; AIDS/HIV			X	X
Urban children and youth			X	X
MDGs and women in urban places				X
Sexualities and queered urban spaces				X
Women in transnational urban families				X

modified from Peake, 2009.

Gendered Fear, Safety, Violence, and Rights to the City

Concerns about identifying and addressing gendered sexual and physical violence galvanized feminist organizing around the world through the 1970s and 1980s and have been central to feminist scholarship on cities. For instance, in India "rape has been the precipitating event that has led to the autonomous women's movement . . . to engage with the law and to forge a collective visible presence in public spaces" (Dutta and Sincar, 2013: 296); the first Reclaim the Night march took place in Brussels in 1976 as part of the International Tribunal on Crimes against Women, and the model quickly diffused to cities throughout Europe, Australia, and North America (Whitzman, Andrew, and Viswarath, 2014); and gendered fear and violence has been identified as one of the main and enduring themes of Anglo-American feminist geography (Bondi and Rose, 2003). With international media attention on the rape culture of North American university campuses, on femicide in Ciudad Juarez in Northern Mexico in the 1990s through 2000s, and on

Table 17.2 Feminist research on gender in the urban global North

Research Foci	1970s	1980s	1990s	2000s+
Gender and urbanization; urban form: city centre versus suburbs; processes of production (paid work) versus reproduction (unpaid work); urban restructuring	X	X	X	X
Labour markets and work-home links, including child care; transportation and access to facilities (as well as from the late 1990s studies of parenting, childhood and children, caregiving and the ethics of care)	X	X	X	X
Urban planning, design, and architecture	X	X	X	X
Women and urban politics; urban social movements; women's political participation (with the introduction in the late 1990s of urban citizenship)	X	X	X	X
Domestic spaces; the home		X	X	X
Urban-based identities of gender, race, ethnicity, and sexuality; urban geographies of patriarchies; social constructions of difference, especially in relation to femininities and masculinities (with the introduction in the late 1990s of issues of racism, whiteness, and transnationalism)		X	X	X
Women's fear and urban places (with the introduction in the 2000s of issues of surveillance)		X	X	X
Urban poverty		X	X	X
Housing: homelessness and gentrification		X	X	X
Immigrant women, First Nations, Aboriginal women in cities; a focus on specific groups of workers such as domestic workers and sex workers			X	X
Lesbian and gay urban geographies; LGBTQ spaces; queer geographies, and later, transgendered geographies			X	X
Embodied urban geographies			X	X
Urban public space (though earlier studies on, for example, women's access to parks and women-only buildings date back to the late 1970s)			X	X
Women's spaces of pleasure/leisure, the female flâneuse			X	X
Urban emotional geographies				X

modified from Peake, 2009.

the gang rape and murder of a young university student, Jyoti Singh Pandey, in Delhi in December 2012, to name just a few examples, gendered violence continues to command attention, though the ways in which it does so have deepened and become more complex.

Gendered violence, its occurrence and reactions to it, is intertwined with the distinction between **public and private**. Identifying and disrupting the work of this binary, in which the feminine is associated with the private sphere and the male citizen with the public, has been central to feminism; as Carole Pateman (1989: 118) states, "it is, ultimately, what the feminist movement is about." The home as a private place has obscured and depoliticized the prevalence of domestic violence and thereby minimized public response to it. Domestic violence, for example, has been routinely ignored in national surveys of gendered fear of crime and experiences of violence. Carolyn Whitzman (2007) identifies only two exceptions: national surveys in Canada and Australia in 1993 and 1996, respectively. In line with feminists' long history of disrupting binaries and reconfiguring conventions of scale, in a classic refusal of the private/public binary Rachel Pain (2014) reconceives domestic violence as a form of everyday **terrorism**—that is, as a form of violence that controls by instilling fear. By reconfiguring domestic violence as terrorism, she aims to unsettle the distinction between international and interpersonal violence and thereby draw public attention to it.

If gendered violence in the home is often hidden or rationalized as a private matter, women are sometimes held responsible for their victimization in public spaces: for being in the wrong place, at the wrong time, wearing the wrong clothing. When a young woman was mobbed and sexually assaulted in 2009 in Mangalore in South India, for example, the founder of a Hindu right-wing group stated, "Whoever has done this has done a good job. Girls going to pubs [sic] is not acceptable" (Kapur, 2012: 2). Police in Ciudad Juarez, Mexico, justified their failure to fully investigate and solve the murders of hundreds of women by casting doubt on these women's morality. Family and friends are often asked of the murdered women (many of whom worked in manufacturing operations called *maquiladora*), "Are you sure that she didn't lead a double life?"—that is, worked as a sex worker (Wright, 2004: 377). As Melissa Wright (2013: 44) notes, "public man as citizen emerged hand in hand with public woman as whore," and the latter designation is evidently reason enough for being killed with impunity.

Given women's tenuous legitimacy in urbanized public spaces, many develop micro-strategies to negotiate their presence (Figure 17.1). These strategies can range from taking up as little space as possible (crossing legs and arms as opposed to "man-spreading," a term coined to describe the expansive bodily postures of men on public transit), to wearing symbolic markers that signify respectability, to navigating the city to avoid unsafe places. Writing about women's negotiation of public space in Mumbai, Shilpa Phadke and her colleagues (2011: 34) argue that markers of respectability "create a bubble of private domestic space around women . . . to ensure that their bodies are 'read right' as being private bodies."

Though a focus of feminist concern for many years, the terms in which gender violence and safety have been—and are—discussed have changed. It is now widely recognized that the category of gender is too blunt to address the issue adequately—it is at once too expansive and too narrow, and a vocabulary of fear and safety is too limiting. We address each of these concerns in turn.

In a thorough review of the international literature on perceptions of crime and safety, Carolyn Whitzman (2007: 2716) concludes that "it is not gender per se, but economic and social powerlessness and exclusion that is the defining factor behind fear of crime." As early as 1990, bell hooks rewrote the geographies of safety and threat from the perspective of an African-American female growing up in the southern United States. She described "home-space" as safe and woman-identified and wrote of her feelings of terror when she passed through low-income white neighbourhoods, not because she was female but because she was an African-American female. Twenty-five years on, Claire Hancock and Anissa Ouamrane (2015) tell a similar story about young immigrant women living in the low-income outer neighbourhoods of Paris. Their geography of security and insecurity is at odds with official maps of safety and risk. Whereas they perceive the neighbourhoods officially defined as security risks to be places of safety, the central and wealthier areas of Paris are for them places of discomfort and fear.

It is not only that geographies of fear vary for different groups of women; it is also the case that particular categories of women are much more vulnerable to **xenophobia** and violence: race, ethnicity, religion, age, disability, sexuality, gender identification, and class are factors that figure into relative risk. Studies of Muslim women wearing *niqabs* or burkas in public spaces in European cities, for instance, reveal the gendered dimensions of **Islamophobia**. Muslim women are vulnerable to a variety of everyday hate crimes, including threats, intimidation, and physical violence. Reporting on research in Malmö, Sweden, Carina Listerborn

Figure 17.1 Woman gives side glance to man-spreading

https://www.flickr.com

(2015) writes that this violence against Muslim women is often not visible because it seems so ordinary, taking the form of pushing, spitting, attempts to pull off the veil, or verbal abuse. The typical offenders are older non-Muslim Swedish women. These experiences of everyday violence are, Listerborn argues, part of the geography of the veil and must be understood within the context of systemic, geopolitical violence. Writing about gender-based violence worldwide, Caroline Moser and Cathy McIlwaine (2014) draw special attention to conditions of poverty and poor urban infrastructure that structure conditions of violence in many cities of the global South. Poor-quality and remote sanitation facilities, for instance, place women in chronic situations of vulnerability. Hence, when the focus on the women's safety movement shifts from the global North to the global South, the expansion and reorientation of priorities—from spatial planning, to poverty reduction, to the provision of urban infrastructure such as housing, water, and sewage treatment—become more apparent (Whitzman, Andrew, and Viswanath, 2014). In very different ways, transgendered persons experience on a day-to-day basis the violent policing of gender in such seemingly non-gendered public spaces as airports, public transit, elevators, and university classrooms (Doan, 2010). Rigidly gendered segregated places like public toilets or bathrooms can be dangerous and violent places for those who do not conform to the gender binary (Browne, 2004). Gendered violence, in other words, is experienced differently by different groups of women (and in some cases men) in different parts of the world, embedded within broad sets of differently configured social, economic, and political relations. It demands globally interlinked but context-specific responses, what Cindi Katz (2001) has conceptualized as **counter-topographies**.

The focus on action and outcomes has led feminists to rethink strategy and in particular to exercise caution around narratives of victimization and fear. Narratives of victimization have been taken up within discourses of paternal protectionism that legitimate and enhance the privatization of urban public space and the neoliberal state's carceral capacities, which systematically target religious minorities, communities of colour, the homeless, and other impoverished and marginalized communities. As Dutta and Sircar (2013: 297) write with respect to the rape and murder of Jyoti Singh Pandey in Delhi (informally known as India's "rape capital") in December 2012, feminist voices were "co-opted by those with a conservative political vision to achieve contrarian ends." Feminist calls to end state apathy towards violence against women, to make public transit safe, to amend rape laws, to make police more responsible, and to stop victim blaming were intermingled with and overwhelmed by demands from the Hindu right wing for retributive justice: the death penalty, chemical castration, and death by stoning of the rapists. In a close reading of the English-language media coverage of this event, Roychowdhury (2013) notes that the murdered woman was celebrated as a modern rights-bearing, professionally successful, and consumer-oriented subject, while the male rapists were portrayed as patriarchal, newly urbanizing migrants to Delhi. That is, representations of violence were played out through portrayals of the old and new India, of those who belong and do not belong in urban space. Roychowdhury (2013: 288–289) cautions that these representations help to "secure Indian women's abilities to be consumers and ready participants in India's neoliberal economy" and do so through "increasingly punitive sanctions against the working class and the disenfranchised"; as such, they pose "real dangers for justice."

Recognizing the limitations of narratives of violence, fear, and safety, Phadke and colleagues (2011) argue for expanding framings of gender and access to public space. In particular, they refuse the masculinization of pleasure and risk-taking in public urban spaces and claim it for women as well. They advocate shifting from a discourse of protectionism to one of rights to access and pleasure. These rights, they argue, subsume rights against violence (through the provision of urban infrastructure with adequate lighting, public transit, and public toilets), along with rights to non-sexist redress if violence should occur. As Elizabeth Wilson (1992) earlier noted, imagining women as risk-takers and agents of pleasure in urban public spaces is profoundly disruptive of the binaries of masculinity and femininity, and of good women and bad women, that structure **heteronormative** conventions of gender and urban space.

Social Reproduction and Production

Theories of change in urban areas often serve to occlude as much as illuminate, and one major realm of investigation that has been under-investigated and misunderstood, even in critical analyses of the urban, has been that of reproduction. Reproduction has biological and social dimensions, the latter incorporating three aspects: the reproduction of labour power, which includes biological reproduction as well as the care, maintenance, and socialization of children and adults throughout their lives and over generations; human reproduction, centring on marriage and kinship and thus issues of the social organization of fertility and sexuality; and the ways in which relations of production are recreated and perpetuated (Kofman, 2017). Although many of the practices on which social reproduction depends are heavily gendered as women's unpaid work (e.g., housework and community mobilization), researchers within urban studies and its cognate disciplines have failed to ask why this is, turning their attention instead to the study of the various modes of provision (such as public versus private) of goods and services. In particular, the provision of goods and services by the state (what Marxist theorist Manuel Castells [1977] referred to as "collective consumption") has been a sustained site of focus. Feminist critiques have indicated that this focus on collective consumption ignores the underpinning of capitalist systems by women's unpaid work in the home and in communities that enable the labour force to be daily and generationally reproduced at virtually no cost to capital.

Urban theory has also ignored women's engagement in urban-based protests, whether these are struggles over social **reproduction of the labour force**, such as rent strikes (labour needs housing to live in), or those of production, such as industrial strikes (Naples, 1992). And yet women have a long history of being centrally engaged in both these types of struggles. Moreover, women in the global South and poor women in the global North have a long history of being present in the labour force, most commonly in the service sector but also in the informal sector in home-based production and sales, including sex work. Middle-class women's increasing entry into the labour force in the latter half of the twentieth century led to women's role in waged work presenting an even more variegated picture than that of men's, although it is the case globally that women's paid work takes place in both formal and informal labour markets; women are invariably paid less than men in commensurate positions; and women are more likely to be in part-time, non-unionized and precarious jobs.

The difficulties women face because of their responsibilities in both social reproduction and production, are commonly exacerbated by the form of the urban environment. The assumption underlying the design of urban settlements has been (and in many places still is) that women are engaged full-time in maintaining and organizing domestic life, while male workers have minimal or no domestic responsibilities and so can work away from the home. Moreover, urban planners have not only ignored women's responsibilities in waged work, but also the fact that women have less access to cars than men. Women's management of their dual roles often requires movement, not only into the city but laterally, across the city, at odd and irregular hours punctuated by trips to shops, daycare, and schools (Figure 17.2). But outside of

Figure 17.2 "Housewives Please Finish Travelling by 4 o'clock": A WWII poster
The National Archives/SSPL/Getty Images.

city cores, services such as affordable and accessible daycare and public transit routes can be limited, the latter inconveniently arranged for women's journeys and often geared only to business hours. Even a cursory knowledge of women's daily activities renders obsolete urban design and urban theory that attempt to understand the urban purely in terms of men's role in waged work.

As the economic recessions of the late 1970s metamorphosed into the social and economic restructuring of the 1980s, new ways of organizing production and social reproduction started to emerge. Women's employment (mostly cheap and flexible) in cities of the global North and South has increased, while male participation in the labour force has declined. The growth of temporary work and "zero hour" contracts, often below the minimum wage, have proliferated as well-paid unionized jobs ("jobs for life") have gone into decline. In response to cuts in public services and to middle-class women's increased participation in the paid labour force, social reproduction is being restructured through the transnational migration of women to be employed as domestic workers in the cities of the global North. Women came to Canada, for example, from the Caribbean to work in the Foreign Domestic Movement Program. In 1992, this was replaced by the Live-in Caregiver Program, which came to fill the growing demand for paid reproductive labour largely by recruiting women from the Philippines. Analyses of the international nature of paid reproductive labour were undertaken (Pratt and Yeoh, 2003), and most recently, in the aftermath of the global financial crisis of 2007–2008, the marketization, privatization, and informalization of social reproduction have led to a renewed interest in the concept and the ways in which households are even more implicated in one another's social reproduction (Kofman, 2017). Indeed, public-sector cuts in social reproduction (such as closure of daycare programs), though more acute in the global South, are affecting large numbers in the global North who also face inadequate access to basic necessities such as food, heating, and shelter. Moreover, the divides between women in cities in the global North, between those with professional, well-paid unionized jobs with pensions and those in low paid, part-time, and (in some cases) home-based jobs with no recourse to benefits, have substantially increased, while in cities in the global South there has been an increase in work in informal sectors, since the formal sectors in many countries have collapsed under the burdens of structural adjustment policies and neoliberal governance. In the twenty-first century, these structural changes have resulted in the feminization of poverty (Chant and McIllwaine, 2016) and increasing gendered urban insecurities in cities everywhere. In the following two case studies, this chapter further explores the gendered themes of social reproduction, production, violence, and everyday lives in Manila and in Georgetown, Guyana.

Disposable Lives, Transnational Families, and Labour Migration in Bagong Barrio, Metro Manila

How is it possible to come to terms with urban lives in Metro Manila, a massive conglomeration of over 12 million inhabitants, the eleventh most populous urban area in the world, home to 4 cities and 13 municipalities, 10 of the most populous in the Philippines? Impossible. And so we enter Bagong Barrio, a neighbourhood of just over 100,000 people located in Caloocan City, a city of roughly 1.5 million located in the northwest of Metro Manila (Figure 17.3).

Various entry points to Bagong Barrio indicate solidity and permanence. The community is reached from EDSA, a major circumferential road in Manila, via wide paved roads flanked by high solid walls and factories. And yet the solidity of the community crumbles on closer examination. In the 1970s, many of the residents were employed in nearby unionized factories, but by the mid-1980s a good number of factories had closed, many unionized workers had lost their permanent jobs, and short-term contractual hiring through agencies had become the norm. Neferti Tadiar (2013: 37) identifies the precarity lived in a place like Bagong Barrio as "lifetimes of disposability." The slow temporality of neoliberalism in the Philippines, she argues, has created the conditions for permanent stagnation, transience, and dislocation, perceived not as "an event" or "immanent fate" but "simply a[n enduring] mode of life."

It is within this context that labour migration—of both men and women—began on a massive scale, creating new relations of social reproduction, new transnational forms of **family** life, and new vulnerabilities to gendered violence. Between 1979 and 2009, more than 30 million Filipinos left the Philippines as labour

Figure 17.3 Bagong Barrio West in Manila, the Philippines

migrants, many of them women working as domestic workers (UN Women, 2013). Migrante International, a migrant organization based in Manila, estimates that over half the residents in Bagong Barrio are, or have worked as, overseas Filipino workers (OFWs) and that the majority of households are highly reliant on remittances for survival. Most of the women work as domestic servants for wealthy families in largely urban places in Saudi Arabia, Malaysia, Hong Kong, Bahrain, Libya, even Canada—the latter one of the most coveted of destinations because of relatively better labour conditions and the possibility for permanent migration. According to the United Nation's International Labour Organization, domestic workers are some of the most likely workers to face abuse and exploitation, and the International Trade Union Confederation estimates that 2.4 million domestic workers face conditions of slavery in the Gulf Cooperation Council countries alone (ITUC, 2014).

Vulnerabilities and violence are lived similarly and differently by different members of the family and across different families, and there is an ethic and politics to insisting on this singularity of experience within what Tadiar (2013: 36) terms "a stagnant surplus population." What follows are the experiences of just one youth in Bagong Barrio, interviewed in August 2014, whose testimony underlines the fact that labour migration is lived in enduring ways by entire families (Pratt, Johnston, and Banta, forthcoming). In the case of Arvin, his mother left him as an infant to work in Saudi Arabia, returning briefly every five years until she was no longer able to do so, having fallen into illegal status as an OFW. Arvin was passed from family member to family member and eventually made his way to Manila to live with relatives there: "I got passed around. When my uncle gets tired of me, I would go to my aunt's. It was like that . . . I had to take care of myself. I didn't experience play when I was young. I didn't experience care." Under the legal age to migrate as an OFW, Arvin nonetheless found overseas work in Saudi Arabia at age 22. Through a series of extraordinary coincidences, he reunited with his mother there, and they lived together as mother and son for two years. And "that was the first and the longest time [that I lived with my mother]. . . . That's where I felt: oh, this is how things are when you have parents who will take care of you. . . . So I was very happy." He described the process of bonding with his mother as an adult: "We will pretend that I'm still young, that I'm still a baby. I will point at food to eat from Jollibee [a fast-food restaurant]. . . . It's like, I want to play myself: I'm still a baby. It was our system. . . . 'Ok, how would you feed me, Ma, now that we are here at Jollibee?' I see that in Jollibee: children, they are fed by their parents. . . . Instead of taking care of someone else. . . . She would show me!" Arvin has committed to caring for his mother in return, and his ambition is to save enough money to build a concrete house for himself and his mother in the province from which they originally came, far from Manila. He plans to build a house strong enough to withstand the onslaught of earthquakes and typhoons commonly experienced in the Philippines. His mother, now 64, is planning to return to the Philippines in the near future. Arvin, in the meantime, left for Jeddah soon after our interview, replacing his mother as the next generation of OFWs, with dreams of sharing the house with his mother sometime in the future.

Arvin's is just one story of a pervasive culture of family separation in the Philippines: it is estimated that at least 9 million children in the Philippines (or 27 per cent of the overall youth population) are growing up with at least one parent working abroad (Parreñas, 2010: 1827). The prevalence of transnational social reproduction is one of the themes that are further taken up in the case study of Georgetown.

"Yuh got to catch yuh hand": Gendered Poverty and Inequality in Georgetown, Guyana

Unlike the sprawling megacity of Manila, Georgetown is the very small capital city of Guyana, with a population of only 235,000 (Census 2012). It is a classic primate city with 31 per cent of the country's population (of 747,884). (Census, 2012). And yet size is no determinant of the complexity of daily life. Guyanese women, for example, are intimately connected with relatives not only within walking distance but also as far afield as the North American cities of New York, Miami, and Toronto, the towns and cities of Venezuela, and the towns and villages of the islands of the Caribbean. The interlocking of these differently scaled interactions and flows of ideas, people, and material resources defines the parameters of the insecurities of daily life for the majority of the city's residents, the working poor. Women's lives in this small city exhibit the same kinds of gendered urban insecurities encountered in much larger cities in the global South, such as Manila. Indeed, domestic violence, labour migration, the creation of new relations of social reproduction and new transnational forms of family life also characterize women's lives in Georgetown, albeit arising from its own historical geographies and experienced differently by the city's classed and racialized (African and Indian) inhabitants.

Although located in South America, Guyana is politically, economically, and socially a part of the Anglo-speaking Caribbean, and Georgetown's location on the Atlantic coast, hemmed in alongside the eastern bank of the Demerara River, speaks to its origins in the colonial geography of the slave trade (Figure 17.4). Based on a flat coastal plain, the city is situated one metre (three feet) below high-tide level, the seawaters held back by a retaining wall (the "seawall") and a system of sluices (known as *kokers*). Emigration combined with a lack of land for building, the high cost of housing, and an urban design that has favoured low-density, low-rise developments, has resulted in a decline in the city's population. Nevertheless, the population of Greater Georgetown is increasing, with growth taking place to the east and south of the city in the form of new housing schemes (Diamond), in squatter and regularized settlements

Figure 17.4 Map of Greater Georgetown, Guyana

(Sophia), as well as in exclusive gated communities (Happy Acres). With the city being the location of CARICOM (Caribbean Community) headquarters—as well as that of foreign embassies, in-country offices of the United Nations, and international development agencies—the cost of accommodation has been pushed up, making

Georgetown an increasingly unaffordable and expensive place to live, with high levels of housing vulnerability (Peake, 2000) (Figure 17.5).

For many of its inhabitants, Georgetown is a city where life is lived on the brink. While up to the 1970s Georgetown was called the "Garden City of the Caribbean," known for its distinctive wooden architecture and its well-laid-out streets, canals, parks, and gardens, its current moniker is that of "garbage city." Georgetown suffered from a lack of capital (Guyana was classified as a **Heavily Indebted Poor Country** [HIPC] until 2012) and government corruption, and resources were withheld from the city because the majority of the previous government's supporters lived outside it. The city fell into disrepair, its decay all too evident; garbage was uncollected, infrastructure was not maintained, and public services, including education and health care, functioned only at a basic level. Climate change and the city's unmaintained drainage infrastructure added to the problem of the increasingly frequent flooding of the city during rainy seasons. The newly minted (in 2015) government is now addressing these issues with much goodwill, albeit with limited human and capital resources.

Daily life for many of the city's inhabitants is a case of living hand to mouth. For women, it is a case of having to "catch yuh hand" not only in terms of finding paid work, but also in terms of securing the time needed for social reproduction, all in a context of normalized relations of domestic violence (Peake, 2000). Days are long, money is tight, and the stress and emotional toll of trying to secure enough to live on is registered in the unenviable record of the country having, according to the World Health Organization (WHO, 2012), the highest rate of suicide in the world.

With an (official) economy based on gold and bauxite mining, agriculture (primarily rice and sugar production), and logging, no sector provides many opportunities for women to be legally and securely employed. The gold, cocaine, and people-smuggling (most commonly called "backtrack") trades that flourish illegally are also predominantly male domains. Stable employment is scarce and symptomatic of a neoliberal economy. The largest private employer in Guyana is Qualfon, a multinational call centre company with offices on the outskirts of Georgetown (Mohabir and Peake, forthcoming) where women are engaged in low-paid, dead-end work. As a mother of a former Qualfon employee described her daughter's call-center job, "It's a chance to get dressed up and get out of the house . . . but you can't move up the ladder" (personal communication). The majority of African women in the city are employed in the precarious informal sector as security guards, domestic workers, sex workers, and street vendors or in low-level civil-service jobs. Indian women are more likely to be employed in the private sector in family businesses, although they also populate the informal sector. Such contexts render recourse to protection from gendered violence in the workplace low.

Racialized conflict and economic uncertainty have meant that since the early 1970s the most common way to earn a living has been to engage in labour migration. Large numbers of women (more so than men) commonly travel illegally ("backtrack") to North America to secure employment, many at first coming to Canada to work in the Foreign Domestic Movement Program. In Georgetown, every family has a story of migration involving its closest members. It is estimated that more Guyanese now live outside the country than within it and the money these women regularly send to support their children until they can send for them to join them has earned Guyana the

Figure 17.5 Georgetown city centre on the banks of the Demerara River

Creative Commons / amanderson2 / Flickr.

pejorative reputation of being a remittance economy. Not only has labour migration led to the transnational family becoming the norm, but social reproduction has also been transformed; it now takes place across extended families in different homes and across different countries. Children commonly grow up with their grandparents, their parents' siblings, their own older siblings, or relatives further removed. Increasing levels of economic insecurity, expressed through migration, have thus spawned increases in social insecurity, in the breaking down of extended family structures, and in the domestic violence that permeates everyday life. This deepening of inequality both within and across social classes is also evident in the city in the growing spatialized divides between the very wealthy, who are increasingly moving into gated communities, and those living in poverty, the working poor, with middle-class residents interspersed among them. These issues of gendered poverty, inequality, and violence are present in the daily lives of women in Canadian cities as well and are explored in the text box below.

Gendered Poverty, Inequality, and Violence in the Canadian Urban Context

Even though many middle-class women in Canada employ women from the global South who have been forced to migrate to care for their children, many of the issues of poverty and inequality faced by women in Manila and Georgetown also face women in Canadian cities, although the extent and degree to which they do varies across and within Canadian urban areas. The previous Conservative government, for example, increased levels of inequality for women in Canada by stopping government funding for a number of organizations that represented women, by voting against pay equity, by voting against a national action plan to end violence against women, and by actively attacking or avoiding efforts to improve the status of women such that Canada dropped in the UN's Gender Inequality Index from fourteenth place in 2006 to twenty-third in 2014 (Hausmann, Tyson, and Zahidi, 2006; UNDP Gender Inequality Index, 2014). The current (Liberal) government, however, is developing initiatives that are starting to address these issues.

While Canadian women are well represented in the labour force, making up over 47 per cent of employees (Statistics Canada, 2013), their employment across Canadian cities presents a varied picture patterned by classed and racialized divides, although arguably it is mainly characterized by the gender wage gap between women and men, which is the most significant factor for women falling into poverty (McInturff with Lockheart, 2015). The wage gap is related to the prevalence of part-time work for women and full-time work for men and to labour-market segmentation, which tends to concentrate women in lower-wage occupations. In cities like Gatineau and Victoria, however, there is a narrow wage gap between women and men, and women in these cities have higher chances of being promoted into senior management positions. These characteristics are due to the presence of large public-sector workplaces that have unionized and robust wage-setting processes and equity regulations to ensure that the employer must keep track of whether or not there are discriminatory gaps in pay and promotion. The public sector also has a high level of transparency when it comes to rates of pay and promotion. The gender gap in income is largest in cities like Calgary, Edmonton, and Kitchener-Cambridge-Waterloo where predominantly "male" industries are located. The construction industry and the mining, oil, and gas sectors are two of the biggest employers in Alberta, and in the province, men hold 88 per cent of the jobs in construction and 76 per cent of the jobs in the extractive sector, with few jobs available to women either within these sectors or outside of them. While employment levels for women in Kitchener-Cambridge-Waterloo are significantly higher than they are on average, the gap in women's and men's representation in the workforce is high, with 72 per cent of men working compared to 64 per cent of women. These figures can be explained by the concentration of jobs in Kitchener-Cambridge-Waterloo in the information technology sector, another male field of employment. This city also has one of the larger wage gaps in urban Canada, with women making only 66 per cent of what men make overall (McInturff with Lockheart, 2015).

Across Canadian cities, white women working full-time continue to bring home 20 per cent less than men, and immigrant women and racialized women, especially Indigenous women, take home even less. Yet stagnant male wages mean that households very much depend on women staying in paid work. But despite women being increasingly engaged in paid work, there has not been a commensurate decline in the number of hours they spend on domestic work: "Women continue to spend nearly twice as many hours as men do looking after the household and taking care of children and other family members" (McInturff with Lockheart, 2015: 7). For many women, whether employed or not, access to publicly provided child and elder care is also severely restricted.

Child care in Canadian cities is provincially regulated, most commonly a licensed and non-licensed patchwork affair, provided in both public and private settings. The previous Conservative federal government spent less than 0.34 per cent of gross domestic product (GDP) on Early Childhood Education and Care (ECEC), a figure, moreover, that is skewed by the fact that 60 per cent of this spending was in Quebec, the only province where a comprehensive child-care policy is in place. However, such a policy is needed across all Canadian cities, as Canadian parents are likely to be working parents: "More than three-quarters of mothers with children under the age of six and even more fathers with children are part of the Canadian labour force" (Macdonald and Friendly, 2014: 6). Daycare fees can play a major role, particularly for women, in the decision to participate in the labour force. Indeed, families rely on daycare in order to be able to work. Although all provinces provide (varying degrees of) subsidies for daycare, the shortage of available spaces means most families have no alternative but to resort to expensive unsubsidized places.

A 2014 study of unsubsidized child-care fees across Canadian cities found that infant child care (compared with toddler and preschooler care) is most expensive and hardest to find in Toronto (costing $1676 a month) and cheapest and easiest to find in the Quebec cities of Gatineau, Laval, Montreal, Longueuil, and Quebec City, where Quebec's generous parental leave, paternity leave, and $7.30-a-day child-care policies keep prices affordable for all families (at $160 a month) (Macdonald and Friendly, 2014). The study also investigated the relationship between child-care fees and women's income, finding that "Brampton is the least affordable city in Canada for child care. In Brampton, fees are worth 36 per cent of a woman's income, the equivalent of four months' worth of work. This is largely due to lower overall incomes for women in Brampton, with Surrey, Windsor, London and Toronto not far behind with child-care fees that absorb 34 per cent to 35 per cent of a woman's income" (Macdonald and Friendly, 2014: 6).

When governments provide access to better-paid parental leave, leave for both parents, and affordable child care, women and their families are better able to access paid work and ensure their financial stability. This is why women in most Quebec cities fare better in achieving a work–life balance. Affordable child care, moreover, is an important issue not only for parents but also for the Canadian economy. It plays a significant role in labour-force participation, particularly for women, a factor that the current Liberal government has recognized in its plans to combine the Universal Child Care Benefit with child tax credits and other forms of social assistance to increase spending on ECEC to 1 per cent of GDP, creating a monthly cheque for all families with a household income below $150,000 and children under 18.

Neither are Canadian cities (nor university campuses) immune to any of the issues that we have

continued

discussed in relation to gendered violence. They have been the locus of both gender violence and innovative strategies to counter it. As noted, in 1993 Canada was one of the only countries to collect, in a national survey, data on all forms of gender violence, including domestic violence. The results were alarming: over half of women respondents had experienced at least one incident of physical or sexual assault since the age of 16, and half of them had been victimized by men they knew (Whitzman, 2007).

So too, Canadian women are no less likely than other women to be blamed for their victimization in public spaces, and as is the case elsewhere, not all women are equally vulnerable to violence. Indigenous women in Canada are particularly vulnerable to extreme violence: between 1980 to 2012 their homicide victimization rate was 4.5 per cent higher than that of all other women in Canada (www.amnesty.ca/blog/missing-and-murdered-indigenous-women-and-girls-understanding-the-numbers), a situation that needs to be understood within the wider context of the enduring systemic violence of colonialism. In March 2015, the UN expert from the Committee on the Elimination of Discrimination against Women concluded that Canada was responsible for a "grave violation" of human rights because of its "protracted failure" to take sufficient action to stop violence against Indigenous women and girls. In the case of many Indigenous women who have gone missing from Canadian cities, this protracted failure reflects the complex layering of imaginative urban and other geographies (Pratt, 2005): that Indigenous women naturally belong on the reserve and thus have left the city on their own volition; that Indigenous women who are sex workers are inherently mobile; and that indigenous women living in central cities are recipients of violence for the very fact of living in inner cities. In Sherene Razack's (2000: 126) phrasing, they are "of the place where murders happen; [their murderers likely are] not."

Canadian women and transgendered persons have organized extensively in reaction to the ubiquity of gender violence and, in some cases, have been remarkably effective in spawning international movements. With respect to violence against transgendered persons, in June 2014 the Vancouver School Board passed a motion pledging the provision of gender-neutral washrooms in all school buildings. Nationally, Bill C-279, which sought to add gender identity to the Canadian Human Rights Act, passed a second reading in the House of Commons in 2012 but was stalled in the Senate and then died when a federal election was called in 2015. It was dubbed "the bathroom bill," and prolonged discussion in the Senate turned in large part around concerns about transgendered persons' use of public washrooms. New legislation was introduced in May 2016, Bill C-16, to include gender identity or expression as prohibited grounds for discrimination within the Canadian Human Rights Act and the Criminal Code. As of 2016 all but four provincial governments have amended their human rights legislation to provide specific protections for transgendered persons.

On the issue of gender violence more generally, in 1989 a non-profit organization affiliated with government, the Metropolitan Toronto Action Committee on Public Violence against Women (METRAC), developed the *Women's Safety Audit Guide* to facilitate urban neighbourhood women's groups' ability to advocate for change. The tool quickly disseminated through Canada and Europe, and then, after the Safer Cities Programme was developed by the United Nations Human Settlements Programme (UN-Habitat) in 1996, through Latin America, Africa, and Australasia (see Whitzman, Andrew, and Viswanath [2014] for an extensive case study of the Toronto Safe City Committee experience from 1989 to 1999). More recently, in 2015, the Twitter hashtag #beenrapedneverreported was started by two Canadian journalists and grew extensively with the stories of millions of survivors worldwide.

Adopting a different form of organizing in response to a police officer's remark in 2011 about combating rape culture through respectable dress, the first SlutWalk was organized in Toronto and attracted several thousand participants to protest against rape culture. In 2011, there were companion SlutWalks in 40 countries across the global North and South (Figure 17.6). The SlutWalk has not been without controversy. *Black Women's Blueprint*, for example, has argued that the SlutWalk is a sign of and reinforces white privilege, given the historic and contemporary representations of black women as sluts. Other feminists assert that it simply circulates rather than problematizes patriarchal fantasies (Miriam, 2012). Still others argue that the SlutWalk has been an effective instance of transnational feminist organizing and that the meaning of the walk has been reconfigured in significant ways in different cities. In Delhi, for instance, it was linked to access to jobs (at night in call centres), and in Latin American cities the Catholic Church has been a central target of criticism, with some participants dressed in costumes of the Catholic Church (e.g., as nuns) (Carr, 2013).

On the issue of endemic violence against Indigenous women, in 2015 the Liberal government responded positively to calls for a national public inquiry into missing and murdered women in Canada from a wide range of organizations, including the Native Women's Association of Canada, Amnesty International, and the United Nations.

Figure 17.6 Edmonton SlutWalk, 2011
Creative Commons / Hugh Lee / Flickr.

Conclusion

The material in this chapter has shown that gendered insecurities both shape and are shaped by processes of urbanization. Regardless of city size or geographic location, women's and men's experiences of the urban are played out through different gendered norms and expectations: in the ways in which they use and give meaning to urban space; in the degree of restrictions placed on their mobility and in levels of spatial exclusion; in their access to basic needs and employment; in their risk of encountering violence; in their exposure to environmental hazards; in ownership of land and property; and in civic disenfranchisement. In short, gendered insecurities influence women's place in the city in terms of their life-chances and well-being, as well as determining the sustainability of city life. Women's agency, however, ensures that the constraints placed on their lives are always being challenged, whether at the level of individual women refusing restrictions on their mobility and thus their ability (the most high-profile case being that of the Pakistani schoolgirl Malala Yousafzai, who survived an assassination attempt after refusing to bow to the Taliban edict that girls in her home town of Mingora in the Swat valley should not go to school) to mass orchestrated and globally implemented endeavours such as the Canadian-initiated SlutWalk. Numerous individual and collective efforts to increase women's autonomy and to seek justice for gendered inequalities take place across towns and cities globally. What we need to ask ourselves is whether the analytical framings used to examine gendered urban subjectivities in the early twenty-first century can still presuppose that the basic binary assumptions underlying knowledge about women and the urban—of public versus private spheres, of mobility versus seclusion, and of good versus bad women—still apply. These inherited framings, derived from interpretations of late nineteenth- and early twentieth-century life in largely European and North American cities, have never been more fluid than at this historical juncture and are ripe for unsettling.

Key Points

- "Women in cities" matters because urbanization is a deeply gendered process; women's and men's experiences of the urban differ.
- Inequality is experienced by women through gendered urban insecurities, which can be understood as having economic, environmental, and social dimensions.
- Three distinct themes through which women's urban lives have been investigated are production and social reproduction; "good" versus "bad" women; and women's bodily experiences of the urban as either fearful or emancipated.
- Concerns to identify and address gendered sexual and physical violence have been central to feminist scholarship on cities.
- The concepts of production and social reproduction offer a lens through which to examine the ways in which gendered, but also sexualized, racialized, and classed, inequalities are lived out.
- The insecurity of women's lives in Manila and Georgetown are interwoven with gendered violence, labour migration, the creation of new relations of social reproduction, and new transnational forms of family life.
- Levels of inequality in women's lives in Canadian cities have been increasing, as measured by access to jobs and pay equity, access to child care and exposure to violence with impunity.

Activities and Questions for Review

1. Why have binary forms of knowledge (i.e., public/private; production/social reproduction; good/bad; and fearful/emancipated) been so common in understanding women's relationship to the urban? Discuss the extent to which these binaries have relevance to your own life and explore reasons why and how women have actively worked to disrupt these boundaries.
2. Work in groups to compare and contrast the ways that the lives of women and men living in Canadian cities are intimately related with women's lives in urban and non-urban places elsewhere.
3. Is it a surprise that it is older, non-Muslim Swedish women who so often exhibit everyday forms of violence towards women wearing niqabs or burkas in public spaces in Sweden? As a class, discuss how we reconcile gender equity with differences among women.
4. Keep a diary for one week of all of the unpaid work performed by different members of your household. Record who does what, where, when, and for how long. In groups, discuss the extent to which you can recognize gender patterns and how you would put a value on this unpaid work.

Acknowledgments

The authors would like to acknowledge the participants in their research and the support of SSHRC funding (Insight grant[s], "Understanding Geographies of Gendered Insecurities in Georgetown, Guyana" and "Migrant Geographies of Politics, Identity and Belonging").

References

Bondi, L., and Rose, D. (2003). Constructing gender, constructing the urban: A review of Anglo-American feminist urban geography. *Gender, Place and Culture*, 10(3): 229–245.

Breughel, I. (1973). Cities, women and social class: A comment. *Antipode*, 5(3): 62–63.

Browne, K. (2004). Genderism and the bathroom problem: (Re)materializing sexed sites and (re)creating sexed bodies. *Gender Place and Culture*, 11(3): 331–346.

Burnett, P. (1973). Social change, the status of women and models of city form and development. *Antipode*, 5(3): 57–62.

Carr, J.L. (2013). The SlutWalk movement: A study in transnational feminist activism. *Journal of Feminist Scholarship*, no. 4: 24–38.

Castells, M. (1977). *The urban question: A Marxist approach.* London: Edward Arnold.

Chant, S. (2013). Cities through a "gender lens": a golden "urban age" for women in the global South? *Environment and Urbanisation*, 25(1): 9–29.

Chant, S., and McIlwaine, C. (2016). *Cities, slums and gender in the global South: Towards a feminised urban future.* London: Routledge.

Doan, P. (2010). The tyranny of gendered spaces—reflections from beyond the gender hierarchy. *Gender Place and Culture*, 17(5): 635–654.

Dorling, D. (2014). *Inequality and the 1%*. London: Verso Books.

Dutta, D., and Sircar, O. (2013). India's winter of discontent: Some feminist dilemmas in the wake of a rape. *Feminist Studies*, 39(1): 293–306.

Hancock, C., and Ouamrane, A. (2015). *Badlands of the "City of Light" Gendered violence and spatial constraints on young banlieusard-e-s in Paris*. Unpublished paper presented at Gendered Rights to the City: Intersections of Identity and Power, Feminist Geography Conference, University of Wisconsin, Milwaukee, April 19–20.

Hausmann, R., Tyson, L., and Zahidi, S. (2006). *The global gender gap report 2006*. Geneva: World Economic Forum.

ITCU (2014). *Facilitating exploitation: A review of labour laws for migrant domestic workers in Gulf Cooperation Council countries*. Retrieved from http://www.ituc-csi.org/IMG/pdf/gcc_legal_and_policy_brief_domestic_workers_final_text_clean_282_29.pdf

Kapur, R. (2012). Pink chaddis and SlutWalk couture: The postcolonial politics of feminism lite. *Feminist Legal Studies*, 20(1): 1–20.

Katz, C. (2001). On the grounds of globalization: A topography for feminist engagement. *Signs*, 24(4): 1213–1228.

Kofman, E. (2017). Feminist methodology. In D. Richardson (Ed.), *The AAG Encyclopedia of Geography*. Oxford, UK: John Wiley and Sons.

Kwan, M.P. (2002). Feminist visualization: Re-envisioning GIS as a method in feminist geographic research. *Annals of the Association of American Geographers*, 92(4): 645–661.

Lefebvre, H. (2003) [1970]. *The urban revolution*. Trans. Robert Bononno. Minneapolis: University of Minnesota Press.

Listerborn, C. (2015). Geographies of the veil: Violent encounters in urban public spaces in Malmo, Sweden. *Social and Cultural Geography*, 16(1): 95–115.

Macdonald, D., and Friendly, M. (2014). *The parent trap: Child care fees in Canada's largest cities*. Ottawa: Canadian Centre for Policy Alternatives. Retrieved from www.policyalternatives.ca

McInturff, K., with Lockheart, C. (2015). *The best and worst places to be a woman in Canada in 2015: The gender gap in Canada's biggest cities*. Ottawa: Canadian Centre for Policy Alternatives. Retrieved from www.policyalternatives.ca

Mackenzie, S. (1989). Women in the city. In R. Peet and N. Thrift (Eds.), *New models in geography: The political economy perspective* (Vol. 2). London: Unwin Hyman, pp. 109–126.

Mackenzie, S., and Rose, D. (1983). Industrial change, the domestic economy and home life. In J. Anderson, S. Duncan, and R. Hudson (Eds.), *Redundant spaces in cities and regions*. London: Academic Press, pp. 155–200.

Miriam, K. (2012). Feminism, neoliberalism, and SlutWalk. *Feminist Studies*, 38(1): 262–266.

Mohabir, N., and Peake, L. (forthcoming). Georgetown, Guyana and El Dorado imaginaries: Postcolonial distance and difference in a small Caribbean city. Submitted to *New West Indian Guide*.

Moser, C., and McIlwaine, C. (2014). New frontiers in twenty-first century urban conflict and violence. *Environment and Urbanization*, 26(2): 331–344.

Naples, N. (1992). Activist mothering: Cross-generational continuity in the community work of women from low-income urban neighborhoods. *Gender and Society*, 6(3): 441–463.

Pain, R. (2014). Everyday terrorism: Connecting domestic violence and global terrorism. *Progress in Human Geography*, 38(4): 531–550.

Parreñas, R. (2010). Transnational mothering: A source of gender conflicts in the family. *North Carolina Law Review*, 88: 1825–1856.

Pateman (1989). *The disorder of women: Democracy, feminism, and political theory*. Stanford, CA: Stanford University Press.

Peake, L. (1998). Living in poverty in Linden, Guyana in the 1990s: Bauxite, the "development of poverty" and household coping mechanisms. In D. McGregor (Ed.), *Resource sustainability and Caribbean development*. Kingston, Jamaica: University of the West Indies Press, pp. 171–194.

Peake, L. (on behalf of Red Thread Women's Development Programme) (2000). *Women researching women: Methodology report and research projects on the study of domestic violence and women's reproductive health in Guyana* (Georgetown, Guyana: Inter-American Development Bank) Project TC-97-07-40-9-GY.

Peake, L. (2009). Urban geography: Gender in the city. In R. Kitchin and N. Thrift (Eds.), *The International Encyclopedia of Human Geography*. London; Elsevier, pp. 320–327.

Peake, L., and Reiker, M. (Eds.) (2013). *Rethinking feminist interventions into the urban*. London and New York: Routledge.

Phadke, S., Khan, S., and Ranade, S. (2011). *Why loiter? Women and risk on Mumbai streets*. New Delhi: Penguin.

Pratt, G. (2005). Abandoned women and spaces of the exception. *Antipode*, 37(5): 1052–1078.

Pratt, G., Johnston, C. and Banta, V. (forthcoming). Filipino migrant stories and trauma in the transnational field. *Emotions, Space and Society*, in press.

Pratt, G., and Yeoh, B. (2003). Transnational (counter) topographies. *Gender, Place and Culture*, 10(2): 159–166.

Razack, S. (2000). Gendered racial violence and spatialized justice: The murder of Pamela George. *Journal of Law and Society/Revue canadienne droit et société*, 15(2): 91–130.

Robinson, J. (2011). Cities in a world of cities: The comparative gesture. *International Journal of Urban and Regional Research*, 35(1): 1–23.

Roychowdhury, P. (2013). "The Delhi gang rape": The making of international causes. *Feminist Studies*, 39(1): 282–292.

Satterthwaite, D. (2008). Cities' contribution to global warming: Notes on the allocation of greenhouse gas emissions. *Environment and Urbanisation*, 20(2): 539–549.

Statistics Canada (2013). Labour Force Survey, Table 282-0087, December 2013. Retrieved from http://www5.statcan.gc.ca/cansim/a26?

Tadiar, N. (2013). Life-times of disposability in global neoliberalism. *Social Text*, 31(2): 19–48.

Tivers, J. (1977). Constraints on spatial activity patterns: Women with young children. *Occasional Paper 6*. Department of Geography, King's College London.

Tonkiss, F. (2005). *Space, the city and social theory*. Cambridge: Polity.

UN–Habitat (2008). *State of the world's cities 2008/09: Harmonious cities*. London: Earthscan.

UNDP Gender Inequality Index (2014). Retrieved from http://hdr.undp.org/en/content/gender-inequality-index-gii

UN Women (2013). *Contributions of migrant domestic workers to sustainable development*. Bangkok: UN Entity for Gender Equality and the Empowerment of Women.

Wekerle, G. (1984). A woman's place is in the city. *Antipode*, 16(3): 11–20.

Whitzman, C. (2007). Stuck at the front door: Gender, fear of crime and the challenge of creating safer space. *Environment and Planning A*, 39: 2715–2732.

Whitzman, C., Andrew, C., and Viswanath, K. (2014). Partnerships for women's safety in the city: "Four legs for a good table." *Environment and Urbanization*, 26(2): 443–456.

WHO (2008). *The global burden of disease 2004 update*. Retrieved from http://www.who.int/healthinfo/global_burden_disease/GBD_report_2004update_full.pdf

WHO (2012). Suicide rates data by country. Global Health Observatory Data Repository. Retrieved from http://apps.who.int/gho/data/node.main.MHSUICIDE?lang=en

Wilson, E. (1992). *The sphinx in the city: Urban life, the control of disorder, and women*. Oxford: Blackwell City Reader.

Woetzel, J., Madgavkar, A., Ellingrud, K., Labaye, E., Devillard, S., Kutcher, E., Manyika, J., Dobbs, R., and Krishna, M. (2015). *The power of parity: How advancing women's equality can add $12 trillion to global growth*. McKinsey Global Institute.

Wright, M. (2004). From protests to politics: Sex work, women's worth and Ciudad Juarez modernity. *Annals of the Association of American Geographers*, 94: 369–386.

Wright, M. (2013). Feminism, urban knowledge and the killing of politics. In L. Peake and M. Reiker (Eds.), *Rethinking feminist interventions into the urban*. London and New York: Routledge, pp. 41–51.

18 Urban Governance, Ethnicity, Race, and Youth

Beverley Mullings and Abdul Alim Habib

Introduction

The United Nations Human Settlements Programme (UN-Habitat) states that half of the 3 billion people currently living on our planet are under the age of 25 and of that number almost half (1.3 billion) are between the ages 12 and 24. Most of these young people live in urban areas, the majority of which are in the global South. With rates of urbanization rapidly increasing, especially in the global South, UN-Habitat projects that as many as 60 per cent of all urban dwellers will be under the age of 18 by 2030. The challenges posed by a fast-growing youth population—the so-called **youth bulge**—are most apparent in cities in the global South, where over 90 per cent of the world's urban growth now occurs (UN-Habitat, 2013). The term "youth" is used in this chapter to describe people in transition from the dependence of childhood to the independence of adulthood. While many policy-makers, for the purposes of data collection, utilize fixed age ranges, it is generally understood that all age categories are context specific. The United Nations and the International Labour Office, for example, both define youth as persons between the ages of 15 and 24 years, while in recognition of the challenges that young people face in transitioning to adulthood, organizations like the African Youth Charter use the age range of 15–35 years (African Union Commission, 2006).

Among youth populations around the world, unemployment is probably the greatest challenge that most face. The International Labour Organization (ILO) estimates that the global youth unemployment rate was just over 13 per cent in 2014 (ILO, 2014), a figure almost three times higher than the global adult unemployment rate. Especially in cities with limited resources, urban managers are increasingly faced with the challenge of meeting the demands of young people for income-generating opportunities. The challenge of unemployment is especially compounded for young people disadvantaged by multiple and overlapping systems of oppression and inequality. Despite the fact that youth are the fastest-growing demographic group in many cities in the global South and the potential lead actors in shaping the futures of their countries, they are rarely included or acknowledged in the processes or institutions that influence how urban space is governed. In fact, as Alain Bertho (2009) notes, youth, and particularly the poorest youth, are no longer considered a social investment for the future, but rather a threat to the present—to be contained. More than it being simply a question of intergenerational difference, young people who are marginalized by poverty and **racism** in many cities around the world are increasingly regarded as a surplus population, as malevolent, criminal, or simply a nuisance. There is a blatant discrepancy between discourses that espouse the need to invest in youth and the actual participation of young people in the programs, policies, and processes being developed to improve quality of life and living standards in the cities where they live. As this chapter demonstrates, whether in Toronto, Montreal, Kingston, or Accra, there is a tendency to perceive youth marginalized by poverty, racism, or ethnic discrimination as a drain on the resources and health of the city.

This chapter explores how social constructions of "race" and ethnicity are implicated in neoliberal forms of urban restructuring in the context of the global South.

Drawing on case studies from Kingston, Jamaica, and Accra, Ghana, the chapter explores how ethno-racial forms of inequality, modes of exclusion, and regimes of violence have come to reassert themselves and even be revitalized in these two cities, which have embraced the market as the primary mechanism for organizing and managing urban space. Focusing on the survival experiences and strategies of young people located at the intersection of overlapping and marginalizing systems of oppression, this chapter invites us to think imaginatively about the ways in which young people potentially offer new modes of creativity and interaction in cities.

Though located on different continents and of vastly different sizes, Kingston, Jamaica, and Accra, Ghana, offer important insights into the ways that the lives of ethno-racially marginalized groups in southern cities are affected by urban strategies that aim to align the functioning of cities with the demands of global markets. Both cities are the product of British imperialism—both started out as coastal cities to facilitate the trade of goods between Europe, Africa, and the Americas, and both later functioned as colonial administrative capitals. The two cities also served as places of departure and arrival in the trafficking of African people to work in the lucrative West Indian sugar cane agro-industry. In addition, given the prominence of West Africa in Jamaican history, many people attribute cultural practices that shape everyday life in the city, such as the dominance of women in informal trade, to the influence of this region on modern-day Jamaica.

For countries that experienced centuries of colonial rule, where ethnic and racial hierarchies structured social, political, and economic relationships, practices, and institutions, de-colonization rarely signaled the end of ethno-racial significations, but instead often marked their recomposition in the face of changing economic, social, and political contexts. An exploration of how emerging modes of governance in each city have affected the most marginalized youth allows us to track the ways in which neoliberal forms of urban governance continue to transform existing ethno-racial significations. The chapter ends with a consideration of urban youth in the Canadian cities of Montreal and Toronto, asking what these four cities can learn from one another. These examples encourage a global view of the growing convergence of the urban experiences of marginalized and racialized youth, both in terms of the hurdles they face and how they respond to them.

Ethno-Racialized Youth and Neoliberalizing Cities in the Global South

Scholars and policy-makers rarely study the marginalizing effects of constructions of race or ethnic difference in cities comprised of predominantly "black" or "brown" populations in the global South and their implications for youth (South Africa is, however, an exception—there is a large literature that examines how racism shapes social interactions in the city, but primarily between the country's white minority and black majority.) This is partly because contemporary understandings of racialization tend to focus on relationships between "whites," "blacks" and/or racialized "others," without acknowledging the existence of ethno-racial formations that retain traces of older colonial systems rooted in hegemonic discourses of **white supremacy** and their impact on patterns of social and urban inequality. Scholars instead have tended to focus on questions of economic inequality, poverty, and gender when addressing the challenges facing young people in cities in the global South (Caldeira, 2014; Chant, 2013; Gough, 2008; Nayak and Kehily, 2013). Only a handful of scholars have systematically examined how racialization and ethnic discrimination shape the lives of young people in cities in the global South (Jeffrey, 2012; Simone, 2010; Vargas and Alves, 2010).

The view that young people embody "the projected dreams, desires and commitments of a society's obligation to the future" (Giroux, 2008: 183) is one that most governments and development agencies embraced in the 1950s. A reflection of the optimism and economic prosperity that characterized the postwar period of the 1950s and 1960s, youth came to be seen as crucial to the course of future development. Especially among newly independent countries in the global South, governments fervently believed that investing in young people, particularly in the areas of education and health, was key to overcoming the levels of impoverishment and inequality that these formerly colonized countries faced. But the large number of youth protests taking place across the globe during this period forced policy-makers to think about youth not simply as objects of policy, but also as agents in the process of social change (Scheper-Hughes and Sargent, 1998). Young people were vibrant participants in anti-colonial struggles during the 1950s and 1960s.

In South Africa, for example, the African National Congress Youth League (ANCYL), founded in 1944, played a key role in shaping the resistance strategies of the African National Congress (ANC). Introducing mass resistance tactics such as the Defiance Campaign, the ANCYL was instrumental in galvanizing support for the ANC and the subsequent liberation struggle. In the 1960s, youth were also instrumental in the protests against state authoritarianism in Brazil, and in Jamaica, youth influenced by the writings of the Guyanese scholar and political activist Walter Rodney played a significant role in bringing black consciousness and black liberation into public conversation.

Outlined in a 1970s document titled *Youth in a Changing World*, UNESCO explained the rationale for this shift in focus, stating that it was imperative to give the young a chance to take part in public affairs because they were a key source of new ideas and new ways of approaching problems (UNESCO, 1977: 31). UNESCO's call for the inclusion of young people in governance processes and its recognition of the importance of incorporating young people as key agents of social justice and change were pioneering in so far as they created spaces for young people's viewpoints to be heard. Many scholars note, however, that despite these earlier insights there has been a steady erosion of the public commitment to nurture, educate, and include young people in national and local projects of development (Scheper-Hughes and Sargent, 1998; Ansell, 2005; Nybell et al., 2013). In fact, as scholars like Henry Giroux (2011) argue, the last 30 years have been little short of an all out war on young people and their capacity to be vanguards of social justice and social change. Giroux identifies the increasing orientation of economies towards the fundamentals of the market as a major catalyst in the shift in the commitment and inclusion of youth in the process of governance. Giroux (2011: 2) argues that for the last 30 years economic and political policy and discourse have been dominated by "a neoliberal dystopian vision that legitimates itself through the largely unchallenged claim that there are no alternatives to a hyper-market-driven society: that economic growth should not be constrained by considerations of social costs or moral responsibility and that democracy and capitalism are virtually synonymous."

The problem of urban poverty that once preoccupied urban managers in most cities in the global South has over the years been superseded by concerns over boosting the wealth-generating capacity of the cities and dismantling many of the social protections and institutions serving the public good that the state once financed. Like cities in the global North, southern cities that embrace the principles of market-driven social transformation in the management of city affairs increasingly seek to direct investments towards elite consumer spaces such as shopping malls, upscale condominiums, and gated communities in a bid to attract further rounds of investment to the city, but market-led strategies like these frequently subordinate urban space in the city to the speculative strategies of private investors, often at the expense of the provision of collective investments oriented towards social needs. In African cities like Johannesburg, Durban, Accra, and Lagos, for example, urban policy-makers are increasingly focusing their energies on "turnkey projects" that will attract foreign architectural and private property development firms seeking to sink investments into the creation of **satellite cities**, with socially polarizing implications (De Boeck, 2013; Watson, 2013).

While most policy-makers see forms of urban governance that are participatory, accountable, transparent, and legal as crucial to the creation of inclusive and equitable cities, these forms of governance are rarely extended to the most impoverished, racialized, or ethnically discriminated-against youth. In many cities that have embraced neoliberal policies, poor **ethno-racialized** youth are subject instead to forms of governance that rely on a politics of abandonment based on the rolling out of policies to make individuals take responsibility for their own means of survival and punishment of those who transgress established social conventions or laws in order to do so. Particularly among countries that were obliged to implement the austerity programs of the International Monetary Fund and World Bank during the 1980s and 1990s, declines in social spending on health and education forced many youth into jobs that were poorly paid, hazardous or often simply outlawed by the state, casual day labour, or household production activities. As the ILO found in its School-to-Work transition (SWTS) household survey of young people aged 15–29 years, in 6 of the 10 countries surveyed, over 60 per cent of young people were neither in the labour force, in education, nor unemployed and, if working, were concentrated in low-quality, irregular, and low-wage informal-sector jobs (ILO, 2013). The World Bank estimates that as many as 43 per cent of women between the ages of 15 and 24 in 2012 were unemployed in comparison to 27 per cent of men of the same age (World Bank, 2015). Combined with the fact that in many parts of the global South political instability or conflict further exposes young people to violence, youth in the global South have precarious lives where opportunities are few and lives can be sometimes short-lived. As poignantly observed by Henry Giroux (2011), too many young people are trapped within a **prison industrial complex**, bearing the brunt of a system that leaves them uneducated and jobless and ultimately offers them one of two available options: poverty or prison.

Global youth statistics elaborate the vulnerabilities that many young people in cities face. As a recent report by the United Nations Population Fund states,

> [t]ens of millions do not go to school, or if they do, they miss even minimum benchmarks for learning. Employment prospects are often dismal, with jobs unavailable or poor in quality, leading to a worsening global youth unemployment crisis. Up to 60 per cent of young people in developing regions are not working or in school or have only irregular employment. (UNFPA, 2014: 31)

While significant regional variations exist in youth unemployment rates (28 per cent and 24 per cent respectively in the Middle East and North Africa versus East Asia's 9.5 per cent and South Asia's 9.3 per cent in 2012 [ILO, 2013]), these variations point to the direct role that national economic and political landscapes play in the lives of young people. For many youth in the global South, the challenge of employment is one of both quantity and quality, and even in places like South Asia where youth unemployment levels are low, the ILO recorded poverty rates among the employed as high as 33.7 per cent in 2010 (ILO, 2013). Being employed is no longer a safeguard against also living in poverty. Particularly in cities where urban population growth is placing a strain on already-overstretched urban infrastructure, it is common for young people, especially the poorest, as well as individuals subordinated by sexism, racism, or ethnic discrimination or homophobia, to be regarded as an impediment to urban transformation. Thus, as scholars like Henry Giroux (2008, 2011) and Mayassoun Sukarieh and Stuart Tannock (2008) note, in the context of the liberalization of markets and the demise of state-led collective welfare provisioning, young people, and in particular poor youth, are increasingly regarded as a **disposable population**—surplus to the requirements of the economic system—rather than as social investments for the future.

The growing concentration of young people in spaces of precarity within neoliberalizing cities is bringing the question of youth governance to the forefront of the development agendas of a growing number of international organizations (UNFPA, 2014; ILO, 2013; World Bank, 2006). But as Mayssoun Sukarieh and Stuart Tannock warn in their critical reading of the World Bank's 2007 *World Development Report* (Sukarich and Tannock, 2008), this shift in attention to youth is not fueled primarily by a desire to empower the young, but rather by a growing concern for the potential political security threat that young people, especially in countries considered to be ungovernable, pose to the global North (Roy, 2008; Fuller, 2003; Herrera, 2010). As the two case studies in this chapter demonstrate, the precarity of the lives of the most marginalized youth when combined with cities' efforts to divest themselves of social responsibility for their welfare creates few opportunities for young people to participate in the life of their cities as full, valued citizens.

Responsibilization versus Containment: Governing Youth in Kingston's Garrison Communities

Slightly larger in land area than the city of Kingston, Ontario (480 sq. km versus 450 sq. km [185 sq mi. versus 173 sq. mi]), with a population approximately three and a half times larger (584,627 versus 558,464 in 2011), the Kingston Metropolitan Region (KMR), Jamaica, encompasses the downtown core and historic city of Port Royal in the parish of Kingston and the business and residential neighbourhoods of the parish of St Andrew (Figure 18.1). These spatial patterns of land use mirror the social divisions that exist between Kingston's uniformly black residents who live in the largely impoverished downtown areas of the city and its "white," "brown," and "black" upper- and middle-class residents who live and work in the uptown neighbourhoods and commercial businesses in St Andrew. To the extent that Jamaican class categories overlap with existing ethno-racial hierarchies that

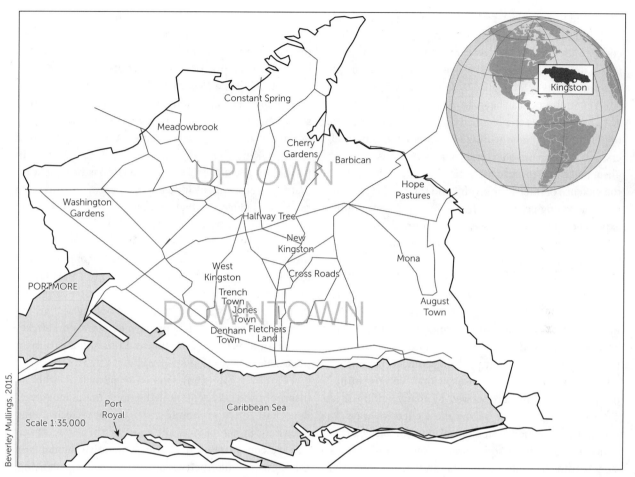

Figure 18.1 Map of the Kingston Metropolitan Area, Jamaica

value proximity to **whiteness**, socio-spatial patterns of settlement and employment in the city also reproduce elements of the ethno-racial divisions that once defined this former plantation economy. Since the 1980s when Jamaica first began to liberalize its economy, the economic gap between these two constituent groups has continued to widen in ways that have further divided the urban landscape into a series of fortressed, gated, and securitized middle-class communities and a series of equally fortressed impoverished garrison communities.

Although 92 per cent of Jamaica's population self-identify as black, such an ascription says little about the multiple meanings of blackness that exist on the island or the **intersectionality** of other markers of identity to racialize poor black bodies in impoverished urban spaces (Rodney, 1969; Gray 2004). While it is impossible to separate the effects of poverty from those of racialization in the material resources available to young people in Jamaican cities, it is possible to discern how "race" functions as a salient factor in the way the poor, young, black, and often male bodies come to be governed in the city (Levy and Chevannes, 2001). Young black women are further marginalized by constructions of gender and sexuality that intersect with constructions of race in ways that portray them as either hypersexualized and immoral or dependent victims of violent and dysfunctional families (Cooper, 1995). Here, blackness and urbanity function as signifiers of threat and distinguish poor black youth as objects of containment and policing in ways reminiscent of the experiences of many black young people in the United States and Canada. Many of Jamaica's poorest youth live in politically aligned and fortressed urban communities popularly known as *garrisons*; these are often governed by gangs and their leaders, popularly known as "Dons," and off-limits to most outsiders (Mullings, 2013). Garrison communities are the spatial expression of Jamaican clientelism and the historical exclusion of the poor from the governing of its cities (Figure 18.2). For young people, garrison communities can be spaces of security from gangs in rival communities or, potentially, places that offer income opportunities to those willing to work for the gang leaders. But these spaces also mark young people as criminal, alien, and threatening to the social order. Caught between the violence of inter-gang rivalry and the violence of state policing, young people who live in garrison communities experience the city as a place of premature death—where the homicide rate exceeds the national average, which at 37 per 100,000 is already one of the highest in the world (Government of Jamaica, 2014).

Unlike middle-class black youth or even poor rural black youth, youth who live in Jamaica's garrison communities experience **urban governance** as containment, hard policing, and fatal violence. Hard policing practices aimed at aggressively pursuing convictions, such as "stop and search," "arrest on suspicion," "**rough riding**," or "paramilitarism," are viewed by many middle-class citizens and politicians as the most effective ways for the police to demonstrate control. But for black urban youth, these strategies are viewed as yet another oppression within an unjust economic and political system. Few politicians or policy-makers seek the views of these young people because they are already regarded as criminal, and for that reason, the neoliberal governance model aimed at facilitating popular democracy through participation in decision-making is rarely extended to them. At best, projects that seek to include these young people do so by entreating them to take responsibility for themselves and encouraging them to develop the formal, legible business practices of the private sector.

Yet, there is a long history of creativity, talent, and entrepreneurialism among youth in Kingston's garrison communities, an example of which is the growth of *dancehall*—a cultural movement that emanates from some of Kingston's poorest communities. Growing in popularity since the 1980s, dancehall has become a lucrative entertainment industry with segments linked to record and video production, sound systems and studio recording services, fashion lines and entertainment promotion (Stolzoff, 2000). In Kingston's poorest communities, dancehall represents one of the few opportunities for racialized young people to make money. Although few manage to rise to the top of the industry as dancehall artists, many secure livelihoods buying and selling dancehall paraphernalia, preparing food and refreshments for street dance events, and offering a variety of dancehall services (Figure 18.3).

However, for young women in particular, who are disproportionately represented among the unemployed, dancehall offers economic opportunities rarely found elsewhere in the city. While most of these opportunities are concentrated in the least lucrative segments of the industry—like petty trading (higglering), dancing, modeling, and a host of other dancehall-related beauty services—a few women have been successful in creating and maintaining small businesses. Despite the innovation and entrepreneurialism that these achievements suggest, young people who earn incomes from

Figure 18.2 A typical "downtown" West Kingston neighbourhood

Figure 18.3 Passa Passa street dance event in West Kingston, Jamaica

dancehall activities are rarely considered to be making a worthwhile contribution to the city. Indeed, insights into forms of creativity, innovation, and resilience that young people routinely employ and that, with state support, could substantially improve quality of life not only in Kingston's poorest neighbourhoods but across the city as a whole are not taken up in urban planning or policy. Recognizing the most marginalized of youth as valued citizens who contribute to the vibrancy and life to the city is a first step towards learning how their everyday survival strategies might provide new ways of thinking about more sustainable and egalitarian urban futures. This issue is also taken up in the study below of Accra, Ghana.

City Modernization, Slum Clearance, and Youth Resilience in Accra

With a population of 4 million people spread over 3245 square kilometres (1252 square miles), the Greater Accra Metropolitan Area (GAMA) is just over half the geographic area and approximately two-thirds of the population of the Greater Toronto Area (GTA) (Figure 18.4). Like Kingston, however, Accra is a city that continues to be shaped by widening divisions between a middle- and upper-class minority eager to embrace the opportunities for commerce, competition, and growth that free markets offer and a youthful impoverished majority who are regarded as a group that is holding back progress in the city. As in Jamaica, colonialism and its aftermath created an uneven social geography in Ghana, bequeathing to Accra a legacy of ethnic differentiations and tensions. Such insecurities and levels of distrust have produced new forms of conflict in the city. From "de-congestion exercises" aimed at removing street vendors from the streets to the demolition and clearance of slum communities, many of Accra's poorest youth are being cast out of the city (discursively and materially) by state officials intent on making the urban landscape attractive to international investors.

Commencing in the early 1980s, Ghana's economic reforms, which entailed among others the liberalization

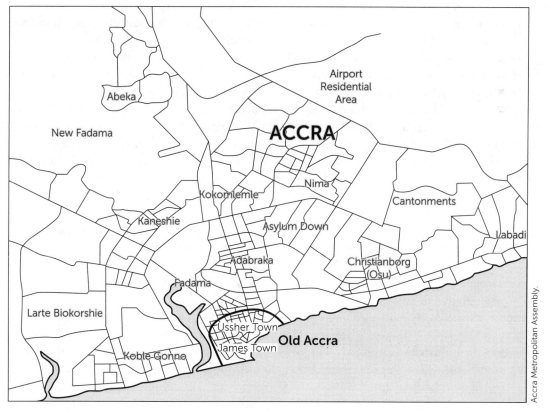

Figure 18.4 Map of Accra, Ghana

of the national economy and the progressive retreat of the state from economic and social responsibilities, amplified the social and spatial inequalities inherited from the colonial and early post-independent eras (Asante and Gyimah-Boadi, 2004; Songsore, 2003). Efforts to decentralize governance only served to deepen existing inequalities when welfare responsibilities that were once assumed by the state shifted to local governments and individuals. In Accra, this dual **responsibilization** further marginalized the poorest ethnic groups, as city governments, embracing "city modernization" policies, now aimed to expel them.

In Accra, it is the indigenous Ga population and ethnic minority groups, mostly from northern Ghana, who are commonly stereotyped as bringing no value to the city. Levels of youth unemployment are especially high among these groups, who constitute the majority of Accra's residents living in informal settlements and slum communities (Agyei-Mensah and Owusu, 2009). In June 2015, in an act that rendered thousands homeless overnight, the city government demolished part of the city's largest slum community, the Old Fadama (Graphic Online, 2015). Pejoratively called "Sodom and Gomorrah," this informal settlement was home for the last 30 years to many of Accra's marginalized ethnic groups, of which over 70 per cent were from the north of the country (Housing the Masses, 2010). Many residents were taken by surprise, most of them awakened by the sheer force of bulldozers, even though officials insisted that they had given at least 24 hours notification (Citifmonline, 2015). The Old Fadama became the symbol of all that was wrong with the city (Afenah, 2012), and the callousness with which the demolition was undertaken highlighted the extent to which the state regarded poor households as a disposable population.

In Ghana, the response to their supposed disposability has been complex, ranging from the self-cultivation of resourcefulness to merely "going with the flow." It is no exaggeration to claim that contemporary Ghanaian youth are incredibly motivated and innovative. Nowhere is this resourcefulness more apparent than in the **creative industries** (specifically in music and movie productions) where, since the early 1990s, youth have created new forms of cultural production that are generating new inflows of capital to the city. Evident in the rise of the movie industry ("Ghallywood") and music industry ("Hip-Life"), many formerly poor youth have been catapulted into continental and international stardom in transnational markets within the diaspora. This resourcefulness also finds expression in the burgeoning informal sector where many of the poorest and often ethnicized youth rely on informal activities such as petty trading, head-porting (using one's head to carry thing, or *kayayee*), and Internet fraud (*sakawa*) for their livelihoods. While some young people have been able to gain access to markets across multiple continents, it is important not to over-romanticize the degree to which these forms of entrepreneurialism are self-empowering. For many, these activities entail significant hardships, and for young women involved in petty trading and head-porting, they routinely include sexual harassment and exploitation (Akuoko, 2013).

Most of the poorest indigenous and ethnically marginalized youth in Accra are obliged to rely on what locals refer to as the *kpakpakpa* movement—the ability to seize any and every opportunity to generate economic resources when no opportunities appear to exist. Spontaneous and innovative, *kpakpakpa* speaks to the ability of the most ethnically marginalized to generate something from activities as mundane as the harvest and sale of fruit from public trees in the city, highlighting not only the innovativeness of the urban poor but also the levels of impoverishment that force many to rely on these forms of income generation. A primary lesson from these conflicting outcomes is that self-aspiration and entrepreneurialism on the part of youth is no substitute for systems of urban governance that would value and seek to support the capacities of the poorest young people.

As critical urban scholars have established, urban youth across the world have been anything but passive with respect to the current forms of economic and political restructurings that marginalize them. In Canada's major cities, for instance, racialized youth have struggled to claim space through forms of cultural production (Teelucksingh, 2006) like Afrofest in Toronto (http://afrofest.ca/2016-festival) and through protests like the Black Lives Matter movement (https://nowtoronto.com/news/meet-the-faces-of-black-lives-matter-tent-city-toronto), often in defiance of city managers and the police who seek to render them invisible in public space.

From Kingston and Accra to Toronto: What Can These Cities Learn from Each Other?

In 2011, nearly 30 per cent of Canadians (10 million people) were under the age of 25 (Ministry of Industry, 2012). Many of these young people live in poverty. A report entitled "The Hidden Epidemic: A Report on Child and Family Poverty in Toronto" (2014) reveals that 29 per cent of Toronto children (145,890) are members of low-income families, an increase of over 10,000 from 2010. Although Toronto is home to five of the richest neighbourhoods in Canada, it also has the highest rates of child poverty of any large Canadian city, 4 per cent higher than Montreal, 5 per cent higher than Winnipeg, and 12 per cent higher than Ottawa (Polanyi et al., 2014). The report reveals that poverty increases significantly by "race" and ethnicity. People of African and Middle-Eastern backgrounds are three times more likely to be living on a low-income than people of European backgrounds.

In Canada, as in Jamaica and Ghana, there is a fraught relationship between racialized and ethnically marginalized young people and the institutions that govern the cities in which they live. Racialized youth in Canada, for example, are on average more likely to be unemployed (23 per cent versus 16 per cent for Canada as a whole in 2006 (National Council of Welfare Reports, 2012), to be involved in precarious income-generating activities, and to live in neighbourhoods with a high degree of violent **crime** (Khenti, 2013). As in Jamaica and Ghana, urban managers and policy-makers in Canada have tended to rely upon explanations that locate violence in the cultural practices and moral shortcomings of young people and the communities in which they live (Anonymous, 2006; Worthington, 2012).

As both victims and perpetrators of violence in the city, underprivileged black, Indigenous, and, increasingly, Muslim youth (and in particular, men) are regularly regarded as a threat to urban stability, especially in cities that have adopted neoliberal modes of urban governance with their emphasis on growth and a heightened role for the private sector in the organization and provision of urban infrastructure and services. In Toronto, for example, relations between the police and the city's black communities have soured over the past decade because of the discriminatory nature of the general intelligence-gathering practice that the police have come to rely on. Known as "carding," this loosely regulated policing strategy has allowed the police to randomly stop and gather intelligence from individuals even if they are not suspected of criminal activity (Rankin and Winsa, 2012).

Carding is a deeply unpopular policy among Toronto's black communities, because young African-Canadians tend to be disproportionately targeted in these operations. According to a *Toronto Star* finding based on information obtained in a freedom of information request, between 2008 and 2011, young men racialized as black were approximately 3.2 times more likely to be carded than whites, and in more affluent neighbourhoods this figure was higher (Rankin and Winsa, 2012). With 1,104,561 names entered into a police database between 2008 and 2011, 23 per cent of which were identified as the names of black people, critics see carding as a form of racial profiling that infringes upon civil liberties and potentially does more harm to targeted communities than good. Carding as a policing practice was suspended in January 2015, but the practice has not officially been brought to a close. While carding is promoted as a form of "community engagement," many young men who are racialized as black view it as an oppressive form of community surveillance that reinforces the idea that they are a threat to the city. The practice of randomly stopping and gathering intelligence from individuals in public spaces—so-called street checks—is a routine part of policing throughout Canada that police argue is invaluable to their data-gathering efforts. Yet, as a practice ultimately aimed at improving quality of life and safety in the city, the policing strategy of carding alienates and further diminishes the ability of racialized young people to exercise their **right to the city**.

continued

Like city managers in Kingston and Accra, city managers in cities like Toronto, Montreal, and Vancouver have been more inclined to treat ethno-racialized youth as problem people than to treat them as people with problems, and thus many of the policies oriented towards youth have focused more on criminalization and incarceration than on employment and governance. While city-supported organizations like the Toronto Youth Cabinet (http://thetyc.ca) or federally registered not-for-profit corporations like the City Youth Council of Toronto (www.toronto.thecyc.ca) provide opportunities for young people to have a voice in the governance of their city, their focus and interests do not always reflect the most pressing issues facing young people and they often focus too narrowly on market-friendly solutions aimed at producing entrepreneurial youth citizens. It is striking, for example, that in light of the deepening tensions between racialized communities and the police over the use of carding, this issue has not been higher on the list of activities and commitments of either group. While initiatives like revenue generation through the placement of local small-business advertisements along dasher boards in the city-owned ice rinks, the creation of an LGBTQ+ youth shelter (www.toronto.thecyc.ca/campaigns), and improvements to the transit system are important and worthy improvements to the life of the city, they do not directly challenge the gnarly issues of precarity, discrimination, or disposability—issues that racialized youth consistently view as primary impediments to their quality of life. While a multitude of grassroots youth organizations do address issues of racism and over-policing, these smaller, often poorly funded groups are rarely as influential in shaping city policy as larger, city-established ones.

There is much that cities like Kingston, Accra, Toronto, and Montreal can learn from each other, for as the case studies in this chapter have demonstrated, it is from the strategies that young people rely upon to eke out a living that new and original forms of innovation, organization, and creativity are emerging. As a number of scholars have suggested, dancehall and Ghallywood's ascent into global youth culture cannot be extricated from the everyday entrepreneurial networks that diasporic youth in cities in the global North operate within (Akrofi, 2013; Galvin, 2014; Garritano, 2013; Power and Hallencreutz, 2002; Stanley Niaah, 2008; Stolzoff, 2000). An example of the importance of the entrepreneurialism of diasporic youth can be seen in the rise and global reach of *sound systems*, an integral part of Jamaican music culture. Sound systems are comprised of groups of disc jockeys, engineers, and MCs who create the quality of sound, produce the music, and determine the mode of performance at music events. As dancehall grew internationally so too did the reputation of particular sound systems in cities in the global North. Some sense of the global reach of sound systems can be gathered from the website Jamworld Movements, which collects and lists the names of past and present sound systems. In 2015, 213 sound systems were listed with operations in countries outside of Jamaica (www.jamworldmovements.com/jwr/sounds/list.htm#EU). Of this number, 49 were located in Canada, of which just over half were in Toronto (25) and just over 30 per cent in Montreal (15).

Without the presence of sound selectors, event promoters, or DVD distributors in cities like Toronto or Montreal, opportunities for the production, distribution, and hybridization of dancehall or Ghallywood cultural innovations would be limited. Diasporic youth have been crucial to the growth and spread of these industries because they function as nodes in these cultural commodity chains, which introduce and circulate new sounds, images, and fashion trends to young people in cities in the global North as well as in the global South. It is ironic that even in cities that expressly seek to attract and promote new forms of innovation and entrepreneurialism, the forms of cultural production and dissemination engaged in by the most marginalized youth are rarely regarded as activities that contribute value to the city. Yet, these are precisely the sort of youth-led activities that with support in the form of infrastructural provision, start-up financing, or access to training would offer racialized and ethnically marginalized youth opportunities for further innovation and employment.

Conclusion

Cities in the global South with high levels of poverty and youthful populations face specific urban challenges. Poverty limits the capacity of young people to acquire the resources they need to become educated, healthy, independent, and responsible urban citizens, while the presence of a poorly educated, unemployed youth population limits the ability of cities to become innovative, attractive locations for investment and wealth creation. These challenges deepen when cities embrace the market as a primary mechanism for urban wealth creation and welfare distribution because they tend to orient urban resources towards sectors and groups with the best prospects for attracting capital to the city, rather than towards their large resident youth population. In Kingston and Accra, the largely unemployed and impoverished youth populations are routinely constructed as a threat to the economic success of the city. In both cities, the idea of threat is exacerbated by the way in which circulating ethnic and racial discourses further devalue the poorest urban youth by casting them as less respectable than and different from the rest of the urban population. With little value placed on their contribution to the life of the city, racialized and ethnically subordinated youth have become a disposable population, excluded from urban governance processes, subject to hard policing, and, in the case of Accra, actually driven out of the city. As these cities orient themselves to the needs of domestic elites and prospective international investors, they are able to see neither the value that the most marginalized youth bring to the city nor their potential for future innovation if adequately supported.

AbdouMaliq Simone argues that the ability of people who have "no predictable means of taking care of themselves, to turn commodities, found objects, resources and bodies into uses previously unimaginable or constrained" (2004: 410) speaks to a level of flexibility and innovation that should be understood in the same way that city planners consider infrastructure. Knowing whom to trust, when to take a risk, when reciprocity is required, how to avoid attention, or when to try something new are all skills and abilities that are rarely viewed as valuable or important to the health of neoliberalizing cities. In both Kingston, Jamaica, and Accra, Ghana, the most marginalized of young people routinely devise creative, collaborative, and often entrepreneurial strategies to generate incomes and investments for their own personal growth. Captured in the idea of *kpakpakpa*, these are some of the skills that the poorest and most marginalized youth identify as essential to their urban lives. Understanding how these calculations and risks help young people to negotiate environments that are hostile to their very presence and creating opportunities for young people to imagine and identify how they could draw upon these assets to create their ideal city offer a mode of urban governance that is more popularly democratic and accountable than is currently the case.

Acknowledging the role played by racialized and marginalized young people in the global spread of youth culture is a step towards recognizing that youth, ordinarily perceived as a threat, bring value to the life of cities. Supporting the transnational networks, modes of cultural production, and entrepreneurial practices in which many young people are already engaged will engender new forms of social innovation that will ultimately bring new resources to cities, whether they are in the global North or South. Whether it is in the form of training programs aimed at deepening skills or in the provision of infrastructure like community-based production studios, the first step towards including racialized or ethnically marginalized youth in the process of urban governance can only begin if they are encouraged to share their vision of the ideal city and the social and economic dispositions and resources needed to achieve it.

Key Points

- A belief in the efficiency and fairness of free markets has become a major organizing force that is shaping urban neoliberal policy and planning in both the global North and the global South and is also influencing the extent to which young people, the poorest in particular, are able to articulate their interest in urban space.
- While most policy-makers see forms of urban governance that are participatory, accountable, transparent, and legal as crucial to the creation of

- inclusive and equitable cities, these forms of governance are rarely extended to the most impoverished, racialized, or ethnically discriminated against youth.
- Ethno-racialized youth are often seen as a surplus population that threatens the future prosperity of the city, the governance of whom tends to take the form of abandonment or containment through hard policing, subjecting them to lives of precarity and the heightened possibility of premature death.
- Racialized black youth in Kingston's garrison communities are viewed by Jamaica's upper and middle classes, as well as by policy-makers, as criminogenic populations that threaten the health of the city because of their proclivity for violence.
- Indigenous and ethnic minority youth in Accra are also perceived as threatening to the city's modernization efforts; their concentration in slum communities in the city is viewed as a blight on the efforts to transform it into the investment gateway to Africa.
- In both cities, marginalized youth are a source of knowledge and value that urban policy-makers rarely appreciate; their efforts to generate incomes with the most minimal of resources and to organize and collaborate with others both locally and transnationally speak of talents that city managers seldom recognize as entrepreneurial and creative, but that nonetheless offer potential insights into alternative spaces and modes of making value in the city.
- Urban racialized youth in Canada also experience cities as alienating spaces, as they are subject to high levels of unemployment, surveillance, and incarceration, with black and indigenous youth, in particular, rarely viewed as contributing value, further diminishing their ability to exercise their right to the city.
- Taking the time to understand how poverty, racialization and ethnic discrimination intersect with other marginalizing markers of identity to exclude young people from having a voice in the governance of their cities, and recognizing value in the levels of innovation and creativity that many youth deploy to survive, offers new ways of thinking about building alternative and socially just urban futures.

Activities and Questions for Review

1. Based on your reading of this chapter and your own experiences growing up, identify how racialization and/or ethnicity play a role in the way that young people in Canadian cities are governed.
2. As a class, debate the extent to which "race" is a useful concept for understanding the marginalization of urban youth in countries where the majority of the population are people of colour.
3. If urban managers want to increase the participation of marginalized youth in the governance of their cities, what would you recommend that they do first?
4. As a member of a youth municipal advisory board, how would you recommend that forms of cultural production (e.g., dancehall, Ghallywood, or Afrofest) be used to improve the social and economic status and political integration of young people?

Acknowledgments

Some of the information presented in this chapter is based upon research funded by the Canada–Latin America and the Caribbean Research Exchange Grants (LACREG) program.

References

Afenah, A. (2012). Engineering a millennium city in Accra, Ghana: The Old Fadama intractable issue. *Urban Forum*, 23(4): 527–540.

African Union Commission. (2006). The African Youth Charter. Retrieved from http://www.carmma.org/reso.urce/african-youth-charter

Agyei-Mensah, S., and Owusu, G. (2009). Segregated by neighborhoods? A portrait of ethnic diversity in the neighborhoods of the Accra Metropolitan Area, Ghana. *Population, space and place*, 16: 499–516.

Akrofi, D.A. (2013). *Assessing brand Ghallywood*. Thesis submitted to the Department of Business Administration, Ashesi University College in partial fulfillment of the requirements for the award of Bachelor of Science degree in Business Administration, Ashesi University College, Ghana.

Akuoko, K.O., Ofori-Dua, K., and Forkuo, J.B. (2013). Women making ends meet: Street hawking in Kumasi, challenges and constraints. *Michigan Sociological Review*, 27. Retrieved from http://www.readperiodicals.com/201310/3234941891.html

Anonymous (2006). The real source of ghetto crime. *National Post*, May 1. Retrieved from http://www.financialpost.com/scripts/story.html?id=004442ca-99e9-47ef-9664-d6240c0c1d16&k=55800

Ansell, N. (2005). *Children, youth, and development*. New York: Routledge.

Asante, R., and Gyimah-Boadi, E. (2004). Ethnic structure, inequality and governance of the public sector in Ghana. UNRISD. Retrieved from http://www.unrisd.org/80256B3C005BCCF9/(httpAuxPages)/8509496C0F316AB1C1256ED900466964/$file/Asante%20(small).pdf

Bertho, A. (2009). *Le temps des émeutes*. Paris: Bayard.

Caldeira, T. (2014). Gender is still the battleground: Youth, cultural production and the making of public space in Sao Paulo. In S. Parnell and S. Oldfield (Eds.), *The Routledge handbook on cities of the global south*. New York: Routledge, pp. 413–428.

Chant, S. (2013). Cities through a "gender lens": A golden "urban age" for women in the global South? *Environment and Urbanization*, 25(1): 9–29.

Citifmonline. (2015). Old Fadama residents were informed ahead of demolition—AMA. June 22. Retrieved from http://citifmonline.com/2015/06/22/old-fadama-residents-were-informed-ahead-of-demolition-ama

Cooper, C. (1995). *Noises in the blood: Orality, gender and the "vulgar" body of Jamaican popular culture*. Durham, NC: Duke University Press.

De Boeck, F. (2013). Challenges of urban growth: Toward an anthropology of urban infrastructure in Africa. In A. Lepik (Ed.), *Afritecture: Building social change*. Munich: Architekturmuseum der TU München, p. 93.

Fuller, G.E. (2003). *The youth factor: The new demographics of the Middle East and the implications for US policy*. Washington, DC: Saban Center for Middle East Policy at the Brookings Institution.

Galvin, A.M. (2014). *Sounds of the citizens: Dancehall and community in Jamaica*. Nashville, TN: Vanderbilt University Press.

Garritano, C. (2013). *African video movies and global desires: A Ghanaian history*. Athens, OH: Ohio University Press.

Giroux, H. (2008). *Disposable youth, racialized memories, and the culture of cruelty*. New York: Taylor & Francis Group.

Giroux, H. (2011). Youth in a suspect society: Coming of age in an era of disposability. *Truthout /Op/Ed*, May 5. Retrieved from http://www.truth-out.org/news/item/923:youth-in-a-suspect-society-coming-of-age-in-an-era-of-disposability

Gough, K.V. (2008). "Moving around": The social and spatial mobility of youth in Lusaka. *Geografiska Annaler: Series B, Human Geography*, 90(3): 243–255.

Government of Jamaica. (2014). *A new approach: National security policy for Jamaica 2013, Towards a secure and prosperous nation*. Ministry Paper # 63. Kingston, Jamaica. Retrieved from http://www.japarliament.gov.jm/index.php/publications/ministry-paper-new/2014-ministry-papers/1286-2014-ministry-paper-63-new-approach-national-security-policy-for-jamaica-2013

Graphic Online (2015). Sodom and Gomorrah demolition exposes illegal connections. June 22. Retrieved from http://graphic.com.gh/news/general-news/44966-portions-of-sodom-gomorrah-razed-down.html

Gray, O. (2004). *Demeaned but empowered: The social power of the urban poor in Jamaica*. Kingston, Jamaica: University of the West Indies Press.

Herrera, L. (2010). Young Egyptian's quest for jobs and justice. In L. Herrera and A. Bayat (Eds.), *Being young and Muslim: New cultural politics in the global south and north*. Oxford: Oxford University Press, pp. 127–143.

Housing the Masses (2010). *Community-led enumeration of Old Fadama Community, Accra—Ghana. Housing the Masses Final Report to People's Dialogue on Human Settlements (People's Dialogue, Ghana)*. Accra, Ghana: Housing the Masses.

ILO (2013). *Global employment trends for youth 2013: A generation at risk*. Geneva: International Labour Office.

ILO (2014). *Global employment trends for youth 2013: Risk of a jobless recovery?* Geneva: International Labour Office.

Jeffrey, C. (2012). Geographies of children and youth II: Global youth agency. *Progress in Human Geography*, 36: 245–253.

Khenti, A.A. (2013). Homicide among young black men in Toronto: An unrecognized public health crisis? *Canadian Journal of Public Health*, 104(1): 12–14.

Levy, H., and Chevannes, B. (2001). *They cry respect: Urban violence and poverty in Jamaica*. Department of Sociology and Social Work, UWI Mona, Kingston, Jamaica. New York: Routledge.

Ministry of Industry (2012). *The Canadian Population in 2011: Age and Sex*. Catalogue no. 98-311-X2011001. Ottawa: Statistics Canada.

Mullings, B. (2013). Neoliberal restructuring, poverty and urban sustainability in Kingston, Jamaica. In I. Voljnovic (Ed.), *Sustainability: A global perspective*. East Lansing: Michigan State University Press, pp. 387–406.

National Council of Welfare Reports (2012). *Poverty profile: Special edition*. Toronto: National Council of Welfare Reports. Retrieved from http://www.esdc.gc.ca/eng/communities/reports/poverty_profile/snapshot.pdf

Nayak, A., and Kehily, M.J. (2013). *Gender, youth and culture: Young masculinities and femininities*. London: Palgrave Macmillan.

Nybell, L.M., Shook, J.J., and Finn, J.L. (2013). *Childhood, youth, and social work in transformation: Implications for policy.* New York: Columbia University Press.

Polanyi, M., Johnston, L., and Khanna, A. (2014). *The hidden epidemic: A report on child and family poverty in Toronto.* Toronto: Social Planning Toronto.

Power, D., and Hallencreutz, D. (2002). Profiting from creativity? The music industry in Stockholm, Sweden and Kingston, Jamaica. *Environment and Planning A, 34*(10): 1833–1854.

Rankin, J., and Winsa, P. (2012). Known to police: Toronto police stop and document black and brown people far more than whites. *Toronto Star.* Retrieved from http://www.thestar.com/news/insight/2012/03/09/known_to_police_toronto_police_stop_and_document_black_and_brown_people_far_more_than_whites.html

Rodney, W. (1969). Message to Afro-Jamaican associations. *Bongo-Man,* January–February: 14–17.

Roy, O. (2008). Al Qaeda in the West as a youth movement: The power of a narrative. CEPS *Policy Briefs,* 1(12): 1–8.

Scheper-Hughes, N., and Sargent, C. (1998). *Small wars: The cultural politics of childhood.* Berkeley: University of California Press.

Simone, A.M. (2004). People as infrastructure: Intersecting fragments in Johannesburg. *Public Culture, 16*: 407–429.

Simone, A.M. (2010). *City life from Jakarta to Dakar: Movement at the crossroads.* London: Routledge.

Songsore J. (2003). *Towards a better understanding of urban change: Urbanization, national development and inequality in Ghana.* Accra: Ghana Universities Press.

Stanley Niaah, S. (2008). Performance geographies from slave ship to ghetto. *Space and Culture, 11*(4): 343–360.

Stolzoff, N.C. (2000). *Wake the town & tell the people: Dancehall culture in Jamaica.* Durham, NC: Duke University Press.

Sukarieh, M., and Tannock, S. (2008). In the best interests of youth or neoliberalism? The World Bank and the New Global Youth Empowerment Project. *Journal of Youth Studies, 11*(3): 301–312.

Teelucksingh, C. (2006). Towards claiming space: Theorizing racialized spaces in Canadian cities. In C. Teelucksingh (Ed.), *Claiming space: Racialization in Canadian cities.* Waterloo, ON: Wilfrid Laurier University Press, pp. 1–18.

UNESCO (1977). *Youth in a changing world: An analysis of UNESCO's youth programme between 1969 and 1977.* Paris: Beugnet, SA, à Paris.

UNFPA (2014). *State of the world's population 2014.* United Nations Population Fund.

UN-Habitat. (2013). *State of urban youth report 2012–2013: Youth in the prosperity of cities.* Nairobi: United Nations Human Settlements Programme.

Vargas, J.C., and Alves, J.A. (2010). Geographies of death: An intersectional analysis of police lethality and the racialized regimes of citizenship in São Paulo. *Ethnic and Racial Studies, 33*(4): 611–636.

Watson V. (2013). *Future African cities: The new postcolonialism.* Public lecture hosted at the Bartlett Development Planning Unit at University College London on March 6. Retrieved from http://www.bartlett.ucl.ac.uk/dpu/pdf-archive/Vanessa_Watson-2013.pdf

World Bank (2006). *Development and the next generation.* Washington, DC: World Bank.

World Bank. (2015). Online database. Retrieved from http://databank.worldbank.org/data/home.aspx

Worthington, P. (2012). You can't stop crime, if you can't say who is doing it. *Huffington Post, Canada.* July 18. Retrieved from http://www.huffingtonpost.ca/peter-worthington/toronto-shooting_b_1683102.html

19 Disabling Cities

Nancy Worth, Laurence Simard-Gagnon, and Vera Chouinard

Introduction

Geographic research on disability and cities is wide-ranging and encompasses the lives of people dealing with disability, physical impairment, and mental illness issues. This chapter focuses on what makes cities more and less disabling for persons with physical and mental health impairments where "disabling" refers to processes of physical and social exclusion arising from physical and social barriers to full participation in city life. It also engages with different ways of understanding disability (i.e., the medical, social, and embodied social models of disability) and the implications of these for whether and how cities need to change. A review of the literature on cities and disability serves to highlight a primary focus on issues of physical impairment and then is followed by an examination of processes shaping the lives of urban residents with mental health impairments.

A global South case study of Georgetown, Guyana, explores some of the ways in which this city disables persons with physical impairments. Since the vast majority (80 per cent) of people with disabilities live in the global South and because most of what we know about disability is based on research conducted in the global North, it is vital to learn more about disabling cities in the global South. This does not mean that valuable insights cannot be gained from studying disability and the city in the global North, and a second case study of visually impaired young people living in Liverpool, in northern England, explores how they negotiate social space in the city and how the aids they use or choose not to use changes how they are treated by others. These two cities were chosen because they are situated very differently in relations of power and oppression in the global order. Taken together, these case studies reveal that while social barriers to inclusion in city life such as poverty are especially severe in cities of the global South, this does not mean that disabled people do not face significant challenges in negotiating social space in cities of the global North. The Canadian context is introduced by an investigation of the use of urban space by persons experiencing mental illness in the neighbourhood of Saint-Roch in Quebec City. This investigation reveals how such persons developed a space for support and social interaction in the Saint-Roch mall and how such spaces have been lost in the course of gentrification in the neighbourhood. This chapter argues that while cities in the global South and North continue to disable persons with physical and mental health impairments, it is important to recognize that disabled people, at least sometimes and in some places, find ways to resist these processes of marginalization in the city.

Disability and the City

Before discussing how geographers and other urban scholars have researched disability and mental health in the city, it is helpful to chart the different terms used to understand disability.

Words Matter: The Power of Definitions Framing the Debates about Disability in the City

This chapter is entitled "Disabling Cities" because it focuses on the experience of "disability" in the city. The words used to talk about disability have powerful meanings. Some, like "lame," "crippled," or "handicapped," are no longer used. Some words, like "impairment" or "deficit," while used in medicine and health, are less often used in the social sciences. These terms map the different ways of thinking about disability: Figures 19.1 and 19.2 show a **medical model** and a **social model of disability**.

In the medical model of disability (Figure 19.1), disability is conceived as an individual's impairment—the problem is with his/her/their body. The social model

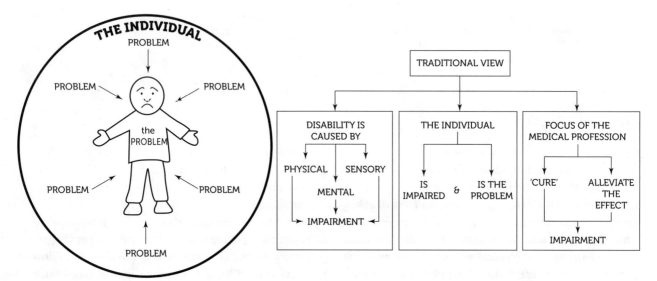

Figure 19.1 The medical model of disability
Democracy, Disability and Society Group/The Thistle Foundation, 2003.

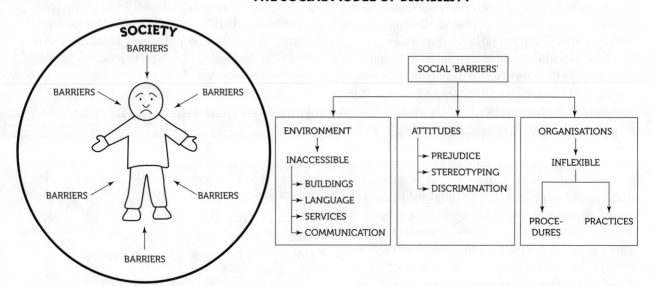

Figure 19.2 The social model of disability
Democracy, Disability and Society Group/The Thistle Foundation, 2003.

of disability conceives of disability very differently: it considers how disability comes from barriers in society that privilege some (able) bodies over others. Here, "the problem" is not the person, but a social world that excludes people with disabilities. These two ways of thinking about disability are important for understanding disability in the city, as they lead to two very different pathways to change. For the medical model, it is up to individuals to overcome their impairment, while the social model demands that society becomes more inclusive and accessible (see Thomas, 2004, for more about definitions).

Geographers and other researchers have recently begun to think about a third way of thinking about disability, a **biopsychosocial model of disability** (Barnes and Mercer, 2010). This model takes parts of the medical model (individual lived experiences of minds and bodies) and the social model (structural or social barriers to participation) and puts them together with the understanding that disability is made up of a complex relation of factors that change over space and time (Worth, 2008). These different models of disability are important because they tell us how researchers have understood disability and the city (for reviews, see Hansen and Philo, 2007; and Imrie and Edwards, 2007).

Much work in geography and urban studies makes use of the biopsychosocial model, or of what some call an embodied social model of disability. Disability in this model is part of our identity (just like gender, ethnicity, and age), and discrimination on the basis of ability is called **ableism**. When assumptions are made about minds and bodies meeting an established norm and anything other than this norm is problematic, this is ableism. Just like sexism or racism, ableism is a social phenomenon that values some bodies over others. In cities there are numerous examples of ableism; for example, the majority of public spaces, buildings, and transportation networks are built to suit an able-bodied user.

Examples of inaccessible (or ableist) urban spaces can be found everywhere—steps to access buildings or a lack of curb cuts, street furniture that impedes the travel of blind people, or a lack of induction loops in shops for the hearing impaired. Geographers have demonstrated that this is not an accident of urban planning, but rather an active process of exclusion: "Disability has distinct spatialities that work to exclude and oppress disabled people. Spaces are currently organised to keep disabled people 'in their place' and 'written' to convey to disabled people that they are 'out of place'" (Kitchin, 1998: 351). For example, the French architect and urban planner Le Corbusier, an influential figure in modern urban design, explicitly stated that he designed buildings and public spaces for tall, strong, masculine, able-bodied, "ideal" bodies (Imrie, 1999). And in contemporary North America and Europe, **accessibility** legislation and building guidelines remain weak. Efforts have been made in the built environment to make spaces accessible to people with disabilities, perhaps by adding ramps or elevators or special assisted entrances—often to meet legislative minimums around accessibility—but even these attempts can powerfully mark disabled people as "out of place"—for example, disabled people might only be able to enter through a back door or with an able-bodied person's help. Even though physical access is sometimes (partially) achieved, it is often without a sense of inclusion or equality.

The final important term to examine is **inclusive design**, a set of principles and practices that goes beyond functional accessibility and aims to create spaces that all users can use and feel comfortable in. Rather than treating disabled people as a separate group that needs to be planned for, inclusive design principles consider the wide diversity of human experience. For example, children and the elderly can have allied access needs, as well as parents with strollers or those who are temporarily injured. According to Imrie and Hall (2001), inclusive design offers a way of thinking through the "**social architecture**" that invades the built environment—ideas and beliefs about the design of our cities that have led to what our cities look like, who can use them, and how. Thinking critically about inclusion means challenging the **stigma** around disability and mental health and understanding that inclusion is about more than the physical design of a space, but also about how our social relationships play out across space.

Geographers, Disability, and the City

Geographers' interests in disability and the city first emerged in the 1980s. Much of this early work was concerned with mental illness and related phenomena, such as community opposition to the siting of mental health facilities (also known as the "Not in My Backyard" syndrome—NIMBY). At this time, it was rare for geographers to consider how persons with physical impairments and/or ill health struggled to access and negotiate urban environments. One exception to this was Kirby, Bowlby, and Swann (1983), who, albeit briefly, considered the mobility issues faced by persons with physical impairments, including the elderly, in the United Kingdom.

It was during the 1990s that geographers began to consider, in detail, how persons with chronic illness and physical impairment were disabled in the city. Isabel Dyck (1995, 1999) explores the shrinking **lifeworlds** of women with multiple sclerosis and the strategies they used to renegotiate a disabled identity in the home, neighbourhood, and workplace. Valentine (1999) explores how physical impairment (spinal cord injury) diminished one man's access to urban places (such as the local pub) that were important in maintaining a working-class masculine identity. In his wide-ranging historical materialist study of disability, Gleeson (1999b) shows how the rise of industrial capitalist cities in the nineteenth century was accompanied by the growing segregation and exclusion of disabled people (e.g., in asylums). He also examines disability and contemporary cities, arguing, for example, that the provision of "community care" was insufficient to ensure that disabled people's rights to fully participate in urban life were realized. Gleeson (1999a) also addresses the question of whether technological innovations (e.g., in **assistive devices**) can overcome the disabling city. He concludes that technology alone cannot "cure" disability or rehabilitate physically impaired bodies. What is needed, he argues, is political-economic, institutional, and cultural change in the forces shaping today's cities so that disabled people's aspirations for a valued place in urban society can be realized. This could include greater access to paid employment and a cultural shift away from regarding persons with impairments as "lesser" in favour of simply being different.

Geographers have also considered how architects' visions of the city and the bodies that move through it have shaped the built environment. As noted above, Imrie (1999) reveals how Le Corbusier envisioned the city as the domain of male able-bodied people and designed the built environment to cater to this ideal or norm of what a city dweller is (e.g., buildings with steps; housing that is inaccessible to wheelchair users). Kitchin (2003) criticizes the architectural profession in Ireland for failing to design accessible urban environments. He argues that access should not be sacrificed because of concerns over such things as cost or aesthetics. Kitchin and Law (2001) point out that design professionals, including urban planners, have failed to ensure that there are accessible public toilets in Irish cities and have thus hindered disabled people's ability to fully participate in urban life. Geographers have also written about the principles and practices of inclusive design as a means of creating more accessible urban environments (Imrie and Hall, 2001).

Geographers' work on disability and the city during the 1990s included efforts to challenge tendencies to assume that disabled people were passive subjects of oppression. Hester Parr (1999), for example, in her study of psychiatric service users in Nottingham, documents instances of resistance to dominant medical treatment practices (e.g., using herbal remedies). Chouinard (1999) discusses disabled women's activism in Canadian cities and internationally, noting that much of the history and geography of it remains to be written.

In the twenty-first century, there have been signs of both continuity and change in geographers' approaches to understanding disability and the city. One sign of continuity is an ongoing interest in disabled people's experiences of marginalization and inclusion in different workplaces. Hall and Wilton (2011), for example, explore the potential of unionized mainstream workplaces, alternative social enterprises, and the arts to provide supportive "working" environments for disabled people. Chouinard (2011) reflects on how university workplaces disable persons with physical and mental health impairments, drawing on her own experiences as a disabled female professor. There has also been ongoing interest in how chronic illness reshapes lives and alters the meanings assigned to places such as the home (Crooks, 2010).

There have also been important signs of change in how geographers understand disability and the city. One such change is a more expansive understanding of disability and the kinds of embodied differences that can be disabling. Longhurst (2010), for example, explores how women who are large or "fat" are disabled in terms of the material environment (e.g., seats and changing rooms that are too small) and also emotionally (e.g., loathing the size difference they embody). People who are of short stature also face challenges. Kruse (2010) reveals that "little people" adapt or "restaturize" private spaces of the home to make them more enabling than the public spaces they have to negotiate. Davidson (2003) helps to put phobias and conditions such as autism (e.g., Davidson and Parr, 2010) on the agenda of geographers concerned with disability and the city. Geographers have also demonstrated an interest in how addiction treatment programs are negotiated in more and less enabling ways. Men with addictions, as Wilton et al. (2014) show, rework their masculine identity in the process of recovery—learning, for example, to be less aggressive in their behaviour, to share their emotions with others, to recognize their interdependence with others, and to no longer associate practices such as drinking alcohol with being a "man."

Geographers' approaches to disability and the city have also become more expansive in terms of the types of interactive spaces that are considered. In particular, there has been growing interest in online interactions in the virtual space of the Internet among groups such as people with mental health impairments, autism, and hearing impairments (Skelton and Valentine, 2010). Findings suggest (as textual interactions are more in line with autistic communication practices) that persons with mental health challenges may find interactions in virtual space enabling in some ways (e.g., in terms of access to information and support) but disabling in others (e.g., when the emotional support demanded of them by other users is experienced as excessive) (Davidson and Parr, 2010).

Almost all of what geographers know about disability and the city is based on research conducted in the global North. Yet the vast majority of disabled people live in the global South. It is important, therefore, that geographers learn more about disability and cities in the global South. Chouinard's (2012, 2014, 2015) work, including the case study of Georgetown, Guyana, in this chapter, begins to address this enormous gap in our understanding of disability and the city. The term "disabled people" is used in this chapter to emphasize, as in the British disability studies tradition, that it is society that disables persons with impairments (as such, it is preferable to the term "people with disabilities" usually used in the North American context).

Mental Health and the City

Understanding the "disabling" actions of the city is incomplete unless attention is also paid to the multiple ways in which city life affects mental and emotional well-being and how city spaces can be disabling or enabling with regard to mental and emotional well-being. Researchers' attention has been directed at how cities foster accessibility, possibilities for inclusion, or, conversely, marginalization and exclusion for people experiencing different types and degrees of emotional distress or madness. It is also important to recognize how city life itself can heighten mental states.

The different terms used to talk about mental states—distress, illness, madness—serve to illustrate the degree of disagreement over terminology—what Parr and Philo (1995: 207) call the "vexed issue of what this thing we call 'madness' actually is." Indeed, the different terms used to discuss madness evoke diverging ideologies through which experiences of mental and emotional difference and/or un-wellness are understood and explained. Thus, the expression "mental illness" usually refers to a medical perspective implying institutional processes of treatment and diagnosis (refer back to Figure 19.1), whereas thinking in terms of mental and/or emotional distress may encompass a wider range of experiences of emotional un-wellness that are not necessarily officially diagnosed and that can be episodic. The terms "madness" and "mad" are used not only to establish a distance from a sole concentration on **medico-psychiatric models of mental illness**, but also to encompass an unlimited range of experiences of mental and emotional distress—for example, from mild depression to major depressive disorders (Mullings and Peake, 2016). Moreover, the terms "mad" and "madness," if they have been long considered negatively, are now being reappropriated by communities of mental health services users (see, for example, Mad Pride Toronto, 2014).

Much like physical illnesses and/or disabilities, experiences of mental distress are intertwined in particular experiences of space that can be constraining and/or exclusionary. Mental and/or emotional distress, in this sense, challenges the conceptual opposition between mind and body by highlighting that, on the one hand, experiences of exclusion and/or difference are embodied—that is, they are lived through the body—and on the other hand, that the ways in which we experience our bodies are shaped by our personal social histories and geographies, including those of exclusion and/or difference. Parr and Philo (1995), for example, discuss how different discourses associated with different types of city spaces (institutional, semi-institutional, and non-institutional) are mobilized by users of mental health services in their construction of complex and sometimes contradictory subjectivities as ill individuals. Moreover, in a very practical sense, mental and emotional distress and/or illness translate into bodily traits, such as particular movements, appearances, reflexes, displays of emotion, and behaviours (Wolch and Philo, 2000). Thus, states of mental and physical illness together shape individual social experiences (Parr, 2008).

It is no surprise then that geographers' interests in disability and the city first developed in the 1980s around community attitudes towards the mentally ill and mental health facilities (see Taylor and Dear, 1981). In 1987, Dear and Wolch published their landmark study entitled *Landscapes of Despair: From Deinstitutionalization to Homelessness*. In this book they linked processes of urban change such as **deinstitutionalization**, the spatial concentration of community mental health facilities in poorer inner-city neighbourhoods, **gentrification**, and the loss of affordable housing to the growing phenomenon of **homelessness** in Canadian and American cities. Across the Atlantic Ocean, in the United Kingdom, Chris Philo (1987) had started to write about the development of urban and rural asylums for the mentally ill in the nineteenth century.

Mental illness was constructed, and became widely understood, as a medical dysfunction during the rapid industrialization of Western cities. As a medicalized condition, the treatment of mental illness involved isolation from the "sane" population and movement into residential institutions—"lunatic" asylums—that were generally located outside of the city. The second half of the twentieth century saw increasing protests against the institutionalization of the mentally ill that, combined with developments in medication allowing for an improved management of certain psychiatric conditions (such as schizophrenia), resulted in large-scale waves of deinstitutionalization from the 1960s onwards.

Deinstitutionalization implied that former inmates were returned to community life and directed towards local mental health services for the provision of mental and social care. It also implied a concentration of mental health service users in inner-city areas, as they followed the location and availability of community services and housing. This meant the development of **post-asylum geographies** through which mental distress quickly became a significant element in the geographies of cities (Wolch and Philo, 2000).

Geographers have been interested in processes of inclusion, exclusion, and marginalization of individuals experiencing mental distress in cities following deinstitutionalization. They have investigated the distribution, location, and access of care and support services (Wolch and Philo, 2000). Of particular interest has been the development of "psychiatric ghettos," which refers to the phenomena of a positive feedback loop through which small-scale mental health centres and housing facilities, as well as populations of mental health services users, become concentrated in a relatively small area (Wong and Stanhope, 2009). Related to this "ghettoization" of mental health services users, geographers have also investigated processes of "drift," that is, the movements of deinstitutionalized or never-institutionalized users through the city following the changing distribution of medical and social care services (Wolch and Philo, 2000).

Post-asylum geographies are also characterized by a particularly virulently stigmatized type of difference (Dear and Wilton, 1997; Parr, 2008). Stigma, an important concept in the geographies of mental distress, refers to particular negative ways of constructing and understanding differences. Stigma occurs when traits are identified and associated with undesirable characteristics (stereotypes) and are used to categorize individuals as belonging to a separate group—"them" versus "us," the "mad" versus the "sane." The persons thus labeled experience loss in status and processes of discrimination and exclusion (Link et al., 2004). The importance of stigma in the lives of individuals experiencing mental distress has led geographers to investigate spatial processes of ostracization—in other words, how mental distress and/or illness is socially articulated as "other," as deviant, and as dangerous—and the practical implications of this ostracization in the day-to-day geographies of urban mental distress.

One phenomenon thus highlighted is the impact of local community mobilization against the establishment of mental health services sites—such as drop-ins, rooming houses, or day centres—in particular neighbourhoods, a phenomenon referred to earlier in the chapter as NIMBY. Geographers have found that NIMBY reactions are especially present in affluent suburban and family-oriented areas, thus contributing to the concentration of mental health services and spaces associated with mental distress in poorer inner-city neighbourhoods (Wilton, 2000).

Alternatively, geographers have been interested in processes or structures that make city spaces more enabling for persons experiencing mental distress. Parr (2008), for example, argues that the "nature work" or gardening done in cities has the potential to promote greater inclusion of persons with mental distress. In particular, she cites the example of the Coach House project in Glasgow, which by creating more aesthetically pleasing public spaces through gardening and other activities, has encouraged surrounding residents to value the work done by persons with mental health impairments.

Geographers are also increasingly concerned with the relationship between neoliberal social and economic policies, financial precarity, and mental distress. The shift towards neoliberal paradigms and models of governance of the last decades throughout the global North has often involved drastic cutbacks in the services and programs used by individuals experiencing mental illness, from access to psychiatric care to government financial assistance (Wilton, 2004). Like canaries in the coal mine, individuals experiencing mental distress, already proportionally poorer and more vulnerable, are among the first to suffer from the dismantling of social safety nets.

The precarity of individuals struggling with mental distress is often related to isolation and difficulties in maintaining regular employment (among other things), which in turn can contribute to a compromised sense of emotional well-being (Wilton, 2004). This translates to a high proportion of people struggling with mental distress relying heavily on government transfers and services for their income (through various social assistance and disability programs) and their housing (e.g., in subsidized rooming houses). Following major cuts in government funding, the increased poverty of mentally ill people has led to new and/or intensified forms of stigma in which stereotypes and emotional reactions associated with mental distress and/or illness combine with those associated with the poor—for example, when limited resources lead individuals to wear clothing that is considered socially undesirable or even inappropriate, marking them as both mad and poor.

Poverty in turn increases both the amount and the steepness of barriers preventing individuals experiencing mental distress from participating in a "normal" social life and urban spaces. Neoliberal cities are characterized by the rarity or absence of open public spaces and the dominance of spaces centred on consumption. Hence, spaces of open and/or cheap consumption, such as malls and fast-food restaurants, often play a critical role in the daily geographies of mentally distressed individuals, particularly those who are poor. They are open longer and more regularly than centres specifically designed to offer mental health care services—and are often much more accessible. Their importance in the lives of individuals experiencing mental distress is symptomatic of the scarcity of mental health care services and of the privatization and marketization of public urban spaces. The inability, or limited ability, to engage in consumption—for example, buying oneself a coffee—in such places enhances experiences of "out-of-placeness" and exclusion (Knowles, 2000).

Poverty can also limit the ability of individuals struggling with mental distress to conform to basic norms of social inclusion, such as maintaining certain standards of personal care and hygiene and regular housing. In urban settings, issues of mental illness and homelessness are inextricably tied to one another, as individuals experiencing mental health issues are more likely to be under-housed or homeless and on the other homelessness, with the associated experiences of intense stress and precarity, can in itself trigger and constitute forms of mental distress.

Neoliberal times are characterized by *both* intensified poverty and exclusion *and* a dominant discourse on the personal responsibility and moral obligation to conform to a norm of economic self-sufficiency. This implies that the associated stigma of mental illness and poverty increasingly leads to experiences and articulations of mental illness as deviancy, threat, and criminality (McGrath and Reavey, 2015; Parr, 1999). It is thus argued that we are currently experiencing a shift in geographies of mental distress from community services back to institutionalization, namely in the shape of the prison system. Meanwhile, contemporary stressors associated with enhanced competition and financial precarity have been associated with increasing rates of mental distress (Sedghi, 2015). Thus it may be that while the mentally ill and/or distressed are being increasingly marginalized and excluded, mental distress is quickly becoming an increasingly common experience. The two case studies that follow from Georgetown, Guyana, and Liverpool, UK, explore some of the key social barriers that serve to disable people with impairments.

Social Barriers to Disabled People's Inclusion in City Life: Georgetown, Guyana

Georgetown is the capital city of Guyana. It is also the country's largest urban centre. According to the 2015 census, the metropolitan city was home to 335,000 people (World Population Review, 2015). Georgetown is located on the coast of the Atlantic Ocean at the mouth of the Demerara River and is the country's main port. It also serves as a retail, administrative, and financial services centre. The city is protected from high tides by a seawall (it lies 1.5 metres [5 feet] below high-tide levels). Owing to the risk of flooding, many older buildings were originally erected on wooden or brick stilts—making them inaccessible to many persons with physical impairments. Today many of the spaces below these buildings have been filled in and put to use (Figure 19.3).

This case study draws on results from 60 interviews with disabled people and 10 interviews with disability activists and service providers in Georgetown, Guyana (Chouinard, 2012). Disabled people are generally acknowledged as being among the "poorest of the world's poor," but the poverty that disabled people face in a global South nation such as Guyana is especially severe. The vast majority of disabled people interviewed struggle to survive on less than USD$1 per day (the equivalent of approximately GYD$200). Poverty acts as a social barrier to inclusion in the life of the city and makes it difficult for disabled people to participate in organizing around disability issues. Disability activists comment on how disability organizations have experienced dwindling membership because people cannot afford the cost of transportation to meetings. Very impoverished living conditions also mean that it is difficult for activists to get disabled people out to rallies in Georgetown—and activists see this as a significant barrier to encouraging government action on disability issues. Poverty also limits disabled people's capacities to access health care and to maintain their health (e.g., because of difficulty affording travel to places offering health care and necessities such as medication or nutritious food). The disabled women and men interviewed adopt a variety of strategies to try to augment their incomes. Some beg for food and money and others do odd jobs such as tiling. Kim, a physically impaired wheelchair user knits chairbacks to sell. Like some of the other disabled people interviewed, however, she finds that people wanted to pay her less for them because she is visibly impaired. Mary, a visually impaired woman, sells small items such as packets of powdered drinks to augment her income. She also sometimes receives money from a relative living abroad (remittances), but this is an unreliable source of income for her.

Violence, particularly male violence against women, is a cause of impairment for some of the women interviewed. And the fact of having been impaired through "chopping violence" (i.e., the use of a cutlass to kill or maim women) in some cases has become a basis for blaming women for the violence they experience. Marta, a woman who lives on the outskirts of Georgetown, lost both forearms to chopping violence carried out by her male partner. Unfortunately, she has found that some people tend to blame her for the violence she experienced (e.g., they argue that she deserved it for being "promiscuous") and that others in her community, as well as members of her family, shun or avoid her as a result. This has limited her opportunities to participate in the life of the city. It is worth noting that contemporary experiences of chopping violence reflect the continued use of cutlasses to harvest sugar cane and a violent colonial past in which black slaves and later indentured servants from India were forced to labour on sugar plantations.

There are also cultural barriers to disabled people's inclusion in life in Georgetown. These barriers include practices such as shutting disabled family members in the home and thus rendering them invisible in public spaces of the city. This invisibility, activists argued, is an important barrier to convincing government officials to act on disability issues. Cultural attitudes that devalue disabled people are also barriers to inclusion and well-being in the city. Lawrence, a man physically impaired by a stroke, related how people in his neighbourhood would call him derogatory names such as "stink nasty" when he was out in public. This name-calling was a way of devaluing his presence in the city and left him feeling isolated and alone.

Figure 19.3 Map of Greater Georgetown, Guyana

Some of the barriers to inclusion and well-being in the city are both economic and cultural in nature. Private mini-bus drivers, for example, frequently fail to pick up passengers who use wheelchairs or other mobility aids. This is because doing so diminishes the number of fares that can be collected (because of space taken up by mobility aids) and also reflects the common practice of devaluing visibly disabled people. This problem was sufficiently widespread to trigger a (largely unsuccessful) "right-to-ride" campaign.

The out-migration of trained medical technicians to countries in the global North and constraints on state funding (e.g., to train replacement workers) have meant that women such as Marta lack access to prosthetic aids that could enable them to more fully participate in life in Georgetown. In Marta's case, a foreign-aid agency provided very basic prosthetic arms on a one-time basis. This, as well as a shortage of prosthetic technicians in the city, meant that when the arms broke they were not repaired and Marta was left without them.

Disabled people in Georgetown face many other barriers to inclusion and well-being in city life. For example, workers who are visibly impaired are paid lower wages than other workers, and disabled family members who are shut in the home are made to feel that there is something "shameful" about being visible in public. The poverty in which so many of them live is especially severe and reflects Guyana's marginalized place in the global capitalist order. So also does the out-migration of trained health professionals and the scarcity of technicians trained to make and fit prosthetic arms or legs. While this case study suggests that social barriers to inclusion in city life are especially severe for disabled people in the global South, disabled people still face significant challenges in negotiating the social space of the city in the global North. The next case study, of visually impaired young people in Liverpool, examines the strategies they use to move about the city and to have their embodied presence read in particular ways by others (e.g., as "independent").

Young People Negotiating Visual Impairment in Liverpool

This case study examines the experiences of visually impaired (VI) young people living in Liverpool, a large coastal city in the northwest of the United Kingdom. Formerly known for the economic activities associated with its port, the city now focuses on aspects of the service sector, including tourism and the knowledge economy. The downtown features many pedestrianized shopping streets close to the main train station, Liverpool Lime Street, with museums and tourist attractions close by in the refurbished Albert Docks (Figure 19.4).

Despite the walkability of the city, for many people with disabilities, going about their everyday lives in cities like Liverpool can expose them to harassment, condescension, or just the uncomfortable feeling of being on display (Morris, 1991). For disabled young people who want to be seen as independent, negative social attitudes about their disability can be especially challenging. Moreover, while disability often makes us think of physical impairment, the focus on sensory impairment in this case study and on mental health in the Canadian text box (see below) demonstrates that mind-body differences are diverse. Rather than asking VI young people how they physically navigate the city, Worth (2013), in this case study, explores how VI young people negotiate the social space of the city—a space that can often feel unwelcoming. VI Young people use a variety of strategies to get around the city; some use white canes, some use a guide dog, and some rely on their friends to get out and about. Critically, the choice and use of different mobility aids—or mobility strategies—also changes how visually impaired young people are treated in public space.

Will is partially sighted and is comfortable getting around on his own. When city streets become crowded and he loses his landmarks, he pulls a symbol cane out of his backpack—this signifier of visual impairment means that people back up and give him space, but it also marks him out as "different," so Will only uses it when he has to (Figure 19.5). James is blind, and is very comfortable navigating the city on his own, using a cane. When people assume he needs help, rather than be upset that his independence is not recognized, James accepts help he does not need, explaining that "there might be somebody else two weeks later that struggles to cross roads and they'd like somebody to offer them their hand. Like it doesn't really bother me, but it's always kind of a thing not to put people off" (Worth, 2013: 581). While this is a kind reaction to a gesture that was kindly meant, it demonstrates how James's capabilities are made invisible by his blindness. For Willow, having a guide dog is useful in several ways—primarily her dog helps her get

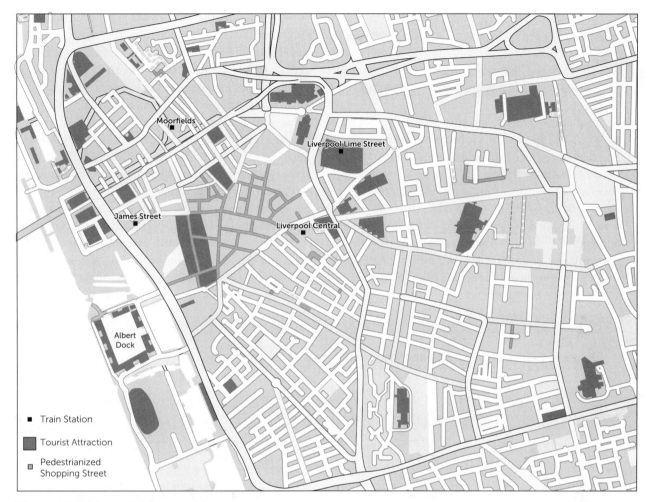

Figure 19.4 Map of Liverpool, UK
OpenStreetMap.

Figure 19.5 Independent mobility: A visually impaired young person travels through the city

around town by herself—but it also has an important social function for her. People she meets on the street, in a café, or on the train to college chat with her about her dog rather than treat her like she was strange or weird. Having a guide dog is a good ice-breaker and helps to create a friendly atmosphere that she values. Finally, Sarah finds that dealing with insults or rude questions as a blind girl in public space is often a difficult experience. Her solution is to travel with her friends rather than on her own, linking arms with her mates rather than using a white cane or a guide dog. She explains:

I don't like the look of [a cane] and I don't like the fact that it makes you, it brings you more attention than other people. And people know straight off that you're a bit "spesh" [special], if you get me.... I don't want to be the odd one out.... I don't really see the benefit. I just think you get bullied, and I'd rather not to be fair. I'd rather trip over a piece of paper than get bullied for being special. (Worth, 2013: 579)

Much to the regret of her parents and her teachers, Sarah is willing to sacrifice her own independence to be more anonymous in public space. Her experience shows how important the social is to a conception of disabling cities.

For Will, James, Willow, and Sarah, how they get around the city is not a question of physical accessibility, although they live in an urban environment with very few features for blind people. (Accessible public spaces for VI people include features such as audible crosswalks that beep during the walk signal, tactical bumps along curb cuts so that VI people know they are entering a roadway, and uncluttered sidewalks). In contrast to the experiences of disabled people in Guyana, for the VI young people involved in this research, the inaccessibility of the built environment is a difficulty that they manage or just get on with. In contrast, dealing with people's reactions to their presence in public space is a much bigger negotiation: it is often very difficult to be recognized as a competent spatial actor because of the negative impact of ableism.

The following section moves to a Canadian context, addressing the state of mental health care funding and service provision in Canadian cities, with particular reference to Quebec City, and reiterating this case study's focus on the city as an often unwelcoming social space.

Disabling Cities in Canada and the Mail Saint-Roch in Quebec City

Emotional and mental instability is a common trait in Canadian lives. Twenty per cent of Canadians will personally experience mental illness in their lifetime, and in this way mental illness affects all Canadian lives, at least indirectly, be it through a family member, a friend, or a colleague (CMHA, 2015). As well, emotional and mental instability represents an important area of government spending in Canada: by 2000 the economic cost nation-wide of mental illnesses to the health care system was estimated to be at least $7.9 billion in 1998—$4.7 billion in care, and $3.2 billion in disability and early death. Moreover, an additional $6.3 billion was spent on uninsured mental health services and time off work for depression and distress that was not treated by the health care system (CMHA, 2015).

Despite its prevalence, emotional and mental instability tends to remain socially unrecognized. For example, although approximately 8 per cent of Canadian adults will experience major depression at some time in their lives, about half of those who feel they have suffered from anxiety or depression will never seek medical help (CMHA, 2015). As discussed above, the stigma attached to such experiences presents an important challenge to inclusion.

In Canada, a large proportion of individuals experiencing mental distress live in poverty and in unstable housing situations (Nelson et al., 2015). As in most of the global North, massive waves of deinstitutionalization occurred throughout Canada starting in the 1960s, leading to concentrations of individuals with mental illnesses in service-dependent neighbourhoods. Vancouver's Downtown Eastside (Boyd and Kerr, 2015), Toronto's South Parkdale (Slater, 2004), Montreal's Downtown and Centre-Sud (Knowles, 2000), and Quebec City's St-Roch (as on page 321) are examples of such neighbourhoods. They are characterized by a relative concentration of mental health community services, and—importantly—by comparatively affordable housing options, such as boarding homes, rooming houses, and blocks of bachelor apartments (Slater, 2004; Boyd and Kerr, 2015).

In these neighbourhoods, the lives of individuals experiencing mental distress weave particular geographies characterized by constant movements between institutional, semi-institutional, and non-institutional places (Parr and Philo, 1995). The latter comprise places where access is relatively open and cheap, such as streets, churches, malls, and fast-food restaurants (Knowles, 2000). The presence of mentally distraught individuals in such places of consumption, which are increasingly replacing open public spaces in Canadian cities, is heavily controlled and repressed by police forces—a circumstance that often leads to additional trauma and strain on these individuals' mental health (Boyd and Kerr, 2015). This heavy policing feeds and is fueled by ideological discourses associating mental illness and poverty with insalubrity, criminality, and violence. In their study of police perception of individuals experiencing mental distress in Vancouver's Downtown Eastside, Boyd and Kerr (2015) demonstrate that such discourses, while feeding general moral panic, promote practices that lead to greater insecurity and exclusion for individuals in the neighbourhood experiencing mental distress.

In a similar vein, Slater (2004) argues that discourses linking mental health issues, single occupancy, criminality, and insalubrity were foundational in gentrification efforts in South Parkdale in Toronto. From the 1950s and 1960s, South Parkdale experienced a series of changes, including the construction of the Gardiner Expressway and the resulting destruction of hundreds of houses; the exodus of the middle and upper class to other urban neighbourhoods or suburbs; and the massive waves of patients discharged from the Queen Street Centre for Addiction and Mental Health, for a long time the largest facility of its kind in Canada and located in the neighbourhood (Slater, 2004). Since the mid-1980s, the city of Toronto has led gentrification efforts in South Parkdale, specifically by tightening regulations around single-occupancy accommodations. The resulting closure of rooming houses and bachelor blocks, while making way for other types of dwelling targeted at families and higher-income individuals, effectively expelled individuals who relied on cheaper accommodation options from the neighborhood. Thus, Slater (2004) argues that the emancipatory action of gentrification, as advocated by the city, must be critically questioned in terms of inclusion and exclusion.

The Saint-Roch Mall, located in downtown Quebec City, is another telling example of a consumption-oriented space whose history, from its construction to its dismantlement, is closely intertwined with histories of neighbourhood changes, urban imaginings, and movements of people experiencing mental and emotional distress (Figure 19.6). The Saint-Roch neighbourhood is centrally located and was once an economic pivot of Quebec City. The decline of the neighbourhood, initiated during the Great Depression in 1929, accelerated in the 1960s with the rise of suburban lifestyles, the increasing importance of the automobile, and the development of shopping centres with convenient parking facilities. By the 1970s, Saint-Roch had experienced a massive exodus of both people and commercial operations. In an attempt to counter this trend, store owners embarked on a venture to put a covering over Saint-Joseph Street, the local commercial artery, in order to create an indoor mall.

However, the development of the Mail Saint-Roch failed in its economic objective and the neighbourhood continued to be increasingly associated with poverty, crime, and marginality. In 2000 the mean household income in Saint-Roch was $18,537, less than half the city's median. Today, approximately half the neighbourhood's population live under

Figure 19.6 Saint-Roch Mall, 1974
Quebec City Archives, all rights reserved.

continued

the poverty line. Saint-Roch residents have poorer health and live shorter lives than the average Quebec City resident (Centre de santé et de services sociaux de la Vieille-Capitale, 2011). In this context, the mall, rather than rekindling local commercial vitality, soon became a haunt for the local poor, including the homeless, the elderly, and individuals experiencing mental distress (Freedman, 2011).

At the same time, the province of Quebec experienced two large waves of deinstitutionalization of mental health care, first in the 1960s and then in the 1970s–1980s. Saint-Roch became an important crossroad for the newly deinstitutionalized patients, thanks in part to the proximity of local mental health service centres. The real estate values in the neighbourhood were lower than in more prosperous areas, and landlords were somewhat more willing to accept individuals experiencing mental health issues, which led to a local concentration of rooming houses. Thus, following deinstitutionalization, about half the individuals with mental health issues in Quebec City congregated in Saint-Roch. These individuals were generally poor, with 57 per cent of them having a yearly income under CAD$12,000 (Centre de santé et de services sociaux de la Vieille-Capitale, 2011). An underground and exploitative industry flourished around these individuals, involving pawning, cashing welfare cheques for a fee, and usury. An important proportion of these individuals also experienced some forms and periods of homelessness.

Lives of deinstitutionalized mentally distraught individuals, particularly those who are homeless, are characterized by unrest and mobility as they navigate networks of resources to meet their basic needs and to socialize on a daily basis. The Mail Saint-Roch represented an anchor in many urban nomadic lives. Designed to foster leisurely consumption, the mall was a place where individuals could loiter while remaining warm and dry and freely access basic resources such as drinking water and washrooms. The mall was also an important place of congregation where individuals with mental health issues would socialize with one another, as well as with other local poor people, especially the elderly (Bourgeois, 2008; Freedman, 2011).

The progressive closure of the mall over the early 2000s meant the loss of a shelter, a landmark, and a central organizing feature in the lives of the Saint-Roch residents, particularly those experiencing mental distress. The dismantlement of the roof over Saint-Joseph Street was part of a wider trend to revitalize Quebec City's downtown and rekindle its attractiveness as both an upscale commercial and a residential area. It involved a process of dilution (Parazelli et al., 2013) meant to break off the concentration of spaces and resources used by mental health services users and the homeless.

Despite the creation of alternative spaces and services, the dismantlement of the mall further marginalized mentally distraught individuals in Saint-Roch by increasing the length and frequency of their movements through the city as they followed the dispersion of resources such as drop-ins or food banks over a wider territory. This nomadism and sense of rootlessness in their lives is enhanced by the increased surveillance and control of public spaces and by the criminalization of "mad" behaviours, such as sleeping on public benches or talking and/or yelling to oneself and/or to passersby, that are included in the gentrification efforts in the neighbourhood (Bourgeois, 2008; Freedman, 2011).

Conclusion

This chapter has examined disability in the city, beginning with definitions of disability and the medical, social, and embodied social models of disability as well as of inclusive design. The chapter then examined work by disability geographers, tracing important trends from early work on mental health to the rise of the subfield of disability geography in the 1990s. Recent work on disability has expanded to include more forms of bodily difference, challenging stereotypes about which bodies are "normal." Next, the chapter's focus shifted to mental health/distress in the city, the reclaiming of the term "mad" and highlighting the inherent connections between mind and body. Weaving geographic literature on mental health into a history of asylums, deinstitutionalization, community living, and NIMBYism, the section emphasized the destructive power of stigma. The case studies examined disability in the city from two perspectives. The first explored the intersection of poverty and disability in Georgetown, Guyana, where an inaccessible built environment and negative cultural attitudes often discourage disabled people from being in public space and taking part in city life. The next case study, on visually impaired young people in the UK, also considered the importance of social inclusion in the city. Even though a space may be physically accessible and disabled people can navigate it, it may still be exclusionary because of the sometimes negative reactions of the public. In the Canadian context, the theme of social inclusion was explored, specifically the intersection of urban poverty and deinstitutionalization through the Mail Saint-Roch in Quebec City. The gentrification of the neighbourhood and the closure of the mall illustrate how cultural attitudes and ableist day-to-day social interactions with people with disabilities and mental health conditions become implicated in city planning.

While research on disability and the city is slowly moving forward in the global North, there remains much work to be done in the global South, where most disabled people live and where hardships can be acute. Just as work needs to be done with different members of disability communities to help us understand their urban experiences, there is a need for research in cities around the world to capture local cultures and show how disability is lived in diverse places. There are three thematic areas where research on disabling cities can be taken forward. The first concerns interrogating the built form of the city, as physically inaccessible spaces undermine other attempts at inclusion. The second future area of work involves scaling up some of the in-depth, qualitative work that is popular in the subfield of disability geography—examining and critiquing policies and practices of exclusion in the city. Finally, in researching the urban more generally, geographers often pay attention to gender, ethnicity, and socio-economic status, yet other forms of difference—especially disability—are unmentioned. Fully understanding citizens and urban life ideally involves embracing the complexity of our identities—disability and mental health are important parts of this intersectional picture.

Key Points

- How disabilities are conceived influences how cities are imagined: thinking of disability as a medical trait emphasizes individual responsibility, and thinking of disability as socially constructed and experienced emphasizes issues of inclusion and accessibility.
- Inclusive design allows ableism to be challenged, as it invites consideration of a wide diversity of human experiences and needs, rather than of particular norms of functionality and associating stigma with individuals who diverge from these norms.
- Geographies of mental, emotional, and physical impairments are closely intertwined with the economic, political, and social processes and phenomena driving city changes (e.g., neoliberal processes of public-service cuts and of privatization of urban spaces has led to reduced access to certain spaces for individuals experiencing physical disabilities and/or mental illness).
- The Georgetown case study reveals how poverty is intimately connected with disability, where a lack of resources limits disabled people's ability to access the city, as well as their ability to advocate for disability rights.
- The Liverpool case study with visually impaired young people demonstrates that disability is very much a social experience and that the way people are treated by others in public space can be more of a barrier than any physical impairment.
- In Canadian cities, the lives of individuals experiencing mental distress are intertwined with histories of deinstitutionalization, neighbourhood changes, and concentrations of mental health resources, with severe forms of exclusion linked to mental health but also to stigma and poverty (e.g., heavy policing, NIMBYism, and gentrification).

Activities and Questions for Review

1. Draw your own version of the medical, social, and embodied social models of disability. What are the differences between these models?
2. In a small group, discuss how ableism is reflected in your journey from home to the university campus and to this classroom.
3. In pairs, discuss how levels of mental and emotional stress can be affected by exposure to the urban built environment and the social relations that animate it.
4. Based on the case studies in this chapter, compare and contrast how disability is differently experienced in cities of the global North and global South.

Acknowledgments

Nancy Worth's research was supported by a Banting Fellowship; Laurence Simard-Gagnon's research was supported by the Fonds de Recherche du Québec—Société et Culture and by the Social Sciences and Humanities Research Council of Canada; and Vera Chouinard's research was supported by a grant from the Social Sciences and Humanities Research Council of Canada.

References

Barnes, C., and Mercer, G. (2010). *Exploring disability*. Cambridge: Polity.

Bourgeois, F. (2008). *La revitalisation du quartier Saint-Roch (ville de Québec) et ses effets sur l'expérience d'exclusion des femmes itinérantes* (MA thesis). Laval, Quebec City, Canada.

Boyd, J., and Kerr, T. (2015). Policing "Vancouver's mental health crisis": A critical discourse analysis. *Critical Public Health*, 26(4): 1–16.

Centre de santé et de services sociaux de la Vieille-Capitale (2011). Plan d'action local en santé publique 2011–2015. Quebec.

Chouinard, V. (1999). Body politics: Disabled women's activism in Canada and beyond. In R. Butler and H. Parr (Eds.), *Mind and body spaces: Geographies of illness, impairment and disability*. London: Routledge, pp. 269–294.

Chouinard, V. (2011). "Like Alice through the looking glass" II: The struggle for accommodation continues. *Resources for Feminist Research*, 33(3–4): 161–178.

Chouinard, V. (2012). Pushing the boundaries of our understanding of disability and violence: Voices from the global South (Guyana). *Disability and Society*, 27(6): 777–792.

Chouinard, V. (2014). Precarious lives in the global South: On being disabled in Guyana. *Antipode*, 46(2): 340–358.

Chouinard, V. (2015). Contesting disabling conditions of life in the global South: Disability activists' and service providers' experiences in Guyana. *Disability and Society*, 30(1): 1–14.

CMHA (Canadian Mental Health Association) (2015). Retrieved from http://www.cmha.ca

Crooks, V.A. (2010). Women's changing experiences of home and life inside it after becoming chronically ill. In V. Chouinard, E. Hall, and R. Wilton (Eds.), *Towards enabling geographies: "Disabled" bodies and minds in society and space*. Surrey, UK: Ashgate, pp. 45–62.

Davidson, J.D. (2003). *Phobic geographies: The phenomenology and spatiality of identity*. Aldershot, UK: Ashgate.

Davidson, J., and Parr, H. (2010). Enabling cultures of (dis)order on line. In V. Chouinard, E. Hall, and R. Wilton (Eds.), *Towards enabling geographies: "Disabled" bodies and minds in society and space*. Surrey, UK: Ashgate, pp. 63–84.

Dear, M., and Wilton, R. (1997). Seeing people differently: The sociospatial construction of disability. *Environment and Planning D: Society and Space*, 15(4): 455–480.

Dyck, I. (1995). Hidden geographies: The changing lifeworlds of women with multiple sclerosis. *Social Science & Medicine*, 40(3): 307–320.

Dyck, I. (1999). Body troubles: Women, the workplace and negotiations of a disabled identity. In R. Butler and H. Parr (Eds.), *Mind and body spaces: Geographies of illness, impairment and disability*. London: Routledge, pp. 117–134.

Freedman, M. (2011). *De la mixité à l'exclusion: Témoignages du nouveau Saint-Roch à Québec* (PhD diss.). Laval, Quebec City, Canada.

Gleeson, B. (1999a). Can technology overcome the disabling city? In R. Butler and H. Parr (Eds.), *Mind and body spaces: Geographies of impairment, illness and disability*. London: Routledge.

Gleeson, B. (1999b). *Geographies of disability*. London: Routledge.

Hall, E., and Wilton, R. (2011). Alternative spaces of "work" and inclusion for disabled people. *Disability & Society*, 26(7): 867–880.

Hansen, N.E., and Philo, C. (2007). The normality of doing things differently: Bodies, spaces and disability geography. *Tijdschrift voor Economische en Sociale Geografie*, 98(4): 493–506.

Imrie, R. (1999). The body, disability and Le Corbusier's conception of the radiant environment. In R. Butler and H. Parr (Eds.), *Mind and body spaces: Geographies of illness, impairment and disability*. London: Routledge, pp. 25–45.

Imrie, R., and Edwards, C. (2007). The geographies of disability: Reflections on the development of a sub-discipline. *Geography Compass*, 1(10): 1–18.

Imrie, R., and Hall, P. (2001). *Inclusive design: Designing and developing accessible environments*. London: Spon Press.

Kirby, A.M., Bowlby, S.R., and Swann, V. (1983). Mobility problems of the disabled. *Cities*, 1(2): 117–119.

Kitchin, R. (1998). "Out of place," "knowing one's place": Space, power and the exclusion of disabled people. *Disability & Society*, 13(3): 343–356.

Kitchin, R. (2003). Architects disable: A challenge to transform. *Building Material*, 10: 8–13.

Kitchin, R., and Law, R. (2001). The socio-spatial construction of (in)accessible public toilets. *Urban Studies*, 38(2): 287–298.

Knowles, C. (2000). Burger King, Dunkin Donuts and community mental health care. *Health and Place*, 6: 213–224.

Kruse, R.J. II (2010). Placing little people: Dwarfism and geographies of everyday life. In V. Chouinard, E. Hall, and R. Wilton (Eds.), *Towards enabling geographies: "Disabled" bodies and minds in society and space*. Surrey, UK: Ashgate, 183–198.

Link, B.G., Yang, L.H., Phelan, J.C., and Collins, P.Y. (2004). Measuring mental illness stigma. *Schizophrenia Bulletin*, 30(3): 511–541.

Longhurst, R. (2010). The disabling affects of fat: The emotional and material geographies of some women who live in Hamilton, New Zealand. In V. Chouinard, E. Hall, and R. Wilton (Eds.), *Towards enabling geographies: "Disabled" bodies and minds in society and space*. Surrey, UK: Ashgate, 199–216.

McGrath, L., and Reavey, P. (2015). Seeking fluid possibility and solid ground: Space and movement in mental health service users' experiences of crisis. *Social Science and Medicine*, 128: 115–125.

Mad Pride Toronto (2014). Retrieved from http://www.madprideto.com

Morris, J. (1991). *Pride against prejudice*. London: Women's Press.

Mullings, B., and Peake, L. (2016). Critical reflections for geography on mental health in the academy. *ACME: An International Journal for Critical Geographies*, 15(2): 253–284.

Nelson, G., Patterson, M., and Maritt, K. (2015). Life changes among homeless persons with mental illness: A longitudinal study of housing first and usual treatment. *Psychiatric Services*, 66(6): 592–597.

Parazelli, M., Bellot, C., Gagné, J., Gagnon, E., and Morin, R. (2013). Les enjeux du partage de l'espace public avec les personnes itinérantes et sa gestion à Montréal et à Québec. Perspectives comparatives et pistes d'actions (No. 2011-PP-144446) (p. 79). Montreal: Fonds de recherche du Québec—Société et culture (FRQSC).

Parr, H. (1999). Delusional geographies: The experiential worlds of people during madness/illness. *Environment and Planning D: Society and Space*, 17: 673–690.

Parr, H. (2008). *Mental health and social space: Towards inclusionary geographies?* RBS-IBG Book Series. Malden, MA: Blackwell Publishing.

Parr, H., and Philo, C. (1995). Mapping mad identities. In S. Pile and N. Thrift (Eds.), *Mapping the subject: Geographies of cultural transformation*. London: Routledge, pp. 199–225.

Philo, C. (1987). Fit localities for an asylum: The historical geography of the nineteenth-century "mad-business" in England as viewed through the pages of the Asylum Journal. *Journal of Historical Geography*, 13(4): 398–415.

Sedghi, A. (2015). What is the state of children's mental health today? *The Guardian*, January 15. Retrieved from http://www.theguardian.com

Skelton, T., and Valentine, G. (2010). "It's my umbilical cord to the world . . . the Internet": D/deaf and hard of hearing people's information and communication practices. In V. Chouinard, E. Hall, and R. Wilton (Eds.), *Towards enabling geographies: "Disabled" bodies and minds in society and space*. Surrey, UK: Ashgate, 85–106.

Slater, T. (2004). Municipally managed gentrification in South Parkdale, Toronto. *Canadian Geographer/Le Géographe canadien*, 48(3): 303–325.

Taylor, S.M., and Dear, M.J. (1981). Scaling community attitudes toward the mentally ill. *Schizophrenia Bulletin*, 7(2): 225–240.

Thomas, C. (2004). How is disability understood? An examination of sociological approaches. *Disability & Society*, 19(6): 569–583.

Valentine, G. (1999). What it means to be a man: The body, masculinities, disability. In R. Butler and H. Parr (Eds.), *Mind and body spaces: Geographies of illness, impairment and disability*. London: Routledge, pp. 163–175.

Wilton, R. (2000). Grounding hierarchies of acceptance: The social construction of disability in NIMBY conflicts. *Urban Geography*, 21(7): 586–608.

Wilton, R. (2004). Putting policy into practice? Poverty and people with serious mental illness. *Social Science and Medicine*, 58(1): 25–39.

Wilton, R., Deverteuil, G., and Evans, J. (2014). "You've got to change who you are": Identity, place, and the remaking of masculinity in drug treatment programs. *Transactions, Institute of British Geographers*, 39(2): 291–303.

Wolch, J., and Philo, C. (2000). From distributions of deviance to definitions of difference: Past and future mental health geographies. *Health and Place*, 6(3): 137–157.

Wong, Y-L.I., and Stanhope, V. (2009). Conceptualizing community: A comparison of neighborhood characteristics of supportive housing for persons with psychiatric and developmental disabilities. *Social Science and Medicine*, 68(8): 1376–1387.

World Population Review (2015). Guyana population. Retrieved from http://worldpopulationreview.com/countries/guyana-population

Worth, N. (2008). The significance of the personal within disability geography. *Area*, 40(3): 306–314.

Worth, N. (2013). Visual impairment in the city: Young people's social strategies for independent mobility. *Urban Studies*, 50(3): 574–586.

20

Cities, Sexualities, and the Queering of Urban Space

David K. Seitz and Natalie Oswin

Introduction

Scholars of sexuality, and especially the geographers among them, have long attested that the sexual is spatial and the spatial is sexual (Bell and Valentine, 1995). They have also shown that there are particular, and particularly important, connections between **sexuality** and city spaces. In short, cities are central sites for the regulation of and resistance to notions of sexual propriety and impropriety, as sexuality is intertwined with numerous aspects of urban life, from reproduction and population growth, to security and the separation of zones of "vice" from "respectable" neighbourhoods, to the emergence of spaces for lesbian, gay, bisexual, transgender, and queer (**LGBTQ**) communities and social movements. Thus, sexuality, sexual identity, and practice have political importance in the lives of a wide range of people in cities, such as LGBTQ-identified people, sex workers, single parents, people on public assistance, migrants, and racialized others. For populations like these, cities can afford a unique platform, offering both the freedom of relative anonymity and the opportunity to gather and organize in larger numbers.

This chapter outlines some of the multiple meanings of the term "**queer**," which is used both as an umbrella term to represent a wide range of identities on the LGBTQ spectrum and as a critical approach to the shifting and exclusive ways that sexual norms work in distinct urban contexts. A queer approach to cities and urban life directs attention to the ways that sexuality is used to mark some urban subjects as worthy of **social inclusion** and others as the bearers of social ills or as problems to be solved. Informed by, but going beyond, LGBTQ experiences, a queer take on the relationship between sexuality and cities invites connections between sexuality and related forms of **difference**, identity, and power, such as race, class, gender, immigration status, religion, ability, and age. Thus, queer theoretical and political perspectives enable urban scholars to analyze the ways in which urban **social exclusion** and structural inequality operate through shifting moral, sexual, and spatial norms. They also help to highlight the dilemmas that urban populations whose sexuality has been marginalized can face with respect to **visibility** and invisibility in cities, showing how visibility can be a double-edged sword.

Following on from this conceptual discussion, we present three case studies—from Chicago, Ciudad Juárez, and Singapore—to showcase some of the ways in which a queer approach can be put to work in relation to diverse urban locales to help shed light on a wide range of urban issues. Chicago is the third-largest city in the United States and one that is both hailed as a haven for LGBTQ people in the American Midwest and decried as a site of stark racial and economic inequality and segregation. For LGBTQ communities in Chicago's Boystown neighbourhood, visibility has enabled the consolidation of certain forms of political clout, beginning in the late twentieth century. But it has also concealed dramatic race and class inequalities and struggles within the LGBTQ community and made the neighbourhood a target of homophobic violence and forms of securitization that welcome some, but not all, LGBTQ people. Second, we look at Singapore. This Southeast Asian city-state's economy was remade over the last couple of decades to shed its reliance on manufacturing-based industry and embrace knowledge-based sectors, and in the process, significant changes in sexual citizenship were made. Most visibly, for Singapore to foster an image of itself as a creative city, official attitudes towards public expressions of homosexuality have been liberalized and an LGBTQ movement has emerged to challenge persistent discrimination in the city-state. Less discussed, Singapore maintains a bifurcated migration regime that invites "foreign talent" and their families to become part of the national family through naturalization, while "foreign workers" have no route to future citizenship and are prohibited from bringing dependants with them as well as from marrying and/or having children locally. Thus, while the bounds of the national family have expanded in Singapore's age of creative urbanism, **heteronormativity** is still alive and well as an exclusionary social force. Third, as the case study of La Zona de la Paz in Ciudad Juárez, Mexico, makes clear, for working-class women participating in informal **sex work** economies, visibility confers both grave danger and possibilities for contestation. The case of Juárez shows that the economic processes of industrialization, deindustrialization, gentrification, and securitization are indelibly shaped by the politics of sexuality and carry profound implications for urban geographies of sexuality. As sex workers have found themselves literally banished from public space near downtown as the city's elites implement a glossier and more high-tech vision for Juárez, visibility has been both a legal tool used to justify police harassment and a weapon that women have reclaimed, defiantly staging their bodies and sexuality in the urban public sphere as a form of resistance. Consideration of both LGBTQ identities and sex work in the Canadian context provides additional geographical layers to these case studies. While the case studies all focus on major population centres, the 2011 Canadian census found that over half of Canadian same-sex couples live outside Canada's three largest cities. In Canada's smaller cities, sexual minorities face similar issues of visibility and invisibility as those in larger cities, but they also face their own geographically distinct dilemmas, as is made clear by the experience of Indigenous sex workers in Regina and queer women in Lethbridge, Alberta.

These case studies, and the discussion that follows on sexual politics in Canadian cities, span cities in the global North and the global South. They demonstrate the relevance of queer theory's view of gender and sexuality as discursive and social forces—rather than private or interior truths—in understanding urban space and life around the globe. They reveal that a queer approach helps us to put LGBTQ urban experiences and politics on the agenda of urban studies, while also helping highlight the manifold forms of violence and exclusion that heteronormative cultural logics perpetuate within cities everywhere, even within urban LGBTQ communities. The case studies track how visibility both empowers and constrains a wide range of urban populations whose intimate and sexual forms locate them outside the bounds of normativity, and how sexual norms are implicated in a wide range of issues (e.g., transnational labour flows, public health, public space, poverty) that are core to the functioning of cities.

Queering Urban Space

"Queer" is a complex term. In popular parlance, it is now used most widely as a simple descriptor for LGBTQ communities and social movements, but its deployment in activist and scholarly circles has grown out of particular lineages and has specific meanings (see Jagose, 1996, for an overview of queer theory's emergence and key tenets). For many decades in the English-speaking global North, queer was generally used as a derogatory slang term for non-heterosexuals. But in these same countries, certain radical HIV/AIDS organizations began to re-appropriate the term in the early 1990s. As groups like ACT UP and Queer Nation fought against the stigma of the widespread labeling of HIV/AIDS as a "gay disease," they advanced a radical, sex-positive, and anti-**assimilationist** politics that they labeled queer rather than gay and lesbian. By re-appropriating queer, these organizations rejected its negative connotations and thereby challenged the notion that there is any putatively normal sexuality. In this usage, a queer identity and politics "embraced literally anyone who refused to play by the rules of **heteropatriarchy**" (Bell and Valentine, 1995: 19).

At around the same time and mostly within the same Anglo-American geographic locales, queer entered the academy as a new mode of theorizing. Here, too, the term took on a radical, politically confrontational tone. Queer theory, as it emerged within literary theory circles in the 1990s, offered a reorientation of scholarly understandings of sexuality. Against prevailing popular and scholarly views of sexual identities as natural, fixed, and biologically determined, queer theory asserts that sexual identities are social constructions that do not pre-exist their worldly (i.e., cultural and linguistic) deployments. It maintains a focus on the plight of sexual minorities while fundamentally challenging the empirical validity and conceptual usefulness of identity categories. As Fran Martin (2003: 25) states, "as 1990s feminist theory did with 'women' and postcolonial theory did with 'race' and 'culture,' queer theory was concerned to disrupt the assumed universality and internal coherence of previous categories of identification in 'gay and lesbian identity.'" The insight that sexualities are performed, that they are something we *do* rather than something we *have*, has formed the basis for the development of a now highly significant interdisciplinary and international literature. By acknowledging both hegemonic heterosexuality and marginal non-heterosexualities as socially, historically, and geographically contingent, work within queer studies has taken on the critical tasks of understanding the ways in which sexual identities are performed and of challenging the myriad processes through which sexual norms become naturalized in a wide range of times and places.

Thus, there are two central meanings of queer. First, it signifies a sexual identity, functioning as an umbrella term that is self-consciously embraced by many of those who fall outside the bounds of "normal" sexuality. Second, it signifies a **poststructuralist** critique of the very notion of sexual identity. Poststructuralist approaches to understanding social life challenge the idea that meaning is fixed, essential, timeless or simply true, emphasizing instead the ways meanings are socially produced and transformed. Poststructuralism thus seeks to uproot and scramble taken-for-granted binary oppositions of meaning, including identity/difference, man/woman, homosexual/heterosexual, and white/black, showing how such binaries reduce prolific diversity into an exclusive either/or model and assume fundamental oppositions between terms that can prove far more murky and fluid in historical experience. For queer theory, it is acknowledged that sexual identity categories come to take on social meaning and thus cannot be abandoned in political struggles over sexuality and social justice. But, while urgently calling attention to and seeking to improve the lives of LGBTQ persons, queer theory pushes beyond liberal frameworks of identity and difference. It challenges the presumed fixity of sexual identities and critiques the ways in which sexual norms are deployed as part of broad structures of governance. In this sense, queer theory provides an analytic and diagnostic tool that facilitates the critical interrogation of the power of sexual norms and of the far-reaching effects of heteronormative cultural logics.

Crucially, this shift from a liberal, identity-politics approach to a poststructuralist queer critique implies that attentiveness to spatial relations might usefully be centralized as we attempt to understand the ways in which sexual norms operate as a social force. With the assertion that the subject comes into being in the world, rather than being born with a fully formed sexual identity, queer theorists effectively re-spatialize our understanding of sexual subjectivity. In short, since sexual identity is not a pre-given, biological essence, but rather a socially and culturally inflected manifestation, space and place very much matter in the determination of sexual experiences and norms. In other words, since sexuality is in significant part a discursive phenomenon and **discourse** changes and mutates across time and space, sexual normativities and non-normativities are geographically and historically dependent. Further, though discourse is unfixed, it is not fully mutable, and sexual normativities take on material force because social constraints limit the expression of sexual identities. Since sexual norms and identities are not easily changeable, being constrained by context and by the ways discourses stick in context, a critical task at hand is to challenge the myriad processes that make the contingent phenomenon of hegemonic heterosexuality seem fixed.

Analyses of sexual politics as they play out in particular urban sites are pivotal to this queer political and scholarly project of destabilizing sexual norms, not least because cities have played a central role in the historical formation of sexual subcultures. Since 1983, when John D'Emilio (1983) noted that the coupling of industrialization and urbanization that separated individuals en masse from their extended families led to the formation of highly gendered public and private spheres and that these urban spaces of same-sex sociality facilitated the creation of gay and lesbian communities and political movements, an abundant literature has emerged within queer studies on how queer politics and culture take place in particular cities. This chapter turns now to a brief overview of the existing queer studies literature, specifically highlighting its geographical parameters and important topical emphases. Note that while space does not permit a detailed discussion here of the ways in which the existing queer studies literature "queers" urban studies in line with the preceding conceptual discussion, the three case studies that follow this section of the chapter demonstrate how a queer theoretical approach can be put to work in the study of particular urban locales.

Cross-disciplinary studies documenting LGBTQ movements and lives in cities of the global North that are popularly known as global gay meccas—such as London, San Francisco, and New York—have been especially plentiful over the last few decades (Abraham, 2009; Chauncey, 1994; Chisholm, 2004; Hanhardt, 2013; Houlbrook, 2005). Further, a wider range of global northern cities are represented in the broader literature, through studies of, for example, LGBTQ residential spaces (Gorman-Murray, 2008, 2012), bars, clubs, "gaybourhoods," and other commercial and community spaces (Bain et al., 2015; Caluya, 2008; N.M. Lewis, 2013; Nash, 2005; Podmore, 2006), and temporary appropriations of public spaces through events such as Pride parades (Browne, 2007; Johnston, 2005; Marston, 2002; Waitt, 2006). Also, in recognition of the fact that cities are not discrete, isolated spaces, many scholars have examined linkages between LGBTQ urban and non-urban or rural spaces (Nash and Gorman-Murray, 2014; Waitt and Stapel, 2011). Going beyond the global North, while it is well acknowledged that queer studies was problematically Western-centric through the first decade or so of its existence, a trickle of work on a more globally representative set of locales began in the late 1990s. Fortunately, that trickle has since become a wave, and there is now a sizable literature on LGBTQ politics and cultures in cities around the globe, including Bangkok (Jackson, 2011), Beijing (Engebretsen, 2014), Beirut (Merabet, 2015), Calcutta (Banerjea, 2015), Cape Town (Tucker, 2009), Hong Kong (Tang, 2011), Manila (Benedicto, 2014), Rio de Janeiro (Hutta, 2013), Singapore (Yue and Zubillaga-Pow, 2012), Tel Aviv (Ritchie, 2015), and many more. Further, smaller cities are now also on the queer studies map everywhere, such as Bloemfontein, South Africa (Visser, 2008), Lethbridge, Canada (Muller-Myrdahl, 2013), and Ulyanosk, Russia (Stella, 2012).

As the field has broadened geographically, it has of course changed. In short, as queer thinking has gone global, queer critique has been put into concerted conversation with postcolonial critique, and some of its founding globalizing assumptions have been strongly shaken. As anthropologists Arnaldo Cruz-Malavé and Martin Manalansan (2002: 6) state, "developmental narrative[s] in which a premodern, pre-political, non-Euro-American queerness must consciously assume the burdens of representing itself to itself and others as 'gay' in order to attain political consciousness, subjectivity, and global modernity" no longer hold as much sway as they once did. At the same time, what we might call this southern, or at least non-Western, turn in queer studies has influenced queer theory in relation to cities of the global North. In other words, as queer scholars are starting to understand and embrace the particular and diverse ways in which LGBTQ subjects are part of urban world-making projects everywhere, we now better understand the ways that sexual politics are always entwined with race, class, and **gender** dynamics. We are also gaining a better understanding of how LGBTQ politics are tied to broader dynamics such as the urban politics of precarity and vulnerability, various modes of inclusion and exclusion, gentrification, austerity, boosterism, migrant worker rights, and more. Thus, we now have significant evidence that LGBTQ individuals, organizers, and movements are key actors in struggles over the right to the city everywhere.

Finally, in contrast to the by now rather large literature on LGBTQ lives and urban space, there is much less work in print that offers queer theoretical analyses of heterosexual experiences in the city or considers urban governance as heteronormative in a broad sense. Scholarship that examines commercial sex in the city develops such a line of enquiry most fully. As Phil Hubbard (2008: 646) notes, such a focus helps to clarify "how heteronormativity is reproduced spatially through the exclusion and containment of commercial sex work away from 'family spaces,'" and a wealth of historical and contemporary studies on sex work and urban red-light districts make important contributions to our understanding of the relationship between sexuality and space (also see Howell, 2004; Hubbard, 2004; Papayanis, 2000; Sanders, 2004). Aside from this literature, however, there remains much room to more fully "queer" the field of urban studies (Oswin, 2008, Seitz, 2015) by interrogating not just the identity politics of homosexuality but also the cultural logics that make particular expressions of heterosexuality seem right (Berlant and Warner, 1998). We now turn to the case studies, which deliberately demonstrate the sorts of critical insights that such a queer approach to both LGBTQ and heterosexual lives in urban spaces can enable.

Rainbows in Boystown: The Dilemmas of LGBTQ Visibility in Chicago

The past several years in the United States have seen significant shifts in public policy and cultural visibility for some LGBTQ subjects and domains of concern. With the end of formal exclusion from military service based on sexual orientation (2011) and gender identity (2016), and the nationwide legalization of same-sex marriage through the Supreme Court (2015), some have hailed the gains of the contemporary moment as indicative of the work that visibility can do to guarantee progress. Yet, more fine-grained attention to LGBTQ life in Chicago, routinely designated as one of, if not, the most segregated cities in the United States, reveals how unevenly the benefits of visibility and progress are distributed in the materiality of urban life. Since the emergence of the so-called Chicago School of urban sociological scholarship in the 1920s and 1930s, the city has long provided a basis for both investigating and modeling urban economic and demographic transformations. Chicago also features prominently in literary and journalistic work that gives voice to the ways in which urban violence and inequality are lived (Addams, 1912; Cisneros, 1984; Sinclair, 1906; Wright, 1940). The city's unique histories of industrialization, central positioning, integration in US infrastructural networks, and rapid transformation by multiple waves of immigration have led to the formation of distinct, highly segregated, and, as will become clear, contested neighbourhood spaces.

Such histories also make Chicago a particularly interesting case study in LGBTQ neighbourhood formation and transformation. As early as the 1950s, social and sexual networks of predominantly white gay men began establishing a social and residential presence in Belmont Harbor, a part of the larger Lakeview neighbourhood by the shores of Lake Michigan on Chicago's north side (Figure 20.1). The neighbourhood, known colloquially as "Boystown," became home to a number of gay social and political institutions and also became a target of the violence of police harassment. In 1970, gay activists chose a route on North Halsted Street on Chicago's north side to hold a Gay Pride parade—a political intervention that became an annual tradition (see Figure 20.1).

In a highly segregated city like Chicago, the formation of a gay enclave not only gave some LGBTQ people a space to congregate, it also yielded political clout (Stewart-Winter, 2016). Strikingly, in Chicago, some city actors have gone out of their way to position themselves as "progressive" on LGBTQ issues in order to cultivate support from LGBTQ voters and districts. In 1991, the city of Chicago itself established a Gay and Lesbian Hall of Fame in the Boystown neighbourhood with the express purpose of recognizing local LGBTQ individuals and institutions. In some respects, Chicago's longtime mayor Richard M. Daley, who served from 1989 to 2011, approached LGBTQ communities as though they were an ethnic or racial community. This approach took concrete form in Daley's efforts in the late 1990s to help ethno-specific neighbourhoods (e.g., Ukrainian Village and Greektown) brand themselves so as to attract visitors and money. Alongside signage in ethnic neighborhoods, in 1998 the city of Chicago erected three 6-metre (20-foot) bronze rainbow pylons (see Figure 20.2) on a stretch of North Halsted Street, meant to designate the presence of the LGBTQ community (Reed, 2003).

Both before and after their installation, the pylons have elicited a range of responses. For some, the pylons represent hard-won and hard-fought recognition and visibility for sexual minorities, who often continue to face discrimination in employment and housing and everyday intimidation and street harassment. Some non-LGBTQ residents felt unfairly lumped in or incorrectly described by the official marking of the neighbourhood as a key site for sexual minorities. Some LGBTQ residents worried about the possibility that marking the neighbourhood would attract homophobic and transphobic violence (Reed, 2003). Others have found the statues playful, imaginative, or simply funny. Indeed, many have jokingly observed that the pylons, with their obvious phallic shape, pay tribute

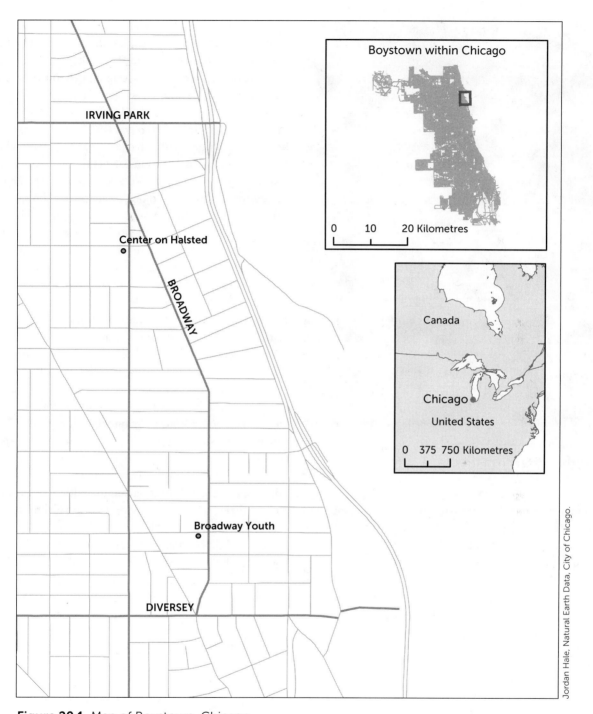

Figure 20.1 Map of Boystown, Chicago

Figure 20.2 Boystown rainbow pylons designating the presence of the LGBTQ community

to men's same-sex desire. However, such a monument to desire among gay men underscores the fact that the neighbourhood caters to gay men, often to the exclusion of lesbians, bisexual, and trans people, and also excludes many gay men who lack the resources or racial privilege to move through Chicago with relative ease (Nast, 2002). Such visible signs of progress and recognition for gay men have been criticized for celebrating the power structure in a city that remains marred by inequality, racism, segregation, inadequate housing, rampant police brutality, and a patriarchal political culture. While the pylons might make the city appear progressive and inclusive, many city actors, including police, continue to act in ways that perpetuate race and class inequality. These urban social problems, critics point out, affect many LGBTQ Chicagoans, including and often particularly working-class LGBTQ people of colour. Thus, while the neighbourhood remains a popular destination for many working-class black and Latino Chicagoans (Daniel-McCarter, 2012), many of whom journey from the city's south side to hang out in Boystown, anxieties about race and class have also led to a backlash from affluent neighbourhood residents and business owners, as in a recent campaign to "Take Back Boystown" (Greene, 2014). The fact that some precarious and street-involved LGBTQ youth engage in sex work (Fishman, 2012) at the edges of Boystown has been cited as a justification for neighbourhood securitization. This slippage—between the mere presence of sex work and a sense of threat to neighbourhood safety—points to the persistence of stigma (Hubbard and Sanders, 2003) around sex work, as some LGBTQ identities and practices are increasingly regarded as normal.

The resulting increased presence of police and private security forces in the neighbourhood—particularly ironic at Pride celebrations meant to honour a struggle for social equality—continues to raise powerful questions about whether the rainbow pylons deliver the belonging and freedom they seem to promise to all LGBTQ Chicagoans, or to just a privileged few. Differences that came to the fore around the "Take Back Boystown" campaign, however, reflect much longer-standing histories of tension and inequality, histories evinced by the work of the Broadway Youth Center (BYC). This organization offers accessible social space and resources around health and housing to racialized, precarious, and street-involved LGBTQ youth in Chicago. BYC's location in Buena Park, a gentrifying neighbourhood just north of Boystown, has led to tension between youth seeking to access its services and neighbouring residents and businesses (Daniel-McCarter, 2012). The trajectory of increasing visibility for some LGBTQ Chicagoans, then, is not a straightforward, linear narrative of progress from concealment to recognition and acceptance. Rather, it is a story in which diverse urban actors struggle over which sexual identities and practices can make legitimate claims on urban spaces and on ordinary (read heterosexual), unimpeded urban life. While dominant urban political and economic actors may embrace some historically marginalized forms of sexual identity and practice, the question of space for other queers remains vigorously contested. We now turn a case study from Southeast Asia that likewise demonstrates the complications of LGBTQ urban visibility and the folly of critical analyses knowledge-based economy. As part of these efforts, the Singapore government has taken on board the now-common contention among urban policy advisors that the creative, highly skilled workers, or "talent," that drive this sort of new economy are attracted to places with "buzz." Phenomenal changes to the urban landscape have been effected as a result. These include infrastructural investments in universities and art centres, the development of "lifestyle quarters," leisure areas and green spaces, and much more (Lim, 2009; Olds, 2007; Tan, 2007; Wong and Bunnell, 2006). Further, this new phase in Singapore's global city project has also entailed a remarkable liberalization of the polity, in part because its government has taken on board the idea put forward by urban policy advisors such as Richard Florida (2002) that urban buzz is best fostered where there is a certain degree of societal openness.

One of the most striking examples of this liberalization has been the change in the Singapore government's stance on homosexuality. Through the 1990s, police raids on gay bars and surveillance of known cruising areas were not uncommon; representations of homosexuality in the public sphere were tightly controlled through strict censorship of literature, film, and theatre; and attempts by a gay and lesbian advocacy organization to register as an official society were flatly denied (Heng, 2001). Since the early 2000s, however, local gay and lesbian film and theatre productions have been produced with limited censorship; gay and lesbian bars and saunas have been allowed to operate without interference; and a large LGBTQ gathering that organizers have named "Pink Dot" has taken place in a public park annually since 2009 (see Figure 20.3). Indeed, so much has changed that Singapore has come to be popularly known as Asia's "new gay capital." Yet, although homosexuality is no longer actively condemned in the city-state, apparently for the sake of the economy, there are strict limits to this official liberalized approach to sexual regulation. Political change in the form of legislative or policy adjustments to facilitate equity for sexual minorities has been deemed beyond the pale, and strictures on LGBTQ community events and activist efforts persist (Chua, 2014; Yue, 2007). This situation has attracted much attention from scholars and other critical commentators, a heartening fact given that the struggle for rights for sexual minorities in Singapore is far from over.

Figure 20.3 "Pink Dot": An LGBTQ gathering in Singapore

that look only for distinctions along the heterosexual-homosexual binary.

Global City Singapore: Heteronormativity and Migration Flows

Singapore emerged from a long period of British colonial rule to become an independent city-state in 1965. Since then, its socio-economic development has been nothing short of extraordinary and it has gained recognition as a leading global city. Maintaining this status in a changing global environment, however, has entailed a drastic economic reorientation in recent years, as state-led development plans have fostered a shift to a post-industrial,

As Oswin (2010, 2014) argues, however, the fight for gay and lesbian rights is only part of the story of **sexual citizenship** in the city-state. Sexual citizenship denotes the contested ways in which dominant discourses of citizenship and belonging normalize some forms of sexuality (including sexual orientation, but not limited to it) over and above others. Exclusion from sexual citizenship, whether as a single parent, migrant worker, and/or LGBTQ person, can bring consequences that are both material and painful. The following brief discussion of two prominent and fairly recent controversies in Singapore illustrates this point.

In 2007, a group of gay and lesbian activists and their allies lobbied for the repeal of section 377A of Singapore's Penal Code, a statute that criminalizes "gross indecency" between men. In his speech to Parliament explaining why the law would stand, Prime Minister Lee Hsien Loong asserted that "the overall society . . . remains conventional, it remains straight" (Lee, 2007: n.p.). As such, he affirmed the city-state as predominantly hetero*sexual*. But if his speech is scrutinized further, it is clear that he did much more than that. He also affirmed the city-state as hetero*normative*. In Lee's (2007: n.p.) words, "The family is the basic building block of our society. It has been so and, by policy, we have reinforced this and we want to keep it so. And by 'family' in Singapore, we mean one man one woman, marrying, having children and bringing up children within that framework of a stable family unit."

Identifying this norm as a fundamental part of Singapore's landscape, he continued: "If we look at the way our Housing and Development Board flats [public housing in which 95 per cent of Singapore's population lives] are, our neighbourhoods, our new towns, they are, by and large, the way Singaporeans live." To understand this statement's far-reaching implications, we need only look to the Housing Development Board's (HDB) well-defined and strictly enforced tenancy regulations. To purchase an HDB flat, the applicant must be at least 21 years of age and "form a proper family nucleus," which is defined as the applicant and fiancé(e); the applicant, spouse, and children (if any); the applicant, the applicant's parents, and siblings (if any); if widowed/divorced, the applicant and children under the applicant's legal custody; and, if orphaned, the applicant and unmarried siblings. Gay and lesbian domestic arrangements are thus not the only ones rendered "queer" by **the state** and its housing program. So are those of unmarried persons, widowed/divorced persons without children, and single parents who have never been married.

The second controversy is tied to citizenship, the final factor in determining eligibility for HDB tenancy beyond age and family composition. Ownership of flats is limited to Singapore citizens and permanent residents, while foreigners legally residing in the city-state may rent or sublet housing flats with one significant exception—foreign construction workers. This group must be housed in dormitories provided by their employers, and it is this fact that led to a national controversy in 2008 when residents of Serangoon Gardens, an affluent private housing estate of semi-detached bungalows, terrace houses, and low-rise condominiums, presented a petition to the government in objection to a plan to convert an unused former school into a dormitory for foreign workers (see Figure 20.4). The concerns raised by the residents centred around assertions that housing workers in the unused school would lead to higher crime rates, an unclean environment, changes to the character of the neighbourhood, and conflicts between Singaporeans and foreign workers because of "cultural differences."

Despite much public criticism over this display of xenophobia in the pages of Singapore's daily newspapers and on the blogs of prominent local social and political commentators, the Ministry of National

Figure 20.4 Foreign-worker dormitory in Serangoon Gardens neighbourhood, Singapore

Development made a series of concessions to accommodate the disgruntled residents. These included the expansion of the "buffer zone" between the dormitory and existing residences, the construction of a separate access road, and requirements that the dormitory management work with the police to implement security measures and that they also provide house rules to minimize disturbances in the neighbourhood. There is much to be said about this conflict over neighbourhood development, but one thread of the debate that was not brought to the fore in local critical responses deserves to be highlighted. The fact that the workers to be housed were all male and without families came up again and again in the public statements made by the residents, especially in response to criticisms that the residents welcome "foreign talent" but not "foreign workers." As one anonymous commentator on the *Straits Times* forum page put it:

> [T]he expats arrive here with their families and they put their children in schools here. Foreign workers are in a "bachelor" state without their family. They are grouped together, single men in dormitories. The situation is entirely different. The connotations emanating from foreign single men living in dorms in an estate which is predominantly family-oriented is only too obvious. (www.straitstimes.com/ST%2BForum/Online%2BStory/STIStory_277226.html)

Thus, although debates over the place of homosexuality in Singapore have made the homosexuality-heterosexuality distinction the most visible aspect of sexual citizenship in the city-state, as attested by the range of exclusions perpetuated by Lee's assertion that Singaporean-ness will remain tied to a specific family form and the Serangoon Gardens residents' judgment of improper domesticity, sexual citizenship in Singapore is clearly about much more than the policing of this binary. The heteronormative logic that underpins Singapore's developmental efforts involves not one norm, but a set of norms. Its practice is animated by ideologies of the sexual in concert with ideologies of race, gender, class, and nationality (to name but some). And it results in the "queering" of a diverse array of subjects, including but by no means limited to those who might identify as LGBTQ.

The Women of La Paz, Ciudad Juárez: Sexual Visibility as Political Intervention

The Chicago and Singapore case studies demonstrated how the rise of visibility for some historically marginalized sexual citizens of the city does not make a uniform guarantee of a right to urban space for all subjects. Some LGBTQ urban citizens may benefit from visibility, while other urban subjects, such as street-involved people, working-class queers, queers of colour, and migrant workers, do not enjoy the same rights to the city because they are cast as improper sexual subjects. In a similar vein, this case study of Ciudad Juárez, Mexico, demonstrates how the provisional embrace of some forms of sex work by urban elites can be unevenly distributed among sex workers, privileging posh red-light districts designed to stimulate urban economic growth over spaces of informal sex work performed by working-class women.

The trajectories of Chicago and other cities in the American Midwest in histories of urban industrialization and deindustrialization are well known and widely discussed by urban geographers and sociologists (e.g., Lewis, 2008; Sugrue, 1996). Yet the significance of Mexico as a key site of industrial production and export processing for US and Canadian firms and the profound impact of international trade agreements on urbanization in Mexico are often overlooked. Located just across the US-Mexico border from El Paso, Texas, Ciudad Juárez, Chihuahua, has long been a city of central economic and political significance in the shifting economic geographies of North America (Figure 20.5). Juárez has been an attractive manufacturing site for American companies because of the significantly lower labour costs in Mexico. Since the North American Free Trade Agreement (NAFTA) came into effect in 1994, Ciudad Juárez has seen an acceleration of this process. *Maquiladoras*—factories in special "free-trade zones" that make it cheap and easy to import materials—have proliferated, manufacturing goods for rapid export. Work in *maquiladoras* is highly gendered, as firms favour women, believing that they are more industrious and better suited to the work of assembling goods, particularly textiles. Women workers' resistance and organizing in confronting poor labour conditions and their defiance of sexist norms around who is fit to be an industrial worker have led to a harrowing pattern of murders

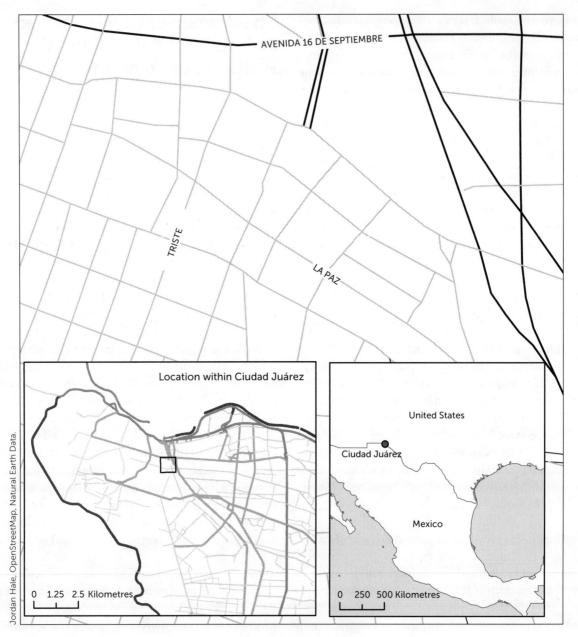

Figure 20.5 Map of la zona de la Paz, Ciudad Juárez, Mexico

and disappearances of women in Ciudad Juárez since the early 1990s. While official government estimates put the number of deaths in the hundreds, Mexico's National Citizen Femicide observatory has documented thousands of murders and disappearances nationwide per year—nearly 4000 from 2012 to 2013 alone, or nearly 6 murders per day. Experts have situated Juárez at the epicentre of what UN officials have called a pandemic (Matloff, 2015). Despite local and international outcry, murders and disappearances have continued with impunity.

It is in this context of changing trade regimes, emerging economic sectors, the transgression of gender norms, labour organizing, and, above all, violence against women—all practised with impunity—that sex work in Ciudad Juárez takes on powerful political significance (Wright, 2004). Urban and state actors have vastly different relationships to sex work and to the spaces that are claimed by sex workers. Power over urban space through the ways in which it is zoned, policed, or commercialized plays a crucial role in the livelihoods, survival, and marginalization of sex workers (Hubbard and Sanders, 2003). Wright (2004) argues that in the 1990s, the dominant vision among urban and corporate elites in Ciudad Juárez began a shift from a city of *maquiladora* industry and women workers to one of high-tech industry and development. This shift in story about Ciudad Juárez meant that the many forms of labour and exchange that working-class women engaged in to survive (including working in *maquilas*, selling drugs, working across the border, or selling sex) have been increasingly stigmatized.

Wright tells the story of La Zona de la Paz (the zone of peace), an area behind the city's Cuauhtemoc market, which was historically an informal market for a wide range of goods, including sex. Because wealthy industrialists and the mayor have begun branding Ciudad Juárez as "Silicon Valley South," sex work performed by low-income women—as opposed to high-end sex work in red-light districts elsewhere in Mexico—has become increasingly reviled. A plan to "systematically correct the vices" in downtown Ciudad Juárez authorized police efforts to remove sex workers from downtown (Wright, 2004: 380) by imposing a ban on women standing outdoors. This ban specifically targeted the working-class and outdoor-market sex work. Yet the women of La Paz have responded by occupying the public square even more visibly, staging both walking and standing protests. In a moment in a city when economic and political elites rework legal and security practices to keep poor women's bodies and sexuality out of sight, claiming space and making sexuality visible prove to be bold and defiant forms of resistance to neoliberal and exclusive political agendas. The case study of Ciudad Juárez thus points to the agency of sex workers in producing urban space (Hubbard and Sanders, 2003). Indeed, sex workers are not simply victims of top-down urban processes of zoning, securitization, or exclusion. They are also urban agents who co-produce and transform urban spaces. Although visibility's consequences are quite mixed for differently situated urban sexual citizens, the persistence of inequality and hierarchy does not foreclose possibilities for both transgression and transformation. As attention to the heterogeneity of forms of sexual citizenship in the Canadian context will make evident in the following section, urban life is characterized by both hierarchy and dynamism that can generate surprising forms of political affinity and solidarity. Moreover, the gaps and interplay between geographical scales (body, neighbourhood, city, nation, globe) can make cities especially inventive and fecund sites of resistance.

Visibility and LGBTQ Issues in Canadian Cities

Recent legislative and judicial changes at the federal scale have both responded to and conditioned shifts in sexual citizenship in Canadian cities. The federal legalization of same-sex marriage in 2005 was a response to provincial and federal court decisions in cases initiated by LGBTQ activists in Toronto, Vancouver, and Montreal. Yet the 2011 Canadian census points to some surprising insights regarding the geographies of LGBTQ people. For instance, while a higher proportion of same-sex couples lived in Canada's three largest cities than opposite-sex couples (45.6 per cent as opposed to 33.4 per cent), over half lived outside them (Statistics Canada, 2015).

That the majority of same-sex couples live outside of Canada's three largest cities suggests that struggles around sexuality are an important facet of the ordinary fabric of many Canadian cities,

continued

regardless of their size. With an urban population of roughly 83,000 and a metropolitan area of 105,000 as of the 2011 census, the city of Lethbridge, Alberta, for instance, may not immediately come to mind when the topic of "sexuality in Canadian cities" comes up. In contrast to Toronto, Montreal, and Vancouver, all of which are home to at least one gay village and dozens of social and political institutions, Lethbridge has no gay bars and only a handful of LGBTQ organizations. Yet, as Muller-Myrdahl (2013) demonstrates, paying attention to the ordinary lives of LGBTQ individuals can reveal a lot about how sexual identity works in southern Alberta's largest city and offer insights into understanding sexuality in other small, conservative cities throughout Canada.

Interviews can be a powerful technique for shedding light on the meanings and identities people attach to urban spaces and to their ordinary lives in cities. While Lethbridge is often (rightly) framed as a hotbed of conservative Christianity, Muller-Myrdahl's interviews with some of the city's queer-identified women residents—both long-time residents and transplants from other cities—tell a rich and complex story about sexuality in the small city. For instance, she spoke to a lesbian parent who, moving from Montreal, chose to challenge her own assumptions about a "small" place. She decided not to expect a frosty reception from neighbours and fellow parents and, to her pleasant surprise, was treated warmly, albeit still as "different." Another Lethbridge resident notes with relief that the city's history of anti-gay protests have become a thing of the past. Yet the interviewees also observe that more and more residential development in the city have taken the form of exclusive gated communities. This observation reminds us that LGBTQ lives in the city are affected by explicit forms of welcome and discrimination on the basis of sexual identity, but also that sexuality connects to other forms of identity and power, such as class, in shaping who can access which parts of the city. Although Lethbridge may not be the most visible queer space because of its size and common assumptions about small cities, its characteristics show that small cities are equally important sites of ordinary struggles around sexuality, race, gender, and class, as well as around the right to survive and thrive in urban space.

Moreover, even in large cities popularly thought of as visible centres of LGBTQ life in Canada, visibility can prove to be desired, controversial, unwanted, or unrealized. The formation of a **gay village** in Toronto in the 1960s—first at the intersection of Yonge Street and Wellesley Street and then a block east at Church Street and Wellesley—and daily life within it have been subject to ongoing debate within LGBTQ communities. In the 1960s and 1970s, gay activists of different political ideologies decried the space as a gay ghetto, arguing that the segregation of gay establishments and residential life kept the rest of Toronto ignorant about sexual minorities and made it easier to forget about the political work of transforming a homophobic society (Nash, 2006). With the increasing mainstreaming of LGBTQ identities (although some are still seen as more mainstream than others), critical perspectives in debates about the meaning and identity of the gay village have challenged contentions that LGBTQ institutions and spaces are no longer necessary (Nash, 2013). In Toronto, claims of a supposed "post-gay" city in which people can live and work and move without giving a care to identity have met with swift response from transgender communities, LGBTQ people of colour, migrants, refugees, and youth—queer subjects for whom the right to move through the city remains under grave threat. As Grundy and Smith (2005) suggest, working at the urban scale has enabled Toronto LGBTQ institutions to address the unique challenges faced by LGBTQ migrants, youth, and racialized people, whose interests have often been overlooked by pan-Canadian organizing for same-sex marriage rights. Examples of the potential value of the urban scale for sexual citizenship are by no means limited to Toronto. Podmore (2001) demonstrates how the sexual and gender heterogeneity of Montreal's Boulevard Saint Laurent commercial strip enables ordinary forms of lesbian sociality, contact, and desire to take place. Moving away from a preoccupation in the literature with visibility and marked, identifiable LGBTQ neighbourhoods, Podmore shows how heterogeneity can in fact facilitate alternative forms of lesbian visibility, community, and intimacy from within a sexually variegated crowd.

The differential consequences of visibility for distinct kinds of sexual citizen are also apparent when one considers transformations in how sex work is governed in Canada at multiple geographical scales. While the 2013 Supreme Court decision in *Canada v. Bedford* ostensibly struck down legislation banning the advertisement of sex work, organizing sex work in a brothel or bawdy house, or practising sex work as a livelihood, federal legislation the following year effectively restored many of those provisions. This quasi-legal, quasi-illegal condition can push sex work further underground, making sex workers more vulnerable to violence on terms that consolidate all the hierarchical relations of difference and power that intersect with sex work. Urban spaces labeled or construed as sites of sex work can mark sex workers as particularly vulnerable to a range of forms of violence, especially for working-class, racialized, and Indigenous sex workers. Sherene Razack's (2002) important study of the brutal murder of Pamela George, a woman from the Salteaux (Ojibway) nation working as a sex worker in Regina, demonstrates how colonialism, racism, and heteropatriarchy and the demarcation of urban space intersect. Razack argues that George's vulnerability to violence was inseparable from her connection to the fringes of Regina, a racialized and sexualized site of sex work that marked those who worked there as disposable in a colonial economy of life, death, and desire.

Thus, as with the cases of LGBTQ neighbourhood formation in Chicago, the limits of liberal inclusion in Singapore for a wide range of sexual outsiders, and the shifting and contested geographies of sex work in Ciudad Juárez, Canadian cities attest to the wide-ranging consequences of visibility and invisibility for urban sexual citizenship. Visibility can take the form of a marked, territorially bound gay ghetto, as in Toronto's Church and Wellesley village, but as the cases of queer parenting in Lethbridge or lesbian social life in Montreal indicate, less obvious forms of visibility can still be an occasion for recognition and community building. While some urban sexual citizens may enjoy the benefits of visibility—or shirk it in a supposedly post-gay era—visibility, recognition, and the right to urban space remain for others a matter of contestation and indeed survival, particularly when sexual citizenship is understood as connected to social relations and urban processes such as racialization, colonialism, segregation, and gentrification.

Conclusion

The case studies investigated in this chapter position cities as complex and unfinished sites of struggle and community formation around sexuality and sexual marginalization. Urban environments enable both the possible freedom of anonymity and the promise of community, but the types of visible community that can form around sexuality and that emerge in cities are not innocent or divorced from the inequalities and exclusions that are so core to contemporary urban life. Thus, while Chicago's rainbow pylons and Singapore's liberalization of its approach to homosexuality seem to promise progress, the benefits of this progress are unevenly distributed among people whose sexualities are stigmatized: working-class and racialized LGBTQ people in Chicago and migrant workers and single parents in Singapore, to name but a few. Likewise, while elite urban actors in Ciudad Juárez may condone glitzy forms of sex work in red-light districts deemed good for neoliberal capitalism as the city lays claim to the status of "Silicon Valley South," the working-class women doing sex work in La Zona de la Paz find their bodies, sexualities, and livelihoods criminalized and scrutinized. In the Canadian context, the visibility and geographical location of sex work on the outskirts of Regina mark sex workers, particularly Indigenous women, for grave vulnerability to colonial and sexual violence, while the (assumed) invisibility of LGBTQ people in Lethbridge, Alberta, is betrayed by important gains in inclusion—at least for more privileged LGBTQ people. At the same time, none of these spaces is devoid of the possibility of contestation and change. "Take Back Boystown" in Chicago led not to an exclusive consensus but to vigorous debate, and institutions like the Broadway Youth Center continue to do the work of reclaiming and recasting space to secure LGBTQ urban inclusion on more racially and economically just terms. The claiming of space by LGBTQ organizations in Singapore keeps debates over sexual citizenship in the limelight in that city. The women of la Paz have boldly challenged their invisibilization, defiantly placing their bodies in public view and daring state authorities to displace them. Finally, as in the case of lesbians in Montreal, visibility and invisibility are themselves relative and subjective. Lesbians might be invisibilized by others or might seem to fade into a more diverse crowd, but simultaneously be visible to one another on terms that enable intimacies, community formation, and political solidarity, under the radar of and in defiance to heteropatriarchy.

Together, the case studies in this chapter show that a queer post-structuralist lens on urban life invites careful attention to how many different people find themselves inside or outside sexual, spatial, and moral norms; how forms of inclusion and exclusion shift across space and time; and how people inhabit and respond critically to the contexts that form them. The ideas and examples presented here are of course just the beginning of a conversation about sexuality and urban life and what it could mean to take a queer approach to urban social movements, policy, history, and transformation. Indeed, much work remains to be done across all those domains to adequately understand and address forms of social inclusion and social exclusion in cities. By suggesting that sexuality is something that we do rather than something that we have, a queer approach argues that all of us are implicated in discourses of sexuality and in forms of power and inequality in cities more broadly. As a result, we all have a stake in understanding the spaces, power relations, and contexts that form us, so that we might transform them.

Key Points

- Though "queer" is often used as an umbrella term for members of the LGBTQ community, a queer theoretical perspective on urban life emphasizes the way that exclusive sexual norms work in tandem with other forms of power (race, class, gender, and more) to marginalize a wide range of people (e.g., LGBTQ people, single parents, sex workers, people on public assistance, migrants). It also highlights how forms of exclusion differ from context to context, which invites place-specific investigation from those who study the urban.
- Scholarly thinking about sexuality and urban space has shifted over time from an initial focus on gay and lesbian spaces, to a concern with the processes that underlie the heterosexualization of space, to a consideration of queer thinking that has increasingly become postcolonial and anti-racist.
- A narrow emphasis on homosexual versus heterosexual spaces in cities misses the ways in which LGBTQ persons are not always and everywhere uniformly excluded from urban spaces, as well as the fact that there are many heterosexualities and that not all expressions of heterosexuality are equally valued within cities—hence the need to analyze cities as hetero*normative*, rather than as simply hetero*sexual* sites.
- Chicago's Boystown illustrates the struggle over which sexual identities and practices can make claims over urban neighbourhood spaces.
- The city-state of Singapore demonstrates how a conservative government has drastically reoriented its stance on homosexuality from condemnation to acceptance in an attempt to develop a creative economy, although sexual citizenship in Singapore is about much more than the policing of the homosexual/heterosexual binary.
- In Ciudad Juárez, sex workers who traverse the conventional boundaries between "public" and "private" use the tactics of both visibility and invisibility to contest their marginalization and claim their rights to urban spaces.
- Paying attention to the ways that different urban communities form around sexuality unsettles grand narratives about large cities as uniform sites of progress and emancipation for people who have experienced sexual marginalization, and about small cities as sites of backwardness and oppression.

Activities and Questions for Review

1. Take turns with a partner to explain in your own words the two meanings of queer as outlined in the "Queering Urban Space" section of the chapter. Together, consider how these meanings could help us to better understand urban lives and social movements.
2. Consider the queer idea that sexuality might describe something we *do* rather than something that we *have*. Evaluate how a shift in attention away from identity and towards power relations might shed light on questions of inequality and justice in the city.
3. Using the case studies as a point of departure, as a class, debate the potential advantages and disadvantages of visibility in urban space for LGBTQ people, sex workers, or other people whose sexual practices are stigmatized.

Acknowledgments

We thank Jordan Hale at the Map and Data Library at the University of Toronto for her intrepid work on the maps in this chapter.

References

Abraham, J. (2009). *Metropolitan lovers: The homosexuality of cities.* Minneapolis: University of Minnesota Press.

Addams, J. (1912). *Twenty years at Hull House.* New York: Macmillan.

Bain, A.L., Payne, W., and Isen, J. (2015). Rendering a neighbourhood queer. *Social and Cultural Geography,* 16(4): 424–443.

Banerjea, N. (2015). Critical urban collaborative ethnographies: Articulating community with Sappho for equality in Kolkata, India. *Gender, Place and Culture,* 22(8): 1058–1072.

Bell, D., and Valentine, G. (1995). Introduction: Orientations. In D. Bell and G. Valentine (Eds.), *Mapping desire: Geographies of sexualities.* London: Routledge, pp. 1–27.

Benedicto, B. (2014). *Under bright lights: Gay Manila and the global scene.* Minneapolis: University of Minnesota Press.

Berlant, L., and Warner, M. (1998). Sex in public. *Critical Inquiry,* 24(2): 547–566.

Browne, K. (2007). A party with politics? (Re)making LGBTQ Pride spaces in Dublin and Brighton. *Social and Cultural Geography,* 8(1): 63–87.

Caluya, G. (2008). "The Rice Steamer": Race, desire and affect in Sydney's gay scene. *Australian Geographer,* 39(3): 283–292.

Chauncey, G. (1994). *Gay New York: Gender, urban culture and the making of the gay male world 1890–1940.* New York: Basic Books.

Chisholm, D. (2004). *Queer constellations: Subcultural space in the wake of the city.* Minneapolis: University of Minnesota.

Chua, L. (2014). *Mobilizing gay Singapore: Rights and resistance in an authoritarian state.* Philadelphia, PA: Temple University Press.

Cisneros, S. (1984). *The house on Mango Street.* Houston: Arte Público Press.

Cruz-Malavé, A., and Manalansan, M. (2002). Introduction: Dissident sexualities/alternative globalisms. In A. Cruz-Malave and M. Manalansan (Eds.), *Queer globalizations: Citizenship and the afterlife of colonialism.* New York: New York University Press, pp. 1–12.

Daniel-McCarter, O. (2012). Us vs them: Gays and the criminalization of queer youth of color in Chicago. *Children's Legal Rights Journal,* 32(1): 5–17.

D'Emilio, J. (1983). Capitalism and gay identity. In A. Snitow, C. Stansell, and S. Thompson (Eds.), *Powers of desire: The politics of sexuality.* New York: Monthly Review Press, pp. 100–113.

Engebretsen, E. (2014). *Queer women in urban China: An ethnography.* New York: Routledge.

Fishman, E. (2012). Pariahs amid the rainbow: Young, queer and homeless in Boystown. *Chicago Reader,* July18. Retrieved from http://www.chicagoreader.com/chicago/boystown-as-hometothe-young-gay-homeless/Content?oid=6990871&showFullText=true

Florida, R. (2002). *The rise of the creative class: And how it's transforming work, leisure, community and everyday life.* New York: Basic Books.

Gorman-Murray, A. (2008). Reconciling self: Gay men and lesbians using domestic materiality for identity management. *Social and Cultural Geography,* 9(3): 283–301.

Gorman-Murray, A. (2012). Queer politics at home: Gay men's management of the public/private boundary. *New Zealand Geographer,* 68(2): 111–120.

Greene, T. (2014). Gay neighborhoods and the rights of the vicarious citizen. *City and Community,* 13(2): 99–118.

Grundy, J., and Smith, M. (2005). The politics of multiscalar citizenship: The case of lesbian and gay organizing in Canada. *Citizenship Studies,* 9(4): 389–404.

Hanhardt, C.B. (2013). *Safe space: Gay neighborhood history and the politics of violence.* Durham, NC: Duke University Press.

Heng, R.H.K. (2001). Tiptoe out of the closet: The before and after of the increasingly visible gay community in Singapore. *Journal of Homosexuality,* 40(3–4): 81–96.

Houlbrook, M. (2005). *Queer London: Perils and pleasures in the sexual metropolis, 1918– 1957.* Chicago: University of Chicago Press.

Howell, P. (2004). Race, space and the regulation of prostitution in colonial Hong Kong. *Urban History, 31*(2): 229–248.

Hubbard, P. (2004). Cleansing the metropolis: Sex work and the politics of zero tolerance. *Urban Studies, 41*(9): 1687–1702.

Hubbard, P. (2008). Here, there, everywhere: The ubiquitous geographies of heteronormativity. *Geography Compass, 2*(3): 640–658.

Hubbard, P., and Sanders, T. (2003). Making space for sex work: Female street prostitution and the production of urban space. *International Journal of Urban and Regional Research, 27*(1): 75–89.

Hutta, J.S. (2013). Beyond the politics of inclusion: Securitization and agential formations in Brazilian LGBT parades. In E.H. Yekani, E. Kilian, and B. Michaelis (Eds.), *Queer futures: Reconsidering normativity, activism and the political.* Farnham, Surrey, UK: Ashgate, pp. 67–82.

Jackson, P. (Ed.) (2011). *Queer Bangkok: 21st century markets, media, and rights.* Hong Kong: Hong Kong University Press.

Jagose, A. (1996). *Queer theory: An introduction.* New York: New York University Press.

Johnston, L. (2005). *Queering tourism: Paradoxical performances of Gay Pride Parades.* New York: Routledge.

Lee, H.L. (2007). Speech to Parliament on reading of penal code (amendment) bill, October 22. Singapore government press release. Ministry of Information, Communications and the Arts, Singapore.

Lewis, N.M. (2013). Ottawa's Le/The Village: Creating a gaybourhood amidst the "death of the village." *Geoforum, 49*(1): 233–242.

Lewis, R. (2008). *Chicago made: Factory networks in the industrial metropolis.* Chicago: University of Chicago Press.

Lim, E.-B. (2009). Performing the global university. *Social Text, 27*(4): 25–44.

Marston, S. (2002). Making difference: Conflict over Irish Identity in the New York City St Patrick's Day Parade. *Political Geography, 21*(3): 373–392.

Martin, F. (2003). *Situating sexualities: Queer representation in Taiwanese fiction, film and public culture.* Hong Kong: Hong Kong University Press.

Matloff, J. (2015). Six women murdered each day as femicide in Mexico nears a pandemic. *Al Jazeera America,* January 4. Retrieved from http://america.aljazeera.com/multimedia/pandemicfemicides.html

Merabet, S. (2015). *Queer Beirut.* Austin: University of Texas Press.

Muller-Myrdahl, T. (2013). Ordinary (small) cities and LGBTQ lives. *ACME: An International E-Journal for Critical Geographies, 12*(2): 279–304.

Nash, C.J. (2005). Contesting identity: Politics of gays and lesbians in Toronto in the 1970s. *Gender, Place and Culture, 12*(1): 113–135.

Nash, C.J. (2006). Toronto's gay village (1969–1982): Plotting the politics of gay identity. *Canadian Geographer, 50*(1): 1–16.

Nash, C.J. (2013). The age of the "Post-mo"? Toronto's gay village and a new generation. *Geoforum, 49*: 243–252.

Nash, C.J., and Gorman-Murray, A. (2014). LGBTQ2s neighbourhoods and "new mobilities": Towards understanding transformations in sexual and gendered urban landscapes. *International Journal of Urban and Regional Research, 38*(3): 756–772.

Nast, H.J. (2002). Queer patriarchies, queer racisms, international. *Antipode, 34*(5): 874–909.

Olds, K. (2007). Global assemblage: Singapore, foreign universities and the construction of a "global education hub." *World Development, 35*(6): 959–975.

Oswin, N. (2008). Critical geographies and the uses of sexuality: Deconstructing queer space. *Progress in Human Geography, 32*(1): 89–103.

Oswin, N. (2010). Sexual tensions in modernizing Singapore: The postcolonial and the intimate. *Environment and Planning D: Society and Space, 28*(1): 128–141.

Oswin, N. (2014). Queer time in global city Singapore: Neoliberal futures and the "freedom to love." *Sexualities, 17*(4): 412–433.

Papayanis, M.A. (2000). Sex and the revanchist city: Zoning out pornography in New York. *Environment and Planning D: Society and Space, 18*(3): 341–353.

Podmore, J. (2001). Lesbians in the crowd: Gender, sexuality and visibility along Montréal's Boul. St-Laurent. *Gender, Place and Culture, 8*(4): 333–355.

Podmore, J. (2006). Gone "underground"? Lesbian visibility and the consolidation of queer space in Montréal. *Social and Cultural Geography, 7*(4): 595–625.

Razack, S.H. (2002). Gendered racial violence and spatialized justice: The murder of Pamela George. In S.H. Razack (Ed.), *Race, space, and the law: Unmapping a white settler society.* Toronto: Between the Lines, pp. 121–156.

Reed, C. (2003). We're from Oz: Marking ethnic and sexual identity in Chicago. *Environment and Planning D: Society and Space, 21*(4): 425–440.

Ritchie, J. (2015). Pinkwashing, homonationalism and Israel-Palestine: The conceits of queer theory and the politics of the ordinary. *Antipode, 47*(3): 616–634.

Sanders, T. (2004). The risks of street prostitution: Punters, police and protestors. *Urban Studies, 41*(9): 1703–1717.

Seitz, D.K. (2015). The trouble with *Flag Wars*: Rethinking sexuality in critical urban theory. *International Journal of Urban and Regional Research, 39*(2): 251–264.

Sinclair, U. (1906). *The jungle.* New York: Doubleday.

Statistics Canada (2015). Same-sex couples and sexual orientation . . . by the numbers. *Statistics Canada,* June 25. Retrieved from http://www.statcan.gc.ca/eng/dai/smr08/2015/smr08_203_2015

Stella, F. (2012). The politics of in/visibility: Carving out queer space in Ul'yanovsk. *Europe-Asia Studies, 64*(10): 1822–1846.

Stewart-Winter, T. (2016). *Queer clout: Chicago and the rise of gay politics.* Philadelphia: University of Pennsylvania Press.

Sugrue, T.P. (1996). *The origins of the urban crisis: Race and inequality in postwar Detroit.* Princeton, NJ: Princeton University Press.

Tan, K.P. (Ed.) (2007). *Renaissance Singapore? Economy, culture and politics.* Singapore: National University of Singapore Press.

Tang, D. T.-S. (2011). *Conditional spaces: Hong Kong lesbian desires and everyday life.* Hong Kong: Hong Kong University Press.

Tucker, A. (2009). *Queer visibilities: Space, identity and interaction in Cape Town.* Malden, MA: Blackwell.

Visser, G. (2008). The homonormalisation of white heterosexual leisure spaces in Bloemfontein, South Africa. *Geoforum, 39*(3): 1347–1361.

Waitt, G. (2006). Boundaries of desire: Becoming sexual through the spaces of Sydney's 2002 Gay Games. *Annals of the Association of American Geographers, 96*(4): 773–787.

Waitt, G., and Stapel, C. (2011). "Fornicating on floats"? The cultural politics of the Sydney Mardi Gras parade beyond the Metropolis. *Leisure Studies, 30*(2): 197–216.

Wong, K.W., and Bunnell, T. (2006). "New economy" discourse and spaces in Singapore: A case study of One-North. *Environment and Planning A, 38*(1): 69–83.

Wright, M.W. (2004). From protests to politics: Sex work, women's worth, and Ciudad Juárez modernity. *Annals of the Association of American Geographers, 94*(2): 369–382.

Wright, R. (1940). *Native son.* New York: Harper and Brothers.

Yue, A. (2007). Creative queer Singapore: The illiberal pragmatics of cultural production. *Gay and Lesbian Issues and Psychology Review, 3*(3): 149–160.

Yue, A., and Zubillaga-Pow, J. (Eds.) (2012). *Queer Singapore: Illiberal citizenship and mediated cultures.* Hong Kong: Hong Kong University Press.

V Urban Infrastructure and Livability

21 Plants, Animals, and Urban Life

Laura Shillington and Alice Hovorka

Introduction

Cities are more than human. Cities are complex entities that comprise numerous ecological processes such as water cycles, soil formation, air circulation, and energy flows, and they provide habitat and spaces of action for diverse plants and animals, including humans. Together, these processes and actors interact to create and shape urban form, function, and everyday life. This means that we as humans do not live alone in cities. We are surrounded by and coexist with trees, shrubs, vegetables, flowers, ground cover, bacteria, insects, reptiles, birds, and mammals. When we consider the urban environment in this way, we cannot help but see that cities are made up of human and non-human, social and natural.

Residents of or visitors to cities around the world can expect to run into **nature** at every turn, and daily life is closely connected to many plants and animals. For those in the global North, urban green spaces offer outdoor activities, rest and relaxation, or even jobs within the lush expanses of, for example, Central Park in New York City, USA, and Stanley Park in Vancouver, Canada. Neighbourhood gardens have long bloomed in cities like Berlin, Germany, or Utrecht, Netherlands, providing city dwellers with exercise and fresh foodstuffs, while local **urban agriculture** projects in Detroit, USA, and Montreal, Canada, provide valuable employment, sustenance, and nutrition in the support of homeless shelters and outreach centres. Urban well-being in the global North is also enhanced through people's companionship with animals: dog parks and cat cafes are increasingly visible, birding serves as an intensive pastime, and wildlife corridors offer a glimpse into "pristine nature" perceived to be outside of city boundaries. Many jobs are connected to urban animals, such as dog-walking, pet retail stores, zoo-keeping, or wildlife caretaking. Even encounters with the not-so-pleasant remind us that we are surrounded by the natural world: pigeons congregate at popular sites such as Trafalgar Square in London, UK, to mingle with and be fed by tourists; raccoons rummage through garbage cans along the streets of Toronto, Canada; and weeds threaten to invade home gardens or modestly maintained parks.

Encounters with nature in cities of the global South are similarly frequent, multifaceted, and closely connected to people's jobs, well-being, and access to productive and recreational green spaces. In Nairobi, Kenya, for example, a short walk around a local homestead can reveal a small-scale backyard dairy that includes five cows and a modest plot of alfalfa for feed, while in Harare, Zimbabwe, the railroad track corridor serves as prime agricultural land for growing maize. Local school programs in Lusaka, Zambia, and Quito, Ecuador, offer students training and hands-on experience with gardening. Urban forests in and around cities of West Africa, including Lagos, Nigeria, and Accra, Ghana, are rich sources of wild foods, medicinal plants, fuel wood, and building materials for people's daily use or for sale at the market; these natural resources may also be used for crafts such as wood carving or grass basket weaving, again to be used at home or sold to generate income. Mexico City is home to one of the largest municipal parks in the world, Bosque de Chapultepec (Chapultepec Forest). For many residents, Chapultepec serves as a weekend escape from the traffic chaos and city pollution. Indeed, it may be the only forest that many Mexican City residents ever see. Urban animals are a common sight in the global South: cows in Delhi, India; elephants in Bangkok, Thailand; horses in Managua, Nicaragua; and donkeys in Addis Ababa, Ethiopia, provide valuable transport of goods and persons on a regular basis. Without these animals, many people would not be able to get to their jobs, feed their families, or move around the city easily. In the neighbourhoods of Johannesburg, South Africa, people rely on domestic dogs for guarding property, as well as for companionship.

The objective of this chapter is to conceptualize cities as **socio-nature**, as well as to acknowledge the myriad of living non-human actors vital to achieving and sustaining urban livability and central to people's **livelihoods**, **food security**, and **right to the city**. We highlight the ways in which plants and animals offer people opportunities to attend to their daily lives in productive and rewarding ways, especially in addressing food insecurity and poverty and achieving their rights to the city as urban dwellers. We also highlight the ways in which these interactions between humans and non-humans in the city are produced and reproduced by relations of power reflecting the socio-economic and political status of various urban actors.

This chapter begins with a discussion of the ways in which scholars and researchers have theorized nature and cities in both the global North and global South. It then highlights the concept of socio-nature, with reference to recent work on the topics of urban agriculture and urban animals, to argue that cities must be viewed as more than human. Specific case studies illustrate how plants and trees in Managua, Nicaragua, and chickens in Gaborone, Botswana, are critical to people's everyday practices and to urban landscapes more broadly. This is followed by a discussion of how plant and animal life matters ecologically and socially in Canadian cities. The chapter concludes by restating the main argument: an enriching and sustainable urban future is premised upon recognizing all human and non-human actors and acknowledging their roles.

Cities and Nature

Humans in many parts of the world have long differentiated themselves from nature, with cities serving as a primary means of entrenching this perspective. Historical urban landscapes throughout the world have reflected a diverse range of human relationships and understandings of nature. The ancient city of Babylon is believed to have had extensive hanging gardens. In the Aztec city of Tenochtitlan, now known as Mexico City, many areas comprised a series of canals around floating artificial islands built for agriculture. These islands, called *chinampas*, are an early example of large-scale urban agriculture. In cities throughout the world, gardens—in addition to other sorts of natures—for both beauty and food have been an integral part of urban life. How such nature has been included or excluded in city-building is rooted in concepts of the city and broader socio-cultural practices. Contemporary concepts of the city and nature in both the global North and global South have been greatly influenced by Western philosophy and science.

The Age of Enlightenment in Europe and North America (1650s to 1780s) prompted an increasing focus on reason, empiricism, and scientific inquiry, leading to technological innovation that enabled people abilities to control and transform nature (Vining, Merrick, and Price, 2008). Advances in scientific knowledge drove the twin forces of industrialization and urbanization to transform human interactions with their natural environments. While the majority of the population in Europe and North America still lived in rural areas until the early twentieth century, rapid urbanization during the Industrial Revolution (1760s to 1850s) gave rise to densely populated cities. Urban inhabitants were brought into closer contact with "bad" natures: bacteria, viruses, and new concoctions of chemical pollutants, such the sulphur and nitrogen in coal smog. The consequences were epidemics of infectious diseases such as cholera, typhoid, and typhus. The increase in epidemics coincided with major advances in scientific knowledge, especially in medicine, which led to the collection of health statistics. The response was a rationalization of urban space, including the sanitary organization and the construction of city-wide public water and sewer systems. This period, from the mid- to late 1800s, gave birth to what is now referred to as "modern urban planning." European, North American, and **colonial cities** in Latin America, Asia, and Africa all became sites of rational organization.

While modern urban planning was concerned with eradicating "bad" nature from the city, it did little to bring "good" nature into the city. Indeed, modern planning emphasized the city as a human space separate from nature, intensifying the idea of nature's sacredness and aiding attempts to protect pristine regions from humanity itself (Vining et al., 2008). Past urban scholarship reflects and reproduces this society-versus-nature dualism. On the one hand, relatively little attention has been paid to our understanding of the natural actors, elements, and processes that shape the form and function of cities. The city is viewed as an object or as part of larger economic and social processes, but is rarely conceived as part of biophysical processes. On the other hand, scholarship that embraces nature in cities often pits humans against the natural environment and focuses on a particular understanding of an ideal nature. Nevertheless, research and ideas generated by the Chicago School, rural–urban connections, urban ecology, and sustainable cities scholars during the twentieth century offer useful insights on the interrelationships between nature and cities.

Early twentieth-century urban scholars were influenced by medical thought from the previous century. Modifying the urban landscape was understood as a key way to improve human health; this involved, in particular, the introduction of large parks to bring wholesome, beneficial nature into the city (but not necessarily integrated into the city as a whole) and the development of middle-class suburbs. These ideas influenced Ebenezer Howard, the founder of the **Garden City movement**. In the UK and USA, drawing on the metaphor of the organism, this movement emphasized the connections between cities and their hinterlands. Howard (1902) argued that human settlement should balance town and country so as to encourage social and urban reform. Specifically, he promoted the idea of garden cities to counter the poverty, overcrowding, inadequate infrastructure, and lack of engagement with nature in industrial cities. Howard believed that new suburban towns based on limited size and surrounded by agricultural land would offer a perfect blend of city and nature.

Expanding Howard's theories, the **Chicago School** of urban sociologists engaged ecological concepts and biological metaphors to explain social, cultural, and spatial patterns of cities (Park, McKenzie, and Burgess, 1925). They used terms like "symbiosis," "competition," "invasion," "succession," "dominance," and "ecological niches" to conceptualize cities as environments similar to those found in nature and governed by evolutionary processes. The **Concentric zone model**, for example, drew heavily on Darwin's theory of survival of the fittest. It explained urban development as emerging through processes of competition among social groups whereby those with access to greater amounts of capital moved outwards from the city centre—through the process of succession—to purchase larger plots of land, while those with less capital concentrated in the central business district. The Chicago School proved foundational in terms of theoretical explanations of urban form and function. Despite reliance on ecological concepts, however, non-human dimensions of the urban landscape held little interest for them and hence their "urban ecology" remained entirely devoid of ecology (Braun, 2005).

The American intellectual Lewis Mumford (1938, 1961) was similarly influenced by ecology. He wrote that the city is a product of the earth and a fact of nature, and argued for city planning to be based on an organic relationship between humans and their living spaces. Along with the landscape architect Ian McHarg, he promoted an ecological view of urban design that embraces hydrology, geology, vegetation, and climate, offering again a means through which to address problems of modern city development (McHarg and Mumford, 1969). In the 1970s, cities became legitimate units of analysis as scientists recognized that humans had significantly transformed nature and that they (and their cities) could no longer remain outside the purview of ecological investigation (Niemelä et al., 2011). Urban ecology emerged with a focus on the scientific study of human-dominated landscapes. Urban ecologists study the ways that human and ecological systems evolve together in urbanizing regions, highlighting especially the patterns, processes, and functions related to climate, hydrology, geomorphology, geochemistry, and biology (Alberti, 2010). Studies often emphasize the impacts of human-driven urbanization, including natural habitat fragmentation, biodiversity loss, and environmental pollution. For example, the study of **urban metabolism** aims to quantify the inputs, outputs, and storage of energy, water, nutrients, materials, and waste for an urban region (Pincetl, Bunjeb, and Holmes, 2012). While urban ecologists advocate for advanced study of coupled human-natural systems, the idea of humans versus nature remains fully entrenched. The city is recognized as borne from nature and yet is viewed simultaneously as "ruining nature" both inside and outside of its boundaries.

While interest in urban ecology and urban nature became popular in Europe and North American, urban scholarship on cities of the global South remained focused primarily on development and introducing Western-style urban planning. Until the 1950s and 1960s, urban planning in much of Latin America, Asia, and Africa was determined by colonial governments, in particular Spain and Britain. During the 1960s in Latin America, for example, there was a widespearad movement to establish urban studies departments in national universities. In addition, regional urban research networks were created, such as the Urban and Regional Development Commission of the Latin American Social Science Council (CLACSO) (Valladares and Prates Coelho, 1995). However, environmental problems and urban nature were not given prominence in urban scholarship until the 1990s with the emergence of global sustainability discourses.

Sustainable cities scholarship gained traction in the 1990s through the examination of problems related to degradation in urban environments and the contributions cities can make to global sustainable development agendas (Haughton and Hunter, 2004). This scholarship brought an international perspective to exploring urban environments, generating research on African, Asian, and Latin American urban contexts and prompting discussion regarding socially and environmentally sustainable development interventions (Drakakis-Smith, 1996; Satterthwaite, 1999; Stren, White, and Whitney, 1992). Specific topical areas included overcrowding, poor sanitation, inadequate housing, and air, water and noise pollution. Notably, scholars in this field recognized and promoted the role of agricultural plants and livestock animals within livelihood strategies of urban dwellers around the world (Mougeot, 2006; Smit, Nasr, and Ratta, 1996). Sustainable cities scholarship calls attention to the importance of nature in helping to produce more livable cities. However, the city in sustainable urban literatures remains divorced from nature; cities are considered "unnatural" places and urbanization the key threat to global ecological sustainability (the nature outside of cities) (Braun, 2005).

Cities as Socio-nature

Viewing cities as separate from nature means that we fail to recognize the ways in which biophysical processes weave themselves into daily urban life and the urban landscape itself. It also means that we fail to recognize the ways in which urban residents socially construct nature, rendering some types of nature as important and valued yet other types of nature as pestilent and hazardous. Failure to acknowledge and embrace nature in its material and discursive form generates serious consequences for present well-being and the future sustainability of cities and their human and non-human residents. For example, the presence of food plants and domesticated animals within city boundaries in many countries of the global South is ignored or condemned by local governments or non-participating residents who view urban agriculture as anti-modernization. Yet this type of nature offers substantial nourishment, food security, income, employment, and empowerment to those involved who may view it as a symbol of necessity, productivity, and success. Without recognition and support, however, agriculture can cause urban environmental problems, such as water pollution from animal waste disposal or soil contamination from overuse of fertilizers, and undermine people's potential to support their households and the broader urban community. More in-depth and holistic investigation of nature-in-cities is thus needed to ensure we promote healthy urban futures.

Recent urban scholarship challenges the view that cities are the antithesis to nature (Braun, 2005). In the late twentieth century, environmental historian William Cronon (1991) and others, including Mike Davis (1998), Alexander Wilson (1992), Nicholas Green (1990), and Andrew Ross (1994), explored the connections between the city and its hinterland, calling attention to how cities have always been constructed in relation to rural areas. Cities are increasingly conceptualized as complex interactive products of human and non-human exchange dynamics formed through relations of power (Hovorka, 2008). Central to this perspective is the concept of socio-nature. This neologism, which combines social and nature together in one word, refers to the natural or ecological conditions and processes that operate together with social conditions or processes whereby existing socio-natural conditions and processes are always the result of intricate transformations of pre-existing configurations that are themselves inherently "natural" and social (Swyngedouw, 1999). Thus nature is "no longer fixed at a distance but emerges within the routine inter-weavings of people, organisms, elements and machines as these configure in the partial, plural and sometimes overlapping time/spaces of everyday living" (Whatmore,

1999: 33). Importantly, nature is viewed as a material and discursive entity: on the one hand, a tree is a physically tangible nature with particular biophysical processes and conditions; on the other hand, a tree is socially constructed nature with particular values, uses, and roles assigned to it according to the people who interact (or not) with it. The tree therefore must be understood as fully socio-nature, an entity whose physical character cannot be separated from its symbolic meaning to humans. Socio-nature is necessarily relational such that cities are co-constituted by humans as well as other entities of nature, including flora, fauna, and environmental processes.

Urban political ecology is an approach to understanding the city and urbanization as the outcome of both social and natural processes (Heynen, 2014). Urban political ecologists have engaged network and hybrid metaphors (drawing on Latour, 1993; and Whatmore, 2002, respectively) to explain the socio-natural processes in cities whereby urbanization occurs in and through vast networks of relationships and within hybrid flows of energy and matter as well as capital, commodities, people, and ideas. Urban political ecologists illustrate people's connections to nature and especially reveal the uneven and contested socio-nature of city landscapes in the global North (Bakker, 2013; Brownlow, 2006; Gandy, 2002; Heynen, Kaika, and Swyngedouw, 2006; Loftus, 2007) and the global South (Hiemstra-van der Horst and Hovorka, 2008; Hovorka, 2006; Myers, 2008; Shillington, 2013; Swyngedouw, 1997). They highlight themes of natural resource access, social marginalization, and environmental conflict among key stakeholders. By seeing the city as a "giant socio-environmental process" whereby the processes of urbanization entail both ecological *and* social transformation, urban political ecology has been able to avoid treating cities as spatially bounded places (Swyngedouw, 2006: 37). In this way, urban political ecology resists strict delineation between city and country or humans and nature. Hence, cities can more easily be viewed as more than human.

Urban Plants and Animals

Viewing the city as socio-nature involves expanding urban scholarship to include living entities besides humans. The topic of urban agriculture is particularly useful in demonstrating how non-humans—in particular plants and animals—create urban space and help humans achieve their livelihoods, food security, and rights to the city. Urban agriculture has long existed in cities, such as the *chinampas* mentioned in the introduction to this chapter. Food production in cities not only provides food and income for households, it also serves as a way for people to establish their place in cities. In this way, urban agriculture is of interest to urban political ecologists because it involves using space in ways that are not considered "urban": the appropriation of space in cities to create urban gardens; the presence of domestic animals in cities; and the production of food as a political act to counter industrial food production. In this way, urban political ecology has explored the ways in which the natures of urban agriculture (plants and animals) challenge taken-for-granted ideas of what counts as urban nature (and urbanity in general).

To this end, scholars have begun to explore the everyday plant ecologies of cities. Plants and trees have always been an important part of urban spaces, both public and private. From urban parks and streets to backyard patios, lawns, and balconies, the city is full of a wide variety of plant and tree life shaping the daily lives of urban inhabitants in multiple ways. Domestic gardens are an important part of creating home (Shillington, 2008). Research has shown the ways in which people attach meanings to gardens and gardening, in particular around gardening as a leisure activity and the garden as a private place, in many cases distinct from the house (Bhatti and Church, 2004; Head and Muir, 2006; Hitchings, 2003; Longhurst, 2006). Home gardens as well as community gardens also serve as important sources of food and income for many urban dwellers, improving household food security (Mougeot, 1996; Redwood, 2012). In addition to contributing to household economies, plants also comprise larger urban and national economies. Robbins' (2007) examination of the North American suburban lawn culture reveals how it has been shaped in part by the industrial chemical economy. Such work shows how human perceptions of nature (e.g., what nature is deemed culturally appropriate for in the city) have produced particular manifestations of urban natures. Other research has focused on how notions of domesticity have shaped our understandings of urban nature (Kaika, 2005). While plants play an integral part in urban cultures and economies, their benefits are not distributed equitably in the city. Access to space to grow food or proximity to urban parks is not the same for all urban dwellers (Gandy, 2002). The unevenness of urban forests and green spaces relates to issues of social justice whereby certain urban populations experience the benefits of plants, trees, and parks while others do not (Pulido, 2000; Heynen, 2003). Such injustices call attention to the problems of dominant plant cultures in urban design and the need to recognize the diversity of plant–human relations in cities.

Urban scholarship has also begun to embrace animals. Indeed bringing animals into urban analysis has been one way to unsettle the presumed human-centric city to recognize that cities are alive with domestic animals, food animals, and wild animals—all of which transform city spaces for their needs and interact with humans and local environments. In **animal geography**, scholars argue that the urban is inherently wrapped up in human–animal relations such that the city itself is characterized by, and thus can be conceptualized as, a product of **transspecies relations** (Wolch et al., 1995). How humans think, feel, and talk about animals will shape their practices and actions towards them, with important consequences regarding the extent to which different species are included or excluded from common sites of human activity. Philo (1995), for example, explores nineteenth-century debates around meat markets and slaughterhouses that facilitated action to exclude livestock animals from cities such as London and Chicago on medical, hygienic, organizational, and moral grounds. Nast (2006) illuminates connections between urbanization and urban life and the emergence of the pet industry giving rise to pet-related consumption of food and care products in contexts as diverse as China, Britain, France, Mexico, and South Korea. Estimates value the industry globally at USD$97million (Companies and Markets, 2013). Further, despite cities being built to accommodate humans and their pursuits, subaltern "animal towns" inevitably emerge with urban growth; animals shape practices of urbanization in key ways; and animals influence the possibilities for human life in cities. For example, Gullo, Lassiter, and Wolch (1998) argue that cougars in and around Santa Monica, California, try to fit humans, who have encroached on their habitat with urban expansion, into preconceived mental maps of prey opportunities. Cougars infiltrating the city often negatively affect people's mobility and sense of security. Yet more often than not cougars construct new sets of distinctions about danger and diet and adapt their behaviours to staying away from human settlements and activities. Yeo and Neo (2010) study human–long-tailed macaque conflicts in Singapore whereby residential development has compromised wildlife corridors and habitat, inciting monkeys to disrupt people's lives through theft and aggression. Monkeys have forced urban residents to change their behaviours to avoid them in particular places and certain time frames. The following two case studies explore other examples of human–plant and human–animal interactions. The first explores plants and trees in Managua, Nicaragua, and the second chickens in the city of Gaborone, Botswana.

Plants and Patios in Managua, Nicaragua

Managua, the capital of Nicaragua, is located on the southern shores of Lake Xolotlán, the second-largest freshwater lake in Central America (Figure 21.1). When the Spanish arrived in 1522, the Nahua indigenous group had a large establishment in what is now the city of Managua. According to Spanish historian Gonzalo Fernández de Oviedo, in 1524 this settlement had approximately 40,000 inhabitants, making it one of the largest human settlements in Central America at that time and on par with London, which in 1530 had an estimated population of 50,000 (Wall, 1996). Managua was officially declared a colonial city in 1819 and in 1852 was named the capital of Nicaragua. At present, the city has a population of 1 million, the largest urban area in the country. Nicaragua is a predominantly urban nation, with approximately 60 per cent of the population living in cities; 20 per cent of the country's population lives in Managua. Most of Managua's growth in both population and land area occurred in the 1980s and early 1990s. The urban landscape is a patchwork of informal settlements in between middle- and upper-class housing, North American style strip malls, and, more recently, gated communities.

Unlike many cities in Latin America (and globally), Managua's skyline remains largely horizontal, with few high-rise buildings and no central business district. The city spreads out to the south towards the Masaya volcano and has within its boundaries four small crater lakes, called *lagunas*. When viewed from above, Managua seems to have a continuous canopy of trees. Indeed,

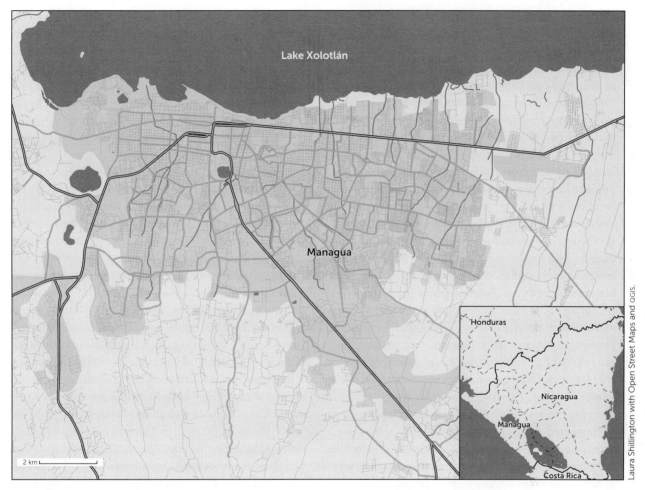

Figure 21.1 Map of Managua, Nicaragua

plants and trees in the patios of houses in Managua are critical in the everyday practices of home (Shillington, 2008). In poor, informal settlements, plants and trees help to create habitable spaces and contribute to household food security. Plants, and especially large shade trees, in patios are a critical part of the materiality of the house. They act as walls and ceilings, making houses more livable, private, and aesthetically pleasing. Plants and trees are especially important in the everyday lives of women, who carry out the majority of domestic work. Kitchens and washing areas are located outside in the patio, sometimes under a small built structure consisting of wooden poles and an aluminium roof, but in most patios large trees provide shade to these areas. The shade trees create microclimates that enable women to carry out their work in relative bodily comfort, critical in a city whose average daily temperature regularly exceeds 35 degrees Celsius. At the same time, the shade trees aid in keeping the inside of houses cool for sleeping at night. Shade is of particular importance for those houses that do not have windows (which allow for air flow to cool the interior) and that are constructed primarily of zinc (corrugated metal walls and roofs, which trap heat inside). Large shade trees also create privacy in the patio, functioning as walls to allow bodies to work and "act naturally" without the gazing eyes of neighbours and passing strangers.

The plants and trees in patios are also part of food security strategies for households. Plants and trees in patios are sources of food and medicine. Fruit trees (mangoes, avocados, oranges, limes, papayas, bananas, plantains, etc.) are the most important source of food. Fruit is also used to make *refresco*, a drink consisting of water, sugar, and fruit. *Refrescos* are a staple in Nicaraguan food culture, served at almost all meals, and are the main source of daily fruit consumption. Other important plants in the patio are medicinal and culinary herbs such as cilantro, basil, and mint. Herbs like cilantro and chiles are common in the patios and used to flavour soups and meats and to make salsas. Such plants are part of the everyday culture of eating. Vegetable crops, though, are rare in patios primarily because they require long periods of sun and the dominance of fruit trees prevents this. Tropical ornamentals, herbs, and medicinal plants thrive in shaded areas and are therefore ideal companions for large shade-producing trees. Large trees are more important for household livelihoods, since they provide fruit and shade, contributing to both food security and livable spaces (Figure 21.2).

Many sustainable development projects in Managua promote urban agriculture as a way to diversify diets, improve nutrition, and create income. These projects seek to establish vegetable gardens in patios but have encountered resistance, especially to plans that reduce the number of trees in patios. The importance of trees to everyday life and culture in Managua illustrates the interconnected complexity of human–nature relations in urban homes. While urban agricultural development projects are seemingly benign and well-meaning, a lack of awareness of the already existing urban socio-natures can produce conflicts. Taking up the challenge to keep trees is a way in which urban inhabitants in Managua fight for their right to produce urban space that is compatible with how they desire to live (Shillington, 2013). The following case study illustrates another such fight about urban agriculture, but the focus of the subject shifts from plants to animals, namely chickens.

Chickens and Entrepreneurship in Gaborone, Botswana

Gaborone, the capital city of Botswana, is located on the Notwane River in the southeast corner of the country, some 50 kilometres (31 miles) from the South African border (Figure 21.3). It was established in 1967, one year after Independence, given its proximity to fresh water and lack of tribal affiliations in the area. The city is Botswana's political capital and the location of the central government and the Southern African Development Community, as well as the economic capital, including the headquarters of Debswana (the national diamond company) and the stock exchange. The urban population is 231,626 (i.e.,10 per cent of Botswana's population) and 421,907 when the surrounding peri-urban villages are included. Gaborone has long reported one of the fastest urbanization rates in sub-Saharan Africa (Hovorka, 2006). Rapid growth has presented challenges with respect to housing and illegal settlements, as well as with employment opportunities with the government of Botswana, a central source of jobs.

In this context, domestic animals, especially chickens, are critical in the everyday lives of people in and around Gaborone through intensive, commercially oriented agricultural production (Hovorka, 2008). Chickens help people generate employment and income, provide food for households and the urban market, acquire entrepreneurial skills, expand social networks, and gain self-confidence. Chickens have especially helped women—both low- and middle-income—establish a foothold in the urban economy, address family needs, and balance multiple roles of employment and in household tasks. Chickens are highly adaptable to women's circumstances in the city because they require little physical space and minimal, albeit regular, care. This allows women, who often face discrimination relative to men in land access on account of legal and economic barriers, to set up chicken houses on small plots of land and care for chickens amid other critical tasks. Chickens have also been catalysts for enhanced social networks

Figure 21.2 Landscape of large fruit trees in patios of an informal barrio in Managua

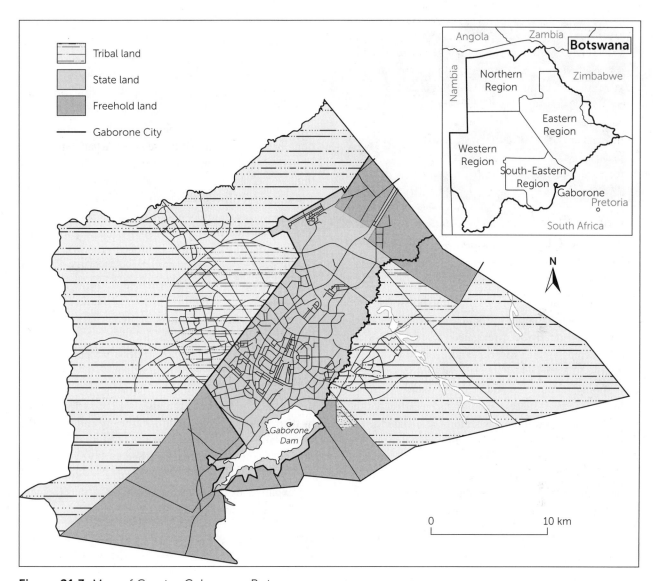

Figure 21.3 Map of Greater Gaborone, Botswana
Alice Hovorka as adapted from Department of Town and Regional Planning, Government of Botswana 2000 database.

and channels of communication in the city. Traditionally individualistic, poultry farmers embraced a collaborative approach to raising chickens in the early 2000s to enhance commercial production: they learned by doing and by sharing their experiences with neighbours, poultry industry suppliers, and government agricultural extension officers. This offered poultry farmers, especially women, greater access to stakeholders—including private- and public-sector employees—to which they would not have previously had access. As commercial poultry farmers, women in Greater Gaborone are lauded entrepreneurs in the community, given their impressive profits and managerial skills, as well as their contributions of broiler meat to the urban market (Figure 21.4).

In turn, the role and significance of chickens in the urban context, especially regarding women's livelihoods and their direct food contributions to the market, have enhanced chicken visibility and status in the city. First, chickens have become agents of influence in urban-planning realms whereby their presence on city land is encouraged and enabled through tribal, state, and freehold land mechanisms. Second, chickens are touted as

Figure 21.4 Female-owned urban poultry farm

a national success story by helping Botswana become self-sufficient in poultry production. Ironically, chickens have traditionally been relegated to subsistence realms, considered "women's work," and deemed of little value relative to cattle, which are revered in Botswana society, economy, and politics and are owned and controlled by men. The urban success of chickens has attracted the attention of men who now wish to be associated with and benefit from them. Men's traditional alignment with cattle does not translate into the urban context given that cattle are deemed out of place in the city and people's interactions with them must take place beyond city limits. This women-chicken pairing is clearly connected to gender relations of power such that rearing chickens, as traditionally "women's work," has allowed female entrepreneurs to benefit despite their subordinate position relative to men. This pairing is also connected to species relations of power such that chickens are able to shape urban life and livability on account of what they are relative to cattle, which may be more highly revered traditionally yet are not suited per se to the urban context.

Greater Gaborone form, function, and daily life is thus shaped by the approximately 3 million chickens and 60 primarily female commercial farmers that have comprised the poultry industry since 2000. Women have benefited from their access to and relations with chickens, having clear contributions to their own livelihoods, food security, and rights to the city. At the same time, urban residents have benefited from the increased broiler meat produced directly for the market, and Botswana as a whole can claim entrepreneurial success and food self-sufficiency through chickens. Ultimately, this Gaborone case study demonstrates that people's ability to thrive and succeed in an urban context is closely connected to the animals they interact with, while at the same time the lives of animals are necessarily affected by their value and roles in human society and within the urban landscape. The roles of urban animals, but also those of plants, as discussed earlier, and the differential opportunities and costs arising from these interactions are explored in the following text box, which includes Canadian examples of interspecies relations.

Plants and Animals Matter in Canadian Cities

Canadian cities are full of plant life. Like the patios in Managua, plants and trees create aesthetically pleasing urban spaces. They line boulevards, river valleys, parks, backyards, and schools. But plants and trees also transform the urban landscape and atmosphere. Urban plants and trees provide habitat for wildlife, such as birds and racoons. Plants and trees also create their own spaces in cities: weeds pop up in sidewalk cracks; tree roots creep into basements; and grasses occupy brownfields. Urban trees also store carbon, moderate the urban climate, and improve air quality. They conserve water and help mitigate the effects of flooding and rainfall runoff. Plants in conjunction with trees also improve quality of life in cities. They reduce noise levels and have been shown to have positive sociopsychological benefits for urban inhabitants, especially in public green spaces and gardens. Urban agriculture in Canadian cities was important in household food security historically and remains so at present. However, the benefit of urban plants and trees in Canadian cities is not equitably distributed. Access to public green spaces is unequal. Some areas of the city have more green spaces and tree cover than others. Some schools have green playgrounds, while others have paved yards. And some houses have more plants and trees. These differences are a result of economic and political processes. In this regard, the socio-natures of cities are uneven. Fighting for equitable access to urban nature is part of the right to the city movements across Canada. For example, there are cities and organizations across Canada that promote urban agriculture. The Toronto Food Policy Council's focus is largely directed towards the creation of urban space and policy for community-supported agriculture as a way to ensure healthy, affordable, sustainable, and culturally acceptable food (tfpc.to). Similarly, the city of Vancouver has a "Food Strategy" that includes programs on growing food (http://vancouver.ca/people-programs/food.aspx). The "Agriculture Urbaine Montréal" network in Montreal (http://agriculturemontreal.com) and the Garden Patch in Saskatoon (http://saskatoonfoodbank.org/gardenpatch) both provide support for individual and community gardeners (Figure 21.5). There are also movements in Canadian cities that seek to improve access to parks and green spaces, such as Park People in Toronto (www.parkpeople.ca).

Canadian cities are also full of animal life. Looking closely, we see food, companion, and wild animals playing diverse roles in urban contexts and interacting with humans in complex ways—good, bad, and otherwise. First, locally grown fresh food is increasingly valued by Canadian urban residents. A pro-backyard chicken movement has emerged over the past decade in numerous cities and towns, including those as diverse as Gatineau, Quebec; Guelph, Ontario; Grand Prairie, Alberta; Whitehorse, Yukon; and Victoria, British Columbia (see Figure 21.6). While supporters tout chickens as an opportunity to reconnect with nature and encourage sustainable and closed-loop food production, detractors highlight possible threats to human public health and chicken welfare. Second, household pets—such as domestic dogs—are ever-present on the Canadian urban landscape. The significance of these animals to city residence is easily garnered by observing packed dog parks and walking trails, stocked retail spaces selling dog accessories, and diverse services offering grooming, sitting, and veterinary care. Lives—both human and dog—are arguably enriched through these intimate relationships, yet such intense trends of pet-related consumption raise questions as to our increasing need to purchase, possess, and control companion animals. Third, those animals we have labeled "wild"—and hence not belonging in cities— are making a nuisance of themselves (e.g., raccoons and coyotes in southern Ontario) or are transgressing municipal boundaries (e.g., bears in northern Alberta). Urban residents and municipalities are debating best management practices for addressing human–animal conflict, including information dissemination on animal behaviour and associated do's and don'ts for living with wildlife.

Figure 21.5 City Farm School solar house and gardens, Montreal, Quebec

Figure 21.6 Urban chickens in Guelph, Ontario

Conclusion

Plants and animals play an important role in cities. They produce urban space, both on their own and in collaboration with humans. Plants and animals contribute to human food security in cities either directly or through income. In this way, they form part of the livelihood strategies of human inhabitants in cities. Plants and animals also contribute to the quality of life in cities. They can produce aesthetically pleasing spaces, which helps to create positive socio-psychological benefits, including stress relief and increased attention span. In addition, plants and animals make spaces such as the house more livable—indeed, they help transform houses into homes. Plants and animals in cities are also agents of change, producing and interacting in urban spaces in ways that are not completely visible or pleasing. Plants occupy spaces on their own, taking over unused land areas such as abandoned industrial sites and growing in lawns and roads. Animals also establish themselves in urban habitats—at times coexisting with humans, at other times challenging the sense of order through their wildness in the city.

We can see the vital roles of plants and animals, with the associated dynamics and outcomes, taking place in our case study cities. Patio plants and trees in Managua and chickens in Gaborone provide urban residents with opportunities to grow food for their households and the market. People's livelihoods are enhanced through their interactions with urban nature. Patios and chickens also offer spaces of enjoyment. Plants and trees create healthy green spaces, while chickens create desired connections with agricultural traditions and status. As agents of change, these non-humans have become established parts of Managua and Gaborone whereby the city landscape and daily life are continually enhanced and shaped by their presence and influence.

These plants and animals are also political. In Managua, plants and trees are used to help residents claim their right to be in the city and to produce space that is of benefit to them. In Gaborone, animals are wrapped up in gender politics, where some animals are considered part of men's work while others, such as chickens, are classified as women's work. Ultimately, plants, trees, and chickens offer avenues for addressing—in these contexts at least—food insecurity, poverty, and rights to productive and rewarding spaces within the city. It is also apparent that plants and animals matter in Canadian cities. Urban plants play a vital ecological role, and they provide opportunities—albeit uneven ones—for food production within city boundaries. Urban animals offer livelihood options and companionship while challenging our ideas about the "proper" place of wildlife.

All told, human and non-human interactions in and across the urban landscape are complex and multifaceted. Yet research and scholarship on plants, animals, and urban life have been relatively slow to emerge; cities have been largely conceptualized as human spaces without authentic consideration given to the ways in which nature matters. Our understanding of the roles, dynamics, and implications of non-humans in cities is limited. This presents a substantial gap in information on how urban life is produced and reproduced through socio-natures, as well as on what relationships and situations this creates and impacts. With the global population—human, plant, and animal—necessarily urbanizing, it is vital that attention be focused on the role of these multiple actors in urban futures and on the opportunities and constraints offered to all within cities.

Key Points

- The city is not just a human space. Rather, the city is a site of socio-nature produced through social, economic, political, and ecological processes and relations of power.
- The city and nature have been understood in different ways. Western thought has commonly understood them as separate, with the city characterized as "dirty" and "bad," and nature as "good" and ideally void of humans.
- Humans' everyday lives are always lived in relation to non-humans (plants, animals).
- Plants and trees are critical to living in the city; in Managua they provide food, shelter, comfort, and possibilities for enacting and demanding the right to the city.

- Animals are also critical to living in the city; in Gaborone they provide food, transport, income, companionship, aesthetics, and the general enhancement of people's livelihoods and well-being.
- The presence of plants, trees, and animals in Canadian cities is determined by social, economic, and political decisions.

Activities and Questions for Review

1. Discuss with a classmate how, if at all, your understanding of "nature" has changed upon reading this chapter.
2. In a small group, explore the way in which the concept of socio-nature is expressed in the two case studies from Managua and Gaborone.
3. In the neighbourhood in which you live, how have plants, trees, and animals shaped the built and social landscape?
4. On a weekly basis, how often and in what ways do you interact with plants, trees, and animals and how does this contribute to your quality of life?

Acknowledgments

Alice Hovorka would like to acknowledge the Government of Botswana, Department of Environmental Science, at the University of Botswana, and research participants supporting her work on gender and (peri-)urban agriculture in Greater Gaborone. Her research was funded by the National Science Foundation, the Social Sciences and Humanities Research Council of Canada, and the AgroPolis Award from the International Development Research Centre. Laura Shillington would like to thank the research participants and FUNDECI in Managua for their time and ongoing collaboration. Her research was funded by an AgroPolis Award from the International Development Research Centre as well as by the Social Sciences and Humanities Research Council of Canada.

References

Alberti, M. (2010). Maintaining ecological integrity and sustaining ecosystem function in urban areas. *Current Opinion in Environmental Sustainability, 2*(3): 178–184.

Bakker, K. (2013). Constructing "public" water: The World Bank, urban water supply, and the biopolitics of development. *Environment and Planning D: Society and Space, 31*(2): 280–300.

Bhatti, M., and Church, A. (2004). Home, the culture of nature and meanings of gardens in late modernity. *Housing Studies, 19*(1): 37–51.

Braun, B. (2005). Environmental issues: Writing a more-than-human urban geography. *Progress in Human Geography, 29*(5): 635–650.

Brownlow, A. (2006). An archaeology of fear and environmental change in Philadelphia. *Geoforum, 37*(2): 227–245.

Companies and Markets (2013). Global pet care market. Retrieved from http://www.companiesandmarkets.com/MarketInsight/Consumer-Goods/Global-Pet-Care-Market/NI8347

Cronon, W. (1991). *Nature's metropolis: Chicago and the Great West.* New York: W.W. Norton.

Davis, M. (1998). *Ecology of fear: Los Angeles and the imagination of disaster.* New York: Metropolitan Books..

Drakakis-Smith, D. (1996). Sustainability, urbanisation and development. *Third World Planning Review, 18*(4): 3.

Gandy, M. (2002). *Concrete and clay: Reworking nature in New York City.* Cambridge, MA: MIT Press.

Green, N. (1990). *Spectacle of nature: Landscape and bourgeois culture in nineteenth-century France.* Manchester, UK: Manchester University Press.

Gullo, A., Lassiter, U., and Wolch, J. (1998). The cougar's tale. In J. Wolch and J. Emel (Eds.), *Animal geographies: Place, politics and identity in the nature-culture borderlands.* London and New York: Verso, pp. 139–149.

Haughton, G., and Hunter, C. (2004). *Sustainable cities.* New York: Routledge.

Head, L., and Muir, P. (2006). Suburban life and the boundaries of nature: Resilience and rupture in Australian backyard gardens. *Transactions of the Institute of British Geographers, 31*(4): 505–524.

Heynen, N. (2003). The scalar production of injustice within the urban forest. *Antipode: A Journal of Radical Geography, 35*(5): 980–999.

Heynen, N. (2014). Urban political ecology I: The urban century. *Progress in Human Geography, 38*(4): 598–604.

Heynen, N.C., Kaika, M., and Swyngedouw, E. (Eds.) (2006). *In the nature of cities: Urban political ecology and the politics of urban metabolism.* New York: Routledge.

Hiemstra-Van der Horst, G.G., and Hovorka, A. (2008). Reassessing the "energy ladder": Household energy use in Maun, Botswana. *Energy Policy, 36*(9): 3333–3344.

Hitchings, R. (2003). People, plants and performance: On actor network theory and the material pleasures of the private garden. *Social and Cultural Geography, 4*(1): 100–113.

Hovorka, A. (2006). The No. 1 Ladies' Poultry Farm: A feminist political ecology of urban agriculture in Botswana. *Gender, Place and Culture, 13*(3): 207–225.

Hovorka, A. (2008). Transspecies urban theory: Chickens in an African city. *Cultural Geographies, 15*(1): 95–117.

Howard, E. (1902). *Garden cities of tomorrow*. London: S. Sonnenschein and Co.

Kaika, M. (2005). *City of flows: Modernity, nature and the city*. New York: Psychology Press.

Latour, B. (1993). *We have never been modern*. Cambridge, MA: Harvard University Press.

Loftus, A. (2007). Working the socionatural relations of the urban waterscape in South Africa. *International Journal of Urban and Regional Research, 31*(1): 41–59.

Longhurst, R. (2006). Plots, plants and paradoxes: Contemporary domestic gardens in Aotearoa/New Zealand. *Social and Cultural Geography, 7*(4): 581–593.

McHarg, I.L., and Mumford, L. (1969). *Design with nature*. New York: American Museum of Natural History.

Mougeot, L.J. (2006). *Growing better cities: Urban agriculture for sustainable development*. Ottawa: International Development Research Centre.

Mumford, L. (1938). *The culture of cities*. New York: Harcourt, Brace and Co.

Mumford, L. (1961). *The city in history: Its origins, its transformations, and its prospects*. New York: Harcourt, Brace and Co.

Myers, G.A. (2008). Peri-urban land reform, political-economic reform, and urban political ecology in Zanzibar. *Urban Geography, 29*(3): 264–288.

Nast, H.J. (2006). Loving . . . whatever: Alienation, neoliberalism and pet-love in the twenty-first century. *acme: An International E-Journal for Critical Geographies, 5*(2): 300–327.

Niemelä, J., Breuste, J.H., Guntenspergen, G., McIntyre, N.E., Elmqvist, T., and James, P. (Eds.) (2011). *Urban ecology: Patterns, processes, and applications*. Oxford: Oxford University Press.

Park, R., McKenzie, R., and Burgess, E. (1925). *The city: Suggestions for the study of human nature in the urban environment*. Chicago: University of Chicago Press.

Philo, C. (1995). Animals, geography, and the city: Notes on inclusions and exclusions. *Environment and Planning D: Society and Space, 13*(6): 655–681.

Pincetl, S., Bunje, P., and Holmes, T. (2012). An expanded urban metabolism method: Toward a systems approach for assessing urban energy processes and causes. *Landscape and Urban Planning, 107*(3): 193–202.

Pulido, L. (2000). Rethinking environmental racism: White privilege and urban development in Southern California. *Annals of the Association of American Geographers, 90*(1): 12–40.

Redwood, M. (Ed.) (2012). *Agriculture in urban planning: Generating livelihoods and food security*. London: Earthscan.

Robbins, P. (2007). *Lawn people: How grasses, weeds, and chemicals make us who we are*. Philadelphia, PA: Temple University Press.

Ross, A. (1994). *The Chicago gangster theory of life: Nature's debt to society*. London: Verso.

Satterthwaite, D. (Ed.) (1999). *The Earthscan reader in sustainable cities* (Vol. 471). London: Earthscan.

Shillington, L. (2008). Being(s) in relation at home: Socio-natures of patio "gardens" in Managua, Nicaragua. *Social and Cultural Geography, 9*(7): 755–776.

Shillington, L.J. (2013). Right to food, right to the city: Household urban agriculture, and socionatural metabolism in Managua, Nicaragua. *Geoforum, 44*(1): 103–111.

Smit, J., Nasr, J., and Ratta, A. (1996). *Urban agriculture: Food, jobs and sustainable cities*. New York: United Nations Development Programme (UNDP).

Stren, R., White, R., and Whitney, J. (1992). *Sustainable cities: Urbanization and the environment in international perspective*. Boulder, CO: Westview Press.

Swyngedouw, E. (1997). Power, nature, and the city: The conquest of water and the political ecology of urbanization in Guayaquil, Ecuador: 1880–1990. *Environment and Planning A, 29*(2): 311–332.

Swyngedouw, E. 1999. Modernity and hybridity: Nature, regeneracionismo, and the production of the Spanish waterscape, 1890–1930. *Annals of the Association of American Geographers, 89*(3): 443–465.

Swyngedouw, E. (2006). Metabolic urbanization: The making of cyborg cities. In N. Heynen, M. Kaika, and E. Swyngedouw (Eds.), *In the nature of cities*. New York: Routledge, pp. 21–40.

Toronto Food Policy Council (2015). Retrieve from http://tfpc.to

Valladares, L., and Prates Coelho, M. (1995). Urban research in Latin America: Towards a research agenda. Discussion Paper Series No. 4, UNESCO Management of Social Transformations (MOST) programme. Retrieved from http://www.unesco.org/most/valleng.htm

Vining, J., Merrick, S.M., and Price, E.A. (2008). The distinction between humans and nature: Human perceptions of connectedness to nature and elements of the natural and unnatural. *Human Ecology Review, 15*(1): 1–11.

Wall, D. (1996). City profile Managua. *Cities, 13*(1): 45–52.

Whatmore, S. (1999). Hybrid geographies, rethinking the "human" in human geography. In D. Massey, J. Allen, and P. Sarre (Eds.), *Human geography today*. Cambridge: Policy Press, pp. 21–40.

Whatmore, S. (2002). *Hybrid geographies: Natures cultures spaces*. London: Sage Publishers.

Wilson, A. (1992). *The culture of nature: North American landscape from Disney to the Exxon Valdez*. Cambridge, MA, and Oxford: Blackwell.

Wolch, J.R., West, K., and Gaines, T.E. (1995). Transspecies urban theory. *Environment and Planning D, 13*(6): 735–760.

Yeo, J.H., and Harvey, N. (2010). Monkey business: Human–animal conflicts in urban Singapore. *Social and Cultural Geography, 11*(7): 681–699.

22

Healthy Cities

Godwin Arku and Richard Sadler

Introduction

Looking back over the 6000-year history of urbanization, one discovers that it is only in the last 160 years that cities have been understood as places that can promote or degrade health. For many centuries, cities served as the focus of commerce, science, religion, politics, and culture. Yet health was of little concern: pre-industrial societies, for example, had little understanding of disease-causing agents or how the built environment itself is a **determinant of health**. In the early twenty-first century, however, the connections between health and urban environments are being vigorously explored.

This chapter begins with an overview of the ways that urban areas have approached health issues in the past in order to frame how "**healthy cities**" are thought about around the world today. Because the world's cities are still challenged by health issues, four key areas of concern are introduced regarding urbanization and health: **communicable disease, non-communicable disease,** poverty and economic inequality, and the provision of urban amenities. The latter two are important because inequalities in economic status and basic service provision directly and indirectly influence health. From here, two case studies are presented—one each from the global North (Flint, Michigan, USA) and global South (Accra, Ghana). These case studies highlight how although some urban health issues vary from the global North to the global South, in other ways each region has very similar challenges. Both studies, for example, show how issues of effective governance can limit the deployment of services to address inequalities in health. The chapter concludes by reflecting upon the linkages to the Canadian context and by reinforcing the assertion that effective governance can help people to lead healthier lives.

While government consideration of cities as health enablers is relatively new to many countries, this chapter argues that effective and socially responsible governance is one of the most important macro-level determinants of the creation of healthy cities. Good governance is essential in providing health-enabling amenities to citizens, addressing poverty-related issues, and promoting activities and policies that limit the spread of both communicable and non-communicable disease. A comparison of the similarities and differences between urban health issues in cities of the global North and South permits a critical reflection upon the challenges and opportunities for building healthy cities.

Historical Overview of Cities and Disease

Thousands of years ago, the earliest cities (in regions such as Mesopotamia, Egypt, the Indus Valley, the Yellow River, and Mesoamerica) brought on the first stage of the **epidemiologic transition**. The emergence of communicable (also known as infectious) disease—caused by increased contact with disease vectors—came as hunter-gatherer cultures moved into closer quarters together with one another and with livestock. For many centuries, cities experienced recurring **epidemics** of infectious disease, including the bubonic plague, smallpox, tuberculosis, and cholera (Armelagos et al., 2005; Nelson et al., 2007).

By the nineteenth century, many cities in Europe were well into the industrialization process. Such development brought many luxuries, but it also created a new range of urban maladies. Not only were people exposed to industrial pollution, but the organization of work, with large numbers of workers living cheek by jowl, brought on large-scale overcrowding that increased the threat of infectious disease. Urban growth was rapidly outpacing advances in public health.

The dominant way of thinking about health in the eighteenth and early nineteenth centuries included the idea that miasma, or "dirty air," was responsible for many health ailments, including infectious disease. As a result, interventions focused on creating environments that promoted open spaces and greenery, but advances in medical treatments and cures had not yet been made. Over this period, **germ theory** displaced this perspective and provided a new way of understanding how pathogens are passed from one host to another. This advance helped reduce the effects of infectious disease substantially, as interventions became informed by this understanding of disease–host transmission.

The mid-nineteenth century saw a major historical moment in the fields that would develop into the study of **epidemiology** and **medical geography**. English physician John Snow demonstrated the connection between place and health by correctly linking a major part of the London cholera outbreak of 1854 to one contaminated drinking-water well (Vinten-Johansen et al., 2003). This is but one example of a breakthrough in spatial thinking around public health issues that we take for granted today, including sanitation services, clean drinking water, and street cleaning. Urban planning philosophies in Europe and North America reflected this orientation, as the **City Beautiful** and **Garden City** movements sought to restore order, harmony, and balance in spatial as well as social terms to cities. While the direct vectors causing disease were not clear, there was by the end of the nineteenth century a major recognition that cities could engender healthy behaviours.

During the early to mid-twentieth century, advances in technology allowed the city to grow both vertically and horizontally. As advances in medicine brought new vaccines for many infectious diseases, the focus of urban development likewise followed a scientific path and became oriented towards rational control of the urban environment. The emphasis of urban planning shifted to concerns such as efficient automobile-oriented transportation, land-use policy that encouraged strict separation of land uses, and urban renewal of poor housing stock. The result was the modern sprawling metropolis and the decline of inner urban areas. Thus, although infectious diseases were on the decline, lifestyle changes associated with modern culture increased the rates of chronic and non-communicable (also known as non-infectious) diseases.

By the late twentieth century, the rapid rise in rates of sedentary behaviour, heart disease, diabetes, and obesity throughout the global North (and more recently in the global South) has brought a new consciousness of the need for cities to not only limit environmental exposure to pollution, but also promote environmental affordance of health-promoting activities. As a result, fields like medical geography, urban planning, and public health have increasingly worked together to address these twenty-first-century health issues. The need for a multi-sectoral perspective on health issues in cities is succinctly underscored by Tulchinsky and Varavikova (2014: 33) when they note that "almost anything a government does or does not do affects the health and well-being of the population, mostly in the spheres of resource allocation, planning, social welfare, public health initiatives and regulation, as well as in taxation, urban planning, and public works policies." This integrative way of looking at health issues in cities is particularly critical in light of ongoing global urbanization trends. Against this background, urban health through effective governance is seen as one of the momentous challenges of the twenty-first century, especially in the global South. This is due to the lack of management capacity and failure to provide basic amenities and infrastructure for the ever-increasing number of people moving to urban areas.

What Are Healthy Cities?

Given the recent recognition of cities as an integral determinant of health, the concept of healthy cities has been pursued in many areas. And even before explicit use of the term, the Alma Ata (Kazakhstan) Declaration (1978) was made to convince all governments of the need for health promotion. Healthy cities are places where people are nurtured in healthy behaviours (including a healthy diet, adequate exercise, and limitations in exposure to environmental pollutants) to maximize quality of life (which includes not only the absence of disease but also the presence of good health). The *Health Promotion Glossary*, created by researchers at the World Health Organization (WHO) for use at a health promotion conference, has defined a healthy city as "one that is continually creating and improving those physical and social environments and expanding those community resources which enable people to mutually support each other in performing all the functions of life and developing to their maximum potential" (WHO, 1998: 13). From the perspective of the WHO, healthy cities aim to create a built environment that affords a good quality of life, meets basic sanitation and hygiene needs, and provides access to health care.

The WHO has been pivotal to the creation of the healthy cities concept and to its promotion in both the global North and South. The WHO's Healthy City Project (HCP) began in Toronto, Canada, in 1984, as part of a conference entitled "Beyond Health Care" (Hancock, 1985). The HCP addresses four key areas: (1) health inequality and poverty; (2) vulnerable groups; (3) participatory governance; and (4) social, economic, and environmental **determinants of health**. Although the meanings attached to the HCP have evolved over the years, the project ultimately seeks to put health on the agenda of decision makers in cities in the global North and global South. The project not only recognizes a strong connection between urban living conditions and health but also takes into account the complex effects of urbanization on health. Cities are seen as the appropriate administrative unit in which resources can be mobilized and action strategies developed to promote health issues, as city managers have the political authority to initiate and implement multi-sectoral approaches to health. Since its creation, the project of healthy cities has become a major long-term international development project and a key avenue for achieving the WHO's strategy of health for all.

Following this project, the Aalborg (Denmark) Charter (1994) and the Zagreb (Croatia) Declaration for Healthy Cities (2008) have focused closely on urban environments as healthy, sustainable places for people to live. This commitment is not limited to European cities. The Alliance for Healthy Cities (2003) has more recently been created by constituent countries throughout east and Southeast Asia, as well as Australia.

Major Urban Health Issues

The evolution of health issues has coincided with the evolution of global cities. Because the cities of the world remain in various stages of human and economic development, however, the health concerns seen across the globe remain very diverse. This section discusses four major health issues that impact the global North and global South to different degrees. First, infectious or communicable disease is addressed as a major issue in the global South. Second, non-infectious or non-communicable disease is considered a health priority in the global North. Third, in both the global North and global South, **poverty** is approached as an important health issue because of the many connections it has with negative health outcomes with regard both to relative poverty within nations and to absolute poverty seen across nations. Last, the provision of amenities has been a response to the need for health-promoting activities in cities, and their inequitable distribution is considered a source of increasing health disparities in many regions. Where effective and participatory governance is lacking, the provision of amenities and the amelioration of poverty have been difficult goals to achieve.

Important for the contemporary context is a disconnect that occurred between planning and public health after the Second World War. In the mid-twentieth century as political orientations refocused to concentrate on business expansion and growth, shifting away from a focus on social-service provision (concomitant with a growth in **neoliberalism**), advances in many fields of health began to lag. Thus, although scientists and public health workers have advanced our knowledge of the determinants of health, we continue to experience urban health issues because of a lack of political coordination and effort.

Communicable Diseases

Globally, communicable diseases continue to constitute the greatest burden among disease categories, especially in urban areas of the global South. These diseases represent an early stage in the epidemiologic transition, during which people are living with increased exposure to one another and to animals. Such diseases were historically not found in pre-urban civilizations and have started to be eradicated in advanced urban centres.

Well-known existing, emerging, and re-emerging diseases like tuberculosis (TB), cholera, meningitis, hepatitis, malaria, dengue/yellow fever, AIDS, Ebola, SARS, and others are bringing suffering and mortality to a global population (Boutayeb, 2006), especially in the global South. Geographically, there are variations in the rates of communicable diseases that reflect the level of economic development, health systems, government policy, institutional capacity, and environmental conditions, among other factors. For instance, avian flu outbreaks among humans have occurred in some Chinese and Southeast Asian cities with large concentrations of poultry operations (Northoff and de Vleeschauwer, 2004). About 95 per cent of all HIV infections occur in the global South, especially in sub-Saharan Africa. TB is among the top 10 causes of global mortality, but the highest rates are found in sub-Saharan Africa and Southeast Asia, particularly in urban areas. As well, the WHO estimates that, each year, malaria claims more than 1 million lives and affects over 300 million people. Similar to most communicable diseases, countries in tropical Africa bear the brunt of malaria.

The persistence of communicable diseases in the global South is largely due to socio-economic and environmental challenges brought about by rapid urbanization, as well as by a legacy of colonialism that has stymied economic growth. In countries where the health system is dysfunctional owing to civil war or political/economic strife, infectious disease is also more prevalent (e.g., the recent widespread occurrence of Ebola in Guinea, Liberia, and Sierra Leone compared to largely unaffected Côte d'Ivoire and Ghana). These disparities in disease rates in bordering countries of West Africa illustrate the fact that effective governance and policy-making can promote the eradication of communicable disease.

The extent of urbanization and the inability of governments to cope with health challenges have weakened the role of urban areas as engines of economic growth and development. Problems such as inadequate housing, severe crowding in slums, extreme poverty, lack of water supply, poor sanitation, transportation problems, and environmental pollution and degradation all contribute to dreadful living conditions for many urban dwellers and have direct implications for the physical and psychological well-being of the urban population. Ultimately, such factors aid the easy spread of communicable diseases (e.g., malaria, diarrhea), especially in highly dense areas in urban centres.

UN-Habitat (2003) estimates that 1 billion people live in slum-like conditions and projects such that by 2030 about 3 billion people, or about 40 per cent of the world's population, will need proper housing and access to basic infrastructure and services like water and sanitation systems. Because of intense crowding and congestion in such urban environments, communicable diseases are spreading faster than ever before. Unsurprisingly, poor people who are trapped in abject poverty because of systematic and systemic barriers to education and stable jobs usually bear the brunt of communicable diseases.

In the global North, although tremendous progress has been made in reducing the incidence of communicable disease (e.g., children are vaccinated against more than 20 diseases), it still remains a threat. The rise of an anti-vaccine culture in North America exists in spite of scientific evidence supporting the safety of vaccines when given properly, and this has created a new concern about outbreaks of previously eradicated diseases like measles and whooping cough (Satcher, 1999; WHO, 2011). Even with the best scientific evidence, emotions and opinions can sometimes lead to inadequate policy responses about health issues or limit the effectiveness of policy implementation. Thus, public health professionals continue to struggle with public health interventions, and officials estimate that 4 million vaccine-preventable deaths occur worldwide each year (WHO, 2011).

The globalization process has also meant an increase in the spread of communicable diseases, a condition that will likely continue as long as such diseases ravage the global South. The mass movement of large numbers of people across geographical boundaries means that no country is completely immune from potentially registering communicable diseases. International migration of people creates opportunities for the spread and establishment of common or novel infectious diseases in various countries, especially countries in the global North, the major recipient of international migrants. For instance, a report in the United Kingdom (Health Protection Agency, 2006) notes that young migrant adults bear a disproportionate burden of infectious disease; approximately 70 per cent of newly diagnosed cases of TB, HIV, and malaria were in patients born outside the UK (EASAC, 2007). As well, the domestic incidence in 2015 of Ebola in the US, UK, and Spain shows how communicable diseases can spread to distant geographical regions.

Non-communicable Diseases

Available evidence suggests that non-communicable diseases have now become a major cause of death in both the global North and global South. In the global North, the increase in non-communicable disease is due to lifestyle changes, including a shift towards diets characterized by calorie-dense, nutrient-poor, value-added food products. The built environment or urban form of many global North metropolitan areas has also become increasingly centred on the automobile as a means of transportation, so the opportunities for leading an active lifestyle in sprawling urban regions have been overlooked in favour of automobility. These twin problems have exacerbated the issue of obesity in countries of the global North.

For many centuries, obesity was considered a disease of the wealthy because food was often difficult to obtain and store in large quantities. But the food products produced by the global food system have created a new culture of plenty in high- and middle-income countries, where it is possible to obtain all kinds of foods year-round, regardless of geographic location (Weis, 2010). Thus, the process of globalization has contributed to the lifestyles that facilitate obesity and obesity-related diseases, as traditional cuisines are infused with or supplanted by less-healthy, value-added food products produced by multinational corporations.

Countries in the global South have also registered increasing rates of non-communicable diseases in recent years, and the vast majority of obese people can now be found in this part of the world (Friedman, 2014; Popkin et al., 2012). The rates are especially high in countries with the highest rates of urbanization. Evidence from Latin America, the Caribbean, the Middle East, and northern Africa suggests that diabetes and obesity rates have been rising rapidly. About 30 per cent of the population in these regions are reported to be overweight (Harpham and Molyneux, 2001). This process has thrown a wrench into the quest for healthier cities, as the very process that brings people to cities and eradicates communicable disease can exacerbate issues with non-communicable diseases.

In densely populated cities, obesity is no longer a disease of the socio-economic elite, as the condition has already crossed over to lower-income groups, especially among women (Monteiro et al., 2004). Some of the highest overweight rates in the world are among the Pacific and Indian Ocean island population. And while the overweight incidence is low in most parts of sub-Saharan Africa and Asia, there is an upward trend in overweight people in urban areas (Monteiro et al., 2004). For example, South Africans living in urban areas have overweight levels comparable to those of North America and exceeding those of many European countries. It is predicted that heart disease will become one of the leading **disease burdens** in most countries in the global South by 2030 (WHO, 1996). Along with these diseases, other non-communicable diseases, such as cardiovascular disease, cancer, coronary heart disease, as well as accidents, have all witnessed increased rates (Harpham and Molyneux, 2001). Estimates are that non-communicable diseases as a whole will account for 80 per cent of the global disease burden, causing 7 out of every 10 deaths in the global South (Boutayeb, 2006).

The reasons for the increased rates of non-communicable disease are complex and relate to lifestyle choices and environmental, socio-economic, and cultural changes. Marketing strategies employed by multinational retail chains, for example, have contributed to an adoption of Western habits. This is especially so in places where such corporate practices and advertising have heavily influenced people to emulate the choices and ethics of those higher in the social rank. In particular, the extension of the global food system to these regions (i.e., McDonald's, Coca Cola) is increasing fat and sugar intake, a potential risk factor for diseases such as diabetes and obesity (Popkin et al., 2012).

In addition, low levels of physical activity in urban areas underpin the obesity problem. For example, across Indian cities, where up to 40 per cent of adults suffer from hypertension, an abnormally high concentration of cholesterol is attributed not only to dietary changes but also to a lack of physical activity (Reddy et al., 2005). In highly urbanized cities of the global South, especially in Latin America (e.g., Rio de Janeiro, Bogotá, Caracas), the worsening incidence of the disease even in preschool children is attributed to poorly built neighbourhood environments and sedentary lifestyles. The problem is compounded by excessive amounts of time spent watching television and playing computer games (Cuevas et al., 2009). This shift is unraveling unfamiliar complexities made evident by the growing incidence of the double burden of under-nutrition and obesity, even in the same household (Popkin, 2009).

Poverty

Although it may not seem readily apparent, the presence of poverty in one's life can have a tremendous impact on health opportunities. Poverty is a health-degrading activity because people who lack resources may be unable to afford medical care, healthy food, or sanitary living conditions, especially in places where governments do not provide effective social services. Furthermore, the stresses of labour-intensive, low-wage jobs, neighbourhood stressors like crime and pollution, as well as poor educational opportunities, can contribute to negative mental and physical health outcomes.

Cities themselves can be effective at eliminating poverty by providing employment opportunities and services, but this means that people constantly move to them to escape rural poverty. Especially in countries where services in urban areas exceed those of rural areas, rural-to-urban migration can be large scale and the demand for services can outpace the government's ability to provide for such needs. Notwithstanding the deficit of services, vast numbers of rural dwellers in countries in the global South continue to migrate to urban areas with the hope of finding economic opportunities in urban areas and a better quality of life. This is, in part, because smallholder farming (which was previously a major livelihood activity) has become an increasingly unproductive and untenable activity. A combination of factors has worked in tandem to push most rural dwellers out of agriculture to seek alternative livelihood in urban areas. These factors include land grabbing by multinational companies (e.g., from countries such as the United Kingdom, the United States, and China) in countries in the global South (including many in sub-Saharan Africa); the rationalization of agriculture; the World Bank's structural adjustment programs, which have resulted in rising costs of input; impacts of climate variability; and neoliberal factors that have exposed poverty-stricken countries to global competition (Pemunta, 2014). Because of the lack of formal employment opportunities, especially for women, the **informal economy** is often the safety net for the vast majority of new migrants to cities (Figure 22.1). According to estimates by the International Labour Organization, the informal economy comprises half to three-quarters of the workforce in the global South (Charmes, 2002).

Figure 22.1 Informal activity in Ekwendeni Trading Centre near the city of Mzuzu, Malawi. Women selling food and household items in open spaces is typical of urban life in the global South

One of the major ways cities can address poverty is through economic development initiatives. Such proposals aim to capitalize on a city's strengths to encourage investment in industry and commerce. Cities in the global South (including, for example, cities in South Africa, Mozambique, and Kenya) have been working to promote small-scale entrepreneurial endeavours in the informal economy, which collectively can help address the challenges of poverty by improving **food security**, thereby improving health status.

In low-income countries, however, an overemphasis on such environmentally friendly projects may have a negative impact on economic development. Within the context of the global economy and environment, this raises the question: what constitutes a healthy city? Must cities sacrifice economic development (and the improvements in health seen from better job prospects) to address environmental issues (and the improvements in health seen from cleaner air, water, and soil), or is there a way to incorporate both? For many concerned with urban-planning theory and practice, the concept of **sustainable urbanization** has gained significant traction as a solution to this dilemma.

Even in countries of the global North the issue of poverty remains a serious public health issue, largely because of the impact that relative poverty can have in creating health disparities. Although twenty-first-century cities in the global North are the sites of the largest concentrations of wealth in the history of civilization, the capitalist economic system under which much of the world operates does not afford an equivalent share of productivity gains to the working classes. In some societies (such as those of northern Europe and Canada), income inequality is less pronounced, but in others, most of the economic gains made during the last 40 years have gone to only the very richest individuals.

Amenities: Opportunities for Food and Physical Activity

The last of the major considerations important for healthy cities is the idea of amenities. The idea that amenities in the urban environment could be health promoters gained traction during the early twentieth century. The environmental conditions of industrial cities were very poor on account of polluting industries. To combat this public health issue, the ideas of City Beautiful, Garden Cities, and other movements were broadly oriented around ensuring that adequate green space was made available to all citizens (e.g., parks as air purifiers).

As cities in the global North have matured, basic amenities have given way to what could be called the "new" environmental determinants of health. Such amenities include the presence of healthy foods, parks and other opportunities for physical activity, schools, jobs, and quality urban design. Taken together, they help constitute the quality of the built environment, which is now known as a major predictor of healthy behaviours. Yet because of the rollback of government services under urban neoliberalism, wealthier communities in the global North have rapidly outpaced poorer communities in their ability to provide health-enhancing activities.

Significant scholarly research has been conducted on health-promoting or -degrading environments as cities around the world pursue interventions to address environmental health determinants (Bonita et al., 2013; Curtis and Jones, 1998; Duhl and Sanchez, 1999; Northridge et al., 2003). This research helps inform "new social and urban planning approaches (e.g., recreational opportunities, access to healthy foods) to reduce the harmful effects of poverty" (Tulchinsky and Varavikova, 2014: 28), which is increasingly important given the increase in non-communicable diseases and the relation of these diseases to poverty.

One idea that has garnered increasing attention is that of **food deserts** (Sadler et al., 2013), as the food system has shifted towards larger-format stores and low-income communities are increasingly found to have inadequate access to healthy foods. These areas also often have high concentrations of junk food in variety stores and fast-food restaurants. The effects of this are that people cannot find the requisite foods for good health. Residents in food deserts have been found to eat less and pay more for healthy foods while simultaneously being exposed to more junk food than other groups.

This concern of social equity in relation to food access in low-income communities can be applied broadly to other amenities, as communities with declining tax bases cut back on community schools, parks maintenance and sports programming, and public safety. These are also the communities least able to invest in new pedestrian and bicycling infrastructure. Contrast this with the rich parks and cycling culture found in cities in countries such as the Netherlands and Denmark.

Even in cultures with high social-service provision and strong economies (such as those of Western and Northern Europe), cities have struggled to find a balance between amenities that are health promoting (such as cycling infrastructure and bike-sharing programs) and the proliferation of fast-food and auto-oriented environments (Figure 22.2).

Unlike in high-income nations in the global North, the idea of amenities as promoters of healthy lifestyles, healthy communities, and sustainable cities is at a rudimentary stage in many nations in the global South. Indeed, in many cities in the global South, policy-makers plan urban expansion and redevelopment without regard for amenities such as public parks, green spaces, and related features or

Figure 22.2 Cycling culture juxtaposed against the provision of fast foods shows the challenge of providing healthy urban amenities, Vienna, Austria

opportunities for physical activity. In part, this disregard is due to a failure to understand the health promotion role of amenities, but also due to the drive for economic development taking centre stage in development policy to the neglect of amenity planning. In many cities in the global South, not only has recent urban redevelopment eradicated or degraded some public parks, but also recent planning policies have failed to preserve natural pockets and corridors, even as development continues to consume green fields (Arku, 2009a). For instance, as demonstrated in the Accra case study below, urban development has virtually consumed the existing green spaces in the city. In Kenya, Nairobi City Park has shrunk from its original 91 hectares (225 acres) to 60 hectares (148 acres). Quite often, parks and open spaces are the first targets for land development when there is a need for an economic development project.

The pressure on urban land is clearly manifested in the mounting urban land conflicts currently prevailing in cities of the global South. The conversion of vacant or idle land is a complex issue, as it not only involves desperately poor urban residents looking for land to occupy, but also members of elite groups who are able to expropriate large areas of urban land for residential, commercial, and office development. The implications of urban land conversion or lack of amenity planning are that there are increasingly fewer recreational spaces for the burgeoning urban population. Because amenities are important health promoters, their disregard in urban planning deprives urban residents and the wider urban economy of the positive contributions such facilities provide. Two case studies are used here to illustrate the foregoing discussion. The first, in Accra, Ghana, showcases the challenges facing rapidly growing urban areas in the global South. The second, in Flint, Michigan, USA, shows that many cities in the global North are also still struggling with basic public health issues (Figure 22.3).

Rapid Urbanization, Economic Development Challenges, and Healthy City Planning in Accra, Ghana

Countries in the global South are currently facing major urban health problems. The problems are not new but have escalated in recent years owing to unprecedented rates of urbanization in these countries. Even though the global South is now going through essentially the same broad process of urbanization that the global North went through over a century ago, the global context in which it is urbanizing means that the specific processes at work create very different health problems and outcomes. One such process involves the global flow of capital, which has transformed some cities in the global South into major economic hubs for the entire country, creating monopoly power and increasing spatial inequality in the national urban system.

Ghana's administrative capital, Accra, is one of the rapidly growing cities in Africa. Spatially, Accra has expanded by more than 500 per cent since the late 1970s. The metropolitan population of 4 million is a twofold increase since the early 1980s. Accra's growth and development have occurred within the context of interaction between global and local forces in the country's political economy; namely, a strong neoliberal agenda of economic rationalization and efficiency influenced economic reforms involving privatization, devolution, and deregulation. As part of this agenda, a special **export processing zone** was created in the metropolis to attract foreign investment. Since the 1990s, Accra has become the headquarters for 94 per cent of all foreign banking institutions established in Ghana and the city has seen modern high-rise office buildings and high-end residential development (Arku, 2009b). Overall, although Accra has long dominated the country's urban system, recent global and local forces have strengthened Accra's position to a historically unparalleled position.

Accra's privileged position has subsequently attracted rural migrants in search of a better life in the city, and the city, like other burgeoning cities in the global South, is grappling with issues that affect people's mental and physical health, such as poor-quality housing, overcrowded streets, and rising poverty levels. This has resulted in informal and squatter settlements where people struggle to survive and diseases are rife. The city authorities have been unable to provide reliable services like water, electricity, and garbage collection. Many people have been unable to secure the basic necessities of life, including housing and food. Only 11 to 20 per cent of the city's residents, for example, have access to adequate house-to-house garbage collection (Afeku, 2005). Thus, although the hope of migrants is to improve their well-being, their living conditions in Accra in most instances end up becoming even worse than their original standard of living in rural villages.

Figure 22.3 Map locating Accra, Ghana, and Flint, Michigan, USA

As elsewhere in the global South, the need for economic development tends to relegate the provision of amenities to the backstage of development planning in the city. Amenities such as urban parks and green spaces are in short supply in Accra, as most existing facilities have been converted to residential, commercial, or retail uses (Figure 22.4). In a recent study involving planning officials on why so few parks exist in Accra, the issue of urban land pressures emerged as the dominant reason, with one official stating that "with so many competing demands for land, public parks are often seen as uneconomic . . . a waste of valuable space" (Arku et al., 2016). Along the same lines, another official commented that "some people see empty spaces as waste. . . . [I]t must be used for more productive purpose" (Arku et al., 2016). These two quotations clearly illustrate the current challenge of building healthy cities in Ghana. While there is limited access to urban parks and green space in Accra, there is a long history of its provision as a public good in North America. The following case study from Flint, Michigan, reveals that while the health issues facing cities in the global North are, in absolute terms, often

Figure 22.4 Original Afrikiko Park in Accra. Urban development pressure led to its conversion to retail uses

less severe than those of the global South, the economic and political changes experienced by North American and European cities have amplified a range of different health concerns in heavily occupied areas.

The Effects of Deindustrialization and Economic Inequality on Health Outcomes in Flint, Michigan, USA

The spread of **deindustrialization** significantly changed the employment structure of manufacturing areas, as their industrial production slowed and jobs either disappeared or transitioned into lower-paid service-sector jobs. Simultaneously, the spread of neoliberalism as a political philosophy meant governments played a much smaller role in managing health and social ills in urban areas. This second case study city of Flint, Michigan, is an extreme example of the negative effects of deindustrialization, neoliberalism, and ineffective governance. In all four areas discussed above (communicable disease, non-communicable disease, poverty, and amenities), Flint is challenged by political and economic changes. The county in which Flint is located ranks towards the bottom of most public health indices in Michigan, including 81st of 82 for quality of life, 78th in length of life, 78th in social and economic factors, and 75th in physical environment (County Health Rankings, 2015). Rates of obesity, diabetes, sexually transmitted diseases, and violence-related deaths are towards the top of these rankings for Michigan. While Flint's situation is exceptionally dire in some neighbourhoods, the city overall broadly exemplifies the processes characteristic of many cities experiencing restructuring in the global North (such as those in Japan, Russia, Germany, and across the North American Rust Belt).

In the mid-twentieth century, Flint (far more than even Detroit) had the lowest industrial diversification of any US city (being heavily dependent on a single employer, the General Motors automotive company) (Rodgers, 1957; Lewis, 1965). Government-sponsored racial segregation was also common practice, with poorer and black populations relegated to less desirable high-pollution areas near factories (Lewis, 1965). These populations were thus more exposed to industrial contaminants and at greater risk for associated health issues like asthma, lead exposure, and cancers. The onset of deindustrialization meant a series of job losses throughout the latter part of the century. These job losses and lower industrial productivity meant a decline in city revenue during the same period in which neoliberalism compelled many governments to roll back social services. The onset of the housing crisis served as a final straw, precipitating large cuts to Flint's county health department (Galewitz, 2011). Even while poverty levels were increasing rapidly throughout the city in the 1970s and 1980s due to the loss of jobs, the government became less equipped to deal with the increasing need for social service provision. Without this safety net, the negative health effects of poverty became amplified, especially among the non-white minority population who were previously disadvantaged by restrictive land-use laws limiting its housing options in segregated, polluted neighbourhoods. Because many residents were unable to pay for health care, medical services were shut down in many neighbourhoods. Other amenities necessary for healthy living closed as well, including grocery stores, banks, and schools. The city rolled back many services, including garbage pickup, police and fire patrols, and parks programming and maintenance. As residents moved out of the city, taxes and rates had to be increased to provide and maintain basic services like water and sewers. Crime also increased as a result of increased poverty, abandonment of homes, and declines in policing levels, making it even more difficult for residents to couct healthy, active lifestyles (Greater Flint Health Coalition, 2012). Recent research on the city has shown how retail demand has continued to shrink and many neighbourhoods are now characterized as food deserts (Sadler et al., 2013). A nearly year-long, semi-permanent satellite camp of the **Occupy Wall Street** movement (Figure 22.5) in 2012 was emblematic of a long-standing concern about the effects of social and economic inequality resulting from deindustrialization (and its attendant health effects).

The consequences of urban disinvestment have created an even more urgent problem: the leaching of lead from water lines into drinking water. As part of a neoliberal governance scheme, the state of Michigan had previously adopted a punitive emergency manager system with broad powers to cut costs in municipalities with declining revenue. In a cost-cutting move while under the direction of an emergency manager, Flint's water system was switched to river water in April 2014. The river water's composition was more naturally corrosive to lead in pipes, and proper corrosion control was not implemented. As a result, lead from pipes and solder began leaching into the drinking water system, and water quality deteriorated. Not long after this change, community members began a campaign to return to the original water source, citing issues with the water's colour, odour, and taste.

Unfortunately, despite significant community activism, it was not until researchers began uncovering elevated water lead levels across the city that higher levels of government took the issue more seriously. An additional research study in September 2015 found that elevated blood lead levels in children nearly doubled within the city as a result of exposure to lead through the water system (Hanna-Attisha et al., 2015). These levels are of concern because lead is a potent neurotoxin that can have long-term health and developmental effects. State officials initially attempted to discredit the study, but relented after continued efforts by community and academic activists.

State and federal government response has included water and filter distribution, educational campaigns, nutrition programs, and proposals for long-term funding to address potential ongoing problems created as a result of this issue. Many charities, non-profits, and community organizations—which had already been providing resources prior to government involvement—have become increasingly active in various ways. What is particularly troublesome, however, is that this environmental injustice most severely affects poor people, further amplifying the challenges they face. This further illustrates the shortcomings of governance that prioritizes the economic bottom line above human health.

Yet even in spite of these challenges, the resiliency of urban areas can be seen in efforts to revitalize the city. In the midst of macroeconomic forces of neoliberalism beyond its control, Flint has over the past few decades adopted a community development philosophy towards urban development. Rather than clearing neighbourhoods of low-income populations to pursue "top-down" economic development (as was the case during the racially motivated slum-clearance era of the 1960s), civic leaders have supported targeted demolition efforts in abandoned areas and a blight-elimination framework served to manage ongoing neighbourhood change (Pruett, 2015). Incidentally, investors have begun to eye Flint as an affordable location where new projects could have a large impact on quality of life for all residents. Significant investments—including many initiated or financed by community-based organizations and faith-based institutions—have been made in pedestrian/cycling infrastructure, community schools, parks maintenance and upgrades, blight elimination, higher education/research, health care and innovation, local food systems, and downtown economic development, and the long decline in population is beginning to slow down (US Census Bureau, 2010). All of these activities

Figure 22.5 The "Occupy Flint" movement encampment sits alongside recent downtown redevelopment (upper left) in Flint, Michigan, USA

may contribute to improving public health indices in the city by strengthening the housing stock, increasing opportunities for healthy eating and physical activity, and improving educational and employment outcomes.

Flint is a reflection of a once-broken planning and economic development system that relied too heavily on private industrial investment for growth and led to significant public health issues, but innovative ways of rethinking the city provide hope for revival. In many ways, although Flint is emblematic of problems facing cities of the global North, its challenges are not dissimilar from those in cities of the global South. This perspective on Flint has been taken by other researchers in the past (Schindler, 2014). This case study therefore serves to show persistent links between health issues in cities globally. Our attention has been focused on cities around the world, with special consideration given to Accra, Ghana, and Flint, USA. While it may be tempting to read these issues only as a problem for these countries, many of the problems are also relevant in the Canadian context. The following text box will make these connections clearer.

Building Healthy Cities in Canada

Canada has benefited from its close association with British town-planning ideals, which have helped create health-promoting environments. The Ottawa Charter for Health Promotion (1986), for example, showed the federal government's commitment to achieving a goal of "Health for All." Action areas of that initiative included building public health policy; creating supportive environments; strengthening community action; developing personal skills; reorienting health-care services towards prevention of illness; and promoting health. These tenets are all reflected in the current worldwide effort to build healthy cities. Although the issues discussed throughout this chapter arise from a range of environmental, political, and economic contexts, they are very much intertwined with the fortunes of the Canadian urban context for a number of reasons.

Foremost, in our globalized world, the issues that affect one nation can have strong effects on others. A major avian flu outbreak in China could adversely affect worldwide poultry prices and international travel and trade more broadly. The vectors through which communicable disease are carried are amplified by the ease of overseas air travel, and this requires new policy approaches to address disease spread. This concern was evident during the 2014 Ebola outbreak, as contentious visa restrictions were put in place for new applicants from affected countries such as Liberia, Sierra Leone, and Guinea (Branswell, 2014).

Canada's cities have become mosaics of many ethnic groups for many reasons. The porous borders of many nations means that immigrants can easily move from one country to another, eventually making their way to Canadian cities, attracted by the high standard of living. Conversely, other countries' failure to help address issues of poverty often translates into civil unrest, which can spur additional migration away from these places. For instance, the recent surge in civil unrest in the Middle East has had a cascading effect on immigrants moving from these nations to places like Southern Europe, the United States, and Canada. Because of the interconnectedness of the global economy, cities necessarily host a large proportion of international migration and trade, providing opportunities for other cultures to benefit from, say, the Canadian urban system and strengthening the domestic economy, which is ultimately one of the strongest predictors of health.

In many ways, European cities have been leading the effort to create cities that embody the tenets of health and engender healthy behaviours. For instance, the increasing popularity of bike-sharing programs in cities like Montreal and Toronto was inspired by the cycling cultures of Amsterdam, Copenhagen, and elsewhere (Béland, 2014). Additional principles about making cities healthy, walkable places have been borrowed from European planning principles. It is through sharing these and other practices shown to improve health that Canadian cities have seen many advances in recent years.

Within Canada, the issues discussed throughout the chapter are also directly relevant. Poverty and the equitable provision of city services remain significant challenges in some Canadian cities, especially the larger metropolises of Vancouver and Montreal where poverty rates have persistently been above 14 per cent (Statistics Canada, 2012). Many cities remain largely automobile-dependent, and for many low-income families accessing health care, healthy foods, or employment is made difficult by the built environment. Furthermore, socio-economic segregation means opportunities for escaping poverty are difficult to find. As above, the largest cities—Montreal, Toronto, and Vancouver (MTV)—have the highest segregation indices, but all Canadian cities experience this issue (Balakrishnan et al., 2005).

The increasing occurrence of non-communicable disease persists as a major public health issue in Canadian cities. While policy-makers have become more aware of the importance of building cities to promote physical activity and healthy eating, Canada's obesity rate has continued to grow to beyond 25 per cent: while the largest cities (MTV) have rates closer to 20 per cent, many other smaller cities and rural areas have rates well above 30 per cent (e.g., Saint John, New Brunswick, at 38 per cent, and Sudbury, Ontario, at 34 per cent) (Statistics Canada, 2014). Thus the poverty-reducing and amenity-enhancing activities being pursued in our cities (seemingly especially in larger cities) are necessary for ensuring that basic health-promoting activities are available for all.

Generally, there is a moral imperative to remain connected to public health challenges across the world. Canadian cities have much to share with those in the global South in terms of building healthy cities, but our cities also have the ability to learn from other cities how to deploy effective urban health strategies. Ultimately, the interconnectedness of our world community means that economic success and stability require a collaborative approach to healthy world cities. Focusing solely on our own cities to the exclusion of international collaboration may increase the disparities in health indicators in the global South versus global North cities.

Conclusion

This chapter has explored cities as places that can promote or degrade health. Historically, there was little understanding about cities as places that could promote health, particularly when compared to our current understanding. Over time, there has been a progressive shift in the way that planners, scholars, residents, and civil organizations think about the role of the city. In particular, the built environment is seen as a determinant of health. Such an understanding of the built environment is reflected in programs like the Healthy Cities Project, spearheaded around the world by the WHO, which recognizes a strong connection between urban living conditions and health.

As countries have urbanized, various social, economic, and environmental challenges have become more prominent. In both the global North and global South, health challenges such as the rising incidence of communicable and non-communicable diseases are well entrenched—albeit at different geographic scales and extents. Poverty is an endemic part of urban life in the global South, although poverty also remains a serious public health issue in cities in the global North. Thus, one key message in this chapter is that cities may be thought of simultaneously as offering opportunities for healthy living and as being places of struggle in terms of achieving healthy living.

As noted in the introductory section of this chapter, cities have historically served as places of innovation and creativity and as centres of commerce, science, and culture. Cities still have the strengths and resources needed to tackle the challenges of building healthy cities. Because the majority of people now live in urban areas, programs that aim to improve the health of cities should be viewed as a set of public health strategies that would benefit most of society. Indeed, the high concentration of people in cities should be viewed as a strength because it presents opportunities for authorities to improve the health of urban residents. The high concentration ensures economies of proximity by reducing the unit costs for the provision of urban services and amenities for healthy living.

Efforts to provide urban services and facilities efficiently will require continued improvements in local governance. Unless governance is radically adapted to the need to be more responsive to local needs, to incorporate participatory governance, and to increase government accountability for health issues, the future will see a continuation of many of the urban health challenges enumerated in this chapter around both communicable (e.g., tuberculosis, malaria, cholera) and non-communicable (e.g., heart disease, diabetes, obesity) diseases. The key question, therefore, is: will we be able to find effective ways to deal with the current urban health problems and build healthy cities? This question is of importance to the Canadian context because, as we saw in the text box, Canada's economic and health fortunes are tied in to the issues being faced around the world.

Key Points

- The emergence of the study of epidemiology in the mid-nineteenth century transformed understandings of the connection between urban places and health.
- Health cities aim to create built environments that afford a good quality of life, meet basic sanitation and hygiene needs, and provide access to health care.
- Four key areas link urbanization and health: communicable disease; non-communicable disease; poverty and economic inequality; and the provision of urban amenities.
- The urbanization of the global South has resulted in momentous urban health challenges with the emergence of new communicable diseases (e.g., Zika) and the re-emergence of others (e.g., tuberculosis), and the continued spread of others, often as the result of overcrowding (e.g., malaria).
- Non-communicable disease has become a major cause of death in the global North and more recently in the global South, and while the underlying factors are complex, they include global economic change, lifestyle choices, environmental factors, socio-economic conditions, and cultural changes.
- In both the global North and global South, it is poor urban residents trapped in abject poverty who bear the brunt of health and health-related issues, given their lack of resources and inability to afford basic necessities of life (e.g., medical care, healthy food, and sanitary living conditions).
- Amenities (e.g., urban green spaces, healthy food, transportation infrastructure, and cultural attractions) are important in promoting healthy lifestyles, healthy communities, and sustainable cities.
- Accra, Ghana, is experiencing rapid urban growth, which creates competition for urban land, including green space, resulting in the prioritization of commercial and residential land uses over public green spaces and a disregard for the health benefits of urban parks.
- In Flint, Michigan, deindustrialization and neoliberalism have combined to worsen economic inequality, and communities continue to experience significant poverty-related public health problems.
- Residents of Canadian cities benefit from public health policies (e.g., promotion of cycling) and a public health-care system, but in a globalized world they are still vulnerable to communicable diseases, while non-communicable diseases also persist as major public health issues (e.g., diabetes). Policies and programs that aim to improve the health of cities should be viewed as a set of public health strategies that will benefit the majority of the world's population.

Activities and Questions for Review

1. What are healthy cities and what are their goals? In pairs, discuss how your future health may benefit from these goals.
2. What particular health challenges are faced by cities grappling with urban decline and whom do they impact upon the most?
3. Imagine that you are on the board of a community-based organization with a mandate to improve neighbourhood health outcomes in a Canadian city. What would your priorities be and why?
4. How are urban health challenges in cities in the global North and global South both similar and different? In groups, discuss how the binary and supposedly opposite categories of "North" and "South" may be inadequate descriptors.

Acknowledgments

We would like to thank Karen Van Kerkoerle of the Department of Geography at Western University for her mapping assistance.

References

Afeku, K. (2005). *Urbanization and flooding in Accra, Ghana*. Doctoral dissertation, Miami University.

Arku, G. (2009a). Rapidly growing African cities need to adopt smart growth policies to solve urban problems. *Urban Forum*, 20(3): 253–270.

Arku, G. (2009b). Housing policy changes in Ghana in the 1990s. *Housing Studies*, 24(2): 261–272.

Arku, G., Yeboah, I., and Nyantakyi-Frimpong, H. (2016). Public parks as an element of urban development: A missing piece in Accra's growth and development. *Local Environment*, 21 (12): 1500–1515. doi: 10.1080/13549839.2016.1140132

Armelagos, G.J., Brown, P.J., and Turner, B. (2005). Evolutionary, historical and political economic perspectives on health and disease. *Social Science & Medicine*, 61(4): 755–765.

Balakrishnan, T.R., Maxim, P., and Jurdi, R. (2005). Residential segregation and socio-economic integration of visible minorities in Canada. *Migration Letters*, 2(2): 126–144.

Béland, D. (2014). Developing sustainable urban transportation: Lesson drawing and the framing of Montreal's bike sharing policy. *International Journal of Sociology and Social Policy*, 34(7/8): 545–558.

Bonita, R., Magnusson, R., Bovet, P., Zhao, D., Malta, D.C., Geneau, R., and Lancet NCD Action Group (2013). Country actions to meet UN commitments on non-communicable diseases: A stepwise approach. *The Lancet*, 381(9866): 575–584.

Boutayeb, A. (2006). The double burden of communicable and noncommunicable diseases in developing countries. *Transactions of the Royal Society of Tropical Medicine and Hygiene*, 100(3): 191–199.

Branswell, H. (2014). "Ebola: Canada suspending visas for residents of outbreak countries." *CBC News*. Retrieved from http://www.cbc.ca/news/politics/ebola-canada-suspending-visas-for-residents-of-outbreak-countries-1.2820090

Charmes, J. (2002). *Women and men in the informal economy: A statistical picture*. International Labour Organization, Employment Sector, Geneva, Switzerland.

County Health Rankings (2015). Retrieved from http://www.countyhealthrankings.org. University of Wisconsin Population Health Institute.

Cuevas, A., Alvarez, V., and Olivos, C. (2009). The emerging obesity problem in Latin America. *Expert Review of Cardiovascular Therapy*, 7(3): 281–288.

Curtis, S., and Jones, I.R. (1998). Is there a place for geography in the analysis of health inequality? *Sociology of Health & Illness*, 20(5): 645–672.

Duhl, L.J., and Sanchez, A.K. (1999). *Healthy cities and the city planning process: A background document on the links between health and urban planning*. World Health Organization Office for Europe, Copenhagen, Denmark.

European Academies Science Advisory Council (EASAC) (2007). *Impact of migration on infectious disease in Europe*. European Academies Science Advisory Council, London, UK.

Friedman, U. (2014). Two-thirds of obese people now live in developing countries. *The Atlantic*, May 29. Retrieved from http://www.theatlantic.com/international/archive/2014/05/two-thirds-of-the-worlds-obese-people-now-live-in-developing-countries/371834

Galewitz, P. (2011). Health services squeezed amid housing bust. *Washington Post*, March 27. Retrieved from https://www.washingtonpost.com/national/health-services-squeezed-amid-housing-bust/2011/03/18/AFpLikkB_story.html

Greater Flint Health Coalition (2012). *2012 Community health needs assessment for the Genesee County/City of Flint Community: An assessment of Genesee County and the City of Flint (Michigan) conducted jointly by Genesys Health System, Hurley Medical Center, McLaren-Flint, and the Greater Flint Health Coalition (GFHC)*. Flint, MI.

Hancock, T. (1985). Beyond health care: From public health policy to healthy public policy. *Canadian Journal of Public Health*, 76(s1): 9–11.

Hanna-Attisha, M., LaChance, J., Sadler, R.C., and Schnepp, A.C. (2015). Elevated blood lead levels in children associated with the Flint drinking water crisis: A spatial analysis of risk and public health response. *American Journal of Public Health*, 106(2): 283–290.

Harpham, T., and Molyneux, C. (2001). Urban health in developing countries: A review. *Progress in Development Studies*, 1(2): 113–137.

Health Protection Agency (2006). *Migrant health: Infectious diseases in non-UK born populations in England, Wales and Northern Ireland*. A baseline report 2006. London: HPA. Retrieved from http://www.hpa.org.uk/web/HPAwebFile/HPAweb C/1201767922096

Lewis, P.F. (1965). Impact of Negro migration on the electoral geography of Flint, Michigan, 1932–1962: A cartographic analysis. *Annals of the Association of American Geographers*, 55(1): 1–25.

Monteiro, C., Moura, E., Conde, W., and Popkin, B. (2004). Socioeconomic status and obesity in adult population of developing countries. *World Health Organization Bulletin*, 82(12): 940–946.

Nelson, K.E., and Masters Williams, C. (2007). Early history of infectious disease: Epidemiology and control of infectious diseases. In K.E. Nelson and C. Masters Williams, (Eds.), *Infectious disease epidemiology: Theory and practice*. Burlington, MA: Jones & Bartlett Learning.

Northoff, E., and de Vleeschauwer, D. (2004). *High geographic concentration of animals may have favoured the spread of avian flu*. Food and Agriculture Organization of the United Nations: Bangkok. Retrieved from http://www.fao.org/Newsroom/en/news/2004/36147/index.html

Northridge, M.E., Sclar, E.D., and Biswas, P. (2003). Sorting out the connections between the built environment and health: A conceptual framework for navigating pathways and planning healthy cities. *Journal of Urban Health*, 80(4): 556–568.

Pemunta, N.V. (2014). New forms of land enclosures: Multinationals and state production of territory in Cameroon. *Studia Universitatis Babes-Bolyai-Sociologia*, 61(2): 35–58.

Popkin, B.M. (2009). Global changes in diet and activity pattern as drivers of the nutrition transition. In S.C. Kalhan, A.M. Prentice, and T. Yajnik (Eds.), *Emerging society: Coexistence of childhood malnutrition and obesity*. Basel: Nestec.

Popkin, B.M., Adair, L.S., and Ng, S.W. (2012). Global nutrition transition and the pandemic of obesity in developing countries. *Nutrition Reviews*, 70(1): 3–21.

Pruett, N. (2015). *Beyond blight: City of Flint comprehensive blight elimination framework*. Report. Flint, MI: City of Flint.

Reddy, S., Shah, B., Varghese, C., and Ramadoss, A. (2005). Responding to the threat of chronic diseases in India. *The Lancet*, 366(9498): 1744–1749.

Rodgers, A. (1957). Some aspects of industrial diversification in the United States. *Economic Geography*, 33(1): 16–30.

Sadler, R.C., Gilliland, J.A., and Arku, G. (2013). A food retail-based intervention on food security and consumption. *International Journal of Environmental Research and Public Health*, 10(8): 3325–3346.

Satcher, D. (1999). *Statement on risk vs benefit of vaccinations*. Given before the House Committee on Government Reform, August 3. Retrieved from http://www.hhs.gov/asl/testify/t990803a.html

Schindler, S. (2014). Understanding urban processes in Flint, Michigan: Approaching "subaltern urbanism" inductively. *International Journal of Urban and Regional Research*, 38(3): 791–804.

Statistics Canada (2012). Income in Canada, 2011. CANSIM Table 202-0804—Persons in low-income families, annual.

Statistics Canada (2014). *Canadian Community Health Survey*. Ottawa: Statistics Canada.

Tulchinsky, T.H., and Varavikova, E. (2014). The new public health. *Public Health Reviews*, 32(1): 25–53.

United Nations Human Settlements Programme (UN-Habitat) (2003). *The challenge of slums: Global report on human settlements 2003*. London: Earthscan.

US Census Bureau (2015). American FactFinder State and County QuickFacts. Retrieved from http://quickfacts.census.gov/qfd/states/26/2629000.html

Vinten-Johansen, P., Brody, H., Paneth, N., Rachman, S., Rip, M., and Zuck, D. (2003). *Cholera, chloroform, and the science of medicine: A Life of John Snow*. New York: Oxford University Press.

Weis, T. (2010). *The global food economy: The battle for the future of farming*. London: Zed Books.

World Health Organization (1996). *Investigating in health research and development*. Geneva: WHO.

World Health Organization (1998). *Health promotion glossary*. Division of Health Promotion, Education and Communications Health Education and Health Promotion Unit. Geneva: WHO.

World Health Organization (2010). *Types of healthy settings*. Retrieved from http://www.who.int/healthy_settings/types/cities/en

World Health Organization (2011). No vaccine for the scaremongers. *Bulletin of the World Health Organization*, 86(6): 417–496. Retrieved from http://www.who.int/bulletin/volumes/86/6/08-030608/en

23 Urban Water Governance

Rebecca McMillan, Sawanya Phakphian, and Amrita Danière

A young girl in New Delhi, India, waits in a long queue at a public water pipe to fill her water buckets. Her household does not have piped water or a latrine. Although an NGO has recently built a public toilet a few blocks away, she can only afford to pay the fee once a day. She also risks violence and harassment, since there is little security in the public bathrooms.

In Johannesburg, South Africa, a group of residents are protesting water cut-offs. Their municipal government recently installed prepaid water meters in an effort to cut administrative costs. When a household can no longer pay, its service is automatically shut off.

In Caracas, Venezuela, a migrant couple from Colombia boasts a piped-water connection directly to their rental property. Yet, water only arrives approximately every 21 days. During the 4 days they have service, they must store enough water for three weeks. They also boil their water, since it becomes contaminated on the liquid's long journey to their hillside residence. Despite this, their government counts them as having an "improved" service.

Introduction

The provision of affordable and safe water in cities has been one of the most important challenges of the twentieth and twenty-first centuries. Water is essential for life and livelihood. Yet, in many cities of the global South, water services, particularly for low-income residents, continue to be undependable, expensive, and potentially unhealthy. As of 2015, 663 million people still lacked access to safe drinking water and 2.4 billion people had no access to improved sanitation (WHO and UNICEF, 2015: 4–5). Providing adequate water is therefore an urgent priority for policy-makers, development practitioners, civil society actors, and scholars alike.

For most urban Canadians, accessing water for drinking and household tasks is as simple as turning on a tap. For many people in the global South, by contrast, life must be organized around procuring water. As the vignettes at the beginning of this chapter illustrate, contamination, high water and sewerage-service prices, and inadequate infrastructure are serious challenges for urbanizing areas in the global South. Climate change will put additional stress on already-deficient water and sanitation systems and further reduce water supplies. It is estimated that by 2030 half of the world's population will be living in high water–stressed areas (UNESCO International Hydrological Programme, 2014: 2).

As is true of many basic services, water access is also highly differentiated along lines of gender, race, citizenship, and class. Water delivery is thus a political and power-laden process (Swyngedouw, 2004). For this reason, policy-makers and development institutions are putting greater emphasis on improving **water governance**, which encompasses all of the processes whereby different groups make decisions, articulate their interests, and mediate their differences in relation to water control, access, and quality. This chapter explores water governance using a framework that distinguishes between public, private, and community modes of water governance and looks at how these modes interact and overlap in practice. It also highlights the need to ground analyses of water delivery in the particulars of local contexts.

The chapter begins with an overview of water supply challenges in the global South followed by a discussion of the merits and limitations of the three main service-delivery models. Consideration is then given to recent water governance reforms in two very different contexts: Pathum Thani in Bangkok, Thailand, where the utility was privatized in 1998; and Caracas, Venezuela, where the public utility operates with community participation. Both cases show how actors' interests and activities interact in practice and also the important role played by communities in struggles for equitable services. The chapter concludes by drawing parallels between governance challenges in cities of the global South and those facing Canadian cities today. Common challenges include securing infrastructure financing and sustainably managing water resources, particularly in the face of climate change.

Water Governance in the Global South

Challenges

In the post-Millennium Development Goal (MDG) era, delivering water to the world's growing urban populations remains a major challenge. Adopted in 2000 by all 189 United Nations (UN) member states, the MDGs comprised eight overarching goals for poverty reduction and development to be achieved by 2015. Goal 7c was a commitment to "halve, by 2015, the proportion of people without sustainable access to safe drinking water and basic sanitation" (WHO and UNICEF, 2015: 2). In a major global achievement, the drinking-water target was met in 2010. However, the world has fallen short of the sanitation target by almost 700 million people (WHO and UNICEF, 2015: 4–5), and with hundreds of millions of people still lacking safe water, there is still a long way to go.

The MDG statistics also understate the gravity of issues concerning water quantity, quality, and access. Due to the difficulty in gathering data on service quality, the WHO-UNICEF Joint Monitoring Programme (JMP), responsible for monitoring progress on the MDG water and sanitation targets, did not directly measure the quality or quantity of water that people accessed. Instead, it used as a proxy the type of supply reported as a household's primary source of drinking water, categorized as *improved* or *unimproved*. Improved sources include not only piped-water connections to households, but also public taps or standpipes, boreholes, protected wells, and protected springs or rainwater. As the JMP's own research has shown, many of these supposedly improved sources are actually contaminated. In addition, where users have to travel for water, it is likely that they are not accessing sufficient quantities for adequate hygiene (Clasen, 2012).

Cities in the global South face three main water-related challenges: ensuring supply and access, providing affordable services, and guaranteeing acceptable water and service quality. At the most basic level, governments struggle to *ensure an adequate supply* of water to industrial, business, and residential users. Urban population growth over the last 40 years has increased demand for water resources, while contamination of surface and groundwater threatens supplies (UN Water, 2007). Climate change will compound these supply challenges by prolonging drought periods and making rainfall more variable and extreme (Sowers, Vengosh, and Weinthal, 2011). The Middle East and North Africa will face the most severe impacts. In Syria, for example, renewable water availability is expected to decrease by approximately 50 per cent below 1997 levels by 2025. Projections are similar for Jordan (Sowers et al., 2011: 603).

Many cities in both the global North and global South are already facing the water-related impacts of climate change. As of 2016, California, USA, is entering its fifth year of a serious drought that has forced urban governments to implement residential water-use restrictions. In 2015, Brazil faced the worst droughts in the country's recorded history. Because of Brazil's dependence on hydroelectricity, residents in major cities, such as Rio de Janeiro and São Paulo, suffered electricity blackouts and Internet cut-offs (Watts, 2015). As this case shows, water shortages also affect other urban systems that rely on water, such as transportation, energy supply, health, food production, and green spaces. Paradoxically, excess water associated with climatic change can also have a negative impact on water access. Flooding damages infrastructure, and sea level rise can cause saltwater to seep into underground aquifers—referred to as *saltwater intrusion*—making the fresh water undrinkable (Loftus, Anton, Philip, and Morchain, 2011).

In addition to resulting from absolute shortages, people's inability to access water may also result from political decisions regarding water distribution (Bakker, 2010; Swyngedouw, 2004). For example, governments frequently prioritize industrial water uses over residential uses. Meanwhile, in many cities in the global South, residential water supply networks frequently neglect poorer neighbourhoods and peri-urban areas (Budds and McGranahan, 2003). Access in **informal** (or **squatter**) **settlements** is particularly problematic, since governments may deny residents services owing to their quasi-legal or illegal land tenure (Crane, Danière, and Harwood, 1997; Jaglin, 2002). Struggles for water services are therefore closely related to struggles for individual or community land tenure security.

The *affordability of water* also influences access. Residents without piped-water connections often obtain water from private water vendors who deliver tap water in tanker trucks or bottles at prices up to 100 times more than are charged to households with connections to the public water network (Budds and McGranahan, 2003: 98). Alternatively, people must line up to fill buckets at a neighbourhood standpipe, often for a fee (WHO and UNICEF, 2014). Struggles around water frequently revolve around the most equitable price structure for water services.

One way to make services more affordable is to subsidize low-income service users, either through direct rebates for the poor or more commonly through subsidized water tariffs. Water subsidies can be incorporated into the price structure of water services in a variety of ways: rising block tariffs (where users are charged a lower per-unit rate for the first consumption "block" or level); social or welfare tariffs (lower charges for low-income households, often at a flat rate); banded charges (lower tariffs for lower-income neighbourhoods); or lifeline tariffs (a designated volume provided free of charge) (Budds and McGranahan, 2003: 109). South Africa's national Free Basic Water Program, introduced in 2001, is an example of a lifeline tariff. National law required utilities to provide households with a free basic supply of at least 6 kilolitres (1585 US liquid gallons) of water per month. Although this amount was subsequently increased to 9 kilolitres (2377 US liquid gallons) in 2009, many experts believe that this quantity is still insufficient for health and hygiene (Nash, 2012).

Initially connecting to a water or sewerage network can also be prohibitively expensive. Like water tariffs, connection costs can be funded through a cross-subsidy—for example, by charging a service improvement fee to all users rather than requiring poor households to cover the entire cost of their connections. Mobilizing financing for sanitary sewerage is particularly challenging. Providing networked sewerage is more expensive than delivering piped water, and it is more difficult to recover costs for sanitation through user fees. Low-income users may be unwilling or unable to spend scarce resources on sanitation (Allen, Hoffman, and Griffiths, 2008). Meanwhile, governments seeking re-election are frequently reluctant to spend public revenues on sewerage, favouring more politically rewarding investments (Krause, 2009). These factors partly explain why progress on sanitation has lagged behind water.

One innovative solution for affordable sewerage was introduced by the NGO-led Orangi Pilot Project (OPP) in Karachi, Pakistan. The OPP succeeded in convincing state authorities to support their condominial sewerage model, which provides improved services at a lower cost to users (Hasan, 2006; Mitlin, 2008). **Condominial sewerage** is a form of simplified sewerage consisting of smaller-diameter, usually plastic, pipes installed at shallower depths than conventional cast-iron sewers. In the OPP model, households finance and participate in the construction of *internal* sewer components (sanitary latrines in houses, underground sewers in lanes, and neighbourhood collector sewers) and the local government pays for and constructs the *external* components (trunk sewers, which collect wastewater from neighbourhood sewers, and treatment plants).

A third key challenge for cities of the global South is ensuring *service quality* (convenience and reliability) and *water quality* (cleanliness and safety). Where service quality is low, collecting water can be a time-consuming burden that tends to fall primarily on women and children (Elmhirst and Resurrection, 2008). Poor water quality and hygiene, meanwhile, cause approximately 88 per cent of diarrheal diseases, which are responsible for 2.2 million deaths annually (Corcoran et al., 2010: 10). A main cause of contaminated water in cities is inadequate wastewater management. Over 2.4 billion people worldwide lack access to improved sanitation, and population growth in urban areas is quickly outpacing past sanitation gains (WHO and UNICEF, 2015: 4–5). Wastewater treatment is also a major challenge. Up to 90 per cent of all wastewater generated in the global South flows untreated into waterways and eventually into fragile coastal ecosystems (Corcoran et al., 2010: 5).

The sustainable management of water resources may receive greater attention under the 2016–2030 **Sustainable Development Goals (SDGs)**, which have succeeded the Millennium Development Goals (MDGs). Unlike the MDGs, the SDGs also include targets aimed at countries in the global North, such as Canada. These include reducing waste and managing resources sustainably, both of which are urgent issues in Canadian cities and globally. The sixth goal of the SDGs is to "ensure access to water and sanitation for all" (UN Sustainable Development, 2016). The goal's targets, to be achieved by 2030, include the following (https://sustainabledevelopment.un.org/sdg6):

6.1 achieve universal and equitable access to safe and affordable drinking water for all;
6.2 achieve access to adequate and equitable sanitation and hygiene for all;
6.3 improve water quality;
6.4 substantially increase water use efficiency;
6.5 implement integrated water resources management at all levels;
6.6 protect and restore water-related ecosystems;
6.a expand international cooperation and capacity-building support to global South countries in water- and sanitation-related activities and programs; and
6.b strengthen the participation of local communities in improving water and sanitation management.

Many of these targets require behavioural changes from water users, support from policy-makers, and the creation of new institutions, which are all questions of governance.

Defining Water Governance

Governance has been defined in a variety of ways, and conceptions have evolved over time. Initially, governance studies focused on stable, collective patterns of formal organization, such as the government ministries, organizations, and laws serving a particular social function (Lauer and Lauer, 2006). Later, the definition of governance was broadened to include both formal and informal institutions, such as the norms and shared rules that affect individuals' behaviours, as well as the processes and practices involved in delivering services (Helmke and Levitsky, 2004).

Water governance has been defined as "the range of political, social, economic and administrative systems that are in place to develop and manage water resources, and the delivery of water services, at different levels of society" (Rogers and Hall, 2003: 16). This definition comprises the interactions among processes, rules, and traditions that determine how people make decisions and share power, articulate their interests, exercise responsibility, mediate their differences, and ensure accountability. Importantly, governance analysis considers who benefits and who loses from such arrangements. Water governance occurs at multiple scales and the water sector is a part of broader social, political, and economic developments. It is thus also affected by decisions in other sectors (Rogers and Hall, 2003). For example, as is discussed in the next section, the global influence of **neoliberalism** in social and economic policy since the 1980s has introduced new priorities into water management, such as an increased focus on private-sector involvement. In the case of Venezuela presented below, pro-poor services were introduced as part of a wider political and economic transformation, the Bolivarian Process.

In sum, governance does not encompass just governments or the structure and lines of decision-making authority, but also the relationships, networks, and processes associated with **civil society** (see, for example, World Bank, 2004). **Urban water governance** thus incorporates public-private partnerships, state-owned networks, and small-scale independent providers such as water vendors and community-run services (Marin, 2009). These actors can be broadly grouped into the categories of *public*, *private*, and *community service providers*, which we discuss next. In practice, however, the boundaries between these categories are quite blurry. A variety of actors usually intersect with each other to manage and regulate water supply within the urban landscape.

Public and Private Networked Service Providers

From the mid-twentieth century onwards, virtually all water utilities in both the global North and global South were run by public (governmental) entities. Partly as a reaction to the poor performance of many of these utilities, the **privatization** of urban water services has been enthusiastically promoted since the 1990s by the major international financial institutions (such as the International Monetary Fund and the World Bank), especially for cities in the global South. Emphasis on private-sector participation is integral to the global shifts towards economic and social policies associated with neoliberalism. Privatization has also met with considerable resistance from communities and labour unions. For example, in the famous Cochabamba "Water War" in 2000, residents of the city of Cochabamba, Bolivia, succeeded in forcing the cancellation of a monopoly concession with the private consortium Aguas del Tunari. Among other grievances, residents were angered that legislation accompanying the deal allowed the company to charge them for water from their own independent wells (see Spronk, 2007). As depicted in the Hollywood film *Even the Rain (También la Lluvia)*, the restrictive new laws would have even outlawed rainwater collection.

Privatization refers to a role for private companies in water provision and management. Full divestiture, where private companies fully own and operate water utilities and infrastructure, is rare. Typically, private-sector participation takes the form of hybrid models or "public-private partnerships" in which the government continues to own the infrastructure but transfers responsibilities to the private sector for a given period through contractual arrangements (Harris, Goldin, and Sneddon, 2013). A recent report on the privatization of water systems in low- and middle-income countries from the World Bank's Public-Private Infrastructure Advisory Facility (PPIAF) notes that the rate of private participation in the water sector has increased by more than 100 per cent since 2000. As of 2012, more than 275 different sponsors were involved in over 500 water projects (Perard, 2012: 9). Taking a longer view, however, the number of new projects per year appears to have peaked in 1997 due to political and environmental problems encountered in the host countries as well as delivery weaknesses within the private sector. Recent estimates suggest that 90 per cent of the world's water users continue to access water from public providers (Spronk, 2010: 157). Moreover, private companies are more interested in working with middle-income than with low-income countries, with much of the new activity concentrated in China (Marin, 2009).

The actors involved in the water market have also changed over time. As of 2001, five major international companies from the global North served 80 per cent of the global South's residents who access water from private providers. Of these, the French company Suez held over a third of the market, followed by SAUR, Veolia, Agbar, and Thames Water. Restrictive prequalification criteria during the earlier rounds of privatization partly explain this concentration, as they precluded smaller actors with less experience. Since 2002, more global South–based companies have succeeded in winning contracts, and as of 2009, they control an estimated 40 per cent of the global South's water market (Marin, 2009: 9). In reality, this proportion may be higher, since this figure does not include China. It is very common for water multinationals such as Veolia or Suez to partner with local companies in joint ventures, as in the failed concession in Cochabamba, Bolivia, as well as in the case of Bangkok, Thailand, discussed below.

Privatization is frequently accompanied by other policy reforms that treat water as a **private good**—for example, the establishment of markets for water rights and the introduction of **full-cost recovery** pricing into service operations. Full-cost recovery means that users pay the entire costs of their service delivery, rather than governments financing services through taxation or cross-subsidies from wealthier users. Economists argue that full cost recovery is more economically and environmentally sustainable since it sends users more accurate information about the cost of the service and can discourage water waste (Jaglin, 2002). While often implemented in the context of privatization, public providers can also implement full-cost recovery.

Privatization and full-cost recovery have been controversial, however, because many people in both the global North and South see water as a **public good** that should be provided to all and paid for through taxation. Water is necessary to sustain life and health and, consequently, holds symbolic and cultural importance. It also delivers public-health and environmental benefits that are greater than those derived by individual users, so many believe that water and sanitation should not be provided solely on the basis of an individual's willingness or ability to pay (Budds and McGranahan, 2003; Hall and Lobina, 2008).

Critics of privatization further argue that markets perform poorly in delivering public goods, an example of **market failure**. Urban water services can be considered a

natural monopoly, which means that it is more economically efficient for one company or entity to provide all of the water in a particular city or region. In the case of natural monopolies, the state is often seen as the best or most legitimate provider of a good or service, because monopoly providers may abuse their market power by overcharging consumers (Budds and McGranahan, 2003). Also, water infrastructure investments yield low returns over long time horizons, making it difficult to attract adequate private investment for infrastructure development.

By contrast, proponents of privatization argue that private companies are inherently more efficient than public providers, since political interference will undermine efficiency and make public utilities captive to special interests—referred to as **state failure** (see Bakker, 2010; Spronk, 2010). Poor public utility management may be exacerbated in cases where government agencies play the role of both regulators and operators, which creates a conflict of interest. Supporters of privatization argue that the competitive pressures of the market will make private providers not only more efficient but also accountable to consumers (World Bank, 2004). Lastly, they believe that the private sector can mobilize much-needed financing for infrastructure improvements (Annez, 2006).

In practice, the outcomes of privatization have been mixed. Several comprehensive scholarly reviews show that privatization has been somewhat successful at improving operational efficiency but much less successful at mobilizing capital finance and extending service coverage to previously unserved areas (Annez, 2006; Marin, 2009). In many cases, companies have been unwilling or unable to extend connections to areas where residents do not have the means or desire to pay rates that would yield a profit (Hall and Lobina, 2007). Because of an inability to mobilize private investment in the water sector, most recent public-private partnerships rely heavily on public financing instead of upon much-anticipated private financing (Marin, 2009: 25).

The literature suggests that privatization contracts have been most successful in situations where governments have been able to establish strong regulatory frameworks with, for example, mandatory targets for extending service coverage and caps on tariff increases (Bakker, 2010: 95). However, many governments lack the capacity for effective private-sector regulation (Annez, 2006), particularly in the contexts where utilities are most in need of reform.

Overall, some scholars and activists suggest that the privatization debate has deflected attention from the fact that in most cities of the global South the barriers that prevent the poor from accessing services—such as poverty and political powerlessness—have tended to persist regardless of whether utilities are publicly or privately owned (Bakker, 2010; Budds and McGranahan, 2003). Moreover, given that public providers continue to supply the vast majority of the world's population, public-sector reform should continue to be a top priority.

Community Models and User Participation

Recently, more emphasis has been placed on how to promote "good governance" by incorporating civil society into service management and delivery (see World Bank, 2004). It is argued that global North models of networked service delivery characteristic of most Canadian cities may be inappropriate in the global South (Pritchett and Woolcock, 2004). Whether privately or publicly operated, utilities have tended to be top-down, centralized, and hierarchical structures that rely on the expertise of bureaucrats and engineers rather than on service users. Emphasizing modern piped-water connections, these strategies fail to account for the ways in which the poor actually access services in the global South (Allen, Hoffman, and Griffiths, 2008). These include the small-scale private water vendors discussed above as well as community-run services. Especially in peri-urban areas, communities may operate their own services such as wells and community toilets, often organized into co-operatives. Some observers advocate supporting these existing practices now by regulating, for example, the prices charged by vendors, rather than by waiting for people to gain formal networked connections (Swyngedouw, 2004). Others have highlighted the benefits of **co-production** arrangements, whereby communities collaborate with professional service providers (public, private, or NGO) in different phases of service planning and delivery, including design, construction, and maintenance, service delivery, monitoring, and collection of user fees (Ostrom, 1996). Proponents argue that participation improves efficiency by facilitating information exchange between providers and users and lowering costs. They also suggest that user involvement in planning and design can ensure that technologies meet user preferences and correspond with what communities are able or willing to pay (Hasan, 2006;

Watson and Jagannathan, 1995: 5), as in the Orangi Pilot Project's sanitation initiative discussed earlier. Scholars also suggest that participation can help empower users, promote accountability, and foster trust between the state and citizens (Mitlin, 2008)—referred to as building **social capital** (Evans, 1996).

Many believe that adaption to climate change requires the involvement of water users in governance. Most conventional approaches to water-service planning rely on quantitative analyses of historical patterns of water supply and demand to plan future infrastructure development. Climate change experts argue that this predict-and-plan approach is ill-suited to conditions of climate change, whose effects are both unpredictable and unprecedented (see Tyler and Moench, 2012). Instead, observers argue that institutions are better able to respond to changes associated with climate change and urbanization when they work closely with service users to explore different future scenarios and plan for different possible courses of action. Less formal institutions such as water committees can also be more flexible in the face of change than conventional, centralized bureaucratic agencies, which can be slower to respond and reluctant to experiment with new approaches (see Pelling, High, Dearing, and Smith, 2008).

In practice, however, participation has taken many different forms (Castro, 2007). Initiatives have ranged from those that enable the poor to participate meaningfully in decision-making processes to those that merely use the poors' voluntary labour to cut service costs or make privatization arrangements more profitable (Jaglin, 2002; Laurie and Crespo, 2007). In some cases, participants are merely "consulted" on pre-existing plans, rather than given a genuine say in how services are delivered or how government funds are allocated. Observers also note that some groups are better equipped to benefit from participation than others, and this can reinforce local inequalities (Watson and Jagannathan, 1995). Finally, it is argued that local communities alone may be unable to manage urban water services sustainably, since many environmental issues transcend the local scale (Bakker, 2010).

There is clearly no single solution to urban water-services challenges. Rather, as McDonald and Ruiters (2012) point out in their global review of service delivery models, the success of governance reforms will depend on many contextual factors, such as the ideological motivations for reforms, their institutional makeup, the capacities of states and civil society, the availability of capital, and environmental conditions. This complexity is apparent in the case studies from Pathum Thani, Thailand, and Caracas, Venezuela.

Water Service Privatization in the Bangkok Metropolitan Area, Thailand

Thailand is a Southeast Asian country of some 67 million people, only 11.6 million (17 per cent) of whom live in urban areas (places of more than 100,000 people). The country's urban population and geography are dominated by the capital, Bangkok. No other city in Thailand has a population of more than 500,000. Bangkok has a long and complicated history with water, particularly because the city once functioned by relying primarily on water transport on a system of intricate canals. These canals have long been filled in and replaced by roads and alleyways to accommodate automobiles and trucks, which has led to very problematic issues of flooding, water pollution, and a lack of equitable water access for urban residents.

As discussed above, while water companies' interest in the world's poorest countries has declined in recent years, they have increasingly seen opportunities in middle-income countries with fast-growing urban markets. Governments of the BRIC countries (Brazil, Russia, India, and China) have only just begun to explore the potential of private-sector participation in national infrastructure provision (Herrera and Post, 2014: 628). Thailand, a middle-income country, also represents a potentially profitable location for a variety of private water corporations.

In the 1990s, Thailand experimented with water privatization as an urban service delivery strategy in response to environmental concerns and economic pressure from the International Monetary Fund (Zaki and Nurul Amin, 2009). A sizable portion of the country's population continues to face difficulties accessing safe drinking water. Despite high levels of access to water—95 per cent of the population with access to water and 96 per cent with access to sewerage in 2010 (World Bank, 2012)—over 57 per cent of the total population remains at risk of contracting water-related diseases as a result of service quality deficiencies (Estache, Goicoechea, and Trujillo, 2006).

Owing to widespread public resistance, however, water privatization in Thailand has been limited to smaller projects (Hewison, 2005), largely on the outskirts of Bangkok, Thailand's capital. The Bangkok Metropolitan Area (BMA) encompasses Bangkok and five rapidly urbanizing surrounding provinces, with a total population of about 9.6 million in 2008 (World Bank, 2015). Thailand's first water-supply privatization scheme was implemented in 1998 in Pathum Thani, one of the provinces immediately to the north of Bangkok (see Figures 23.1 and 23.2), with a population of approximately 900,000 inhabitants.

Water-service privatization was implemented by the Provincial Waterworks Authority (PWA), the agency responsible for water supply and distribution in most of Thailand. The privatization followed a Build Operate Own and Transfer (BOOT) approach, which means that the concessionaire builds, owns, and operates the infrastructure for a set period (in this case 25 years) before ownership is transferred back to the government authority (Zaki and Nurul Amin, 2009: 2310). In Pathum Thani, the contract was awarded through

Figure 23.2 Pathumthani Water Company

a competitive bidding process to a joint venture consortium led by the UK's Thames Water International. The consortium, in turn, set up a company called the Pathumthani Water Supply Company Ltd (PTW). The PTW is still operating and provides water to much of the province, even as the population of this urbanizing area has grown by some 10 per cent over the past decade (see Figure 23.3).

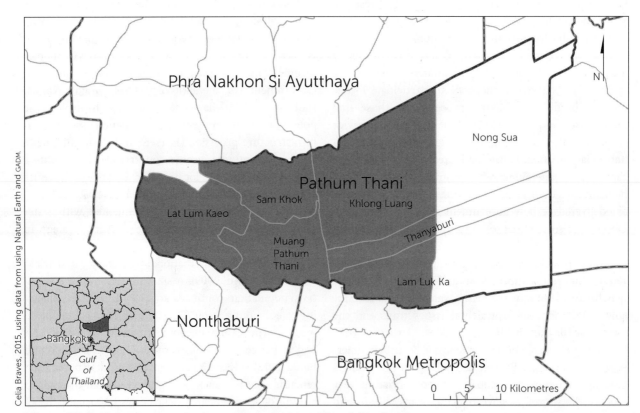

Figure 23.1 Map of water service areas in Pathum Thani, Thailand

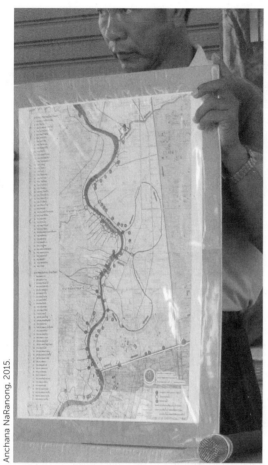

Figure 23.3 A utility official explains the water system to community members in Pathum Thani

Under the 1998 agreement, the newly created PTW was to extract and treat water according to quality standards set by the Provincial Waterworks Authority, as well as maintain and operate the reservoirs, bulk-main pipelines, and local distribution pipelines (Aekaraj, 2013). The PWA agreed to pay for a minimum daily amount of water from the consortium, regardless of the actual amount delivered to users. In turn, the PTW was to increase production and expand the network according to defined targets. The PWA acted as the regulator and retained control over billing, fee collection, and pricing (Zaki and Nurul Amin, 2009). The government agreed to pay a flat rate to the private company for the water but charged consumers different rates based on income level, which allowed for a cross-subsidy from wealthier to poorer users.

How, then, did the utility perform according to criteria of access, affordability, and quality? On the basis of household surveys, Zaki and Nurul Amin (2009) identified significant improvements in both water access and water quality for the poor and the residents of informal settlements. Access was defined in terms of connections to people's homes. Water-quality indicators were drinkability, clarity, and turbidity (quantity of particles), while service quality was measured in terms of reliability, supply, pressure, and responsiveness. The authors note that connection costs and monthly charges did increase as a result of privatization but that most respondents expressed satisfaction with the service (Zaki and Nurul Amin, 2009: 2323).

What can be learned from this case? Public resistance to private water supplies slowed the privatization process in Thailand and required the central government to retain a strong role in management and regulation, such as regulating quality and tariffs and setting benchmarks for increasing production and service coverage. This helped mitigate potentially adverse social consequences. The relative success of this case, moreover, was enabled by the high capacity of the PWA and the fact that the region had the potential to be financially lucrative for the private sector given the large and growing population.

Although strong regulation may help to mitigate some of the potential risks of private-sector involvement in some cases, many critics of privatization (e.g., the Council of Canadians) maintain that water should be a *commons*: a resource that is collectively managed rather than one that is a source of profit for private actors (Barlow, n.d.). This has led to an increased interest in alternatives to both privatization and conventional top-down public systems. Venezuela, a middle-income country in South America, offers one such example.

State-Community Co-production in Caracas, Venezuela

Venezuela offers a markedly different context than Thailand. In Venezuela, water reforms have accompanied a broader political and economic shift in the country away from neoliberalism, referred to as the Bolivarian Process. The "process began in 1998 with the election of popular leftist" President Hugo Chávez (1999–2013) and has continued under the government of President Nicolás Maduro. With support from an array of civil society actors, the political project attempts to reverse

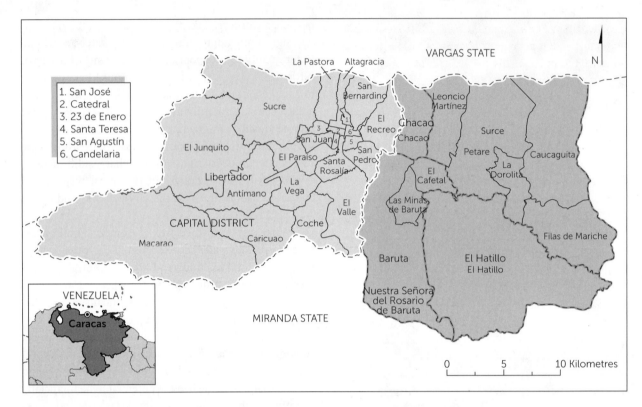

Figure 23.4 Map of municipalities and parish boundaries of Caracas, Venezuela
Cecilia Braves, 2015, using data from OpenStreetMap contributors; Baruta parish boundaries are not to scale.

a history of poverty and inequality by promoting economic redistribution and citizen participation in all areas of social and economic life, including the water sector. Thanks to increased infrastructure spending and community participation, the country reached the MDG for water and sanitation in 2005, 10 years ahead of schedule.

Venezuela is a highly urbanized country, with approximately 93.9 per cent of its population living in urban areas (UN Data, 2014). The country's capital region, the Caracas Metropolitan Area (see Figure 23.4), is home to approximately 6 million people, about a fifth of the country's population (Brillembourg and Klumpner, 2010: 121). A public water utility, Hidrocapital, is responsible for delivering services in the Capital District and neighbouring Vargas and Miranda states.

Caracas residents have long experienced water-service deficiencies, particularly in the city's populous hillside *barrios* ("slums"), which are home to the majority of the city's low-income residents, as well as to those that self-identify as Afro-descendant or *mestizo* ("mixed race"). Considered to be illegal because of their lack of formal land tenure, the barrios were historically neglected by the state. One of this chapter's co-authors has been researching the technical water committees' effects on governance and service equity. In 2012, research was conducted in Antímano (see Figures 23.4 and 23.5), a series of barrios in the west of Caracas

Figure 23.5 The barrios of Antímano in the west of Caracas

(see McMillan, Spronk, and Caswell, 2014). Despite being home to 150,000 people, Antímano did not even appear on city maps in the 1990s.

Since Chávez's election in 1998, the national government has sought to reverse structural discrimination against the urban poor by providing land title and a variety of social services, such as subsidized groceries, health care, and educational programs. However, the "technical water committees" (*mesas técnicas de agua*, MTAs) actually predate this period. They were first piloted by the municipal government in two Caracas neighbourhoods in the early 1990s. Based on their success and on input from civil society, the Chávez administration decided to extend the MTA model across Caracas and then nationwide.

The MTAs can be considered a case of state-community co-production. To form a neighbourhood MTA, residents work with their local public utility to map and conduct a census of the neighbourhood, diagnose service issues, and plan and implement infrastructure projects. Because residents installed many of their water pipes themselves, community self-mapping provides utility staff with much-needed information about the local infrastructure. At regular public meetings, residents and utility staff discuss service issues and follow up on work (see Figure 23.6). In contrast, before the 1990s, utility officials reportedly never set foot in the barrios (Interview with Victor Díaz, Hidrocapital community coordinator, August 20, 2012).

Owing to state investment in infrastructure, as well as over 1500 community-led infrastructure projects nation-wide (Mesas técnicas de agua, 2011), access to water and sanitation has improved significantly. Between 1998 and 2013, the proportion of households with piped-water service increased from 80 per cent to 96 per cent. Over the same period, sewerage coverage increased from 62 per cent to 85 per cent (Instituto Nacional de Estadística, 2012). Water users do not pay for a connection, and roughly 35 per cent of water users are not billed (often due to service deficiencies), while 16 per cent are considered "social users"—low-income users who should receive public support—which means that they pay only 20 per cent of the full tariff (WSP-LAC, 2007: 198).

Information sharing in the MTAs has improved certain service-quality measures, such as reliability, water pressure, and utility responsiveness. Since the barrios are situated well above the city's reservoirs, water must be pumped long distances. In many hillside settlements, there is insufficient pressure to deliver water to all neighbourhoods simultaneously, so many households receive piped water only periodically and must store water for the intervening period. Antímano residents explained that, prior to the MTAs, the water would sometimes arrive unexpectedly at 3:00 a.m. and they would have to sacrifice sleep to fill up containers (focus group, November 22, 2012). Now, the engineers work with residents to deliver water according to a predetermined schedule, with continuous follow-up to verify it is arriving on time and with adequate pressure (see Figure 23.7). Nonetheless, the periods between water deliveries continue to be long in some places.

Figure 23.6 Public MTA meeting in Antímano, Caracas

Figure 23.7 Residents of Barriolita, Caracas, explain their water problems to Hidrocapital engineer Daniel Pereira

Sanitation has also improved, as water committees have replaced aging sewerage pipes and installed pipes where none had existed. However, serious challenges remain with respect to water quality and wastewater management. Namely, 80 per cent of Caracas's wastewater continues to be dumped untreated into the Guaire River, the city's main waterway (Interamerican Development Bank, 2012: 3).

The MTAs also have little voice when it comes to such city-wide issues. From an ecological perspective, this is problematic. The **watershed** is considered the most ecologically appropriate scale for management because it brings together all users that share the same water source. Yet, the MTAs usually operate at a much smaller neighbourhood scale. They thus may be unable or unwilling to address issues that transcend neighbourhood limits, such as wastewater treatment.

Overall, the Venezuelan case shows how participation can make services more responsive to the needs of marginalized communities and improve efficiency. Part of the MTAs' success is owed to strong government spending on infrastructure and human resources in low-income areas. Thus, Venezuela's context differs from neoliberal contexts where participation has been implemented as a cost-cutting measure. In addition, the capacity of residents to participate has been enhanced through a variety of free training programs and workshops. Many residents explained that they have gained knowledge and confidence through their work with the water committees. Like many fellow MTA leaders, Florencia Gutiérrez first became involved in her community through the committees: "Before, the community wasn't organized, and it wasn't participative either. The community resigned itself with waiting for elections to come, when they [politicians seeking election] would give you what you needed to fix your house: cement and other materials" (Interview with Florencia Gutiérrez, December 6, 2012).

Now, residents are demanding services as a basic right. However, in a context of increasing economic and political flux such as existed in Venezuela in 2016, the sustainability of the MTAs and other participatory reforms is an open question. In December 2015, in a major political shift, Maduro's government lost its majority in the National Assembly to a coalition of centre-right and right-wing parties. The MTAs have been closely aligned with the governments of Chávez, and Maduro, and it is presently unclear whether they will outlast a change in government. Despite these challenges, service provision "alternatives" such as the MTAs will likely receive greater attention in the future. In improving water governance, Canadian cities, as the following text box explores, can learn from the global South and vice versa. There are shared challenges of aging infrastructure, especially given that the provision of water is recognized as a basic human right and a municipal responsibility.

Urban Water Governance Challenges in Canada

Urban water governance is of crucial concern in Canada. Although countries of the global North usually have in place well-established public water authorities or boards, water quality, access to water, and price of water are still matters of political and economic concern. In Canada, water delivery is a municipal responsibility. Water-related legislation is set provincially, following national water-quality guidelines. As in many countries in the global South, mobilizing financing for infrastructure is a major challenge for Canadian cities. Most water-related infrastructure was constructed in the twentieth century when Canada's urban areas were first developing, so many pipes and treatment plants are now reaching the end of their life cycle. It is estimated that Canadian municipalities currently need to spend $31 billion to repair existing infrastructure and an additional $56.6 billion for new infrastructure (Environment Canada, 2012a: 1).

Infrastructure is typically funded through a combination of sources, such as taxes, development levies, and user fees. The average Canadian household pays approximately $37.55 per month for water service, with considerable variation across provinces (Environment Canada, 2012a: 8). As of 2009, 72.1 per cent of Canadian households have water meters and thus pay according to the volume of water they use. The most common type of volumetric rate in Canada is the constant unit rate, where customers are charged a uniform amount per unit of water used (Environment Canada, 2012a: 4–5), as compared with a block tariff discussed above. There is considerable evidence to suggest that households in jurisdictions with volumetric pricing consume less water than those in jurisdictions with flat-rate tariffs.

In the face of financing challenges, it is likely that Canadian municipalities will explore ways to provide services at a lower cost. Hamilton, Ontario (a city of about 500,000 people on Lake Ontario, known for its steel-manufacturing past), for example, entered into a contract in 1994 with a local private water-supply company, the Philip Utilities Management Corporation, with the hope of delivering water more efficiently within the city. The corporation had responsibility for the maintenance and operation of Hamilton's water and wastewater treatment plant (Pigeon, 2012: 75). At the time, it was one of the largest public-private partnerships in North America. Unfortunately, the partnership was soon revealed to be problematic. While the corporation made profits, service levels reportedly decreased and the city did not make any money from the arrangement. Pressure on the city's municipal council from citizen organizations and some local politicians led to the contract not being renewed in 2004. In essence, the city opted to "re-municipalize" the water and wastewater treatment services.

Another major concern in Canadian cities, as in many cities in the global South, is wastewater management. As of 2009, only 78.7 per cent of Canadians connected to sanitary sewers had their wastewater treated at a secondary level or higher (Environment Canada, 2012b: 11). This means that almost a quarter of Canadian municipalities only use primary treatment or have no treatment at all. Primary wastewater treatment removes solid or floating matter, while secondary treatment also removes dissolved organic matter. In 2012, Canada introduced legislation requiring municipalities to introduce secondary treatment by 2040. By contrast, the United States has had similar legislation in place for almost 40 years (Proskow, James, and Rowney, 2013). Meanwhile, some major Canadian cities are deficient even in primary wastewater treatment. Victoria, British Columbia, for example, continues to pump raw sewage directly into the Juan de Fuca Strait (Meissner, 2014). The same was true of Halifax, Nova Scotia, until a major infrastructure project was completed in 2011 (CBC News, 2014).

In the years ahead, water shortages will be another major consideration for Canadian cities. Despite the fact that Canada is one of the most well-endowed places on earth in terms of freshwater resources, many of these resources are geographically distant from population centres. While 84 per cent of Canadians live in the southernmost part of the country, 60 per cent of water resources flow north (Environment Canada, 2012a: 1). Canada is also likely to look to water as a source of income as water resources continue to become more limited and more contaminated in other parts of the world. There is and will continue to be pressure on Canada to provide water to water-stressed countries. There is a growing consensus that water will become as valuable in the twenty-second century as petroleum is today. How much Canada decides to sell or to provide and at what price are key governance questions.

Conclusion

Access to water supply and sanitation infrastructure is essential for urban sustainability and health. However, in the global South, economic growth and urbanization are outpacing existing infrastructure. Many municipal services continue to be poorly managed and weakly financed, and climate change and contamination threaten water availability in cities of the global South and North alike. These economic and environmental pressures force municipal governments to consider ways to improve water governance.

Three modes of governance are state, private, and community management. As this chapter has shown, the lines between these models are blurry and actors frequently overlap. For example, in the Thailand case, the government used privatization strategically as part of its broader effort to bring a more commercial orientation to water services while mitigating potential negative social effects, such as runaway tariff increases. In Venezuela, government-run utilities rely heavily on citizen participation to deliver services effectively. This chapter has also argued that the local context will shape the relative success and outcomes of reforms. In Thailand, citizen resistance played a key role in putting pressure on the government to slow the implementation of privatization. In the Canadian context, in Hamilton, Ontario, an overly ambitious public-private partnership failed and services were eventually remunicipalized. And in Venezuela, unlike the participatory reforms implemented in contexts of neoliberal austerity, participation was accompanied by significant increases in public infrastructure spending in poor areas.

Overall, it is likely that confronting the water challenges of access, affordability, and quality will defy technical solutions or ideal-type models. Water access is a political process, and how well a reform works in a given city will depend on many factors, including environmental conditions, the relative capacities and influence of government and civil-society actors, and the broader political economic context. A key concern as we go forward should be finding ways to support the existing efforts of poor and marginalized groups to achieve equitable services. This includes involving existing community groups in developing solutions that meet their needs, such as the Orangi Pilot Project in Pakistan and the neighbourhood water committees in Venezuela. Steps must be taken to ensure that those participating have a real voice in decision-making. **Participatory budgeting**, which has been implemented successfully in some Brazilian cities, is one way that community voices might be included in infrastructure planning. With participatory budgeting, residents have the opportunity to vote on funding priorities at regular public meetings. Research suggests that public participation in budgeting, when accompanied by sufficient resources, can make decision-making processes more transparent and ensure that decisions taken are in the public interest (see, for example, de Sousa Santos, 1998). Some municipalities, such as Cochabamba, Bolivia, have also experimented with community and labour union representation on water utility boards (see Spronk, 2007).

For such efforts to be effective, however, new sources of financing will need to be mobilized for water and sanitation in both Canada and the global South, either through changes to water-pricing structures (e.g., increasing water charges for industrial users) or other forms of taxation. Cross-subsidies will continue to be essential to ensure that water remains affordable for low-income users. In addition, alternative, lower-cost technologies such as condominial sewerage will likely play a role. This is particularly true for sanitation, given the large gap in sanitation financing in both urban and rural areas. However, low-cost technologies must be acceptable to their users and not come at the expense of service quality or environmental sustainability.

Finally, a key global concern will be ensuring that governments and communities have the capacity and flexibility to confront the water-related impacts of climate change. Water-service planning must take into account likely climate change impacts and urbanization trends, and relevant actors must be involved in determining possible courses of action as we plan for an uncertain future.

Key Points

- Many cities in the global South struggle to deliver high-quality water services to all residents at an affordable price.
- Governance refers to the processes of decision-making, structured by institutions (rules or norms), whereby different groups articulate their interests, while water governance includes governmental actors, as well as civil society—community groups, NGOs, and the private sector.
- The major challenges facing water governance in the global South are adequate water supply and access (affordability), and acceptable water quality (cleanliness and safety) and service quality (convenience and reliability).
- The sixth of the Sustainable Development Goals is to ensure access to water and sanitation for all.
- Urban water governance incorporates public-private partnerships, state-owned networks, and

- small-scale independent providers; these three can be grouped into categories of water-service delivery models that include public, private, and community.
- Proponents of privatization argue that state delivery is inefficient because of political interference ("state failure"), while opponents of privatization suggest that markets will not equitably allocate water ("market failure"), and so it should be provided by the state at subsidized prices.
- In Pathum Thani, Bangkok, Thailand, water delivery was privatized in 1998 and was accompanied by strong government regulations, which helped to ensure that services remained affordable and of acceptable quality.
- In Caracas, Venezuela, community water committees work with the public water company to oversee service delivery, execute infrastructure projects, and improve information sharing between the utility and residents, opening spaces for citizen empowerment.
- Canadian cities face water challenges, including aging infrastructure, inadequate wastewater treatment, and political tensions over the potential to profit from vast water resources while other countries in the world face water stress.

Activities and Questions for Review

1. Compare water issues in Canadian cities with those faced by communities in the global South. What are the commonalities and differences?
2. What does governance mean, and how does it differ from government? In small groups, brainstorm the variety of actors involved in water governance at the municipal, provincial, and federal government levels in Canada.
3. As a class, debate the merits of public versus private urban water management and service delivery. Should the private sector be involved in the management and delivery of water?
4. If you were the president of the World Bank, what policies would you promote for improving water governance in the urban global South and what obstacles might you encounter?

Acknowledgments

The authors would like to thank the Social Sciences and Humanities Research Council of Canada (SSHRC), which has supported this research.

References

Aekaraj, S. (2013). Thailand's water sector: Overview and limitations. In P.O. Lee (Ed.), *Water issues in Southeast Asia: Present trends and future directions*. Singapore: Institute of Southeast Asian Studies.

Allen, A., Hoffman, P., and Griffiths, H. (2008). *Moving down the ladder: Governance and sanitation that works for the urban poor*. Paper presented at IRC Symposium on Urban Sanitation, Delft, Netherlands.

Annez, P.C. (2006). *Urban infrastructure finance from private operators: What have we learned from recent experience?* World Bank Policy Research Working Paper 4045. Washington, DC: World Bank.

Bakker, K.J. (2010). *Privatizing water: Governance failure and the world's urban water crisis*. Ithaca, NY: Cornell University Press.

Barlow, M. (n.d.). *Our water commons: Toward a new fresh water narrative*. Ottawa, ON: Council of Canadians.

Brillembourg, A., and Klumpner, H. (2010). Rules of engagement: Caracas and the informal city. In F. Hernandez, P. Kellett, and L.K. Allen (Eds.), *Rethinking the informal city: Critical perspectives from Latin America*. New York: Berghahn Books, pp. 119–136.

Budds, J., and McGranahan, G. (2003). Are the debates on water privatization missing the point? Experiences from Africa, Asia and Latin America. *Environment and Urbanization*, 15(2): 87–114.

Castro, J.E. (2007). Poverty and citizenship: Sociological perspectives on water services and public-private participation. *Geoforum*, 38(5): 756–771.

CBC News (2014). Untreated sewage still flowing into Halifax Harbour. *CBC News*, December 17. Retrieved from www.cbc.ca

Clasen, T.F. (2012). Millennium Development Goals water target claim exaggerates achievement. *Tropical Medicine and International Health*, 17(10): 1178–1180.

Corcoran, E., Nellemann, C., Baker, E., Bos, R., Osborn, D., and Savelli, H. (Eds.) (2010). *Sick water? The central role of wastewater management in sustainable development: A rapid response assessment*. United Nations Environment Programme, UN-HABITAT, GRID-Arendal.

Crane, R., Danière, A., and Harwood, S. (1997). The contribution of environmental amenities to low-income housing: A comparative study of Bangkok and Jakarta. *Urban Studies*, 34(9): 1495–1512.

de Sousa Santos, B. (1998). Participatory budgeting in Porto Alegre: Toward a redistributive democracy. *Politics and Society*, 26(4): 461–510.

Elmhirst, R., and Resurreccion, B.P. (2008). Gender, environment and natural resource management: New dimensions, new debates. In R. Elmhirst and B.P. Resurreccion (Eds.), *Gender and natural resource management: Livelihoods, mobility and intervention*. London, UK: Earthscan, pp. 3–20.

Environment Canada (2012a). *2011 municipal water pricing report*. Cat. No. En11-3/20011E-PDF.

Environment Canada (2012b). *2011 municipal water use report*. Cat. No. En11-2/2009E-pdf.

Estache, A., Goicoechea, A., and Trujillo, L. (2006). *Utilities reforms and corruptions in developing countries*. World Bank Policy Research Working Paper No. WPS 4081. Washington, DC: World Bank.

Evans, P. (1996). Government action, social capital, and development: Reviewing the evidence on synergy. *World Development*, 24(6): 1119–1132.

Hall, D., and Lobina, E. (2007). Profitability and the poor: Corporate strategies, innovation and sustainability. *Geoforum*, 38(5): 772–785.

Hall, D., and Lobina, E. (2008). *Sewerage works: Public investment in sewers saves lives*. Retrieved from http://www.psiru.org/reports/2008-03-W-sewers.pdf

Harris, L.M., Goldin, J., and Sneddon, C. (2013). Introduction. In L. Harris, J. Goldin, and C. Sneddon (Eds.), *Contemporary water governance in the global South: Scarcity, marketization and participation*. London, UK: Routledge, pp. 1–10.

Hasan, A. (2006). Orangi Pilot Project: The expansion of work beyond Orangi and the mapping of informal settlements and infrastructure. *Environment and Urbanization*, 18(2): 451–480.

Helmke, G., and Levitsky, S. (2004). Informal institutions and comparative politics: A research agenda. *Perspectives on Politics*, 2(4): 725–740.

Herrera, V., and Post, A.E. (2014). Can developing countries both decentralize and depoliticize urban water services? Evaluating the legacy of the 1990s reform wave. *World Development*, 64: 621–641.

Hewison, K. (2005). Neo-liberalism and domestic capital: The political outcomes of the economic crisis in Thailand. *Journal of Development Studies*, 41(2): 310–330.

Instituto Nacional de Estadística (2012). *Indicadores básicos de salud, 1990–2011*. Retrieved from INE website: www.ine.gov.ve

Interamerican Development Bank (2012). *Bolivarian Republic of Venezuela, Rio Guaire Sanitation Loan Proposal*. VE-1037.

Jaglin, S. (2002). The right to water versus cost recovery: Participation, urban water supply and the poor in sub-Saharan Africa. *Environment and Urbanization*, 14(1): 231–245.

Kochasawasdi, P. (2002). Role of the private sector in water management in Thailand: Case study of the Pathumthani Waterworks Co., Ltd. In M. Chantawong, P. Boonkrob, C. Chunjai, and P. Kochasawasdi (Eds.), *Water privatization in Thailand*. Bangkok: Foundation for Ecological Recovery, pp. 38–48.

Krause, M. (2009). *The political economy of water and sanitation*. New York: Routledge.

Lauer, R.H., and Lauer, J.E. (2006). *Social problems and the quality of life* (10th ed.). New York: McGraw Hill.

Laurie, N., and Crespo, C. (2007). Deconstructing the best case scenario: Lessons from water politics in La Paz–El Alto, Bolivia. *Geoforum*, 38(5): 841–854.

Loftus, A.-C., Anton, B., Philip, R., and Morchain, D. (2011). *Adapting urban water systems to climate change: A handbook for decision makers at the local level*. Freiburg, Germany: ICLEI European Secretariat.

McDonald, D., and Ruiters, G. (Eds.) (2012). *Alternatives to privatization: Public options for essential services in the global South*. London, UK: Routledge.

McMillan, R., Spronk, S., and Caswell, C. (2014). Popular participation, equity, and coproduction of water and sanitation services in Caracas, Venezuela. *Water International*, 39(2): 201–215.

Marin, P. (2009). *Public-private partnerships for urban water utilities: A review of experiences in developing countries*. Washington, DC: World Bank/PPIAF.

Meissner, D. (2014). Victoria sewer dispute hits the fan as Washington State urges B.C. intervene. *Globe and Mail*, June 11. Retrieved from www.theglobeandmail.com

Mesas técnicas de agua han recibido más de Bs. 500 milliones (2011). *Correo del Orinoco*. Retrieved August 11 from www.correodelorinoco.gob.ve

Mitlin, D. (2008). With and beyond the state—Co-production as a route to political influence, power and transformation for grassroots organizations. *Environment and Urbanization*, 20(2): 339–360.

Nash, F. (2012). Participation and passive revolution: The reproduction of neoliberal water governance mechanisms in Durban, South Africa. *Antipode*, 45(1): 101–120.

Ostrom, E. (1996). Crossing the great divide: Coproduction, synergy, and development. *World Development*, 24(6): 1073–1087.

Pathum Thani Water Company Limited (2010). Service area. Retrieved from http://www.ptw.co.th/aboutus_servicearea_en.php

Pelling, M., High, C., Dearing J., and Smith, D. (2008). Shadow spaces for social learning: A relational understanding of adaptive capacity to climate change within organisations. *Environment and Planning A*, 40(4): 867–884.

Perard, E. (2012). *Private sector participation in water infrastructure: Review of the last 20 years and the way forward*. Retrieved from www.ppiaf.org

Pigeon, M. (2012). Who takes the risks? Water remunicipalisation in Hamilton, Canada. In M. Pigeon, D.A. McDonald, O. Hoedeman, and S. Kishimoto (Eds.), *Remunicipalisation: Putting water back in public hands*. Washington, DC: Transnational Institute.

Pritchett, L., and Woolcock, M. (2004). Solutions when the solution is the problem: Arraying the disarray in development. *World Development*, 32(2): 191–212.

Proskow, J., James, H., and Rowney, M. (2013). Canada's method of wastewater treatment "backwards." *Global News*, May 13. Retrieved from www.globalnews.ca

Rogers, P., and Hall, A. (2003). *Effective water governance*. TEC Report No. 7. Stockholm, Sweden: Global Water Partnership.

Sowers, J., Vengosh, A., and Weinthal, E. (2011). Climate change, water resources, and the politics of adaptation in the Middle East and North Africa. *Climatic Change, 104*(3): 599–627.

Spronk, S. (2007). Roots of resistance to urban water privatization in Bolivia: The "new working class," the crisis of neoliberalism and public services. *International Labour and Working-Class History, 71*: 8–28.

Spronk, S. (2010). Water and sanitation utilities in the global South: Recentering the debate on "efficiency." *Review of Radical Political Economics, 42*(2): 156–174.

Swyngedouw, E. (2004). *Social power and the urbanization of water: Flows of power*. Oxford: Oxford University Press.

Tyler, S., and Moench, M. (2012). A framework for urban climate resilience. *Climate and Development, 4*(4): 311–326.

UN Data (2014). Venezuela country profile. Retrieved from http://data.un.org/CountryProfile.aspx?crName=Venezuela%20(Bolivarian%20Republic%20of)

UNESCO International Hydrological Programme (2014). Water in the post-2015 development agenda and sustainable development goals. Discussion paper. Retrieved from www.unesdoc.unesco.org

UN Sustainable Development (2016). Sustainable Development Goals. Retrieved from http://www.un.org/sustainabledevelopment

UN Water (2007). *Coping with water scarcity: Challenge of the twenty-first century*. Retrieved from http://www.fao.org/nr/water/docs/escarcity.pdf

Watson, G., and Jagannathan, N.V. (1995). *Participation in water and sanitation*. Environment Development Papers Participation Series, No. 002. Washington, DC: World Bank.

Watts, J. (2015). Brazil's worst drought in history prompts protests and blackouts. *The Guardian*, January 23. Retrieved from www.theguardian.com

WHO and UNICEF (2014). *Progress on drinking water and sanitation—Joint Monitoring Programme update 2014*. Retrieved from www.who.int

WHO and UNICEF (2015). *Progress on drinking water and sanitation—2015 update and MDG assessment*. Retrieved from www.wssinfo.org

World Bank (2004). *Making services work for the poor: World Development Report 2004*. Washington, DC: World Bank.

World Bank (2012). Improved water source (% of population with access). Retrieved from http://data.worldbank.org/indicator/SH.H2O.SAFE.ZS

World Bank (2015). Urbanization in Thailand is dominated by Bangkok urban area. Retrieved January 26 from http://www.worldbank.org/en/news

WSP-LAC (2007). *Saneamiento para el desarrollo ¿ Cómo estamos en 21 países de América Latina y el Caribe?* Retrieved from www.wsp.org

Zaki, S., and Nurul Amin, A.T.M. (2009). Does basic services privatisation benefit the urban poor? Some evidence from water supply privatisation in Thailand. *Urban Studies, 46*(11): 2301–2327.

24

Delivering and Managing Waste and Sanitation Services in Cities

Carrie L. Mitchell, Kate Parizeau, and Virginia Maclaren

Introduction

We live in a predominately urban world, and this has implications for sanitation and waste management, particularly in cities in the global South. In 2014, the United Nations estimated that 54 per cent of the world's population lived in cities, or approximately 3.8 billion people worldwide (UN, 2014). But urbanization trends are not equal across the globe; middle- and large-sized cities in Africa and Asia are expected to experience faster rates of urbanization than those in other regions in the coming decades (UN, 2014). This rapid urban expansion, in terms of both the overall number of urban residents and the expanding geographical area that is required to house new residents, means that the provision of waste and sanitation services often cannot keep pace with the rapid expansion of cities.

For many urban residents and governments around the world, sanitation and waste are serious urban-planning, environmental, and public health issues. Indeed, according to the World Health Organization (2014), a full 2.5 billion people lack access to improved sanitation. In other words, out of a global population of 7 billion, over a third of the people on this planet do not have access to a flush toilet connected to a sewer or septic system. The situation is similar for waste: municipal collection of garbage is a luxury that a large percentage of the global population does not have. These trends have serious health and environmental implications for cities and their citizens in the global South, particularly for low-income residents who typically lack access to these basic urban services. In Ethiopia, for example, urban growth in the capital city, Addis Ababa, has occurred rapidly owing to national-level political-economic changes in the early 1990s. At first glance, changes to the city are impressive: new roads, hotels, and high-rise buildings now mark the urban landscape. Yet, approximately 40 per cent of the urban population lives below the poverty line of $1 per day, and 42 per cent lack access to a toilet (Keller, 2012).

In some cities in the global South, the state of poor waste and sanitation services is increasingly in conflict with municipal attempts to present an image of a world-class, modern, and hygienic city (Banerjee-Guha, 2009; Gidwani and Reddy, 2011; Mitchell, 2008, 2009; Parizeau, 2015b). Governments may try to "clean up" the city for affluent residents, tourists, and foreign investors by focusing funds for waste and sanitation services in high-profile neighbourhoods and by cracking down on behaviours inconsistent with the brand of the modern city, such as informal recycling, squatting in settlements without water and sanitation services, practising open defecation, and littering. Such efforts may also focus on the privatization of city services (including waste and sanitation services) in an attempt to expand service delivery without burdening the municipal budget.

This chapter examines how access, delivery, and management of sanitation and waste services in cities in the global South are connected to broader patterns of urbanization and globalization. It explains how economic, spatial, and social inequality affect access to and delivery of waste and sanitation services. In the sections that follow, trends in waste and sanitation services in the global South are explored. To better explain waste management, attention is directed to the life cycle of waste—from generation and composition to disposal. The discussion then moves on to current global trends in the provision of sanitation and common themes of sanitation and waste, including health and environmental impacts; social inequity in provision of services; and service delivery mechanisms. Briefly, consideration is given to the growing problem of electronic waste, or e-waste, and what happens to technology that is rendered out of date or obsolete when a new model is introduced. In the case studies provided, specific examples are given of research on sanitation and waste delivery and management: a case study from Accra, Ghana, reveals the globalized nature of e-waste; and a case study of Delhi, India, showcases sanitation and gender-based violence concerns in relation to access to basic urban services; namely, toilets. Both studies emphasize issues of social and economic inequality, which are then illuminated in the Canadian context with reference to a discussion of waste management and informal recycling in Vancouver. Taken together, the chapter explores the ways that researchers have been working in and with cities to create more sustainable infrastructure and livelihood solutions. The chapter concludes with a discussion of future trends in the provision of urban services in the global South.

Trends in Waste Management

The amount of **solid waste** produced in urban areas is closely related to a city's level of economic activity, the presence of industry, total population, and the wealth of its residents. Waste generation tends to increase with income: as people earn higher incomes, they can afford to consume more. The World Bank has estimated that a person living in a city will generate 1.2 kilograms of solid waste per day (Hoornweg and Bhada-Tata, 2012). However, this is just an average; people living in more affluent countries tend to throw out more. For example, Canadians currently discard an average of 2.33 kilograms of solid waste per day (Hoornweg and Bhada-Tata, 2012). That may not seem like too much on an individual basis—merely the by-products of our urban lifestyle. Collectively, however, this amounts to a lot of solid waste.

Research shows that the composition of waste we generate differs between cities in the global North and global South. In the global North, about one-third of the waste stream by weight is paper, one-third is organic (e.g., food waste, yard waste), and the remainder consists of plastics (over 10 per cent and growing), metal, glass, rubber, leather, textiles, and miscellaneous materials. In the global South, the waste stream is dominated by organic waste (up to two-thirds) with smaller percentages of paper, plastics (about 10 per cent each), and other materials (Hoornweg and Bhada-Tata, 2012). These percentages primarily reflect the fact that there is considerably less packaging in the global South. However, again, we expect to see changes in the composition of waste in the global South in the coming decades, as more packaged materials are introduced into the marketplace and more people can afford to purchase them.

In the global South, the percentage of households that have access to waste collection services is often spatially varied. Residents in affluent, centrally located neighbourhoods are much more likely to receive waste collection service than residents residing in the outskirts of a city. For example, in Siem Reap, Cambodia, only about 50 per cent of the city receives municipally provided waste collection services (Parizeau et al., 2006). The municipality cannot afford to provide full waste collection coverage and chooses to focus its efforts on central areas of town frequented by tourists and more affluent residential neighbourhoods.

Households without waste collection service often sell or give recyclable materials to **informal waste collectors**—people who buy and/or collect recyclable materials but are not employed by municipal waste management departments (Figures 24.1 and 24.2). Households may also carry waste to a city-owned container, which may be quite far from their home, or they may deal with it themselves

Figure 24.1 Informal recyclers in Buenos Aires, Argentina

Figure 24.2 Informal recycler in Hanoi, Vietnam

by burying it, burning it, or throwing it into public spaces or water bodies. Even if they do manage to bring the waste to a container, the container itself may be overflowing because the municipality responsible for collecting the containers may not collect the containers on a regular basis.

Many low-income residents in cities in the global South live in what are known as *informal settlements*, or communities without legal claim to the land on which they reside. Informal settlements are usually very crowded, and the residents do not typically pay into a city's tax base. In Karachi, Pakistan, for example, it is estimated that 62 per cent of the population live in informal settlements that take up only 8 per cent of the city's land area (Pervaiz et al., 2008). In practice this means that even if municipal governments could afford to provide waste collection service throughout the city, it is logistically difficult to collect waste from inside densely populated informal settlements. Withholding waste collection services from certain informal squatter settlements may also be a political decision that allows the municipality to deny the legitimacy of these settlements.

Although municipalities in the global South seldom provide recycling collection services, most cities have a thriving informal recycling sector, using materials supplied by informal recyclers who buy and/or pick out valuable waste materials from the waste stream. Informal recyclers include people who go door-to-door and buy recyclable items from households and businesses (primarily metal, plastic, paper, and glass) and people who sort through garbage in trash bins, at waste collection points, or in landfills to pull out valuable items (Mitchell, 2008).

In contrast to the global North where municipalities organize and pay for recycling collection, informal recyclers in the global South provide a "free" service (i.e., they are not paid by the municipality) that is essentially a subsidy by the poor to the rest of the city (UNHSP, 2010). Their work reduces the quantities of waste requiring collection by the municipality while also contributing to the environmental sustainability of the city. For example, in Hanoi, Vietnam, it is estimated that informal recyclers reuse and recycle about one-quarter of the city's waste (World Bank, 2004) (Figure 24.2). In many cities in the global South, informal recyclers are rural migrants, or people from lower socio-economic classes (Mitchell, 2008, 2009; Parizeau, 2015a); social inequity is a common feature of both formal and informal waste management services. Informal recycling, while useful in terms of diverting usable materials from a landfill and in creating and/or supplementing livelihoods, also has serious health implications for some who engage in the activity.

Despite the fact that in most Southern cities two-thirds of the waste generated is organic, there are very few organized **composting** programs in the global South. Most of the composting programs that do exist have been started by non-governmental organizations and tend to be subsidized, since income from selling compost rarely covers the cost of production. **Incineration**, which is often relied upon in the global North, is in many cases not a viable option in Southern cities because of high capital costs associated with the technology. As the technical and financial capacity of countries of the global South increases, disposal is shifting from highly polluting **open dumping** to controlled dumping in **sanitary landfills**. Open dumps have no measures to manage leachate or methane gas production and no formal siting requirements (Figure 24.3). Landfills have more stringent siting requirements and often include measures to collect and treat leachate and methane gas.

Despite the fact that landfilling waste is an improvement over open dumping, best practice in waste management in both the global South and the global North is, first, source reduction, followed by reuse of waste materials, composting of organic material, and, finally, recycling of materials that cannot be reused in their present form. Many cities in the global North use expensive technology to achieve higher waste diversion rates, and some cities in the global South are following suit. However, effective management of waste, whether

Figure 24.3 Open dump outside of Vientiane, Lao PDR

in the global North or global South, is less about technology and more about changing people's behaviour when it comes to waste generation and disposal.

Sanitation Trends in the Global South

The quality of **sanitation** in a city is closely correlated to overall human development. Cross-country studies have shown that the method used to dispose of excreta, or human waste, is one of the strongest determinants of a child's survival rate in the global South (UN Water, 2014). Indeed, the transition from unimproved (e.g., open defecation, pit latrines without a slab or platform, and hanging or bucket latrines) to improved sanitation facilities that ensure separation of human excreta from human contact (e.g., pit latrines with a slab, composting, or flush toilets) reduces child mortality by about one-third (UN Water, 2014). It is estimated that 80 per cent of diseases in some parts of the global South are caused by unsafe water and poor sanitation (UN Water, 2014).

However, much like waste collection, access to sanitation services is not evenly distributed throughout regions, countries, or cities. In many cases, rural areas, informal settlements in cities, newer settlement areas on outskirts of cities, and otherwise marginalized groups have reduced-access sewage management systems. Relative wealth of residents is also a determinant of access to sanitation. For example, in Mozambique open defecation, or defecating outside and not into a designated toilet or latrine, rates vary from 13 per cent among the rural rich (the richest 20 per cent of the rural population) to 96 per cent among the rural poor (the poorest 20 per cent of the rural population) (WHO and UNICEF, 2013). In cities, similar inequality exists. The *Millennium Development Goals Report* (UN, 2011) notes that sanitation improvements often bypass the poor, documenting how improvements in parts of Southern Asia between 1995 and 2008 disproportionately benefited affluent residents, with little improvement in sanitation coverage for the poorest 40 per cent of households (most of whom continued to practise open defecation). As such, it is important to recognize that global figures on access to improved sanitation mask spatial and socio-economic inequality within countries and cities.

At the Millennium Summit, held in September 2000, world leaders adopted what have become known as the Millennium Development Goals (MDGs), a set of time-bound, quantifiable, global targets to reduce extreme poverty. Goal 7 (target 10) of the MDG document focuses on improvements in access to safe drinking water and basic sanitation. The goal set forth by world leaders in 2000 was to halve, by 2015, the proportion of people without sustainable access to safe drinking water and improved sanitation using 1990 global sanitation data as a baseline. At that time, it was estimated that 51 per cent of the global population lacked access to improved sanitation (WHO and UNICEF, 2013), and the goal was to increase this figure to 75 per cent by 2015.

The World Health Organization (2014) estimates that there are still over 2.5 billion people who do not have access to improved sanitation facilities, mostly in sub-Saharan Africa and South Asia, and about 1 billion people still practise open defecation. While the practice of open defecation is on the decline in most of the world (from 24 per cent in 1990 to 15 per cent in 2011), rates remain high in South Asia (39 per cent) and are still on the increase in sub-Saharan Africa (WHO and UNICEF, 2013). Within densely inhabited informal settlements, open defecation is often the only option for residents because there are either too few public toilets or maintenance is so poor that they are unusable. Also, toilets require water and electricity to function, and these urban services are often unreliable and/or expensive for low-income urban residents. In Chennai, India, a study on the state of shared toilets in the city found that of the 57 toilets they counted, only 14 were usable. Unusable toilets did not have doors, lights, and/or water. Some of the toilets were deemed unusable because of blockages, while others were boarded up, locked, or completely dilapidated (Transparent Chennai, 2011).

Maintenance of shared toilets is sometimes difficult for city governments, as many shared toilets are owned and maintained by private caretakers. In the Chennai-based study, researchers found that individuals owned and/or maintained 26 of the 57 toilets. Similar private ownership of shared toilets exists in other cities. In Jakarta, Indonesia, for example, individuals own and maintain shower and toilet stalls in informal settlements and "pay per use" is a common practice. Fees in Chennai to use a toilet (2013 figures) range from Rs. 1 to Rs. 5 (1–10 cents) per use. In Jakarta, user fees amount to the equivalent of approximately 44 cents per day. This may not seem like a lot of money, but if you earn $1 to $2 per day, as many urban poor do, spending money to urinate, defecate, and shower constitutes a significant proportion of your daily wage.

Another challenging issue with shared toilets is inconsistent and/or inconvenient hours of operation. Very few toilets are available on demand: in Chennai only 12 of the 57 toilets surveyed were open 24 hours a day, and most were open between specific hours only (from 3 a.m. to 6 a.m. and from 6 p.m. to 11 p.m.). The lack of maintenance, the limited availability, and the cost to use toilet/shower facilities are rarely considered in the global literature on sanitation and often come at the cost of personal health, safety, and dignity. For example, the lack of toilets in schools (and thus a lack of private space to attend to hygienic needs) is one reason why adolescent girls in the global South may miss school or drop out altogether. Rarely discussed are the needs of menstruating women, pregnant women, elderly men and women, and disabled men and women (Greed, 2003). Moreover, using shared toilets or practising open defecation at night puts women and girls at risk of sexual assault.

Efforts are underway to deliver more sustainable and equitable sanitation services around the world. For example, the World Toilet Organization, a global non-profit organization, was founded in 2001 to improve toilet and sanitation conditions worldwide. Multiple international organizations also work on improved and sustainable sanitation across the globe. Recognition of some of the challenges noted above is a key factor in providing improved sanitation across the global South.

Health and Environmental Challenges Associated with Solid and Human Waste and Sanitation

Disposal of human and solid waste is a significant environmental challenge for cities in the global South. Waste and sewage wastes (also known as **black water**—excreta, urine, and fecal sludge) have multiple impacts on human health and our environments. Improper disposal of waste can contribute to air pollution (burning waste), water pollution (dumping waste in waterways or leachate seeping out of low-quality landfills), and burying waste can introduce toxins to land (Figure 24.4). Sewage wastes can also introduce pathogens (e.g., bacteria, amoeba, viruses, and parasites) into land and waterways, especially when black water wastes are not properly treated before they are disposed of.

When reusable and recyclable materials are not fully diverted from waste streams, there is increased pressure for resource extraction in order to maintain consumption patterns. Improper solid and sewage waste management

Figure 24.4 Indiscriminate dumping of waste in canal, Jakarta, Indonesia

leads to lost materials and nutrients that cannot be easily recovered. These environmental problems impact upon both the global North and the global South.

The environmental impacts of waste have health implications for people who come into close contact with them. Waste workers, and particularly informal recyclers, report multiple health problems associated with waste contact, including exposure to chemical hazards and infections, injuries from heavy lifting and touching dangerous items in the trash, and emotional stresses connected to the stigma of being a waste worker (Binion and Gutberlet, 2012). Contact with the pathogens in black water wastes can cause a number of infections and diseases, including diarrhea. The World Health Organization estimates that 280,000 diarrheal deaths per year can be attributed to inadequate sanitation, and suggests that many other sanitation-related deaths are harder to observe and quantify (WHO, 2014).

In the global North, the focus is often on a **green sustainability agenda** of environmental issues like biodiversity loss and resource depletion. In contrast, the prevalence of environmental health threats in cities in the global South (including unsanitary living conditions, pollution, and the accumulation of wastes and waste water) lead to a different type of environmentalism in these communities that can be characterized as a **brown environmental health agenda**. In other words, there is a close relationship between environment and health in cities that face challenges in managing waste, black water, and other urban pollutants (McGranahan and Satterthwaite, 2000). The environmental inequalities associated with sanitation and waste management are particularly pronounced in the global South.

Social Inequality

Sometimes low-income urban residents are represented as being "ignorant" of the health, safety, and environmental problems associated with their ad hoc sanitation and waste practices, but this is an unfair characterization. In many cases, low-income urbanites are well aware of the varied social and environmental impacts of open defecation, burning waste, and littering, but they have few alternatives. For many, the problem is not their unwillingness to participate in or pay for such services, but rather their *inability* to pay for expensive hygiene and waste services or to fund infrastructure projects on their own. It is not uncommon for development institutions and non-governmental organizations to implement waste and sanitation programs that require low-income residents to "volunteer" their time in community-based schemes and pay for waste, water, and sanitation services that middle- and high-income urbanites receive at little or no cost; middle- and high-income urbanites receive services provided by the city, or they can afford to pay for services from private-sector service providers. Hygiene education programs promoted by development institutions may also seem insulting and even culturally inappropriate to low-income urbanites. Providing appropriate sanitation and waste infrastructure for the urban poor is therefore a matter of equity and dignity, in addition to being a health, safety, and environmental issue.

Many informal and formal workers in the waste and sanitation sectors face stigma and discrimination because they are associated with so-called dirty jobs. This stigma is compounded when it is vulnerable groups who are pushed into these types of work. Informal waste workers, for example, often earn far less money than formal waste collectors, and these informal workers often face labour exclusions (e.g., factors that prevent them from taking on formal work: household and child-care responsibilities, health problems, addictions, and other mental health issues). In Buenos Aires, Argentina, informal recyclers discussed how years of economic crisis made it difficult to find stable work. In some cases, children dropped out of school in order to help support their low-income families through informal recycling work, thus missing out on educational opportunities for themselves (Parizeau, 2015a). Women often earn less than men who do the same job thanks to social biases about the types of material women should handle or where they should be allowed to work. In Hanoi, Vietnam, for example, women earned 61 per cent of what men earned on average. In the case of Hanoi, this income inequality can be explained by the social stigma against women collecting the more lucrative e-waste (Mitchell, 2008).

Sometimes formal waste and sanitation systems perpetuate inequalities. For example, in many Indian cities, "manual scavenging" of dried excrement from sewers, latrines, and railroad tracks (a common site for open defecation) is an unhealthy and low-paid job exclusively reserved for Dalit or "untouchable" people, who occupy the lowest place in the Indian caste system. In the early 2000s in Cape Town, South Africa, unemployed black women were encouraged to volunteer their time as public street sweepers. They routinely provided unpaid litter cleanup and public hygiene services to the community in the hope that it might lead them to a paid job (Miraftab, 2004). Social inequality is an inherent feature of formal and informal waste and sanitation work, and it will likely remain as such until the social stigma associated with solid and human waste is tackled.

The case studies that follow introduce two critical issues related to the management and delivery of waste and sanitation services: the globalization of waste management and access to sanitation and gender-based violence. In the case of e-waste in Accra, Ghana, the increasingly globalized nature of our waste is explored with a focus on the management of electronic waste. In the case of Delhi, India, the way in which social and economic inequality challenges safe access to basic urban services, such as access to a toilet, is examined. Both of these case studies illustrate the complex and dynamic nature of managing and delivering waste and sanitation services.

Electronic Waste in Accra, Ghana

Accra is the capital of and largest city in Ghana, a country located along the coast of West Africa (Figure 24.5). The population of Accra, as of 2012, was about 2 million people in the city proper and about 4 million people in the Greater Accra Metropolitan Area (GAMA) (World Atlas, 2015).

Agbogbloshie marketplace in Accra is the electronic waste (e-waste) recycling and reuse hub for Ghana, and it is also the site of a large local food and vegetable market (Figure 24.6). **E-waste** arrives here from Europe, from neighbouring countries, and from within Ghana itself. It includes used or surplus computers, cellphones, monitors, televisions, and other electronic devices. Some of the equipment is still in good working order, some will require refurbishment, and some is only suitable for dismantling and recycling. E-waste recycling recovers valuable resources such as gold, silver, aluminum, and copper, but it also releases harmful toxics such as mercury, cadmium, lead, polyvinyl chloride (PVCs) plastics, and brominated flame retardants (BFRs). E-waste recycling is hazardous work in the absence of stringent health and safety precautions.

As a site of e-waste trade and transformation, Agbogbloshie illustrates both sides of the dominant narratives about e-waste trade. One narrative sees international e-waste flows as "toxic trade," while the other sees it as "digital development" (Pickren, 2014). The first narrative claims that most e-waste from the global North ends up in the global South where recycling and disposal occur with little or no regulation, resulting in serious consequences for the environment and for the health of those who do the recycling. Is this what can be seen in Agbogbloshie? Yes and no. Unregulated recycling is clearly present. One of the main recycling techniques used at the site is open burning of plastic-coated wiring and cables to recover the copper. Studies have found significant chemical contamination of soils and the presence of toxic metals on-site and in a nearby lagoon (Brigden et al., 2008). On the other hand, claims about the origin of e-waste arriving in Agbogbloshie are not so clear-cut. Grant and Oteng-Ababio (2012) found that while most used computers arriving in Ghana come from Europe and smaller percentages from North America and Asia, by far the most rapid growth in used computer imports is from within Africa (Grant and Oteng-Ababio, 2012). The growth in internal trade is consistent with Lepawsky's (2015) study of the global e-waste trade, which found that there has been a sharp increase in the flow of e-waste among countries of the global South from 1996 to 2012. Lepawsky also looked at e-waste exports from countries of the global South and found an even greater increase in e-waste movement from the global South to processing facilities in the global North. Another contradiction of the toxic trade narrative is that most e-waste sent to Ghana is being repaired, refurbished, and resold for reuse rather than being dismantled for recycling or disposal (Amoyaw-Osei et al., 2011).

The digital development narrative focuses on how trade in e-waste destined for reuse provides affordable access to information and communication technologies (ICTs) for residents of the global South (Pickren, 2014). There is truth in this narrative as it applies to Accra, but it is more nuanced than just being a narrative about access. Grant and Oteng-Ababio (2012) found that the demand for ICT in Ghana and the availability of used electronic equipment and parts at Agbogbloshie have stimulated an extensive refurbishment sector in the city. It employs 10,000–15,000 workers earning an average of about USD\$7/day, which is more than three times the country's official minimum wage. In fact, the entire e-waste sector provides a livelihood for tens of thousands of people, almost all of them informal workers. E-waste collectors, those who collect e-waste from households and businesses in the city, require the least skills and have the lowest income, earning an average of USD\$3.50

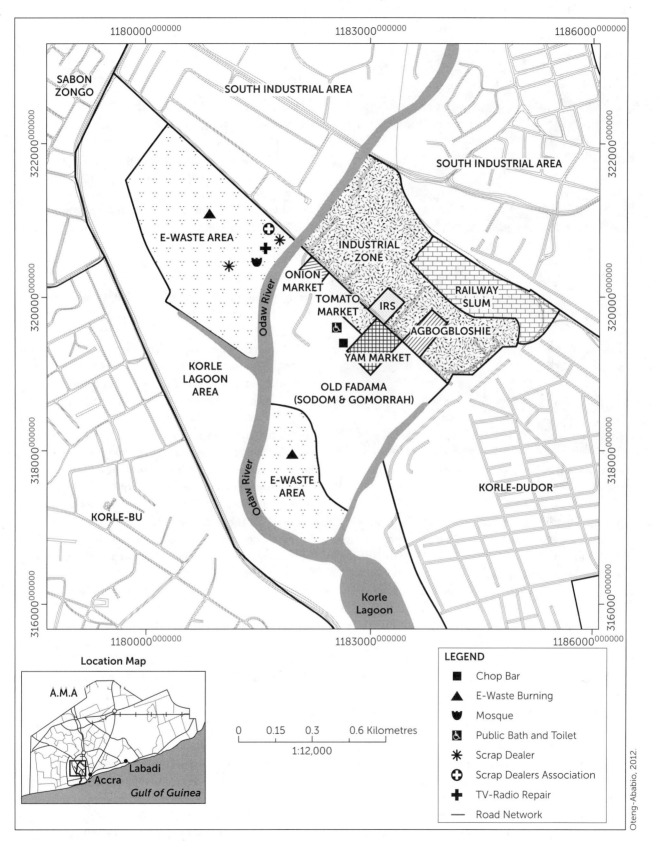

Figure 24.5 Map of Agbogbloshie e-waste recycling in Accra, Ghana

Figure 24.6 E-waste incineration at the Agbogbloshie metal scrapyard in Accra

per day. Dismantlers and recyclers earn about $8 per day (Grant and Oteng-Ababio, 2012) but are exposed to high levels of toxic pollution on a daily basis. Many informal workers are youths (under 15 years old) who work on a part-time basis and earn considerably less money.

This brief case study both supports and contradicts elements of the dominant narratives in e-waste trade while adding to our understanding of the contribution of e-waste to the urban economy. The case study also suggests that Agbogbloshie is an interesting example of both green sustainability and brown environmental health. Repairing and refurbishing electronic equipment so that it can be used again diverts from disposal what would otherwise be waste and conserves resources. Recycling at Agbogbloshie also conserves resources, but with much higher cost to the environment. Working with e-waste provides a livelihood for many, but can also have significant health consequences. Health concerns related to gender-based violence and safety are featured in a case study of Delhi, India, in relation to the public provision of toilets.

Sanitation Infrastructure and Women's and Girls' Safety in Delhi's Low-Income Resettlement Communities

Delhi, India, is a vibrant, bustling megacity in South Asia (Figure 24.7). With a population of approximately 11 million people, it is India's second-largest city (after Mumbai). It is also home to India's national government and is positioning itself to become a "world-class city" (The Hindu, 2015).

But not all Indian citizens have equal rights to access this emerging world-class city or its basic urban services. The Master Plan of Delhi (MPD) is the guiding document for city planners. The most recent MPD, published in 2001, stipulates how land is zoned in the city, allocating certain areas to residential, commercial, and industrial purposes. The MPD also includes plans for water, sanitation, and solid waste management throughout the city. Over the course of the last few decades, the city government has evicted thousands of residents for living on land zoned, or rezoned, for non-residential purposes in central Delhi (Khosla and Dhar, 2013). The result has been the growth of "unauthorized" resettlement colonies, many located in the outskirts of the city where land is less valuable and zoning restrictions less stringently enforced. In the lead-up to the Commonwealth Games in 2010, efforts to evict so called squatter settlements from central Delhi accelerated in an attempt to portray the city as an emerging star on the global stage.

Access to basic urban services in Delhi, and in many cities around the world, is linked to where in the city residents live and to the type of tenure arrangement they have acquired. There are eight different types of settlements in Delhi, ranging from "unauthorized" and "slum designated" colonies to planned residential communities. Indeed, the majority of the city's residents (76 per cent as of the 2008/2009 Economic Survey of Delhi) reside in unplanned and/or unauthorized settlements in and on the outskirts of the city. For those living in these unauthorized and/or unplanned settlements (typically the cities' urban poor), access to basic urban services is a daily challenge. Minimum standards have been set in Delhi's National Slum Policy for the urban poor; these include one water-supply tap for every 150 people; one latrine for every 20–50 people; and open sewer drains

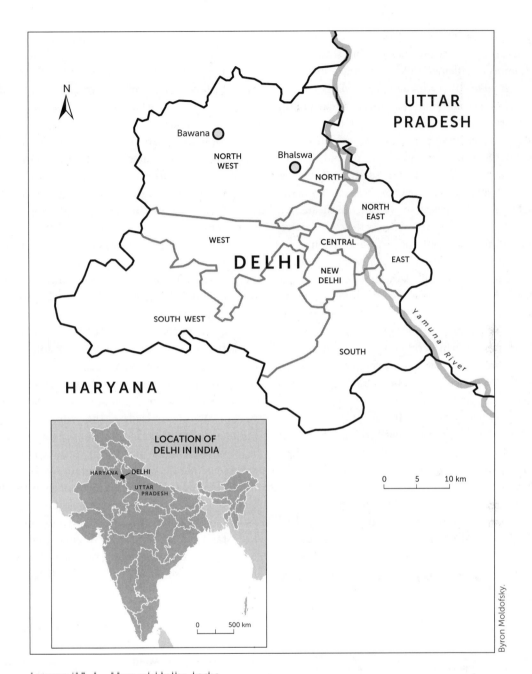

Figure 24.1   Map of Delhi, India

with normal outflow to avoid the accumulation of stagnant wastewater. Electricity, which is needed to power water pumps for toilets, is supplied by a variety of private distribution companies.

In practice, however, these minimum standards are often not met, resulting in criticism from citizens, civil-society groups, and the Indian government itself. For example, following a report on the conditions of community toilet complexes in an area of Delhi called Jhuggi-Jhompri (Sheikh, 2008), the Delhi High Court ordered the city government to take immediate action to make the toilet complexes safe and functional (WICI and Jagori, 2011).

How can a city government with a population of 11 million residents, the majority of whom are living in unplanned and unauthorized settlements, ensure the safety and functionality of sanitation and sewerage service across the city? This was one of the questions a team of Indian and Canadian researchers posed in a study of two Jhuggi-Jhompri resettlement colonies on the outskirts of Delhi from 2009 to 2011 (WICI and Jagori, 2011).

The first study community, Bawana, is situated in the northwest corner of Delhi. Starting in 2004, thousands of evicted slum dwellers from central Delhi were relocated to this site, about 35 kilometres away. Plots of land were assigned in the form of a licence for five, seven, or nine years. At this point, original licences have expired, and residents' tenure rights are again unclear. In practice, this means that residents are unsure about what services they will receive, but they are afraid to ask for fear of further eviction.

The second study community, Bhalswa, is located in northeast Delhi, next to the Bhalswa landfill. Most people who reside in this community were evicted from the north and east of Delhi starting in 2000. Today, the approximate population of Bhalswa is 25,000. In both communities, provision of basic urban services is inadequate. Community taps rarely yield water; toilet complexes are often closed because of lack of water and electricity supply; and open sewers are typically filled with uncollected garbage, which results in sewage overflow (Figure 24.8).

This lack of access to basic urban services affects men and boys and women and girls differently. Girls interviewed in Bhalswa, for example, reported that there was no toilet allocated specifically for girls at their school. Instead, they would use the boys' toilet in shifts, so there would be no overlap. There was no disposal option for menstrual waste, so girls would often not go to school when they were menstruating. Women and girls in both Bawana and Bhalswa reported instances of sexual harassment when they attempted to use toilet complexes in their communities. In both communities, men and boys loitered and played around the toilet complexes. Some women and girls reported men and boys peeking in at them when they were using the toilet; others reported hearing lewd songs and comments. Power failures not only affected the functioning of the toilets themselves, but resulted in women and girls reporting a decreased sense of safety as they tried to return to their homes with inadequate street lights. Indeed, many women reported instances of sexual assault, including rape, during their attempts to urinate and defecate.

The findings from this project illustrate the connection between inadequate urban infrastructure and women's and girls' safety in the city. However, finding appropriate solutions was more complex. One important tool that researchers used to catalyze communication and action was the Women's Safety Audit. In Bawana and Bhalswa, researchers conducted "safety audit walks" with members of the community, including women, girls, boys and men, police, municipal officials, and other relevant community leaders. The primary goal of these walks was simple: to identify and document the social and environmental factors that contributed to, or inhibited, a sense of safety in their community. The end result was community maps identifying "hot spots" for further intervention. A secondary and more difficult goal was to have different groups in these two communities better understand the experiences of women and girls in the city and take concrete action to change current social and cultural mores. Researchers report, however, that the combination of the women's safety audit walks, subsequent conversations among the residents, and follow-up discussions with public toilet caretakers has resulted in enhanced awareness and understanding of the behaviours that make women and girls feel uncomfortable and unsafe. But researchers also note that changing attitudes and behaviour is a long-term project, one that cannot be fully achieved in a short-term research project.

By the end of the two-year project, residents in Bawana had developed a new, more gender-sensitive design for community toilets. In Bhalswa, for the first time in 10 years, the community had a system of solid waste management. A vacant area of this community, previously used as a dumping ground, was cleaned up and developed into a park. Perhaps most importantly,

Figure 24.8 Woman collecting water at community tap, outskirts of Delhi, India

the police have increased their presence in the community, and at the time of final reporting, women and girls informed researchers that they were experiencing less harassment and felt safer.

Overall, this case study illustrates the importance of understanding complex social issues in the provision of sanitation services in cities. Sanitation infrastructure in cities could be improved with thoughtful interventions that target the particular needs of users. Sanitation waste, however, is only one dimension that needs to be considered in the development of urban-planning solutions to waste management systems. It is also imperative to address the uneven social and environmental impacts of waste management. The following text box examines the social marginalizations and health hazards of informal recycling work in Canadian cities.

Waste Management in Canada, Informal Recycling in Vancouver

Canadians dispose of nearly 13 million tonnes (14.3 million tons) of waste each year, and this number is on the rise (Conference Board of Canada, 2013). Most of this waste, approximately two-thirds, is disposed of in one of Canada's more than 10,000 landfills. The remainder of our municipal solid waste is recycled or composted. As global patterns of wealth distribution shift, we can expect to see increases in the per capita waste generation rates in cities around the world. If waste generation rates and disposal patterns follow trends similar to those in Canada, we will experience a global garbage crisis.

In Canada, residential waste management services are coordinated by municipal governments and are paid for through the municipal tax base. Industrial, commercial, and institutional sectors may use elements of the residential waste management system, but they often must arrange and pay for the private collection of their waste outside of the municipal system. The laws and policies that regulate waste management are primarily mandated at the provincial level, although there are also federal environmental regulations that pertain to waste management planning (such as the transboundary movement of hazardous wastes).

Formal waste management systems in Canadian municipalities usually include separated garbage and recycling streams and, increasingly, an organic composting stream. In many Canadian cities, informal recycling is a conspicuous part of the waste management system, particularly in places that have deposit-return systems (like the deposit that is paid on alcohol bottles in Ontario and all beverage containers in British Columbia). In many cities, it is not uncommon to see people sorting through blue bins or garbage containers, collecting large quantities of items that they can return for the deposit money. Municipal bylaws often prohibit such practices, although authorities do not consistently enforce these regulations for low-income recyclers (Parizeau and Lepawsky, 2015).

In addition to providing income for recyclers, informal collecting can have environmental benefits: it takes more energy to recycle bottles that end up in the blue recycling bin than it does to reuse those bottles when they are returned for a deposit. In Vancouver, BC, informal recyclers work throughout the city to collect beverage containers and other items that can be reused or resold. Many of these informal workers are prevented from holding full-time jobs because they face labour exclusions such as disabilities, addiction or other mental health concerns, language barriers, or migrant status that does not allow for traditional work activities. Informal recycling offers an important income-earning opportunity under these circumstances. There are also connections between informal recycling and large-scale economic change. Tremblay et al. (2010), for example, describe how informal recycling activities increased when British Columbia made cuts to social assistance rates.

Similar to informal recyclers in the global South, these workers experienced health problems from touching waste and hazardous materials found in the waste stream. Because of the stigma of "dirtiness"

continued

associated with informal recycling, many of these workers reported facing discrimination and social exclusion from members of the public and from social-service providers (including health-care providers). Stigma can cause psychological stress and can also prevent recyclers from accessing the services they need to remain healthy (Wittmer, 2014).

One key initiative that addresses the economic and social problems confronting informal recyclers in Vancouver is the United We Can depot. This social enterprise was started by and for informal recyclers and provides a place where they can sell their collected containers for a fair price. United We Can has also provided diverse services and opportunities for informal recyclers since it was founded in 1995 (including bicycle repairs and opportunities for recyclers to pick up recyclables at public events or from commercial partners). United We Can is one example of how informal recyclers can work together and with the community to build social networks and address some of the social and economic challenges that many of these workers face (Dale and Newman, 2006; Tremblay et al., 2010).

The example of informal recycling in Vancouver highlights how social inequality in Canada can be reflected in waste management systems: those who carry out low-paid and relatively hazardous informal recycling work tend to have lower socio-economic status, and their waste work can cause further social marginalization and discrimination. Technical solutions may improve elements of our waste management system, but addressing the uneven environmental burdens posed by waste in Canadian cities will also require social interventions.

Conclusion

In this chapter, we introduced key themes in the current waste and sanitation literature in the global South. Urbanization patterns are shifting, and rapid intra-national migration is increasingly a key issue facing large cities. At the same time, globalization means that cities are more connected to each other than ever before and they are increasingly competing for foreign investment. The result is spatial inequality in the delivery of services: residents in central, high-profile neighbourhoods often have better access to waste and sanitation (and water, electricity, and transport) than those living on the periphery of cities or in informal settlements. We also see social inequity in terms of who does the "dirty work" of cleaning up the city and people's ability to access services. Gender, class, caste, and race hierarchies, for example, are perpetuated in unequal access to, and delivery of, urban services in the city. Low-income women without access to private toilets face sexual harassment and assault in Indian cities; in African cities, race and poverty intersect in waste workers' experience of gaining employment opportunities. Globally, e-waste workers toil to extract valuable material from our old phones, computers, and other e-waste, often resulting in negative health and environmental impacts. Overall, it is clear that sustainable and equitable provision of sanitation and waste management services requires more than simply technical solutions. Consideration of the particular social challenges of service delivery in cities is critical to designing and implementing effective sanitation and waste management programs.

Key Points

- Urbanization and globalization are intimately linked to access to, and provision of, urban services such as waste and sanitation.
- Waste management in the global South differs from cities' waste management efforts in the global North with respect to the amount generated, its composition, collection methods, and how it is finally disposed of.
- Despite global efforts, lack of "improved" sanitation in the global South is still a significant issue.
- Inadequate waste and sanitation services in the global South result in serious health and

- environmental impacts; in some cities, they can also catalyze violence against women.
- Social and spatial inequity is a key feature of access to, and delivery of, urban services in many cities in the global South.
- Electronic waste in Accra, Ghana (and other places around the world), can be understood as both a toxic trade and digital development. Regardless, the practice of e-waste recycling has significant implications for workers' health.
- The delivery of sanitation services in cities like Delhi, India, is not just about installing pipes and taps; effective delivery of sanitation services requires understanding, and planning for, social and spatial inequalities in terms of access to services and gendered violence.
- Modern waste management systems in Canadian cities and other parts of the world cannot be built around technical solutions alone but must also consider the social and economic challenges faced by waste workers and the recipients of waste services.

Activities and Questions for Review

1. If you had nowhere to dispose of your household garbage for (a) a period of two weeks and (b) a period of two months, what impact would this have upon life in your household?
2. In a small group, discuss methods of informal waste collection in your city with particular emphasis on the people involved and the work being performed.
3. In pairs, discuss how often you buy a new phone or laptop and what happens to your old communications devices. What are the e-waste consequences of your electronics purchasing habits?
4. Taking into consideration the case study about sanitation in Delhi, India, think about your own experience of toilet access on the university campus. How does your access and sense of personal safety differ from those of the women and girls in Delhi?

Acknowledgments

Thank you to the International Development Research Centre for funding the research discussed in the case study on Delhi, India. Thank you to Jagori and Women in Cities International for conducting the research on this critical issue. Thank you to the Social Sciences and Humanities Research Council for funding the research discussed in the case study on Vancouver, Canada.

References

Amoyaw-Osei, Y., Agyekum, O.O., Pwamang, J., Mueller, E., Fasko, R., and Schleup, M. (2011). *Ghana e-waste country assessment*. Prepared for the Secretariat of the Basel Convention. Retrieved from http://ewasteguide.info/Amoyaw-Osei_2011_GreenAd-Empa

Banerjee-Guha, S. (2009). Neoliberalizing the "urban": New geographies of power and injustice in Indian cities. *Economic and Political Weekly, 44*(22): 95–107.

Binion, E., and Gutberlet, J. (2012). The effects of handling solid waste on the well-being of informal and organized recyclers: A review of the literature. *International Journal of Occupational and Environmental Health, 18*(1): 43–52.

Brigden, K., Labunska, I., Santillo, D., and Johnston, P. (2008). *Chemical contamination at e-waste recycling and disposal sites in Accra and Korforidua, Ghana*. Greenpeace International, Amsterdam. Retrieved from http://www.greenpeace.org/international/en/publications/reports/chemical-contamination-at-e-wa

Conference Board of Canada (2013). Municipal Waste Generation. Retrieved from http://www.conferenceboard.ca/hcp/details/environment/municipal-waste-generation.aspx

Dale, A., and Newman, L. (2006). Sustainable community development, networks and resilience. *Environments Journal, 34*(2): 17–27.

Gidwani, V., and Reddy, R. (2011). The afterlives of waste: Notes from India for a minor history of capitalist surplus. *Antipode, 43*(5): 1625–1658.

Grant, R., and Oteng-Ababio, M. (2012). Mapping the invisible and real "African" economy: Urban e-waste circuitry. *Urban Geography, 33*(1): 1–21.

Greed, C. (2003). *Inclusive urban design: Public toilets*. Oxford, UK: Architectural Press.

Hindu, The (2015). BJP promises to make Delhi a world class city. Retrieved from http://www.thehindu.com/elections/delhi2015/vision-document-bjp-promises-to-make-delhi-a-world-class-city/article6852230.ece

Hoornweg, D., and Bhada-Tata, P. (2012). What a waste: A global review of solid waste management. Retrieved from http://siteresources.worldbank.org/INTURBANDEVELOPMENT/Resources/336387-1334852610766/What_a_Waste2012_Final.pdf

Keller, E.J. (2012). The delivery of public goods in a rapidly expanding African city: Financing policies in Addis Ababa, Ethiopia. *Journal of International Affairs*. Retrieved from http://jia.sipa.columbia.edu/online-articles/delivery-public-goods-rapidly-expanding-african-city-financing-policies-addis-ababa-ethiopia

Khosla, P., and Dhar, S. (2013). Safe access to basic infrastructure: More than pipes and taps. In *Building inclusive cities: Women's safety and the right to the city*. New York: Earthscan.

Lepawsky, J. (2015). The changing geography of global trade in electronic discards: Time to rethink the e-waste problem. *Geographical Journal*, 181(2): 147–159.

McGranahan, G., and Satterthwaite, D. (2000). Environmental health or ecological sustainability? Reconciling the brown and green agendas in urban development. In C. Pugh (Ed.), *Sustainable cities in developing countries*. London: Earthscan, pp. 73–90.

Miraftab, F. (2004). Making neo-liberal governance: The disempowering work of empowerment. *International Planning Studies*, 9(4): 239–259.

Mitchell, C.L. (2008). Altered landscapes, altered livelihoods: The shifting experience of informal waste collecting during Hanoi's urban transition. *Geoforum*, 39(6): 2019–2029.

Mitchell, C.L. (2009). Trading trash in the transition: Economic restructuring, urban spatial transformation, and the boom and bust of Hanoi's informal waste trade. *Environment and Planning A*, 41(11): 2633–2650.

Oteng-Ababio, M. (2012). Electronic waste management in Ghana—Issues and practices. In S. Curkovic (Ed.), *Sustainable development—Authoritative and leading edge content for environmental management*. Rijeka, Croatia: InTech, pp. 149–166.

Parizeau, K. (2015a). When assets are vulnerabilities: An assessment of informal recyclers' livelihood strategies in Buenos Aires, Argentina. *World Development*, 67: 161–173.

Parizeau, K. (2015b). Re-representing the city: Waste and public space in Buenos Aires, Argentina in the late 2000s. *Environment and Planning A*, 47(2): 284–299.

Parizeau, K., and Lepawsky, J. (2015). Legal orderings of waste in built spaces. *International Journal of Law in the Built Environment*, 7(1): 21–38.

Parizeau, K., Maclaren, V., and Chanthy, L. (2006). Waste characterization as an element of waste management planning: Lessons learned from a study in Siem Reap, Cambodia. *Resources, Conservation, and Recycling*, 48(2): 110–128.

Pervaiz, A., Rahman, P., and Hasan, A. (2008). *Lessons from Karachi: The role of demonstration, documentation, mapping and relationship building in advocacy for improved urban sanitation and water services*. Retrieved from http://pubs.iied.org/pdfs/10560IIED.pdf

Pickren, G. (2014). Political ecologies of electronic waste: Uncertainty and legitimacy in the governance of e-waste geographies. *Environment and Planning A*, 46(1): 26–45.

Sheikh, S. (2008). Public toilets in Delhi: An emphasis on the facilities for women in slum/resettlement areas. Retrieved from http://ccs.in/internship_papers/2008/Public-toilets-in-Delh-192.pdf

Transparent Chennai (2011). Public toilets in Chennai. Retrieved from http://www.transparentchennai.com/wp-content/uploads/downloads/2013/07/Public%20Toilets%20%20In%20Chennai.pdf

Tremblay, C., Gutberlet, J., and Peredo, A. (2010). United we can: Resource recovery, place and social enterprise. *Resources, Conservation and Recycling*, 54(7): 422–428.

UN (United Nations) (2011). *The Millennium Development Goals Report 2011*. Retrieved from www.un.org/millenniumgoals/reports.shtml

UN (United Nations) (2014). World's population increasingly urban with more than half living in urban areas. Retrieved from http://www.un.org/en/development/desa/news/population/world-urbanization-prospects-2014.html

UNHSP (United Nations Human Settlement Programme) (2010). *Solid waste management in the world's cities*. London: Earthscan. Retrieved from http://mirror.unhabitat.org/pmss/listItemDetails.aspx?publicationID=2918

UN Water (United Nations Water) (2014). Access to Sanitation. Retrieved from http://www.un.org/waterforlifedecade/sanitation.shtml

WHO (World Health Organization) (2014). *Preventing diarrhoea through better water, sanitation and hygiene: Exposures and impacts in low- and middle-income countries*. Retrieved from www.who.int/water_sanitation_health/gbd_poor_water/en

WHO (World Health Organization) and UNICEF (United Nations International Children's Emergency Fund) (2013). *Progress on sanitation and infrastructure: 2013 update*. Retrieved from http://apps.who.int/iris/bitstream/10665/81245/1/9789241505390_eng.pdf

WICI (Women in Cities International) and Jagori (2011). *Gender and essential services in low-income communities: Report on the findings of the Action Research Project Women's Rights and Access to Water and Sanitation in Asian Cities*. Retrieved from http://www.idrc.ca/Documents/105524-Gender-and-Essential-Services-in-Low-Income-Communities-Final-Technical-Report.pdf

Wittmer, J. (2014). *Environmental governance, urban change, and health: An investigation of informal recyclers' perspectives on well-being in Vancouver, BC*. MA thesis, University of Guelph. Retrieved from https://dspace.lib.uoguelph.ca/xmlui/handle/10214/8288

World Atlas (2015). Where is Accra, Ghana? Retrieved from http://www.worldatlas.com/af/gh/aa/where-is-accra.html

World Bank, MONRE, CIDA (2004). *Vietnam environmental monitor 2004: Solid waste*. Retrieved from http://www-wds.worldbank.org/external/default/WDSContentServer/WDSP/IB/2005/07/28/000012009_20050728112421/Rendered/PDF/331510rev0PAPER0VN0Env0Monitor02004.pdf

25 Global Convergence and Divergence in Urban Transportation

Craig Townsend

Introduction

Cities are concentrations of people and infrastructure that together enable many types of social interactions, including shopping, socializing, studying, and working. **Infrastructure networks**, including physical "bundles" of transport, telecommunications, energy transmission, water, and wastewater, are components of most human settlements, but vary in quality and quantity and in terms of access (Graham and Marvin, 2001). Transportation infrastructure, such as freeways, container ports, airports, and railway stations, links each city with territories beyond its boundaries. In addition to this inter-urban transport infrastructure, cities and their adjacent suburbs are traversed by infrastructure networks that include everything from roads and subways to aerial cable cars, such as the Roosevelt Island Tramway in New York City or Metrocable in Medellín, Colombia. All of these form the base of **transport systems** used by metropolitan residents on a daily basis. Transport systems include locations or "nodes" where movements and people come together, infrastructure networks that make movement possible, and demand for movement resulting from people (or things) being needed or desired in another place (Rodrigue et al., 2009). A growing stock of urban transport infrastructure has enabled the expansion of urban agglomerations or metropolitan areas to sizes of territories and populations that are larger than ever before; during the last 25 years, 16 urban agglomerations in the global South have grown to more than 20 million inhabitants (UN, 2014).

The levels of concentration and types of people and activities vary greatly within and between metropolitan areas, as do the amount and types of transportation infrastructure. For most of human history, people walked, rode animals, or rode in carriages pulled by other humans or domesticated animals (Schaeffer and Sclar, 1975). The invention of steam-powered railways in the early 1800s in England, as well as further technological innovations that led to electric railways and automobiles with internal combustion engines in Germany and the United States in the late 1800s, enabled movement in vehicles travelling at higher speeds. During the twentieth century, the use of automobiles and the building of cities compatible with automobile use grew in the global North, particularly in the United States. In the early 2000s in the global South, the use of walking for urban transport remained high, but other forms of transport technologies were being adopted at very rapid rates in many places. In 2009 there were more than 1 billion motor vehicles (including cars, trucks, buses, and motorcycles) in the world, and projections were for a doubling to 2 billion by the year 2020 (Sperling and Gordon, 2009).

This chapter describes some general features of urban transport systems, with an emphasis on conditions in the global South. Because automobile use has slowed in much of the global North while increasing rapidly in parts of the global South, experiences may be converging. However, the fast pace of transport system **motorization** (which often includes extraordinarily high motorcycle use), the high human health impacts of motorization, the expansion of informal transport as a gap filler, and the flows of foreign investment in transport infrastructure may be new phenomena, and they are described in the sections that follow. This chapter details some of the diversity of experiences with the aid of case studies from Tokyo and Singapore. The Tokyo case study focuses on different modes of transport provision in a global city in which the state has played a significant role in decision-making, resulting in high levels of public transit investment and use. Singapore has also witnessed significant transformation of and government investment in its urban transport system, which is now recognized as one of the most technologically advanced in the world. Significantly less public or private investment, however, has been made in public transportation in Canadian cities, resulting in a high modal share for the automobile. A further contrast between Canadian cities and those in the global South is the scale and persistence of informal transport in the latter, especially the growth in use of motorcycles, trends without parallel in Canada.

Rapid Motorization

The prevalence of urban transport technologies varies between metropolitan areas and has changed over time. Within metropolitan areas of the global North, walking and transit are commonplace in central areas of cities (also referred to as downtowns, Central Business Districts, or inner cities) in comparison with outer suburbs, where automobiles dominate. Measured as metropolitan averages, automobile use now varies from a high of 97 per cent (an average share of total distance travelled for all daily trips) in sprawling Denver to 54 per cent in compact Vienna (Kenworthy, 2014). Canada's Vancouver, which like Denver and Vienna has a population of about 2.5 million metropolitan residents, is similar to the former, with approximately 89 per cent of the average daily distance travelled by car. However, transit use per capita in Vancouver has been slowly but steadily rising since the year 2000 and is now approaching levels found in the larger and more transit-oriented Montreal and Toronto metropolitan areas.

Among large metropolitan areas in the global North, widespread availability of automobiles and public transit offers possibilities for significantly higher mobility. People make most trips using these transport modes, which tend to vary inversely: when the use of one is high, the use of the other is low. **Public transit** (also referred to as public transportation, or just transit) refers to vehicles that offer regularly scheduled journeys that are open to all passengers (usually in exchange for a fare) and usually carrying multiple passengers whose trips may have different origins, destinations, and purposes (Walker, 2012). Public transit fleets typically comprise standard buses of 10–12 (33–39 feet) metres and 35–45 seats that are powered by diesel engines, or railway carriages propelled by electricity or pulled by diesel locomotives. In the global North, use of transit almost always ranks second to automobile use. In exceptionally low density US metropolitan areas such as Birmingham, Alabama, or Oklahoma City, the share of transit use for the trip to work is less than 1 per cent, and the use of walking or bicycling amounts to less than 2 per cent. Among Canada's large metropolitan areas in 2011, the average **modal share** of transit for the trip to work varied from a low of 11 per cent in Edmonton to a high of 23 per cent in Toronto (see text box for more discussion). High levels of urban mobility in Canada are made possible by extensive infrastructure networks.

In contrast, urban transportation systems in the global South lack both the spaces dedicated to private motor vehicles or public transit and depend far more on muscle power than on fuel-burning engines; people use their feet to walk or to pedal bicycles for most daily trips. These non-motorized or "active" modes formerly prevailed in the global North but were supplanted over time by automobiles, buses, and trains. While the modal shares characterizing urban transport systems have been highly stable in most cities of the global North for almost half a century, rapid motorization (the shift from walking and pedaling to the use of cars and motorcycles) has been occurring in many cities of the global South. The process has been particularly pronounced in countries experiencing growth in household incomes. For example, in Shanghai, between 1995 and 2009, bicycling declined from 40 per cent to 14 per cent of all trips, while private motor vehicle use climbed from 8 per cent to 20 per cent (Zhao et al., 2013).

As in the global North, the increasing use of automobiles and motorcycles in the global South has been spurred by rising household incomes and falling costs of motor vehicle production (Axhausen, 2008). City planning and government investments in transport infrastructure have also influenced change. Until the 1980s, cities in the former Union of Soviet Socialist Republics (USSR) and the Eastern Bloc states had lacked mass automobile ownership and relied heavily on bicycling, walking, and railways, including street railways that had been removed from most cities in the capitalist world in the 1940s and 1950s. Since the implementation of pro-market reforms and economic liberalization during the latter decades of the twentieth century in much of Asia and Europe, global motorization accelerated (Acharya and Morichi, 2007). This massive growth in motor vehicle use had major environmental impacts, and some researchers advocated mitigation of these impacts through changing car designs and transport fuels (Sperling and Gordon, 2009), while others emphasized accommodating a greater share of future travel on public transit (Grescoe, 2012).

In addition to expanding roads to accommodate more privately owned motor vehicles, governments and private firms have upgraded existing transit and constructed new rail rapid-transit infrastructure in global South cities, including Bangkok (Thailand), Beijing and Shanghai (China), Delhi and Mumbai (India), Dubai (UAE), Mexico City, and São Paulo (Brazil). In many of these cities and in some bus-only cities, such as Bogota (Colombia), Havana (Cuba), and Ho Chi Minh City (Vietnam), governments have invested in new buses and created bus lanes that enable higher operating speeds. However, in cities where the transport system has undergone rapid motorization, infrastructure expansion has still lagged demands for the use of that infrastructure, resulting in conditions of severe crowding on public transit and traffic congestion on roads. Long and unpredictable traffic jams have occurred throughout the day and even during the night in many cities of the global South. Researchers have characterized cities in the global North as "automobile dependent" (Newman and Kenworthy, 1999) and cities in the global South as "saturated" with automobiles, buses, and other motorized vehicles (Kenworthy and Townsend, 2007).

Under these circumstances, many households purchase cheap motorcycles, which run on noisy two-stroke engines burning a mixture of oil and gasoline. These motorized two-wheelers provide affordable mobility that enables greater accessibility, particularly to households that have limited capacity to purchase automobiles. Most of the metropolitan areas with extraordinarily high shares of motorcycle use are in South and Southeast Asia. Some notable examples include Ahemdabad and Pune in India and Hanoi and Ho Chi Minh City in Vietnam; in 2004 in the latter city, an exceptionally high 66 per cent of trips were made by motorcycle and 16 per cent by walking (JICA, 2004). In general, the growth in motorcycle ownership and use has outpaced car ownership and use, raising the future possibility of high motorcycle use being "built in" to the form and functioning of these cities in a way not seen before (EMBARQ, 2014). This raises concerns because while motorcycles provide mobility, they also impose substantial social costs such as human health problems.

Human Health Impacts of Automobile and Motorcycle Use

While the mobility provided by motor vehicles and public transit facilitates increased social interaction and wealth generation, it also incurs negative side effects, including harm to human health. The most immediate of these harmful effects result from exposure to air pollution produced by motor vehicles and collisions or crashes involving fast-moving motorized vehicles. The combustion of fuels for transportation results in **air pollution**, defined as air-borne substances at concentrations, durations, and frequencies that adversely affect human health or the environment (Murray and McGranahan, 2003). The most common pollutants include particulates, carbon monoxide, sulphur oxides, ozone and other photochemical oxidants (often referred to as "smog"), nitrogen oxides, toxic compounds, and a variety of volatile organic compounds. Extensive research has demonstrated clear links between these pollutants and adverse effects on human health.

For instance, in Boston, where exposure to air pollution produced mainly by automobiles is not particularly intense, medical researchers linked particulates with reduced lung functioning that led to premature deaths of elderly people (Lepeule et al., 2014). In the global North, most air pollution is produced by cars, but human exposure levels are reasonably low because the population is relatively dispersed. However, there are still always some instances, under these conditions, when central city residents adjacent to major roadways experience high levels of exposure to pollutants. This seems inequitable when those populations are socio-economically disadvantaged, have few choices of residential location, and also drive relatively little or not at all (Sider et al., 2015). Governments in the global North identify critical levels of air contaminants and sometimes take actions to legislate or formulate public policies that may decrease exposure. In the cities of the global South, and particularly middle-income cities, exposure to air pollution produced by combustion of fuels for electricity generation, transportation, and industrial processes can be particularly severe.

In addition to the negative impacts resulting from air pollution, motorization exacts a heavy human health toll from collisions and crashes. Deaths and injuries result immediately when heavy vehicles moving at high speeds collide with other objects, including vehicles or pedestrians. These impacts are found wherever there are motorized vehicles, but they are particularly intense in the global South. Impacts from collisions increase with the quantity and speed of travel, but are reduced in highly motorized cities of the global North to varying extents by actions such as discouraging people from operating vehicles while under the effects of alcohol and regulating speed in areas where there is potential for interaction with slower and more vulnerable road users. According to the United Nations' World Health Organization (WHO), legislation that can mitigate risk factors (speed, drunk driving, motorcycle helmets, seatbelts, and child restraints) is lacking in many countries experiencing rapid motorization.

Collisions between automobiles, motorcycles, and people accounted for more than 1 million deaths in 2013 (WHO, 2013). Most of these deaths (90 per cent) occurred in the global South and involved pedestrians being hit by automobiles and motorcycles; as a result, the WHO (2013) has recommended protecting pedestrians and cyclists from high-speed traffic. Rates of deaths and injuries are particularly high in countries with high rates of motorcycle use. Southeast Asian and South Asian governments reported 75,331 fatalities of riders of two- and three-wheeled vehicles in 2007, and this excluded pedestrian fatalities resulting from collisions with motorized two- and three-wheeled vehicles (WHO, 2009). Motorized two-wheeled vehicles require much less space than automobiles and are an adaptive response to a lack of road infrastructure, but their manoeuvrability then often infringes upon pedestrians (Figure 25.1). Thus, while motorcycles provide mobility benefits to individuals or households, they also impose large social costs; as well, they degrade the quality of streets and sidewalks, which are places used by low-income households living in small and crowded dwellings to socialize and carry out their livelihoods.

Figure 25.1 Motorcycle-covered sidewalk in Ho Chi Minh City, Vietnam, 2004

Informal Transport

By walking or pedaling more, people increase their mobility. However, these non-motorized modes require human effort that could be used for other purposes, such as working for a wage or caring for other household members, and there are limits to the distances that can be travelled. If automobiles, roads, and public transit are available and affordable, demands for more mobility can be met without significant additional physical exertion. However, in much of the global South, transport networks are highly insufficient. In response, many households pool resources to purchase motorcycles, but even these cannot serve all of their needs because it is difficult and dangerous to carry more than one person or bags and packages on a small motorcycle, although it is common in some places to see large amounts of goods and several people being transported by a single motorcycle.

In places where small-scale profit-seeking enterprise is possible, individual entrepreneurs purchase vehicles (usually small buses of 5–7 metres [16–23 feet] and 12–20 seats), although occasionally cars or even motorized two- and three-wheelers, and sell rides to destinations for fares usually negotiated between the drivers and passengers. Most owners only operate one vehicle, which they also drive. Because the owner-operators of these services, which do not completely meet the definition of "transit," lack official sanction or a license to operate the service (or in some cases, even the license to drive), the mode is referred to as **informal transport** (Cervero and Golub,

2007). Usually they lack fixed schedules and prices and operate in contravention of laws regulating the use of motor vehicles and running businesses, so they are actually illegal. These "in-between" modes of transport go by various local names, such as "tap-tap" (converted vans and pickup trucks) in Haiti, "tuk-tuk" (motorized three-wheeled vehicles) in Thailand, "jeepney" (converted military jeeps) in the Philippines, and "baby taxis" (motorized three-wheel vehicles) in Bangladesh (Figure 25.2). The vehicles are usually old and in poor condition because of a lack of money to invest in better vehicles or maintenance.

In cities where the state does not provide much in terms of public goods such as roads or transit, the informal transport system can fill gaps not covered by the formal public transit system or can offer lower waiting times and higher travel speeds. In some cases they "poach" customers from the formal public transit system by offering a more comfortable, less-crowded, and more direct route with fewer stops for a slightly higher fee. In other cases the relationship is more complementary. For instance, in Bangkok and Mexico City, informal transport operators serve streets where passing is difficult for full-size buses, and they pick up and drop off passengers at stations on urban rail rapid-transit lines and suburban areas where the formal bus routes have not been extended. In Mexico City, these services are provided by mini-buses called *collectivos*, while in Bangkok they are provided by mini-buses and approximately 100,000 motorcycle taxis offering rides to passengers who sit behind the drivers. These informal transit services provide employment (mainly to men) who have migrated from rural areas and lack the formal education or skills to participate in manufacturing or service jobs requiring higher levels of literacy or numeracy. In other cities, such as Lagos, Nigeria, the entire public transit system is actually informal, with 75,000 old and polluting buses owned and managed by a large number of individual entrepreneurs (World Bank, 2011).

While it provides benefits, there are also social costs associated with informal transport. Many park their vehicles and wait for passengers on crowded streets and contribute to increased congestion and difficulties for people selling, socializing, or walking on streets. And because those vehicles offer only low to medium passenger capacity and are in poor operating condition, they contribute much higher levels of air pollution per passenger carried than do large buses, which burn less fuel per passenger carried (provided that passenger loading is moderate to high). The lack of regulation can also lead to dangerous driving practices, such as unlicensed drivers racing to pick up or deliver passengers or taking stimulants to stay awake in order to work longer hours.

Informal transport is not completely unknown in the global North. In Canada and the United States between 1914 and 1929, large sedans referred to as "jitneys" (in the early 1900s "jitney" was slang for a five-cent coin, which was a typical fare) competed illegally to carry passengers who otherwise would have used privately owned and operated street railways (Davis, 1989). They were eventually banned, and over time public authorities took over the provision of buses and railways. Ad hoc taxi services like Uber that have emerged in the 2010s may represent the return of informal transport to the global North. The long-term implications of these emergent modes of urban transport are still unclear, and there is disagreement concerning whether they will contribute to substantial shifts in either automobile or transit use.

To some neoliberal critics of public provision of transit, the public takeover and operation of trains and buses was a mistake that should be avoided in the global South. The Washington, DC–based World Bank, which since the 1980s has been promoting liberalization and privatization in the global South where it makes loans for infrastructure, has sought to limit government involvement to regulating the private operators of informal transport vehicles or encouraging their replacement

Figure 25.2 Motorized three-wheel taxis in Bangkok, Thailand, 2011

by competing firms providing bus services. The World Bank has often counselled global South governments against investing in urban passenger railway systems (including subways or "metros"), which require large capital expenditures.

Foreign Investment in Urban Transport Infrastructure

Investment in transport infrastructure by governments or private enterprises from outside a nation-state has a long history. During the European colonial period from the 1500s until the mid-1900s, Europe's governments and companies invested in transport infrastructure such as roads and railways in colonies in order to facilitate resource extraction and establish settlements where market exchanges could take place. They also provided consumer markets for the export of products from colonized territories. For example, Montreal was founded as a trading post in New France in 1642, and it later became an important city largely because it offered a port that could accommodate the largest ships at the time. When it was part of the British Empire, Montreal served as a terminus for numerous railway lines. Goods were transferred between ships on the St Lawrence River and railways, including one that crossed the Dominion of Canada to the Pacific Ocean. In the early 1900s, "[m]uch was made of the British Empire's connections through Canada" as part of a British-controlled route from Britain to Australia (Hayes, 2010: 107). In postcolonial times, governments and private firms have continued to invest in other places according to their commercial and political interests, although they are dealing with sovereign governments and firms based within those independent nation-states.

In addition, there are now supra-national institutions that provide loans and grants of capital and expertise to governments in the global South. Following the Second World War, the United Nations created the World Bank in order to invest in transport and other infrastructure in war-torn Europe. Beginning in the early 1970s, the World Bank shifted towards providing large loans to governments in the global South (Goldman, 2005). The selection of places where the World Bank loaned capital reflected the global geopolitical interests of the US, and the types of projects were aligned with the politics of infrastructure in the US. The World Bank (sometimes in conjunction with the American government) has provided the needed expertise for planning and engineering the construction of national highways for passenger and truck transport and building railways for carrying freight. In terms of public transport, one of the World Bank's main policies since the early 1970s has been to encourage the creation of bus-only lanes and the upgrading of bus services by private companies. With few exceptions, the World Bank has avoided making loans for the construction of urban rail rapid-transit infrastructure.

However, the US government and the World Bank are not the only powerful transnational actors in planning and constructing urban transport infrastructure. The US and the World Bank face competition in providing loans to governments in the global South, in particular from Japan, and increasingly, from the People's Republic of China (Krauss and Bradsher, 2015). Following periods of rapid economic expansion in Japan (from the late 1950s until the early 1990s) and China (from the early 2000s until the present), the governments of those countries have made loans and investments for urban transport infrastructure in the global South. The governments of Japan and China are more directly involved than the US government through tight coordination between large private corporations and government agencies in Japan or through direct state ownership of enterprises in China. They provide training, expertise, and financial loans, some of which may be provided in exchanges for other resources, such as access to natural resources or manufacturing and construction contracts for firms. The conduits for these funds influence the types of systems that get built.

The Chinese government is increasingly providing funding for transport and other infrastructure, particularly in Asia and Africa. These investments are tied to the procurement of equipment and expertise from firms in China and have been directed towards countries that possess natural resources (including crude oil) that are needed as inputs for China's industrial economy. In terms of urban transport equipment, buses manufactured in China can be found throughout the global South and particularly in cities such as Havana Cuba, and Tehran, Iran, where companies of the US or its allies are prohibited from doing business. Until recently, China's state-owned firms have not invested in public transit infrastructure (other than buses), although this changed recently in Addis Ababa (Ethiopia) where a 34-kilometre

(21-mile) light rail line was built by state-owned and state-linked firms from China (*Railway Gazette*, 2015). In many ways, the recent actions of China and state-linked companies resemble those undertaken by Japan in the 1950s. During that period, investors from Japan went to countries in East and Southeast Asia to extract raw materials for the home market; then, in the 1960s and 1970s, investors shifted labour- and capital-intensive industries such as textiles, chemicals, and steel into those countries (Hatch and Yamamura, 1996). In the 1980s, firms from Japan began using countries in Southeast Asia as a platform for export-oriented production, in some ways like the US-based firms that manufactured automobiles in Canada or other consumer goods in Mexico. In the latter cases, much of the production was for consumption in the US, while in the Southeast Asian example much of the production was for export to "third" countries in the global North. Much of this investment was coordinated from Tokyo, the world's most populous urban agglomeration.

A Railway-Oriented Influence on the Global South: Tokyo

With 33 million inhabitants, equivalent to the population of Canada, Tokyo comprises a well-defined area called the Tokyo Metropolis, home to 13 million residents, which is governed by the Tokyo Metropolitan Government (Figure 25.3). The larger Tokyo region has many definitions, including the Tokyo Major Metropolitan Area; however, this larger region lacks a unified government authority or planning agency (Okata and Murayama,

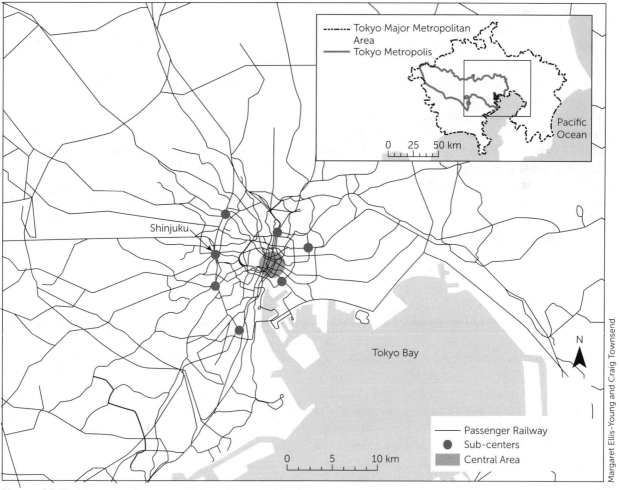

Figure 25.3 Map of railways in Tokyo Metropolis, Japan

2010). Tokyo's population grew very rapidly between the 1920s and the 1980s (although growth was interrupted during the Second World War), as Japan experienced rapid rural–urban transition and industrialization. By the 1980s, Tokyo, along with London, New York, and Paris, was one of the world's top-tier global cities. Tokyo is the headquarters for some of the world's most powerful financial and specialized service firms, as well as some of the largest automobile and railway manufacturers. The government and business activity of Japan (population 127 million in 2013) is highly concentrated in Tokyo, which is a "center of command functions" in Japan's national economy, and the national government, which is based in Tokyo, "has played a crucial role in developing Tokyo's capabilities as a global city" (Sassen, 2001: 163).

In comparison with that of London, New York, and Paris, Tokyo's urban transport system stands out for having a higher share of public transit use, and of that transit use, the share of railway use is higher than that of bus use (Focas, 1998). Considering all modes across the wider metropolitan area, the use of transport in Tokyo is balanced between around one-third non-motorized transport (including a significant share for bicycles), one-third transit (most by railway), and one-third by private motor vehicles (Kumagai, 2000). Within the Tokyo Metropolis, railways dominate the modal shares with 50 per cent of all trips (Chorus and Bertolini, 2011). Commuting between home and workplace in Tokyo is highly oriented towards a central area and a number of sub-centres, many of which are at points of intersection between radial suburban railway lines and a circular inner railway called the Yamanote loop (Figure 25.3). Within the loop there is an extensive subway system operated by two entities: the Tokyo Metropolitan Government and a private company. The view in Figure 25.4 is from the Tokyo Metropolitan Government headquarters located at Shinjuku, an important sub-centre.

Tokyo also has freeways designed to enable rapid car movement through the metropolitan area. Many of these freeways, sometimes referred to as expressways, are elevated through highly built up areas, although the overall level is much lower than in metropolitan Paris and New York, and comparable with London. In comparison with the other three global cities, the **population density** (the amount of people residing on a unit of land area) is the highest overall and much higher in the suburbs, although New York and Paris have higher-density

Figure 25.4 Dense Tokyo suburbs, view facing southeast from Shinjuku, 2013

central areas than Tokyo (Focas, 1998). This dense suburban concentration of mixed-use development followed railway construction without much government regulation until the 1960s, when higher levels of land-use regulation were introduced and public corporations began to build new residential areas along railway lines (Okata and Murayama, 2010). In terms of urban transport and land use, Tokyo lacks the public provision of extensive street (and even sidewalk) networks found in suburbs, and the land uses are more mixed and on small plots of land.

The transport and land-use system of Tokyo has been produced by a higher level of private-sector activity and less government regulation than in London, New York, and Paris; however, these have also been influenced by a higher degree of intervention by the national government than in those other three global cities and by most other cities in the global North (Sorensen, 2002). Cervero (1998: 181) describes Tokyo as an "entrepreneurial transit metropolis" in which railway conglomerates make money by building railways together with new communities. However, this private activity involves close coordination between higher levels of government. The pattern of public-private relations in Tokyo's urban transport and land-use system resemble those in Japan more generally and have been described as different from those in liberal democracies or welfare states. Japan has been referred to as a "capitalist developmental state" in which "the state's role in the economy is shared with the private sector, and both the public and private sectors have perfected the means to make the market work for developmental goals" (Johnson, 1982: viii).

Because Tokyo's transport system uses much fewer (per person and per unit of economic output) fossil fuels and produces fewer emissions than those in many cities in the global North, some environmentalists and planners have studied it as a case of a sustainable transport system. Like the inner parts of London, Paris, and New York, Tokyo's built environment is transit oriented, defined as "moderate and high-density housing, along with complementary public uses, jobs, retail and services, . . . concentrated in mixed-use developments at strategic points along the regional transit system" (Calthorpe, 1993: 41). Indeed, Tokyo provides an example of **transit-oriented development** writ large.

In addition to admirers from the global North, people from the global South have looked at Tokyo as an example of possibilities that could be realized, particularly in places where land and crude oil are in short supply. The use of Tokyo as an example that could be emulated has been facilitated by the Japanese government, which provides overseas development assistance in the form of grants and loans of experts, technology, and money to countries in the global South for urban transport. In particular, the government devotes substantial resources to support planning and construction of urban transport infrastructure in parts of the global South where private firms based in Japan are likely to benefit from higher demand for urban transport equipment, construction equipment and services, as well as in offshore manufacturing and service industries. A large number of the automobiles produced by Japanese companies for export to markets around the world are now manufactured in Southeast Asia, where firms from Japan have established a network of industrial clusters (Hatch and Yamamura, 1996).

One example of the relationship between Japan's government and private-sector involvement in the global South can be found in Bangkok, Thailand's capital, which had well over 10 million inhabitants by the early 1990s, making it one of Southeast Asia's largest metropolitan areas. Firms from Japan have exported goods to Thailand, established regional offices, and operated manufacturing plants in Bangkok for decades. By the late 1980s Bangkok was notorious for high levels of traffic congestion and air pollution resulting from rapid motorization, a lack of infrastructure (particularly public transit), and a lack of coordinated government responses (Daniere, 1995). In the late 1980s the Japan International Cooperation Agency (JICA), a government agency providing assistance to governments in the global South, studied and prepared a plan for improvements to road transport in Bangkok, following this with a study of railway improvements in the early 1990s. In the late 1990s another organization, the government-linked Japan Railway Technical Service, participated in the preparation of a rail mass-transit master plan for Bangkok, an urban redevelopment master plan prepared by JICA, and then training programs organized by JICA in the early 2010s. In the early 2000s construction companies from Japan worked with construction companies from Thailand to build a 20-kilometre (12-mile) subway line, partially financed with a low-interest loan from Japan, that began operations in 2004 (Figure 25.5). A more recent project involved the financing and construction by firms from Japan (in joint ventures with Bangkok-based firms) of a 23-kilometre (14-mile) elevated rail rapid-transit line serving Bangkok's northern suburbs that began operations in 2016 (Reuters, 2007). For the first time in Thailand, this project also involved the use of railway cars designed and manufactured in Yokohama, which is part of the Tokyo metropolitan region.

While Bangkok's urban transport system retains a level of informal transport and other characteristics that make it distinct from Tokyo, ongoing changes resemble those experienced in Tokyo decades ago. Similar processes are underway in other cities of the global South;

Figure 25.5 Subway station exit (top left), Bangkok, Thailand, 2011

public and private institutions from Japan have participated in the rapid construction of an urban rail rapid-transit system in New Delhi (DMRC, 2015), and an initial urban rail rapid-transit system began construction in Jakarta in 2013 (Otto, 2014). Singapore constitutes another example in which rapid developments in transportation have been made over the last half-century.

Transformation of Transportation: Singapore

Singapore is a city-state of about 5.5 million residents (1.6 million of whom are temporary foreign workers) located on an island nestled between Indonesia and Malaysia (Figure 25.6). In the early 1970s, this former British colony had an urban transport system that was unremarkable and displayed the same types of problems (e.g., insufficient infrastructure, unreliable informal transport, pollution from motor vehicles, and road congestion) as were found throughout the global South. However, since that time, Singapore's urban transport system has been transformed into one of the most technologically advanced, innovative, and productive systems in the world. Singapore has become known for a high quality of public transit, integrated with urban development, charging high prices for road use and vehicle ownership (through information technology) to minimize traffic congestion, and the coordination of the transport system by one government agency (May, 2004). That this has been accomplished in less than

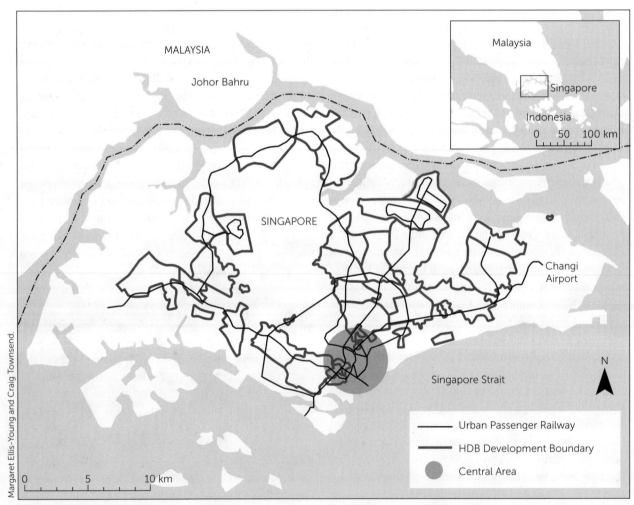

Figure 25.6 Map of Singapore's urban passenger railways and HDB public housing

50 years makes it even more remarkable. Singapore has become a showcase city, and the government's Land Transport Authority even has a permanent exhibition to educate curious visitors seeking to learn about the transport system.

Singapore's urban transport changes have been intertwined with the larger transformation of the settlement from a small colonial outpost to a medium-sized, wealthy (but highly unequal) metropolis, a hub for air travel and shipping, and a global financial centre. In 1965 Singapore's gross domestic product (GDP) per capita was lower than that of Mexico or Venezuela, which were then, and still are, middle-income countries in the global South. By 2015 Singapore's GDP per capita was equal to that of Canada, one of the world's highest-income countries. Singapore's rapid ascendance into the global ranks has been made possible by urban transport system improvements that make it highly productive and accessible for multinational firms. In comparison with other Southeast Asian cities like Bangkok and Jakarta, Singapore's system works exceptionally well and offers high accessibility to opportunities with low negative impacts. According to the WHO (2013), Singapore in 2009 had 3.8 road traffic fatalities per 10,000 persons, in comparison with 6.5 per 10,000 persons in Canada in 2010, and 19.9 per 10,000 persons in Thailand (WHO, 2013). While Singapore sometimes experiences high levels of air pollution, it is most frequently the result of activities relating to industrial production, power generation, and forest fires occurring in neighbouring countries.

Singapore's transformation was orchestrated by a national government controlled by one political party. Since national independence in the 1960s until the present, this party has permitted little organized opposition or dissent. Furthermore, Singapore's government dominates land, property, and economy to an extent "that has few parallels elsewhere in the world" (Shatkin, 2014: 116). As a result, the government has been able to enact policies that have been proposed but are typically not implemented in other cities in the global South or the global North. In many cities, governments face opposition from groups who would be made worse off by changes, but in Singapore an authoritarian state has pursued highly coordinated transport development strategies (including prohibitively high charges on the use of cars and motorcycles) as part of a strategy to transform the city-state. Singapore has become a model city to many people (Chua Beng, 2011), including environmentalists and planners seeking to encourage public housing coordinated with rail rapid transit, free-market economists seeking to implement real-time pricing of road use, and authoritarian leaders such as those of the People's Republic of China and the United Arab Emirates, where forms of rapid capitalist economic development are underway but governments are interested in harnessing forces for economic liberalization in ways that will support rather than undermine their political legitimacy.

Until the early 1970s, Singapore's bus services were largely private operations that had many of the characteristics of informal transport. Under the careful guidance of the government, the many small private operations were amalgamated into two private operations. In addition to consolidating bus operations, the government invested in rail rapid-transit infrastructure that connected the central business district, which houses national and multinational corporate offices and key employment locations (including one of the world's busiest container ports and one of the world's largest airports), with primarily residential areas called "new towns." These "towns" mainly comprise constellations of residential towers, planned and built by the Housing Development Board (HDB), a government enterprise that plans and builds affordable apartments that are then leased to citizens. Paradoxically for a high-income nation, over 80 per cent of the population lives in this housing (Lam and Toan, 2006). All of the HDB developments are served by railway infrastructure, which is continuously being expanded and which connects these housing areas with the central area and Changi Airport where many jobs are located (Figures 25.6 and 25.7).

Singapore also stands out for its system of electronically operated "congestion pricing" (referred to as Electronic Road Pricing; see Figure 25.8); variable tolls are charged for road use based on levels of traffic (Olszewski, 2007). The idea of reducing traffic congestion through charging variable prices for the use of roads was first formally proposed in the United Kingdom at the beginning of the 1960s. However, the system was not tried until the 1970s in Singapore, when the government charged cars for entering the central area during peak travel hours. Later, the number of new car registrations was managed through a quota system that required people to bid on a fixed number of vehicle registrations that had to be obtained before people could buy

Figure 25.7 View north towards Singapore's high-rise suburbs from the central area, 2009

Figure 25.8 Electronic road pricing gantry, Singapore, 2009

automobiles (Chin and Smith, 1997). More recently, the system of requiring payment for entry to the central area was expanded to cover much of the arterial road network and to use remote sensing equipment to electronically charge drivers. As a result of these actions, Singapore has a much higher level of transit use and a lower level of motor vehicle use than other Southeast Asian cities like Bangkok or Kuala Lumpur, which, though they have less productive economies and proportionately smaller middle classes, have higher levels of private motor vehicle ownership and use. To some observers, while the comparison is not particularly illuminating, since Singapore is a "contained" city that lacks the complexity of a larger country with multiple cities, non-urban urban areas, and rural hinterlands (Shatkin, 2014), it nonetheless illustrates possibilities and has become a reference point for discussions of urban transport.

The case of Singapore (along with the case of Tokyo) suggests that the particularities of urban transport systems, even among cities with high average incomes, can differ in significant ways. The text box illustrates some of the distinctive features of urban transport in Canada's cities; they differ from more transit-oriented metropolitan areas such as Singapore and Tokyo, as well as from global South cities where levels of motorized two-wheel, three-wheel, and informal transport are more prevalent. In addition, urban transport systems vary between cities and suburbs within Canada.

Impacts of Automobile Dependence and Globalization of Urban Transport Infrastructure Provision in Canada

Rapid motorization took place in Canada between the 1930s and 1970s and then grew slowly or declined. Continued growth in automobile share of travel predominated in the more dispersed and low-density sprawling suburbs, while the declines were most prevalent in the central cores and inner-city residential districts of Canada's largest metropolitan areas. Shares of transit were higher in cities that were larger and denser; the highest transit shares were in Canada's five largest metropolitan areas. However, transit shares were also relatively high (above 10 per cent of trips to work) in Halifax, Quebec City, Victoria, and Winnipeg, which are small to mid-sized metropolitan areas. Transit shares in some other small to mid-sized metropolitan areas, such as Kelowna, Moncton, Trois Rivières, and Windsor, were under 3.5 per cent. An overall leveling off or decline in driving to work has been observed in most countries of the post-industrial global North (Jones, 2014; Millard-Ball and Schipper, 2011). At present, the main issues facing Canada include the long-term impacts of a very high modal share for automobiles and increased foreign investment in Canada's urban transport infrastructure, along with increasing investment abroad by firms based in Canada.

Variation in levels of automobile use and in whether, or even how, they could be changed have been emotionally debated (Walks, 2015). Concerns over human health impacts of urban transport and high automobile use have grown in recent decades, even as levels of motor vehicle use have stabilized. Early concerns were with air pollution and collisions, which had immediate or short-term effects, as is the case now in the global South. However, other concerns have become the long-term impacts of motorization. These include high levels of fossil fuel use and the contribution this makes to human-induced climate change. Most Canadians enjoy a high quality of life associated with high levels of accessibility to people and activities. Many Canadians worry that altering urban transport systems may disadvantage segments of the population. In addition, many Canadians express dissatisfaction with fluctuating gasoline prices, increases in fees and taxes on automobile or transit use, and the implementation of user fees on new road infrastructure.

An emerging contentious issue confronting planners and politicians is what to do with large elevated freeway structures built in the 1950s and 1960s in the central areas and inner suburbs of Canada's cities. Many of these structures require replacement, and this raises the question of whether infrastructure that we now know exposes people to air pollution and produces greenhouse gas emissions should be rebuilt as before, removed without replacement, or replaced with ground-level infrastructure. In Montreal, for example, a massive elevated freeway interchange (the Turcot Interchange) is being reconstructed at ground level and will carry a similar number of cars as the present structure. In 2015 Toronto's mayor decided (apparently against the preferences of the city's planning staff for removal) to build a hybrid ground-level arterial road and elevated freeway to replace a section of the aging Gardiner Expressway.

An increasingly popular approach, one supported by many local governments, planners, and developers, is to build transit-oriented development around rail rapid-transit stations. This approach was followed, to an extent, following the opening of initial rail rapid-transit lines in Toronto (in 1954), Montreal (in 1966), and Vancouver (in 1986). In all three cities, intensification of commercial and residential development was supported in the central areas, while in Toronto and Vancouver there were also efforts to encourage transit-oriented development in suburban areas such as Scarborough and Burnaby. In Vancouver, restrictions on using agricultural land for suburban development have encouraged densification that began in the 1970s, particularly around rail rapid-transit (SkyTrain) stations. By the early 2010s virtually every local government in Canada had plans for transit-oriented development, along with improvements to transit. While the change to urban built environments and infrastructure is slow, the phenomenon of densification around transit (usually in the form of high-rise buildings) is increasingly visible in Canada and around the globe (Figure 25.9).

Figure 25.9 Transit-oriented development in suburban Vancouver, 2013

continued

Canada's history as a nation-state began with a cross-country railway investment by firms with strong links to Britain. More recently, Canada has become both a source and a site for foreign direct investment in urban transport infrastructure in an emerging global system with an increasingly multinational character. Some companies based in Canada have been highly active in selling transport hardware around the world, including in the global South. In addition, Canada has become a site for foreign companies to establish manufacturing for local markets and export; the province of Ontario in particular has been the site for offshore manufacturing of automobiles by foreign companies. Increasingly, foreign firms are active in building and financing transport infrastructure in Canada, as governments seek to avoid tax increases and shift to user-pay or deferring payments into the future. For example, two companies based in Madrid, Spain (ACCIONA and ACS), led a consortium that designed, built, and financed a 42-kilometre (26-mile) suburban "bypass" freeway through Montreal's southwestern suburbs, and the two firms also have a contract to operate the system for 30 years (ACCIONA, 2012). More recently, foreign firms (from Italy, Spain, and the US) have, together with Canada-based firms, designed, constructed, and in some cases financed (and will eventually operate and maintain) light rail transit projects in Kitchener-Waterloo, Ottawa, Toronto, and Vancouver. The Kitchener-Waterloo system, called "Ion," is being built by firms that were involved with a similar project in Gold Coast City, Australia, a few years prior. The Ion project consortium (called Grandlinq) includes Paris-based Keolis, a private firm that operates public transport in many cities in Europe. The involvement of companies such as Keolis in light rail projects is indicative of increasing transnationalization of investment in the urban transport sector. Such public-private arrangements have sometimes proven to be politically contentious, with segments of the population (including trade unions in particular) opposed to the private finance and operation of transport infrastructure. They suggest growth in the complexity and new patterns of foreign involvement in constructing urban transport infrastructure networks.

Conclusion

This chapter began by identifying four urban transport issues of global significance in the early twenty-first century. It suggested that populations in very different cities have experienced similar changes, but there may be some significant differences. Rapid motorization and the decline of muscle power in getting around cities in much of the global South represent convergence. This is evident in cities like Shanghai where walking and cycling have declined as automobile and transit use have increased. However, the experience of people in many cities in the global South may be different owing to the widespread diffusion of motorcycles and automobile use at relatively low average-income levels. The costs of private motorized transport are now much more affordable than at any time in human history. As a result, cities like Ahemdabad and Ho Chi Minh City have a level of motorcycle use that has never been seen before. And cities like Lagos, which have completely informal transport systems based around minibuses, also represent divergence. Informal transport is an adaptive response to a lack of public transit capacity and an example of people making technologies work, even when the public institutions are missing. The lack of these institutions in places where formal transit systems are virtually absent, exacerbates the negative impacts of those systems.

The governments of the world's three largest economies (China, Japan, and the US) have sought to invest or to steer investments towards different types of urban transport infrastructure in the global South. The example of Tokyo demonstrates that a high level of wealth and quality of life, together with a metropolitan population of exceptionally large size, is possible with lower levels of automobile use than found in cities like those in Canada. Singapore is a rather unique case that demonstrates the implementation of a number of transport policies (including congestion pricing, auctioning a restricted number of licences for new automobiles, and public provision of housing well-integrated with rail rapid transit).

The case studies, including the involvement of Japanese organizations based in Bangkok and other cities, suggest that there are new ways in which cities in the global North and South are linked through innovations in urban transport. The emergence of China as a source of funds in the global South, for example through the financing of a light rail line in Addis Ababa, one of the world's poorest cities, suggests that the future may not be like the past. Firms building transport infrastructure or equipment are increasingly globalized, and there is more diversity in the sources of equipment and infrastructure than in the past.

The growth of informal transport and the greater number of motorcycles in the global South, together with investments by foreign firms from a variety of countries, suggest that experiences in the twenty-first century may be substantially different from those in the global North in the twentieth century. While it is unlikely that many cities in the world will reach the extraordinarily high level of automobile use found in most cities in the US and to a lesser extent in Canada, the potential for future growth in travel by motor vehicles (and transit) is huge. The short- and long-term ramifications for human health will also be huge.

Key Points

- Urban transportation has been changing rapidly since the Industrial Revolution, and accelerated change has been occurring in the global South over recent decades.
- Automobiles are a dominant mode of urban transportation in the global North, and rapid motorization of urban transport systems is currently underway in the global South.
- As a result of rapid motorization, human health impacts of automobile and motorcycle use are on the rise globally; short-term impacts include illnesses caused by air pollution and injuries resulting from collisions.
- Informal transport is prevalent in cities of the global South where government provision of public transit is insufficient to meet needs.
- Governments of the countries with the largest economies (particularly the US, Japan, and China) seek to provide assistance through investment in urban transport infrastructure in the global South.
- Tokyo's transport and land use are railway oriented and based on close interaction between public and private sectors; the Japanese government has encouraged governments in the global South to pursue similar policies, in part through loans for railway projects.
- Singapore's transport system has been transformed in a relatively short time from one typical of those in the global South to one that is considered a good model by people elsewhere for a variety of reasons.
- Canada's urban transport systems are characterized by very high modal shares for the automobile, and transit-oriented development has been pursued by governments as a means of reducing or mitigating the negative impacts.

Activities and Questions for Review

1. In pairs, discuss how you commonly travel between your home and the university. How typical is your mode of transport relative to other journeys to work in (a) your household and (b) your neighbourhood?
2. In your province or territory's largest city, what does the physical evidence (e.g., infrastructure, street layouts, building design, and railway tracks) of past urban transport systems reveal about the relationship between changing transportation technologies and urban form?
3. In your own words, explain to the classmate beside you what informal transport is. When and in what forms has informal transport existed in Canadian cities?
4. How has the sharing economy (e.g., Uber or car-sharing programs like Zip Car, car2go, and AutoShare) affected transportation use in Canadian cities?

References

ACCIONA (2012). ACCIONA and ACS inaugurate Montreal's A-30. Retrieved from http://www.acciona.com/news/acciona-and-acs-inaugurate-montreals-a-30

Acharya, S.R., and Morichi, S. (2007). Motorization and role of mass rapid transit in East Asian megacities. *IATSS Research*, 31(2): 6–16.

Axhausen, K. (2008). Accessibility: Long-term perspectives. *Journal of Transport and Land Use*, 1(2): 5–22.

Calthorpe, P. (1993). *The next American metropolis: Ecology, community, and the American dream*. New York: Princeton Architectural Press.

Cervero, R. (1998). *The transit metropolis: A global inquiry*. Washington, DC: Island Press.

Cervero, R., and Golub, A. (2007). Informal transport: A global perspective. *Transport Policy*, 14(6): 445–457.

Chin, A., and Smith, P. (1997). Automobile ownership and government policy: The economics of Singapore's vehicle quota scheme. *Transportation Research Part A: Policy and Practice*, 31(2): 129–140.

Chorus, P., and Bertolini, L. (2011). An application of the node place model to explore the spatial development dynamics of station areas in Tokyo. *Journal of Transport and Land Use*, 4(1): 45–58.

Chua Beng, H. (2011). Singapore as model: Planning innovations, knowledge experts. In A. Roy and A. Ong, *Worlding cities: Asian experiments and the art of being global*. Chichester, UK: Wiley-Blackwell, pp. 29–54.

Daniere, A.G. (1995). Transportation planning and implementation in cities of the Third World: The case of Bangkok. *Environment and Planning C: Government and Policy*, 13: 25–45.

Davis, D. (1989). Competition's moment: The jitney-bus and corporate capitalism in the Canadian city, 1914–29. *Urban History Review/Revue d'histoire urbaine*, 18(2): 103–122.

DMRC (Delhi Metro Rail Corporation Ltd.) (2015). *Annual report 2014–2015*. New Delhi: DMRC.

EMBARQ (2014). Working paper: Motorized two-wheelers in Indian Cities: A case study of the City of Pune. Retrieved from http://spotidoc.com/doc/123032/motorized-two-wheelers-in-indian-cities

Focas, C. (Ed.) (1998). *The four world cities transport study*. London, UK: London Research Centre.

Goldman, M. (2005). *Imperial nature: The World Bank and struggles for social justice in the age of globalization*. New Haven: Yale University Press.

Graham, S., and Marvin, S. (2001). *Splintering urbanism: Networked infrastructures, technological mobilities and the urban condition*. New York: Routledge.

Grescoe, T. (2012). *Straphanger: Saving our cities and ourselves from the automobile*. Toronto: HarperCollins Publishers.

Hatch, W., and Yamamura, K. (1996). *Asia in Japan's embrace: Building a regional production alliance*. Cambridge: Cambridge University Press.

Hayes, D. (2010). *Historical atlas of the North American railroad*. Vancouver: Douglas & McIntyre.

JICA. (2004). *The Study on Urban Transport Master Plan and Feasibility Study in Ho Chi Minh Metropolitan Area (houtrans): Final report*. Vol. 1: *Summary*. Japan International Cooperation Agency (JICA), Ministry of Transport, Socialist Republic of Vietnam (MOT), Ho Chi Minh City People's Committee (HCMC-PC).

Johnson, C. (1982). *MITI and the Japanese miracle: The growth of industrial policy, 1925–1975*. Stanford, CA: Stanford University Press.

Jones, P. (2014). The evolution of urban mobility: The interplay of academic and policy perspectives. *IATSS Research*, 38(1): 7–13.

Kenworthy, J. (2014). Total daily mobility patterns and their policy implications for forty-three global cities in 1995 and 2005. *World Transport Policy and Practice*, 20(1): 41–55.

Kenworthy, J., and Townsend, C. (2007). A comparative perspective on urban transport and emerging environmental problems in middle-income cities. In G. McGranahan and P.J. Marcotullio (Eds.) (2007), *Scaling urban environmental challenges: From local to global and back*. London: Earthscan, pp. 206–234.

Krauss, C., and Bradsher, K. (2015). China's global ambitions, with loans and strings attached. *New York Times*, July 24.

Kumagai, M. (2000). *Automobile dependence in Japanese cities: An international comparative review with implications for urban policy*. BSc (Hons.), Murdoch University, Western Australia.

Lam, S.H., and Toan, T.D. (2006). Land transport policy and public transport in Singapore. *Transportation*, 33(2): 171–188.

Lepeule, J., Litonjua, A.A., Coull, B., Koutrakis, P., Sparrow, D., Vokonas, P.S., and Schwartz, J. (2014). Long-term effects of traffic particles on lung function decline in the elderly. *American Journal of Respiratory and Critical Care Medicine*, 190(5): 542–548.

May, A.D. (2004). Singapore: The development of a world class transport system. *Transport Reviews*, 24(1): 79–101.

Millard-Ball, A., and Schipper, L. (2011). Are we reaching peak travel? Trends in passenger transport in eight industrialized countries. *Transport Reviews*, 31(3): 357–378.

Murray, F., and McGranahan, G. (2003). Air pollution and health in developing countries—The context. In G. McGranahan and F. Murray (Eds.), *Air pollution and health in rapidly developing countries*. London: Earthscan, pp. 1–20.

Newman, P., and Kenworthy, J. (1999). *Sustainability and cities: Overcoming automobile dependence*. Washington, DC: Island Press.

Okata, J., and Murayama, A. (2010). Tokyo's urban growth, urban form and sustainability. In A. Sorensen and J. Okata (Eds.), *Megacities: Urban form, governance, and sustainability*. Tokyo: Springer, pp. 15–41.

Olszewski, P.S. (2007). Singapore motorisation restraint and its implications on travel behaviour and urban sustainability. *Transportation*, 34(3): 319–335.

Otto, B. (2014). When will Jakarta's MRT be finished? New answer coming in September. *Wall Street Journal*, July 2. Retrieved from http://blogs.wsj.com/indonesiarealtime/2014/07/20/when-will-jakartas-mrt-be-finished-new-answer-coming-in-september

Railway Gazette (2015). Addis Ababa light rail opens. Retrieved from http://www.railwaygazette.com/news/urban/single-view/view/addis-ababa-light-rail-opens.html

Reuters. (2007). Japan's JBIC lends $554 million to Thai rail project. Retrieved from http://in.reuters.com/article/thailand-jbic-idINBKK2880520071203

Rodrigue, J.P., Comtois, C., and Slack, B. (2009). *The geography of transport systems*. New York: Routledge.

Sassen, S. (2001). *The global city: New York, London, Tokyo*. 2nd ed. Princeton NJ: Princeton University Press.

Schaeffer, K.H., and Sclar, E. (1975). *Access for all: Transportation and urban growth*. New York: Columbia University Press.

Shatkin, G. (2014). Reinterpreting the meaning of the "Singapore Model": State capitalism and urban planning. *International Journal of Urban and Regional Research*, 38(1): 116–137.

Sider, T., Hatzopoulou, M., Eluru, N., Goulet-Langlois, G., and Manaugh, K. (2015). Smog and socioeconomics: An evaluation of equity in traffic-related air pollution generation and exposure. *Environment and Planning B: Planning and Design*, 42(5): 870–887.

Sorensen, A. (2002). *The making of urban Japan: Cities and planning from Edo to the twenty-first century*. New York: Routledge.

Sperling, D., and Gordon, D. (2009). *Two billion cars: Driving toward sustainability*. New York: Oxford University Press.

United Nations (2014). *World urbanization prospects: The 2014 revision: Highlights*. New York: United Nations Department of Economic and Social Affairs.

Walker, J. (2012). *Human transit: How clearer thinking about public transit can enrich our communities and our lives*. Washington, DC: Island Press.

Walks, A. (Ed.) (2015). *The urban political economy and ecology of automobility: Driving cities, driving inequality, driving politics*. New York: Routledge.

WHO (2009). *Global status report on road safety: Time for action*. Geneva: World Health Organization.

WHO (2013). *Global status report on road safety 2013: Supporting a decade of action*. Geneva: World Health Organization.

World Bank (2011). *Implementation completion and results report (IDA-37200 and IDA-37201) on a credit in the amount of SDR 109 million (USD$150 million equivalent) to the Federal Republic of Nigeria for a Lagos Urban Transport Project*. Abuja: Sustainable Development Department.

Zhao, Z., Zhao, J., and Shen, Q. (2013). Has transportation demand of Shanghai, China, passed its peak growth? *Transportation Research Record: Journal of the Transportation Research Board*, 2394: 85–92.

26

Conclusion: Envisioning Global Urban Futures

Alison L. Bain and Linda Peake

Introduction

Urbanization in the twenty-first century is one of the world's most complex and important societal challenges. This textbook has sought to address some of the breadth and complexities of urbanization around the world and to examine some of the many lived realities of urban places. Invariably, in such an assemblage of urban knowledge there are silences and omissions, both geographical and analytical, as well as differing philosophical vantage points. Nevertheless, this textbook emphasizes the importance of understanding cities as material phenomena, embedded within broader webs of power that are manifested spatially, linking urban places to rural hinterlands, urban regions, nation-states and other cities, both within and beyond their own national context. It is from understanding the political economies and ecologies of these linkages and how the lives of urban residents—their movements across space, their practices of everyday life, their hopes, dreams, and desires—play out within them that we can learn about global urban futures.

This concluding chapter speculates about these futures and reflects critically upon pressing urban issues and challenges. Discussion begins with attention to the utopian tradition of envisioning alternative urban realities; a long history of aspiring to both imagine and build better cities has been a source of inspiration for the present (Cooper, 2014). Not only can the utopian imagination be a powerful way to foster intellectual curiosity about the urban, but the urban can also be a "privileged site . . . for actualizing new, alternative forms of citizenship and belonging" (Firth, 2012: 89). Consideration of how such alternative urban futures can be achieved in times of **crisis** necessitates a discussion of fundamental challenges facing cities today: the rapid rate of urban growth in the global South in relatively poor postcolonial countries; the persistent spread of urban neoliberalism as the dominant mode of urban development and regulation; the rise of inequality, rooted in a context of over-production and over-consumption; and the detrimental impact of urbanization and climate change on urban environments. These challenges have provoked different urban policy responses, and the final section of this chapter reflects upon the circulation of dominant policy discourses that aim to produce "green," "intelligent," and "resilient" cities. The chapter concludes with an examination of how any future urban trajectories must grapple with a fundamental question about how regulatory bodies at different spatial scales—municipal, regional, national, supranational, and global—intersect and interact as a new era of global urban policy is being ushered in.

Utopian Imaginaries: From Urban Form to Urban Politics

The concept of **utopia** dates from the book of the same name, *Utopia,* written by Thomas More, six centuries ago, in 1516. It was written during a period of turbulent change and power struggles in England when King Henry VIII was attempting to divorce the country from the rule of the Catholic Church. More used the term "utopia" to refer to an imaginary society that had its own political constitution (Beauchesne and Santos, 2011). Traditionally, "utopia" is an ambiguous word with a double meaning that describes "somewhere good" and yet is also "nowhere," temporally or spatially, in the present world. Utopias are perhaps best defined functionally "as articulating critique and dissatisfaction (with the present frame or *status quo*) as well as desire, for something better" (Firth, 2012: 90). Utopias can be fictional, playful, or experimental articulations of the imagination. Indeed, there is a long tradition of imagining utopias and dystopias in fiction, particularly in science fiction (Boscaljon, 2014). Utopias are a way of expressing many different hopes and desires; one tradition of utopianism is a desire for freedom, for example, while another is a desire for order. As with *Utopia*, they are also ways of presaging real change, through attempts to put these imaginings into practice, either through designs for and the building of utopias or through attempts to reorder socio-spatialities.

In many respects, utopia is a kind of "social dreaming" that involves the "expression of aspirations for a desired way of being, or a future state of society, by an individual or a group" (Levitas in Frith, 2012: 93). Such dreaming is a manifestation of a longing and hope for a better future that may be more socially just or politically progressive. But while it can become a way for some groups to transcend the obstacles and frustrations of the present, it can also be interpreted as a privilege that not everyone has the time or the opportunity or the will to engage with. Moreover, while one person's dreaming may represent a utopian future for them, for others it may constitute a nightmare—a dystopia—something David Harvey (2000) recognizes in the appendix to his book *Spaces of Hope*, in which he describes a utopian future he calls "edilia." In this future, people do not live in families but in "hearths" that consist of groups of 20–30 adults with their children, made up of "pradashas" of individuals who bond together for the purpose of raising children. Members of hearths eat, work, and make collective decisions together. In the hearths, relationships are not bound by marriage—a recognition that reimagining space means reimagining relations of power (cooking, cleaning, and chores, for example, are shared on a rota system and sexual favours are exchanged for points in a computerized market). Ten or so hearths group together to form "neighbourhoods," the scale at which educational and health-care functions are provided and where urban agriculture and gardening have a prominent role to play. Approximately 20 neighbourhoods combine together to form an "edilia" (of about 60,000 people), and about 20 to 50 edilias come together to form the largest contiguous political unit, called a "regiona" (of at most 3 million people). Each regiona is also a bioregion that aims to be as self-sufficient as possible for its inhabitants. This spatialized form of organization may not seem to be very different from what already exists in a number of places and at a number of scales. Nevertheless, Harvey radically departs from current forms of organization in that regionas combine to form "nationas," which are not spatially contiguous or ethno-culturally homogeneous but rather combine regionas in temperate, tropical, subtropical, and sub-arctic parts of the world, brought together for the purposes of trade and barter. Moreover, nationas are not permanent features of the geopolitical map. They are expected to dissolve and reform periodically.

How we imagine and reimagine cities and the networks in which they reside works, in part, to define the scope and contours of what is intellectually and experientially possible in urban policy, planning, design, and practice (Pinder, 2002). In the early nineteenth century, for example, in the context of the Industrial Revolution, there were attempts to design new living and working conditions that focused on treating industrial workers with dignity and respect. These efforts to establish experimental communities based on social change were soon, however, to be pejoratively referred to by some as a form of "utopian socialism" because they were divorced from the class struggle considered essential for socialism to emerge. The social reformer Robert Owen pioneered one such effort when he developed the textile settlement of New Lanark, Scotland, which housed approximately 2500 people. Owen introduced a reduction in working hours to eight hours a day, schools for children, and renovated housing for workers. In 1824 he also set up a short-lived community of approximately 1000 people called New Harmony in Indiana, United States. Although Owen's material efforts may not be considered socialist, they led to the development of the co-operative movement as well as the idea that human beings have the free will to organize themselves into different kinds of societies.

In Europe and North America, the turn of the twentieth century was another particularly influential time for envisioning the possibilities of cities (and for China, later in the twentieth century, cf. Hua, 2009). One British social and urban reformer, influenced by co-operative socialism, was Ebenezer Howard, who wrote *Tomorrow:*

A Peaceful Path to Real Reform, which was significantly revised in 1902 as *Garden Cities of Tomorrow*. His book described a utopian city in which people lived harmoniously together with nature. It not only offered a vision of towns free of slums and enjoying the benefits of both town (e.g., employment opportunities and good wages) and country (e.g., natural beauty, fresh air, and low rents), but also led to the formation of exemplars such as Letchworth Garden City and Welwyn Garden City as well as to the founding of the British Town and Country Planning Association. The Swiss-French founder of modern architecture, Le Corbusier, also influenced utopian urbanisms, although from a right-wing perspective. In 1935 he published his utopian ideas in a book, *La Ville Radieuse* (*The Radiant City*), in which he outlined his plan for a well-ordered, linear, high-rise city composed of "machines for living" that were based upon blocks of housing in the abstract shape of the (male) human body interspersed with abundant green spaces. Although he tried to influence many authorities, including Mussolini in Italy and the Vichy government in France, by propagandizing his ideal city plans through international tours and lectures, his plans were not applied until 1949. This first application was in Chandigarh, India, and it was also to prove very informative in the planning and construction of the new capital of Brazil, Brasilia, in 1960 (Hubbard, 2006).

Although the urban planning legacies of Howard and Le Corbusier can still be felt in the present-day form and organization of cities around the world (Pinder, 2005), both have been critiqued for using abstract principles to produce sanitized and controlled urban spaces that inadequately account for the lived realities of everyday life. Brasilia was the inspiration for Danish architect and planner Jan Gehl's (2010) critique of much modernist urban planning as fixated on a bird's-eye view and neglectful of the human scale—what he termed the "Brasilia Syndrome." While Brasilia might have embodied the hopes and desires of modernity and used the most advanced technology to construct an ideal and perfect city, it is a city of vast spaces that in his estimation is over-designed with little left for its citizens to do or shape. Pinder (2005: 85) argues that both Howard's pre-modern and Le Corbusier's modern utopias were "dreams of urban order." Both believed that they could achieve spatial and social order in the city by constructing carefully regulated and structured urban space. But the utopian project need not just be about a quest for perfect spatial and social order. To shut out disorder and informality, Pinder (2005) asserts, limits the possibilities for imagining alternative futures, reminding us that utopian visions of the city are many and contested.

People continue to be inspired to imagine future alternative urban realities and to make real changes in the urban present. For the political scientist Iris Marion Young (1990: 6, 256), "ideals," with respect to the production, management, and experience of urban space, "are a crucial step in emancipatory politics" because "they offer standpoints from which to criticize the given and inspiration for imagining alternatives." Utopian visions can be considered expressions of such ideals that have the capacity to motivate and to drive human ingenuity. David Harvey (2000: 195) is optimistic that we are in "precisely" such a utopian moment—"a time and place in the ceaseless human endeavour to change the world, when alternative visions, no matter how fantastic, provide the grist for shaping powerful political forces for change." He writes about an **urban actor** that he calls the "**insurgent architect**," who he insists is not a professional architect but a metaphorical figure who represents human agency—the power and capacities that all humans have to actively critique and question the way in which urban spaces are conceived, designed, and used. Inspired by feminist and postcolonial theory, the critical urban planner Leonie Sandercock (1998) envisions utopia as a social project about "living together in difference"—a project that supports communication through both dialogue and conflict and is open to radical change to social and spatial relations. If utopianism can focus on socio-spatial process rather than particular spatialized urban forms, it has, David Pinder reminds us, the potential to be a vibrant source of emancipatory urban politics. Utopian urban imaginaries, then, extend an invitation to engage with the world of the urban differently—to understand and to challenge the networks of power and the many intersecting forms of social injustice that cross-cut urban development, urban policy and planning, urban forms, urban lives, and urban infrastructure.

Urban Challenges "In a Fierce New World"

Urban change, utopian or real, may not only be imagined, it may also emanate from crises. For many critical urban scholars in this time of financial and economic crises, "a fierce new world has been born—full of shaken premises, complicated contradictions, serious fractures, severe backlash, broken promises and uncertain outcomes for the world's peoples" (Sen and Durano, 2014: 4–5). This is a world marked by the outbreak of wars, faith-based crises, human rights abuses, **corruption**, food shortages, climate change, and natural disasters, all of which have fundamental implications for access to the basic resources of **social reproduction** (for human survival and well-being) such as housing, water, food, education, health care, and security within cities in the present and into the future. In this fierce new world, "life and death, and . . . hope and harm, and . . . endurance and exhaustion" are all unequally distributed, impacting upon social groups and bodies differently (Povinelli, 2011: 3). In this textbook, we have identified four fundamental and interrelated challenges to urban futures.

The first is the rapid rate of urban growth due to population increase and migration to urban areas in the global South in relatively poor postcolonial countries, which under current projections will continue on the same path until at least the mid-century (UN-DESA-PD, 2015). Migration scholarship has shown that extreme poverty, lack of economic opportunities, civil unrest, and political uncertainty are all push factors that impel people to leave areas of origin, whether legally (through immigration or asylum seeking) or illegally (through human smuggling and human trafficking), on missions of survival (Jac-Kucharski, 2012). Whether moving from rural to urban areas or from one urban place to a larger one, these migrants are adding another layer to those at the very bottom, resulting in the urbanization of poverty in the twenty-first century (Craig and Porter, 2006). As Harvey (2014: 60) has noted, "[t]he massive forced and unforced migrations of people now taking place in the world . . . will have as much if not greater significance in shaping urbanization in the 21st century as the powerful dynamic of unrestrained capital mobility and accumulation." For some, such as the Marxist scholar Mike Davis (2006), this growth presages a pessimistic and dystopian urban future—a *Planet of Slums*. The geographer Kevin Archer (2013: 193) agrees when he states, "[s]uch dense concentrations of destitute people surely represent[s] a ticking time bomb both in social and environmental terms." Persistently dismal slum-like living conditions for the majority of the world's population can only contribute to escalating social unrest in the future. Others, such as Saunders (2011) and Amin and Thrift (2002), offer a much more positive and optimistic vision of technologies, footholds, networks, and mobilities, while the postcolonial scholar AbdouMaliq Simone (2004) stresses the ingenuity and innovation of those faced with urban squalor. What these differing views of urban futures mark is the need to continuously develop new insights into experiences of everyday life in urban areas of the global South. It is also imperative to develop new conceptualizations of these experiences of urban places in ways that can effectively speak to related local understandings of territory, land, spatialities, density, propinquity, urban social life, femininity, masculinity, and biological reproduction.

A second challenge is the corrosive persistent spread of **urban neoliberalism** as the dominant mode of urban development and regulation. In its most basic form, neoliberalism is characterized by the privatization and financialization of markets and the dismantling of social contracts as the role of the state is redefined in this period of austerity. The market has become the regulative principle of the state, such that urban policies and decisions are increasingly designed and evaluated primarily on the basis of the criteria of profitability and efficiency (Paton, 2014). Urban citizens are also increasingly disciplined to act responsibly by making decisions that support entrepreneurialism. In such an entrepreneurial regulatory framework, where cities are run more like businesses and local governments rely on risk taking, inventiveness, self-promotion, and profit-maximization, cities around the world have increasingly come to resemble one another in their drive to produce urban space for progressively more affluent users (Bain, 2015). One trope in particular has proved to be virtually global in its travel and deployment, that of an Americanized urban development paradigm rooted in a lifestyle of car-dependence, hyper-consumption, privatized forms of suburban governance, and spatial seclusion behind gates and walls. Such an urban development paradigm destroys nature, produces more waste, and socially fractures cities along ethnic and class lines (Graham and Marvin, 2001). Aiding the globalization of this **splintering urbanism** paradigm (Graham and Marvin, 2001) has been the transformation of urban territorial governance and regulation from within nation-state framings to "broader institutional and territorial frameworks through which urbanization processes are being managed at every spatial scale" (Brenner and Schmid, 2015: 153). City networks are becoming increasingly complex and intertwined at different scales with the emergence of new jurisdictional arrangements and regulatory frameworks. Such transformations in governance entangle the urban in complex and uneven ways into regional, national, and global scales, demanding urban analysis that is contextualized within the extra-urban flows of a neoliberalized world economy.

A third and related challenge is that of increasing levels of inequality. In the late period of neoliberalism, the increasingly unregulated nature of capital alongside economic growth in the post-financial crisis period is not only causing production and consumption to approach unsustainable levels, but is also resulting in increased inequality. This is manifested through large increases in wealth and income for the top 1 per cent and further austerity policies for the poor (Sen and Durano, 2014; Dorling, 2014), whose lives are lived out in conditions of precarity, marked by transience and dislocation, and who are considered replaceable, and, ultimately, disposable (Beall and Fox, 2009; Tadiar, 2009). These inequalities are also marked out through increasing levels of socio-spatial polarization within cities. The "fortressed" city in which the elite lives, developed through gentrified areas, gated enclaves, securitized buildings, and the continual privatization of public space, often exists in close proximity to the marginalized places of the urban poor. McIntyre and Nast (2011) have coined the term "bio(necro)polis" to speak to the entanglement of these polarities of the elite and the poor in many urban places. Ways of thinking about urban futures that do not speak to urban places only in terms of commodities, commodification, and profit-making are required to recognize alternative ways forward. Critical urban scholars have much to offer here through an engagement with grassroots urban-based protests, from Occupy, SlutWalks, Black Lives Matter, austerity protests, living-wage campaigns, migrant rights movements, struggles around sexuality, and everyday struggles for housing, food, and work, that are already rethinking socio-spatial forms and processes in ways that recognize and benefit the urban poor.

The fourth challenge is that urbanization is the most extreme collision of humanity and nature. Continued rapid urbanization in combination with the negative effects of climate change (e.g., increases in temperature, heavier precipitation, severe droughts, dangerous floods, and rising sea levels) and extreme weather events holds dire consequences for the continued existence of diverse plant and animal life and for environmental justice. The large scale of urbanization not only has serious social and economic consequences for its human and non-human inhabitants, but also places cities at the forefront of political arenas devising solutions to address ecological conservation and sustainability (Ackerman, 2014), raising questions about the relationship of human urban residents to non-human nature. For some environmental activists, cities are the most ecologically viable form of settlement (Meyer, 2013), while for others they are ecological culprits. There are certainly grounds for thinking quite negatively about cities, but some scholars argue that it is increasing affluence and the production and consumption associated with economic growth that leave an **ecological footprint** and not cities (Wackernagel et al., 2006). Although the concentration of people and activities in cities does produce high levels of environmental stress, that same concentration also affords important opportunities for strategies to reduce environmental degradation. Consequently, cities are sites with the potential to reduce environmental damage and resource depletion. Indeed, it is cities as opposed to national governments that are leading the way in measuring carbon, cutting CO_2 emissions and air pollution, and developing renewable energy. The challenge for critical urban analysis is being able to take into account the deleterious effects of climate change and of social and economic stresses on urban environments in ways that go beyond seeing the city as a bounded, coherent unit and that recognize the fluidity of the boundaries between so-called natural and built environments.

At their foundations, then, cities around the world face similar fundamental challenges in a context of urban expansion, neoliberal economies, processes of inequality, and environmental stress—how to obtain the energy to feed, hydrate, shelter, care for, educate, employ, and move residents from place to place and how to manage the waste that is the by-product of these basic human activities without disproportionately negatively impacting upon existing environmental insecurities, already marginalized social groups, and exacerbating social and political conflict. The provision of such critical urban services as water, electricity, sanitation, garbage, housing, and transportation demands the investment of significant public resources in construction, maintenance, and management costs. With dwindling, if non-existent, city tax bases to draw upon, fundamental questions remain about how in today's neoliberal, globalized context to both meet and finance the basic needs of urban residents in the present and into the future (Archer, 2013). The cities of the future, within both the global South and global North, will need to be realized within very real financial, technical, and infrastructural constraints. Despite these daunting constraints and the complex social and environmental issues that they are entwined with, we remain cautiously optimistic about the collaborative problem-solving capacities of urban citizens to be active participants in the creation of innovative and inclusive urban futures (Knox, 2014). In the following section, this chapter considers how city "experts" and citizens are addressing these urban futures.

Discourses on Urban Futures: Enabling Green, Intelligent, and Resilient Cities?

Perhaps the most widely engaged with discourse about urban futures (and the urban present) is that of sustainability. The idea of **urban sustainability** is that cities and towns have a minimal environmental footprint (which decreases inputs of energy, water and food, and waste outputs of heat, air pollution, and water pollution) enabling them to meet the needs of the present without sacrificing the ability of future generations to meet their own needs. Although there are many different ways of achieving sustainability, the green agenda of sustainable urban development focuses on conservation, biodiversity, and the protection of natural resources. It attends to the long-term ecological effects that affect future generations, and it has important implications for the management of urban environments. In **green cities** (places like Curitiba, Brazil; Masdar City, Abu Dhabi, UAE; Portland, Oregon, USA; Freiberg, Germany; and Stockholm, Sweden) citizens, politicians, and urban planners prioritize the development of more "resilient infrastructures, institutions, and behaviors to help them face the problems associated with climate change" (Mayer, 2014: 212).

Commitment to reductions in greenhouse gas emissions and urban planning and urban policy leadership to support changes in urban lifestyles aligned with more sustainable ways of life are essential components of the green city movement. Through compact, green urban design, particularly at the scale of the neighbourhood, a city's built environment can be adapted to conserve energy and green space. Sustainable building materials and products, new building standards, sustainable transportation modes (e.g., walking, cycling, and public transit), and use of renewable energies can all reduce the ecological footprint of cities. But "a green city is only as sustainable as its residents' lifestyles" (Mayer, 2014: 218). Food security concerns and the locavore movement have advocated for the importance of de-globalizing and re-localizing urban food supplies (Bedore, 2010), and there has been a corresponding increase in urban agriculture and gardening in cities. Another important urban lifestyle shift is the increase in car-sharing and bicycle-sharing programs in cities of the global North and global South (Cervero, 2009; Hass-Klau, 2015; Pucher and Buehler, 2008). Embedded in all of these lifestyle shifts is a requirement that people everywhere take greater ownership of the cities in which they live—what can be termed active **urban citizenship**. About more than paying taxes and voting, active urban citizenship involves deliberate and considerate participation in the everyday life of the city with the goal of collaborating to take ownership over urban problems and the development of solutions to them. Based on the concept of enablement whereby municipal governments create contexts within which citizens and institutions can connect, in this framework the city is reimagined as a hub of innovation through collaboration and co-operation (Camponeschi, 2009).

It behooves us to remember that urban sustainability is not always linked to progressive discourses. While several scholars have made important critical interventions to link issues of urban sustainability to those of uneven spatial development, neoliberalization, and struggles for environmental justice (Atkinson, 2009; Young, Wood and Keil, 2011) for others, urban sustainability is more a technical than political discourse, an issue for which technological and rational "fixes" can be found. It also behooves us to remember that in many cities in the global South, sustainability is not on the agenda, whether it be for reasons of competing agendas, lack of qualified personnel, a lack of political will, or lack of financial capability. For example, while the preservation or creation of green space is a priority for some cities, for others it is low on their agendas, especially when such land can raise capital if given over to more lucrative and profit-maximizing commercial or residential development. There are also very powerful forces, such as the carbon economy with its investments in certain types of urban development (e.g., large infrastructure projects and suburban sprawl), that mitigate against planning for sustainability and environmental equity.

A newly emerging discourse, since the 2000s heavily associated with cities, is that of electronic data and particularly that which is known as **big data**. In the twenty-first century, significant advances in communication and computational technologies have permitted the collection and analysis of new kinds of big data about cities. Big data refers to massive data sets that are characterized by huge volumes of diverse, fine-grained, interlocking data produced on a dynamic basis (some updated hourly), much of which are spatially and temporally referenced and are reshaping how geographic knowledge is produced, business conducted, and governance enacted (Kitchin, 2016). Big data are not without negative issues, however. For example, as Kitchin (2016) reports, much big data are mined and produced by private corporations (e.g., mobile phone operators, application developers, social media companies, retailers, and security firms) who are not obligated to share that data. There are also technical concerns with big data with respect to how clean (error-free), objective (bias-free), and consistent (few discrepancies) the data are. The surveillance function of big data in particular is a major cause for concern to civil liberties groups raising ethical questions about infringements on privacy and other human rights (Koskela, 2002; Ruppert, 2006). The ability of big data to engage in urban social engineering by increasing socio-spatial inequalities (e.g., enabling differential access to services), posing security concerns, such as identity theft, and enabling "control creep" whereby data generated for one purpose are used for another is another major issue (Batty, 2014; Kitchin, 2016). Finally, the use of big data has been linked to neo-positivist understandings of cities in which the use of sophisticated techno-scientific methods replaces politics for understanding and "managing" flows and relations both within and between cities, thus encouraging an understanding of cities as bounded and inherently knowable entities.

The dystopian aspects of big data are somewhat alleviated by the emergence of **open data**. Some information that was previously withheld from the public and hidden in administrative, infrastructure, and service systems has been made freely available as open data, allowing citizens to leverage collaborative expertise and insight and to take a more active role in shaping urban futures. For example, "[c]itizens are not only creating mobile apps that promote smarter ways of conducting various functions, but are also building online platforms so as to source problems and solutions from their fellow dwellers" (Desouza, 2014: 228). In the **information age**, the widespread adoption of location-aware mobile technologies and the increasing availability of Internet connectivity have provided users with location-specific, real-time information that can strengthen connections to local surroundings and embed social knowledge in the fabric of the city through public authoring projects. Mobile technologies are reshaping urban spaces into hybrid physical and virtual worlds that allow those people who can afford smartphones, tablets, and laptops to filter, control, and manage their relationships with cities and the people within them (de Souza and Frith, 2012). The ability of citizens and urban managers to synthesize data from different sources (e.g., objects, actors, and events) and generate actionable new knowledge that can guide real-time decision-making suggests a future of increasingly "intelligent" and resilient cities.

The discourse of **urban resilience** is increasingly a contested one. In contemporary "risk society," urban resilience refers to the quest to preserve the material and social fabric of cities (Beck, 1992). It addresses the "bouncebackability" capacity of cities to adapt to stress and to recover swiftly from threats by minimizing urban disorder. And it speaks to a recognition of the need for cities to prepare for severe weather events, the spread of diseases (e.g., SARS, the HIV virus, the Zika virus), political violence, and the protection of environmental resources. It could be argued that resilience has always been at the core of urbanism—it is what has allowed cities to adapt and be rebuilt through history. While urban resilience can be actively planned, it is important to remember that "cities are resilient to different extents and in different ways and have different periods of recovery" (Savitch, 2008: 152). While "local resilience is a long-term and inherent condition of cities," cities can also foster it by "the strength of their social fabric, the dynamism of their economies, and the optimism of their citizenry" (Savitch 2008: 174, 159). It is people, after all, who hold and carry the promise of city life forward.

Critics of resilience argue that it can serve to normalize market-based governance—as the state pulls back from providing welfare and public goods, it is citizens who have to bear the brunt of "being resilient" (Slater, 2014), with the expectation that they will form social networks to help develop adaptive and transformative capacity, helping to manage threats and conditions of uncertainty. Thus, critics claim, it is neoliberal urbanism itself that is proving to be resilient. Another discursive formation of resilience has developed post 9/11, as "security-induced notions of resilience have become a core idea round which threat-response policy is increasingly produced and mobilized at a variety of spatial scales" (Coafee et al., 2009: 4). Post 9/11, cities and nations throughout the world have had to dramatically and proactively rethink systems of defence and emergency management in response to terrorism. As Savitch (2008: 171) explains, "[t]errorists attempt to use the city's own strength against itself, forcing it to implode"; they seek to "turn a city upside down by exploiting its freedoms, its openness, its interdependencies, and its very magnitude." Counter-terrorism measures are increasingly framed as strategies of urban resilience and **military urbanism**. Building designs, traffic management strategies, official urban plans, social policies for diverse neighbourhoods and cities, all now have the potential to be regulated by national security agendas.

Finally, a further, and at this time unknowable, factor has recently been thrown into the mix of discourses on urban futures—a new interest in global urban policy by supra-municipal bodies, such as UN-Habitat. Regional and global organizations are increasingly adopting an interest in cities as they come to contain more and more of the world's population, potentially making them not only one element of a sustainable future, but the potential tipping point into unsustainability. Prior to 2015, global development policy on cities was virtually non-existent. **Sustainable Development Goal 11,** the first global goal in relation to cities, and Habitat III's New Urban Agenda are creating a new potential force in urbanization, the impact of which has not yet been felt. It remains to be seen whether and how these global urban policy agendas will be implemented and interact with urbanization, and how they may reconfigure inequalities and rights to the city, issues that lie at the heart of democratic urban transformations.

Conclusion

We hope this textbook has inspired you to become part of a new generation of urban scholars, to be intellectually curious about the inner workings and multiple sociospatial configurations of cities, and to become a politicized urban citizen. We invite you to continue to look at cities (in terms of both urban development and people's experiences of the urban) with an analytical eye—to gaze in different directions, to focus on the all-too-often neglected realities of cities beyond the global North (Parnell and Oldfield, 2014), and to think ever more critically about the production, use, and governance of urban space and the pressing urban challenges that face us in the twenty-first century. Together, we all have active roles to play in building the urban knowledge needed to envision and realize just and inclusive urban futures.

Key Points

- We can think of cities no longer as bounded and self-contained entities but as nodes within circulating bundles of flows, characterized by increasing connectedness through flows of people, capital, and ideas.
- **Urban utopias,** whether fictional or "real," speak to attempts to imagine new spatialities in which issues of democracy, equality, and justice are addressed.
- Consideration of how alternative urban futures can be achieved in times of crises necessitates addressing the challenges posed by the rapid rate of urban growth in the global South in relatively poor postcolonial countries, by the persistent spread of urban neoliberalism as the dominant mode of urban development and regulation, by the rise of inequality, and by the detrimental effects of stress, including climate change, on urban environments.
- Contested discourses that dominate discussions of urban futures relate to urban sustainability, electronic data, and resilience.
- We remain cautiously optimistic about the collaborative problem-solving capacities of urban citizens to be active participants in the creation of innovative and inclusive urban futures.

Activities and Questions for Review

1. Present to a group of your peers three ideals that you would use to design your own urban utopia. How might your utopia differ in socio-spatial form from existing urban places?
2. As a class, identify which concepts, common in twenty-first-century urban theory (e.g., resilience, precarity, and inequality), you find most helpful in thinking through urban futures.
3. Do you think the Canadian urban future overall is positive or negative? On what factors and discourses do you base your reasoning?
4. If you think back across the chapters from this textbook that you have read, which topics or issues have most galvanized your urban geographical imagination?

References

Ackerman, D. (2014). *The human age: The world shaped by us*. New York: Harper Collins.

Amin, A., and Thrift, N. (2002). *Cities: Reimagining the urban*. Cambridge and Oxford: Polity Press in association with Blackwell.

Archer, K. (2013). *The city: The basics*. New York: Routledge.

Atkinson, A. (2009). Cities after oil—One more time. *City, 13*(4): 493–498.

Bain, A.L. (2015). Re-imaging, re-elevating, and re-placing the urban: The cultural transformation of Canadian inner cities. In P. Filion, M. Moos, T. Vinodrai, and R. Walker (Eds.), *Canadian cities in transition: Perspectives for an urban age* (5th ed.). Don Mills, ON: Oxford University Press, pp. 244–257.

Batty, M. (2014). *The new science of cities*. Cambridge, MA: MIT Press.

Beall, J., and Fox, S. (2009). *Cities and development*. London and New York: Routledge.

Beauchesne, K., and Santos, A. (Eds.) (2011). *The utopian impulse in Latin America*. New York: Palgrave Macmillan.

Beck, U. (1992). *Risk society: Towards a new modernity*. London: SAGE.

Bedore, M. (2010). Just urban food systems: A new direction for food access and urban social justice. *Geography Compass, 4*(9): 1418–1432.

Boscaljon, D. (Ed.) (2014). *Hope and the longing for utopia: Futures and illusions in theology and narrative*. Eugene: Pickwick Publications.

Brenner, N., and Schmid, C. (2015). Towards a new epistemology of the urban? *City, 19*(2/3): 151–182.

Camponeschi, C. (2009). *The enabling city: Place-based creative problem-solving and the power of the everyday*. Unpublished MES MRP. York University, Toronto.

Cervero, R., et al. (2009). Influences of built environments on walking and cycling: Lessons from Bogotá. *Journal of Sustainable Transportation, 3*(4): 203–226.

Coafee, J., Wood, M., and Rogers, P. (Eds.) (2009). *The everyday resilience of the city: How cities respond to terrorism and disaster*. New York: Palgrave Macmillan.

Cooper, D. (2014). *Everyday utopias: The conceptual life of promising spaces*. Durham, NC: Duke University Press.

Craig, D., and Porter, D. (2006). *Development beyond neoliberalism? Governance, poverty reduction and political economy*. London: Routledge.

Davis, M. (2006). *Planet of slums*. London: Verso.

de Souza, A., and Frith, J. (2012). *Mobile interfaces in public spaces: Locational privacy, control, and urban sociability*. London: Routledge.

Desouza, K.C. (2014). The intelligent city. In P. Knox (Ed.), *Atlas of cities*. Princeton, NJ: Princeton University Press, pp. 228–243.

Dorling, D. (2104). *Inequality and the 1%*. London: Verso Books.

Firth, R. (2012). Transgressing urban utopianism: Autonomy and active desire. *Geografiska Annaler: Series B, 94*(2): 89–106.

Gehl, J. (2010). *Cities for people*. Washington: Island Press.

Graham, S., and Marvin, S. (2001). *Splintering urbanism: Networked infrastructures, technological mobilities, and the urban condition*. London and New York: Routledge.

Harvey, D. (2000). *Spaces of hope*. Los Angeles: University of California Press.

Harvey, D. (2014). Cities or urbanization? In N. Brenner (Ed.), *Implosions/Explosions: Towards a study of planetary urbanization*. Berlin: Jovis Verlag, pp. 52–66.

Hass-Klau, C. (2015). *The pedestrian and the city*. London: Routledge.

Howard, E. (1902). *Garden cities of tomorrow*. London: S. Sonnenschein & Co.

Howard, E. ([1898] 2010) *To-morrow: A peaceful path to real reform*. Cambridge: Cambridge University Press.

Hua, S. (2009). *Chinese utopianism: A comparative study of reformist thoughts with Japan and Russia, 1898–1997*. Stanford, CA: Stanford University Press.

Hubbard, P. (2006). *City*. London and New York: Routledge.

Jac-Kucharski, A. (2012). The determinants of human trafficking: A US case study. *International Migration, 50*(6): 150–165.

Kitchin, R. (2106). Big data. In D. Richardson (Ed.), *The aag encyclopedia of geography*. Oxford: John Wiley and Sons.

Knox, P. (Ed.) (2014). *Atlas of cities*. Princeton, NJ: Princeton University Press.

Koskela, H. (2002). Video surveillance, gender, and the safety of public urban space. *Urban Geography, 23*(3): 257–278.

Le Corbusier (1935). *La Ville Radieuse*. Boulogne: Editions de l'Architecture d'Aujourd'hui.

McIntyre, M., and Nast, H. (2011). Bio(necro)polis: Marx, surplus populations, and the spatial dialectics of reproduction and "race." *Antipode, 43*(5): 1465–1488.

Mayer, H. (2014). The green city. In P. Knox (Ed.), *Atlas of cities*. Princeton, NJ: Princeton University Press, pp. 211–225.

Meyer, W. (2013). *The environmental advantages of cities*. Cambridge, MA: MIT Press.

Parnell, S., and Oldfield, S. (Eds.) (2014). *The Routledge handbook on cities of the global South*. London and New York: Routledge.

Paton, K. (2014). *Gentrification: A working-class perspective*. Surrey, UK: Ashgate.

Phillips, R. (2014). Space for curiosity. *Progress in Human Geography, 38*(4): 493–512.

Pinder, D. (2002). In defence of utopian urbanism: Imagining cities after "the end of utopia." *Geografiska Annaler: Series B, 84*(3/4): 229–241.

Pinder, D. (2005). *Visions of the city: Utopianism, power, and politics in twentieth-century urbanism*. London and New York: Routledge.

Povinelli, E.A. (2011). *Economies of abandonment: Social belonging and endurance in late liberalism*. Durham, NC: Duke University Press.

Pucher, J., and Buehler, R. (2008). Making cycling irresistible: Lessons from the Netherlands, Denmark, Germany. *Transport Reviews, 28*(4): 495–528.

Ruppert, E. (2006). Rights to public space: Regulatory reconfigurations of liberty. *Urban Geography, 27*(3): 271–292.

Sandercock, L. (1998). *Towards Cosmopolis: Planning for multicultural cities*. London: Wiley.

Saunders, D. (2011). *Arrival cities: The final migration and our next world*. Toronto: Vintage.

Savitch, H.V. (2008). *Cities in a time of terror: Space, terror, and local resistance*. New York: M.E. Savage.

Sen, G., and Durano, M. (Eds.) (2014). *The remaking of social contracts: Feminists in a fierce new world*. London: Zed Books.

Simone, A.M. (2004). *For the city yet to come: Changing African life in four cities*. Durham, NC: Duke University Press.

Slater, T. (2014). *The resilience of neoliberal urbanism*. Open Democracy. Retrieved from https://www.opendemocracy.net/opensecurity/tom-slater/resilience-of-neoliberal-urbanism

Tadiar, N.X.M. (2009). *Things fall away: Philippine historical experience and the makings of globalization*. Durham, NC: Duke University Press.

UN-DESA-PD (United Nations, Department of Economic and Social Affairs, Population Division) (2015). *World urbanization prospects: The 2014 revision (st/esa/ser.a/366)*.

Vanolo, A. (2013). Alternative capitalism and creative economy: The case of Christiania. *International Journal of Urban and Regional Research, 37*(5): 1785–1798.

Wackernagel, M., et al. (2006). The ecological footprint of cities and regions. *Environment and Urbanization, 18*(1): 103–112.

Young, D., Wood, P., and Keil, R. (Eds.) (2011). *In-between infrastructure: Urban connectivity in an age of vulnerability*. Kelowna, BC: Praxis (e)Press.

Glossary of Key Terms

Ableism is discrimination or prejudice against people with disabilities. It includes ideas, practices, institutions, and social relations that presume ablebodiedness and, as a consequence, constructs persons with disabilities as marginalized and largely invisible "others."

Accessibility is a practice (and often a legal protection) that aims to mitigate discrimination against people with disabilities. While often focusing on physical access to buildings, accessibility also involves making reasonable accommodations in employment and standards to ensure that people with disabilities can use goods and services.

Accumulation is the primary goal of capitalism and is the process by which capital is reproduced through the continual reinvestment of surplus value (or profit) that results from the appropriation of labour power by capital. Surplus value is the difference between a worker's wages (exchange value) and the value of the goods and services labour produces (use value). *See* **Accumulation by dispossession**, **Labour**, **Capitalism**.

Accumulation by dispossession describes a process through which low-income communities are dispossessed of valuable land and resources that will in turn generate capital gain for the state or the market. The dispossession takes place largely through urban development and gentrification projects but can equally take on a rural character such as when farmers are dispossessed of their land because of the industrialization of agriculture. *See* **Accumulation**, **Capitalism**.

Adaptation (to climate) refers to policy measures that reduce an area's vulnerability to extreme weather associated with climate change.

Adaptive reuse is the process of using older, redundant, or abandoned buildings or sites for new purposes. It offers a method for preserving historic land and heritage buildings but also represents contested spaces of revitalization, gentrification, and façadism. *See* **Brownfield development**.

Agglomeration economies are the savings or benefits that firms accrue by clustering in the same place or region as other firms engaged in similar types of business. These benefits might, for example, be due to localized supplies of skilled labour, essential infrastructure, or knowledge transfer within the sector.

Air pollution comprises air-borne substances at concentrations, durations, and frequencies that adversely affect human health or the environment.

Alienation is a state in which individuals are estranged from or indifferent to their urban material world and/or urban community in which they are living.

Animal geography is a sub-discipline that explores human–animal relationships spatially and temporally to understand human valuing of animals, the roles animals play and the spaces they occupy, and the outcomes of varying encounters between human and non-human animals.

Anonymity refers to the condition of being anonymous. City life affords many members of oppressed groups the freedom to discover and live out their identities anonymously in the absence of opposition from those who may know them in other places, although some have argued that in a digitally networked world and an increasingly securitized urban environment, anonymity is much less possible to attain.

Apartheid refers to a policy of spatial separation of racialized groups. The most well-known system of apartheid, which removed millions of black and coloured South Africans from urban areas, existed in South Africa between 1948 and 1994 and had its roots in the apartheid practices of Canada's Indian Act of 1876.

Arab Spring refers to the democratic uprisings leading to demonstrations, protests, riots, and civil wars in the Arab world that began on December 18, 2010, in Tunisia and spread throughout the countries of the Middle East, leading to regime changes in Egypt, Libya, Iraq, and the Yemen, with ongoing wars in Syria and Palestine. It will be many years before the consequences of these transformations become clear.

Artistic precariat, drawing on Guy Standing's definition of precariat (a hybrid term that combines precarious and proletariat), refers to cultural workers (e.g., artists, dancers, musicians, writers, performers, etc.) who are underclass members of the creative class and lack labour-related security.

Assimilationism is a form of political thinking in marginalized or minoritized communities that prioritizes seeking recognition from and fitting into the values and frameworks of dominant or "mainstream" social groups and institutions as the best means of survival or flourishing for a minoritized group. In the space of the city, assimilationism often takes the form of integration, though integration can have many other aims and political values attached to it.

Assistive devices are devices used to enable movement for persons with physical/mobility impairments (e.g., wheelchairs, scooters, canes), hearing loss (e.g., hearing aids), and way-finding challenges (e.g., white canes for persons with visual impairments).

Asylum-seekers, according to the United Nations Human Commission on Refugees, are individuals who claim refugee status but whose claims have yet to be definitively evaluated by a host state.

Austerity refers to government policies that seek to reduce budget deficits may include spending cuts, reducing or freezing labour costs, tax increases, privatization, and a reconfiguring of public services and the welfare state or any combination thereof. *See* **Austerity urbanism**.

Austerity urbanism describes how urban authorities are embracing policies that seek to reduce budgets and include cutting back on essential services, reducing or freezing labour costs, privatizing of public services, and the welfare state. Austerity policies and discourses as a way of life have become a key feature of the urban condition under neoliberal forms of capitalism, especially since the financial crisis of 2007–2008. See **Austerity**.

Behaviouralism is a paradigm (a framework for thinking about and viewing the world) within the discipline of human geography that when applied to urban geography sought to overcome the shortcomings of spatial analysis by studying human spatial behaviour in cities using the scientific method. It strove to explain spatial patterns of behaviour in terms of the mental processes and decision-making that individuals use to codify and to respond to the built environment. See **Spatial analysis**.

Bid-rent models of urban land show how the trade-off between accessibility to the greatest concentration of jobs and markets (often assumed to be the central business district, or CBD), and the need for space on behalf of bidders, results in differences in the price of land and the types of users who end up occupying urban space. These models explain why high-density and high-intensity uses are more likely to bid the highest for accessible space, and why users who need more spaces (family residences, manufacturing, etc.) often bid the most for land near the fringe of the city. See **Central Business District**.

Big data is a broad term for massive and complex data sets that differ from other data sets with respect to volume, velocity, and variety. Big data, its proponents argue, enable new forms of knowledge that produce disruptive innovations with respect to how business is conducted and governance enacted leading to debates concerning the socio-spatial implications of big data technologies for urban systems and everyday life.

Biopsychosocial model of disability brings together the focus on external barriers in the social model, while also recognizing that the lived experience of a disabled mind/body can be challenging. See **Medical model of disability, Social model of disability.**

Black water is waste water that is produced from excreta, urine, and fecal sludge that have multiple impacts on human health and the environment. Contact with the pathogens in black water wastes can cause a number of infections and diseases, including diarrhea.

Bodies have been the focus of much academic attention that emphasizes their social construction as much as their biological formation and speaks to the relationship between subjectivity, corporeality, and identity. Although naturalized accounts of embodiment (and of the hierarchical and binary divide of "male" and "female" bodies) have been resisted, the distinctive materiality of embodiment has nonetheless been maintained.

Brain circulation is a form of emigration in which skilled workers depart from a country for the purpose of pursuing educational or work opportunities elsewhere but return, resulting in a circular movement of skilled labour between nations. See **Brain drain**.

Brain drain describes the emigration of skilled labour for the purpose of pursuing educational or work opportunities elsewhere, resulting in a loss of intellectual capital and domestic skilled professionals from the nation of origin. See **Brain circulation**.

Brown environmental health agenda is a framing of environmental issues focused on concern for the quality of human habitats, (e.g., sanitation, waste, air quality, housing quality); proponents of this agenda are often urbanists concerned with cities in the global South.

Brownfield development describes the development of land previously used for industrial purposes or some commercial uses that may have been contaminated with hazardous waste or pollution but once cleaned up is considered fit (usually) for commercial development.

Bubble economy refers to major periods of speculation-induced asset inflation, such as the 1929 stock market boom in the US or the "dot-com" bubble of the late 1990s. In Japan this term refers to the period of the late 1980s when both real estate and stocks saw huge but temporary increases in value.

Built environment is the collection of transportation, housing, economic, and cultural infrastructure constructed by humans and in which humans interact in their daily lives. It is intimately related to the walkability and the quality of life in cities.

Capitalism is the most extensive economic system to have existed whose primary aim is capital accumulation, achieved though private ownership of the means of production and a system of salary or wage labour for the majority of workers. It can take various forms in practice, including free market, welfare, or state capitalism. See **Accumulation, Division of labour, Labour, Labour market, Middle class, Mode of production, Reproduction of the labour force, Uneven development, Working class**.

Capital switching, theorized by David Harvey in the 1970s, refers to the tendency of capital to "switch" between the primary circuit investments (manufacturing and commodity production) and the secondary circuit (urban land development) once the return on investment in the current circuit declines. Switches of capital investment into the secondary circuit are responsible for the many residential building booms that pepper the histories of cities.

Caste is a hierarchical system of social stratification mandated by Hinduism through which an individual's social identity and location are marked; separation in marriage and occupational interdependence are the key processes by which caste distinction is institutionalized.

Central business district (CBD) refers to the concentrated area of commercial and business activities located at the historic centre of each city. The CBD is typically the location demanding the highest bid rent per unit of land. See **Bid-rent models**.

Chicago School refers to a small group of male scholars who founded the Department of Sociology at the University of Chicago in the early twentieth century. Using a variety of approaches, they drew on the city of Chicago as a laboratory for their studies, and although much critiqued for some of their essentialist assumptions about social life, their work has proved incredibly enduring and still threads though much urban theory over a century later. See **Community, Concentric zone model**.

Citizenship is a legal status that recognizes a person's formal relationship with a sovereign nation, guaranteeing the person her/his/their civil, political, and social rights. Alternatively, citizenship may also be understood as a practice enacted through illegal occupation of vacant lands and informal access to basic services, challenging the liberal-democratic logic of citizenship. See **Digital citizenship, Indigenous-inclusive citizenship, Sexual citizenship, Urban citizenship**.

City Beautiful movement is a movement within urban planning during the late nineteenth and early twentieth centuries that aimed to address beautification and the monumentality of cities to engender civic virtue, increase access to green space, and increase the level of public health in cities.

City-building is a summary term for a constellation of policies and processes undertaken by a range of urban actors with the intention of giving physical and social shape to urban places. See **Urban actors**.

Cities as systems is an approach to urban geography that studies cities as complex social, economic, ecological, and political systems. It investigates the inner workings and socio-spatial processes that operate within cities to influence the locational arrangements of humans, activities, and institutions (e.g., transportation infrastructure, retail structure, and neighbourhood change).

Civic life refers to participation in matters concerning a particular group and/or community in city and national life. See **Civil society**.

Civil society is a domain of private associations and institutional networks in which the struggle of social, political, and cultural hegemony is achieved. In capitalist societies, it is considered to mediate between the economy and the state.

Class mobility is a form of social and economic mobility in which a person's status in social and economic hierarchies moves upward or downward in terms of the prevailing socio-economic hierarchies in place.

Climate change is a transformation in the statistical distribution of weather patterns when that change lasts for an extended period of time (i.e., decades to millions of years). It is caused by factors such as biotic processes, variations in solar radiation received by the earth, plate tectonics, and volcanic eruptions, although the current period of global warming has been caused by human activities, mainly industrialization and the resulting release of greenhouse gases into the atmosphere. See **Climate justice, Climate policy mitigation, Greenhouse gas, Urban Anthropocene**.

Climate justice refers to a fair distribution of responsibility where the wealthy and privileged expend effort on reducing their greenhouse gas emissions and on helping to fund adaptation measures to reduce the vulnerability of the poor and racialized, who bear far less responsibility for greenhouse gas emissions. See **Climate change**.

Climate policy mitigation refers to policy measures that reduce the emissions of greenhouse gases. See **Climate change, Greenhouse gas**.

Closed-circuit television (CCTV) refers to the use of digital and video technologies for surveillance purposes by private firms (e.g., banks, casinos) and public institutions (e.g., police and fire departments) so as to prevent unwanted activity in urban space. Studies question CCTV's effectiveness as a crime control tool suggesting that property and vehicle crime may be reduced but there is no evidence of a reduction in violent crime; instead crime is often displaced to areas not under surveillance and the root social causes of crime are not addressed. See **Crime**.

Colonial cities developed in societies colonized during the early expansion of the capitalist world system. Their primary function was as administrative nodes in the extraction of value from the colony back to the metropole.

Commoning practices are the messily negotiated relational ways in which the living urban commons (as a physical resource and/or a bounded plot of urban space like a community garden or a do-it-yourself space), past and present, are produced in practical ways with other people in specific situations. This process of sharing is a form of togetherness that can help to create "publics" and "communities." See **Urban commons**.

Communicable diseases are infectious diseases caused by micro-organisms that can be spread by humans. They include malaria, tuberculosis, and many tropical diseases. See **Epidemic, Non-communicable disease**.

Community commonly refers to a group of people with common ties and social networks who may live in the same area (a neighbourhood) or have the same interests or values, as depicted by the Chicago School. Communities can also be defined as local groupings of people who engage in face-to-face interaction, but globalization and technological developments have also led to the development of virtual communities and "imagined communities" that may never meet or "see" each other. See **Chicago School, Gemeinschaft, Gesellschaft, and Neighbourhood**.

Composting is the biological process of breaking up organic material.

Concentric zone model was developed by members of the Chicago School based on their research in Chicago to explain the social geography of the city. Modern cities were assumed to expand outwards from their downtown (CBD) area by a process of "invasion" and "succession" whereby residents moved into a series of adjacent zones: first into the zone in transition, then into a working-class residential zone, followed by a residential zone, until they finally reached the commuter zone of the suburbs. See **Central business district, Chicago School**.

Condominial sewerage is a form of simplified sewerage consisting of smaller-diameter, usually plastic pipes installed at shallower depths than conventional cast-iron sewers.

Condominiums are housing units that are owned or leased typically as townhouses, apartment flats, or sections of a house. Condominium owners are typically responsible for the care and maintenance of their entire property and pay "condo" fees towards the upkeep of common property (e.g., common grounds, passageways, and amenities like security, pools, saunas, gymnasia, squash courts, and party rooms).

Congestion charge is a policy that levies a charge on all vehicles (sometimes with certain exceptions) that enter a designated zone. The idea is to disincentivize driving in congested areas while raising funds to pay for improved public transit.

Consumption-based carbon accounting is a way of attributing responsibility for greenhouse emissions that tracks the full lifecycle emissions of any good or service, including manufacturing, transportation, and end use, and assigns responsibilities for all the emissions in that process to the end consumer.

Conurbation usually refers to a metropolitan-scale grouping of smaller and larger cities in a particular region.

Co-production refers to the involvement of multiple organizations in service planning and delivery, usually where NGOs or service users co-operate with professional providers.

Corruption is a form of dishonest, unethical, or illegal conduct by a person entrusted with a position of authority, often for personal gain. It can occur on different scales, such as small favours between a small number of people (petty corruption), corruption that affects the government on a large scale (grand corruption), and corruption that is so prevalent that it is part of the everyday structure of society (systemic or political corruption).

Counter-publics denote the political practice of the possibility of formations and contestations of alternative or subaltern representations—what Nancy Fraser describes as parallel "discursive scenes" that allow members of subordinated social groups (e.g., women, workers, people of colour, LGBTQ people) to put new topics on the public agenda and to talk about them in new and different ways. Current critical urban analyses emphasize the role that urban and digital spaces play in mediating various dimensions of counter-publics. *See* **Subalterns**.

Counter-topographies is a concept Cindi Katz devised as a means of recognizing the historical and geographical specificities of particular places while also inferring their analytic connections in relation to specific material social practices that serve as the basis for a feminist politics of connection. Its underlying intent is to suggest the importance of situated knowledge for a gendered oppositional politics.

Counterurbanization refers to population de-concentration from larger urban areas to rural and suburban areas. It is often the result of populations seeking cheaper land or housing and more pleasant rural landscapes at a distance from urban areas, which can contribute to rural gentrification and shrinking urban regions.

Creative class is a concept, devised by Richard Florida, that includes a broad swathe of professionals who are considered to be the iconic citizens of the knowledge-based economy (e.g., lawyers, engineers, scientists, doctors, architects, designers, and artists) and who supposedly share similar approaches to complex problem solving and a common work ethos that values individuality, difference, and merit. City branding initiatives cater to this class and seek to provide an urban environment that offers 3Ts (technology, talent, and tolerance). *See* **Creative city theory, Creative industries**.

Creative city theory deploys cultural planning and policy to favour a flexible creative capital model of urban development that privileges knowledge, creativity, and commodified difference as a means of civic renewal. *See* **Creative class, Creative industries**.

Creative industries are activities that are based on individual creativity, skill, and talent with the potential to create wealth and jobs through developing intellectual property. *See* **Creative city theory, Creative class**.

Crime denotes an illegal act for which the enactor is punishable by law, usually through prosecution by the state. There is no singular definition of a crime, and it is commonly considered a category that has been created through law.

Crisis is a concept social scientists have been using for decades as a way of interpreting various periods of profound social and economic change in the urban experience under capitalism. As a system capitalism has been viewed as inherently crisis-prone and certain cities have become emblematic symbols of boom (and bust) throughout history.

Critical geopolitics refers to an understanding that the arenas of the geopolitical are not just "out there" waiting to be discovered but rather are socially constructed through human geopolitical praxis, as political interests carve up of the earth's surface and its peoples into (primarily) states and nations. Critical geopolitics, in conjunction with feminist geopolitics, has been useful in bringing into view the nationalist, masculinist, chauvinistic, and racist ideologies that underlie the myriad violent conflicts that divide and conquer, resulting in the drawing up of the world of nation-states that today dominate the global map. *See* **Geopolitics, Urban geopolitics**.

Cultural capital, as defined by Pierre Bourdieu, is the symbolic power derived from the accumulation of manners, credentials, knowledge, and skills that are acquired through education and upbringing. There are three kinds: embodied (learned bodily movements that show class or manners); objectified (the choice and consumption of things such as art, objects, food, dress, and buildings); and institutionalized (educational degrees and academic titles).

Cultural economy captures the idea that culture and the economy are inseparable and mutually constitutive, meaning that economies are shaped by culture and culture is shaped by economies. The term can also refer to a specific urban economic system that is dominated by cultural industries such as film, art, media, fashion, food, and advertising.

(A) **De-differentiated city** is one in which housing, amenities and facilities, and workplaces are evenly distributed without inequalities between neighbourhoods. Social-class differences are likewise invisible in the landscape of a de-differentiated city or are substantially muted (there are no discernible "rich" or "disadvantaged" neighbourhoods, for example).

Deep democracy refers to practices of self-organization and mutual interdependency among the urban poor—practices that deepen democracy beyond liberal forms like the franchise or basic civil liberties to encompass expanded access to resources and livelihoods in the city.

Deindustrialization is the economic transition marked by a decline in heavy manufacturing activities and a shift to service-sector jobs. In many economies, it has been accompanied by neoliberalization and meant a decline in quality of life for former industrial communities.

Deinstitutionalization describes the release of institutionalized individuals from institutional care (e.g., psychiatric hospitals) into the community for social service, health, and housing support.

Democratic racism is a concept that encompasses a variety of discourses that grow out of the collision of democratic liberalism and discourses such as blaming the victim of racism rather than addressing the institutionalized racism prevalent throughout the media, legal system, education, and social services, among other institutions. See **Racism**.

Demographic transition refers to changes in birth and death rates in the population of a country. Improvements in nutrition and medical care bring down mortality rates, even as high birth rates remain, leading to a high rate of population growth. Eventually, fertility and birth rates decline and the population stabilizes or may even decline.

Densification refers to processes whereby an area's residential density (e.g., the number of residents per square kilometre) increases over time, typically as a result of public policy and/or development practices. Residential density, associated with relatively low greenhouse gas emissions, should also be distinguished from density of buildings, which may or may not contain many residents.

Desakota is a term coined by geographer Terry McGee, based on the Indonesian *desa* ("village") and *kota* ("city"). It refers to the pattern of urban growth seen in densely populated, lowland, rice-producing areas of Asia, where urban and industrial landscapes exist alongside features that are considered rural: farms and fisheries, non-cash barter economies, and rural planning and administrative structures. See **Peri-urban**.

Determinants of health are the social, economic, environmental (natural and built), and biological characteristics that can impact upon health behaviours and outcomes. It contrasts with earlier conceptions of health, which held that biological characteristics were the dominant determinant.

Developmentalism is a discourse through which most cities in the global South are assessed as lacking in qualities of cityness and, in terms of urban theory, can therefore only be understood by their difference from cities in the global North, which form the objects of urban theorizing.

Difference expresses the ways in which social groups are always defined in relation to one another and, in this sense, is best understood as relational connectedness rather than essential or timeless opposition.

Digital citizenship refers to the relationship between politics, political subjects, and virtual space in the contemporary era of new socio-technical arrangements. See **Citizenship**.

Discourse is a term used by the French philosopher Michel Foucault to describe the ways that power operates through the organization of knowledge; discourses produce inherent norms and set parameters for how identities, social practices, spaces, social problems and forms of agency are understood. Discourses are also a subtle form of power, because they work by setting inherent limits on the possible, rather than by directly prohibiting subjects from acting in certain ways, and while discourses make certain norms and parameters seem natural and inevitable across an entire social field, they also make it possible for people to resist and transform them from within.

Disease burden is the impact of a disease or health problem in terms of financial cost, mortality, morbidity, and other factors.

Displacement is the forcing out, directly or indirectly, of long-established and working-class residents of a neighbourhood typically by landlords seeking higher rents or higher-income residents, reducing local housing affordability. It is typically associated with redevelopment and gentrification. See **Dispossession**, **Gentrification**.

Disposable populations refers to groups marginalized by attributes such as race, gender, sexuality, and class who are perceived as having little or no value in a given social or economic system and therefore are consigned to fend for themselves.

Dispossession (in the city) is a process by which land and houses belonging to low-income communities in the city are expropriated to make way for urban development projects. See **Displacement**, **Gentrification**.

Distress is an important concept in urban public finance and can be measured to determine the fiscal and overall social and economic health and well-being of a municipality. Indicators used include population, income, poverty rates, employment rates, and other characteristics of well-being such as social inclusion and citizen engagement.

Division of labour refers to the specialized undertakings of various tasks. Feminism emphasizes the unequal distribution of domestic work between men and women as a result of gendered discourses. See **Labour**, **New international division of labour**.

Drones are aerial vehicles that are usually armed with missiles but have no on-board human pilot. They have been used in attacks in wars since the 1980s, especially by the United States, which has increased its use of drone strikes against urban targets in foreign countries and elsewhere as part of its War on Terror.

Ecological footprint measures human demand on ecosystems, estimating the amount of land and water area necessary to supply the resources a human population consumes and the waste it produces. Although calculable at any scale, cities, owing to population concentration, have large ecological footprints and have become the major foci for footprint reduction.

Economy is the wealth and resources of a city, region, or nation and a framework for managing the production and consumption of goods and services that now also operates on the global scale.

Edge cities exist on the outer reaches of metropolitan areas with all the facilities and services that would be expected in downtown areas.

Emergency urbanism refers to processes of urbanization that occur during emergencies, particularly large-scale humanitarian crises, such as the establishment of large areas of temporary accommodation. See **Urbanism**.

Encounter speaks to the everyday interactions between urban residents of different socio-cultural backgrounds. Streets and public parks are quintessential urban public spaces of superficial encounters and chance social contact, which often comply with norms of social civility but may offer little opportunity to learn respect for difference, or solidarity with, the "Other." See **Public space**.

Entrepreneurial city is focused on initiating rather than simply managing the effects of economic development. This includes attempts by local authorities to engage in city marketing campaigns, investment-friendly strategies, and highly speculative economic development projects to compete with other cities for inward investment. See **Entrepreneurial urbanism**.

Entrepreneurial urbanism describes how entrepreneurial practices in many cities (e.g., city marketing, investment-friendly strategies, and highly speculative economic development approaches) have become a way of life for many urban places in the context of neoliberalism. City authorities have reoriented themselves away from state priorities to manage growth and towards those that promote the construction of spectacular places in order to attract and compete for inward investment. See **Entrepreneurial city, Urbanism**.

Epidemic is a widespread occurrence of an infectious disease in a community or area at a particular time. See **Communicable disease**.

Epidemiologic transition is the shifts in disease burden brought on by advances in human development. They include the advent of infectious diseases in early urban cultures and the transition from infectious to non-infectious diseases in advanced urban cultures. See **Disease burden, Epidemiology**.

Epidemiology is the branch of medicine science that deals with the incidence, distribution, and control of disease in large populations and the application of this study to the control of diseases and other health-related problems. See **Disease burden, Epidemiologic transition**.

Ethnoburb is a suburb that contains a majority, or large minority, of people with a common ethnic background. It typically includes many first-generation immigrants. See **Suburb**.

Ethnography is a research methodology that draws on mixed qualitative methods to generate fine-grained accounts of everyday life in specific places emphasizing the experience and standpoint of research participants. Typical methods include open-ended interviews and participant or non-participant observation.

Ethno-racialized refers to a socially constructed feature of groups of people who are singled out for unique treatment on the basis of real or imagined cultural and physical characteristics. See **Racism**.

Everyday life refers to mundane daily rituals oriented to commodity consumption on the one hand and socialistic modes of relating on the other—the latter of which are conceptualized, particularly by Lefebvre in the literature on the right to the city, as a potential site of radical transformative politics. See **Right to the city**.

E-waste, or electronic waste, includes used or surplus computers, cellphones, monitors, televisions, and other electronic devices, some of which is are still in good working order or may require refurbishment, and some of which is are only suitable for dismantling and recycling. E-waste recycling recovers valuable resources such as gold, silver, aluminum and copper, but it also releases harmful toxics and is hazardous work in the absence of stringent health and safety precautions.

Export processing zone is an administratively defined area within which firms are given incentives such as lower taxes, regulatory exemptions, and subsidized utilities. These zones are typically set up by countries in the global South to attract foreign investment and generate employment, but working conditions in such zones have often been criticized. See **Offshoring**.

Extended metropolitan region (EMR) is a term that is used to differentiate an administratively delimited metropolitan region from the actual extent of urbanization, particularly when the former is smaller than the latter. This term proves useful when comparing agglomerations across countries that might have different criteria for bounding metropolitan areas, although the criteria for delimiting the EMR can differ across researchers and organizations.

Family is a group of people related by blood ties across generations who may or may not live together. See **Household**.

Feminism is a global political movement that seeks to challenge the subordinate position of women and other marginalized groups. Never a unifying political movement, it includes, among others, liberal, socialist, postmodern, postcolonial, and anti-racist positionings while also sometimes depicted as having "stages" that encompass first-, second-, third-, and fourth-wave feminism. See **Second-wave feminism**.

FIRE is an acronym standing for "finance, insurance and real estate," sectors recognized as the key industries associated with so-called global city economies. These sectors are emblematic of the rise of service-based urban economies and, in particular, of producer service economies in major urban centres.

Flâneur is a term made popular by Walter Benjamin to describe a person (usually a white male) who idles his time away in the city streets "just looking," in search of hidden truths and pleasure. Although a marginal figure for Benjamin, the flâneur has come to have a surprising influence on thinking about subjects in the city, both as a signifier for cruising and as an exemplar for modern subjectivity. See **Psychogeography.**

Floor area ratio (FAR) is also called floor space ratio and represents the ratio of a building's total floor area to the size of the piece of land upon which it is built.

Food desert describes a food-insecure area where residents face economic and geographic barriers to accessing fresh, affordable, and healthy foods. These are places with limited transit options and high rates of carless residents, where a lack of supermarkets puts residents in a position to purchase foods of inferior quality and freshness at small markets or fringe food retailers at inflated prices. See **Food security**.

Food security is generally defined as a situation whereby individuals have at all times the physical, social, and economic access to sufficient, safe, and nutritious food that meets dietary and cultural food preferences. It is distinguished from food sovereignty, which expands security to include the right to define how food is produced (the food system) and is linked to the rights of small-scale community farmers as opposed to the contemporary industrial food system. *See* **Food desert.**

Foreclosure refers to a process by which the collateral—the property—used to secure a mortgage is repossessed and ownership is transferred to the lender or an assignee of the lender. It can take between 40 days to over a year, and once foreclosed, the property is put up for sale. *See* **Mortgage, Mortgage-backed securities.**

Foreign direct investment (FDI) refers to flows of investment from "offshore" companies into either existing assets or new ventures of a given country. Ownership in such cases is held by those living in a separate country, and often profits are repatriated back to the home country where the corporation is based.

Full-cost recovery describes the financing of service delivery entirely through user fees.

Garden City movement is a movement within urban planning during the late nineteenth and early twentieth centuries that aimed to create master-planned hub-and-spoke format communities surrounded by greenbelts, with the goal being to increase access to green space.

Gated communities are residential areas of cities with protective measures such as barriers, fences, gates, and private security guards designed to exclude social groups deemed undesirable and dangerous.

Gateway city is one that serves as a central entry and exit point for flows of people, commodities, investments, and employment. Most gateway cities represent important nodes in the world economy, as they offer key destinations and spaces where global activities "touch down" and diffuse across national and regional areas.

Gay village is a term used to describe urban enclaves that see a concentration of social activity, political organizing, residential settlement, and/or commercial activity by people who identify as part of LGBTQ communities. Gay villages depend on complex configurations of visibility and invisibility in order to enable LGBTQ people to congregate while avoiding the increased risk of homophobic and transphobic violence that greater visibility can incur, and yet gay villages are themselves often profoundly hierarchical sites, in many instances giving way to gentrification and privileging some members of the LGBTQ communities (often white, middle-class gay men) while attenuating social, economic, and sexual membership for many others. *See* **LGBTQ, Queer.**

Gemeinschaft is a German word developed by the sociologist Ferdinand Tönnies to describe community groupings based on feelings of togetherness and on mutual bonds of kinship. *See* **Community, Gesellschaft.**

Gender refers to the socially constructed, as opposed to biological, differences between men and women. Although commonly used in relation to women and femininity, it also refers to men and masculinity; current understandings of gender have a focus on performativity—that is, as something that people "do," embodied acts and gestures that when repeated come to take on the appearance of the "real." *See* **Patriarchy, Reproduction of the labour force, Social reproduction.**

Gentrification is a process of capital investment, neighbourhood renewal, and rebuilding aimed at attracting more-affluent people into formerly deteriorating urban areas. It is important to note that this process, in which older, often working-class houses in an inner city are renovated by more-affluent residents, displaces poorer residents. *See* **Displacement, Dispossession, Green gentrification, Rent gap, Stage model of gentrification.**

Geopolitics is a field of study that developed in the early twentieth century and is concerned with investigating the impacts of geography, both physical and human, on international relations and the resultant territories of nation-states. It assumes there is a supposedly objective and determinate relationship between geographic territory and global political interests that can lead to "laws" being unearthed, such as Mackinder's Heartland of the World Island, Mahan's theory of sea power, and Ratzel's *Lebensraum*. *See* **Critical geopolitics.**

Germ theory is a theory of disease transmission that emerged in the late nineteenth century, which postulated that diseases were passed on by micro-organisms. Prior to the advent of this theory, pseudo-scientific theories of "miasma" dominated public health discussions.

Gesellschaft is a German word developed by the sociologist Ferdinand Tönnies to describe community groupings based on instrumental, rational, impersonal, and voluntary forms of social interaction. *See* **Community, Gemeinschaft.**

Ghost estates is the local term given to the unfinished estates that were left standing in Dublin and elsewhere in Ireland after the global financial crisis popped the Irish housing bubble. In many cases, construction sites remained frozen in time, equipment often unmoved since the day the crisis hit.

Global cities, also called "world cities," are commonly seen as leading metropolitan regions in a country that boasts a strong articulation with the global economy, have large foreign or immigrant populations, and are command centres of the globalized economy.

Globalization is the process of increasing economic, demographic, and cultural integration and interaction between people, cultures, companies, and governments in a global context, resulting in greater cultural exchange, international trade, and heightened global interconnectivity.

Governance, in contrast to government, usually refers to a process of governing through a mix of modalities (e.g., institutional [state], private [the market], and civil society actors) that can be understood to be in various competitive or collaborative relationships with each other. The rise of governance has been associated with the neoliberalization of democratic societies.

Gray spaces are spaces of inbetween-ness often associated with displacement in which individuals are neither bona fida residents of a location nor temporary guests.

Green cities are urban places where citizens, politicians, and urban planners prioritize the development of resilient infrastructures, institutions, and behaviours to address the problems associated with climate change. Commitments are given to reduce greenhouse gas emissions, urban planning and urban policy leadership is provided to support changes in urban lifestyles aligned with more sustainable ways of life, and alternative economies are supported that are low-carbon, resource-efficient, and socially inclusive. See **Green sustainability agenda**.

Green gentrification refers to social displacement of poor residents as a result of neighbourhood improvements associated with environmental amenities, like parks and trees, or the reduction of environmental harms, like local pollution. See **Gentrification**.

Green sustainability agenda is an interpretation of sustainable development that emphasizes environmental sustainability. It focuses on conservation, biodiversity, and the protection of natural resources, with particular attention paid to the delayed, dispersed, and long-time ecological effects that affect future generations, and its proponents are often environmentalists based in the global North. See **Green cities**.

Greenhouse gas is a gas that, when dispersed into the atmosphere, traps heat on the planet's surface. In order of prominence, the three most important greenhouse gases are carbon dioxide, methane, and nitrous oxide. See **Climate change**, **Production-based carbon accounting**.

Growth poles refer to the idea long promoted by economic development planners that peripheral economic development is best supported by focusing investment in a few locations to promote economic synergies, instead of by widely spreading investment geographically.

Habitat III is shorthand for a major global summit, formally known as the United Nations Conference on Housing and Sustainable Urban Development, held in Quito, Ecuador, in October 2016. It is the third in a series that began in 1976 to reinvigorate global political commitment to the sustainable development of towns, cities, and other human settlements, both rural and urban; this reinvigoration is being referred to as the "New Urban Agenda" that will set a new global strategy around urbanization for the next two decades. See **New Urban Age**.

Healthy cities are places where people are nurtured in healthy behaviours (including a healthy diet, adequate exercise, and limitations in exposure to environmental pollutants) to maximize quality of life (which includes not only the absence of disease but also the presence of good health).

Heavily Indebted Poor Country (HIPC) is an initiative started in 1996 and initially composed of a group of 39 countries in the global South with high levels of poverty and debt considered to be unsustainable. They are eligible for special monetary assistance through debt relief and low-interest loans from the World Bank and International Monetary Fund.

Heroic informality is the celebration of informality as deliberatively resistant to planning and regulation. It also praises the entrepreneurial nature of low-income urban dwellers and their ability to survive despite the odds against them. See **Informal settlement, Informal urbanism, Informal urbanization, Informality, Slum, Squatter settlement**.

Heteronormativity is a shorthand term for normative heterosexuality, which refers to a virtually all-encompassing regime in which an individual is sexually attracted only to a person of the opposite sex, with this assumed to be the natural and universal norm or way of being human. Heteronormative logics simultaneously render pathological, odd, or queer any intimacy or embodiment that falls outside them. See **Heteropatriarchy, Queer, Sexual citizenship, Sexuality**.

Heteropatriarchy is the social, cultural, and political condition in which male dominance and heteronormativity carry and reinforce one another. It tends to rely on essentialist, binaristic, and naturalized understandings of appropriate gender and sexual roles in organizing divisions of labour and distributing agency, vulnerability, and autonomy, and takes material form in the city by normalizing some forms of inequality and intimacy while rendering others as private, frivolous, or apolitical. See **Heteronormativity, Queer, Sexuality, Sexual citizenship**.

Hidden austerity is a term used to describe much of Canada's response to government austerity policies. Despite the evidence that Canadian governments at all levels have implemented austerity policies, much of it is hidden in public discourse, as Canadians still map a national imaginary of collectiveness onto urban political decisions that in reality signal a fading of redistribution policies in many parts of the country.

Homelessness describes the situation of an individual or family without stable, permanent, appropriate housing, or the immediate prospect, means, and ability of acquiring it. It thus describes a range of housing and shelter circumstances, with people being without any shelter at one end and being insecurely housed at the other, the result of systemic or societal barriers, a lack of affordable and appropriate housing, the individual/household's financial, mental, cognitive, behavioural, or physical challenges, and/or racism and discrimination.

HOPE VI is a program, funded by the US federal government, that identified and demolished selected social housing communities and replaced them with new "mixed" tenure communities, often based on new-urbanist design principles.

Household is a group of people, who may cross generations, who may or may not be related to each other, and who live together in the same dwelling(s). See **Family**.

Housing bubble refers to a rapid increase of market-housing prices, often outside alignment with "market fundamentals," such that an overvaluation of housing occurs relative to these fundamentals, which include the rents that such housing would provide on the rental market and the incomes out of which households pay for housing. It is colloquially said that a housing bubble only appears after it bursts, and one cannot know for sure if there has been a bubble until after the fact. See **Housing market**.

Housing market refers to the institutions and procedures that bring together housing supply and demand (e.g., buyers and sellers, renters and landlords, builders and consumers) for the purposes of exchanging resources. See **Housing bubble**.

Housing stock involves the total of individual residential dwelling units currently occupied or available for occupancy, which includes both market and non-market housing units. *See* **Housing system**.

Housing system involves interrelationships between all of the actors (individual and corporate), housing units, and institutions that regulate the production and consumption of housing. Housing system is a broader term than housing market or housing stock and includes methods and structures of housing-based finance, as well as government housing policy. *See* **Housing stock**.

Human agency refers to the capacity to act independently. Such actions occur in relation to other agents, both human and non-human, and social structures (e.g., social class, religion, gender, sexuality, and ethnicity), which set limits on agency, although the nature of the relationship between structure and agency is contested.

Humanism is an approach to studying urban geography that views people as purposeful agents of change in cities. The aim is to understand human social behaviour using methodologies that privilege subjective perceptions and experiences of cities and the human meanings, values, and significance associated with them.

Hyperbuilding is a form of spectacular, state-led urban development that often symbolizes the state's expression of both economic and political sovereignty—a gesture that is typically a claim to power both on the world stage and locally. These projects are sometimes a cornerstone of speculative real estate markets.

Incineration refers to the process of taking waste materials and turning them into ash, flue gas, or heat.

Inclusive design is a design theory that seeks to create spaces that diverse groups can use and enjoy. Consultation with different user groups is often a key part of the process, where heterogeneity is valued.

Incremental urbanization describes forms of urbanization that happen by the accretion of many small steps. The pace of this form of urbanization can be gradual or rapid, but it usually produces urban space through numerous small, cumulative, and independent actions that are undertaken by many different people over time.

Indigeneity is a term used to describe the basis for Indigenous rights claims and mobilization to advance the political and cultural project of embedding Indigenous self-determination in society's institutions at all levels. *See* **Indigenous peoples**.

Indigenous density is a correction to the prevalent focus on how Indigenous communities may be "different" from the settler cultural mainstream. Indigenous "density" relates to an understanding of self-determination that is not constantly focused on trying to demonstrate how an Indigenous approach would be different from a non-Indigenous approach: self-determination requires no such rationale, and indeed Indigenous approaches may incorporate strategies consistent with the settler mainstream, in whole or in part, when suitable to the circumstances. *See* **Indigenous-inclusive citizenship**.

Indigenous-inclusive citizenship hinges on the notion that Indigenous self-determination, an Indigenous group right, is essential to common national citizenship within a settler nation-state, thereby invalidating the simplistic view that universal citizenship presents. Universal citizenship, on the other hand, centres on the common identity that is imagined to attach to all individual citizens equally within a nation-state framework and does not adequately account for the prior occupancy, treaties, and group rights of Indigenous peoples. *See* **Indigenous density**.

Indigenous peoples is a term encompassing descendants of the original inhabitants of settler countries around the world, including Aboriginal peoples in Canada. "Aboriginal" is the legal term used to refer to Canada's Indigenous peoples in the Constitution Act of 1982—the First Nations, Métis, and Inuit peoples. *See* **Indigeneity**, **Reconciliation**, **Urban reserve**.

Indigenous planning is an approach to community planning by Indigenous peoples built on cultural identity, place, kinship, stewardship, and world view, which informs goals, processes, and techniques used to connect past with present while planning for the future.

Indigenous urbanism refers to the participation in, production and re-territorialization of urban space using Indigenous approaches, and the enjoyment of a good urban life by Indigenous peoples. *See* **Urbanism**.

Industrial Revolution is the era, beginning in the late 1700s, of rapid change in the production technologies that resulted in hitherto unseen productivity and economic growth in Europe.

Inequality is a structural power relation, a condition of being unequal, and refers to the widening of economic and social divisions in contemporary society. It is commonly referred to economically in relation to inequalities in income, wealth, or resources and socially in relation to differences in caste, class, ethnicity, gender, race, sexual orientation, and other axes of difference as a result of unequal opportunities and rewards for social positions or statuses within a group or society.

Informal economy is economic activity that falls outside governmental regulation or taxation. Also referred to as the black market or shadow economy, it forms a major sector of the economy in the countries of the global South.

Informality arises when economic actions that are not in themselves criminal in some way evade or challenge state regulation. An example includes construction that violates zoning or building regulations. *See* **Heroic informality**, **Informal urbanism**, **Slum**, **Squatter settlement**.

Informal settlement is a neighbourhood that has developed without state oversight or approval for which residents typically lack formal land title. *See* **Heroic informality**, **Informal urbanism**, **Informal urbanization**, **Slum**, **Squatter settlement**.

Informal transport is urban transport provided to passengers for a fee by private entrepreneurs lacking official recognition, often using small motor vehicles in contravention of laws.

Informal urbanism takes place in many forms, although in most cases it emerges as a result of the absence, insufficiency, or unaffordability of housing options for low-income populations in formal urban spaces, and in many countries, processes of informal urbanism constitute the dominant force of urbanization. *See* **Heroic informality, Informality, Informal settlement, Informal urbanization, Occupancy urbanism, Slum, Squatter settlement, Urbanism.**

Informal urbanization is the process of urbanization occurring independently from formal frameworks that comply with official rules and regulations. *See* **Informal urbanism, Urbanization.**

Informal waste collectors are people who pick or buy waste but who do not work for a city's municipal or contracted waste-collection service.

Information age (also known as the computer or digital age) is the period characterized by the shift from traditional industry that the Industrial Revolution initiated to an economy based on the computerization of information. Its onset is associated with the digital revolution of the late 1950s when the adoption of computers and the use of the Internet created a knowledge-based society in which technology is inherent to (an increasingly paperless) daily life and social organization.

Infrastructure refers to the facilities and services that are needed for the functioning of a society. It includes physical facilities such as transportation, water supply, sewage treatment, Internet connectivity, and electrical power, but also various institutions such as emergency services, health-care systems and education. *See* **Infrastructure networks.**

Infrastructure networks are physical "bundles" of transport, telecommunications, energy transmission, water, and wastewater—infrastructure that enables human settlements. *See* **Infrastructure.**

Inhabitance expresses the everyday experience of living in—or inhabiting—the city and forms the basis for claims to the right to the city. *See* **Everyday life, Right to the city.**

Instant urbanization refers to processes of urbanization that appear to occur very rapidly and often involve the development of large parcels of land using significant investments in capital and labour to produce ambitious development projects. *See* **Spectacular urbanism, Urban mega-events.**

Insurgent architect is a metaphorical urban actor who is not a professional architect but rather represents for David Harvey the power and capacities that humans have to critique the built environment and to make changes to it. The word "insurgent" means to rise in active revolt against established authority.

Internally displaced persons (IDP) are those who have been forced to flee their homes because of persecution, war, natural disaster, or other reasons for forced displacement. They differ from refugees in that they are people who have not crossed an international frontier. *See* **Refugees.**

International Monetary Fund (IMF) is an international organization of 188 countries, established in 1945, with the goal of promoting international monetary cooperation, facilitating international trade, promoting high employment and sustainable economic growth, and reducing poverty worldwide. It makes loans available to members experiencing balance of payments difficulties; however, this practice has been widely criticized for its market-oriented approaches that impose austerity, privatization, and deregulation programs across the global South.

Intersectionality refers to the multiple axes of structural difference, such as race, class, gender, age, and sexuality, expressed in individuals and groups that have the potential to oppress. It is the lived intersection of these social structures that affects the ways in which lives are lived among oppressed groups.

Islamophobia refers to anti-Muslim sentiment based on fear of Muslims, which intensified after 9/11 2001. *See* **Racism, Xenophobia.**

(The) Just city strain of planning theory advocates a redistribution of wealth and opportunity, guided by principles of humanity and justice, in a manner that operates within, rather than transcends, a fundamentally capitalist mode of accumulation.

Keynesian-inspired policy was formulated by the British economist John Maynard Keynes and was widespread in the global North during the latter part of the Great Depression, the Second World War, and the postwar economic expansion (1945–1973), losing its influence in the mid-1970s with the advent of neoliberalization. Keynes argued that private-sector decisions often lead to inefficient macroeconomic outcomes that require active policy responses from the public sector, especially policy actions by the central banks and government; thus, Keynesian economies promote a mixed economy based primarily on the private sector but with an active role for government intervention, especially during recessionary times.

Labour is the aggregate of all human physical and mental effort used in the creation of goods and services. Labour is also a primary factor of production in economic systems, and Marxist economics theorizes that the gap between the value a worker produces and her/his wage is a form of unpaid labour, known as surplus value (or profit). *See* **Capitalism, division of labour, Labour market, New international Division of labour, Reproduction of the labour force.**

Labour market is a socially constructed, political-geographic institution, typically governed by state regulations that facilitates the buying and selling of labour power, in most contexts in classed, gendered, and racialized ways. *See* **Capitalism, Labour.**

Land-banking is the process of buying up large parcels of land many years in advance of their being developed. Land may be banked by a municipal government in order to control the planning of the future expansion of the city, or it may also be banked by private-sector developers as a way of enhancing the profitability of future development (the assumption being that the value of the land at the time of development will be much greater than when it was purchased many years in the past). *See* **Land banks.**

Land banks in the United States constitute public authorities in large urban centres that hold the titles of once-vacant or blighted properties, maintain a civic inventory of land, and "bank" available properties to be sold back into use and to contribute to the tax base. Developed over the last three decades, this form of urban land governance positions "urban blight" as a community asset for development and revitalization. See **Land-banking**.

Land grabbing refers to the political power exerted by national and transnational capital to appropriate land from local communities for the purposes of privatization and commodification. In the process, local communities become alienated from their land and lose access to resources on which they depend on for their livelihoods.

Land title is conferred upon those who legally purchase ownership rights to land. There are different forms of land title, but if one has title to land, one has certain legal rights that one can exercise with respect to that land.

Leapfrog development, also known as discontinuous development, occurs when development takes place on sites that are separated from the main urban area by vacant land. It is a common feature of urban sprawl, and it also refers specifically to development beyond growth-inhibiting conservation areas such as greenbelts.

LGBTQ is an umbrella term for lesbian, gay, bisexual, transgender, and queer communities. Although widely accepted, there remains robust debate among different sexual and gender minorities about whether the term is adequately inclusive of all identities or what a better term to describe people on a spectrum of sexual and gender diversity might look like. See **Gay village**, **Queer**.

Lifeworld is a term coined by the German philosopher Edmund Husserl that refers to the sum total of physical surroundings and everyday experiences.

Livelihood refers to the capacity and means through which one secures basic human needs or pursues related tasks. It is often defined in terms of activities focused on subsistence tasks (e.g., producing food, collecting water and fuel wood, securing shelter and clothing) or income-generating tasks (e.g., employment).

Living wage is the amount of wages a worker requires to afford a standard of living that is deemed be acceptable to the majority of the population. It is set above the minimum wage, a legal requirement in many countries, but often only allows people to live above the poverty line.

Lofts are large, open areas typically found in factories, warehouses, or other large buildings, commonly in downtown or waterfront locations, that have been converted into living and/or working spaces.

Loneliness is a complex and usually unpleasant emotional response to isolation or lack of companionship. It typically includes anxious feelings about a lack of connectedness with other people, both in the present and extending into the future; as such, loneliness can be felt even when one is surrounded by other people (such as in a crowd in a city).

Market failure refers to a situation where the self-interested economic pursuit of profit for a few results in the inability of the market to effectively or efficiently meet the collective basic goods and services needs of its citizens. It includes, for example, the market's failure to deliver public goods such as water.

Market socialism is the term used in Vietnam (and China) to describe a market-oriented economy—one that has, for example, privately owned enterprises and prices set by the market—operated by a socialist state. The term acknowledges that socialist states are more commonly associated with centrally planned economies, which often involve state-determined price-setting, subsidies and rationing, and a partial or full state monopoly on production and distribution of all goods.

Median asking rent is a measure used to describe the cost of rental housing "on the ground." It refers to the mid-price point of all properties in a given area currently for rent. It does not include the rental cost of currently occupied properties. See **Median monthly gross rent**.

Median monthly gross rent is a measure used to describe the mid-price point of all monthly rents plus utility costs for a given area. It includes both properties that are occupied and those that are on the housing market. See **Median asking rent**.

Medical geography is a branch of geography concerned with the effects of space and place on health behaviours and outcomes that is closely related to health geography and spatial epidemiology.

Medical model of disability conceives of disability as an individual's impairment, implying that the problem is with his/her/their body. See **Biopsychosocial model of disability**, **Social model of disability**.

Medico-psychiatric models of mental illness understand mental illness/distress as individual pathologies rather than as embodied social experiences.

Megacities are exceptionally large metropolitan areas, commonly defined as those exceeding 10 million people. The term may imply that urban growth has defied rational planning. See **Metacities**.

Megalopolis is a term coined by urban geographers to describe very large urban areas that have grown to encompass what were previously distinct conurbations. These areas are not continuously built up, but the whole area is understood to function as an integrated economic unit. See **Megaregions**, **Metacities**.

Megaregions refer to extensive agglomerations of urban regions with increasing economic and infrastructural ties with each other. See **Megalopolis**.

Metacities, also known as hypercities, are exceedingly large metropolitan areas, agglomerations with populations of 20 million or more. Sometimes the term "metacity" is used to refer to the pattern described by the term "megalopolis." See **Megacities**, **Megalopolis**.

Middle class is a widely used term to describe a socio-economic group positioned between the upper and working classes. The middle class often includes individuals in professional and business employment categories, families living in urban and suburban areas, and key consumers in contemporary society. See **Working class**.

Migration is the movement of people (migrants) out of the country in which they live (emigration) to migrate to another country (immigration) in which they are not a member or citizen, with the aim of living there permanently or temporarily. Although migration can take place within countries, it is commonly understood in relation to the crossing of national boundaries. *See* **Rural–urban migration**.

Militarism refers to a militarized system and set of ideas that a society adopts to sustain peace and/or prepare for war. *See* **Militarization, Military urbanism**.

Militarization is a process by which something becomes controlled by or derives its value either from the military as an institution or from militaristic criteria. An analysis of militarization therefore takes us beyond the actions of militaries and into complex social relations produced through and in relation to militaristic institutions and ideas. *See* **Militarism**.

Military urbanism is a response to a fear of crime, interpersonal violence, and terrorism in cities that uses urban design, urban planning, and policing in a comprehensive security effort to produce defensible space and technological surveillance at the city-wide level. By foregrounding security as a motif of urban planning and design and hardening the city in different ways with walls, gates, security guards, and cameras, the city is treated as a fortress. *See* **Urbanism, Closed-circuit television**.

Modal share is the share of all trips carried out by a specific mode, usually divided into public transit, private motor vehicle, walking, and cycling.

Mode of production refers to the ways in which human beings produce and reproduce their means of subsistence. In any given mode of production (e.g., feudalism, capitalism, socialism, or communism), people enter in a definite relationship with instruments of production (i.e., tools, machines, and buildings), subjects of production (i.e., raw materials), and each other through their labour. *See* **Capitalism**.

Modernism is a philosophy in urban planning and architecture that became prominent from the 1920s onward and is driven by a desire to advance the social project of universal progress. Modernism seeks to correct urban problems and to improve social life through interventions in urban form.

Modernist urbanism is a way of living (imagined and/or actually existing) that confronts urban problems by breaking with the past and traditional ways of living in cities. *See* **Urbanism**.

Modernization is model of development that seeks to transform traditional, rural, agrarian societies into secular, urban, industrial ones. It is based on the premise that the Western industrial model of social transformation is one to which all other societies should aspire.

Mortgage is a loan for the purchase of real property that is secured by collateral contained in the property. Mortgage terms range widely with regard to amortization periods, loan-to-value ratios, minimum down payments, and household income requirements. *See* **Foreclosure, Mortgage-backed securities**.

Mortgage-backed securities (MBS) involve the pooling of many mortgages together under a single security, often being seen as a way of reducing overall risk in a pool of loans. The packaging of mortgages into MBS and the sales of these on the secondary market allow banks to take the mortgages off their books and make new loans. *See* **Foreclosure, Mortgage**.

Motorization is the process of change from movement requiring human (or animal) power to vehicles (automobiles, buses, trucks, and motorcycles), which have fuel-burning engines.

Multi-level governance climate model has as its centre a concept for understanding urban climate politics that focuses on how key policy processes occur outside city government proper and are instead shaped by (a) relations with other levels of government and (b) "horizontal" policy networks, including other city and transnational policy groups.

National urban systems can be defined as the network of cities within a particular country that are linked together by flows of people, goods, money, and information. The fundamental conditions for such networks of cities are the transportation and communications systems that connect cities together.

Natural monopoly refers to a situation where it is more economically efficient when a single enterprise produces the entire output for a given market (e.g., urban water delivery).

Nature, in the broadest sense, is the natural (geological and atmospheric) and physical (including microbes, plant and animal life) world or universe and includes human life itself. It was traditionally understood as separate (and secondary) from (and untouched by) human life but is now critically understood as a social and cultural construction inherently tied to human life. *See* **Socio-nature.**

Neighbourhood, often described as the building block of cities and protected within many official plans, is a specific geographical area within an urban place that may or may not contain social communities and/or social networks. *See* **Community**.

Neoliberal governance refers to a set of government practices that are based on the ideology that the competitive free market is the most efficient way of organizing society and the economy in general. This set of practices became particularly widespread in the 1980s and 1990s in the global North where many governments privatized some services and outsourced others and in the global South many governments were forced to adopt Structural Adjustment Programs. *See* **Neoliberalism, Structural Adjustment Programs**.

Neoliberalism is a political philosophy increasingly popular during the late twentieth century based on privatization, fiscal austerity, deregulation and privatization of business, free trade, and cutbacks in government spending on social programs. It is a modified form of the laissez-faire economic liberalism of the 1970s and 1980s, which confers authority to free markets, capitalism, and private rather than public forms of investment and control. *See* **Neoliberal governance, Neoliberal urbanism, Urban neoliberalism**.

Neoliberal urbanism refers to the ongoing state-, culture-, and finance-led deconstruction and reconstitution of urban space in ways that are deemed neoliberal (see above) and which, since the 1980s, scholars have utilized to theorize an assortment of concepts (e.g., gentrification, post-Fordism, deindustrialization, re-industrialization, revanchism, global city formation, urban entrepreneurialism, and socio-spatial polarization). In Canada, critical scholars have critiqued neoliberal urbanism for its disproportionate impact on racialized and socially vulnerable populations through the remaking of social contracts of previously public services. *See* **Neoliberal governance, Neoliberalism, Urban neoliberalism.**

New Institutionalism is a broad research agenda in the social sciences that focuses on the institutional structures that shape economic, social, and political processes and outcomes and how they change over time. Institutions are commonly defined as the shared norms and understandings, laws and regulations, and standard operating practices in any given society.

New international division of labour (NIDL) is a spatial division of labour that has appeared with the geographic reorganization of production as a result of globalization since the early 1970s. It refers to the spatial shift of manufacturing industries from advanced capitalist countries to the global South that occurred when the process of production was no longer confined to national economies. *See* **Division of labour, Labour.**

New Urban Age is a discourse promulgated by the United Nations to indicate that we live in a world in which over 50 per cent of the world's population lives in an urban settlement. This tipping point into the urban is estimated to have happened at some point in 2007. *See* **Habitat III.**

Non-communicable diseases are diseases that are not non-transferrable between people, including many chronic diseases such as Alzheimer's, cancers, and diabetes. They are more common among populations in the global North. *See* **Communicable disease.**

Occupancy urbanism is the appropriation of surplus real estate with little regard for land titles and hegemonic master planning. *See* **Urbanism, Informal urbanism.**

Occupy Wall Street is a grassroots social movement that emerged on September 17, 2011, in Zuccotti Park in New York City. The movement quickly grew to a worldwide social movement against social and economic inequality, spurred on when the US government prioritized bailing out the banking sector immediately after the 2007–2008 financial crisis at the expense of homeowners and more disadvantaged people affected by the subprime mortgage crisis and subsequent recession.

Offshoring is a process of relocating some or all parts of a business operation (either production or service related) to another, typically lower-wage, country. Offshoring may take the form of *production offshoring*, typified in recent years by the relocation of manufacturing from industrialized countries to lower-wage countries like Bangladesh, China, and Vietnam, or *services offshoring*, which relies on information and communications technology to allow services such as sales and customer support to be rendered remotely. *See* **Export processing zone.**

Open data is information that was previously withheld from the public and hidden in administrative, infrastructure, and service systems but has been made freely available to citizens.

Open dumping is an uncovered waste disposal site that does not have any environmental controls in place to protect the surrounding environment and people.

Ordinary cities are those that are not exceptional in size or significance and that commonly attract less than their share of scholarly attention. A term originating with Jennifer Robinson, it is used by postcolonial scholars to critique a global hierarchy of cities that has largely ignored cities of the global South.

Over-urbanization is an outdated term that referred to rates of urban growth that exceeded what was expected, given a country's level of industrialization and wealth. This was based on a now-discredited notion of urbanization having a direct and deterministic relationship with industrialization and economic growth, an idea based on the experience of urbanization in Europe and North America from the Industrial Revolution onward but was found to be poorly suited for understanding urbanization in the global South.

Palimpsest, when applied to cities, refers to how the built environment of a city is a record of the social, economic, and often political visions that constructed it, where it is often possible to read several layers of historical change in those visions through the newer layers of more recent ideas about social and economic systems and urban planning.

Participatory budgeting is a way that communities' voices might be included in infrastructure planning by providing residents with the opportunity to vote on municipal funding priorities at regular public meetings.

Patriarchy constitutes a form of governance through hierarchical power relations that legitimize, perpetuate, and privilege masculinity over femininity, and permits the domination, oppression, and exploitation of women by men, although this is not to deny the agency of women to bargain with patriarchy. In conjunction with capitalism and racism, patriarchy has constituted a major form of socio-spatial exclusion. *See* **Gender.**

Peri-urban areas are characterized by the intermingling of urban and rural features. In the Asian context, peri-urban areas produce a particular landscape that is referred to as a desakota. *See* **Desakota.**

Pirate settlement is a residential area that contravenes government regulations. This may be because it is not zoned for development or for residential use; it is unserviced, where bylaws require all new settlements to be serviced; or its methods or materials of construction contravene building regulations. *See* **Informal settlement, Informal urbanism, Informal urbanization, Informality, Pirate urbanism, Slum, Squatter settlement.**

Pirate urbanism is the illegal siphoning of inaccessible urban infrastructures that are then made available at low cost. *See* **Informality, Informal settlement, Informal urbanization, Pirate settlement, Slum, Squatter settlement, Urbanism.**

Place is understood as the human transformation of space through the ascription of socially produced associations with particular locales. Urban places are physically embedded in the built environment in neighbourhoods and come into being through the reiterative social practices of inhabitants.

Place-branding refers to a broad set of activities that are aimed to make a place more attractive to a range of groups who will then invest or spend money in it. With the rise of neoliberalism, place-marketing has become an activity in which city governments have invested.

Place annihilation is a term coined by the geographer Kenneth Hewitt to refer to the destruction resulting from war-time area bombings of resident civilians, their communities and neighbourhoods, and any major features of their urban environment and civil ecology.

Planetary urbanization is a contested theory popularized by Neil Brenner and Christian Schmid that postulates that processes of urbanization are now planetary in scale. Although not all places may look urban, they are now incorporated into capitalist urbanization such that there is no longer an urban "outside."

Policy mobilities describe the intercity learning that occurs through the transfer of policies between policy-makers who work to identify and circulate successful urban policy models and to adapt them to local conditions.

Political society is a site of negotiation and conflict opened up as a result of attempts by the poor (who are excluded from the legal and liberal-democratic sphere of civil society) to channel their demands to the state.

Polity refers to the constitution or framework by which an organized unit is governed in a systematic fashion.

Poor doors are separate entrances for lower-income tenants that are built in new residential buildings that have mixed-income (lower- and higher-income) residents.

Population density is the number of people residing on a unit area of land.

Positionality highlights how people see the world from different social, cultural, and geographical locations, with those dimensions most commonly assumed to be of influence being gender, race, class, and sexuality. Geographers have highlighted how positionality is itself both contextual and mercurial, while awareness of one's positionality has been most commonly explored via the methodological tool of self-reflexivity.

Positivism is a philosophy of science that adheres to the scientific method of investigation by using hypothesis testing, statistical inference, and theory construction. This approach dominated the spatial analysis school of urban geography in the 1950s and contributed to models that helped one understand the spatial patterning of settlements (e.g., Central Place Theory). *See* **Spatial analysis**.

Post-asylum geographies emerged after large-scale waves of deinstitutionalization from residential asylums after the 1960s. They imply an understanding of the interweaving of urban geographies and geographies of formerly institutionalized individuals within communities.

Postcolonialism refers both to an intellectual direction (as in postcolonial theory) and to the period of time after colonialism in the mid-twentieth century when colonial countries started to become independent. Postcolonial studies analyze the politics of knowledge production in relation to matters that constitute the postcolonial identity of a decolonized people while rebalancing the power relationships between colonists and colonial subjects by establishing the intellectual spaces for subaltern peoples to speak for themselves.

Post-industrial cities, often referred to as "Rust Belt" cities in the North American context, are commonly characterized by processes of industrial decline, a decreasing manufacturing workforce, and depopulation, all of which contribute to weak urban economies and the slow deterioration of the urban landscape. These are places often reconfigured for middle-class consumption, in which the process of abandonment that characterizes contemporary capitalist and neoliberal state organization has contributed to uneven socio-economic development, income inequality, and social polarization.

Postmodernism is a philosophy that has been influential in the arts and social sciences for its critique of "grand theory" and universal truths. In urban geography, a postmodern perspective emphasizes difference, uniqueness, and individuality, asserting that urban phenomena need to be understood from diverse points of view (e.g., through the intersectional lenses of class, sexuality, ethnicity, age, racialization, and disability). *See* **Intersectionality**.

Post-socialist cities are urban places located in countries that were once, but are no longer, socialist. The built legacies of their socialist era (e.g., large districts of prefabricated apartments) often present challenges to present-day policy-makers. *See* **Socialist cities**.

Poststructuralism refers to a development in social theory popularized since the late 1960s that eschews a totalizing epistemology in favour of understandings that go beyond structures and hierarchical binaries to emphasize plurality of meaning and instability of the concepts that structuralism uses to define society. Its main features include foregrounding discourse, the constructedness of all knowledge and behaviour, and an anti-foundationalist stance that decentres the human subject (as having a coherent identity). *See* **Structuralism**.

Post-suburbanization is the process by which suburbanization matures from a singular process of producing new, low-density urban peripheries to a more complex, multifaceted process that produces variable morphologies and land uses. *See* **Suburbanization**.

Poverty is a systemic condition in which the poor are not able to meet basic subsistence needs, including food and shelter. A significant aspect of poverty also has to do with the stigmatization of poor people and poor people's own perception of their relative status and standard of living in society. *See* **Precarity**, **Urbanization of poverty**.

Precarity refers to living and working conditions that render people vulnerable and excluded with no certainty of security, safety, and employment. Like poverty, it is shaped by structural conditions such as financial crisis, climate change, or war. *See* **Poverty**.

Primate city refers to a pattern of urbanization seen in many countries around the world, where the largest city is more than two times the size of the next-largest city and exerts economic, social, and political dominance. Primacy is often historically rooted in the era of European colonization, when these cities played the role of ports and colonial administrative centres.

Prison industrial complex refers to the relationship between carceral spaces, practices of imprisonment, and a punitive state. Geographers have discussed the relationship between prisons and the city in the context of the racism of "hyperincarceration," particularly in the United States, that incorporates the swing from the social to the penal management of poverty, with a punitive revamping of public policy to address urban marginality through containment, establishing a single carceral continuum between the ghetto and the prison.

Privatization refers to the transfer of public infrastructure to the private sector (full divestiture) or more commonly to private-sector participation in service delivery.

Private good is a good whose benefits accrue to individual owners or users and where it is possible to exclude others from enjoying these benefits. *See* **Public good**.

Production-based carbon accounting is a method of attributing responsibility for greenhouse gas emissions that focuses on what is emitted by activities within a certain jurisdiction, in all possible ways, and that ignores any emissions outside the jurisdiction that may have occurred before the good or service arrived. *See* **Greenhouse gas**.

Provincial city is a regional service centre that is urban owing to its size and service function, but provincial in terms of its place in the urban hierarchy of the nation in which it is located.

Psychogeography explores the relationship between environments and landscapes and the emotions and behaviour of individuals. It is an approach that emphasizes playfulness and drifting (*dérive*) around urban environments, most commonly discussed in relation to the flâneur or the French Situationists influenced by Guy Debord but also applied to anyone walking for pleasure. *See* **Flâneur**.

Public address denotes the political practice of the possibility of formations and contestations of representations. To speak publicly and to address a public requires a public sphere where a public can be communicated with; but, at the same time, a public sphere cannot exist, if people do not address it. *See* **Public sphere**.

Public and private are discursively constructed, contested categories that define the boundaries between households, market economies, the state, and political participation. The binary is central to the ideologies and practices of separate spheres.

Public good is a product that has social benefits that are greater than those that accrue to an individual. In economics, it is a good from which it is impossible to exclude other users. *See* **Private good**.

Public space refers to an idealized understanding of social space considered essential to citizens formulating their right to the city, albeit with restricted access and with some people having more rights than others. *See* **Public space**, **Right to the city**.

Public sphere is an area of social life where individuals can come together to freely discuss and identify societal problems (usually in public space) and through that discussion influence political action. It is a realm for the production and circulation of discourses, separate from the state and market relations, in which public opinion is formed. *See* **Public address**, **Public space**.

Public transit, also called public transportation or transit, refers to vehicles offering regularly scheduled journeys that are open to all passengers (usually in exchange for a fare) and usually carrying multiple passengers whose trips may have different origins, destinations, and purposes.

Quality of life refers to the current social, cultural, political, and economic livability of cities. Livable cities indices, rank cities on an annual survey of living conditions (e.g., education, health care, housing, recreation, transportation, and natural environment), and the top cities have changed little over time—they tend to be mid-sized, to be in wealthier countries, and to have low population densities.

Queer is an umbrella term used to describe LGBTQ and other sexual and gender minority communities and to assert a resistant or anti-assimilationist political identity and set of commitments. It is also a theoretical lens, influenced by post-structuralist, postcolonial, and feminist theories, that approaches sexual identity as a performance (related in space and time to other forms of power like gender, class, and race) rather than as an identity or inner truth. *See* **Gay village**, LGBTQ, **Sexual citizenship**, **Sexuality**.

Racism is the belief that people can be categorized according to (arbitrary) biological categories that determine one's positioning in a hierarchical order of inferiority and superiority, with lightness of skin colour often used as the primary biological marker to determine a raced identity. The contextual specificity indicates, however, that skin colour or other biological markers are never stable bodily signifiers, but this is not to deny that racism has real effects on people's lives and that it is invariably deployed to augment oppressive relations of power. *See* **Democratic racism**, **Islamophobia**, **Xenophobia**.

Rank-size rule describes the consistent empirical finding that when city sizes within countries are ranked in order of population, there is a statistical regularity in the relationship between the rank of the city and the population of each city as a fraction of the population of the largest city. Despite many conforming examples, no theoretical explanation has ever adequately described why this is so.

Reconciliation refers specifically to the process of healing and redistribution between Indigenous and non-Indigenous peoples to repair damage from the abusive colonial relationship. Truth and reconciliation commissions can help national communities bear witness to the deplorable human rights abuses against Indigenous peoples (i.e., the truth). *See* **Indigenous peoples**.

Redlining is a racially discriminating process, originating in the practices of the United States Federal Housing Administration (FHA), by which lenders classify certain neighbourhoods using four different colours to represent the neighbourhood's development trajectory and by extension the level of credit worthiness. Red was the colour used for the category of neighbourhoods that the FHA thought were highest risk. *See* **Mortgages**, **Reverse redlining**.

Refugees are people who have crossed an international frontier because of persecution, war, natural disaster, or other reasons for forced displacement from their countries of origin. *See* **Internally Displaced Persons**.

Rent burden is a measure of housing affordability based on the proportion of household income devoted to rent. The generally accepted benchmark for rent-burdened households or individuals is when 30 per cent or more of income (monthly or yearly) is devoted to rent. *See* **Median asking rent, Median monthly gross rent**.

Rent gap refers to the process whereby a gap emerges between the value of a property (ground rent) and its potential value should it be redeveloped. The rent gap describes, for example, what happens (in theory) when property values in former working-class districts become gentrified. *See* **Gentrification**.

Representation refers to the complex web of language and texts that people use to make sense of a world that cannot speak for itself. Urban scholars continue to produce representations of the urban that are used, modified, or rejected by students, other scholars, and the public according to how persuasive and coherently reasoned they are.

Reproduction of the labour force refers to the activities and processes, conducted by both paid and unpaid labour, that ensure that the labour force can "go to work." The essential elements for Marxists have been conceptualized as those of "collective consumption," namely the provision of goods and services by the state, such as education and housing, while feminist scholars have argued that it is women's unpaid daily and generational labour in housework that is essential to ensure that the labour force can "go to work." *See* **Capitalism, Gender, Labour, Social reproduction**.

Resilience is a controversial and complex term. Disputes aside, it is broadly understood to refer to the capacity of communities to recover quickly from a disaster.

Resource towns are relatively small and often isolated settlements that have grown around a resource-based industry (often a single industry in a given place) and its transportation requirements. Examples include mining towns, forest industry towns, and fishing villages.

Responsibilization refers to processes through which subjects are rendered individually responsible for functions that previously would have been collectively carried out through institutions like the state or would not have been recognized as a responsibility at all.

Revanchist urbanism is a term that derives from the French word for revenge (*revanche*) and it was introduced by Neil Smith to address the violent and aggressive urban politics of 1990s New York. In order to create spaces of privilege, consumption, and mass surveillance, policies were introduced against the most vulnerable populations, including homeless and racialized youth, and laws were introduced against begging, panhandling, and sleeping on sidewalks. *See* **Urbanism**.

Reverse redlining, also known as "yellowlining" or "greenlining," is a process in which neighbourhoods are identified and targeted by lenders for (often aggressive) marketing of loans, often those with high levels of equity due to previous house price appreciation but lower incomes and older populations. In the United States, many neighbourhoods that received such aggressive marketing were the same ones that were previously redlined in the early postwar period. *See* **Mortgages, Redlining**.

Right to the city is a slogan, a movement, and a concept first introduced by French sociologist Henri Lefebvre in 1968 that advocates for the rights of a city's inhabitants to urban life, broadly constituted as the right to live, work, and play. This notion of rights encompasses not only the legal rights within a liberal-democratic framework, but also social and cultural rights, and while it is concerned with the rights of individuals and communities to access urban resources (e.g., education, transportation, housing, water), it is much more about the right to *collectively produce* urban space through reshaping the processes of urbanization. *See* **Everyday life, Inhabitance, Public space**.

Rough riding is an aggressive driving tactic used by the Jamaican police to "rough suspects up." Prisoners subjected to a rough ride are handcuffed and placed without a seatbelt in a police van, which when driven erratically throws them violently about.

Rural–urban migration is the movement of people from rural areas to urban centres, which significantly contributes to urban population growth. This form of migration is catalyzed by push (e.g., war and conflict, natural disasters, climate change, land-use change, poverty, and unemployment) and pull factors (e.g., the promises of better employment opportunities, higher income, and increased access to infrastructures and facilities). *See* **Migration**.

Rust Belt cities refer to previously industrialized cities in the northeast United States, like Gary, Indiana, Detroit, Michigan, and Milwaukee, Wisconsin, that have experienced population loss, rising crime rates, loss of union jobs, particularly in manufacturing, white flight to the suburbs, and a generally declining urban environment. The decline of these cities is part of the process of deindustrialization and uneven development and was linked to the rise of the Sun Belt cities. *See* **Sun Belt cities**.

Sanitary landfill is a waste disposal site that attempts to protect the environment by burying waste underground in a lined pit until it has degraded.

Sanitation refers to the method used to dispose of excreta, or human waste, and in a city it is closely correlated to overall human development. Basic sanitation includes open defecation, pit latrines without a slab or platform, and hanging or bucket latrines, while more improved sanitation facilities include the separation of human excreta from human contact (e.g., pit latrines with a slab, composting, or flush toilets).

Satellite cities are cities located close to, but independent of, a larger metropolitan centre.

Scale is traditionally and most commonly understood as how maps represent the earth's surface, but geographers also treat scale as a conceptual arrangement of space. Questions have arisen, however, over whether scale is in fact an ontological bedrock of social space or merely an epistemological framework that we impute to space to help provide order and meaning. *See* **Space**.

Scientific method is a process of knowledge production, originally used in the natural sciences, based on empirical or measurable evidence that can be subjected to observation, measurement, and the formulation, testing (via experiments), and modification of hypotheses to formulate laws and theories. It has been critiqued for its association with positivism and its assumptions of universality and objectivity, not least by the scholar Donna Haraway, who claims all knowledge is partial, embodied, and situated. See **Positivism**.

Second-wave feminism is usually referred to as a period that covered the early 1960s to the mid-1980s in both the global South and the global North. It addressed a wide range of issues, including sexuality, family, the workplace, reproductive rights, domestic violence, changes in the law to benefit women, and inequalities but has been critiqued for its exclusion of the voices of marginalized women—women of colour, working-class women, women from the global South—and for its primary focus on engaging with issues of patriarchy and of women's inequality in relation to men, as opposed to other oppressive systems of inequality, such as racism and class. See **Feminism**.

Settler colonialism occurs when a political entity and its associated population seek to conquer another territory and its population and exploit them for their resources, while settling in the colonized territory. Australia, Canada, Israel, New Zealand, South Africa, and the United States are all examples of settler countries. See **Settler state**.

Settler state refers to the general state apparatus derived from colonial traditions in countries like Australia, Canada, Israel, New Zealand, South Africa, and the United States, among many others, that has been superimposed, often by force and violence, onto Indigenous lands and community structures. See **Settler colonialism**, **State**.

Sexual citizenship describes the implicit and explicit ways that forms of social, emotional, and legal membership idealize some modes of organizing erotic life, intimacy, kinship, and social reproduction over others. People whose experience of sexuality—in its intersections with race, class, gender, and other social relations—distances them from the ideal sexual citizen may contest their exclusion both by demanding inclusion in citizenship as it is and/or seeking to fundamentally transform the norms of citizenship. See **Citizenship**, **Heteronormativity**, **Heteropatriarchy**, **Queer**, **Sexuality**.

Sexuality is the capacity of humans to have erotic experiences and responses, usually expressed through a person's sexual orientation, practices, and/or fantasies. Sexuality may be experienced and expressed in a variety of ways, usually referred to as straight (or cisgender or heterosexual), queer (or gay or lesbian), bisexual, or asexual, although these categories are fluid in nature. See **Heteronormativity, Heteropatriarchy, Queer**.

Sex work is a broad term designed to encompass many forms of intimate and sexual labour (stripping, escorting, performance of various sexual acts, and more). Sex work can be performed by people who have any sexual or gender identity, under conditions of legality, ambiguous legality, or illegality depending on when, where, and between whom a transaction and/or relationship takes place.

Shrinking cities refer to urban agglomerations experiencing sustained population decline, which in turn might be caused by aging populations, closure and/or relocation of major employers, conflict, or disasters.

Slum, or slum settlement, is a value-laden term often used loosely to mean an informal settlement that is heavily populated and characterized by housing that varies by standard but commonly lacks reliable sanitation services, a supply of clean water, reliable electricity, timely law enforcement, and other basic services. The term is almost always used pejoratively, being part of a public discourse in which slum dwellers (who are disproportionately migrant, ethnic, or racial minorities) are denigrated. See **Informal settlement**, **Squatting settlement**.

Smart city is one in which information and communication technologies (ICT) are used to enable a more efficient and networked city. However, it is a concept for which there is little agreement, with many pointing to its ambiguity and its use as a promotional label and how it could variously apply to technologies, urban forms, and policies, which are invariably applied in a top-down manner.

Social architecture refers to the ideas and beliefs about the design of cities that have led to what cities look like, who can use them, and how.

Social capital refers to the social norms and networks that enable people to act collectively, including in relationships of trust.

Social exclusion describes experiences at the periphery of the social world; whether through urban segregation and polarization, practices of stigmatization and shaming, or the casual denial of basic political rights and material entitlements, social exclusion describes patterned, structural ways of differentiating social insiders from social outsiders, and of relegating some people and practices to the undesirable edges of an environment. People experience social exclusion when their embodiments, identities, social practices or membership in a minoritized group are marked as abnormal, potentially dangerous, disposable or trivial. See **Social inclusion**.

Social inclusion is a way of describing the experiences of enfranchisement, access to the forms of material redistribution, and political rights that come with membership in a society, as well as more ineffable feelings of solidarity and belonging. Social inclusion thus requires the absence of material deprivation, but it also goes beyond this; it entails being able to move through the social world with relative ordinariness, reasonably unimpeded, without having to denigrate, apologize for, or minimize one's embodiment, identity, social practices, or membership in a minoritized group. See **Social exclusion**.

Socialism is a way of organizing economic activity based on collective ownership rather than on private or individual ownership as in capitalism. The organization of society is also different under socialism: instead of being divided into different and unequal social classes (as is the case in capitalist society), socialist society is made up of equals who act in the collective interest, although actually existing socialism has taken many different forms. See **Socialist cities**.

Socialist cities are fashioned to reflect socialism's desired transformation of social relations and to redress the pre-modern urban problems understood as failings of capitalist urbanization (e.g., cities that were dirty, crowded, and dark). Socialist cities are, moreover, an instrument in achieving that transformation, built or reformed to a design which emphasizes urban efficiency and de-differentiation, and opened up with large public spaces in which the urban masses could gather. *See* **Post-socialist cities, Socialism, Swedish model**.

Social mixing refers to an urban policy agenda that aims to increase neighbourhood and tenure diversity in urban redevelopment by deliberately seeking to mix tenants and/or owners from different socio-economic backgrounds.

Social mobility is the movement of people between categorical groups or stratifications within society denoting a change in status.

Social model of disability conceives of disability as derived from barriers in society that privilege some (able) bodies of others. In this model "the problem" is with a social world composed of places and attitudes that excludes people with disabilities. *See* **Biopsychosocial model of disability, Medical model of disability.**

Social networking sites enable users to create public or private accounts on a given online platform through which they can interact with each other in various verbal and visual forms.

Social polarization refers to the increasing gap between low- and high-income groups within an urban area. This gap can be physical (i.e., two groups occupy different territories) and/or cultural (i.e. their ways of living are quite markedly different). The two groups have completely different lived experiences of the same city and little or no knowledge or understanding of how the other group lives.

Social reproduction refers to the practices and structures that enable people to live (and live together) and that enable systems of production to function over time, for example, from the provision of meals, education, and health to the presence of love and affection. *See* **Gender, Reproduction of the labour force.**

Socio-nature is a neologism that combines social and nature together in one word, to refer to the indissoluble connections between what we call nature and what we call society. It encourages a non-dualist way of thinking and reflects the complexity and fundamental intertwining of nature and social. Also referred to and/or written as socionature(s), social nature(s), socialnature(s), natureculture(s), socio-ecological(ies). *See* **Nature, Urban Political Ecology.**

Socio-spatialities denote the spatial-temporal production of social relations. They signify how spaces can be enabling and/or restrictive for social relations. *See* **Space**.

Socio-spatial segregation is the separation of social groups into particular spaces. For example, people of a particular ethnicity might be forced (either socially, economically, or politically) to reside in a specific neighbourhood in a city.

Solid waste is discarded solid materials from household, commercial, industrial and/or institutional sources.

Space is a socially produced (and contested) term central to geographic knowledge production that has been understood as absolute, relative, relational, abstract, processual, more than-representational, structured, and experienced. As a power relation it can be thought of as a shorthand term for the stretching out of social relations. *See* **Scale, Socio-spatialities, Spatiality.**

Spatial analysis is a theoretical and methodological approach to research that utilizes quantitative methods. It was popular in geography from the 1960s, and although it fell out of favour with many for its supposed association with positivism, it is enjoying a resurgence though the use of both GIS and big data. *See* **Big data, Positivism.**

Spatial fix refers to the ways in which a capitalist system finds new ways of generating profit by seeking out new geographical sources of labour, resources, or demand.

Spatiality is a condition of relationship between people and things mediated through space, indicating that space and society are mutually produced. Theorists such as Doreen Massey have insisted that spatialities be considered in relation to time, as time-space, or as spatiotemporalities. *See* **Space**.

Spectacular urbanism relates broadly to ways of being or relating in cities that rely on expressions of a "global" sense of place. These relations can often involve grand architectural gestures or spectacular events that create new meanings of place. *See* **Urban integrated megaprojects, Urbanism, Urban mega-events.**

Splintering urbanism describes how the contemporary urban landscape of advanced transportation and communication infrastructure fragments the social and material fabric of the city between those who have access and those who do not have access to this infrastructure.

Sprawl is suburban development that is extensive and/or of especially low density. *See* **Suburb**.

Squatter settlement is a residential area to which the inhabitants, usually the poor, have no legal claim to ownership or use, and no formal access to basic services such as water and electricity. *See* **Informal Settlement, Slum, Squatting.**

Squatting occurs when people settle an area for which they lack the legal right to occupy the land. In already built-up areas, squatting refers to the occupation of empty buildings and houses for reuse, usually residential. *See* **Informal Settlement, Squatter settlement.**

Stage model of gentrification describes four stages of neighbourhood gentrification: first, there is small-scale renovation of centrally located working-class housing and/or empty warehouses; second, an enlarging group of housing consumers displaces local residents as real estate agents and small-scale speculators renovate houses for resale or rental; third, developers renovate old structures and build new housing; and fourth, the presence of upper-class residents encourages the construction of condominiums, specialized retail, and commercial services. *See* **Gentrification**.

(The) State refers to a set of institutions, including government and the law, that exercise control over and governance of a particular territory. States exist at various scales, such as nation-states, city-states (e.g., Singapore, the Vatican), and local states (i.e., urban levels of governance). See **Settler state**, **State failure**.

State failure refers to a state's inability to effectively achieve its policy goals, typically attributed to corruption, political interference, and a lack of performance-based incentives. See **State**.

Stigma is the marginalization of a group of individuals based on a specific trait that is used to distinguish and separate the group from wider society. This term only applies when this process is accompanied by some form of disenfranchisement of the group thus demarcated.

Stranger refers to someone who is not known to you. The trope of the stranger, which speaks to social distance and difference, features prominently in foundational urban theory and is making a comeback in critical urban studies as a result of increased levels of migration and xenophobia. See **Xenophobia**.

(A) Street is usually a public thoroughfare that in residential areas would house homes in which people live. It is also a scale at which urban ways of life have been popularly studied in terms of how people interact with and use public space—pedestrians, flâneurs, crowds, for example. See **Flâneur**.

Structural adjustment programs (SAPs) were first imposed in the late 1970s and consist of loans provided by the International Monetary Fund and the World Bank to countries, usually in the global South, that experienced economic crises. SAPs have been heavily critiqued for the criteria attached to loans that required borrowing countries to implement free-market policies that led to privatization of public assets and increases in austerity, as social programs were inevitably cut. See **Neoliberal governance**.

Structuralism is a twentieth-century intellectual movement, rooted in modernism, that attempts to analyze a specific field as a complex system of interrelated parts. It maintains that scientific research can reveal the underlying universal and general rules governing human existence by focusing on the relationship between cause and effect. In the 1970s, however, it came under increasing critique for being too deterministic and ahistorical and for its lack of attention to agency. See **Human agency**, **Poststructuralism**.

Subalterns are disenfranchised population groups that are marginalized within established social structures and political hierarchies to the extent that they are deemed to not even have a legitimate "voice" with which to speak about their own marginality.

Subaltern urbanism is a discourse that challenges the dominant apocalyptic and dystopian narratives about megacities. It humanizes the urban poor by focusing on their lived experiences and political agency. See **Urbanism**.

Subjectivity refers to the fluid and dynamic nature of individual identity formation based on various axes of power (e.g., gender, sexuality, class, ethnicity, age).

Subprime mortgage is a loan offered to borrowers with "less than prime" credit scores, justified as an innovation that "completes the market" and allows lower-income households and those with poorer credit scores to become homeowners. However, many subprime mortgages involved predatory lending and disproportionately impacted upon racialized communities, especially African Americans and immigrant Hispanics. See **Mortgages**, **Subprime mortgage crisis**.

Subprime mortgage crisis was initially a global banking crisis that coincided with the global economic recession of 2007–2008, triggered by a corrupt mortgage lending system that passed on risk to others, which led to the erosion of lending standards. When the credit crisis hit, risk was highly concentrated among several highly levered financial institutions, but then it was quickly passed down to high-risk borrowers who saw the rapid decline in their home prices lead to mortgage delinquencies and home foreclosures. See **Subprime mortgage**.

Suburb originally referred to a recently developed area, commonly residential and of lower density, located at or near the edge of the urban area. It can also be industrial or mixed-use in nature, be the location of factories, offices, and warehouses, as well as major infrastructure facilities, and have selectively high densities such as in tower neighbourhoods in Canadian, Chinese or European cities. See **Ethnoburb**, **Sprawl**, **Suburbanism**, **Suburbanization**.

Suburbanism refers to a suburban way of life. See **Sprawl**, **Suburb**, **Suburbanization**.

Suburbanization is the process by which urban areas expand at their periphery by creating suburbs. See **Post-suburbanization**, **Suburb**, **Suburbanism**.

Sun Belt cities describe those cities in the American South and southwest that grew rapidly after WWII because of the growth in retirement communities and high rates of immigration and house "new" industries like oil, information technology, aerospace, defence, and film and media. Typically, Sun Belt cities are in Florida (Orlando), Arizona (Phoenix), Nevada (Las Vegas), and California (Los Angeles). See **Rust Belt cities**.

Surveillance is the monitoring of activities and behaviours of individuals via various technologies. Surveillance practices are often introduced, and justified, as safety and security measures. See **Closed-circuit television**.

Sustainable Development Goals (SDGs), whose adoption began in September 2015, were drawn up by the United Nations to determine global policy on reducing inequality and levels of poverty. The SDGs differ from the Millennium Development Goals (MDGs) that were in place between 2000 and 2015: in that they are more comprehensive (there are 17 of them as opposed to 8); they apply to countries of the global North as opposed to just the global South; and the SDG framework addresses key systemic barriers to sustainable development that the MDGs neglected, such as inequality, unsustainable consumption patterns, environmental degradation, and weak institutional capacity. See **Sustainable Development Goal 11**.

Sustainable Development Goal 11 is the first-ever statement of a global urban development goal. It aims to "[m]ake cities and human settlements inclusive, safe, resilient and sustainable." See **Sustainable Development Goals**.

Sustainable urbanization is a concept that espouses a balance between economic growth and environmental impacts in a way that can be sustained in the long term for the needs of future generations. See **Urbanization, Urban sustainability**.

Sweat equity in housing refers to the increase in value, or ownership interest, that is created in one's home as a direct result of work (e.g., do-it-yourself housing construction, plumbing, electrical) by the owner(s).

(The) Swedish model is based on the "social contract" achieved between three previously antagonistic parties—labour, capital, and the state—in which labour would be organized into trade unions that agreed to avoid labour disruptions in exchange for generous pay from employers (capital) and a range of generous social services from the state, including health, education, housing, child-care, and pension benefits. See **Socialist cities**.

Systems of cities refers to an approach to studying urban phenomena by considering the linkages and interdependencies that connect cities to one another within regional, national, and global urban systems through flows of spatial interaction.

Tacit knowledge refers to information and understanding, or a skill, that cannot easily be reduced to a codified version or explanation. It is usually something that has to be learned and absorbed through demonstration or experience.

Tea Party is a political movement that emerged in 2009 after Barack Obama's federal election. The Tea Party is known for its conservative policies and is active in the Republican Party, where it pushes for less government intervention and lower taxes.

Technopoles refer to a concept, much discussed in the 1980s, based on the model of "Silicon Valley" near San Francisco. According to the concept, it makes sense to promote clusters of high-tech industries, universities, and venture capitalists in order to promote self-sustaining economic synergies.

Terrorism is organized violence that targets civilians with the intent of generating fear, most typically for political purposes.

Transit-oriented development is moderate and high-density housing, along with complementary public uses, jobs, retail, and services concentrated in mixed-use developments at strategic points along the regional transit system.

Transnationalism describes having multiple connections to or across nations. It is the result of processes of globalization, including increased migration, economic linkages between countries, and flows of people, goods, information, and finance, which have led to increasing linkages and more-complex ties between places.

Transport systems comprise locations or "nodes" where movements of people and goods converge, networks that include infrastructure (such as streets or railways), and movement that results from people (or things) being located in one place but needed or desired in another place.

Transspecies relations refer to the mutually constituted interaction between two (or more) species. For example, humans influence possibilities for animal life on the basis of how they think, feel, and talk about animals, as well as on their subsequent actions towards animals; the opposite also holds true whereby animals influence possibilities for human life with important consequences regarding how different species are in/excluded from particular contexts.

Uneven development is a Marxist concept that challenges the conventional neoclassical assumption that over time economic growth will tend to equalize levels of wealth and development around the world, as capital seeks out profitable locations to invest in. Instead, the uneven development hypothesis suggests that geographical differences in levels of development are inherent in the way contemporary capitalism operates in the endless search for profit and that economic growth routinely produces concentrations of wealth in some locations and underdevelopment in others. See **Capitalism**.

Urban actors are all of the individuals, institutions, and organizations involved in city-building. These can include ordinary citizens, businesses, government agencies, and non-profit organizations that may act alone, in concert with other actors, or in opposition to each other.

Urban agriculture describes a range of activities, including beekeeping, vegetable, fruit or herb gardens, hen rearing, or aquaponic fish production, that all result in the growth of edible goods by using urban land and other locations, such as rooftops, backyards, balconies, and repurposed industrial buildings. Urban agricultural goods are produced by individuals for household consumption, as well as by faith-based or non-profit and public organizations for distribution through emergency food networks, urban farmers' markets, farm-to-foodbanks, and soup kitchen partnerships.

Urban Anthropocene describes the current geologic age marked by human-induced global warming, habitat destruction, species extinction, and ocean acidification, in which people living in cities, which are considered to have the greatest potential to ensure sustainable lifestyles, are considered to be the key to determining the fate of earth's life forms. How the relationship between urbanism and the Anthropocene will evolve is difficult to anticipate and will in part be determined by developments in lower middle-income countries, where the pace of urbanization is fastest. See **Climate change**.

Urban citizenship highlights the importance of cities in promoting *practices* of citizenship, particularly the engagement and participation of diverse groups in political, social, and economic life. It is also a term used to describe how people in cities can claim rights and responsibilities related less to their official status as citizens of a country, but more to their status as inhabitants of a particular urban place, often expressed in the form of public protests on city streets and public spaces rather than in voting. See **Citizenship**.

Urban commons are beyond private and public property; they are a hybrid assemblage of material (e.g., land, natural resources, built environment) and immaterial (e.g., ideas, images, languages, codes, affections) resources and things that belong to everyone. They are counter-hegemonic and insurgent sociospatial practices, based on appropriation, use value, and spatial experimentation that are underwritten by informal socio-economic practices of reciprocity, redistribution, and subsistence.

Urban entrepreneurialism broadly refers to approaches to urban governance that focus on fostering economic growth and cost savings. As a term that was originally derived from research focusing on cities in the global North, it is often used to denote a shift from governance activities that focused on municipal management and local service provision (often called managerialism) towards economy-boosting activities (e.g., place-branding, hosting mega-events to bring investment to the city, lowering municipal tax rates to encourage a pro-business climate in the city, or promoting public-private partnerships in the development process). See **Urban imagineering**, **Place-branding**.

Urban geopolitics as a field of study refers to the explicit attempt to connect the historically distinct fields of urban studies and international relations in order to understand how urbanization and geopolitics are co-constituted. See **Critical geopolitics**.

Urban governance is the day-to-day management of a city. This involves the processes used to influence the short-term and long-term agenda of a city's development and the enactment of policies and decisions concerning public life and economic and social development.

Urban heat island is the effect on temperatures generated by the presence of concrete and asphalt, which absorb sunlight, and the relative lack of vegetation, which cools the air. In general, urban areas are hotter than surrounding regions, and those areas where the proportion of concrete to vegetation is highest tend to suffer the most from the urban heat island effect.

Urban imagineering fuses the words "imagine" and "engineer," and refers to the construction of a new civic image for a neighbourhood or city. Place-branding exercises are used to replace negative perceptions of a neighbourhood or city as a place of disinvestment, decay, crime, and poverty, with positive images of growth, vitality, and prosperity. See **Place-branding**, **Urban entrepreneurialism**.

Urban inhabitant is an urban resident whose right to the city, in a Lefebvrian conception, should not be defined by citizenship status, permanent residency, or any other such formal, legal status, but rather by the practice of inhabiting the city in everyday life.

Urban integrated megaprojects are those that transform a large parcel of land, often under the direction of a single (and frequently private) developer. Typically requiring exceptional forms of state intervention into established development protocols, the objective of these projects is increasingly to create new channels for profit and investment opportunities by transforming the built environment. See **Spectacular urbanism**.

Urban mega-events are large-scale events of great social, cultural, and/or political significance taking place in cities (e.g., the Olympics, the Pan Am Games, or world expositions), that are often closely aligned with powerful national, colonial, imperial, or economic interests. These events involve staging dramatic forms of spectacle, which often entail the temporary intensification of global tourism, security, commerce, and logistics, as well as significant modifications to the built form and functioning of the host city. See **Instant urbanization**, **Spectacular urbanism**.

Urban metabolism refers to the flow or circulation of material and energy through the city (inputs and outputs). In urban political ecology, urban metabolism is used as a framework for analysing the transformation of social and natural elements into particular socio-natural urban forms and relations. See **Urban political ecology**.

Urban neoliberalism is a policy orientation away from state regulation and management of urban processes towards governance mechanisms that seek to attune all aspects of urban life to the economic concerns of free markets. See **Neoliberal urbanism**.

Urban policy is a broad term used to describe the kinds of government frameworks developed for allocating resources to meet needs. The definition can include both explicit urban policies (e.g., park and land-use allocation) as well as implicit urban policies, usually developed at more senior government levels, such as federal regulation of financial institutions or federal highway development spending that affects the development of urban areas.

Urban political ecology (UPE) is an approach to understanding the city as socio-natural—as a product of interconnected economic, social, cultural, and ecological processes, which are shaped by uneven power relations. UPE emerged out of traditional political ecology, urban ecology (the study of ecological processes in cities), and urban metabolism (analysis of the ecologic and economic flows and circulations through the city to reveal how social, political-economic processes create uneven urban environments). See **Socio-nature**, **Urban metabolism**.

Urban primacy describes cases where the largest city in some countries is much larger than that of the next largest, often more than three times as large. This has been described as a product of colonization, export-oriented economic growth, and self-reinforcing patterns of growth. See **Primate city**.

Urban reserve is a satellite reserve to a principal reserve (which is generally outside the urban area), sometimes nearby and sometimes as many as hundreds of kilometres away. The designation of reserve status, under the Canadian federal Additions to Reserves policy, is given to a parcel(s) of urban land purchased by a First Nation on a willing-buyer from willing-seller basis, often to house commercial enterprises and generate income. See **Indigenous peoples**.

Urban resilience refers to the quest to preserve the material and social fabric of cities. It addresses the "bouncebackability" capacity of cities to adapt to stress and to recover swiftly from threats by minimizing urban disorder.

Urban revolution references a transformative politics that can come about not only through organized movements and spectacular, often violent, expressions of resistance, but also through more gradual processes rooted in socialistic modes of everyday life. See **Urban social movements**.

Urban social movements refer to the organized and coordinated activities of social actors that aim to change the existing structure of urban environments. Insisting on instituting a participatory model of decision-making, the participants of these movements usually seek recognition as a key part of urban inhabitation and transformation. See **Urban revolution**.

Urban sustainability is the idea that cities and towns have a minimal environmental footprint (which minimizes inputs of energy, water, and food and waste outputs of heat, air pollution, and water pollution) that enables them to meet the needs of the present without sacrificing the ability of future generations to meet their own needs. See **Sustainable urbanism**.

Urban transition describes the movement of people from rural to urban areas and the increasing percentage of people living in cities. It began to take place in Europe and North America in the late nineteenth and early twentieth centuries and was closely linked to the economic development and demographic growth of the Industrial Revolution, which spawned industrial cities.

Urban utopia refers to an attempt to realize socio-spatial processes or an alternative urban form that speaks to social justice. There have been many attempts to realize urban utopias in fiction and in reality, such as Charlotte Perkins Gilman's imagined community of Herland and Robert Owen's intentional community of New Harmony, Indiana. See **Utopia**.

Urban village is usually taken to refer to a self-contained urban development typically characterized by medium- to high-density housing and mixed land use, with public transit and public space. Urban villages are assumed to reduce the need to travel long distances and to reduce the subsequent reliance on private transport, but the complexity of modern urban life reduces this likelihood and, moreover, the term has also been used merely as a marketing device for downtown neighbourhoods.

Urban water governance describes the range of political, social, economic, and administrative systems established to manage water resources and the delivery of water services in an urban area as well as the relationships, networks, and processes associated with civil society. It involves the activities of public-private partnerships, state-owned networks, and small-scale independent providers such as water vendors and community-run services.

Urbanism is a term with a diversity of meanings, but it generally relates to ways of *being* in the city, such as urban residents' practices of living and interacting with each other or the production and reproduction of place-based imaginaries among those who live within and outside of the city. See **Emergency urbanism, Entrepreneurial urbanism, Indigenous urbanism, Informal urbanism, Military urbanism, Modernist urbanism, Neoliberal urbanism, Occupancy urbanism, Pirate urbanism, Revanchist urbanism, Spectacular urbanism, Subaltern urbanism**.

Urbanization refers to a broad set of processes taking place to transform space; to material, economic, and demographic growth and development; to socio-material transformations in the built environment; and to the connections—economic, ecological, social, or cultural—that settlements are forging with places and processes well beyond their own borders. See **Informal urbanization**.

Urbanization of poverty refers to evidence for two interrelated processes: an ongoing shift of the global demographics of poverty, which until recently has been predominantly rural, to urban areas; and a rate of urbanization for populations below defined poverty lines that are faster than the average for the total population. See **Poverty**.

Utopia is an imagined place or state of things in which everything is highly desirable or perfect but exists nowhere. See **Urban utopia**.

Varieties of capitalism is a new institutionalist research agenda that seeks to understand why, despite increasing global integration, different countries maintain distinct policy and institutional structures and very different economic patterns over long periods. This failure to converge on a common pattern is usually explained as a product of path-dependent institutions that are interlinked, making changes to one set of institutions difficult without changing others as well.

Violence is a structural form of oppression against individuals or groups of people that includes any act, including the threat of acts like assault, that result in physical, sexual, or psychological harm or suffering and pose a threat to the life, body, emotional well-being or liberty of a person or persons.

Visibility is a political goal that many urban social movements both seek and regard with ambivalence, one that can take the form of recognition of a marginalized or minoritized group by state actors or by the social field more broadly conceived. Visibility also raises important questions about what narratives or subject positions within that group are privileged or expected from the perspective of the dominant order.

War is a political label to characterize specific forms of violence in specific contexts as either legitimate or on a larger scale than other expressions of violence.

Water governance encompasses all of the processes whereby different groups make decisions, articulate their interests, and mediate their differences in relation to access to and quality of water.

Watershed refers to the area of land that encompasses all of the water that drains towards the same end point, such as a river or lake (also a catchment or drainage basin).

White flight describes the large-scale migration in the United States of white people of mostly Anglo-European origin that started in the mid-twentieth century from racially mixed city-regions to more racially homogenous suburban and ex-urban areas.

Whiteness is a social construction that normalizes and confers privileges on individuals and groups on the basis of real or imagined meanings attached to white skin. See **White supremacy**.

Whitestream society refers to a powerful set of perceptions in settler societies according to which whiteness is seen as neutral and objective, and societal institutions—descended mostly from white European society—are perceived to treat everyone "equally." See **Settler colonialism**.

White supremacy is a historically based and institutionally embedded racial ideology that constructs white peoples and nations of the European continent as superior to continents, nations, and peoples of colour that are racialized as not white. Emerging during the period of European colonial expansion, this ideology generally serves as a justification for the maintenance and defence of unequal systems of wealth, power, and privilege. *See* **Whiteness**.

Working class is a socio-economic group that typically consists of people who are employed for wages, especially in service, manual, or industrial work, and who have a relatively low degree of autonomy within the workplace. *See* **Middle class**.

World city is a term coined by Peter Hall in the mid-1960s and defined by John Friedman in seven theses that linked urbanization processes to global economic forces. World cities have a disproportionate concentration of businesses and command centres that organize the global economy, and a common synonym for them is "global city." *See* **Global cities**.

Worlding, originally coined by Heidegger to mean "being in the world," has been taken up by Ananya Roy to explain how cities and their inhabitants in the global South are positioned within a global power regime in which they are understood primarily only through reference to dominant global North perspectives.

Xenophobia is the fear of that which is perceived to be foreign or strange. Applied to the study of cities, it has been used to refer to fear of strangers, new arrivals in cities, particularly immigrants, or refugees or asylum-seekers of a different racialized, ethnic, or religious group to the dominant group. *See* **Stranger**.

Youth bulge is a demographic profile in which a disproportionate percentage of a country's population are between the ages of 15 and 24 years old.

Zero-tolerance policies enforce punishment without compromise for any transgressions of given rules and regulations. These policies do not pay attention to the particularities of different transgressions; a grave violation and a petty infraction should be punished with the same severity.

Zoning is a common feature of urban land-use planning and refers to the practice by municipal authorities of demarcating areas of cities for specific uses (e.g., residential, industrial, commercial). Restrictions commonly apply not only to land use but also to the density, set-backs, design, and forms of buildings within zones.

Index

Note: Page numbers in italics indicate selected figures and photographs.

Aalborg Charter (European sustainable cities), 363
ableism, 311, 320
Aboriginal Australians, in Sydney: "Block" community of, 139–41; early railway work by, 139–40; housing crisis faced by, 140–1; housing proposal of, 141; later planning/development by, 141–3. *See also entries below;* "Block (The)"
Aboriginal Housing Association Grant Program (Australia), 141
Aboriginal Housing Committee (Sydney), 141
Aboriginal Housing Company (Sydney), 139–43, 147; background of, 139–41; formation/early challenges of, 141; later redevelopment project of, 141–3. *See also* "Block (The)"; Pemulwuy Redevelopment Project
Academy of Visual Arts (Leipzig), 112
accessibility, 311
Accra, Ghana, 296, 297; e-waste centre in, 400–2; health issues in, 368–70; slum demolition in, 301, 302; urban forest near, 347; youth resilience/creativity in, 301–3
accumulation, of wealth, 190
accumulation by dispossession, 262
Acharya Tole (Kathmandu squatter settlement), 261, 265–70; neighbouring property owners and, 267–9; planned demolition of, 267–8; previous dispossession/resistance in, 267–8; proof of inhabitance/right to compensation in, 269–70; rebuilding process in, *269*
Acholi (people), of Uganda, 234; in IDP camp, 233–5
Adalet ve Kalkinma Partisi (AKP; Turkish governing party), 215, 216, 218
adaptation (climate policy), 157, 159
adaptive reuse, 174, 179, 181
ADO (women's human rights organization), 235, *235*
Afghanistan, Canadian mission in, 238
African National Congress Youth League (ANCYL), 297
African Youth Charter, 295
Agbogbloshie e-waste centre (Accra), 400–2
agglomeration economies, 8, 25–6, 43–4
Agricultural Redevelopment and Rehabilitation Act (1961), 47
air pollution, 398, 419, 430, 431; traffic-related, 159, 411–12, 413, 417, 421
Al-Baqa'a Refugee Camp, Jordan, 235–8; as considered non-urban, 233, 237; as symbol of Palestine, 236–7; temporality of, 231–2, 233, 236–7
Alexander the Great, 210
Algonquins: Lansdowne Park revitalization project and, 148–50, *149–50*
alienation, 8
Alliance for Healthy Cities, 363
Alliance of Citizens of Valenzuela and Caloocan (Manila), 32
Alma Ata Declaration (health care), 362
amenities (food/physical activity), 361, 367–8, 369
animal geography, 352

animals, 347–8, 350, 352; in Canadian cities, 357; on Gaborone poultry farms, 354–6
anonymity, 1, 320, 326
apartheid, 151
Appadurai, Arjun, 230, 264–5, 267
Arab Spring, 212, 214
Arendt, Hannah, 213
artistic precariat, 110
assimilationism, 327
assistive devices, 312
asylum-seekers, 239, 429
Auckland, New Zealand, Indigenous planning in, 143–6. *See also* Ngāti Whātua Ōrākei, *and entry following*
austerity, 89, 297, 329, 390
austerity urbanism, 89–100; Canadian experience of, 97–9; crisis/distress and, 90–2; municipal bankruptcies and, 92–7; neoliberalism and, 90, 91–2, 98, 179, 180, 429–30. *See also* Detroit, as post-industrial city; Stockton, California, bankruptcy of
Awere IDP camp, Uganda, 233–5; humanitarian relief in, 234–5; "protected villages" of, 234; temporality of, 231–2, 233; women's activism in, 235

Bagong Barrio (Manila): transnational families/labourers of, 283–5
Bai Bang Paper Mill (Vietnam), 123–4
Bangkok Metropolitan Area: water privatization in, 378, 383–5
bankruptcies: Japanese financial crisis and, 46; Spanish mortgage debtors and, 200
bankruptcy, municipal: Canada and, 97–8; of Detroit, 91, 95–7, 108; of Stockton, 92–5
Barcelona, housing bubble in, 199–200; foreign investors and, 200; mortgage debtors and, 200
Barlow Report, on industrial population in UK, 39
Bedford case, on sex work in Canada, 339
behaviouralism, 12–13
Beijing, 7, 21, 23, 58, 73, 176, 329, 411
Bell, Daniel: *The Coming of Post-industrial Society*, 104
Bellear, Bob and Kaye, 141
Benjamin, Rich, 177
Benjamin, Walter, 9
Berlin, 2, 121–2, 124, 174, 347
bicycles. *See* cycling
bid-rent models, of land use, 191
big data, 28, 432
Bing, David, 108
biopsychosocial model of disability, 311
Black Lives Matter, 214, 303, 430
black water, 398–9
"Block (The)" (Redfern, Sydney), 139–43; as "Black Heart" of Australia, 143; Black Power movement and, 140; Black Theatre of,

140; community services in, 140; as contested space, 140–3; redevelopment program for, 141–3; social tensions/rioting in, 141, 142; university involvement/assistance in, 140, 141–2. *See also* Aboriginal Australians, in Sydney; Aboriginal Housing Company; Pemulwuy Redevelopment Project
body, 228; disability/mental health and, 309–16, 318–20; privileging vs "othering" of, 211–15, 238–9, 299, 303–5, 310–13; sexuality and, 326–40
Bonifacio Global City (Taguig, Metro Manila), 31
Bosque de Chapultepec (Mexico City), 347–8
Boulanger, Eugene, 222
Bourne, Larry S., 37
Boystown (Chicago), 326, 330–3
brain circulation, 247
brain drain, 247
Brasilia, 2, 428
Brenner, Neil, and Christian Schmid, 6, 11
Broadway Youth Center (BYC, Chicago), 332
Brown, Michael, 219
brown environmental health agenda, 399
brownfield development, 110, 112; in Canada, 114, 115, 186
bubble economy: in Japan, 45–7. *See also* housing bubbles
Buck-Morss, Susan: *Dreamworld and Catastrophe: The Passing of Mass Utopia in East and West*, 133
built environment, 12, 57–8, 71; disability and, 309–23; as exclusionary, 177–8, 181; as green/eco-friendly, 430, 431–3; health and, 361–74; megaprojects/spectacles and, 72–6, 83; planning of, 142–3; as transit-oriented, 165, 410, 417, 420, 421
Bulkeley, Harriet, and Michelle Betsill, 159
Burgess, Ernest, 8–9
Burj Khalifa (Dubai), *73*, 83
Bush, George W., 195

Cabbagetown area (Toronto), 132, 184
Calgary, 33, 65, 99, 136, 164–5, 185, 270, 288
Canada Mortgage Bonds (CMBs), 201
Canadian Border Services Agency (CBSA), 239
Canadian Charter of Rights and Freedoms, 238
Canadian Human Rights Act, 290
capitalism, 9; cities and, 10–11, 27, 262–4; crises of, 90–2; patriarchy and, 211, 277; varieties of, 38–9
capital switching, 194–5
Caracas, Venezuela, 377; state-community water governance in, 378, 385–8
caste, 252, 399
Castells, Manuel, 10–11, 282
central business district (CBD), 9, 191, 349, 410
Central (later Canada) Mortgage and Housing Corporation (CMHC), 201
Central Park (New York), 347
Centre for Addiction and Mental Health (Toronto), 321
C40 Cities Climate Leadership Group, 159, 160, 162, 165
Chandigarh, India, as base for prospective student migrants, 246, 249–54; in context of student travel, 246–9; high costs of applications/English-language training in, 252–3; housing/amenities in, 253; immigration industry in, 252; men's/women's social life in, 253–4; modernist design/legacy of, 249–51, 428; population of, 251; as provincial city, 245–6, 248; Punjabi migration culture and, 247–9; Sikh community/culture and, 247, 252, 254; "un-Indian"/global environment of, 251–2. *See also* transnational student migrants, from India
Chatterjee, Partha, 264
Chávez, Hugo, 385, 387, 388
Chicago, 2, 7, 9, 104, 160, 335, 352; LGBTQ community in, 326, 330–3
Chicago School, 8–9, 10, 330, 349
child care, 277, 282–3; in Canada, 98, 165, 239, 289; Indigenous, 142, 239; in socialist cities, 125, 126, 127, 130
Childe, V. Gordon, 6
China, 3, 4, 6, 177, 204–5, 255, 352, 366, 372, 411, 419, 427; emissions in, 160; private water companies in, 381, 383; public transit investments by, 414–15; rapid urbanization in, 7, 19–20, 21, 23, 58–9, 70, 75; rural gentrification in, 176; rural-urban migration in, 24, 25, 75–6; socialist urbanism in, 121, 122, 124, 125
China Central Television (CCTV) building (Beijing), 73
Chinese Immigration Act (1923), 238
cholera, 348, 361, 362, 364
CIAM (Congrès Internationaux d'Architecture Moderne), 123, 124, 130
cities: ancient, 6–7; capitalism and, 10–11, 27, 262–4; colonial, 348; de-differentiated, 122, 133; edge, 52; entrepreneurial, 98; gateway, 25, 185; green, 431; healthy, 361, 362–3, 372–3; of industrial era, 7; as megacities, 7, 23, 53–5; as metacities, 7, 27; ordinary, 11; primate, 29–30, 45, 285; provincial, 245–6; satellite, 297; shrinking, 2, 20, 23, 29; smart, 28; as systems, 12. *See also* global cities; post-industrial cities; socialist cities; world cities
citizenship, 209, 261; geopolitics and, 229, 231, 238–9; Indigenous-inclusive, 137, 148; of Singapore's foreign workers, 327; transnational student migrants and, 246–7, 248; urban poor and, 262, 264–5, 266, 268, 272. *See also* sexual citizenship; urban citizenship
City Beautiful movement, 362, 367
city-building: informal, 66–7; nature and, 348; neoliberal, 120; socialist/post-socialist, 120–33; spectacular/instant, 74–6
Ciudad Juárez, Mexico: femicide in, 278, 280, 335–6; sex work in, 280, 327, 335–7
civic life, 91, 209
civil society, 120, 215, 264–5, 380; water governance and, 382, 383, 385–8
class mobility, 245
climate change, 155–67; coastal/floodplain cities and, 28–9, 30, 54, 157–8; displacement caused by, 79, 158; extreme weather and, 28–9, 30, 155–6, 157–9, 162–4, 165, 287; greenhouse gas emissions and, 58, 155, 156–7, 160–2, 164–6, 421, 431; as ongoing challenge, 426, 429, 430–1; politics/community activism and, 32, 159–60, 163–4; rising sea levels and, 28, 157–8, 162–3, 378–9; water/sanitation systems and, 158, 377, 378–9, 383
climate justice, 160

climate policy mitigation, 155
Clos, Joan, 178
closed-circuit television (CCTV), 177, 215
Cochabamba, Bolivia: "Water War" in, 381
Coleman, Michael, 93
"collective consumption," 10–11, 282
collective housing: in Hanoi, 125, 126–9, 131
colonial cities, 348
commoning practices, 11
communicable diseases, 361–2, 363–4, 372
communities, 8; capacity-building in, 221; everyday life and, 263–4; Indigenous land ownership by, 138, 143–4, 146, 151; of occupancy urbanism, 83, 271; resilience/creativity of, 106, 107, 110. *See also* gated communities
composting, 221, 396, 397, 405
Comprehensive National Development Plan (CNDP) system, of Japan, 45
Concentric Zone Model, 9, 349
condominial sewage, 379, 390
condominiums, 58, 74, 173, 176, 191, 198, 297; in Canada, 133, 184, 185, 186, 272; in global South, 29, 129, 266, 334; "poor doors" in, 181–4
congestion charge, 159
consumption-based carbon accounting, 156–7
Continuous Passage Act (1908), 238
conurbations, 23, 53
coordinated market economies (CMEs), 38–9
co-production: of water services, 382, 385–8
corruption, 429; in development/construction industries, 59; municipal/state, 91, 97, 287
counter-publics, 212–13, 222
counter-topographies, 281
counterurbanization, 176
creative city theory, 103, 105–7
creative class, 26, 271; and LGBTQ community, 333; in post-industrial city, 103, 105–7, 109–10, 115
creative industries, 104, 321; of ethno-racialized youth, 300–1, 302
crime: gated communities and, 177, 185; in post-industrial cities, 103, 109, 370–1; in poor/racialized areas, 141, 186, 219, 272, 303, 321, 366; xenophobic fears of, 334. *See also* police; women, violence against
crisis, 90–1. *See also* global financial crisis (2007–8); subprime mortgage crisis
Crisp, Jeff, 231
critical geopolitics, 228
Cronon, William, 350
cultural capital, 105–6
cultural economy, 96–7, 106, 115
cycling, 371, 410, 412, 416; benefits of, 58, 156, 372, 431; bicycle lanes for, 106, 162, 165, 166; bike-sharing and, 367, 372, 431

Daley, Richard M., 330
dancehall culture, of Jamaica, *300*, 300–1
David Pranteau Aboriginal Children's Village (Vancouver), 148
Davis, Mike, 55, 157, 350, 429
daycare. *See* child care
De Blasio, Bill, 183
Debord, Guy, 263
de-differentiated cities, 122, 133
deep democracy, 264–5, 267
deindustrialization, 370
deinstitutionalization, 314
Deis, Bob, 94
Delhi, 7, 348, 411; gang rape/murder in, 279, 281; as megacity, 23, 54, 245; safe access to sanitation infrastructure in, 377, 402–5; safety of women/girls in, 281, 291, 377, 402–5
democratic racism, 136
demographic transition, 24–5
densification, 56, 58, 156, 159, 164, 165, 421
Department of Regional Economic Expansion (DREE), 47
"desakota," 21, 24
Designated Countries of Origin (DCOs), 239
determinants of health, 361, 363
Detroit, as post-industrial city: auto industry in, 95; bankruptcy of, 95–7; cultural initiatives in, 96–7, 109–10; house expropriations/auction sales in, 108–9; land redevelopment in, 107–9; Occupy movement in, 97; urban agriculture in, *109–10*, 109–10
Detroit Future City project, 108, 110
Detroit Garden Resource Program, 109
Detroit Institute of Art, 96–7
Detroit Land Bank Authority, 107, 108–9
developmentalism, 11
Dharavi (informal settlement in Mumbai), 80
diabetes, 58, 362, 365, 370
difference, 326
digital citizenship, 209, 214
disability, 309–23; body and, 309–16, 318–20; gentrification and, 309, 314, 321, 322; geographers and, 311–13; mental health and, 313–16; models of, *310*, 310–11; stigma of, 311, 314–15, 320
disability case studies: Georgetown (Guyana), 309, 316–18; Liverpool, 309, 318–20; Quebec City, 309, 320–2
discourse, 89, 328
disease, in cities, 348, 361–2; communicable, 361–2, 363–4, 372; non-communicable, 361, 362, 365, 373. *See also* cholera; HIV/AIDS
disease burdens, 365
displacement: challenges/resistance to, 80, 141, 144, 159, 268; gentrification and, 173, 174–6, 178–9, 184, 193, 221, 271; of Indigenous people, 140–1, 143–4, 239; land grabbing and, 79, 80; of urban poor, 81–2, 215, 261, 265–72; war and, 79, 219, 227–38, 239
disposable populations, 298, 302
dispossession, 79, 261; accumulation by, 262; of farmers, 32; foreclosure and, 108–9; of Indigenous people, 136–51; of Palestinians, 235–6
distress (urban public finance), 91
division of labour, 8, 211. *See also* new international division of labour
drones, 230

droughts, 29, 158, 378, 430
Dubai: as example of spectacular urbanism/instant urbanization, 70–1, 76–8; development financing in, 74, 76, 77; exploitation of labour in, 74–8; megaprojects in, 74, 83; rapid transit in, 411
Dublin, housing bubble in, 196–8; "ghost estates" and, 197–8, *197–8*

Eastcliffe Retirement Village (built on Ngāti Whātua Ōrākei land, Auckland), 144
Ebola, 364, 372
ecological footprint, 156–7, 165, 276, 430, 431
economies: bubble, 45–7; coordinated vs liberal market, 38–9, 48; cultural, 96–7, 106, 115; informal, 70, 366; knowledge-based, 104, 105, 271, 318, 327, 333
edge cities, 52
Edo, Japan (later Tokyo), 41–2, 43
emergency urbanism, 232, 235
emissions. *See* greenhouse gases (GHGs), emissions from
encounters, 1, 9, 209, 211–12, 214, 265
Enloe, Cynthia, 229
entrepreneurial cities, 98
entrepreneurial urbanism, 92–3
epidemics, 361
epidemiologic transition, 361
epidemiology, 362
Erdogan, Recep Tayyip, 217, 218
ethnoburbs, 57
ethnography, 261, 265
ethno-racialized youth, 295–306; in Accra, 301–3; in Kingston, 298–301; in neoliberalizing cities of global South, 296–8; resilience/creativity of, 300–1, 302; in Toronto and Montreal, 303–5
European Union (EU), 91, 112, 113, 114, 196–7, 199
Eveleigh Railway, workshops/yards of (Sydney), 139–40
everyday life: lived reality of, 428; participation in, 431; plants/animals and, 347, 354; of poor, 79, 260, 261–70, 272; public spaces and, 209, 210–11; right to the city and, 10, 213–14; in socialist cities, 125–6, 130; violence in, 229, 288; war/militarization and, 229, 230, 233
e-waste, 399; in Accra, 400–2
export processing zones, 11; in Accra, 368; in Ciudad Juárez, 335–7; in Manila, 29
expressways/freeways, 61, 114, 321, 416, 421, 422
Extell Development Company: New York condominium project of, 181–3, *182*
extended metropolitan region, 29
extreme weather: climate change and, 28–9, 30, 155–6, 157–9, 162–4, 165, 287

families, 8; daycare for, 289; of disabled, in Georgetown, 316, 318; homelessness/poverty of, 270–1, 303, 373; Indigenous, housing for, 142, 144–5, 148; of Punjabi student migrants, 247–9, 251–4; Singapore heteronormativity and, 334–5; in suburbs, 52, 57, 58, 165, 314; transnational, 29, 31, 271, 283–5, 287–8; utopian vision of, 427

Federal Housing Administration, US (FHA), 191–2, 193
Federal National Mortgage Association, US ("Fannie Mae"), 192
femicide, in Ciudad Juárez, 278, 280, 335–6
feminism, 13; on public sphere concept, 212–13; second-wave, 277; urban activism and, 221–2, 290–1; urban geopolitics and, 228–9; urbanist research of, 277, 278, 279. *See also* women, in cities/urban areas; women, violence against
FIRE (finance, insurance and real estate), 74, 179
First Nations, 136, 239; Idle No More movement of, 222; missing/murdered women of, 291; reserve system and, 148, 239, 240; on urban reserves, 147–8, 150. *See also* indigeneity, urban, *and entries following*
flâneur, 9
Flint, Michigan: health issues in, 370–2
flooding, 28, 32, 158, 197, 287, 356, 378, 430; human negligence/intervention and, 81–2, 144; protective infrastructure and, 125. *See also entry below*
flooding, in specific cities: Bangkok, 383; Georgetown (Guyana), 287, 316; Manila, 30; New York, 162–4; Okahu Bay (Auckland), 144
floor area ratio (FAR), 182–3
Florida, Richard: on creative class, 105–6, 333; on technology/talent/tolerance, 255
food deserts, 367, 371
food security, 348, 366
Ford, Rob, 166
Fordism, 107; Detroit as symbol of, 90, 95; Keynesianism and, 98, 130
foreclosures, 91, 108–9, 200, 203, 204; class/race and, 195–6
foreign direct investment (FDI), 196, 422
Foreign Domestic Movement Program (Canada), 283, 287
Fort McMurray, Alberta: forest fires in, 33
Fraser, Nancy, 212–13
full-cost recovery, 381

Gaborone, Botswana: urban poultry farms in, 354–6
Gans, Herbert, 57
Garden City movement, 349, 362, 367
Gardiner Expressway (Toronto), 83, 321, 421
gated communities, 60, 173, 177–8, 185, 338, 430; in Canada, 185; outside North America, 58, 176, 179, 250, 286, 288, 297, 299, 352
gateway cities, 25, 185
gay village, in Toronto, 338, 339
Gehl, Jan, 428
Gemeinschaft, 8
gender, 276, 329
gentrification, 173, 174–6, 193; in Canada, 174, 184–6, 272; creative class and, 106, 113–14; disability and, 309, 314, 321, 322; in London, 174, *175*, 179–81, 187; opposition to, 113–14, 180–1; stage model of, 174–5
geopolitics, 228–9; critical, 228. *See also entry below;* urban geopolitics; war and militarization
geopolitics, Canadian, issues of, 238–9; asylum-seekers, 239; Indigenous displacement, 239; monitoring of port workers,

238–9; peacekeeping mythology, 238; racialized immigration policies, 238–9; resource company conflicts, 238
George, Pamela, 339
Georgetown, Guyana: disabled in, 309, 316–18; gendered poverty/inequality in, 285–8
German Democratic Republic (GDR), 111, 121–2, 123–5
germ theory, 362
Gesellschaft, 8
Gezi Park (Istanbul): as contested space, 217–18; as demolished/replaced by building complex, 215, *216–17*. *See also entry below*
Gezi Park protests, 215–18; as creating public space through virtual space, 218; expansion of, 216–17; government response to, 217; groups/constituencies involved in, 218; legacy of, 218; police crackdown on, 216, 217; social media use in, 215–16, 218
"ghost estates," in Dublin area, 197–8, *197–8*
Glass, Ruth, 174, 175
global cities, 7, 52, 55–6, 65, 74, 232, 363; London, 7, 26, 55, 179, 416; New York, 7, 26, 55, 181–2, 416; Paris, 416; Singapore, 31, 333–5; Tokyo, 7, 45, 49, 55, 416
global financial crisis (2007–8), 33, 264, 283, 430; housing markets and, 89, 91, 92, 190, 194–200; municipal bankruptcy/austerity and, 89–98; as predictable, 195; US government/Federal Reserve response to, 195
globalization, 36; cities/suburbs and, 52–67; climate change/sustainability and, 156–7, 431–2; disease and, 364, 365, 372; e-waste and, 399, 400–2; gentrification and, 173, 175–6, 187; of housing crisis, 190–205; industrial cities and, 103; national urban systems and, 36–7, 38, 40, 42; public space/protest and, 209–23; spectacular urbanism and, 72–8; of transportation investment, 422; urban challenges posed by, 428–31. *See also* transnationalism; transnational student migrants, from India
Golden Horseshoe region (Southern Ontario), 104, 115, 165
Gordon, Jessica, 222
Gottmann, Jean, 43
governance, 89, 204. *See also* urban governance; urban water governance
graffiti, as form of protest/resistance: in Brixton (London), *180*; of Gezi Park protests (Istanbul), 218, 223; in Leipzig, 113–14
Graham, Stephen, 227, 230, 240
Gray, Freddie, 219
Greece, 45, 89, 91; agora of, 210
Green, Nicholas, 350
green cities, 431
green gentrification, 166
greenhouse gases (GHGs), emissions from, 58, 155, 431; Canadian opportunities to cut, 164–5, 421; production-based vs consumption-based, 156–7; São Paulo's efforts to cut, 160–2
green sustainability agenda, 399
"gray spaces," 232
growth poles, 45

Habermas, Jürgen, 212–13
Habitat III, 7, 433
Halifax, 83, 98, 99, 165, 420; Africville district of, 132; wastewater disposal in, 389

Hall, Peter, 55
Hamilton, Ontario, 66, 104, 115, 211, 256, 389
Hanoi, 123, 124; gendered income inequality in, 399; informal recyclers in, 396; motorcycle use in, 411; volunteer labour in, 125. *See also entry below;* Vietnam
Hanoi, collective housing estates in, 125, 126–9; design/amenities of, 127; poor quality/degradation of, 127, 129; residents' ownership of units in, 127–8; as slated for demolition/redevelopment, 129, 131
Harvey, David, 10, 74, 429; on accumulation by dispossession, 262; on capital switching, 194; on Swedish welfare state, 130; on utopian communities, 427, 428
health issues, 361–74; background/history of, 348, 361–2; communicable diseases, 361–2, 363–4, 372; neoliberalism, 363, 366, 367, 368, 370–1; non-communicable diseases, 361, 362, 365, 373; poverty/economic inequality, 361, 363, 366–7, 370–2; rapid urbanization, 368–70; urban amenities, 361, 367–8, 369
healthy cities, 361; in Canada, 372–3; definition of, 362–3
Healthy City Project (WHO), 363
heat waves, 28, 158, 165
Heavily Indebted Poor Countries (HIPCs), 287
heroic informality, 80
heteronormativity, 327
heteropatriarchy, 327
Hewitt, Kenneth, 229
hidden austerity, 99
HIV/AIDS, 327, 364, 432
Hobson, William, 143
Ho Chi Minh, 128, *128*
homelessness, 46, 281, 347; in Canada, 83, 270–1, 322; displacement and, 79, 141, 215, 302; mental illness and, 314, 315, 322; occupancy urbanism and, 83, 271
hooks, bell, 280
HOPE VI program (US), 193
Hoskins, Rau, 139
household, 12
housing bubbles: in Ireland/Dublin, 196–8; in Spain/Barcelona, 199–200; in US, 92–5, 195–6
housing markets, 190; Canadian, 201–4; evolution of, 192–4; falling house prices and, *194*, 194–6; gentrification and, 174–6, 179–81, 193; housing systems and, 190–2; informal, 79; mixed projects and, 178, 193; racism of, 108–9, 191–2, 193, 195–6; securitization/volatility of, 193–4; subprime mortgage debacle and, 89, 91, 92, 194–204
Housing New Zealand Corporation, 144
housing stock, 190; gentrification and, 174–5, 179–80, 184–6; renewal/strengthening of, 362, 372; varieties of, 190–1. *See also* condominiums; public housing; rental housing
housing stock, in specific cities: Detroit, 107–9; Leipzig, 111–12; Sydney, 140–1
housing systems, 190–2, 193
house prices, in Canada, 33, 165, 201–2; (1996–2013), 202, *202*; in selected national markets (2000–14), *194*, 194–6
Howard, Ebenezer, 349, 427–8
Hulchanski, David, 99, 184

human agency, 13, 428
humanism, 13
Hurricane Sandy, 155–6, 157, 162–4; housing issues of, 163–4, *183*
hyperbuilding, 73

Ibadan, Nigeria: Oke-Foko settlement in, 70–1, 81–2
Idle No More, 222
immigration: to Canada, 238–9; to Canadian cities, 25, 32, 65, 185, 271–2
incineration, 396; of e-waste, *402*
Inclusionary Housing Program (IHP), of New York, 182–4
inclusive design, 311
incremental urbanization, 72; informal urbanism and, 78–9, 83
India, 3, 4, 23, 24–5, 60, 76, 228, 365, 383; gendered violence in, 279, 280, 281, 291, 377, 402–5; informal settlements in, 79, 80; IT industry in, 26, 28; land grabbing/dispossession in, 79, 80; motorcycle use in, 411; prospective student migrants in, 246–54
Indian Act (1876), 239
indigeneity, urban, 137–8. *See also entries below*
Indigenous and Northern Affairs Canada, 148
Indigenous density, 137, 143, 146
Indigenous-inclusive citizenship, 137; in Canada, 148
Indigenous peoples, 136–51; coexistence of, with settler state, 136–9; colonial violence/cultural erasure and, 136, 138–9, 143–4; inclusive citizenship of, 137, 148; "savagery" of, 137; truth/reconciliation process and, 150–1; urban planning by, 138–46. *See also* Aboriginal Australians, in Sydney; Aboriginal Housing Company; First Nations; Māori; Ngāti Whātua Ōrākei, *and entry following*
Indigenous planning, 138–9
Indigenous urbanism, in Canadian cities, 147–51; Lansdowne Park revitalization, 148–50; social/affordable housing, 148; urban reserves, 147–8, 150
Industrial Revolution, 7, 21, 210, 348, 427
inequality, 7, 8, 10, 28, 163–4, 173, 204, 272, 330; health and, 363, 370–2; income, 83, 99, 130, 176, 367; mixed communities and, 174, 178, 181–6; Occupy movement vs, 92, 264, 271, 371, 430; poverty and, 80, 260, 263–4, 276, 285–8, 296, 361; racial/class, 92, 195–6, 326, 332; social, 399–400, 406; water/sanitation access and, 158, 394–5, 397, 399–400
informal economy, 70, 366
informality, 59–60, 65–6; heroic, 80; importance of, 428; reframing of, 79–81; spectacular urbanism and, 79; subaltern urbanism and, 79–80, 264–5; in Tehran, as extensive, 63–4
informal settlements, 5, 8, 56, 232; in Canada, 65–6, 83; demolition of, 267–8, 301, 302; framing/reframing of, 78–81; governance and, 59–60; as heroic, 80; as incremental, 78–9; plants/trees in, 353; rural-urban migration and, 25, 66, 78–9; sanitation in, 397–8; self-built homes in, 59, 64, 78; as types of suburbs, 63–4; water access/quality in, 379, 385. *See also entry below;* "slums"; squatter settlements
informal settlements (specific): Acharya Tole (Kathmandu), 261, 265–70; in Karachi, 396; in Managua, 352–3; Oke-Foko (Ibadan, Nigeria), 70–1, 81–2; Old Fadama (Accra), 302
informal transport, 412–14

informal urbanism, 71, 72, 84, 232; in Canada, 83; informal urbanization and, 81–2; incremental urbanization and, 78–9, 83; types of, 79–80
informal urbanization, 56; of IDPs, 235; informal urbanism and, 81–2
informal waste collectors/recyclers, 395–6, 399–400, 402; in Vancouver, 405–6
information age, 432
infrastructure, 5, 12, 23, 26–9; costs of, 57, 98, 115; green, 108, 158; Japan's investments in, 39–40, 43–5, 48; lack of access to, 80, 82, 158, 165, 364; marketization/financing of, 74, 77; of refugee camps, 231, 235, 237; for urban mega-events, 83. *See also* sanitation; transportation, urban; urban water governance; waste management; water, access to
infrastructure networks, 409
inhabitance, 261, 262; in Kathmandu squatter settlement, 268, 269–70
instant urbanization, 70, 72
Insured Mortgage Purchase Program (IMPP; Canada), 201
insurgent architects, 428
internally displaced persons (IDPs), 231; in Ugandan camp, 233–5
International English Language Testing System (IELTS), 248, *249*, 252, 253
International Labour Organization (ILO), 284, 295; School-to-Work transition survey of, 297, 298
International Monetary Fund (IMF), 89, 197, 297, 381, 383
intersectionality, 299
Iraq, US-led war in, 238
Isin, Engin, and Evelyn Ruppert, 214, 218
Islamophobia, 280–1
Istanbul, 215. *See also* Gezi Park (Istanbul), *and entry following*

Jacobs, Jane, 6, 114, 209
Jakarta, 7, 28, 58, 263, 398, 418, 419
James, Colin (Col), 141
Japan, 36–7, 40–8; coordinated market economy of, 38–9; core-periphery tensions/imbalance in, 43–6; feudal city-system in, 40–2; imperial restoration/modernization in, 42; land price control attempts/financial crisis in, 46–7; national development plans of, 45, 48; Pacific Belt region of, 43–5, *44*; pollution crisis in, 43; population growth/concentration in, 40, 41–2, 43, 45; railway system of, 42–3, *44*, 48, 415–18; rapid postwar economic growth of, 43; state interventions in urban system of, 39–40, 44–8
Japan Housing Loan Corporation, 192
jitneys, 413
Johnson, Boris, 179
Jojola, Ted, 138
Jordan, 378. *See also* Al-Baqa'a Refugee Camp, Jordan
just city, 264

Kassab, Gilberto, 160, 162
Kathmandu: squatter settlement in, 261, 265–70. *See also* Acharya Tole

Kathmandu Valley Development Authority (KVDA), 266, 267, 269
Katz, Cindi, 228–9, 230
Keolis (Paris public transport firm), 422
Keynesian-inspired policy, 89, 92, 195; Fordism and, 98, 130
Kingston, Jamaica, ethno-racialized youth in, 298–301; as caught between gangs and police, 299–300; dancehall culture of, 300–1; diasporic music culture of, 304–5
knowledge-based economy, 104, 105, 271, 318, 327, 333
Kony, Joseph, 234
Koolhaas, Rem, 73

labour: division of, 8, 211; new international division of, 12, 55, 156, 277; volunteer, 125
labour, migrant: in China, 25, 75–6; in Dubai, 74–8; transnationalism of, 283–5, 287–8, 327
labour markets, 13, 197, 200; construction, 74–8; women and, 11, 211, 277, 281–8. *See also* social reproduction
labour unions, 197, 381, 390
Lagoons megaproject (Dubai), 74, *74*
Lagos, Nigeria: incremental growth of, 70
land-banking, 125
land banks, 107; in Detroit, 107, 108–9
landfills, 27, 48, 404, 405; low-quality, 398; sanitary, 396
land grabbing, 79, 80
land title, 190; of collective housing units in Hanoi, 127–8; as imposed on Indigenous people, 143–4, 146, 151; as provided to urban poor in Venezuela, 387
Lansdowne Park (Ottawa), as revitalized by city and Algonquin community, 148–50; Aberdeen Square basket-weave paving, 148, *149*; culturally significant landscaping, 148; teaching circle, 149, *150*
Lavo Project (Montreal development), 186
League of California Cities, 93
leapfrog development, 56, 58
Le Corbusier: CIAM and, 123, 124; as designer of Chandigarh, 249–51, 428; disability design and, 311, 312; Radiant City concept and, 428
Lee Hsien Loong, 334, 335
Lee, Spike, 174
Lefebvre, Henri, 9–10, 11, 52, 213–14, 261, 262–5
Leipzig, as post-industrial city, 110–14; history of, 110–11; "imagineering" of, 111–14; new media cluster in, 112; refurbishing/protection of buildings in, 111–12; resistance to gentrification in, 113–14
Lethbridge, Alberta: LGBTQ community in, 327, 329, 338, 339
LGBTQ community, 326, 327–30; in Canadian cities, 327, 337–9; in Chicago, 326, 330–3; in Singapore, 326–7, 333–5
Liberal Democratic Party (LDP), of Japan, 44, 45, 46
liberal market economies (LMEs), 38, 48
lifeworld, 312
Lindenau-Plagwitz creative cluster (Leipzig), 113; guardian house project in, *112*; Spinnerei arts venue in, 113, *113*
Liu Gong Li, China, 19, 20, 24
Live-in Caregiver Program (Canada), 283
livelihood, 348
Liverpool: visually impaired in, 309, 318–20
Livingstone, Ken, 159
living wage, 110
lofts, 176; in London, 179–81; in New York, 176. *See also entry below*
lofts, in London: Honour Oak Lofts (East Dulwich), 179, 180, *180*; Notting Hill Lofts, 179–80, *180*; Sanctuary (Clapham Junction), 179, 180, *180*; Village Lofts (Battersea), 179, 180
London, 2, 39, 41, 253, 256, 329, 347, 352; adaptive re-use/loft housing in, 179–81; cholera outbreak of (1854), 362; emissions in, 157, 160; financial sector in, 26, 55, 90, 230; gentrification in, 174, *175*, 179–81, 187; as global city, 7, 26, 55, 179, 416; "poor doors" in, 178, 183; public housing in, 192; recent development in, 179; transportation in, 159, 416, 417
loneliness, 1, 8
Lord's Resistance Army (Uganda), 232, 233–5
Los Angeles: as post-suburban megacity, 61–3
Lu'ma Native Housing Society (Vancouver), 148

McAdam, Sylvia, 222
McGee, Terry, 21
McHarg, Ian, 349
Mackenzie, Suzanne, 276, 277
McLean, Sheelah, 222
Maduro, Nicolás, 385, 388
Mah, Alice: *Industrial Ruination, Community, and Place*, 104
Managua, Nicaragua: plants and patios in, 352–4
Manila, 29–32; concentration of wealth in, 29–30; export processing zones of, 29; flooding in, 30, 32; IT industry in, 29; luxury developments in, 31; metropolitan region of, 20, 29; poor transit system of, 32; as primate city, 29–30; remittances to, 29, 31; slums/marginalized communities of, 30–2; transnational families/labourers of, 29, 31, 283–5; urban planning activism in, 32
Māori, 136, 139; colonial dispossession of, 143, 146; loss of communal land ownership by, 143–4, 146; resistance/protest by, 144, 146; urban planning by, 144–5, 146. *See also* Ngāti Whātua Ōrākei, *and entry following*
maquiladoras: in Ciudad Juárez, 335–7
Marcuse, Peter, 262, 263
Marineland Animal Defense, 221–2
Markelius, Sven, 123, 130
market failure, 381
market socialism, 125
Martin, Trayvon, 177, 219
Matunga, Hirini, 138, 139
median asking rent, 183
median monthly gross rent, 183
medical geography, 362
medical model of disability, 310–11
medico-psychiatric models of mental illness, 313
megacities, 7, 23; emergence of, 53–5
megalopolises, 7, 23–4, 43, 53; Manila, 29–32; Tokaido, 43, *44*; US northeastern seaboard, 43
megaregions, 7, 52, 53, 56

mental illness, 313–16; homelessness and, 314, 315, 322; models/terminology of, 313; neoliberalism and, 315; NIMBYism and, 311–12, 314–15; post-asylum geographies of, 314; poverty/precarity and, 315, 320
metacities 7, 27
Metropolitan Toronto Action Committee on Public Violence against Women (METRAC): *Women's Safety Audit Guide*, 290
Michigan Municipal League, 97
Michigan Right to Farm Act (1981), 109
middle class, 27–8, 79, 105; flight to suburbs by, 95, 107; gentrification/mixed communities and, 174–5, 176, 178–81, 184–5, 186, 193; suburban, 57, 349; transnational student migrants and, 246, 248, 250, 251–4; women of, 211, 282–3, 288
migration. *See* immigration; rural-urban migration; transnationalism; transnational student migrants, from India
militarism, 229
militarization, 229. *See also* war and militarization
military urbanism, 227–40, 433
Millennium Development Goals (MDGs), 378, 380, 386, 397
Miller, David, 165–6
Minghine, Anthony, 97
Ministry of State for Urban Affairs (MSUA), 47
mixed communities, 173, 178, 193; in Canada, 185–6
modal share, of transportation types, 410, 416, 421
mode of production, 210–11
modernism, 120, 122, 123; of Brasilia, 428; of Chandigarh, 249–51; of Vällingby, 126, 129–31
modernist urbanism, 122, 428
modernization: Accra's attempts at, 301–3; of post-Edo Japan, 36, 41–3; São Paulo's intrusive attempt at, 162; social mobility and, in Chandigarh, 256; of Soviet state-building, 126
Montreal: gentrification in, 184, 185; household debt/disposable income in, *203*, 203–4; lesbian community in, 338, 339; mixed communities in, 186; urban agriculture in, 347, 357
More, Thomas: *Utopia*, 426–7
Morris Justice Project (New York), 218–21
mortgage debt: in Montreal, *203*, 203–4; in Spain, as unremovable by bankruptcy, 200
mortgages, 190, 192–4; in Canada, 201–4; default/foreclosure crisis and, 195–6; securitization of, 193–5, 201; subprime, 89, 91, 107, 194–6
mortgage-backed securities (MBS), 193–5, 201
motorcycles, 409, 410–11, 419; health impacts of, 411–12; as taxis, 413
motorization, 409; fast pace of, 410–11; health/environmental impacts of, 411–12, 417, 421
multi-level governance climate model, 159
Mumford, Lewis, 6, 55, 209, 349
Munro, Jenny, 142
Museveni, Yoweri, 234
Muskeg Lake Cree Nation urban reserve (Saskatoon), *147*, 147–8

National Resistance Army/Movement (Uganda), 234
national urban systems, 36–49; attempts to shape, 39–40, 44–8; Canadian, 47–8; Japanese, 36–7, 40–8; "varieties of capitalism" approach to, 38–40. *See also* Japan

natural monopoly: urban water services as, 381–2
nature, in cities, 347–58; history of, 348–50; plants/animals and, 351–2; socio-nature and, 350–1. *See also* animals; parks and green spaces; plants and trees; urban agriculture
Nehru, Jawaharlal, 249
neighbourhoods, 12; city vs suburban, 57; climate change and, 158–9, 164, 165; destruction/redevelopment of, 46, 108–10, 132–3, 371; gentrification and, 106, 113–14, 173, 174–6, 179–81, 184–6, 187, 193, 314; income inequality and, 99, 122, 271–2; LGBTQ, 326, 330–3, 338–9; rebranding of, 104–7; redlining of, 191–2, 193; social mixing in, 173, 178; sustainability of, 431; utopian vision of, 427
neoliberalism, 90, 120, 283, 300, 371; austerity urbanism and, 90, 91–2, 98, 179, 180, 429–30; as health issue, 363, 366, 367, 368, 370–1; mental illness and, 315. *See also* privatization
neoliberal urbanism, 92; culture/creativity and, 103, 106–7; ethno-racialized youth and, 296–8; resilience of, 432–3; as revanchist, 215
New Institutionalism, 38
new international division of labour (NIDL), 12, 55, 156, 277
New Leipzig School of modern art, 112–13
New Urban Age thesis, 2; contesting of, 5–6; Los Angeles and, 62
New Urban Agenda (Habitat III), 7, 433
New York: as global city, 7, 26, 55, 181–2, 416; after Hurricane Sandy, 162–4, *183*; lofts in, 176; Morris Justice Project in, 218–21; Occupy Wall Street in, 92, 264; "poor doors" in, 174, 178, 181–4
Ngāti Whātua Ōrākei (Māori *hapū*/sub-tribe), 143–6, 147; eviction of/occupation protest by, 144; joint management agreement of, 145, 146; planning process of, 144–5, 146. *See also entry below*; Māori
Ngāti Whātua Ōrākei, land of (Ōrākei Block): as broken up into individual titles, 143–4, 146; evictions from/destruction of buildings on, 144; retirement village built on, 144; as returned to community ownership, 144, 146; stewardship/future use planning of, 144–5, 146
NIMBY ("Not in My Backyard") syndrome, 311–12, 314–15
non-communicable diseases, 361, 362, 365, 373
NORR (Canadian architectural firm), 83
North, Douglass, 38
North American Free Trade Agreement (NAFTA), 335

Obama, Barack, 92
obesity, 28, 58, 362, 365, 370, 373
Obote, Milton, 234
occupancy urbanism, 71, 80, 83, 271
Occupy movement, 11, 212, 214, 222, 271, 430
Occupy Wall Street, 92, 264; Detroit offshoot of, 97; Flint offshoot of, 371
offshoring, 28, 29
Oke-Foko (Ibadan, Nigeria): informal urbanization/urbanism of, 70–1, 81–2
Old Fadama (informal settlement in Accra): demolition of, 302, *302*
One Riverside Park (luxury New York condominium), 181–3, *182*

Ontario, 33, 48, 83, 357, 405; auto industry in, 422; Golden Horseshoe of, 104, 115, 165; green belt in, 65; international students in, 255–6; municipal operating expenses in, 98–9
open data, 432
open dumping, 396, 397
Ōrākei Act (New Zealand, 1991), 144
Ōrākei Block. *See* Ngāti Whātua Ōrākei, land of
Orangi Pilot Project (Karachi), 379, 383, 390
ordinary cities, 11
Orwell, George, 213
Ottawa: Lansdowne Park revitalization in, 148–50, *149–50*
Ottawa Charter for Health Promotion, 372
over-urbanization, 30–1
Owen, Robert, 427

Pacific Belt region, of Japan, 43–5, *44*
Palestinian refugees, of Arab-Israeli wars, 235–6. *See also* Al-Baqa'a Refugee Camp, Jordan
palimpsests, 9, 62, 122
Pandey, Jyoti Singh, 279, 281
Papakāinga planning process, of Māori, 144–5, 146
Paris, 2, 7, 9, 41, 72, 417; gentrification in, 174; as global city, 416; immigrant women in, 280; Lefebvre and, 9–10, 262; suburbs in, 57, 64
Park, Robert, 8–9
parks and green spaces, 347–8, 349, 351–2; in Canada, 356–7; loss of, 215–18, 368, 369
Parr, Hester, 312, 315; and Chris Philo, 313–14
participatory budgeting, 390
Pathum Thani, Thailand: water privatization in, 378, 383–5
Pathumthani Water Supply Company, 384–5
patriarchy, 52; capitalism and, 211, 277
Peck, Jamie, 90, 91–2, 99
Pemulwuy Redevelopment Project (Sydney), 141–3; affordable housing in, 142–3; municipal challenge to, 142, 143; state approval of, 142. *See also* Aboriginal Housing Company; "Block (The)"
People's Climate March (New York), 160
peri-urban regions, 24, 32; water services to, 379, 382
Persian Gulf region: real estate/infrastructure development funds in, 77
Philippines. *See* Manila
Philo, Chris, 314, 352
physical activity amenities, 367–8, 369
pirate settlements, 59
pirate urbanism, 80
Pitts, Angela, 141
place, 12, 228, 230
place annihilation, 229
place-branding: of post-industrial cities, 104–5, 106–7, 116; spectacular urbanism and, 73
Places to Grow Act (Ontario), 165
planetary urbanization, 11
plants and trees, 347–8, 351–2; in affluent neighbourhoods, 158–9, 165; in Canadian cities, 356–7; in Managua patios, 352–4; urban planning and, 348–50

Platform of Mortgage Victims (Spain), 200
police, 27; activists/protestors and, 140–1, 142, 144, 215–16, 271, 303; African Americans and, 218–21; brutality of, 215, 219, 300, 332; carding by, in Toronto, 303–4; cuts to services by, 93, 94, 95–6, 370; drone use by, 230; feminist response to comments by, 221, 290–1; gendered violence and, 222, 280, 281, 404–5; harassment by, 327, 330; LGBTQ community and, 330, 332, 333; mentally ill and, 321
policy mobilities, 105–7
political society, 264–5
polity, 213
Pol Pot, 124
"poor doors," 178; in New York buildings, 174, 178, 181–4
population density, 5, 21, 416
positionality, 13
positivism, 12
post-asylum geographies, 314
postcolonialism, 13
post-industrial cities, 103–16; aesthetics of, 104; analysis of, 104; in Canada, 114–15; challenges of, 104–5; creative/cultural remaking of, 105–7, 109–10, 111–14; housing in, 108–9, 111–12; land redevelopment in, 107–10; "meds and eds" approach to, 104; tensions between groups in, 104–6, 110, 113–14; urban agriculture in, 106, 109–10. *See also* Detroit, as post-industrial city; Leipzig, as post-industrial city
postmodernism, 13
post-socialist cities, 121. *See also* socialist cities
poststructuralism, 13, 328
post-suburbanization, 52, 60, 65. *See also* Los Angeles
poultry farming, urban, 354–6, 357
poverty: in Canadian cities, 373; everyday life/right to the city and, 79, 260, 261–70, 272; of families, 303, 373; gendered, 285–8; health and, 363, 366–7; inequality and, 80, 260, 263–4, 276, 285–8, 296, 361; urbanization of, 27–8, 429
precarity, 283, 329; mental distress and, 315; poverty and, 263, 272, 430; of youths' lives, 298, 304
Pride parades/celebrations, 329, 330, 332
primate cities, 29–30, 45, 285
prison industrial complex, 298
privatization: austerity/neoliberalism and, and, 89, 92, 96, 196, 368, 413–14, 429–30; gated communities and, 173, 177–8, 185, 429; of public housing, 193; of public services, 196, 394; of public space, 11, 210, 214–15, 281, 315, 430; of rural land, 79; of suburbs, 60, 62, 429; of urban planning, 73; of urban water services, 96, 120, 380–2, 383–5, 389
private good, 381
production-based carbon accounting, 156
protest, 209–23; in Canada, 221–23; digital/social media, 209–10, 214, 221–23; graffiti as form of, 113–14, *180*, 218, 223; in public spaces, 209, 212. *See also* urban social movements
provincial cities, 245–6. *See also* Chandigarh, India, as base for prospective student migrants
Provincial Waterworks Authority (Thailand), 384–5
psychogeographies, 1
public address, 209, 212–13

public and private, women's negotiation of, 279–80, 290
public good, 381
public health: history of, 362
public housing, 65, 144, 190–1, 193; in New York, 163–4; in Toronto, 132, 185–6
public spaces, 209; in ancient world, 210; civic life/culture and, 209, 211–12; decline/loss of, 214–15, 221; digital sphere/social media and, 209–10, 214, 221–3; gendering of, 210–11; global urbanization and, 211–12; Habermas on, 212, 213; revanchist urbanism and, 215; scholarly debate over, 212–13; as sites of protest/resistance, 209, 212; urban citizenship and, 213–14. *See also* Gezi Park (Istanbul), *and entry following*; Morris Justice Project (New York); protest; urban social movements
public sphere, 127; sexuality in, 327, 333; urban citizenship and, 210–13
public transit, 410, 411; average modal share of, 410; in Canada, 420, 421; funding of, 413–14; informal transport vs, 412–14; in Singapore, 418, 419–20; in Tokyo, 416–18, 420. *See also* rapid transit
Public Works Act (New Zealand), 144
Punjab, India: colonial history of, 247, 253; migration culture of, 247–9; prospective overseas students in, 246–54. *See also* Chandigarh, India, as base for prospective student migrants
Purcell, Mark, 213, 262

Qualfon (call centre in Georgetown, Guyana), 287
quality of life, 28, 45, 90; environmental, 159, 160, 162, 165; for ethno-racialized youth, 295, 301, 304; health and, 363, 366, 370, 371; Indigenous, 146, 151; in mixed communities, 186; plants/trees and, 356; in post-industrial cities, 103, 105, 115; transportation and, 421
Quebec City: Saint-Roch neighbourhood/mall of, 309, 320–2
"queer" (term), 326, 327–30
queering, of urban space, 326, 327–30; in Canadian cities, 327, 337–9; in Chicago, 326, 330–3; in Singapore, 326–7, 333–5

racism: of Canadian immigration policies, 238–9; democratic, 136; of housing markets, 108–9, 191–2, 193, 195–6; toward Indigenous people, 136–46, 150–1, 239, 339; of police, 140–1, 142, 144, 218–21, 300, 303–4; toward youth, 295–306
railways, 409, 410; Aboriginal employees of, in Australia, 139–40; in Canada, 414, 422; funding of, 413–14; in Los Angeles, 61–2; in Singapore, 419; in Tokyo/Japan, 42–3, 44, 48, 415–18
railways, street (streetcars/trams), 57, 410–11, 413
rainfall: damaging effects of, 32, 158, *237*, 356, 378
rank-size rule, 37, 38
rapid transit, 27, 32; bus, 165; rail, 27, 411, 413, 414, 417–18, 419, 421. *See also* public transit
Rauch, Neo, 113
Razack, Sherene, 239, 290, 339
reconciliation, between settler and Indigenous peoples, 150–1
recycling, informal, 405–6
Redfern (Sydney neighbourhood). *See* Aboriginal Housing Company; "Block (The)"
Redfern-Waterloo Authority (RWA): Built Environment Plan of, 142, 143

redlining, 191–2, 193
refugees: in camps/urbanized settings, 227, 229, 231–3; in Canada, 185, 238–9, 338; as living in public spaces, 215; Palestinian, in Al-Baqa'a camp (Jordan), 235–8
Regent Park (Toronto public housing project), 132, 185–6
rental housing, 190, 193; as lost, in London, 175; illegal basement units as, in Canada, 83; "poor door" access to, in New York, 174, 178, 181–4
rent burden, 183
rent gap, 10
representations, 13, 104–6
reproduction of the labour force, 11, 282. *See also* social reproduction
resilience, 158, 431–2
resource towns, 33
responsibilization, of ethno-racialized youth: in Accra, 301–3; in Kingston, 298–301
revanchist urbanism, 215
reverse redlining, 192
right to the city: in Canada, 272; constraints on, 215; core principles of, 262; defence of, by squatter community, 265–70; everyday life and, 10, 213–14; of LGBTQ community, 327–30; nature and, 348, 357; of poor, 79, 260, 261–70, 272; postcolonial scholarship on, 261, 264–5; of racialized youth, 304; urban revolution and, 261, 262, 263–4; of women, 221–2. *See also* urban citizenship
Robinson, Jennifer, 11
Rodney, Walter, 297
Rose, Damaris, 277
Ross, Andrew, 350
"rough riding," 300
Rowell-Sirois Commission (Royal Commission on Dominion-Provincial Relations), 47
Roy, Ananya, 11–12, 33
rural-urban migration, 2, 25, 66, 78–9; in China, 19, 24, 25, 75–6; in Japan, 42, 43, 416; to Kathmandu, 266; to Manila, 29, 30; to Tehran, 64
"Rust Belt" cities, 2, 89, 104, 115, 370

Saint-Roch (Quebec City): neighbourhood/mall of, 309, 320–2
Sanguinetti, Michael, 221, 290
sanitary landfills, 396
sanitation, 81–2, 158, 281, 350, 364, 394–406; in global South, 397–8; health/environmental challenges of, 398–9; historical improvements to, 57–8, 362; safe access to, by women/girls, 281, 377, 402–5; social inequality and, 399–400; urban water governance and, 377, 378–83. *See also* sewers; toilets, access to; waste management, *and entries following*
São Paulo, 7, 54, 58, 378, 411; efforts to cut emissions in, 160–2
Saskatchewan and Manitoba Treaty Land Entitlement (TLE) Framework Agreement, 148
Saskatoon, 33, 222; food gardens in, 357; urban reserves in, *147*, 147–8
Sassen, Saskia, 55, 213
satellite cities, 297
Saunders, Doug, 19, 429

scale, 228, 230; of modernist urban planning, 428; of spectacular urbanist projects, 72–3
scientific method, 12
sea levels, rising, 28, 157–8, 430; of Hurricane Sandy, 157, 162–3; saltwater intrusion and, 378–9
second-wave feminism, 277
Serbia and Kosovo: NATO bombing of, 238
settler colonialism, 227
settler states, 136–7; Indigenous peoples and, 136–51. *See also* Aboriginal Australians, in Sydney; Aboriginal Housing Company; Indigenous peoples; Māori; Ngāti Whātua Ōrākei, *and entry following*
sewers, 348, 379, 383, 387, 402–3; condominial, 379, 390; problems associated with, 19, 59, 144, 237, 379–80, 388, 389, 394, 399, 404
sexual citizenship, 334–5, 337–9
sexuality, and urban space, 326, 327–30; in Canadian cities, 327, 337–9; in Chicago, 326, 330–3; in Ciudad Juárez, 327, 335–7; in Singapore, 326–7, 333–5
sex work, 211, 218, 287; in Boystown (Chicago), 332; Canadian case law on, 339; in Ciudad Juárez, 280, 327, 335–7; by Indigenous women, 290, 327, 339; spatial separation of, 329
Shanghai, 7, 23, 54, 58, 160, 176, 187, 410, 411
shrinking cities, 2, 20, 23, 29. *See also* urban decline
Sikh community/culture, in Punjab, 247, 252, 254
Simmel, Georg, 8, 209
Simone, AbdouMaliq, 12, 261, 263, 264–5, 270, 305, 429
Singapore: foreign workers' housing in, 334–5; as global city, 31, 333–5; heteronormativity of, 327, 334–5; Housing Development Board of, 334; LGBTQ community in, 326–7, 333–5; transportation in, 410, 418–20
"slums," 27, 58, 158, 267, 429; clearance/demolition of, 301, 302, 371; everyday life in, 263, 265; global population living in, 260, 364; poverty/informality of, 59, 78; refugee camps as resembling, 232. *See also entry below;* informal settlements
"slums" (specific): in Caracas, 386; in Delhi, 402–3, 404; in Kathmandu, *267*; in Manila, 30–2; Oke-Foko (Ibadan, Nigeria), 70–1, 81–2; Old Fadama (Accra), 302; in Rio de Janeiro, 25
SlutWalks, 221–2, 290–1, 430
smart cities, 28
Smith, Neil, 10, 175, 215
Snow, John, 362
Snyder, Rick, 95
social architecture, 311
social capital, 246, 383
social inclusion/social exclusion, 326
socialism, 120–1
socialist cities, 120–33; building of, 122–6; incarnations of, 121–2; relevance of, for Canada, 132–3. *See also entry below;* Hanoi, *and entry following;* Vällingby, Sweden
socialist urbanism, 121–2; apartment blocks of, 123, 124, 125, 130–1; architecture/design of, 122, 123; CIAM's influence on, 123, 124, 130; in Europe and Asia, 123–4; everyday life and, 125–6; prefabricated construction of, 124–5; urban planning and, 124; volunteer labour of, 125
social mixing, 173, 178

social mobility, 9, 109; in Punjab, 248, 251–2, 256
social model of disability, 310–11
social movements. *See* urban social movements
social networking sites, 209, 214, 223
social polarization, 132
social reproduction, 8, 105, 272, 276, 281–3, 429; vs capitalist production, 175, 210–11; transnationalism of, 283–8; women and, 11, 211, 277, 281–3, 285–8
socio-nature, 348
socio-spatialities, 209; of transnational student migration, 251–3; utopian reordering of, 427
socio-spatial segregation, 178
solid waste: amounts generated, 28, 395, 398; disposal of, 156, 398–9; management of, in Delhi, 402, 404
Somalia: Canadian forces scandal in, 238
sound systems (Jamaican music culture), 304–5
South Africa, 148, 247, 296, 297, 348, 365, 366, 377, 379, 399–400
Soviet Union, 55; city-building in, 121; collapse of, 59, 121; shrinking cities of, 2, 23; socialist urbanism of, 121, 122, 123–6; transportation in, 410–11
space, 10, 12. *See also* parks and green spaces; public spaces
spatial analysis, 12–13
spatial fix, 26, 28
spatiality, 12
Special Economic Zones (SEZs), in India, 79
spectacular urbanism, 71–2; in Canada, 71, 83; in Dubai, 70–1, 74–8; entrepreneurialism and, 72–3, 74; low-cost/exploitable labour and, 74–8; megaprojects/architecture of, 73–4, 83; place-making and, 72–3, 83. *See also* Dubai
splintering urbanism, 429–30
sprawl, 43, 46, 52, 56–7, 65, 157, 160, 164–5, 362, 365, 420, 432
squatter settlements: in global South, 78, 191, 192, 368, 379, 396, 402; in Kathmandu, 261, 265–70; regulation of, 60. *See also* informal settlements
squatting, 59, 64, 141, 394; in Canada, 65, 83, 271; in post-industrial cities, 109, 113–14
stage model of gentrification, 174–5
Stanley Park (Vancouver), 347
state, 334
state failure, 382
Stewart, Patrick (Nisga'a architect), 148
stigma: of disability/mental illness, 311, 314–15, 320; of HIV/AIDS, 327; of post-industrial landscape, 106; of sex work, 332, 337; of waste/sanitation work, 399–400, 405–6
Stockton, California, bankruptcy of, 92–5; entrepreneurial urbanism and, 92–3; public sector employees and, 93–4, 95
strangers, 8, 192, 209, 211–12, 213, 353
streets, 12, 209; dancehall events on, *300*, 300–1; disabled/visually impaired and, 311, 318–20; police-youth interactions on, 218–21, 303–4; residents' claim to, 268; zero-tolerance policies and, 215
student migrants: in Canada, 254–6. *See also* Chandigarh, India, as base for prospective student migrants; transnational student migrants, from India
structural adjustment programs (SAPs), 89, 277, 366

structuralism, 13
subalterns, 264–5
subaltern urbanism, 79–80, 264–5
subjectivity, 8
subprime mortgage crisis, 89, 91, 107, 194–6
subprime mortgages, 195–6
suburbanisms, 60
suburbanization, 52, 56–61; in Canada, 64, 65–6; socialist rejection of, 123
suburbs, 56–61; in Canada, 64, 65–6; development/governance of, 59–60; ethnic, 57; in global North and South, 58–9; lifestyle in, and consequences of, 57–8; meaning of, 60–1; sprawl and, 56–7. *See also* Los Angeles; Tehran
subways, 219, 229, 230, 414; in Bangkok, 417; in New York, 162, 164; in Stockholm/Vällingby, 130–1; in Tokyo, 416; in Toronto, 27
Sudbury, Ontario, 115, 373
sukumbasi (Nepali squatter communities), 266–70, 272. *See also* Acharya Tole
"Sun Belt" cities, 89, 92–5
Suplicy, Marta, 160
surveillance, 8, 214, 220, 333; big data and, 432; carding as, 304; gentrification and, 185, 322; technologies of, 177, 187, 214, 215
Sustainable Development Goals (SDGs), 7, 380; Goal #11, for cities, 7, 433
sustainable urbanization, 366
sweat equity, 174
Swedish model of urbanism, 126, 129–31
Sydney, Australia, Indigenous urban planning in, 139–43. *See also* Aboriginal Australians, in Sydney, *and entries following*
systems of cities, 12

tacit knowledge, 25–6
Task Force on Housing and Urban Development (Canada), 115
Tea Party movement (US), 92
technical water committees (MTAs), of Caracas, 387–8
technopoles, 44
Tehran, 63–4; informal settlements of, 63–4; social geography of, 63
terrorism: domestic violence as, 279–80; post-9/11 responses to, 433
Test of English as a Foreign Language (TOEFL), 248
Thames Water (UK), 381, 384
Thatcher, Margaret, 193, 264
toilets, access to, 82, 394, 397–8, 402–5; community/shared, 382, 397–8, 403–5; for disabled, 312, 398; fees for, 377, 398; gendered violence and, 281, 377, 398, 404–5
Tokaido megalopolis, Japan, 43, *44*
Tokugawa Shoganate (Edo period), in Japan, 40–2, 43
Tokyo: in feudal era, 41–2, 43; as global city, 7, 45, 49, 55, 416; population growth/state investment in, 42–6; transportation in, 42–3, 410, 415–18
Tönnies, Ferdinand, 8
Toronto, 7, 48, 64, 65, 83, 99, 115, 222, 239, 270, 285, 289, 303, 305, 347, 373; bike-sharing in, 372; carbon efficiency of, 164, 165–6; Food Policy Council of, 357; gentrification in, 184–5; Healthy City Project and, 363; house prices in, 201–2; immigration to, 25, 32, 65, 185, 271–2; international students in, 253, 255–6; LGBTQ community in, 337, 338, 339; mixed communities in, 185–6; Pan Am Games in, 83; police carding in, 303–4; public housing in, 65, 132, 185–6; SlutWalks in, 221–2, 290–1; socio-spatial inequality in, 271–2; South Parkdale area of, 320–1; Tent City in, 83, 271; three sub-cities of, as distinguished by income, 184; transportation in, 27, 165–6, 271–2, 410, 421, 422
Toronto Community Housing Corporation (TCHC), 185–6
Toronto Youth Cabinet, 304
Tory, John, 166, 421
transit-oriented development, 417; in Canada, 165, 410, 420, 421
transnationalism: in Canada, 65; of culture, 302, 306; of families/households, 190, 281–8; of migrants/migrant workers, 245–57, 283–8, 327; national urban systems and, 36–49; of prospective overseas students, 246–54; of social movements/activism, 222, 291; of urban investment, 196, 414, 422; of urban megaprojects, 73, 74–8, 79, 83
transnational student migrants, from India, 246–54; in context of student travel, 246–9; high costs incurred by families of, 252–3; English-language proficiency of, 248, 253; loss of student visas by, 253; Punjabi migration culture and, 247–9; Sikh community/culture and, 247, 252, 254; social life of, 253–4; women as, 248, 252, 253–4. *See also* Chandigarh, India, as base for prospective student migrants
transportation, urban, 409–23; in Canada, 410, 420–1; congestion charges and, 159; evolution of, 409; expressways and, 61, 114, 321, 416, 421, 422; fast-paced motorization of, 410–11; foreign investment in, 414–15, 421; health/environmental impacts of, 159, 411–12, 413, 417, 421; informal, 412–14; systems of, 409; transit-oriented development and, 165, 410, 415–20, 421; vs walking, 58, 156, 318, 367, 371, 372–3, 409, 410, 411. *See also* cycling; motorcycles; public transit; railways; rapid transit; subways
transportation, urban, case studies of: Singapore, 410, 418–20; Tokyo, 42–3, 410, 415–18
transport systems, 409
transspecies relations, 352
Treaty of Waitangi (New Zealand), 143; tribunal of, 144, 146, 148
truth and reconciliation commissions, 150–1
Tsuu T'ina (First Nation), 136
tuberculosis, 361, 364
Tuperiri (Māori chief), 143, 144

Uber, 413
Uganda, 4. *See also* Awere IDP camp, Uganda
uneven development, 10, 20, 26–7, 37
United Nations Children's Fund (UNICEF), 378
United Nations Committee on the Elimination of Discrimination against Women, 290
United Nations Educational, Scientific and Cultural Organization (UNESCO), 261; *Youth in a Changing World*, 297
United Nations High Commission on Refugees (UNHCR), 231, 232
United Nations Human Settlements Programme (UN-Habitat), 2, 78, 178, 261, 290, 295, 364, 433

United Nations Relief and Works Agency (UNRWA), 235–6, 237
universities: international students at, 254–6
University of Sydney, 140, 141–2
urban actors, 10, 132, 265, 332, 348, 428
urban agriculture, 106, 347, 350–1, 427, 431; in Aztec Mexico, 348, 351; in Botswana, 354–6; in Canada, 347, 357, *357*; in Detroit, 109–10, 347; in Managua, 354
Urban Anthropocene, 7–8
urban areas: key variables used to define, 5, 20, 21
urban citizenship, 10, 95–7, 209, 213–14, 221, 431; as digital, 209, 214. *See also* right to the city
urban commons, 1, 11
urban decline, 20, 23, 28; in Canada, 33, 114–15; drivers of, 24, 26–7, 33. *See also* post-industrial cities
urban entrepreneurialism, 72–3
urban geopolitics, 227; Canadian, 238–9; feminism and, 228–9; of forced displacement, 231–3; geopolitics and, 228–9; of militarization, 228–33; of war/militarization, 229–31. *See also* geopolitics, *and entry following*; war and militarization
urban governance: of ethno-racialized youth, 295–306; under neoliberalism, 90, 120, 283, 300, 371; under socialism, 125–6
urban growth, 19–33; in Canada, 32–3; challenges/opportunities of, 27–9; demographic change and, 24–5; economic drivers of, 25–7; patterns of, 20–4; rural-urban migration and, 25. *See also* Manila
urban heat island effect, 158
urban imagineering, 73; of Leipzig, 110–14
urban inhabitants, 10, 262
urban integrated megaprojects, 73
urbanism, 1, 71–2; austerity, 89–100; emergency, 232, 235; entrepreneurial, 92–3; Indigenous (Sydney/Auckland), 136–46; Indigenous (Canadian), 147–51; informal, 71, 72, 78–83, 232; military, 227–40, 433; modernist, 122, 428; neoliberal, 92, 103, 106–7, 215, 432–3; occupancy, 71, 80, 83; pirate, 80; revanchist, 215; socialist, 120–33; spectacular, 70–8, 83; splintering, 429–30; subaltern, 79–80, 264–5
urbanization, 1–13, 71–2; contemporary theory of, 10–12; foundational theory of, 8–10; future challenges/opportunities of, 426–33; geographical approach to, 12–13; global development of, 6–8; growth/decline of, 19–33; incremental, 72, 78–9, 83; informal, 56, 81–2, 235; instant, 70, 72; level of, 2, 4, 5–6, 20, 21; lived experiences of, 19; over-, 30–1; planetary, 11; of poverty, 27–8, 429; process of, 2, 6–8, 12–13, 19, 21; rapid, 368–70; in rapidly transforming world, 2–4; rate of, 2–3, 4, 21; sustainable, 366; urban age thesis and, 2, 5–6, 61; utopian visions of, 426–8. *See also entry below*; urban growth
urbanization (1950–2050), statistics on/trends in, 20–4; average annual rate of population change, 4; global urbanization (map/graphs), 22; population increase, 3; population milestones, 5
urban mega-events, 83
urban metabolism, 349
urban neoliberalism, 175
urban planning, 348–50; climate change and, 155–67; Indigenous peoples/settler states and, 136–51; in post-industrial cities, 103–16; public health and, 362; in socialist/post-socialist cities, 120–34; transnational actors in, 73; women's involvement in, 82
urban policy, 7–8; austerity as, 89–100; Canada's lack of, 47–8; climate change and, 155–67; geopolitics and, 238–9; social mixing as, 178; sustainability as, 431–3
urban political ecology (UPE), 351
urban primacy, 37
urban reserves, *147*, 147–8, 150
urban resilience, 158, 432–3
urban revolution: in city/suburb development, 52–67; right to the city and, 261, 262, 263–4
urban social movements, 11, 214; Arab Spring, 212, 214; Black Lives Matter, 214, 303, 430; Gezi Park Protests, 215–18; Idle No More, 222; Marineland Animal Defense, 221–2; Morris Justice Project, 218–21; Occupy, 11, 212, 214, 222, 271, 430; Occupy Wall Street, 92, 264, 371; SlutWalks, 221–2, 290–1, 430
urban sustainability, 431–2
urban transition, 6, 7, 8; in global South, 3, 7
urban utopias, 433
urban villages, 8
urban water governance, 377–90; in Canada, 388–9; community models of, 382–3; co-production arrangements of, 382, 385–8; definition of, 380; in global South, 378–80; privatization of, 380–2, 383–5, 389; sanitation and, 377, 378–83; user participation in, 383
utopia, concept of, 426–7
utopian imaginaries, of urban life, 133, 426–8, 433
Vällingby, Sweden, 129–31; as ABC town, 130; in context of social welfare state, 129–30; design/amenities of, 130; modernism of, 130; population of, 131; recognition of, as unique/livable place, 131
Vancouver, 7, 21, 48, 64, 65, 83, 99, 222, 270, 290, 304, 347, 373; Downtown Eastside of, 320–1; EcoDensity Charter of, 165; food strategy of, 357; gentrification in, 174, 184, 185; house prices in, 165, 201–2; immigration to, 32, 185; Indigenous social/affordable housing in, 148; informal recycling in, 405–6; LGBTQ community in, 337–8; Public Space Network of, 221; transportation in, 410, 421, 422
varieties of capitalism (VOC), 38–9
violence: of colonialism, 138–9; displacement/dispossession and, 233–8, 267–8; ethno-racial, 177, 219, 280–1, 296–304; homophobic/transphobic, 326, 327, 330; war/militarization and, 229–35, 238. *See also* women, violence against
Victoria, BC, 185, 288, 357, 420; wastewater disposal in, 389
Vietnam, 120, 121, 122, 124; ambivalence/mistrust of urbanism in, 124; collective housing in, 125, 126–9, 131; Cultured Family campaign in, 125, 129; everyday life in, 125; history of conflict in, 126; market socialism in, 125, 126; socialist network support of, 123–4
Viet-Xo Friendship Palace (Hanoi), 123
visibility of sexuality, 326; in Canadian cities, 327, 337–9; in Chicago, 326, 330–3; in Ciudad Juárez, 327, 335–7; in Singapore, 326–7, 333–5

Waitangi Tribunal (New Zealand), 144, 146, 148
war, 227, 229

war and militarization, 227–40; displacement caused by, 231–3; emergency urbanism of, 232, 235; geopolitics of, 229–31; IDP/refugee camps, as urban forms of, 233–8. *See also* Al-Baqa'a Refugee Camp, Jordan; Awere IDP camp, Uganda

Warner, Michael: *Publics and Counterpublics*, 212–13

"War on Terror," 232

waste management, 394–406; global trends in, 395–7; social inequality and, 399–400. *See also entry below*

waste management, types of: composting, 221, 396, 397, 405; informal collection/recycling, 395–6, 399–400, 402, 405–6; open dumping, 396; sanitary landfills, 396

wastewater, management/treatment of, 379–80, 388, 389, 402–3. *See also* sewers

water, access to, 377–90; affordability of, 379; in Canada, 388–9; climate change and, 158, 377, 378–9, 383; governance and, 377–83; politics of, 377, 379; pollution and, 371; public vs private, 96, 120, 380–2, 383–5, 389; quality of, 379–80; sewerage and, 379, 389; sources of, 378, 388, 389; sustainability of, 380, 381, 383. *See also* urban water governance

water governance, 377–8, 380. *See also* urban water governance

water pollution: in Flint, 371

watershed, 388

Wekerle, Gerda, 277

Weston, Ontario (Toronto), 271–2

white flight, 95

whiteness, 136, 239, 298–9

"whitestream" society, 136

white supremacy, 296

Whitlam, Gough, 141, 143

Wilson, Alexander, 350

Wilson, Nina, 222

Wirth, Louis, 9, 57, 209

women, in cities/urban areas, 276–91; activism by, 235; bodily experiences of, 277, 280; in Canada, 288–91; child care for, 98, 277, 289; condominium ownership by, 184; as construction workers, 76; feminist research on, 277, 278, 279; as poultry farmers, 354–6; public/private binary and, 279–80, 290; racialization of, 280; safe access to sanitation infrastructure by, 281, 377, 402–5; societal/moral constraints on, 277, 280, *282*, 282–3, 290–1; transnational families/households of, 190, 281–8; as transnational migrant workers, 283–5, 287–8; waged/unwaged work of, 11, 210–11, 277, 281–3, 285–9; xenophobia toward, 280–1. *See also entry below*

women, violence against, 239, 276, 278–81, 288, 290–1, 303, 316, 337, 339; in Canada, 221–2, 290–1; in Ciudad Juárez, 278, 280, 335–6; in Delhi, 279, 281, 291, 377, 402–5

working class, 165, 211, 252, 281, 312, 366–7; of LGBTQ community, 332, 335; in post-industrial cities, 104–7; sex work by, 327, 335–7, 339; urban planning and, 132, 162

working-class areas, 9, 211; gentrification of, 173, 174, 176; housing bubble and, 200, 204; suburban, 57

World Bank, 56, 297–8, 366, 381, 395, 413–14

world cities, 53, 62, 179, 187, 373; emergence of, 55–6

World Health Organization (WHO), 287, 363, 378, 394, 397, 399, 412

"worlding," 11–12

World Toilet Organization, 398

World Urbanization Prospects (UN report), 20–4

xenophobia, 280–1, 334

Yellow Quill First Nation, urban reserve of (Saskatoon), 147, *147*

Young, Iris Marion, 213

"youth bulge," 24, 28, 295

Zagreb Declaration for Healthy Cities, 363

zero-tolerance policies, 215, 219

Zimmerman, George, 177

Zona de la Paz (Ciudad Juárez), 327, *336*, 337

zoning, 57, 59, 78, 79, 192, 211, 337, 402; inclusionary, 181–4